The Million Person Study of Low-Dose Radiation Health Effects

This book presents original research findings of *The Million Person Study of Low-Dose Radiation Health Effects (MPS)*, the largest and most comprehensive epidemiologic study of its kind to investigate the health effects of low-level chronic radiation exposure on American workers and veterans throughout the 20th century.

Since the early 1900s, epidemiologists have studied the consequences of radiation exposures, yet the health effects of low levels received gradually over time remain unresolved. This uncertainty comes at a time when the public and workers are experiencing ever-increasing levels of radiation exposure from advances in medical radiation imaging techniques (e.g., CT scans), frequent flying at high altitudes, and environmental and occupational exposures. The MPS is providing answers by studying 30 radiation-exposed U.S. populations, including workers at nuclear power plants, radiologists, workers at former Manhattan Project sites, nuclear submariners, nuclear weapons test participants (atomic veterans), industrial radiographers, and radium dial painters. Ongoing for more than 20 years and coordinated by the National Council on Radiation Protection and Measurements, Vanderbilt University Medical Center and Memorial Sloan Kettering Cancer Center, the MPS is a national effort supported by the Department of Energy, National Aeronautics and Space Administration, U.S. Navy, Defense Threat Reduction Agency, Nuclear Regulatory Commission, Centers for Disease Control and Prevention, and the Environmental Protection Agency.

Unparalleled in scope and quality, the MPS provides an understanding of low-dose health effects that is desperately needed for decision-makers and the radiation protection community as society continues to increase the uses of radiation technologies. Individual chapters were originally published in the *International Journal of Radiation Biology.*

John D. Boice Jr is past President of NCRP and Professor of Epidemiology at Vanderbilt University, Nashville, USA. He served on the Main Commission of the International Commission on Radiological Protection and on the United Nations Scientific Committee on the Effects of Atomic Radiation. He directs the *Million Person Study of Low-Dose Health Effects.*

André Bouville is a leading expert on radiation dose reconstruction. He was Head of the Dosimetry Unit of the Radiation Epidemiology Branch at NCI until he retired in 2010. He chaired NCRP SC 6-9 on the dosimetry for the Million Person Study, producing NCRP Report No. 178.

Lawrence T. Dauer is Attending Physicist specializing in radiation protection at Memorial Sloan Kettering Cancer Center in the Departments of Medical Physics and Radiology. He is a Council and former Board member of the NCRP, served on ICRP Committee 3, Protection in Medicine, and is Scientific Director for the Million Person Study.

Ashley P. Golden is Senior Director of ORISE Health Studies at Oak Ridge Associated Universities, USA, where she conducts multidisciplinary projects in occupational epidemiology, biostatistics, radiation exposure and dosimetry, medical surveillance, and environmental assessments.

Richard Wakeford is Honorary Professor in Epidemiology in the Centre for Occupational and Environmental Health at The University of Manchester, UK, where he specializes in radiation epidemiology. Professor Wakeford has served on UNSCEAR, ICRP, NCRP, UK, and EU committees throughout his career.

[illegible]

[illegible]

[illegible]

[illegible]

John D. Boice Jr. is past president [illegible]

[illegible]

[illegible]

[illegible]

[illegible]

The Million Person Study of Low-Dose Radiation Health Effects

Edited by

John D. Boice Jr., André Bouville, Lawrence T. Dauer, Ashley P. Golden and Richard Wakeford

CRC Press
Taylor & Francis Group
Boca Raton London New York

CRC Press is an imprint of the
Taylor & Francis Group, an **informa** business

First edition published 2025
by CRC Press
4 Park Square, Milton Park, Abingdon, Oxon, OX14 4RN

and by CRC Press
2385 NW Executive Center Drive, Suite 320, Boca Raton FL 33431

CRC Press is an imprint of Informa UK Limited

British Library Cataloguing-in-Publication Data
A catalogue record for this book is available from the British Library

ISBN13: 978-1-032-60717-7 (hbk)
ISBN13: 978-1-032-60719-1 (pbk)
ISBN13: 978-1-003-46028-2 (ebk)

DOI: 10.1201/9781003460282

Typeset in Minion Pro
by codeMantra

Publisher's Note
The publisher accepts responsibility for any inconsistencies that may have arisen during the conversion of this book from journal articles to book chapters, namely the inclusion of journal terminology.

Disclaimer
Every effort has been made to contact copyright holders for their permission to reprint material in this book. The publishers would be grateful to hear from any copyright holder who is not here acknowledged and will undertake to rectify any errors or omissions in future editions of this book.

Contents

Citation Information

The following chapters in this book were originally published in the *International Journal of Radiation Biology*, volume 98, issue 4 (2022), except Chapter 20 and Chapter 21, which were published in volume 99, issue 2 (2023). When citing this material, please use the original page numbering for each article, as follows:

Chapter 1
Introduction to the U.S. Million Person Study of health effects from low-level exposure to radiation
John D. Boice Jr, André Bouville, Lawrence T. Dauer, Ashley P. Golden, and Richard Wakeford
International Journal of Radiation Biology, volume 98, issue 4 (2022), pp. 529–532

Chapter 2
Radiation in the workplace – an opportunity for substantial epidemiological evidence
Richard Wakeford
International Journal of Radiation Biology, volume 98, issue 4 (2022), pp. 533–536

Chapter 3
The Million Person Study, whence it came and why
John D. Boice Jr, Sarah S. Cohen, Michael T. Mumma, and Elizabeth D. Ellis
International Journal of Radiation Biology, volume 98, issue 4 (2022), pp. 537–550

Chapter 4
The Million Person Study relevance to space exploration and Mars
John D. Boice Jr
International Journal of Radiation Biology, volume 98, issue 4 (2022), pp. 551–559

Chapter 5
Historical perspective on the Department of Energy mortality studies: focus on the collection and storage of individual worker data
Elizabeth D. Ellis, David Girardi, Ashley P. Golden, Phil Phillip W. Wallace, Joyce Phillips, and Donna L. Cragle
International Journal of Radiation Biology, volume 98, issue 4 (2022), pp. 560–567

Chapter 6
50 Years of the Radiation Exposure Information and Reporting System and importance to the Million Person Study
Derek A. Hagemeyer, Gregory P. Nichols, Michael T. Mumma, John D. Boice Jr, and Terry A. Brock
International Journal of Radiation Biology, volume 98, issue 4 (2022), pp. 568–571

Chapter 7
Evaluation of statistical modeling approaches for epidemiologic studies of low-dose radiation health effects
Ashley P. Golden, Sarah S. Cohen, Heidi Chen, Elizabeth D. Ellis, and John D. Boice Jr
International Journal of Radiation Biology, volume 98, issue 4 (2019), pp. 572–581

Chapter 8

Obtaining vital status and cause of death on a million persons
Michael T. Mumma, Sarah S. Cohen, Jennifer L. Sirko, Elizabeth D. Ellis, and John D. Boice Jr
International Journal of Radiation Biology, volume 98, issue 4 (2022), pp. 580–586

Chapter 9

Validating the use of census data on education as a measure of socioeconomic status in an occupational cohort
Sarah S. Cohen, Michael T. Mumma, Elizabeth D. Ellis, and John D. Boice Jr
International Journal of Radiation Biology, volume 98, issue 4 (2022), pp. 587–592

Chapter 10

Cohort profile – MSK radiation workers: a feasibility study to establish a deceased worker sub-cohort as part of a multicenter medical radiation worker component in the million person study of low-dose radiation health effects
Lawrence T. Dauer, Meghan Woods, Daniel Miodownik, Brian Serencsits, Brian Quinn, Michael Bellamy, R. Craig Yoder, Xiaolin Liang, John D. Boice Jr, and Jonine Bernstein
International Journal of Radiation Biology, volume 98, issue 4 (2022), pp. 593–599

Chapter 11

Dosimetry and uncertainty approaches for the million person study of low-dose radiation health effects: overview of the recommendations in NCRP Report No. 178
Lawrence T. Dauer, André Bouville, Richard E. Toohey, John D. Boice Jr, Harold L. Beck, Keith F. Eckerman, Derek A. Hagemeyer, Richard W. Leggett, Michael T. Mumma, Bruce Napier, Kathy H. Pryor, Marvin Rosenstein, David A. Schauer, Sami Sherbini, Daniel O. Stram, James L. Thompson, John E. Till, R. Craig Yoder, and Cary Zeitlin
International Journal of Radiation Biology, volume 98, issue 4 (2022), pp. 600–609

Chapter 12

Dosimetry associated with veterans who participated in nuclear weapons testing
John E. Till, Harold L. Beck, Jill W. Aanenson , Helen A. Grogan, H. Justin Mohler, S. Shawn Mohler, and Paul G. Voillequé
International Journal of Radiation Biology, volume 98, issue 4 (2022), pp. 610–618

Chapter 13

Dosimetry for the study of medical radiation workers with a focus on the mean absorbed dose to the lung, brain and other organs
R. Craig Yoder, Lawrence T. Dauer, Stephen Balter, John D Boice Jr, Helen A. Grogan, Michael T. Mumma, Christopher N. Passmore, Lawrence N. Rothenberg, and Richard J. Vetter
International Journal of Radiation Biology, volume 98, issue 4 (2022), pp. 619–630

Chapter 14

MPS dose reconstruction for internal emitters: some site-specific issues and approaches
Richard W. Leggett, Keith F. Eckerman, and Michael Bellamy
International Journal of Radiation Biology, volume 98, issue 4 (2022), pp. 631–643

Chapter 15

Potential improvements in brain dose estimates for internal emitters
Richard W. Leggett, Sergei Y. Tolmachev, and John D. Boice Jr
International Journal of Radiation Biology, volume 98, issue 4 (2022), pp. 644–656

Chapter 16

Mortality from leukemia, cancer and heart disease among U.S. nuclear power plant workers, 1957–2011
John D. Boice Jr., Sarah S. Cohen, Michael T. Mumma, Derek A. Hagemeyer, Heidi Chen, Ashley P. Golden, R. Craig Yoder, and Lawrence T. Dauer
International Journal of Radiation Biology, volume 98, issue 4 (2022), pp. 657–678

Chapter 17

Mortality among U.S. military participants at eight aboveground nuclear weapons test series
John D. Boice Jr, Sarah S. Cohen, Michael T. Mumma, Heidi Chen, Ashley P. Golden, and John E. Till
International Journal of Radiation Biology, volume 98, issue 4 (2022), pp. 679–700

Chapter 18

Updated mortality analysis of the Mallinckrodt uranium processing workers, 1942–2012
Ashley P. Golden, Elizabeth D. Ellis, Sarah S. Cohen, Michael T. Mumma, Richard W. Leggett, Phillip W. Wallace, David J. Girardi, Janice P. Watkins, Roy E. Shore, and John D. Boice Jr
International Journal of Radiation Biology, volume 98, issue 4 (2022), pp. 701–721

Chapter 19

Mortality among workers at the Los Alamos National Laboratory, 1943–2017
John D. Boice Jr, Sarah S. Cohen, Michael T. Mumma, Ashley P. Golden, Sara C. Howard, David J. Girardi, Elizabeth D. Ellis, Michael Bellamy, Lawrence T. Dauer, Caleigh Samuels, Keith F. Eckerman, and Richard W. Leggett
International Journal of Radiation Biology, volume 98, issue 4 (2022), pp. 722–749

Chapter 20

Mortality among Tennessee Eastman Corporation (TEC) uranium processing workers
John D. Boice, Jr, Sarah S. Cohen, Michael T. Mumma, Ashley P. Golden, Sara C. Howard, David J. Girardi, Elizabeth D. Ellis, Michael B. Bellamy, Lawrence T. Dauer, Keith F. Eckerman, and Richard W. Leggett
International Journal of Radiation Biology, volume 99, issue 2 (2023), pp. 208–228

Chapter 21

Mortality among medical radiation workers in the United States, 1965–2016
John D. Boice, Jr., Sarah S. Cohen, Michael T. Mumma, Sara C. Howard, R. Craig Yoder, and Lawrence T. Dauer
International Journal of Radiation Biology, volume 99, issue 2 (2023), pp. 183–207

Chapter 22

Radium dial workers: back to the future
Nicole E. Martinez, Derek W. Jokisch, Lawrence T. Dauer, Keith F. Eckerman, Ronald E. Goans, John D. Brockman, Sergey Y. Tolmachev, Maia Avtandilashvili, Michael T. Mumma, John D. Boice Jr, and Richard W. Leggett
International Journal of Radiation Biology, volume 98, issue 4 (2022), pp. 750–768

Chapter 23

Sex-specific lung cancer risk among radiation workers in the Million Person Study and among TB-fluoroscopy patients
John D. Boice Jr, Elizabeth D. Ellis, Ashley P. Golden, Lydia B. Zablotska, Michael T. Mumma, and Sarah S. Cohen
International Journal of Radiation Biology, volume 98, issue 4 (2022), pp. 769–780

Chapter 24

Asbestos exposure and mesothelioma mortality among atomic veterans
John E. Till, Harold L. Beck, John D. Boice Jr, H. Justin Mohler, Michael T. Mumma, Jill W. Aanenson, and Helen A. Grogan
International Journal of Radiation Biology, volume 98, issue 4 (2022), pp. 781–785

Chapter 25

Mesothelioma mortality within two radiation monitored occupational cohorts
Michael T. Mumma, Jennifer L. Sirko, John D. Boice Jr, and William J. Blot
International Journal of Radiation Biology, volume 98, issue 4 (2022), pp. 786–794

Chapter 26

A million persons, a million dreams: a vision for a National Center for Radiation Epidemiology and Biology
John D. Boice Jr, Brian Quinn, Isaf Al-Nabulsi, Armin Ansari, Paul K. Blake, Steve R. Blattnig, Emily A. Caffrey, Sarah S. Cohen, Ashley P. Golden, Kathryn D. Held, Derek W. Jokisch, Richard W. Leggett, Michael T. Mumma, Caleigh Samuels, John E. Till, Sergei Y. Tolmachev, R. Craig Yoder, Joey Y. Zhou, and Lawrence T. Dauer
International Journal of Radiation Biology, volume 98, issue 4 (2022), pp. 795–821

For any permission-related enquiries please visit:
http://www.tandfonline.com/page/help/permissions

Notes on Contributors

Jill W. Aanenson, M.S., has over two decades of experience in environmental risk assessment, including emphasis on contaminant transport and modeling, uncertainty estimation and statistics, environmental monitoring, environmental database development, and exposure and risk assessment, with a special interest in radiation exposures in dentistry.

Isaf Al-Nabulsi Served as Acting Director at the Office of Domestic and International Health Studies within the Office of Health and Safety and Office of Environment, Health, Safety and Security at the U.S. Department of Energy (DOE). Dr. Al-Nabulsi was responsible for managing and coordinating day-to-day activities associated with domestic and international health studies, including the Million Person Study.

Armin Ansari was Radiological Assessment Team Lead at the Centers for Disease Control and Prevention (CDC) serving as subject matter expert in CDC's radiation emergency preparedness and response activities since 2002. He is a fellow and past President of the Health Physics Society and Adjunct Associate Professor of Nuclear and Radiological Engineering at Georgia Institute of Technology, Atlanta, USA. He serves on the National Council on Radiation Protection and Measurements (NCRP), provides consultancy to the International Atomic Energy Agency (IAEA), and serves as member of the United States delegation to the United Nations Scientific Committee on the Effects of Atomic Radiation (UNSCEAR). Dr Ansari is currently Director, Center for Science and Technology, EPA.

Maia Avtandilashvili is Assistant Research Professor at the U.S. Transuranium and Uranium Registries (USTUR) at College of Pharmacy and Pharmaceutical Sciences at Washington State University, USA. She has over 15 years of experience in internal radiation dosimetry and biokinetic modeling of actinides. Dr. Avtandilashvili serves on NCRP scientific committee (SC) 6-12 'Development of Models for Brain Dosimetry for Internally Deposited Radionuclides' and is a member of the European Radiation Dosimetry Group (EURADOS) Working Group 7 on Internal Dosimetry.

Stephen Balter is Professor of Clinical Radiology (physics) at Columbia University, USA. He is an international authority on all aspects of medical fluoroscopy. He is a member of the National Council on Radiation Protection and Measurements and served as Chair of NCRP Report-168 – Radiation Dose Management for Fluoroscopically-Guided Interventional Medical Procedures.

Harold L. Beck, B.S, is a leading expert on radiation dose reconstruction with more than 50 years of experience in radiation dosimetry, radiation metrology, and dose reconstruction uncertainty. He served for over 36 years in various scientific and management positions in the Department of Energy Environmental Measurements Laboratory.

Michael Bellamy, Ph.D., is Assistant Attending at Memorial Sloan Kettering Cancer Center. His research is currently concentrated on the development of accurate dosimetric assessments of organ and effective dose for both occupational and clinically exposed populations, as well as understanding fundamental micro and macro effects of radiation on the body.

Jonine Bernstein, Ph.D., is an attending epidemiologist at Memorial Sloan Kettering Cancer Center in the Department of Epidemiology & Biostatistics. She is a Council and Board member of the NCRP and the Principle Investigator of the NCI-funded international 24-center Women's Environmental, Cancer, and Radiation Epidemiology Study.

Paul K. Blake was a civilian health physicist with the Defense Threat Reduction Agency (DTRA), Fort Belvoir, USA. DTRA safeguards America and its allies from weapons of mass destruction by providing capabilities to reduce, eliminate, and counter the threat and mitigate its effects. He co-led the Nuclear Test Personnel Review Program, which confirms participation and reconstructs radiation doses for U.S. atomic veterans.

Steve Blattnig is currently at NASA Langley Research Center and has been working on a wide variety of different aspects of space radiation research for the last 20 years. He is one of the primary developers of the NASA Standard for Models and Simulations, NASASTD-7009. More recently, his focus has been on the development of probabilistic risk methodology and radiation biology modeling for effects including acute radiation syndrome, cancer, cardiovascular disease, and degenerative central nervous system diseases. He was the project manager for the space radiation transport and measurement project and was the PI of the space radiation risk assessment project.

William J. Blot is Professor of Medicine at Vanderbilt University School of Medicine, Nashville, USA, and Chief Executive Officer of the International Epidemiology Institute (IEI). Prior to co-founding IEI in 1994, he directed research into the causes of cancer for 20 years at the National Cancer Institute. He directed the NIH-funded Southern Community Cohort Study, one of the largest epidemiologic studies yet undertaken to assess reasons for racial disparities in cancer incidence and mortality.

John D. Boice Jr is Director of Science and past President of NCRP, and Professor of Epidemiology at Vanderbilt University, Nashville, USA. He served on the Main Commission of the International Commission on Radiological Protection and on the United Nations Scientific Committee on the Effects of Atomic Radiation. He directs the *Million Person Study of Low-Dose Health Effects*.

André Bouville is a leading expert on radiation dose reconstruction. He was Head of the Dosimetry Unit of the Radiation Epidemiology Branch at NCI until he retired in 2010. He chaired NCRP SC 6-9 on the dosimetry for the Million Person Study, producing NCRP Report No. 178.

Terry A. Brock is the current NRC senior project manager for the Radiation Exposure Information and Reporting System (REIRS), and the agency's participation in the One Million U.S. Radiation Workers and Veterans health study. He has also served as the project manager for the State-of-the-Art Reactor Consequence Analyses study and on the Risk Task Group that explored risk-informing the radioactive materials arena.

John D. Brockman is an Associate Professor in the Department of Chemistry at the University of Missouri. He researches problems in diverse fields that benefit from radio-analytical techniques. Currently, his research group is working on problems in pre-and post-detonation nuclear forensic analysis, trace element epidemiology, and nuclear engineering.

Emily A. Caffrey is President of Radian Scientific, which supported Risk Assessment Corporation in independent environmental dose and risk assessments. Her research includes environmental dose assessment and computational dosimetry methods. Her expertise is in statistical methods and uncertainty analysis, source term reconstruction and development, and nuclear engineering.

Heidi Chen is Research Assistant Professor of Biostatistics in the Vanderbilt Center for Quantitative Sciences. Her Ph.D. is in Biomathematics from North Carolina State University, USA, and her research focuses on applying mathematical, statistical, and computational methods to fit models for biological systems. Since joining the Department of Biostatistics, Vanderbilt University Medical Center in 2003, her research has involved identifying biomarkers and building models to predict disease progression and clinical outcome for cancer patients. Heidi has been a statistical collaborator on the Million Person Study of Low-Dose Health Effects since 2013.

Sarah S. Cohen is Director of Public Health Research at DLH Corporation, USA, and has directed observational research studies in the areas of pharmacoepidemiology, nutritional epidemiology, and occupational epidemiology as well as leads large data management projects and statistical analyses. She is also Adjunct Assistant Research Professor of Medicine in the Department of Medicine at Vanderbilt University School of Medicine, Nashville, USA. She has been a collaborator on the Million Person Study of Low-Dose Health Effects for nearly 20 years, providing analytic support as well as co-authoring numerous publications.

Donna L. Cragle is an occupational epidemiologist at Oak Ridge Associated Universities (ORAU), USA. She has studied the health effects of radiation and chemical exposures related to work in the nuclear industry for over 35 years. She is past Senior Vice President for Health, Energy, and Environment at ORAU.

Lawrence T. Dauer is Attending Physicist specializing in radiation protection at Memorial Sloan Kettering Cancer Center in the Departments of Medical Physics and Radiology. He is a Council and former Board member of the NCRP, served on ICRP Committee 3, Protection in Medicine, and is Scientific Director for the Million Person Study.

Keith F. Eckerman, Ph.D., is a leading expert in computational and phantom dosimetry, served on ICRP Committee 2, and is Emeritus Member of the ICRP. He has received several scientific achievement awards, including the 39th Lauriston Taylor Lecturer for the NCRP. He is currently a retired staff member of the Environmental Sciences Division of the ORNL.

Elizabeth D. Ellis worked in the Health, Energy and the Environment Program at Oak Ridge Associated Universities, USA. Dr. Ellis conducted research in Public Health, Occupational Health, and Epidemiology. A major project is contributing to the Million Person Study. She is also active in human subject protection in research.

David Girardi is scientific programme analyst at Oak Ridge Associated Universities, USA. He has over 15 years programming experience working with epidemiologists and statisticians and has been working with the Million Person Study for over five years. He is a member of the Project Management Institute and has been a certified Project Management Professional (PMP) for over five years.

Ronald E Goans is Senior Medical Advisor with MJW Corporation and Senior Medical/Scientific Advisor with the Radiation Emergency Assistance Center/Training Site (REAC/TS). He received his PhD in radiation physics from the University of Tennessee in 1974, his MD from the George Washington University School of Medicine in 1983, and the MPH from the Tulane School of Public Health and Tropical Medicine in 2000. He has worked in the field of nuclear physics and radiation effects since 1966.

Ashley P. Golden is Senior Director of ORISE Health Studies at Oak Ridge Associated Universities, USA, where she conducts multidisciplinary projects in occupational epidemiology, biostatistics, radiation exposure and dosimetry, medical surveillance, and environmental assessments.

Helen A. Grogan is President of Cascade Scientific Inc with more than 35 years of experience in radioecology, dose reconstruction, and the assessment of radioactive and non-radioactive hazardous wastes. She's been a U.S. delegate to UNSCEAR since 2014.

Derek A. Hagemeyer is Associate Director for Human Health & Environment at Oak Ridge Associated Universities, USA. He is responsible for program operations associated with human subject health and protection surveillance as well as the independent environmental assessment and verification under the DOE Oak Ridge Institute for Science and Education (ORISE), USA. He serves as the principal investigator for the DOE Radiation Exposure Monitoring System (REMS) and the U.S. Nuclear Regulatory Commission (NRC) Radiation Exposure Information and Reporting System (REIRS) projects. He is currently serving as Vice Chairman of the international Information System on Occupational Exposure (ISOE) Working Group on Data Analysis (WGDA) to provide support in the collection and analysis of ISOE's international database of radiation exposure information. ISOE is jointly sponsored by the Organisation for Economic Co-operation and Development (OECD) and the International Atomic Energy Agency (IAEA).

Sara C. Howard is an epidemiologist at Oak Ridge Associated Universities, USA. She has a master's degree in Epidemiology and is currently pursuing a Ph.D. in Epidemiology with a focus on chronic diseases and occupational exposure. She has been a collaborator on the Million Person Study of Low-Dose Health Effects for nearly three years.

Kathryn D. Held is past President of the NCRP and an Associate Radiation Biologist and Associate Professor, Department of Radiation Oncology, Massachusetts General Hospital/Harvard Medical School. She has served on review panels for numerous federal agencies and other organizations such as the Radiological Society of North America and Brookhaven National Laboratory. She is a past President of the Radiation Research Society and a member of the Board of the Radiation Research Foundation.

Derek W. Jokisch is Professor of Physics and Chair of the Department of Physics and Engineering at Francis Marion University, Florence, USA. He holds a Joint Faculty Appointment at Oak Ridge National Laboratory (ORNL) within the Center for Radiation Protection Knowledge (CRPK) and is a member of ICRP Committee 2 on Doses from Radiation Exposure and is a member of the U.S. Scientific Review Group for the Department of Energy's Russian Health Studies program.

Richard W. Leggett, Ph.D., is a research scientist in the Environmental Sciences Division at Oak Ridge National Laboratory (ORNL), USA. His main research is in physiological systems modeling, with primary applications to the biokinetics and dosimetry of radionuclides and radiation risk analysis. He is a member of ICRP Committee 2 and the Task Group on Internal Dosimetry.

Xiaolin Liang is a senior program analyst at Memorial Sloan Kettering Cancer Center in the Department of Epidemiology & Biostatistics. Her primary collaboration is the Women's Environmental, Cancer, and Radiation Epidemiology Study group and has served as an informatics specialist at the coordinating center of the study since its inception.

Nicole E. Martinez, Certified Health Physicist, is Associate Professor in the Department of Environmental Engineering and Earth Sciences at Clemson University, USA. She also holds a Joint Faculty Appointment at Oak Ridge National Laboratory (ORNL) within the Center for Radiation Protection Knowledge (CRPK). While serving in the U.S. Navy, she was a nuclear power instructor and radiation health officer. Her current research focuses on dosimetric modeling and the behavior and effects of radiological contaminants in the environment. She serves on ICRP Committee 4, Application of the Commission's Recommendations.

Daniel Miodownik is a physicist in the Department of Medical Physics at Memorial Sloan Kettering Cancer Center. He currently manages the dosimetry database for the Center and ongoing assessments of worker exposure scenarios.

H. Justin Mohler, M.S., has 20 years of experience in various aspects of environmental health physics, including sampling and measurement, data analysis and interpretation, transport modeling and exposure and risk assessment, as well as extensive experience with database design and management.

S. Shawn Mohler, M.S., has almost 20 years of experience in risk assessment and dose reconstruction projects, particularly related to exposure analysis, exposure scenario development, technical editing and review, and technical graphic design. Shawn comes from an occupational health and safety background with an M.S. in Environmental Health specializing in industrial hygiene.

Michael T. Mumma is Director of Information Technology at the International Epidemiology Institute and the International Epidemiology Field Station at Vanderbilt University Medical Center, USA. He has over 20 years of experience in data analysis and conducting epidemiologic investigations.

Bruce Napier, M.S., is a staff scientist in Environmental Analysis and Engineering Group at PNNL. He is an expert in the areas of radiation dose reconstruction, computer modeling, environmental analysis, and human health risk analysis. Mr. Napier is Scientific Vice President and a past member of the Board of Directors of NCRP.

Gregory P. Nichols is President of GP Nichols & Company, LLC. He has nearly 20 years of experience spanning clinical laboratory medicine, emergency medical care, and occupational/environmental health research. He has supported projects for a variety of government agencies including the Department of Energy and the Nuclear Regulatory Commission.

Christopher N. Passmore is a certified health physicist with over 33 years of experience in dosimetry and a former lead U.S. Department of Energy Laboratory Accreditation Program (DOELAP) assessor and a current National Voluntary Laboratory Accreditation Program (NVLAP) assessor for NIST in ionizing radiation dosimetry. He is Chief Technology Officer (CTO) for Radiation Detection Company and still provides dosimetry consulting services through Passmore Dosimetry.

Joyce Phillips is a senior scientific programmer analyst at Oak Ridge Associated Universities, USA. She serves as database administrator with over 30 years of experience providing technical support and maintaining data integrity and accuracy of the Center for Epidemiological Research historical data.

Kathy H. Pryor, M.S., is currently on the NCRP Board of Directors and is Scientific Vice President of Program Area Committee 2. Ms. Pryor was Chief Health Physicist at PNNL with previous experience in nuclear power and universities. Ms. Pryor served as President of the HPS and Chair of the ABHP and is President of the AAHP.

Brian Quinn is Technical Lead Physicist in the Department of Medical Physics at Memorial Sloan Kettering Cancer Center. He continues to research both patient and occupational dosimetry issues, concentrating on more accurately estimating organ doses.

Marvin Rosenstein, Ph.D., is an NCRP consultant. He was Director of Office of Health Physics at the Center for Devices and Radiological Health, US FDA, and previously served 33 years as Commissioned Officer in the US PHS. He is Distinguished Emeritus Member of NCRP and Emeritus Member of ICRP Committee 3, Protection in Medicine.

Lawrence N. Rothenberg is Attending Physicist Emeritus at Memorial Sloan-Kettering Cancer Center (MSKCC). He directed the Diagnostic X-Ray Quality Assurance Laboratory for MSKCC and New York Presbyterian Weill Cornell Medical Center for over 35 years until January 2007 when he was appointed Member Emeritus of MSKCC.

Caleigh Samuels is Radiation Dosimetry Lead in the Center for Radiation Protection Knowledge at Oak Ridge National Laboratory, USA. Her research in radiation dosimetry includes element-specific biokinetic model development, Monte Carlo applications in dosimetric and detector response modeling, development of radiological risk assessment tools, and dose reconstruction for internal emitters. She collaborates on the Million Person Study and is currently focusing on organ dose reconstructions following plutonium intakes at the Rocky Flats nuclear facility.

David A. Schauer, Sc.D., is Executive Secretary of the ICRU. He served as an officer in the U.S. Navy and on the faculty of the Radiology and Radiological Sciences Department at the Uniformed Services University of the Health Sciences, USA. He previously served as NCRP Executive Director and is Executive Director Emeritus.

Brian Serencsits is Lead Health Physicist in the Department of Medical Physics at Memorial Sloan Kettering Cancer Center. He currently supports the medical health physics aspects of laboratory and clinical uses of radiation and radioactive material for the Center as well as ongoing training programs.

Sami Sherbini, Ph.D., is a senior level advisor for health physics at the U.S. NRC, where he has held progressively more technical positions in providing expertise to support rulemaking in radiation protection, nuclear medicine, and other radioactive materials regulation. Prior to the NRC, he worked in the commercial nuclear industry, including Three Mile Island.

Roy E. Shore is Professor Emeritus at New York University Langone School of Medicine, USA, and Chief of Research at Radiation Effects Research Foundation, Hiroshima, Japan (retired). His areas of interests are radiation epidemiology of cancer and non-cancer, low doses, and low dose rates. He has served as a member of the ICRP and NCRP and on UNSCEAR task groups.

Jennifer L. Sirko is an epidemiologist with over 10 years of experience in occupational and public health research. She is currently a lead technical assessor/evaluator at MITRE, USA.

Daniel O. Stram is Professor in the Department of Preventive Medicine at the Keck School of Medicine of the University of Southern California, USA. He served as a postdoctoral fellow in the Biostatistics Department of the Harvard School of Public Health and was a research associate at RERF, Hiroshima, Japan.

James L. Thompson, B.S., initially served as a nuclear propulsion plant operator in the U.S. Navy, then became a health physicist working as a radioactive materials inspector for the Texas Department of Health at Bureau of Radiation Control, and currently is Senior Health Physicist with the US NRC.

John E. Till, Ph.D., is the founder and President of Risk Assessment Corporation with more than 40 years of experience in environmental dosimetry. He received the E.O Lawrence award in 1995 and delivered the L.S. Taylor lecture in 2013. He also served in the U.S. Navy Nuclear Submarine Program, retiring as a Rear Admiral in 1999.

Sergey Y. Tolmachev is Associate Research Professor in the College of Pharmacy and Pharmaceuticals Sciences (COP) at Washington State University, USA, where he directs the U.S. Transuranium and Uranium Registries Research Center and the National Human Radiobiology Tissue Repository. His current research interests are in the area of radiation protection, including radionuclide biokinetic modeling, actinide decorporation treatment, retrospective biodosimetry, and study of radioactive nanoparticles as well as application of mass-spectrometric techniques for stable element determination in the human body.

Richard E. Toohey, Ph.D., was a physicist with over 45 years of experience in health physics, especially internal radiation dosimetry, dose reconstruction, and radiological emergency response. He recently retired from Oak Ridge Associated Universities, USA, where he served as Director of the Radiation Internal Dose Information Center and held other leadership roles. Dr Toohey died Nov 2023.

Richard J. Vetter is Professor Emeritus and former Radiation Safety Officer for the Mayo Clinic and Medical School, Rochester, USA.

Paul G. Voillequé, M.S., is an expert in historical dose reconstruction whose broad interests range from personnel protection and internal dosimetry to the behavior of radionuclides within nuclear facilities and in the environment. He was a member of the Subcommittee on DTRA Dose Reconstruction Procedures of the Veterans' Board on Dose Reconstruction.

Richard Wakeford is Honorary Professor in Epidemiology in the Centre for Occupational and Environmental Health at The University of Manchester, UK, where he specializes in radiation epidemiology. Professor Wakeford has served on UNSCEAR, ICRP, NCRP, UK, and EU committees throughout his career.

Phillip W. Wallace was Information Technology Manager, supporting epidemiological research. He semi-retired in 2016 after 32 years in this position and contributed 10 years to the MPS. He was responsible for managing the databases at Oak Ridge Associated Universities (ORAU), USA, containing DOE rosters, work history information, radiation exposure data, and vital status. He was also one of the original developers of the DOE Comprehensive Epidemiologic Data Resource in 1990.

Janice Watkins is a biostatistician with extensive experience working with data from a wide variety of areas including occupational epidemiology, radiation dosimetry and dose reconstruction, workplace violence, and drugs/alcohol abuse. She worked extensively on the Mallinckrodt cohort study. She recently retired from Oak Ridge Associated Universities, USA, after 16 years of employment.

Meghan Woods is Research Project Manager in the Department of Epidemiology & Biostatistics at Memorial Sloan Kettering Cancer Center. She currently manages data collection for two multi-center studies: the Women's Environmental, Cancer, and Radiation Epidemiology Study and the MRI Background Parenchymal Enhancement as a Risk Factor for Breast Cancer: the IMAGINg and Epidemiology Study.

R. Craig Yoder, Ph.D, directed Landauer's technical activities relating to radiation dosimetry, particularly for applications in radiation protection for over 30 years. An internationally known expert in radiation monitoring, he led the development of optically stimulated luminescence. He currently is using his historical knowledge to advise the MPS epidemiologists regarding the methods to translate personal monitoring information into mean absorbed doses to various organs. He is a member of NCRP and former President of the Council on Ionizing Radiation Measurements and Standards.

Lydia B. Zablotska is a Professor in the Department of Epidemiology and Biostatistics in the School of Medicine at the University of California, San Francisco (UCSF), where she serves as the Leader of the Occupational and Environmental Epidemiology Area of Concentration. She is the Principal Investigator of the NCI-funded study to update dosimetry and conduct lifespan mortality and incidence follow-up of the Canadian Fluoroscopy Cohort Study.

Cary Zeitlin, Ph.D., is a research scientist at Leidos Innovations Corporation, working in the Space Radiation Analysis Group at NASA Johnson Space Center, with their programs in space radiation physics and biology. He performs operations and data analysis for the Radiation Assessment Detectors aboard the Curiosity Mars rover and the International Space Station.

Joey Y Zhou is a senior epidemiologist in the DOE Office of Health and Safety under the Office of Environment, Health, Safety and Security. He is past DOE program manager for the agency's participation in the Study of One Million U.S. Radiation Workers and Veterans. He also serves as the program manager for the United States Transuranium and Uranium Registries, the Russian Health Studies Program and the Radiation Emergency Assistance Center/Training Site (REAC/TS).

Introductions

Introduction to the U.S. Million Person Study of health effects from low-level exposure to radiation

John D. Boice, André Bouville, Lawrence T. Dauer, Ashley P. Golden, and Richard Wakeford

The epidemiologic study of one million U.S. radiation workers and veterans on health effects following low-level radiation exposure (Boice, Cohen, et al. 2019), or the Million Person Study (MPS[1]), has been underway in some form for more than a quarter of a century. The MPS was designed to examine health effects after chronic exposure to low dose-rates of radiation, in contrast to the brief exposure at a high dose-rate experienced by the Japanese atomic bomb survivors. The study will provide important scientific evidence needed for sound radiation protection policy and recommendations (NCRP 2018a; Boice, Held, et al. 2019). This special issue consists of 26 articles, including this introduction and an editorial (Wakeford 2021). The aim for this special issue is to present a comprehensive overview of the MPS with regard to: its conceptual development and historical perspectives, methodological approaches for both epidemiology and dosimetry, the first publications of quantitative results to date, as well as a summary of the first international virtual symposium with key stakeholders and researchers on the MPS that also casts a vision for the future.

Articles grouped as 'Reviews' specifically address the importance of the MPS and its history (Boice, Cohen et al. 2019), including the relevance to space exploration and Mars (Boice 2019). In addition, one article discusses the historical perspective of the U.S. Department of Energy (DOE) mortality studies with an emphasis on collection and storage of individual worker data (Ellis et al. 2019) while another provides a five-decade overview of the Nuclear Regulatory Commission's (NRC) Radiation Exposure Information and Reporting System (REIRS) database focusing on the importance of this 'registry' of radiation workers to the MPS (Hagemeyer et al 2018). The Comprehensive Epidemiologic Data Resource (CEDR) was created in 1990 to allow researchers to access data from the U.S. DOE epidemiological studies program (US DOE 2020). CEDR became a major source of data used to extend the follow up of U.S. DOE worker cohorts for the MPS. All MPS studies are planned to be uploaded to CEDR once they are completed and published, and not just the U.S. DOE worker cohorts. Currently, the studies of Mound, Rocketdyne and Mallinckrodt workers are available on CEDR, the atomic veteran and the Los Alamos National Laboratory studies shortly, and most of the others published in this special issue within a year.

Articles within the 'Methods' section identify certain key epidemiologic methods utilized for the MPS, including the evaluation of statistical modeling approaches for such studies of low-level radiation health effects (Golden et al. 2019), the essential methods of obtaining vital status and cause of death on a million persons (Mumma, Cohen, et al. 2018), and the validation of census data on education as a measure of socioeconomic status (Cohen et al. 2019). A fourth article in this section provides a profile for a pilot sub-cohort of a multicenter medical radiation worker component of the MPS, including early worker data prior to the 1960s (Dauer et al. 2019).

Exquisite dosimetry is the key to quality epidemiology and several articles are grouped within the 'Dosimetry' section. The first summarizes the dosimetry methods recommended by the National Council on Radiation Protection and Measurements (NCRP) in Report No. 178 (NCRP 2018b) emphasizing the dosimetry and uncertainty approaches for the MPS (Dauer et al. 2018). An article delineating the specific dosimetry associated with individual veterans who participated in nuclear weapons testing emphasizes the importance of an in-depth understanding of exposure scenarios (Till et al. 2014, Till, Beck, Aanenson, et al. 2018; Beck et al. 2017), while another addresses approaches associated with dose reconstruction for internal emitters (Leggett et al. 2018). Dosimetry for medical radiation workers also is particularly unique and was given specific development attention with a focus on absorbed doses to the lung, brain, heart, colon and red bone marrow (Yoder et al. 2019). Largely based on this work and additional development, the NCRP published general guidance in NCRP Commentary 30 (NCRP 2020; Yoder et al. 2021) specifically directed toward the MPS for using personal monitoring records to derive organ doses for medical radiation workers. The final article in this section identifies potential improvements in brain dose estimates for internal emitters as relevant to possible cognitive impairment following high-LET exposures (Leggett et al. 2019). The concepts discussed in this paper helped form the basis of a forthcoming NCRP Commentary on the development of kinetic and anatomical models for brain dosimetry for internally deposited radionuclides, developed by NCRP Scientific Committee SC 6-12 (NCRP 2021).

Perhaps the *pièce de résistance* of this special issue are 10 articles, grouped within the 'Original Articles' section, that provide the first quantitative publication of results from epidemiologic analyses of specific and wide-ranging MPS cohorts. Six articles provide results on mortality (including leukemia, cancer, heart disease, Parkinson's disease, and other diseases) among large cohorts of the MPS, including: U.S. nuclear power plant workers (Boice, Cohen, Mumma, Hagemeyer, et al. 2021), medical radiation workers (Boice, Cohen, Mumma, Howard, et al. 2021), U.S. military participants (atomic veterans) at eight aboveground nuclear weapons test series (Boice et al. 2020), Mallinckrodt uranium processing workers (Golden et al. 2019), Los Alamos National Laboratory workers (Boice et al. 2021a), and Tennessee Eastman Corporation uranium processing workers (Boice et al. 2021b) including both male and female employees. The MPS has re-opened the epidemiologic study of the extremely important and influential study of the radium dial workers (Martinez et al. 2021), complemented with a new NCRP Scientific Committee 6-13 on *Methods and Models for Estimating Organ Doses from Intakes of Radium*. Quantitative information on female and male lung cancer risks is provided among workers in the MPS and TB-fluoroscopy cohorts (Boice, Ellis, et al. 2019). In addition to radiation, the MPS evaluated asbestos exposure and mesothelioma and asbestosis mortality among radiation-monitored occupational cohorts including the atomic veterans (Till, Beck, Boice, et al. 2018; Boice et al. 2020), industrial radiographers and nuclear power plant workers (Mumma, Sirko, et al. 2018; Boice, Cohen, Mumma, Hagemeyer, et al. 2021).

The 'Meeting Report' section represents the finale of the special issue. A comprehensive paper (Boice, Quinn et al. 2021) summarizes the first international virtual MPS symposium wherein key stakeholders and researchers discussed the importance of the MPS, its relevance for radiation protection, what has been done, current emerging results, plans for the future, as well as a vigorous question and answer dialogue with the over 300 participants. It is anticipated that a powerful evaluation of radiation effects at low doses and low dose rates will be achieved when all MPS cohorts are completed and pooled, i.e. combined to provide substantial statistical power to evaluate health effects following low-level exposures. A continued follow-up of the 34 individual cohorts comprising the MPS is planned for the next several decades. Incidence data on the occurrence of disease, in addition to mortality data, are forthcoming as is a comprehensive approach to obtain individual information on tobacco use via linkages with the Centers for Medicare and Medicaid Services (CMS) nationwide databases (Boice, Quinn et al. 2021). A National Center for Radiation Epidemiology and Biology is envisioned.

Note

1. Also known as the Million U.S. Workers and Veterans Study (MWS), and earlier as the Atomic and Nuclear Energy Worker (ANEW) Study (Boice 2012).

Disclosure statement

RW is a consultant to the UK Compensation Scheme for Radiation-Linked Diseases (CSRLD). Otherwise, no potential conflict of interest was reported by the author(s).

Funding

NCRP acknowledges the financial support of the MPS-related work presented in this special issue from U.S. DOE [Grants # DE-SC0008944, DE-SC0014664, DE-AU0000042 and DE-AU0000046], NASA [Grants # NNX15AU88G, 80NSSC17M0016 and 80NSSC19M0161], U.S. Navy [Grant # N00024-17-C-4322], the National Cancer Institute [U01 CA137026], and the CDC [Grants # 5EU1EH000989 and 5NUE1EH001315].

ORCID

John D. Boice Jr. http://orcid.org/0000-0002-8755-1299
André Bouville http://orcid.org/0000-0002-9090-2862
Lawrence T. Dauer http://orcid.org/0000-0002-5629-8462
Ashley P. Golden http://orcid.org/0000-0002-9430-7702
Richard Wakeford http://orcid.org/0000-0002-2934-0987

References

Beck HL, Till JE, Grogan HA, Aanenson JW, Mohler HJ, Mohler SS, Voillequé PG. 2017. Red bone marrow and male breast doses for a cohort of atomic veterans. Radiat Res. 187(2):221–228.

Boice JD Jr. 2012. A study of one million U.S. radiation workers and veterans a new NCRP initiative (DOE Grant Awarded September 2012). Health Phys News. XL:7–10; [accessed 2021 Sep 21]. https://ncrponline.org/the-boice-report/.

Boice JD Jr, Cohen SS, Mumma MT, Ellis ED. 2019. The Million Person Study, whence it came and why. Int J Radiat Biol. DOI:10.1080/09553002.2019.1589015

Boice JD Jr, Cohen SS, Mumma MT, Golden AP, Howard S, Girardi DD, Ellis ED, Bellamy M, Dauer LT, Eckerman KF, et al. 2021b. Mortality among Tennessee Eastman Corporation uranium processing workers, 1943–2017. Int J Radiat Biol.; submitted for publication.

Boice JD Jr, Cohen SS, Mumma MT, Golden AP, Howard SC, Girardi DJ, Dupree Ellis ED, Bellamy M, Dauer LT, Samuels C, et al. 2021a. Mortality among workers at the Los Alamos National Laboratory. Int J Radiat Biol. DOI:10.1080/09553002.2021.1917784

Boice JD Jr, Cohen SS, Mumma MT, Hagemeyer DA, Chen H, Golden AP, Yoder RC, Dauer LT. 2021. Mortality from leukemia, cancer and heart disease among U.S. nuclear power plant workers, 1957–2011. Int J Radiat Biol.; in press.

Boice JD Jr, Cohen SS, Mumma MT, Howard SC, Yoder RC, Dauer LT. 2021. Mortality among medical radiation workers in the United States, 1965–2016. Int J Radiat Biol. in press.

Boice JD Jr, Ellis ED, Golden AP, Zablotska LB, Mumma MT, Cohen SS. 2019. Sex-specific lung cancer risk among radiation workers in the Million Person Study and among TB fluoroscopy patients. Int J Radiat Biol. DOI:10.1080/09553002.2018.1547441

Boice JD Jr, Quinn B, Al-Nabulsi I, Ansari A, Blake PK, Blattnig SR, Caffrey EA, Cohen SS, Golden AP, Held KD, et al. 2021. A million persons, a million dreams: a vision for a National Center of Radiation Epidemiology and Biology. Int J Radiat Biol.; in press.

Boice JD Jr. 2019. Relevance of the Million Person Study to research needs for NASA and space exploration. Int J Radiat Biol. DOI:10.1080/09553002.2019.1589020

Boice JD, Held KD, Shore RE. 2019. Radiation epidemiology and health effects following low-level radiation exposure. J Radiol Prot. 39(4): S14–S27.

Boice JD, Cohen SS, Mumma MT, Chen H, Golden AP, Beck HL, Till JE. 2020. Mortality among U.S. military participants at eight above-ground nuclear weapons test series. Int J Radiat Biol. DOI:10.1080/09553002.2020.1787543

Cohen SS, Mumma MT, Ellis ED, Boice JD Jr 2019. Validating the use of census data on education as a measure of socioeconomic status in an occupational cohort. Int J Radiat Biol. DOI:10.1080/09553002.2018.1549758

Dauer LT, Bouville A, Toohey RE, Boice JD, Beck HL, Eckerman KF, Hagemeyer D, Leggett RW, Mumma MT, Napier B, et al. 2018. Dosimetry and uncertainty approaches for the million-worker study of radiation workers and veterans: overview of the recommendations in NCRP Report No. 178. Int J Radiat Biol. DOI:10.1080/09553002.2018.1536299

Dauer LT, Woods M, Miodownik D, Serencsits B, Quinn B, Bellamy M, Yoder C, Liang X, Boice JD Jr, Bernstein J. 2019. Cohort profile – MSK radiation workers: a feasibility study to establish a deceased worker sub-cohort as part of a multicenter medical radiation worker component in the Million Person Study of low-dose radiation health effects. Int J Radiat Biol. DOI:10.1080/09553002.2019.1587194

Ellis ED, Girardi D, Golden AP, Wallace PW, Phillips J, Cragle DL. 2019. Historical perspective on the Department of Energy mortality studies: focus on the collection and storage of individual worker data. Int J Radiat Biol. DOI:10.1080/09553002.2018.1547851

Golden AP, Cohen SS, Chen H, Ellis ED, Boice JD Jr 2019. Evaluation of statistical modeling approaches for epidemiologic studies of low-dose radiation health effects. Int J Radiat Biol. DOI:10.1080/09553002.2018.1554924

Golden AP, Ellis ED, Cohen SS, Mumma MT, Leggett RW, Wallace PW, Girardi D, Watkins JP, Shore R, Boice JD Jr. 2019. Updated mortality analysis of the Mallinckrodt uranium processing workers. Int J Radiat Biol. DOI:10.1080/09553002.2019.1569773

Hagemeyer D, Nichols G, Mumma MT, Boice JD, Brock TA. 2018. 50 years of the radiation exposure information and reporting system. Int J Radiat Biol. DOI: 10.1080/09553002.2018.1540896

Leggett RW, Eckerman KF, Bellamy M. 2018. MPS dose reconstruction for internal emitters: some site-specific issues and approaches. Int J Radiat Biol. DOI:10.1080/09553002.2018.1558302

Leggett RW, Tolmachev SY, Boice JD Jr. 2019. Potential improvements in brain dose estimates for internal emitters. Int J Radiat Biol. DOI: 10.1080/09553002.2018.1554923

Martinez NE, Jokisch DW, Dauer LT, Eckerman KF, Goans RE, Brockman JD, Tolmachev SY, Avtandilashvili M, Mumma MT, Boice JD Jr, et al. 2021. Radium dial workers: back to the future. Int J Radiat Biol. DOI:10.1080/09553002.2021.1917785

Mumma MT, Cohen SS, Sirko JL, Ellis ED, Boice JD Jr. 2018. Obtaining vital status and cause of death on a million persons. Int J Radiat Biol. DOI:10.1080/09553002.2018.1539884

Mumma MT, Sirko JL, Boice JD Jr, Blot WJ. 2018. Mesothelioma mortality within two radiation monitored occupational cohorts. Int J Radiat Biol. DOI:10.1080/09553002.2019.1642540

[NCRP] National Council on Radiation Protection and Measurements. 2018a. Implications of recent epidemiologic studies for the linear-nonthreshold model and radiation protection. NCRP commentary 27. Bethesda (MD): NCRP.

[NCRP] National Council on Radiation Protection and Measurements. 2018b. Deriving organ doses and their uncertainty for epidemiologic studies (with a focus on the one million U.S. workers and veteran study of low-dose radiation health effects). NCRP report 178. Bethesda (MD): NCRP.

[NCRP] National Council on Radiation Protection and Measurements. 2020. Using personal monitoring data to derive organ doses for medical radiation workers, with a focus on lung. NCRP commentary 30. Bethesda (MD): NCRP.

[NCRP] National Council on Radiation Protection and Measurements. 2021. Development of kinetic and anatomical models for brain dosimetry for internally deposited radionuclides. NCRP Draft SC 6-12. Bethesda (MD): NCRP; [accessed 2021 May 18]. https://ncrponline.org/program-areas/sc-6-12/.

Till JE, Beck HL, Aanenson JW, Grogan HA, Mohler HJ, Mohler SS, Voillequé PG. 2014. Military participants at U.S. atmospheric nuclear weapons testing-methodology for estimating dose and uncertainty. Radiat Res. 181(5):471–484.

Till JE, Beck HL, Aanenson JW, Grogan HA, Mohler HJ, Mohler SS, Voillequé PG. 2018. Dosimetry associated with veterans who participated in nuclear weapons testing. Int J Radiat Biol. DOI:10.1080/09553002.2018.1551639

Till J, Beck H, Boice JD Jr, Mohler H, Mumma M, Aanenson J, Grogan H. 2018. Asbestos exposure and mesothelioma mortality among atomic veterans. Int J Radiat Biol. DOI:10.1080/09553002.2018.1551641

[US DOE] US Department of Energy. 2020. Comprehensive Epidemiologic Data Resource: CEDR epidemiologic databases. Oak Ridge (TN): Oak Ridge Institute for Science and Education; [accessed 2021 Aug 25]. https://oriseapps.orau.gov/cedr/.

Wakeford R. 2021. Radiation in the workplace - an opportunity for substantial epidemiological evidence. Int J Radiat Biol.; submitted for publication.

Yoder RC, Dauer L, Balter S, Boice JD Jr, Grogan H, Mumma M, Passmore CN, Rothenberg LN, Vetter RJ. 2019. Dosimetry for the study of medical radiation workers with a focus on the mean absorbed dose to the lung, brain and other organs. Int J Radiat Biol. DOI:10.1080/09553002.2018.1549756

Yoder C, Balter S, Boice JD Jr, Grogan H, Mumma M, Rothenberg LN, Passmore C, Vetter RJ, Dauer LT. 2021. Using personal monitoring data to derive organ doses for medical radiation workers in the Million Person Study – considerations regarding NCRP Commentary no. 30. J Radiol Prot. 41(1):118–128.

Radiation in the workplace – an opportunity for substantial epidemiological evidence

Richard Wakeford

Studies of the long-term health experience of the Japanese survivors of the atomic bombings of Hiroshima and Nagasaki are highly impressive investigations of the late effects of exposure to ionizing radiation, which continue today and still contribute significant evidence some seven decades after the bombings (Ozasa et al. 2019). It is upon the cancer risk estimates derived from the LifeSpan Study (LSS) of a large cohort of atomic bomb survivors that the present system of radiological protection recommended by the International Commission on Radiological Protection is primarily (but not solely) based (ICRP 2007). Nonetheless, no single epidemiological study, however impressive, will provide a sufficient basis for the framework of radiological protection because there will be aspects of protection that will not be covered. For example, the Japanese atomic bomb survivors provide no direct evidence on the effects of alpha radiation, and this must be gleaned from other sources, such as the studies of underground hard-rock miners exposed to radon and its decay products. A further source of uncertainty is that the atomic bomb survivors are a Japanese population briefly exposed at a high dose-rate in the mid-twentieth century, which raises questions about applicability of risk estimates derived from the survivors to those chronically exposed to low dose-rates and who experience different background health risks, such as a twenty-first century population in North America. These questions point to the evidential value of studies of those exposed to radiation in the course of their employment.

The information that could potentially be provided by radiation workers was recognized half a century ago when efforts were made to establish cohorts of nuclear industry workers. Unfortunately, an early study of the Hanford workforce in the USA that apparently showed substantially higher risks of cancer than predicted by the standard risk models of the time (Mancuso et al. 1977) did not provide a shining example of how such studies should be conducted: it was heavily criticized for its seriously flawed methods, e.g. by Hutchison et al. (1979), and the findings were not confirmed by analyses using an appropriate methodology, e.g. by Darby and Reissland (1981). Despite this controversy, efforts continued to assemble cohorts of nuclear workers in North America (Gilbert and Marks 1979; Checkoway et al. 1985; Wilkinson et al. 1987; Gilbert et al. 1989; Gribbin et al. 1993) and Europe (Beral et al. 1985; Smith and Douglas 1986; Beral et al. 1988). It was recognized that with the low levels of radiation exposure experienced occupationally, especially in more recent years, large numbers of workers would have to be included in studies to generate sufficient statistical power to produce meaningful results in terms of risk estimates, which led to combined studies of workers from a number of installations: from three nuclear sites in the USA (Gilbert et al. 1989, 1993) and a variety of sites in the UK (Kendall et al. 1992; Carpenter et al. 1994).

The desire to improve the power of worker studies led to an international collaborative project coordinated by the International Agency for Research on Cancer (IARC) to pool the data for nuclear workers from three countries: the USA, UK and Canada (IARC 1994; Cardis et al. 1995), essentially based on data from previous studies in these countries (Gilbert et al. 1993; Gribbin et al. 1993; Carpenter et al. 1994). The IARC-coordinated international collaboration was expanded, first to include workers from 15 countries (Cardis et al. 2005, 2007) and then, most recently, to include nuclear industry workers from the USA, UK and France in the International Nuclear Workers Study (INWORKS) (Leuraud et al. 2015; Richardson et al. 2015). Some aspects of these studies were problematical (Wakeford 2009; Ashmore et al. 2010; Wakeford 2014), but the risk estimates obtained were generally reassuring when compared with equivalent estimates derived from the LSS, although the estimates provided by the worker studies are still comparatively imprecise and statistical uncertainties would be reduced by more extensive data.

Somewhat disappointing, perhaps, is that a more substantial database of radiation workers from the USA has not been built upon the original efforts made in the late-1980s and early-1990s. In the first multi-site study in the USA, the three Department of Energy (DOE) sites of Hanford, Oak Ridge and Rocky Flats were included (Gilbert et al. 1989, 1993), and it was workers from these three sites who contributed to the first international study (IARC 1994). A decade later, the US component of the 15-country study consisted of workers from the three DOE sites of Hanford, Oak Ridge and Idaho Falls, and the combined workforces from 15 nuclear power stations (Cardis et al. 2005). A further decade later, the US sites included in INWORKS were the four DOE sites of Hanford, Oak Ridge, Idaho Falls and Savannah River, and the Portsmouth Naval Shipyard (Schubauer-Berigan et al. 2015), which cannot realistically be regarded as approaching comprehensive coverage of radiation workers in the USA, and is in contrast to the national databases in the UK (Muirhead et al. 2009) and France (Metz-Flamant et al. 2013) that contributed to INWORKS. This is unfortunate because the USA was the country hosting many of the early nuclear sites, and radiation exposures at some of these facilities were often at levels that would not be regarded as acceptable today, with thousands of workers accumulating doses exceeding 100 mGy (received over periods of years). Moreover, many of these early workers have now died, so potential information on radiation-related late effects has grown significantly in recent years, inviting powerful analyses of large datasets to provide meaningful risk estimates of adverse health effects.

In this Special Issue of *International Journal of Radiation Biology*, papers are published describing the Million Person Study (MPS) of US radiation workers and nuclear weapons test veterans, with some of these papers presenting results from initial analyses of the various component groups of the MPS (Boice et al. 2021). The MPS promises to provide a large and comprehensive database of those exposed occupationally to radiation in the USA, which is a welcome advance in increasing the power of studies of radiation workers, especially with the inclusion of substantial numbers of workers who have accumulated doses of a few hundred mGy during their employment. The MPS will make a significant contribution to the understanding of the risks to health arising from a series of many low doses received at a low dose-rate over extended periods.

However, it should not be thought that, even after nearly half a century of research, issues requiring resolution are absent from radiation worker studies, and these do need to be addressed. Nuclear worker studies have largely examined the impact of exposure to penetrating photon radiation from external sources because doses are comparatively easy to measure and available in records. Doses from intakes of radionuclides are not so easy to estimate, and have not been so widely used in worker studies, although there are exceptions (Gillies et al. 2017), and it is encouraging that significant efforts are being made in component groups of the MPS, such as the Los Alamos workforce (Boice et al. 2021), to assess internal doses from intakes of radionuclides. Even so, there are challenges in reconstructing internal doses from early monitoring data that have to be met. For example, at Sellafield in the UK, early urinalysis data have been found to be inadequate to derive plutonium doses sufficiently reliable for epidemiological purposes, and job-exposure matrices have had to be constructed instead (Riddell et al. 2019). A further problem is posed by neutron doses, which have been excluded from some worker studies, including INWORKS. Again, it is the early years of nuclear operations that present most difficulties because personal neutron monitoring was a rarity in these times (the neutron only having been discovered in 1932) often leading to an absence of neutron dose records and a need for dose reconstruction – some of these unrecorded neutron doses could have been large (Kite and Anderson 1996; Fix et al. 1997). Then there are other sources of radiation exposure that need to be considered, such as X-ray exposures during the annual medical examinations that were often required for employment as a radiation worker.

Although the photon doses that have been used in INWORKS are generally the most reliable in terms of accuracy, it is, once again, early records of individual photon doses that can be questionable. This was recognized at the outset of the nuclear worker studies, and an impressive review of dosimetry was conducted for the purposes of the first international study (Fix et al. 1997), followed by dosimetry reviews for the 15-country study (Thierry-Chef et al. 2007) and INWORKS (Thierry-Chef et al. 2015). Nonetheless, there remain aspects of early photon doses that need to be properly investigated, together with any impact upon the findings of epidemiological studies of workers, before full confidence can be had in the risk estimates produced by these studies (Wakeford 2018, 2021). Of particular concern are the upward adjustments made to early (before about 1960) annual recorded external doses at (at least, some) DOE facilities for the purposes of the US worker compensation program (Merwin et al. 2008) because some of these revisions are substantial – and it is notable that these adjustments were not deemed necessary for recorded external doses received in later years. I have argued elsewhere (Wakeford 2018, 2021) that the factors behind these upward adjustments need to be examined carefully to determine whether revisions to external doses from the early years of operations at nuclear facilities need to be made beyond any that have already been incorporated into the doses used in epidemiological studies. Undoubtedly, the dose revisions that are made for the purposes of worker compensation will be generous to the claimant (possibly substantially so), but the need to upwardly adjust early external doses, as outlined by Merwin et al. (2008), demands a close examination of the reasons behind these revisions to determine what might be required for the purposes of epidemiology.

As succinctly pointed out by Ellis et al. (2018) in the title of an earlier MPS paper: 'dosimetry is key to good epidemiology'. Considerable effort has been expended in the MPS to reconstruct organ/tissue-specific absorbed doses from available dose records (Bouville et al. 2015; Dauer et al. 2018; NCRP 2018, 2020; Yoder et al. 2021). It may be that those factors that led to the upward adjustments of early external doses presented by Merwin et al. (2008) are ultra-conservative, being made in the context of worker compensation, but it would be reassuring to conclude that the dosimetry assessments incorporate these factors to the extent necessary for epidemiological purposes. I have noted elsewhere (Wakeford 2018, 2021) that, on the face of it, the temporal distributions of annual external doses for the early years of operations at the DOE sites of Hanford and Savannah River look rather puzzling, but that there could be reasonable explanations. This is not a trivial issue because an underestimation of cumulative doses, particularly toward the higher end of the cumulative dose distribution, could lead to the slope of the dose-response being overestimated (Wakeford 2021). This may be something of nothing, but it would be satisfying to see this issue closed.

The MPS is a welcome advance in collecting, collating, and analyzing data for radiation workers in the USA, the country where the first large nuclear facilities were operated. This will be an important source of information on the health effects of low-level exposure to radiation, and knowledge will grow with increasing length of follow-up of these early workers. Let us hope that the USA will finally achieve a comprehensive database of radiation workers, equivalent to those in the UK and France.

Disclosure statement

The author is a member of the Technical Working Party of the UK Compensation Scheme for Radiation-Linked Diseases (CSRLD, http://www.csrld.org.uk) and receives a consultancy fee for this membership.

Funding

The author(s) reported there is no funding associated with the work featured in this article.

ORCID

Richard Wakeford http://orcid.org/0000-0002-2934-0987

References

Ashmore JP, Gentner NE, Osborne RV. 2010. Incomplete data on the Canadian cohort may have affected the results of the study by the International Agency for Research on Cancer on the radiogenic cancer risk among nuclear industry workers in 15 countries. J Radiol Prot. 30(2):121–129.

Beral V, Inskip H, Fraser P, Booth M, Coleman D, Rose G. 1985. Mortality of employees of the United Kingdom Atomic Energy Authority, 1946–1979. Br Med J (Clin Res Ed). 291(6493):440–447.

Beral V, Fraser P, Carpenter L, Booth M, Brown A, Rose G. 1988. Mortality of employees of the Atomic Weapons Establishment, 1951–82. BMJ. 297(6651):757–770.

Boice JD Jr, Bouville A, Dauer LT, Golden AP, Wakeford R. 2021. Introduction to the special issue on the US Million Person Study of health effects from low-level exposure to radiation. Int J Radiat Biol. 1–4. [online ahead of print]. doi: 10.1080/09553002.2021.1989906

Boice JD Jr, Cohen SS, Mumma MT, Golden AP, Howard SC, Girardi DJ, Ellis ED, Bellamy MB, Dauer LT, Samuels C, et al. 2021. Mortality among workers at the Los Alamos National Laboratory, 1943–2017. Int J Radiat Biol. 1–28. [online ahead of print]. doi: 10.1080/09553002.2021.1989906

Bouville A, Toohey RE, Boice JD Jr, Beck HL, Dauer LT, Eckerman KF, Hagemeyer D, Leggett RW, Mumma MT, Napier B, et al. 2015. Dose reconstruction for the million worker study: status and guidelines. Health Phys. 108(2):206–220.

Cardis E, Gilbert ES, Carpenter L, Howe G, Kato I, Armstrong BK, Beral V, Cowper G, Douglas A, Fix J, et al. 1995. Effects of low doses and low dose rates of external ionizing radiation: cancer mortality among nuclear industry workers in three countries. Radiat Res. 142(2):117–132.

Cardis E, Vrijheid M, Blettner M, Gilbert E, Hakama M, Hill C, Howe G, Kaldor J, Muirhead CR, Schubauer-Berigan M, et al. 2007. The 15-Country Collaborative Study of Cancer Risk among Radiation Workers in the Nuclear Industry: estimates of radiation-related cancer risks. Radiat Res. 167(4):396–416.

Cardis E, Vrijheid M, Blettner M, Gilbert E, Hakama M, Hill C, Howe G, Kaldor J, Muirhead CR, Schubauer-Berigan M, et al. 2005. Risk of cancer after low doses of ionising radiation: retrospective cohort study in 15 countries. BMJ. 331(7508):77.

Carpenter L, Higgins C, Douglas A, Fraser P, Beral V, Smith P. 1994. Combined analysis of mortality in three United Kingdom nuclear industry workforces, 1946–1988. Radiat Res. 138(2):224–238.

Checkoway H, Mathew RM, Shy CM, Watson JE Jr, Tankersley WG, Wolf SH, Smith JC, Fry SA. 1985. Radiation, work experience, and cause specific mortality among workers at an energy research laboratory. Br J Ind Med. 42(8):525–533.

Darby SC, Reissland JA. 1981. Low levels of ionizing radiation and cancer—are we underestimating the risk? J R Statist Soc A. 144(3): 298–315.

Dauer LT, Bouville A, Toohey RE, Boice JD Jr, Beck HL, Eckerman KF, Hagemeyer D, Leggett RW, Mumma MT, Napier B, et al. 2018. Dosimetry and uncertainty approaches for the million person study of low-dose radiation health effects: overview of the recommendations in NCRP Report No. 178. Int J Radiat Biol. 1–10. [online ahead of print]. doi: 10.1080/09553002.2018.1536299

Ellis ED, Boice JD Jr, Golden AP, Girardi DJ, Cohen SS, Mumma MT, Shore RE, Leggett RW, Kerr GD. 2018. Dosimetry is key to good epidemiology: workers at Mallinckrodt chemical works had seven different source exposures. Health Phys. 114(4):386–397.

Fix JJ, Salmon L, Cowper G, Cardis E. 1997. A retrospective evaluation of the dosimetry employed in an international combined epidemiological study. Radiat Prot Dosim. 74(1):39–53.

Gilbert ES, Marks S. 1979. An analysis of the mortality of workers in a nuclear facility. Radiat Res. 79(1):122–148.

Gilbert ES, Cragle DL, Wiggs LD. 1993. Updated analyses of combined mortality data for workers at the Hanford Site, Oak Ridge National Laboratory, and Rocky Flats Weapons Plant. Radiat Res. 136(3): 408–421.

Gilbert ES, Petersen GR, Buchanan JA. 1989. Mortality of workers at the Hanford site: 1945-1981. Health Phys. 56(1):11–25.

Gilbert ES, Fry SA, Wiggs LD, Voelz GL, Cragle DL, Petersen GR. 1989. Analyses of combined mortality data on workers at the Hanford Site, Oak Ridge National Laboratory, and Rocky Flats Nuclear Weapons Plant. Radiat Res. 120(1):19–35.

Gillies M, Kuznetsova I, Sokolnikov M, Haylock R, O'Hagan J, Tsareva Y, Labutina E. 2017. Lung cancer risk from plutonium: a pooled analysis of the Mayak and Sellafield worker cohorts. Radiat Res. 188(6):645–660.

Gribbin MA, Weeks JL, Howe GR. 1993. Cancer mortality (1956–1985) among male employees of Atomic Energy of Canada Limited with respect to occupational exposure to external low-linear-energy-transfer ionizing radiation. Radiat Res. 133(3):375–380.

Hutchison GB, MacMahon B, Jablon S, Land CE. 1979. Review of report by Mancuso, Stewart and Kneale of radiation exposure of Hanford workers. Health Phys. 37(2):207–220.

IARC. 1994. Direct estimates of cancer mortality due to low doses of ionising radiation: an international study. IARC Study Group on Cancer Risk among Nuclear Industry Workers. Lancet. 344: 1039–1043.

ICRP. 2007. The 2007 Recommendations of the International Commission on Radiological Protection. ICRP publication 103. Ann ICRP. 37:1–332.

Kendall GM, Muirhead CR, MacGibbon BH, O'Hagan JA, Conquest AJ, Goodill AA, Butland BK, Fell TP, Jackson DA, Webb MA. 1992. Mortality and occupational exposure to radiation: first analysis of the National Registry for Radiation Workers. BMJ. 304(6821): 220–225.

Kite AV, Anderson RW. 1996. An overview of retrospective occupational dosimetry at BNFL. In: Duftschmid KE, editor. IRPA9: 1996 International Congress on Radiation Protection. Proceedings. Vol. 3. Vienna: IRPA.

Leuraud K, Richardson DB, Cardis E, Daniels RD, Gillies M, O'Hagan JA, Hamra GB, Haylock R, Laurier D, Moissonnier M, et al. 2015. Ionising radiation and risk of death from leukaemia and lymphoma in radiation-monitored workers (INWORKS): an international cohort study. Lancet Haematol. 2(7):e276–e281.

Mancuso TF, Stewart A, Kneale G. 1977. Radiation exposures of Hanford workers dying from cancer and other causes. Health Phys. 33:369–385.

Merwin SE, Smith MH, Winslow RC, McCartney KA, Fix JJ, Taulbee TD, Macievic GL. 2008. External dose reconstruction under Part B of the energy employees compensation act. Health Phys. 95(1): 95–106.

Metz-Flamant C, Laurent O, Samson E, Caër-Lorho S, Acker A, Hubert D, Richardson DB, Laurier D. 2013. Mortality associated with chronic external radiation exposure in the French combined cohort of nuclear workers. Occup Environ Med. 70(9):630–638.

Muirhead CR, O'Hagan JA, Haylock RG, Phillipson MA, Willcock T, Berridge GL, Zhang W. 2009. Mortality and cancer incidence following occupational radiation exposure: third analysis of the National Registry for Radiation Workers. Br J Cancer. 100(1):206–212.

NCRP. 2018. NCRP Report No. 178. Deriving organ doses and their uncertainty for epidemiologic studies (with a focus on the One Million U.S. Workers and Veterans Study of low-dose radiation health effects). Bethesda (MD): National Council on Radiation Protection and Measurements.

NCRP. 2020. NCRP Commentary No. 30. Using personal monitoring data to derive organ doses for medical radiation workers, with a focus on lung. Bethesda (MD): National Council on Radiation Protection and Measurements.

Ozasa K, Cullings HM, Ohishi W, Hida A, Grant EJ. 2019. Epidemiological studies of atomic bomb radiation at the Radiation Effects Research Foundation. Int J Radiat Biol. 95(7):879–891.

Richardson DB, Cardis E, Daniels RD, Gillies M, O'Hagan JA, Hamra GB, Haylock R, Laurier D, Leuraud K, Moissonnier M, et al. 2015. Risk of cancer from occupational exposure to ionising radiation: retrospective cohort study of workers in France, the United Kingdom, and the United States (INWORKS). BMJ. 351:h5359.

Riddell A, Wakeford R, Liu H, O'Hagan J, MacGregor D, Agius R, Wilson C, Peace M, de Vocht F. 2019. Building a job-exposure matrix for early plutonium workers at the Sellafield nuclear site, United Kingdom. J Radiol Prot. 39(2):620–634.

Schubauer-Berigan MK, Daniels RD, Bertke SJ, Tseng CY, Richardson DB. 2015. Cancer Mortality through 2005 among a Pooled Cohort of U.S. Nuclear Workers Exposed to External Ionizing Radiation. Radiat Res. 183(6):620–631.

Smith PG, Douglas AJ. 1986. Mortality of workers at the Sellafield plant of British Nuclear Fuels. Br Med J (Clin Res Ed). 293(6551): 845–854.

Thierry-Chef I, Richardson DB, Daniels RD, Gillies M, Hamra GB, Haylock R, Kesminiene A, Laurier D, Leuraud K, Moissonnier M, et al. 2015. Dose estimation for a study of nuclear workers in France, the United Kingdom and the United States of America: methods for the International Nuclear Workers Study (INWORKS). Radiat Res. 183(6):632–642.

Thierry-Chef I, Marshall M, Fix JJ, Bermann F, Gilbert ES, Hacker C, Heinmiller B, Murray W, Pearce MS, Utterback D, et al. 2007. The 15-Country Collaborative Study of Cancer Risk among Radiation Workers in the Nuclear Industry: study of errors in dosimetry. Radiat Res. 167(4):380–395.

Wakeford R. 2014. Nuclear worker studies: promise and pitfalls. Br J Cancer. 110(1):1–3.

Wakeford R. 2018. The growing importance of radiation worker studies. Br J Cancer. 119(5):527–529.

Wakeford R. 2009. Radiation in the workplace-a review of studies of the risks of occupational exposure to ionising radiation. J Radiol Prot. 29(2A):A61–79.

Wakeford R. 2021. Overview of epidemiological studies of nuclear workers: opportunities, expectations, and limitations. J Radiol Prot. 41:1075–1092..

Wilkinson GS, Tietjen GL, Wiggs LD, Galke WA, Acquavella JF, Reyes M, Voelz GL, Waxweiler RJ. 1987. Mortality among plutonium and other radiation workers at a plutonium weapons facility. Am J Epidemiol. 125(2):231–250.

Yoder RC, Balter S, Boice JD Jr,Grogan H, Mumma M, Rothenberg LN, Passmore C, Vetter RJ, Dauer LT. 2021. Using personal monitoring data to derive organ doses for medical radiation workers in the Million Person Study – considerations regarding NCRP Commentary No. 30. J Radiol Prot. 41(1):118–128.

Richard Wakeford

Reviews

The Million Person Study, whence it came and why

John D. Boice, Sarah S. Cohen, Michael T. Mumma, and Elizabeth D. Ellis

ABSTRACT

Purpose: The study of low dose and low-dose rate exposure is of immeasurable value in understanding the possible range of health effects from prolonged exposures to radiation. The Million Person Study (MPS) of low-dose health effects was designed to evaluate radiation risks among healthy American workers and veterans who are more representative of today's populations than are the Japanese atomic bomb survivors exposed briefly to high-dose radiation in 1945. A million persons were needed for statistical reasons to evaluate low-dose and dose-rate effects, rare cancers, intakes of radioactive elements, and differences in risks between women and men.

Methods and materials: The MPS consists of five categories of workers and veterans exposed to radiation from 1939 to the present. The U.S. Department of Energy (DOE) Health and Mortality study began over 40 years ago and is the source of ~360,000 workers. Over 25 years ago, the National Cancer Institute (NCI) collaborated with the U.S. Nuclear Regulatory Commission (NRC) to effectively create a cohort of nuclear power plant workers (~150,000) and industrial radiographers (~130,000). For over 30 years, the Department of Defense (DoD) collected data on aboveground nuclear weapons test participants (~115,000). At the request of NCI in 1978, Landauer, Inc., (Glenwood, IL) saved their dosimetry databases which became the source of a cohort of ~250,000 medical and other workers.

Results: Overall, 29 individual cohorts comprise the MPS of which 21 have been or are under active study (~810,000 persons). The remaining eight cohorts (~190,000 persons) will be studied as resources become available. The MPS is a national effort with critical support from the NRC, DOE, National Aeronautics and Space Administration (NASA), DoD, NCI, the Centers for Disease Control and Prevention (CDC), the Environmental Protection Agency (EPA), Landauer, Inc., and national laboratories.

Conclusions: The MPS is designed to address the major unanswered question in radiation risk understanding: What is the level of health effects when exposure is gradual over time and not delivered briefly. The MPS will provide scientific understandings of prolonged exposure which will improve guidelines to protect workers and the public; improve compensation schemes for workers, veterans and the public; provide guidance for policy and decision makers; and provide evidence for or against the continued use of the linear nonthreshold dose-response model in radiation protection.

Introduction

The Million Person Study (MPS) of low-dose health effects of radiation-exposed workers and veterans was designed to evaluate the health consequences of exposure to radiation received gradually over time and not acutely (Boice 2015). The MPS (Boice 2012a; NCRP 2018a) was conceived over 25 years ago following a National Cancer Institute (NCI) request to the U.S. Nuclear Regulatory Commission (NRC) to consider creating a radiation worker registry by modifying the NRC requirement as to how licensees report radiation doses received by individual workers (Muirhead et al. 1996; Boice 2012a; Hagemeyer et al. 2018). Over time, the MPS (Boice 2015; NCRP 2018a) was formed with the inclusion of 115,000 atomic veterans who participated in above-ground nuclear weapons testing (Boice 2012b; Till et al. 2014; Till, Beck, Aanenson, et al. 2018; Till, Beck, Boice et al. 2018; Beck et al. 2017; Boice, 2014b, 2014c; Caldwell et al. 2016), 360,000 Department of Energy (DOE) workers employed during and after the Manhattan Project (Boice, Cohen, et al. 2006, 2014; Boice 2013a; Ellis and Girardi et al. 2018), 250,000 early radiologists and other medical workers (Yoder et al. 2018), 150,000 nuclear power plant (NPP) workers (Boice 2013b, 2016a; Golden et al. 2018; Boice et al. 2018; NCRP 2018b; Boice, Ellis et al. 2019), and 130,000 industrial radiographers (NCRP 2018a, 2018b; Boice, Ellis et al. 2019). The follow-up started as early as 1939 for industrial radiographers.

The overriding goal of the MPS is to estimate the risk of organ-specific cancers from radiation doses received at low dose rates over the course of years, for which there is much interest and much uncertainty (Jacob et al. 2009). These risk estimates from healthy American workers and veterans are deemed more valid for American workers and populations than those from the atomic bomb survivor study which involved a brief, acute exposure in a Japanese population in 1945 with a profile of cancer types and other health conditions that differ from the U.S. population. Initial MPS publications stressed the importance of dosimetry for the DOE facility workers (Leggett et al. 2005; Boice, Cohen, et al. 2006; Ellis, Boice, et al. 2018), the atomic veterans (Till et al. 2014; Till, Beck, Aanenson, et al. 2018; Beck et al. 2017), and for all cohorts studied (Boice, 2013c, 2014b, 2016a; Bouville et al. 2015; NCRP 2018a). It is a credit to the Atomic Energy Commission (now the DOE) for having industrial hygiene programs in the 1940s for worker protection and the foresight to archive the comprehensive records that could be used for research applicable to problems faced today and will be faced tomorrow (Ellis, Girardi, et al. 2018). Statistical issues of the uncertainties in estimating radiation doses and risks are being addressed (NCRP 2012, 2018a; UNSCEAR 2014; Stram et al. 2015).

The MPS is poised to address the public health issues on radiation facing the nation today. Ever-increasing population exposures are accompanied by public concerns about medical exposures (e.g. computed tomography (CT) imaging), nuclear waste and facility cleanups, nuclear power generation, certain occupations (e.g. interventional radiologists), increased air travel and cosmic-ray exposures, technologically enhanced naturally occurring radioactive materials (TENORM) in the environment from hydraulic fracturing, reactor accidents (e.g. at the Daiichi facilities in Fukushima, Japan; O'Brien 2012), and the possibility of terrorist events with dirty bombs or improvised nuclear devises. The MPS will inform regulations (ICRP 2007; NCRP 2018c), improve methods of dose reconstruction, provide emergency response guidance, address medical radiation uses, help set clean up levels after a major nuclear incident, support TENORM recommendations, contribute to compensation schemes (NRC/NA 2003; Neton et al. 2008; Neton 2014), guide occupational and environmental limits, and support radiation guidance for astronauts embarking to Mars (Boice 2017a, 2017c, 2019; NCRP 2014, 2016). Further, differences between men and women in their response to radiation can be fully examined (Boice, Ellis et al. 2019), and noncancer outcomes such as cardiovascular disease (NCRP 2018b) and neurological disorders (NCRP 2016; Boice 2017c) can be examined comprehensively.

Publications to date have included evaluations of Rocketdyne (Atomics International) workers (Boice et al. 2011), Mound workers (Boice et al. 2014; Boice 2017a; NCRP 2018b; Boice, Ellis et al. 2019), Mallinckrodt workers (Boice et al. 2018; Ellis, Boice, et al. 2018; Boice, Ellis et al. 2019; Golden et al. 2019), atomic veterans (Caldwell et al. 2016; Boice et al. 2017a; Till, Beck, Aanenson, et al. 2018; NCRP 2018b; Boice, Ellis et al. 2019), nuclear power plant workers (Golden et al. 2018; NCRP 2018b; Boice, Ellis et al. 2019; Boice et al. 2019a), and industrial radiographers (NCRP 2018b; Boice, Ellis et al. 2019; Boice et al. 2019b). The necessary statistical power to obtain precise risk estimates over a range of low dose exposures, particularly for many cancer sites that are relatively rare, will come from the combination of the study cohorts, similar to the large-scale INWORKS study of nuclear workers from three countries: France, the United Kingdom and the U.S.A (Laurier et al. 2017). The earlier MPS publications, while informative, were not sufficiently powerful to address radiation risks in the low-dose domain or evaluate the consistency or lack of consistency with the LNT model[1] used for radiation protection (NCRP 2001, 2018b, 2018c; ICRP 2007; Boice 2017b; Shore et al. 2018). In contrast, the great statistical power, precision, and validity of the risk estimates within the complete MPS will be due to: the large numbers, the broad dose distribution, standard methods for population selection, thorough vital status tracing (Mumma et al. 2018), and comprehensive dose reconstruction applied across all exposed cohorts (Dauer, Bouville, et al. 2018; NCRP 2018a). The dose distribution is broader than for most occupational studies, with more workers with cumulative doses >50 mGy than reported among atomic bomb survivors. Brief descriptions of the five major components of the MPS are provided below: DOE workers, nuclear weapons test participants,

Table 1. Summary of the epidemiologic cohorts and main sources of exposure for the five major categories of workers within the Million Person Study (as of April 2017)[a].

MPS Cohort	Main sources of exposure
NPP workers first employed from 1957 through 1984 (~145,000)	External: gamma-ray-emitting fission products (e.g. ^{137}Cs) and activation products (e.g. ^{58}Co, ^{60}Co); in some cases, neutrons Internal: few documented cases of ingestion or inhalation of radionuclides
Industrial radiographers/nondestructive testing (~126,000)	External: 192Ir, 60Co and 75Se systems; x-ray units Internal: none
Nuclear weapons test participants (~115,000 atomic veterans)	External: gamma-ray-emitting fission products (e.g. 140Ba-140La, 95Zr-95Nb) Internal: some intakes of radionuclides (e.g. 131I, 133I; 239Pu)
DOE workers (~360,000, many last followed 15–20 years ago)	External: gamma-ray-emitting radioactive materials; in some cases, neutrons Internal: intakes of uranium and plutonium isotopes (e.g. Rocketdyne workers); intakes of ^{210}Po, ^{238}Pu, ^{3}H (tritium) (e.g. Mound workers); intakes of ^{235}U and ^{226}Ra (e.g. Mallinckrodt workers)
Medical workers (~170,000) (includes early radiologists, radiologic technologists, nuclear medicine technologists, radiation oncologists, cardiologists, interventional radiologists, and other medical specialties with potential for relatively high doses)	External: diagnostic x-ray production devices, x-ray or gamma-ray teletherapy units, ^{99m}Tc and other medical radiopharmaceuticals, brachytherapy sources such as radium Internal: possible inadvertent intakes of radionuclides may occur during nuclear-medicine procedures

[a]Adapted from Table 2.1 in NCRP (2018a). Note that because of the dynamic nature of epidemiologic research for the Million Person Study over the years, the population numbers throughout may differ but were current at the time of the specific publication or report.

nuclear power plant workers, industrial radiographers and medical, and other radiation workers (Table 1).

The five major components of the MPS include a total of 29 individual cohorts. Some of the individual cohorts, such as the study of nuclear weapons test participants, include multiple sub-cohorts. The status of the 29 cohorts under study is summarized in Table 2. Each study cohort is placed into one of five categories: completed ($n = 127{,}000$), completed with additional analyses ongoing ($n = 271{,}000$), ongoing and to complete in 1–3 years ($n = 193{,}000$), recently begun and to complete in 2–3 years ($n = 70{,}000$), to begin in stepwise approach as resources become available ($n = 122{,}000$). Our goal is to complete all cohort studies within the next 3–4 years. If resources are not sufficient, the first mortality follow-up for the MPS will include at a minimum ~810,000 workers and veterans, and the second mortality follow-up would include all 29 individual cohorts and over 1,000,000 persons.

Study populations of workers and veterans

The sources of population and dosimetry data and how they were incorporated into the MPS are described in detail in NCRP Report 178 (2018a) and elsewhere (Till et al. 2014; Hagemeyer et al. 2018; Dauer, Bouville, et al. 2018; Ellis, Boice, et al. 2018; Yoder et al. 2018). Briefly, the DOE Health and Mortality study began over 40 years and is the source of ~360,000 DOE workers (Ellis, Girardi, et al. 2018). Over 25 years ago, the NCI collaborated with the U.S. NRC to change their reporting requirements and effectively created a database to study nuclear power plant workers ($n = \sim150{,}000$) and industrial radiographers ($n = \sim130{,}000$) (Hagemeyer et al. 2018). For over 30 years, the Department of Defense (DoD) collected data that facilitated studies of aboveground nuclear weapons test participants (~115,000) (Till et al. 2014). At the request of NCI in 1978, Landauer, Inc. saved their dosimetry databases which became the source to study ~170,000 medical and other workers (Yoder et al. 2018).

DOE workers

The U.S.A. was one of the first countries to develop both nuclear weapons and nuclear power reactors to generate electricity and to fuel military ships and submarines. Within the MPS, epidemiologic studies are being conducted or are planned of ~360,000 workers employed by DOE and its contractors since 1940. While there are many DOE facilities, the MPS focuses on the larger and previously studied DOE facilities for both efficiency and cost savings. The MPS has followed the same classification scheme used by the DOE Health and Mortality Studies program of defining facilities as to uranium or plutonium activities (Shy 1982; Lushbaugh et al. 1983; Fry et al. 1985; CEDR 2018; Ellis, Girardi, et al. 2018). About 400,000 workers in the DOE worker database could conceivably be studied; the MPS focuses on ~360,000 workers: ~240,000 uranium workers (Table 3), and ~120,000 plutonium workers (Table 4). Sites that had not been comprehensively studied are excluded, such as the Sandia National Laboratory, for which a follow-up would be both challenging and expensive. Because of unique exposure or worker characteristics, some facilities not previously studied were purposely included: the Tennessee Eastman Corporation (TEC) female workers employed in Oak Ridge

Table 2. Status of the epidemiologic cohorts under study in the Million Person Study as of October 2018 and updated.

Cohort	Relevant references	Approximate number of workers
Completed		
Rocketdyne	Boice et al. (2011)	5801
Mound	Boice et al. (2014), (2017), NCRP (2018b), Boice, Ellis et al. (2019)	4101
Mallinckrodt	Golden et al. (2019)	2514
Atomic Veterans (8 Series)	Caldwell et al. (2016), NCRP 2018b; Boice, Ellis et al. 2019; Boice, Mumma, et al. (2019)	115,000
Completed with additional analyses ongoing		
Nuclear power plant workers	Golden et al. (2018), NCRP (2018b); Boice, Ellis et al. (2019); Boice et al. (2019a)	145,000
Industrial radiographers	NCRP (2018b), Boice, Ellis et al. (2019), Boice et al. (2019b)	126,000
Ongoing to complete in 1–3 years		
Middlesex		700
Los Alamos National Laboratory		15,727
Fernald		6000
Medical radiation workers		170,000
Recently begun		
Rocky Flats		17,000
Linde		1500
Hanford		32,643
Tennessee Eastman Corporation (Y-12) – women		20,423
Planned to begin shortly		
Oak Ridge Gaseous Diffusion Plant (K-25)		49,749
Portsmouth Gaseous Diffusion Plant		9215
Planned to begin as resources become available		
Tennessee Eastman Corporation (Y-12) – men		41,107
Savannah River Site		18,883
Oak Ridge National Laboratory, X-10 Graphite Reactor		28,528
Union Carbide Corporation Nuclear Division, Oak Ridge (Y-12)		26,059
Paducah		6526
Pantex		1031
Total planned		122,134
Overall total		1,028,301

Table 3. Prior studies of workers at DOE uranium facilities[a].

Worker Cohort	Number in previous studies	Relevant publications	Last follow-up	Total dead	Comment	Total in database
Oak Ridge segment	106,020	Frome et al. (1997)	2005	27,982	All Oak Ridge	
K-25	35,712	Dupree et al. (1994)	1989	12,848	White males	49,794
X-10	6348	Gilbert, Cragle, et al. (1993), Schubauer-Berigan et al. (2015)	1984	1246	Recent follow-up combined	28,528
Y-12 (TEC)[b] (males)	18,348	Polednak and Frome (1981)	1977	5394	White males	41,107
Y-12 (TEC) (females)	N/A	None, first time studied	N/A	15,429	Females, white and black; never studied, in process	20,423
Y-12 (Union Carbide Corporation Nuclear Division)	8116	Loomis and Wolf (1996)	1990	1861	Electromagnetic enrichment	26,059
Portsmouth Gaseous Diffusion	9215	NIOSH (2001), Yiin et al. (2017)	1991	1275	Enriched uranium	9308
Paducah	6759	Chan et al. (2010)	2003	1638	Gaseous diffusion	5731
Manhattan Engineer District (uranium)						
Niagara Frontier						
Harshaw, ElectroMet, Bethlehem	N/A	None	N/A	N/A	N/A	1144
Linde	995	Dupree et al. (1987)	1979	429	Employed 1943–1949	1551
Middlesex	N/A	None, first time studied	N/A	N/A	Never studied, in process	670
Fernald Feed Materials Production Center	4014	Cragle et al. (1995), Ritz (1999), Silver et al. (2013)	1989	1064	White males	7337
Mallinckrodt Chemical Works	2514	Dupree-Ellis et al. (2000)	1993	1013	White males, in process	3272
Savannah River Site (uranium processing)	18,883	Cragle et al. (1988), Richardson et al. (2007)	2002	5096	Internal exposure to uranium, tritium	21,509
Pantex (weapons assembly)	4668	Acquavella et al. (1985), Silver et al. (2005)	1995	1031	Minimum uranium, possible control	12,670
Rocketdyne (Atomics International)	5801	Boice, Cohen, et al. (2006), Boice, Leggett, et al. (2006), Boice et al. (2011)	1999	1468	Comprehensive dosimetry	46,970
Sum of unique workers				62,379		240,000[c]

[a]Adapted from Table 4.1 in NCRP (2018a).
[b]TEC: Tennessee Eastman Corporation
[c]Because workers may be in more than one cohort, the individual sites sum to more than the total which is for unique workers; 240,000 is an approximate figure.

1943–1947 [the so-called girls of the atomic city (Kiernan 2013)] and the workers at the Middlesex Sampling Plant (Middlesex, New Jersey) with unique exposures to pitchblende dust (Eisenbud 1975).

Atomic veterans

Approximately 250,000 U.S. military personnel participated in aboveground nuclear weapons testing between 1945 and 1962. Previous studies were conducted on participants at seven of the test series. An eighth series, TRINITY, was added to these previous seven making up the Eight Series Study. The study population was obtained from databases within the DoD Nuclear Test Personnel Review (NTPR) program (NRC/NA 2003) coupled with epidemiologic research investigations conducted by the Medical Follow Up Agency (Johnson et al. 1996, 1997; Thaul et al. 2000) and the Veterans Affairs (VA) (Watanabe et al. 1995; Dalager et al. 2000). These test series were selected because of the availability of high-quality records of personnel and radiation dosimetry and indications that radiation exposures were higher than at other tests series. Notable scientists in the study include J. Robert Oppenheimer, Enrico Fermi, and Hans Bethe, who were at the first nuclear test (TRINITY) in 1945 at Alamogordo, New Mexico.

There were 98 weapon detonations (or shots) within the eight test series (DOE 2000). The estimation of radiation was facilitated by the Nuclear Test Review Information System (*NuTRIS*) maintained by the U.S. Defense Threat Reduction Agency (DTRA), DoD which included film badge readings (Till et al. 2014; Till, Beck, Aanenson, et al. 2018; Beck et al. 2017). Table 5 shows the number of participants at each of the eight nuclear weapons test series, the number from each military service, and other demographic characteristics.

The tracing and determination of vital status among these veterans were challenging because the social security number was not used as the military identification number until 1968 and thus was not readily available. We relied heavily on mortality linkages based on military identification number and name and date of birth. The Department of VA BIRLS system (Beneficiary Identification Record Location Subsystem) was a primary source of vital status information (Fisher et al. 1995), supplemented with other sources such as the Social Security Administration records and the National Death Index (NDI) (Mumma et al. 2018). The VA Patient Treatment File was also of value (Fisher et al. 1995; Boyko et al. 2000).

Publications to date include a second follow-up, with collaborators at the Centers for Disease Control and Prevention (CDC), of the 1979 SMOKY shot within the PLUMBBOB test series where an excess of leukemia had led to future studies and compensation programs (Caldwell et al. 1983, 2016); detailed dose reconstructions (Till et al. 2014; Till, Beck, Aanenson, et al. 2018; Beck et al. 2017); an evaluation

Table 4. Studies of workers at DOE plutonium facilities[a].

Worker cohort	Number in previous studies	Relevant Publications	Last follow-up	Number dead	Total in database[b]
Hanford	32,643	Peterson et al. (1990), Gilbert, Cragle, et al. (1993), Gilbert, Omohundro, et al. (1993), Wing et al. (2004), Wing and Richardson (2005)	1986	9452	56,688
Los Alamos	15,727	Gilbert, Cragle, et al. (1993), Wiggs et al. (1994), Wilkinson et al. (1987)	1990	3196	23,288
Mound	4101	Boice et al. (2014), Wiggs, Cox-DeVore, Voelz, et al. (1991), Wiggs, Cox-DeVore CA, Wilkinson (1991); Boice et al. (2017), NCRP (2018b); Boice, Ellis et al. 2019)	2009	3686	7270
Rocky Flats	5413	Brown et al. (2004), Gilbert, Cragle, et al. (1993)	1979	409	53,033
Sandia National Lab	N/A	None	N/A	N/A	24,685
Total	–	–	–	55,000[c]	120,000

[a]Adapted from Table 4.2 in NCRP (2018a).
[b]Because workers may be in more than one cohort, the individual sites sum to more than the total which is for unique workers; 120,000 is an approximate number.
[c]Number found to have died during the pilot and previous investigations.

of mesothelioma among sailors at the Pacific Proving Grounds linked to work in boiler rooms and other areas with high likelihood of asbestos exposures (Till, Beck, Boice, et al. 2018); and an evaluation of ischemic heart disease (Boice 2017a; NCRP 2018b).

Also, not all exposures were low; e.g. exposures greater than 300 mSv occurred during the 1954 Bravo test at Operation Castle, where the wind blew the wrong way, and personnel stationed at Rongerik Atoll were exposed to direct fallout before they were evacuated (DTRA 2015). A special study was conducted by the NCI that focused on the living high-dose servicemen (>200 mSv) at the Bravo test or the Nevada test site to compare three forms of dose estimation: (1) historical personnel monitoring data; (2) dose reconstruction based on detailed scenarios; and (3) two chromosome aberration assays and a telomere length assay (Simon et al. 2019).

Rarely can an occupational study address personal medical x-ray procedures, such as CT that are so frequent today (although much less so in the years covered by this study). We were able to link to the electronic records of the VA medical care system and evaluated radiological procedures for all veterans with leukemia and lung cancer and a sample of the cohort (Fisher et al. 1995; Boyko et al. 2000). Preliminary findings indicated that the frequency of CT examinations was the same between leukemia cases with the cohort sample and did not differ by years prior to death. Atomic veterans who died of lung cancer, however, had more CT examinations than the cohort sample, but the increased numbers were entirely due to procedures performed within 1–2 years of death, indicating that they were conducted for the purposes of diagnosis and treatment planning.

A unique population of 37,000 military personnel who participated in the underground nuclear weapons tests from 1962 to 1992 could be studied where the potential for exposure includes radioactive dust and gas venting such as occurred at the Baneberry test in 1970. Cancer incidence and noncancer diseases could also be evaluated for all atomic veterans who reach the age of 65 and were eligible for Medicare (Rector et al. 2004). A feasibility effort was conducted but at this time there are no plans to initiate a follow-up of this population.

The study of atomic veterans provides a service to the veterans and their families in obtaining sound information on the possible health risks related to service to our country. It also is relevant to understanding the consequences of possible terrorist attacks with dirty bombs or improvised nuclear devices, since the nuclear exposures in these circumstances are similar to those from weapons testing. It is the relevant population for evaluating compensation claims and radiation risks among atomic veterans.

Nuclear power plant workers

By law, NRC is required to record radiation dosimetry information from its licensees to assure that they are in compliance with existing regulations. These data have been collected since the late 1950s, but it was not until 1994 that the reporting requirements changed and it was possible to make these data available for epidemiologic study (Hagemeyer et al. 2018; NCRP 2018a). In 1986, Drs. Gilbert Beebe and John Boice, both at the NCI, composed a letter that was sent by NCI Director Vincent DeVita, MD, to NRC Chairman Lando Zech requesting the changing of the reporting requirements in order to develop a registry that would be valuable for health studies. In 1991, a positive response was received from NRC Chairman Kenneth Carr to NCI Director Samuel Broder. In 1994, as part of the implementation of the 1991 revisions of 10 CFR 20 'Standards for Protection Against Radiation,' the NRC began requiring annual radiation exposure records for every monitored worker (NRC 1992). In 1994 Dr. Boice urged the voluntary reporting of additional occupational radiation exposure data so that a high-quality radiation worker registry might be created for epidemiologic studies. In 1994 NRC (1994) sent a 'generic letter' to licensed utilities requesting that they report voluntarily the career doses of current and past employees; the utilities responded favorably, and a registry of radiation workers suitable for epidemiologic study was created. Twenty years later, the REIRS registry (ORISE 2011; REIRS 2018) became the cornerstone for identifying the ~150,000 early nuclear utility workers (1957–1984) being studied today as part of the MPS (Boice, 2013b,

2016a, 2017a; Boice et al. 2018, 2019a; Dauer, Bouville, et al. 2018; NCRP 2018a).

In 1957 the town of Moorpark, California, became the first in the country to be powered (briefly) by electricity produced from a nuclear reactor. Later that year, the Shippingport Atomic Power Plant became the first commercial nuclear power plant in the U.S.A. There have been over 600,000 nuclear power plant workers over the years, of which ~150,000 were selected for study on the basis of NRC REIRS files, supplemented by dosimetry records from Landauer, Inc. Feasibility studies were conducted utilizing NRC, Landauer, Inc., and utility records (Jablon and Boice 1993; Muirhead et al. 1996), but the current REIRS database now supersedes all previous approaches to conduct research studies (NCRP 2018a, 2018b; Boice, Ellis et al. 2019). Additional records from Landauer include 2880 rolls of microfilm of early workers (1954–1979) which have been used in validation studies (see below) and could be tapped to expand and enhance the early worker data if resources became available.

Early regulations allowed workers to receive up to 30 mSv per quarter (120 mSv y^{-1}), and although rare, some workers did receive high cumulative exposures (Hall et al. 2009). A broad distribution of cumulative radiation doses found that 28% of the workers in the current study have cumulative doses >50 mSv which will enhance the statistical power of the study to detect any associated increases in cancer or other causes of mortality should they have occurred (NCRP 2018a). A study of current workers in the nuclear industry would not be informative because the individual doses are so low; indeed they are lower than doses received from ubiquitous natural background radiation and lower than those received from medical radiation exposures. From 1973 to 2006, the average annual measurable dose per worker was reduced from 6.6 to 1.4 mSv (Blevins and Andersen 2011).

To verify the accuracy of the reported film badge doses for workers monitored 1950–1979, 27 workers were selected, and their film dosimeters (commercial product name: Gardray) were retrieved from archive storage in salt mines in Hutchinson, Kansas (Underground Vaults & Storage, Salt Mine). Twenty-seven dosimeters (and control dosimeters which were stored alongside the worker film badges) were reanalyzed using current Macbeth TD 931 reader and compared with their past recorded film badge doses. The actual reports were found by searching 2880 rolls of microfilm from 1954 to 1979. Overall, for film badge readings in the MPS sample, the accuracy of the historical measurement data and dose calculation to current readings was within 10% thus verifying the accuracy of the reported doses used for early workers understudy (Kirr et al. 2016; Kirr and Passmore 2016).

Industrial radiographers

Industrial radiographic nondestructive testing typically utilizes ^{192}Ir and ^{60}Co sources (Bouville et al. 2015; NRC 2017; NCRP 2018a), and thus radiographers receive external irradiation, generally in an anterior-posterior geometry. Similar to the cohort of nuclear power plant workers described above, information on annual recorded dose was collected for ~130,000 industrial radiographers within the NRC Radiation Exposure Information and Reporting System (REIRS 2018) and the Landauer, Inc. dosimetry database (NRC 1992, 1994, 2018; Muirhead et al. 1996; Anzenberg et al. 2010). The average cumulative recorded dose for these industrial radiographers is ~20 mGy, with 10% receiving a cumulative recorded dose >50 mGy (NCRP 2018a). Over 32,000 of the industrial radiographers are currently known to have worked in naval shipyards and analyses suggest a high risk for mesothelioma likely attributable to asbestos exposure (Mumma et al. 2019). Approaches to follow-up, dose reconstruction, mortality ascertainment, and analyses were equivalent to those for the study of NPP workers. A unique feature of the MPS being utilized in the NPP and industrial radiographer studies is the use of census block group data for education as a proxy measure of socioeconomic status which is included in the statistical models to control for potential confounding by smoking or other lifestyle factors (Cohen et al. 2018). Preliminary analyses for leukemia, other than chronic lymphocytic leukemia, and lung cancer have been conducted (NCRP 2018b; Boice, Ellis et al. 2019; Boice et al. 2019b). The industrial radiographers include nearly 200 leukemia cases, in comparison with the 98 cases among adult male atomic bomb survivors.

Medical radiation workers

The Medical Radiation Workers in the MPS include radiologists, radiologic technologists, radiopharmaceutical workers, interventionalists, cardiologists, nuclear medicine workers, physicists, and nurses (Bouville et al. 2015; NCRP 2018a; Yoder et al. 2018). Approximately 170,000 persons were selected from an eligible population of 1.71 million medical workers identified within the Landauer, Inc. dosimetry database. The database was created in the 1980s to assist with an epidemiology study of radiological technologists being conducted by the Radiation Epidemiology Branch of the NCI (Boice et al. 1992; Simon et al. 2006). The database contains annual doses for the years 1977 and later; cumulative doses up to 1977 for workers who had coverage with Landauer before 1977; and annual doses for the years 1960–1976 can be obtained through manual examination of copies of dosimetry reports for many workers available on 2880 rolls of microfilm. The estimation of radiation doses among medical workers over the years has been challenging but improving (Linet et al. 2010). NCRP has convened Scientific Committee 6–11 to comprehensively address the dosimetry issues associated with the medical radiation workers in the MPS (Yoder et al. 2018; NCRP 2018a).

The medical radiation workers represent the largest occupational cohort in the MPS and include large numbers of women that will allow a statistically powerful evaluation of any sex-specific radiation risks. Studies of Japanese atomic bomb survivors report an approximately three times greater radiation risk of lung cancer among females compared with males (Ozasa et al. 2012; NCRP 2014; Cahoon et al. 2017; Boice 2019; Boice, Ellis, et al. 2019). The National Aeronautics and Space Administration (NASA) provides

Table 5. Number of nuclear weapons test participants at each of the eight test series by military service. These tests involved 98 bomb detonations[a].

Test series	Year	Test site	No. of tests	Air force	Army	Marine corps	Navy	Total
CROSSROADS	1946	Pacific	2	0	3395	551	39,188	43,134
GREENHOUSE	1951	Pacific	4	2442	1548	70	3854	7914
UPSHOT-KNOTHOLE	1953	Nevada	11	2175	13,401	2256	886	18,718
CASTLE	1954	Pacific	6	2763	1644	306	11,918	16,631
REDWING	1956	Pacific	17	2976	1708	250	6993	11,927
PLUMBBOB	1957	Nevada	23	2216	7052	2,120	601	11,989
HARDTACK	1968	Nevada	34	3476	1535	187	9487	14,685
TRINITY	1945	New Mexico	1		726			726
Total	–	–	98	16,048	31,009	5740	72,927	125,724

[a]Seven military cohorts (test series) had been carefully constructed and studied by the Medical Follow-Up Agency (MFUA) (Robinette et al. 1985; Johnson et al. 1996, 1997; Thaul et al. 2000) and the U.S. Department of Veterans Affairs (VA) (Watanabe et al. 1995; Dalager et al. 2000). These cohorts, along with the TRINTY test, formed the basis of the study population.

support for the medical worker study because 1) they are interested in protracted rather than acute radiation and 2) since the NASA protection standards for astronauts are based on individual lifetime risk projections, sex-specific differences limit the time women can spend in space, including on long-term missions to Mars (NCRP 2014; Boice 2016b, 2017a, 2017c, 2019; Boice et al. 2018; Boice, Ellis, et al. 2019).

Landauer, Inc. has provided radiation dosimetry services since 1953 for U.S. medical, nuclear, and other facilities. A computerized database of millions of workers is maintained which formed the source of the medical radiation worker cohort. An initial cohort of 1.71 million workers first monitored before 1995 was assembled. To reduce this size to a more cost-efficient design without losing statistical power, a 50% random sample of workers with cumulative recorded doses between 10 and 50 mSv and a 2% random sample of workers with doses less than 10 mSv were taken. All workers having cumulative recorded doses equal to or greater than 50 mSv were selected (Table 6). The broad dose distribution is seen in Table 6 with 36% ($n \sim 62{,}000$) of the cohort having cumulative recorded doses over 50 mSv, 16% ($n \sim 27{,}000$) over 100 mSv, and 4% ($n \sim 6{,}600$) over 250 mSv. For comparison, there were 18,443 Japanese survivors of the atomic bombings with exposures >100 mSv. The large numbers and broad range of doses in the MPS indicate substantial statistical power to detect any late-occurring health effects and to distinguish any differences associated with radiation exposures experienced gradually over time compared with acute exposures for individual cancer sites.

Other – U.S. Navy submariners

Navy personnel who served on nuclear submarines are the most important population currently not included in the MPS. A third follow-up could be conducted of 85,498 nuclear submariners who were previously studied by Yale University (Charpentier et al. 1993) and New York University (Shore et al. 2001; Friedman-Jimenez 2003a, 2003b; Mueller et al. 2018). In this study, Navy enlisted men who served on nuclear submarines sometime between 1969 and 1982 were followed through 1995. The average lifetime badge reading was 5.2 mSv with 1.4% over 50 mSv. Most sailors were alive at last follow-up and only 3263 (3.8%) had died. The follow-up was too short to provide a powerful evaluation of possible effects. An extended follow-up of 22 years, through 2018, would increase the statistical ability to detect a radiation effect, had there been one. To further increase statistical power and to improve the study design, the study could be expanded to include submariners who served prior to 1969 (the Nautilus was commissioned in 1954) and to include officers. To date, over 100,000 submariners have been identified who served from 1954 through 1984. Between 10-15% were officers. Discussions are ongoing with U.S. Navy personnel concerning permission to conduct a new follow-up of the submariner population and, if permitted, we would include it within the MPS program of studies.

National effort

The DOE had for many years supported low-dose radiation research (Hall et al. 2009). The DOE funded an important pilot study that proved the feasibility of conducting the MPS (Boice 2012a); it provided additional support with interagency collaborations (NASA, NRC, Environmental Protection Agency (EPA)) and has recently renewed its support for the DOE worker studies with a 5-year grant to the NCRP. Further, the DOE provided access to the databases essential for study conduct. The vision to conduct a study of the early U.S. radiation workers began 30 years ago with requests to the U.S. NRC to change its reporting requirements to facilitate possible future study of licensee workers. These changes were made and the NRC remains one of the major supporters of the study. NASA also is a strong supporter of the MPS, first through an interagency agreement with DOE and now directly to National Council on Radiation Protection and Measurements (NCRP). Over the years, the CDC continued to provide foundational support essential for the study conduct. The U.S. EPA provided support through the DOE interagency agreement as well as indirect support via the Oak Ridge National Laboratory (ORNL). The NCI awarded a five-year grant to Vanderbilt University for the study of atomic veterans, with support and cooperation from the U.S. DoD, the Defense Threat Reduction Agency (DTRA), and VA. Recently the Naval Sea Systems Command awarded a five year contract for research focusing on naval personnel. National laboratory collaborators include ORNL, the Los Alamos National Laboratory (LANL) as well as the Oak Ridge Institute for Science and Education (ORISE). Other academic involvement comes from Vanderbilt University, the University of Southern California, Harvard University, and the Fred Hutchison Cancer Research Center. Private-sector contributors and collaborators are from the

International Epidemiology Institute, the Risk Assessment Corporation, Landauer, Inc., and Oak Ridge Associated Universities (ORAU).

The importance of the MPS and the continued funding for research in the U.S.A. was covered in the recent U.S. Government Accountability Office (GAO) Report on Low Dose Radiation: Interagency Collaboration on Planning Research Could Improve Information on Health Effects (GAO 2017; Davis et al. 2019). This Report highlighted the need for federal funding of low-level radiation research, extensively described the funding history of the MPS, and reported discussions with various government agencies on the value of ongoing MPS research to their missions. Congress recently supported the MPS by directing the DOE to provide research funds for the MPS of workers and veterans in recent appropriation bills.

The scientific understandings obtained also will be used in setting standards to protect workers and the public, and in improving compensation schemes for workers (Neton 2014), veterans (NRC/NA 2003) and civilians who were on-site participants at above-ground nuclear weapons tests, down-winders who lived in proximity to the weapons testing, and underground miners who worked in uranium mines (NA/NRC 2005).

In brief, for claimants requesting compensation for cancer and diseases they believe were linked to their prior radiation exposure, extensive dose reconstructions are often required. The estimated radiation doses and characteristics of each claimant are then evaluated as to whether or not the dose was the contributing cause, i.e. more likely than not. The procedures to estimate individual risk and a so-called 'probability of causation' (more correctly 'assigned share') were developed by law to maximize the likelihood of an award. However, the major source of information on cancer or disease risks used in these computations comes from the study of Japanese atomic bomb survivors. This database is based on a single exposure in 1945 to an Asian population and the risks are then modified, often appreciably, to be made appropriate for a U.S. population of healthy workers or civilians. The numbers of cancers for specific sites, when broken down by sex, age, and follow-up after exposure, can be quite small with risk coefficients having great uncertainty. All compensation schemes could be improved by the results from the MPS because the numbers will be much larger than from the atomic bomb survivors so the estimates will have more precision; they are more representative of healthy Americans than the Japanese population; and they do not require challenging procedures that attempt to convert a single exposure to a Japanese population with different disease rates and make it appropriate to American populations with chronic or prolonged exposures. The MPS also will be able to provide precise estimates for rare cancer sites that cannot be determined among the Japanese atomic bomb survivor population except by assumption. The MPS then will provide a high-quality assessment of cancer risks among healthy men and women that could be incorporated into the 'probability of causation' computations and thus contribute to enhancing the quality of information needed for these three different, but similar, compensation programs for workers and veterans and civilians who were exposed to radiation in the service of their country. Congress has passed laws that address these compensation issues.

Dosimetry

Good dosimetry is the key to good epidemiology (NCRP 2018b, UNSCEAR 2018). NCRP Report 178 (NCRP 2018a) entitled '*Deriving organ doses and their uncertainty for epidemiologic studies* (*With a focus on the one million U.S. workers and veterans study of low-dose radiation health effects*' guides the overall dose reconstruction process. The approach to estimating occupational doses received by individuals has also been described in several publications: Rocketdyne workers (Leggett et al. 2005; Boice, Cohen, et al. 2006, 2011); Mound workers (Boice et al. 2014); atomic veterans (Till et al. 2014; Till, Beck, Aanenson, et al. 2018; Caldwell et al. 2016; Beck et al. 2017); Mallinckrodt workers (Ellis, Boice, et al. 2018; Golden et al. 2019); and all workers (Bouville et al. 2015; NCRP 2018a). Six manuscripts in this special issue are focused on the complexities of dose reconstructions: all workers (Dauer, Bouville, et al. 2018); atomic veterans (Till, Beck, Aanenson, et al. 2018); internal emitters (Leggett, Eckerman, et al. 2018); brain dosimetry (Leggett, Tolmachev, et al. 2018); and medical radiation workers (Yoder et al. 2018; Dauer, Woods, et al. 2019). Dosimetry improvements are to continue, with more emphasis on handling the uncertainties and possible biases in studies (UNSCEAR 2008, 2014, 2018; Boice 2010; NCRP 2012) and in measurements and dose estimation, especially in the early years of monitoring (Gilbert et al. 1996; Gilbert 1998; Wakeford 2018; NCRP 2018a).

Table 7 provides a comparison between an early combined dose distribution of all MPS workers as of about 5 years ago with that of the Japanese atomic bomb survivors (Ozasa et al. 2012). The numbers of workers in the MPS study cohorts and the radiation dose distributions change during active follow-up as new personnel information is incorporated from employment or other records and as dose information is obtained and refined. For example, dose

Table 6. Distribution of the cumulative recorded doses of 168,565 medical radiation workers as of April 2015 (NCRP 2018a)[a,b].

Cumulative recorded dose (mSv)[c]	Number of workers[d]	Percentage of workers
<10	29,902[d]	17.7
10–<25	50,464[d]	30.0
25–<50	26,673[d]	15.8
50–<100	34,410	20.4
100–<250	20,512	12.1
250–<500	4842	2.87
500–<1000	1247	0.74
≥1000	515	0.31
All	168,565	100.0

[a]Adapted from Table B.5 in NCRP (2018a).
[b]This distribution was based on the information available at the time NCRP Report No. 178 (2018a) was prepared and has changed somewhat during the course of the study.
[c]The doses are expressed in terms of $H_p(10)$.
[d]The initial selection of workers eligible for study was those first monitored before 1995 ($n = 1,710,951$). To reduce the population to a more manageable and cost-effective size, without reducing statistical power, a 2% sample of the ~1,500,000 eligible workers with <10 mSv, and a 50% sample of the ~160,000 eligible workers with 10–49 mSv were selected for study.

Table 7. Distribution of the accumulated recorded doses, for the Million Person Study (MPS)[a] and for the study of Japanese atomic bomb survivors (Ozasa et al. 2012).

Cumulative recorded dose[b,c] (mSv)	Number of MPS workers and veterans	Cumulative number starting >50 mSv[d]	Number of atomic bomb survivors	Cumulative number starting >50 mSv
<10	6,874,268	–	43,361	–
10–	510,719	–	18,215	–
50–	104,395	104,395	6894	6894
100–	77,670	182,065	12,330	19,224
500–	4121	186,186	3424	22,648
≥1000	1217	187,403	2387	25,035
All	7,572,390	–	86,611	–

[a]Extracted from the NRC REIRS, DOE REMS, DoD NTP, historical DOE databases, and Landauer, Inc. dosimetry databases. All subjects >50 mSv were initially included for study. The numbers of low-dose subjects were so large, that only samples were taken which reduced the total population study to about a million workers and veterans.

[b]Dose categories for the MPS are expressed in terms of $H_p(10)$ primarily low-LET radiation. Absorbed doses from intakes of radionuclides are being estimated and not included in this table.

[c]Dose categories for the Japanese atomic bomb survivors are weighted absorbed colon dose. The absorbed dose from neutrons was weighted by a factor of 10.

[d]Cumulative doses from 50 mSv to indicate the differences in study size for relatively high-dose subjects.

reconstructions for individual workers will change to reflect the incorporation of organ doses from intakes of radionuclides, adding doses from work at other facilities, and converting personal dose equivalent [Hp(10)] (the film badge or personnel dosimeter measurement) to organ dose. The doses for the MPS were initially expressed in terms of personal dose equivalent [Hp(10)] and thus not strictly comparable with the absorbed dose to colon displayed for the atomic bomb survivors. On the other hand, the MPS tabulations in Table 7 do not include doses received at other facilities (see for example the influence of such doses on cumulative individual dose in Boice, Leggett, et al. 2006); nor does it include the doses from intakes of radionuclides, which can be substantial for some organs in several studies (Boice, Leggett, et al. 2006, 2014). Nonetheless, the distribution supports the tenet that the MPS has sufficient numbers and a broad dose distribution to provide new information on the effects of radiation when dose is delivered in a chronic or fractionated manner over a period of years.

Estimates of risks from gradual exposure over years will be substantially improved by the MPS (Boice 2014a). Radiation risk estimates will be precise having 20 times the number of adults than the atomic bomb survivor study (86,000 total and 54,000 adults), as broad a dose range (>100 mSv), as long a follow-up (>70 years), and based on U.S. radiation workers and veterans for whom the follow-up for vital status and death information and dose reconstruction (external and internal) are done in the same manner for all cohorts within the MPS (Mumma et al. 2018; Dauer, Bouville, et al. 2018; NCRP 2018a).

Summary

The MPS of U.S. radiation workers and veterans has been ongoing for over 15 years. The MPS will fill a critically important gap in knowledge by answering the question: what are the health consequences of exposure to radiation received gradually over time? The size of the MPS and range of radiation doses will provide substantial understanding as to the adequacy of the LNT model as used today in radiation protection (not for risk assessment) (ICRP 2007; Boice 2017b; NCRP 2018b; Shore et al. 2018). Approximately 810,000 workers and veterans within 21 of 29 MPS cohorts are currently under study to fill this need, with the remaining 8 cohorts to be incorporated once funds become available. The MPS is relevant today to federal agencies (GAO 2017; Davis et al. 2019) and the public: NRC will obtain information on the worker populations they regulate and set standards for; DoD and the DTRA will learn the extent to which atomic veterans have developed diseases related to their service for the nation; EPA has interest in accurate risks related to environmental exposures, especially to radionuclides; CDC's mission includes emergency response, environmental and medical exposures; and NASA requires more precise information on risks following gradual exposures and high-LET exposures to provide protection guidance for astronauts on long-duration missions such as to Mars. The scientific understandings obtained will be used in setting standards to protect workers and the public, in improving compensation schemes for workers, veterans, and the public, and in providing guidance for policy and decision makers.

Note

1. LNT refers to the linear-nonthreshold (LNT) dose-response model as used in radiation protection. Its purpose is to give regulators an approach to manage the potential risk of low-dose radiation exposures where radiation epidemiology is incapable of providing clear or convincing evidence or risk. It is inappropriate to use the LNT model in risk assessment for individuals or populations since it incorporates many assumptions and judgments that are not related to radiation risk (ICRP 2007; Boice 2017b; Shore et al. 2018; NCRP 2018b).

Disclosure statement

The authors report no conflicts of interest. The authors alone are responsible for the content and writing of the article.

Funding

This work was supported in part by a grant from the U.S. Nuclear Regulatory Commission (NRC-HQ-60-14-G-0011), grants from the Centers for Disease Control and Prevention (5UE1EH000989, 5NUE1EH001315), grants from the National Aeronautics and Space Administration (NNX15AU88G, 80NSSC17M0016), and grants from the U.S. Department of Energy (Grant No. DE-SC0008944 awarded to the National Council on Radiation Protection and Measurements, which included interagency support from the U.S. Nuclear Regulatory Commission, the U.S. Environmental Protection Agency, and the National Aeronautics and Space Administration; and more recent Grants No. DE-AU0000042 and DE-AU0000046). Contract support also was recently received from the Naval Sea Systems Command (N00024-17-C-4322). Further, contract support was received by Oak Ridge National Laboratory from the Office of Radiation and Indoor Air, U.S. Environmental Protection Agency, under Interagency Agreement DOE No. 1824 S581-A1, under contract No. DE-AC05-00OR22725 with UT-Battelle; and contract support was received by Oak Ridge Associated Universities from the U.S. Department of Energy under contract No. DE-SC0014664. The Radiation Nuclear Countermeasures Program of the National Institute of Allergy and Infectious Diseases (NIAID) provided support to the NCI under NIAID agreement #Y2-Al-5077.

ORCID

John D. Boice http://orcid.org/0000-0002-8755-1299
Sarah S. Cohen http://orcid.org/0000-0003-0421-1983
Michael T. Mumma http://orcid.org/0000-0001-7506-8710
Elizabeth D. Ellis http://orcid.org/0000-0003-2871-1587

References

Acquavella JF, Wiggs LD, Waxweiler RJ, Macdonell DG, Tietjen GL, Wilkinson GS. 1985. Mortality among workers at the Pantex weapons facility. Health Phys. 48:735–746.

Anzenberg V, Lewis DE, Dickson ED, Bush-Goddard SP. 2010. The U.S. nuclear regulatory commission radiation exposure information reporting system (REIRS). Radiat Res. 173:254–255.

Beck HL, Till JE, Grogan HA, Aanenson JW, Mohler HJ, Mohler SS, Voillequé PG. 2017. Red bone marrow and male breast doses for a cohort of atomic veterans. Radiat Res. 187:221–228.

Blevins MR, Andersen RL. 2011. Radiation protection at U.S. nuclear power plants-today and tomorrow. Health Phys. 100:35–38.

Boice JD, Jr. 2010. Uncertainties in studies of low statistical power. J Radiol Prot. 30:115–120.

Boice JD, Jr. 2012a. A study of one million U.S. Radiation Workers and Veterans a New NCRP Initiative (DOE Grant Awarded September 2012). Health Phys News. XL:7–10. [accessed 2018 Sep 8]. https://ncrponline.org/the-boice-report/.

Boice JD, Jr. 2012b. Study of atomic veterans who participated at U.S. aboveground atmospheric nuclear weapons tests, 1945–1962. Health Phys News. XL:6–9. [accessed 2018 Sep 8]. https://ncrponline.org/the-boice-report/.

Boice JD, Jr. 2013a. DOE worker studies. Health Phys News. XLI: 22–23. [accessed 2018 Sep 8]. https://ncrponline.org/the-boice-report/.

Boice JD, Jr. 2013b. Study of U.S. nuclear power plant workers. Health Phys News. XLI:11–12. [accessed 2018 Sep 10]. http://ncrponline.org/wp-content/themes/ncrp/PDFs/BOICE-HPnews/8-NPP-JAN2013.pdf.

Boice JD, Jr. 2013c. NCRP dosimetry committee for the million worker study: Oak Ridge, Tennessee, April 2013. Health Phys News. XLI: 15–17. [accessed 2018 Sep 9]. https://ncrponline.org/the-boice-report/.

Boice JD, Jr. 2014a. The importance of radiation worker studies (editorial). J Radiol Prot. 34:E7–E12.

Boice JD, Jr. 2014b. The million worker study comes to oak ridge. Health Phys News. 42:16–17. [accessed 2018 Sep 9]. http://ncrponline.org/wp-content/themes/ncrp/PDFs/BOICE-HPnews/26_MWS_OakRidge_July2014.pdf.

Boice JD, Jr. 2014c. Atomic bombs, asbestos, and healthy warriors. Health Physics News. XLII:21–23. [accessed 2018 Sep 9]. https://ncrponline.org/the-boice-report/.

Boice JD, Jr. 2015. The million worker study - from whence it came. Health Phys News. 43:22–23. [accessed 2018 Sep 9]. http://ncrponline.org/wp-content/themes/ncrp/PDFs/BOICE-HPnews/37_MWS_June2015.pdf.

Boice JD, Jr. 2016a. From oak ridge to Indian point studying workers at nuclear power plants. Health Phys News 44:23–24. [accessed 2018 Sep 8]. http://ncrponline.org/wp-content/themes/ncrp/PDFs/BOICE-HPnews/45-From-Oak-Ridge-to-Indian-Point-Feb-2016.pdf.

Boice JD, Jr. 2016b. Mars matters! Health Phys News. 44:32–33. [accessed 2018 Sep 9]. http://ncrponline.org/wp-content/themes/ncrp/PDFs/BOICE-HPnews/46_Mars_Matters_Mar2016.pdf.

Boice JD, Jr. 2017a. Space: the final frontier-research relevant to mars. Health Phys. 112:392–397.

Boice JD, Jr. 2017b. The linear nonthreshold (LNT) model as used in radiation protection: an NCRP update. Int J Radiat Biol. 93: 1079–1092.

Boice JD, Jr. 2017c. The Martian Brain. Health Phys News. 45:23–24. [accessed 2019 Mar 29]. https://ncrponline.org/wp-content/themes/ncrp/PDFs/BOICE-HPnews/55_MartianBrain_Feb2017.pdf

Boice JD, Jr. 2019. Relevance of the million person study to research needs for NASA and space exploration. Int J Radiat Biol. DOI: 10.1080/09553002.

Boice JD, Jr., Cohen SS, Mumma MT, Dupree Ellis E, Eckerman KF, Leggett RW, Boecker BB, Brill AB, Henderson BE. 2006. Mortality among radiation workers at Rocketdyne (Atomics International), 1948–1999. Radiat Res. 166:98–115.

Boice JD, Jr., Cohen SS, Mumma MT, Ellis ED, Eckerman KF, Leggett RW, Boecker BB, Brill AB, Henderson BE. 2011. Updated mortality analysis of radiation workers at Rocketdyne (Atomics International), 1948–2008. Radiat Res. 176:244–258.

Boice JD, Jr., Cohen SS, Mumma MT, Ellis ED, Cragle DL, Eckerman KF, Wallace PW, Chadda B, Sonderman JS, Wiggs LD, et al. 2014. Mortality among mound workers exposed to polonium-210 and other sources of radiation, 1944–1979. Radiat Res. 181:208–228.

Boice JD, Jr., Ellis ED, Golden AP, Girardi DJ, Cohen SS, Chen H, Mumma MT, Shore RE, Leggett RW. 2018. The past informs the future: an overview of the million worker study and the Mallinckrodt Chemical Works cohort. Health Phys. 114:381–385.

Boice JD, Jr., Leggett RW, Ellis ED, Wallace PW, Mumma M, Cohen SS, Brill AB, Chadda B, Boecker BB, Yoder RC, et al. 2006. A comprehensive dose reconstruction methodology for former Rocketdyne/Atomics International radiation workers. Health Phys. 90:409–430.

Boice JD, Jr., Mandel JS, Doody MM, Yoder RC, McGowan R. 1992. A health survey of radiologic technologists. Cancer. 69:586–598.

Boice JD, Jr., Mumma M, Chen H, Cohen SS, Till JE, Beck HL. 2019. Mortality among US military participants at eight above-ground nuclear weapons test series. Int J Radiat Biol. (Submitted).

Boice JD, Jr., Cohen SS, Mumma MT, Hagemeyer D, Chen H, Golden AP, Yoder RC, Dauer LT. 2019a. Leukemia among early nuclear power plant workers employed between 1957 and 1984 in the United States. Int J Radiat Biol. (Submitted).

Boice JD, Jr., Cohen SS, Mumma MT, Hagemeyer D, Chen H, Golden AP, Yoder RC, Dauer LT. 2019b. Leukemia among industrial radiographers employed between 1939–2011 in the United States. Int J Radiat Biol. (Submitted).

Boice JD, Jr., Ellis ED, Golden AP, Zablotska LB, Mumma MT, Cohen SS. 2019. Sex-specific lung cancer risk among radiation workers in the Million Person Study and among TB-fluoroscopy patients. Int J Radiat Biol. DOI: 10.1080/09553002.2018.1547441

Bouville A, Toohey RE, Boice JD, Jr., Beck HL, Dauer LT, Eckerman KF, Hagemeyer D, Leggett RW, Mumma MT, Napier B, et al. 2015. Dose reconstruction for the million worker study: status and guidelines. Health Phys. 108:206–220.

Boyko EJ, Koepsell TD, Gaziano JM, Horner RD, Feussner JR. 2000. U.S. Department of veterans affairs medical care system as a resource to epidemiologists. Am J Epidemiol. 151:307–314.

Brown SC, Schonbeck MF, McClure D, Barón AE, Navidi WC, Byers T, Ruttenber AJ. 2004. Lung cancer and internal lung doses among plutonium workers at the rocky flats plant: a case-control study. Am J Epidemiol. 160:163–172.

Cahoon EK, Preston DL, Pierce DA, Grant E, Brenner AV, Mabuchi K, Utada M, Ozasa K. 2017. Lung, laryngeal and other respiratory cancer incidence among Japanese atomic bomb survivors: an updated analysis from 1958 through 2009. Radiat Res. 187:538–548.

Caldwell GG, Kelley D, Zack M, Falk H, Heath CW, Jr. 1983. Mortality and cancer frequency among military nuclear test (SMOKY) participants, 1957 through 1979. JAMA. 250:620–624.

Caldwell GG, Zack M, Mumma M, Falk H, Heath CW, Till JE, Chen H, Boice JD, Jr. 2016. Mortality among military participants at the 1957 PLUMBBOB nuclear weapons test series and from leukemia among participants at the SMOKY test. J Radiol Prot. 36:474–489.

Chan C, Hughes TS, Muldoon S, Aldrich T, Rice C, Hornung R, Brion G, Tollerud DJ. 2010. Mortality patterns among Paducah Gaseous diffusion plant workers. J Occup Environ Med. 52:725–732.

Charpentier P, Ostfeld AM, Hadjimichael OC, Hester R. 1993. The mortality of US nuclear submariners, 1969–1982. J Occup Med. 35: 501–509.

Cohen SS, Mumma MT, Ellis ED, Boice JD, Jr. 2018. Validating the use of census data on education as a measure of socioeconomic status in an occupational cohort. Int J Rad Biol. DOI:10.1080/09553002.

[CEDR] Comprehensive Epidemiologic Data Resource. 2018. U.S. Department of energy. [accessed 2018 August 31]. https://apps.orau.gov/cedr/#.W4l4WLgpBPY.

Cragle DL, McLain RW, Qualters JR, Hickey JL, Wilkinson GS, Tankersley WG, Lushbaugh CC. 1988. Mortality among workers at a nuclear fuels production facility. Am J Ind Med. 14:379–401.

Cragle DL, Watkins JP, Ingle JN, Robertson-Demers K, Tankersley WG, West CM. 1995. Mortality among a cohort of white male workers at a uranium processing plant: Fernald feed materials production center, 1951–1989. Oak Ridge (TN): Oak Ridge Institute for Science and Education.

Dalager NA, Kang HK, Mahan CM. 2000. Cancer mortality among the highest exposed US atmospheric nuclear test participants. J Occup Environ Med. 42:798–805.

Dauer LT, Bouville A, Toohey RE, Boice JD, Jr., Beck HL, Eckerman KF, Hagemeyer D, Leggett RW, Mumma MT, Napier B, et al. 2018. Dosimetry and uncertainty approaches for the million-worker study of radiation workers and veterans: overview of the recommendations in NCRP report no. 178. Rev Int J Rad Biol. 19:1–10. DOI:10.1080/09553002

Dauer LT, Woods M, Miodownik D, Serencsits B, Quinn B, Bellamy M, Yoder C, Liang X, Boice JD, Jr., Bernstein J. 2019. Cohort profile - MSKCC radiation workers: a pilot sub-cohort of a multicenter medical radiation worker component of the million person study of low-dose radiation health effects. Int J Rad Biol. 27:1–21. DOI: 10.1080/09553002.

Davis J, Dewji SA, Abelquist E, Hertel N. 2019. Synopsis of the Oak Ridge radiation protection research needs workshop. Health Phys. 116:69–80.

[DTRA] Defense Threat Reduction Agency. 2015. Operation CASTLE fact sheet. May 2015. [accessed 2018 Sep 7]. http://www.dtra.mil/Portals/61/Documents/NTPR/1-Fact_Sheets/CASTLE%20-%202017.pdf?ver = 2017-02-07-081707-913.

Dupree EA, Cragle DL, McLain RW, Crawford-Brown DJ, Teta MJ. 1987. Mortality among workers at a uranium processing facility, the Linde Air Products Company Ceramics Plant, 1943–1949. Scand J Work Environ Health. 13:100–107.

Dupree EA, Wells SM, Watkins JP, Wallace PW, Davis NC. 1994. Mortality among workers between 1945 and 1984 at a uranium gaseous diffusion facility. Oak Ridge (TN): Oak Ridge Institute for Science and Education.

Dupree-Ellis E, Watkins J, Ingle JN, Phillips J. 2000. External radiation exposure and mortality in a cohort of uranium processing workers. Am J Epidemiol. 152:91–95.

Eisenbud M. 1975. Early occupational exposure experience with uranium processing. Presented at the Conference on Occupational Health Experience with Uranium, Arlington, Virginia, April 28, 1975. ERDA-93. Washington (DC): Energy Research and Development Administration. [accessed 2018 September 9]. http://www.osti.gov/scitech/servlets/purl/4082757.

Ellis ED, Boice JD, Jr., Golden AP, Girardi DJ, Cohen SS, Mumma MT, Shore RE, Leggett RW, Kerr GD. 2018. Dosimetry is key to good epidemiology: workers at Mallinckrodt chemical works had seven different source exposures. Health Phys. 114:386–397.

Ellis ED, Girardi D, Golden AP, Wallace PW, Phillips J, Cragle DL. 2018. Historical perspective on the department of energy mortality studies: focus on the collection and storage of individual worker data. Int J Radiat Biol. 29:1–23. DOI:10.1080/09553002.2018.1547851.

Fisher SB, Weber L, Goldberg J, Davis F. 1995. Mortality ascertainment in the veteran population: alternatives to the national death index. Am J Epidemiol. 141:242–250.

Friedman-Jimenez G, Kato I, Afanasyeva Y, Shore R. 2003a. Relationship between low doses of ionizing radiation and mortality of nuclear submariners in the U.S. NAVY. Abstract 039. Am J Epidemiol. 157:S1–S106.

Friedman-Jimenez G, Kato I, Afanasyeva Y, Shore R. 2003b. Mortality of nuclear submariners in the U.S. Navy. Abstract 317. Am J Epidemiol.157:S1–S106.

Frome EL, Cragle DL, Watkins JP, Wing S, Shy CM, Tankersley WG, West CM. 1997. A mortality study of employees of the nuclear industry in oak ridge, Tennessee. Radiat Res. 148:64–80.

Fry SA, Lushbaugh CC, Shy CM, Cragle DL, Checkoway H, Blum S, Carpenter AV, Dupree EA, Frome EL, Groer PG, et al. 1985. The U.S. department of energy health and mortality study: the oak ridge studies. Epidemiological studies of some populations exposed to ionizing radiation, AECL-8360. Pinawa, Manitoba: Energy of Canada Limited. [accessed 2018 Aug 30]. http://www.iaea.org/inis/collection/NCLCollectionStore/_Public/20/008/20008287.pdf.

Gilbert ES. 1998. Accounting for errors in dose estimates used in studies of workers exposed to external radiation. Health Phys. 74:22–29.

Gilbert ES, Cragle DL, Wiggs LD. 1993. Updated analyses of combined mortality data for workers at the Hanford site, oak ridge national laboratory, and Rocky flats weapons plant. Radiat Res. 136:408–421.

Gilbert ES, Omohundro E, Buchanan JA, Holter NA. 1993. Mortality of workers at the Hanford site: 1945–1986. Health Phys. 64:577–590.

Gilbert ES, Fix JJ, Baumgartner WV. 1996. An approach to evaluating bias and uncertainty in estimates of external dose obtained from personal dosimeters. Health Phys. 70:336–345.

Golden AP, Cohen SS, Chen H, Ellis ED, Boice JD, Jr. 2018. Evaluation of statistical modeling approaches for epidemiologic studies of low-dose radiation health effects. Int J Radiat Biol. 30: 1–26. DOI: 10.1080/09553002.2018.1554924.

Golden AP, Ellis ED, Cohen SS, Mumma MT, Leggett RW, Wallace PW, Girardi DJ, Watkins JP, Shore RE, Boice JD, Jr. 2019. Updated mortality analysis of the Mallinckrodt uranium processing workers, 1942–2012. Int J Radiat Biol. 17:1–21. DOI:10.1080/09553002.2019.1569773.

Hagemeyer D, Nichols G, Mumma MT, Boice JD, Jr., Brock TA. 2018. 50 years of the radiation exposure information and reporting system and importance to the million person study. Int J Radiat Biol. 25: 1–14. DOI: 10.1080/09553002.2018.1540896.

Hall EJ, Metting N, Puskin J, Ron E. 2009. Low dose radiation epidemiology: what can it tell us? Radiat Res. 172:134–138.

[ICRP] International Commission on Radiological Protection. 2007. The 2007 recommendations of the international commission on radiological protection. ICRP publication 103. Ann ICRP. 37(2–4): 1–332.

Jablon S, Boice JD, Jr. 1993. Mortality among workers at a nuclear power plant in the United States. Cancer Causes Control. 4:427–430.

Jacob P, Rühm W, Walsh L, Blettner M, Hammer G, Zeeb H. 2009. Is cancer risk of radiation workers larger than expected? Occup Environ Med. 66:789–796.

Johnson JC, Thaul S, Page WF, Crawford H. 1996. Mortality of veteran participants in the CROSSROADS nuclear test. Washington (DC): National Academies Press.

Johnson JC, Thaul S, Page WF, Crawford H. 1997. Mortality of veteran participants in the CROSSROADS nuclear test. Health Phys. 73: 187–189.

Kiernan D. 2013. The girls of atomic city: the untold story of the women who helped win World War II. New York (NY): Simon and Schuster.

Kirr M, Passmore C. 2016. Retrospective review of dosimeter film processing. Poster P06.13. IRPA 2016. 50 years of practicing radiation protection. May 9–13; Cape Town, South Africa.

Kirr M, Passmore C, Koperski B, Moscatel M, Zhang R. 2016. Dosimeter archeology - Digging up the past. 61st WAM-B.1. Annual H PS Meeting; July 20; Spokane, Washington. http://hpschapters.org/2016AM/program/singlesession.php3?sessid = WAM-B

Laurier D, Richardson DB, Cardis E, Daniels RD, Gillies M, O'Hagan J, Hamra GB, Haylock R, Leuraud K, Moissonnier M, et al. 2017. The international nuclear workers study (Inworks): a collaborative epidemiological study to improve knowledge about health effects of protracted low-dose exposure. Radiat Prot Dosimetry. 173: 21–25.

Leggett RW, Eckerman KF, Boice JD, Jr. 2005. A respiratory model for uranium aluminide based on occupational data. J Radiol Prot. 25: 405–416.

Leggett RW, Tolmachev SY, Boice JD, Jr. 2018. Potential improvements in brain dose estimates for internal emitters. Int J Rad Biol. 4:1–54. DOI: 10.1080/09553002.2018.1554923.

Leggett RW, Eckerman KF, Bellamy M. 2018. MPS dose reconstruction for internal emitters: some site-specific issues and approaches. Int J Radiat Biol. 18:1–13. DOI: 10.1080/09553002.2018.1558302.

Linet MS, Kim KP, Miller DL, Kleinerman RA, Simon SL, Berrington de Gonzalez A. 2010. Historical review of occupational exposures and cancer risks in medical radiation workers. Radiat Res. 174: 793–808.

Loomis DP, Wolf SH. 1996. Mortality of workers at a nuclear materials production plant at Oak Ridge, Tennessee, 1947–1990. Am J Ind Med. 29:131–141.

Lushbaugh CC, Fry SA, Shy C, Frome EL. 1983. The DOE health and mortality study at oak ridge. Health physics society: epidemiology applied to health physics. Proceedings of the 16th Midyear Topical Symposium. U.S. Department of Energy Report CONF-830101, Albuquerque (NM): DOE. [accessed 2018 Sep 5]. https://inis.iaea.org/collection/NCLCollectionStore/_Public/15/013/15013337.pdf.

Mueller TJ, Weishar TM, Hallworth JM, Bonamer DG. 2018. Occupational radiation exposure from U.S. Naval Nuclear Plants and their support facilities. Washington (DC): Naval Nuclear Propulsion Program, NT-18-2. [accessed 2018 Sep 4]. https://www.energy.gov/sites/prod/files/2018/07/f53/NT-18-2.pdf.

Muirhead CR, Boice JD, Jr., Raddatz CT, Yoder RC. 1996. Comparison of dose histories for U.S. nuclear power plant workers, based on records held by a major dosimetry service company and on the NRC REIRS database. Health Phys. 70:645–650.

Mumma MT, Cohen SS, Sirko JL, Ellis ED, Boice JD, Jr. 2018. Obtaining vital status and cause of death on a million persons. Int J Radiat Biol. 9:1–21. DOI: 10.1080/09553002.2018.1539884.

Mumma MT, Sirko JL, Boice JD, Jr., Blot WJ. 2019. Mesothelioma mortality within two radiation monitored occupational cohorts. Int J Radiat Biol. (Submitted).

[NCRP] National Council on Radiation Protection and Measurements. 2001. Evaluation of the linear-nonthreshold dose-response model for ionizing radiation. NCRP report 136. Bethesda (MD): NCRP.

[NCRP] National Council on Radiation Protection and Measurements. 2012. Uncertainties in the estimation of radiation risks and probability of disease causation. NCRP report 171. Bethesda (MD): NCRP.

[NCRP] National Council on Radiation Protection and Measurements. 2014. Radiation protection for space activities: supplement to previous recommendations. NCRP commentary 23. Bethesda (MD): NCRP.

[NCRP] National Council on Radiation Protection and Measurements (NCRP). 2016. Potential for central nervous system effects from radiation exposure during space activities. Phase 1: overview. NCRP Commentary 25. Bethesda MD: NCRP.

[NCRP] National Council on Radiation Protection and Measurements. 2018a. Deriving organ doses and their uncertainty for epidemiologic studies (With a focus on the one million U.S. workers and veterans study of low-dose radiation health effects. NCRP report 178. Bethesda (MD): NCRP.

[NCRP] National Council on Radiation Protection and Measurements. 2018b. Implications of recent epidemiologic studies for the linear nonthreshold model and radiation protection. NCRP commentary 27. Bethesda (MD): NCRP.

[NCRP] National Council on Radiation Protection and Measurements. 2018c. Management of exposure to ionizing radiation: radiation protection guidance for the United States. NCRP report 180. Bethesda (MD): NCRP.

[NIOSH] National Institute for Occupational Safety and Health. 2001. Final report: mortality patterns among uranium enrichment workers at the Portsmouth gaseous diffusion plant, Piketon, Ohio. Atlanta (GA): Centers for Disease Control and Prevention. [accessed 2019 January 31]. https://www.cdc.gov/niosh/oerp/pdfs/2001-133g5-2.pdf.

[NRC/NA] National Research Council of the National Academies. 2003. A review of the dose reconstruction program of the defense threat reduction agency. Committee to review the dose reconstruction program of the defense threat reduction agency. Board on radiation effects research, division on earth and life studies. Washington (DC): National Academies Press. [accessed 2018 Sep 3]. https://www.nap.edu/catalog/10697/a-review-of-the-dose-reconstruction-program-of-the-defense-threat-reduction-agency.

[NRC/NA] National Research Council of the National Academies of Science. 2005. Assessment of the Scientific Information for the Radiation Exposure Screening and Education Program. Washington (DC): National Academies Press. [accessed March 24, 2019]. https://www.nap.edu/download/11279.

Neton JW, Howard J, Elliott LJ. 2008. Radiation dose reconstruction program of the national institute for occupational safety and health: overview. Health Phys. 95:6–13.

Neton JW. 2014. Characterization of exposures to workers covered under the U.S. energy employees compensation act. Health Phys. 106:249–258.

O'Brien M. 2012. Fukushima after the meltdown. PBS newshour special report. March 15, 2012. [accessed 2018 August 30]. http://www.pbs.org/newshour/multimedia/fukushima/#.UHLFGK5siSr.

[ORISE] Oak Ridge Institute for Science and Education. 2011. Establishing a U.S. nuclear regulatory commission worker registry for epidemiologic investigations. A feasibility study using the Radiation Exposure Information and Reporting System. Submitted to the Nuclear Regulatory Commission. [accessed 2018 Sep 8].

Ozasa K, Shimizu Y, Suyama A, Kasagi F, Soda M, Grant EJ, Sakata R, Sugiyama H, Kodama K. 2012. Studies of the mortality of atomic bomb survivors, report 14, 1950–2003: an overview of cancer and noncancer diseases. Radiat Res. 177:229–243.

Peterson GR, Gilbert ES, Buchanan JA, Stevens RG. 1990. A case-cohort study of lung cancer, ionizing radiation, and tobacco smoking among males at the Hanford Site. Health Phys. 58:3–11.

Polednak AP, Frome EL. 1981. Mortality among men employed between 1943 and 1947 at a uranium-processing plant. J Occup Med. 23:169–178.

[REIRS] Radiation Exposure Information and Reporting System. 2018. Nuclear regulatory commission. accessed 2018 August 30]. http://www.nrc.gov/reading-rm/doc-collections/gils/rad-exp.html.

Rector TS, Wickstrom SL, Shah M, Thomas Greeenlee N, Rheault P, Rogowski J, Freedman V, Adams J, Escarce JJ. 2004. Specificity and sensitivity of claims-based algorithms for identifying members of Medicare + Choice health plans that have chronic medical conditions. Health Serv Res. 39:1839–1857.

Richardson DB, Wing S, Wolf S. 2007. Mortality among workers at the Savannah river site. Am J Ind Med. 50:881–891.

Ritz B. 1999. Radiation exposure and cancer mortality in uranium processing workers. Epidemiology. 10:531–538.

Robinette CD, Jablon S, Preston TL. 1985. Mortality of nuclear weapons test participants. Washington (DC): National Academies Press.

Schubauer-Berigan MK, Daniels RD, Bertke SJ, Tseng CY, Richardson DB. 2015. Cancer mortality through 2005 among a pooled cohort of U.S. nuclear workers exposed to external ionizing radiation. Radiat Res. 183:620–631.

Shore RE, Beck HL, Boice JD, Jr., Caffrey EA, Davis S, Grogan HA, Mettler FA, Jr., Preston RJ, Till JE, Wakeford R, et al. 2018. Implications of recent epidemiologic studies for the linear nonthreshold model and radiation protection. J Radiol Prot. 38:1217–1233.

Shore RE, Friedman-Jimenez G, Kato I. 2001. Final report of mortality study of naval nuclear submariners. Navy Contract No. N66604-96-C-0260. New York (NY): New York University School of Medicine.

Shy CM. 1982. Research objectives and university collaboration in the health and mortality study of Department of Energy contract facilities. Hearing before the Subcommittee on Investigations and Oversight of the Committee on Science and Technology, U.S. House of Representatives, Ninety-Seventh Congress, Second Session, May 19, 1982. Washington (DC): Office of Scientific and Technical Information. [accessed 2019 Mar 28]. https://www.osti.gov/biblio/6311127-review-department-energy-health-mortality-study-hearing-before-subcommittee-investigations-oversight-committee-science-technology-us-house-representatives-ninety-seventh-congress-second-session-may.

Silver SR, Anderson-Mahoney P, Burphy J, Hiratzka S, Schubauer-Berigan MK, Waters KM. 2005. Mortality update for the Pantex Weapons Facility: final report. Publication No. 2005–124. Cincinnati (OH): National Institute of Occupational Safety and Health. [accessed 2018 Sep 10]. http://www.cdc.gov/niosh/docs/2005-124/pdfs/2005-124.pdf.

Silver SR, Bertke SJ, Hein MJ, Daniels RD, Fleming DA, Anderson JL, Pinney SM, Hornung RW, Tseng CY. 2013. Mortality and ionising radiation exposures among workers employed at the Fernald Feed Materials Production Center (1951–1985). Occup Environ Med. 70:453–463.

Simon SL, Weinstock RM, Doody MM, Neton J, Wenzl T, Stewart P, Mohan AK, Yoder RC, Hauptmann M, Freedman DM, et al. 2006. Estimating historical radiation doses to a cohort of U.S. radiologic technologists. Rad Res. 166:174–192.

Simon SL, Bailey SM, Beck HL, Boice JD, Bouville A, Brill AB, Cornforth MN, Inskip PD, McKenna MJ, Mumma MT, et al. 2019. Estimation of radiation doses to U.S. military test participants from nuclear testing: a comparison of historical film-badge measurements, dose reconstruction, and retrospective biodosimetry. Radiat Res. DOI:10.1667/RR15247.1.

Stram DO, Preston DL, Sokolnikov M, Napier B, Kopecky KJ, Boice J, Beck H, Till J, Bouville A. 2015. Shared dosimetry error in epidemiological dose-response analyses. PLoS One. 10:e0119418.

Thaul S, Page WF, Crawford H, O'Maonaigh H. 2000. The five series study: mortality of military participants in U.S. nuclear weapons tests. Washington (DC): National Academies Press. [accessed 2018 August 30]. http://www.nap.edu/catalog.php?record_id¼9697.

Till JE, Beck HL, Aanenson JW, Grogan HA, Mohler HJ, Mohler SS, Voillequé PG. 2014. Military participants at U.S. Atmospheric nuclear weapons testing-methodology for estimating dose and uncertainty. Radiat Res. 181:471–484.

Till JE, Beck HL, Aanenson JW, Grogan HA, Mohler HJ, Mohler SS, Voillequé PG. 2018. Dosimetry associated with veterans who participated in nuclear weapons testing. Int J Radiat Biol. 4:1–30. DOI: 10.1080/09553002.2018.1551639.

Till JE, Beck HL, Boice JD, Jr., Mohler HJ, Mumma MT, Aanenson JW, Grogan HA. 2018. Asbestos exposure and mesothelioma mortality among atomic veterans. Int J Rad Biol. 4:1–15. DOI: 10.1080/09553002.2018.1551641.

[UNSCEAR] United Nations Scientific Committee on the Effects of Atomic Radiation. 2008. Effects of ionizing radiation. UNSCEAR 2006 report. Report to the general assembly. Vol. 1. Scientific annex A. Epidemiological studies of radiation and cancer. Publication E.08.IX.6. New York (NY): UNSCEAR.

[UNSCEAR] United Nations Scientific Committee on the Effects of Atomic Radiation. 2014. UNSCEAR 2012 report to the general assembly with scientific annexes. Scientific annex B. Uncertainties in risk estimates for radiation-induced cancer. New York (NY): UNSCEAR.

[UNSCEAR] United Nations Scientific Committee on the Effects of Atomic Radiation. 2018. Sources, effects and risks of ionizing radiation. UNSCEAR 2017 report. Report to the general assembly, with scientific annexes. Scientific annex A. Principles and criteria for ensuring the quality of the Committee's reviews of epidemiological studies of radiation exposure. Publication E.18.IX.1. New York (NY): UNSCEAR.

[DOE] U.S. Department of Energy. 2000. United States nuclear tests July 1945 through September 1992. DOE/NV - 209-REV 15. Las Vegas, NV: U.S. department of energy Nevada operations office. [accessed 2018 September 3]. https://www.nnss.gov/docs/docs_LibraryPublications/DOE_NV-209_Rev16.pdf.

[GAO] U.S. Government Accountability Office. 2017. Los-dose radiation: interagency collaboration on planning research could improve information on health effects. GAO-17-546. Washington (DC): GAO. [accessed 2018 September 8]. https://www.gao.gov/products/GAO-17-546A.

[NRC] U.S. Nuclear Regulatory Commission. 1992. Instructions for recording and reporting occupational radiation exposure data. Regulatory guide (RG) 8.37. Washington (DC): NRC. [accessed 2018 September 8]. https://www.orau.org/ptp/PTP%20Library/library/NRC/Reguide/08-007.pdf.

[NRC] U.S. Nuclear Regulatory Commission. 1994. Voluntary reporting of additional occupational radiation of exposure data (Generic Letter 94-04). [accessed 2018 September 8]. https://www.nrc.gov/docs/ML1204/ML12047A243.pdf.

[NRC] U.S. Nuclear Regulatory Commission. 2017. Industrial radiography licensee toolkit. Washington (DC): NRC. [accessed 2018 Mar 30]. https://www.nrc.gov/materials/miau/industrial-uses/rad-toolkit. html.

[NRC/NA] National Research Council of the National Academies of Science. 2005. Assessment of the scientific information for the radiation exposure screening and education program. Washington (DC): National Academies Press. [accessed 2019 Mar 24]. https://www.nap.edu/download/11279.

[NRC] U.S. Nuclear Regulatory Commission. 2018. Radiation exposure information and reporting system (REIRS). Washington (DC): NRC. [accessed 2018 Sep 8]. http://www.nrc.gov/reading-rm/doc-collections/gils/rad-exp.html.

Wakeford R. 2018. The growing importance of radiation worker studies. Br J Cancer. 119:527–529.

Watanabe KK, Kang HK, Dalager NA. 1995. Cancer mortality risk among military participants of a 1958 atmospheric nuclear weapons test. Am J Public Health. 85:523–527.

Wiggs LD, Cox-DeVore CA, Voelz GL. 1991. Mortality among a cohort of workers monitored for 210Po exposure: 1944–1972. Health Phys. 61:71–76.

Wiggs LD, Cox-DeVore CA, Wilkinson GS, Reyes M. 1991. Mortality among workers exposed to external ionizing radiation at a nuclear facility in Ohio. J Occup Med. 33:632–637.

Wiggs LD, Johnson ER, Cox-DeVore CA, Voelz GL. 1994. Mortality through 1990 among white male workers at the Los Alamos National Laboratory: considering exposures to plutonium and external ionizing radiation. Health Phys. 67:577–588.

Wilkinson GS, Tietjen GL, Wiggs LD, Galke WA, Acquavella JF, Reyes M, Voelz GL, Waxweiler RJ. 1987. among plutonium and other radiation workers at a plutonium weapons facility. Am J Epidemiol. 125:231–250.

Wing S, Richardson D, Wolf S, Mihlan G. 2004. Plutonium-related work and cause-specific mortality at the United States department of energy Hanford site. Am J Ind Med. 45:153–164.

Wing S, Richardson DB. 2005. Age at exposure to ionising radiation and cancer mortality among Hanford workers: follow up through 1994. Occup Environ Med. 62:465–472.

Yiin JH, Anderson JL, Daniels RD, Bertke SJ, Fleming DA, Tollerud DJ, Tseng CY, Chen PH, Waters KM. 2017. Mortality in a combined cohort of uranium enrichment workers. Am J Ind Med. 60:96–108.

Yoder RC, Dauer L, Balter S, Boice JD, Jr., Grogan H, Mumma M, Passmore CN, Rothenberg LN, Vetter RJ. 2018. Dosimetry for the study of medical radiation workers with a focus on the mean absorbed dose to the lung, brain and other organs. Int J Radiat Biol. DOI:10.1080/09553002.2018.1549756

The Million Person Study relevance to space exploration and Mars

John D. Boice

ABSTRACT
Understanding the health consequences of exposure to radiation received gradually over time is critically needed. The National Aeronautics and Space Administration (NASA) bases its safety standards on the acute exposures received by Japanese atomic bomb survivors. Such a brief exposure differs appreciably from the chronic radiation received during a two to three year mission to Mars. NASA also applies an individual risk-based system for radiation protection that accounts for age, sex, smoking history, and individual life styles. Because the Japanese life span study (LSS) reports women to be at 2 to 3 times greater lifetime risk of developing cancer than men, female astronauts are allowed less time in space. Another concern is the potential behavioral and cognitive impairments from galactic cosmic radiation (GCR) impinging on the nervous system that might jeopardize the mission, and, possibly, lead to dementia later in life. GCR are high-velocity heavy ions traveling through space. There are no human circumstances/analogs similar to GCR that can provide direct information on the possible effects of such high-LET exposure to brain tissue. The MPS provides a more representative group (healthy men and women) for risk estimates than the 1945 Japanese population exposed briefly to the atomic bombs. The permissible career exposure limit set by NASA for each astronaut is a 3% risk of exposure-induced death (REID) from cancer at a 95% confidence level to account for uncertainties in risk projections. Because the MPS is 10 times larger than the LSS, the 95% confidence levels will be narrower and thus allow more time in space, all things being equal. Sex-specific differences in radiation risk can be examined more fully in the MPS with over 250,000 women compared with about 32,000 women in the LSS. Non-cancer outcomes such as neurological disorders also can be evaluated following low-dose rate exposures to high-LET alpha particles. Workers at several nuclear facilities had intakes of radionuclides, such as plutonium, that exposed brain tissue to alpha particles (Helium nuclei) for life. Such workers are being evaluated for mortality from dementia and other motor neuron diseases; can be evaluated for clinically diagnosed incidences of these conditions; and, though challenging, could be interviewed and ask to take cognition tests. Ischemic heart disease is also under study. The MPS, thus, provides another line of human inquiry to assist in decision-making and policy guidance for space missions beyond earth orbit.

Introduction

I'd like to die on Mars, just not on impact – Elon Musk

Much is known about the health effects of brief high-dose exposures such as those received by patients during the treatment of benign and malignant conditions and by survivors of the atomic bombings in Japan (UNSCEAR 2008; Ozasa et al. 2012, 2018; Grant et al. 2017). However, the concerns today are related to the health consequences from prolonged low-dose exposure received over a period of years (Shore et al. 2017). The Million Person Study of Low-Dose Health Effects (MPS) was designed to precisely estimate radiation risks from chronic exposures (NCRP 2018a; Boice et al. 2019a) and to address issues of the appropriateness of the linear non-threshold dose-response model as used in radiation protection (Boice 2017a; Shore et al. 2018, 2019; NCRP 2018b). Because of the diversity of exposure circumstances within the MPS (Boice et al. 2006, 2019a; Bouville et al. 2015; Ellis et al. 2018; Leggett et al. 2018a, 2018b; Dauer et al. 2018; NCRP 2018a), the MPS has the ability to address specific issues of interest to the National Aeronautics and Space Administration (NASA), and its system of radiation protection for astronauts (NCRP 2000, 2006, 2014, 2016; NA/NRC 2012; Cucinotta et al. 2013; Cucinotta 2015; NASA 2015). The MPS populations include 170,000 medical radiation workers, 140,000 nuclear power plant workers, 130,000 industrial radiographers, 114,000 atomic veterans who participated in above-ground atmospheric weapons tests, and 360,000 workers at US Department of Energy (DOE) nuclear facilities. The large study size and broad range of doses, coupled with a very long follow-up period and consistent methodological approaches, indicates the substantial quality and statistical ability to quantify and clarify the health risks from exposures that are received gradually over time.

Because of the relatively high cumulative doses possible during a mission to Mars and return, cosmic radiation is arguably the limiting factor in going to Mars (Shay et al. 2011; Zeitlin et al. 2013) in addition to the many stresses that flight crews would have to deal with that affect behavioral health, performance and cognition, e.g. sleep and circadian rhythm disturbances, CO_2 build-up, microgravity, isolation and confinement, mission duration, workload and psychosocial adaptation, including team coordination, cooperation, and communication (NCRP 2014, 2016; NASA 2015). Astronauts on a Mars mission would receive approximately 200 mGy from the exposure to high-energy protons but perhaps more importantly, a comparable dose from the heavy ions of high atomic number and high energy (HZE particles) traveling at high speed. The NASA-defined effective dose (Cucinotta et al. 2013, 2013; NA/NRC 2012; Zeitlin et al. 2013; Norbury et al. 2016; Simonsen and Zeitlin 2017) would be about 900–1200 mSv received at a rate of about 1 mSv d^{-1} which would be similar to receiving a whole-body CT examination every 7 or 8 days. Overviews are available describing the many aspects of the MPS that are relevant to NASA, space exploration and, perhaps someday, even colonization (Davies and Schulze-Makuch 2011; Boice 2014a, 2014b, 2015a, 2015b, 2016, 2017b; Bouville et al. 2015; Boice et al. 2018a; NCRP 2018a; see also Davies and Schulze-Makuch 2011) and will be summarized here.

NASA is unique in applying an individual risk-based system for radiation protection that considers an individual's lifetime radiation risk of cancer mortality that accounts for the astronaut's age, sex, smoking history, and other risk factors as known (NA/NRC 2012; Cucinotta et al. 2013; NCRP 2014; NASA 2015). This is in contrast to the system of dose limits for occupational exposures used by terrestrial-based organizations which is based on age-averaged, sex-averaged risk estimates, subjective tissue-weighting factors and other considerations for computing effective dose and demonstrating compliance for regulatory purposes (ICRP 2007). NASA sets a permissible career exposure limit of a 3% lifetime Risk of Exposure-Induced Death (REID) from cancer at a 95% confidence level to account for uncertainties in risk projection assumptions (NCRP 2006; NA/NRC 2012; Cucinotta et al. 2013; NASA 2015). Once these limits are exceeded, allowable time in space is reduced. Current estimates are based on the Japanese atomic bomb survivor life span study (LSS) of acute exposures among 86,000 survivors, and about 54,000 adults (Ozasa et al. 2012). There is some evidence from epidemiologic studies that prolonged exposures over a period of years may result in a lower estimate of radiation-related cancer risk than seen following the acute exposure received by Japanese bomb survivors (Shore et al. 2017). The MPS will reduce the statistical uncertainty in the LSS risk estimates, narrow the 95% confidence (credibility) interval, and thus allow more time in space for flight crew even if the lifetime risk estimates turn out to be similar to those from the atomic bomb survivor study.

Because of the sex-specific differences in radiation-induced cancer, female flight crew have reduced allowable time in space due to their higher risk of radiation-induced cancer based on the Japanese LSS (NCRP 2014). This difference is not only because the much higher female radiation-risk of breast and ovarian cancer than the male radiation-risk of testicular and prostate cancer, but is related in large part to a comparatively high risk of lung cancer, about 2–3 times higher, in women that observed for men. The MPS has nearly 250,000 women which will provide a very strong evaluation of any risk differences by sex for all cancers including lung cancer in particular. The ongoing study of 170,000 medical radiation workers is designed to address sex-specific differences in risk focusing on lung cancer dosimetry and lung cancer risks (Yoder et al. 2018; Boice et al. 2019b).

One major difference between the study of Japanese atomic bomb survivors exposed in 1945 and occupationally exposed workers is the dose rate, i.e. the time over which the exposures were delivered (Shore et al. 2017). The Japanese LSS can only address brief exposures that occurred in a very short period of time, whereas worker exposures occurred over many years if not decades. The second major difference is that healthy American workers are more representative of today's workers, the public, and flight crews than the 1945 population of Japanese atomic bomb survivors.

Animal experiences following brief exposure to high-velocity heavy ions (HZE particles) simulating GCR exposures in space indicate detrimental early and late effects of space irradiation on behavior and cognitive performance (NCRP 2016). These studies have raised concern about possible effects on astronauts that might impair performance so that the mission would not be completed or, conceivably, there might be a risk of Alzheimer's or dementia years after the voyage is over (Cherry et al. 2012; Cucinotta et al. 2014; Limoli 2015; NCRP 2016). There is no analog of human studies or circumstances that can approximate these heavy-ion (HZE) exposures to brain tissue. If fact, there is no evidence in human studies that low-LET radiation is associated with dementia or Alzheimer's disease (Yamada et al. 2009). However, worker cohorts at several DOE facilities (e.g. Mound, Rocketdyne, Mallinckrodt, Middlesex, Los Alamos National Laboratory, Rocky Flats, and Hanford) had intakes of radionuclides such as polonium, plutonium, radium, uranium, and americium which provide alpha particle dose to the soft tissue within the brain and not from depositions in bone (Leggett et al. 2018a; Boice et al. 2018b).

Although mortality is a weak indicator of behavioral issues, large numbers of deaths due to dementia, Alzheimer's disease, Parkinson's disease, and motor neuron disease are being evaluated with regard to brain tissue dose from high-LET alpha particles coupled with low-LET gamma radiation (Boice et al. 2014; Boice 2017b). More to the point, however, ongoing studies of workers at nuclear facilities with high intakes of radionuclides could incorporate cognition tests similar to those astronauts take. Clinically based diagnoses of dementia, Alzheimer's and other motor neuron conditions, including behavioral and cognition impairment, also could be evaluated within Medicare or other nursing record files (Rector et al. 2004; CMS 2018a, 2018b).

The study of workers with intakes or radionuclides that expose brain tissue to alpha particles has the following strengths: (1) humans and not rodents are being studied; (2) the exposure occurs over many years (the radionuclides have very long half-lives and continually expose brain tissue for years after intake) and not over minutes or weeks as in the rodent experiments; (3) the radionuclides emit alpha particles (He nuclei) which constitute about 14% of the GCR flux (though of lower energy); (4) the working environment would include many years of concomitant exposure to gamma rays (low-LET radiation) which would contribute about 90% or more to the brain dose which can be compared with the 85% contribution of protons (low-LET radiation) to the GCR flux (Zeitlin et al. 2013; Norbury et al. 2016) and (5) standardized tests of cognitive function (Basner et al. 2015; CMS 2018b) are available or could be administered, and dose-response analyses made, i.e. the test scores could be contrasted with alpha-particle brain dose as estimated from extensive urinary excretion data on individual workers. The alpha-particle dose would be computed from intake up until the time of diagnosis and/or of testings. The weakness of this study is that HZE particles are not being evaluated. While HZE particles are of high energy and would interact with neural tissue differently than deposited radionuclides emitting alpha particles, the approximately uniform distribution of alpha particle dose to brain tissue might be a similarity.

A previous analysis of heart mortality among Apollo astronauts claimed an elevated rate of heart disease (Delp et al. 2016), but was severely limited (Cucinotta et al. 2016) and not borne out with subsequent study using a more valid methodology (Elgart et al. 2018). However, it remains uncertain whether low doses (<500 mGy) cause ischemic heart disease (IHD) (NCRP 2018a; Shore et al. 2019) and whether cumulative exposures to chronic low-dose low-LET or high-LET galactic cosmic rays should be considered a detriment of space travel (Cucinotta et al. 2015; NASA 2015). The size of the MPS will allow comprehensive evaluations of non-malignant conditions such as IHD, including both low-LET and high-LET exposures (Boice 2017b; Golden et al. 2018, 2019; NCRP 2018a).

Overview of relevant MPS cohorts

People will visit Mars, they will settle mars, and we should because it's cool – Jeff Bezo

The various cohorts within the MPS are of such diverse nature that they can contribute to research efforts that are relevant to the NASA radiation protection program (Table 1). The combination of all cohorts will provide the most precise estimates of radiation risk following low-dose-rate exposures over years yet possible. This precision will provide guidance as to the relevancy of the continued use of the Japanese LSS study of acute exposures for use in model projections that are used to set individual guidance limits. The large number of women in the MPS, nearly a quarter of a million, will provide evidence as to any difference of sex-specific radiation risks, particularly for lung cancer, following chronic exposures. The major cohorts within the MPS to evaluate sex-specific differences include the medical radiation workers, nuclear power plant workers, industrial radiographers, as well as the workers at Mound, Los Alamos National Laboratory, Rocky Flats, Hanford and other facilities. The DOE workers with intakes of radionuclides, such as plutonium, will provide estimates of radiation dose from high-LET particles to various organs such as the lung, brain and heart which can be used in the evaluation of organ-specific conditions, e.g. lung cancer, as well as the possibility of looking at any association with dementia and Alzheimer's disease and, conceivably, cognitive behavioral impairments.

Anticipated contributions

You can see a lot just by observing – Yogi Berra

The MPS is so large, so complete, with such comprehensive dosimetry and long follow-up that it will make critically important contributions regarding the level of risk when radiation is received over a period of years and not briefly (GAO 2017; Davis et al. 2019) (Table 1). Gradual exposures to low doses at low-dose rates are the type of exposures experienced by workers, the public, and flight crew. For NASA and the decisions that will have to be made with regard to radiation protection for astronauts on long missions beyond Earth orbit, the MPS will provide precise estimates of cancer risk following prolonged exposure that can be incorporated into lifetime risk projections; the MPS will provide precise estimates of sex-specific radiation risks for lung and other cancers that can also be incorporated into NASA lifetime risk models for radiation guidance for individual astronauts; the MPS will provide risk estimates following chronic exposure to brain tissue from high-LET radiation that might be relevant, in certain ways, as a human analog to GCR experienced in space. There is interest in integrating the MPS epidemiology with biologically based models to improve risk estimates at low doses (Preston 2015; Rühm et al. 2017). Finally, the MPS will contribute knowledge regarding the risks of non-malignant diseases such as ischemic heart disease and central nervous system disorders and whether such conditions might be relevant to adding to models of detriment following radiation exposures received at low doses and at low-dose rates.

Better risk estimates for cancer following prolonged exposure

There's nothing I would like more than to watch a manned Mars landing – Andy Weir

The MPS is so large that the estimates of radiation risk following prolonged or chronic exposures over a period of years will be very precise (Table 1). NASA uses an individual risk-based system for radiation protection based on the lifetime risk estimates considering the radiation risk and the astronaut's characteristics such as age, sex, smoking history,

Table 1. Relevance of the Million Person Study (MPS) research to NASA's missions in space.

Issue for NASA	MPS relevant research
Dose rate	
Radiation Protection Guidance is based primarily on the acute (brief) exposures received by Japanese atomic bomb survivors in 1945.	MPS workers accumulated radiation exposures over a period of years and thus at a dose rate more similar to what astronauts would receive during a 2–3 year mission to Mars.
Representativeness	
The Japanese survivors lived in a war-torn country and faced malnutrition, infections, and deprivations after the War.	The MPS consists of healthy male and female workers and thus more similar to healthy astronauts.
Time in space	
Radiation Protection Guidance is based on the 95% upper confidence level about a 3% increased lifetime risk of cancer death from the Japanese LSS.	The MPS is 10 times larger than the Japanese LSS. Statistical variability will be reduced, 95% confidence levels narrowed, and more time in space allowed, all things being equal.
Time in space for women	
The Japanese life span study (LSS) finds women to be at 2–3 times greater lifetime risk of dying from cancer than men, particularly for lung cancer, and thus women are allowed less time in space than men.	The Japanese data include ~32,000 adult women and ~21,000 adult men. The MPS has 250,000 adult women; and a cohort of 170,000 medical radiation workers (1/2 females) is designed to address differences in sex-specific lung cancer risks.
High-let effects on brain and other tissues	
Galactic Cosmic Radiation (GCR) consists of heavy ions traveling through space at high energy. Animal studies at high-dose rate (about a minute) indicate a decrease in cognitive abilities and even development of Alzheimer's disease.	There are no human analogues or circumstances similar to GCR exposures in space. MPS workers at DOE facilities comes close: intakes of radionuclides, e.g. plutonium, continually exposed brain tissue to alpha particles (Helium nuclei) for life. Human brain tissue exposed to low-dose-rate high-LET radiation is being evaluated with large numbers and long follow-up. All organs can be similarly evaluated.
Dementia, Alzheimer's other CNS	
Modified mouse strains indicate an early-onset of Alzheimer's disease after heavy ion exposures.	The MPS will evaluate deaths from dementia, Alzheimer's, Parkinson's and other motor neuron diseases following brain exposure to alpha particles from plutonium, polonium, radium, uranium, and americium.
Cognition	
Animal studies of high-dose rate exposures to energetic heavy ions (GCRs) find a decrease in measures of cognitive function. The concern is whether a similar decline might affect astronauts to such an extent that the mission would be in jeopardy.	Thousands of workers with alpha particle exposures to brain tissue could be evaluated for the occurrence of dementia and cognitive impairment, including neuropsychological test scores, via Medicare records. Further, though challenging, large numbers of workers could be asked to take cognition tests similar to those taken by astronauts.
Ischemic heart disease (IHD)	
It is uncertain whether low doses (<500 mGy) cause ISH and whether cumulative exposures to chronic low-dose low-LET or high-LET galactic cosmic rays should be considered a detriment of space travel.	The MPS has substantial statistical power to evaluate the risk of ischemic heart disease (IHD) related to radiation doses below 300–500 mGy.
Uncommon cancer sites	
Uncommon or infrequent cancer sites within the Japanese LSS cannot be evaluated because of small numbers.	The MPS is so large, that all cancer sites (by sex) can be evaluated and incorporated explicitly into lifetime risk projection models.

and more. Because there is uncertainty with regard to the radiation risk from space radiation as well as the uncertainties of converting an acute radiation risk from a 1945 Japanese population to an American cohort of astronauts, NASA sets their limitation guidance based on the 95% upper confidence (or credibility) level about the permissible career exposure limit of a 3% lifetime Risk of Exposure-Induced Death (REID) (NCRP 2014; NASA 2015). The MPS will reduce the statistical uncertainty in the risk estimates being used for lifetime cancer risk projections; will provide relevant estimates of risk based on chronic prolonged exposures which eliminate the need for making adjustments based on dose rate considerations (Shore et al. 2017); will alleviate the need to adjust for differences in underlying background cancer rates between an Asian population and a western population, thus removing this contribution to uncertainty; and will provide sex-specific radiation-risk estimates for uncommon and infrequent cancer sites not possible within the Japanese LSS because of small numbers. Reducing these measures of uncertainty, would reduce the 95% confidence (credibility) level and thus allow more time in space given current radiation protection approaches, all things being equal. That is, that the risk levels from the MPS turn out to be the same or less than those of the atomic bomb survivors which appears to be the case for radiation-induced lung cancer (Boice et al. 2019b).

Better estimates of sex-specific radiation risk for lung and other cancers

> If you don't know where you're going, you might not get there
> – Yogi Berra

Lung cancer is the largest potential cancer risk from space travel (Shay et al. 2011). This risk is not entirely from gamma radiation, but from high-energy protons from solar flares (which occur 5 to 10 times per year) and from GCR (the HZE particles such as Fe). It is unclear whether sex-specific lung cancer risks differ appreciably between women and men. Based on the Japanese LSS data (Ozasa et al. 2012; NCRP 2014; Cahoon et al. 2017), women are about 2–3 times more sensitive to radiation than men, due in part to the differences in the sensitivity of the sex organs, i.e. cancers of the female breast and ovaries are increased after radiation exposure, whereas cancers of the testis and prostate are not. But the major difference is due to a 2–3 times greater risk of lung cancer among women than men. This difference limits the time female astronauts can spend in space as well as their potential departure on long-term missions such as to Mars (Table 1). The Japanese data, however, are based on relatively small numbers, only about 32,000 adult women and about 21,000 adult men. The MPS has 250,000 adult females under study so that precise comparisons with the 750,000 adult males in the study will be possible (Boice et al. 2019b).

In addition to the ongoing comparisons between male and female risks for lung cancer in the ongoing studies mentioned (Boice et al. 2019b), the definitive comparison will be with the ongoing study of 170,000 medical radiation workers of which 65,000 are men and 65,000 are women. The dose distribution is broad and organ doses are being comprehensively being determined (Yoder et al. 2018). These data may influence the current models used by NASA to project lifetime estimates of cancer risk and thus modify the amount of time that astronauts are allowed in space taking into account the most accurate and relevant estimates of sex-specific risks. Further, several of the studies will provide information on high-LET exposures to the lung and other organs from the intake of radionuclides for men and women which also should contribute to the understanding of any sex-specific differences.

A new study will be of 21,000 female workers who worked from 1943 to 1947 at the Tennessee Eastman Corporation in Oak Ridge, TN. These women had never been studied and are mentioned in the novel *The Girls of the Atomic City* (Kiernan 2013). They worked in the facility that eventually became known as Y-12 and many operated the 1052 calutrons, i.e. mass spectrometers that used electromagnetic fields to separate uranium-235 from natural uranium for the World War II effort (specifically for the 'Little Boy' weapon dropped on Hiroshima).

Risk estimates following chronic exposure to brain tissue from high-LET radiation

> It's not going to do any good to land on Mars if we're stupid – Ray Bradbury

High-LET particles in space pose a great mystery since it is unknown what human health effects might result from these exotic high energy and high atomic number (Z) ions (HZE particles). The challenging questions concern their biological effectiveness to cause cancer, non-cancers, and cognitive impairment and behavior changes. Rodent experimental models have raised these important issues but are not entirely satisfactory. The MPS attempts to address some of these issues in studies of humans with brain exposure to helium nuclei (alpha particles) as another line of evidence supporting or not some of the experimental results.

The radiation exposures in space are reasonably well-characterized but are extremely complex and vary over time (Cucinotta et al. 2013; NA/NRC 2012; Zeitlin et al. 2013; Borak et al. 2014; Norbury et al. 2016; Simonsen and Zeitlin 2017; NCRP 2018a). The flux of galactic cosmic rays (GCRs) 'consists of about 85% protons… and about 14% helium ions. The remainder of the flux consists of heavier ions referred to as "HZE particles" [high (H) atomic number (Z) and high energy (E)]' (Zeitlin et al. 2013). Further, the absorbed dose depends not only on the changing space radiation environment, but also on mission time, shielding characteristics of the space capsule and of structures on Mars, and any radionuclide power sources brought on board. Depending on the assumptions made regarding exposure scenarios for a 30 month roundtrip journey to Mars, including a 6 month stay on the surface, the absorbed dose from all sources of exposure might range from 250 to 400 mGy or about 1–1.2 Sv (NASA 'effective dose' or 'dose equivalent') (NA/NRC 2012; Cucinotta et al. 2013; Zeitlin et al. 2013; Norbury et al. 2016).

The exposure scenarios and assumptions made include: solar particle events (SPEs)—mainly high-energy protons ($Z = 1$, 130–180 mGy); GCRs: He ($Z = 2$, 45–70 mGy); $Z =$ 3–9 particles (20–40 mGy); $Z > 10$ particles (30–40 mGy), and neutrons and other particles (20–30 mGy); the thickness of the space capsule; secondary radiation, e.g. neutrons and pions, from proton and GCR fragments after impacts with the space capsule; shielding characteristics of structures on the surface of Mars; any nuclear isotope power sources transported with the astronauts; and the 'quality factors' selected for the menagerie of space radiations (the fluence of low and high Z particles from atomic numbers 1–28 and their energy spectra) to multiply times the absorbed does to obtain a dose equivalent measure in mSv [i.e. a NASA defined 'effective dose' that applies weighting factors that are not the ICRP tissue weighting factors (ICRP 2007)], and the extent of solar activities during the 30 month journey and sojourn (Cucinotta et al. 2013; Zeitlin et al. 2013; NASA 2015). The experimental simulations of the radiation environments in outer space and on Mars is to say the least exceedingly challenging (Norbury et al. 2016).

The MPS can provide evidence of high-LET health effects associated with continuous, low-dose-rate exposures to human brains following intakes of alpha-particle-emitting radionuclides. Ingested or inhaled intakes of radionuclides that decay by emitting alpha particles can deposit in the brain and expose brain tissue for many years (Leggett et al. 2018a; Boice et al. 2018b). Such exposure is more uniform across brain tissue and differs from alpha particle exposure that might occur from bone-seeking radionuclides that deposit in the cranium. Currently, an NCRP committee with colleagues at the Oak Ridge National Laboratory are developing biokinetic models for estimating brain dose directly following the intake of radionuclides and not by using the current indirect approach that considers dose to all soft tissue and apportions the dose to brain based on weight or other considerations (Leggett et al. 2018a).

Postmortem radiochemical analyses of tissues have shown that intakes of radium deposit in the brain and result in alpha particle exposure to brain tissue (Schlenker et al. 1982). Similarly, radiochemical analyses have detected brain depositions for radionuclides such as plutonium (James et al. 2003, 2007), americium (McInroy et al. 1985, 1991), uranium (Avtandilashvili et al. 2015; Kathren and Tolmachev 2015), and polonium (Nathwani et al. 2016). Thorium has been detected in brain tissue of a patient injected with Thorotrast, a colloidal radiographic contrast medium containing thorium dioxide (McInroy et al. 1992). The generosity of radiation workers and patients made this remarkable research possible by donating their bodies for scientific research to the United States Transuranium and Uranium Registries (USTUR) (Filipy and Russell 2003; Tolmachev et al. 2011; Tolmachev and McComish 2016; USTUR 2018). In addition to postmortem radiochemical analysis of tissues, location of radionuclide deposition within the brain is to be examined by synchrotron-based X-ray fluorescent microscopy using the Advanced Photon Source at Argonne National Laboratory (Paunesku et al. 2012; Chen et al. 2015).

Limitations to these evaluations include the different dose distributions within the brain and the energy depositions from alpha particle emitters compared with that of high-energy particles and heavy ions (HZE particles) experienced in space (Cucinotta et al. 2013; Zeitlin et al. 2013; Norbury et al. 2016; Boice 2017b). Nonetheless, there is no other human analog/circumstance that provides information on possible health risks from high-LET radiation to brain tissue, recognizing the limitations (Boice et al. 2018a). The mouse models provide valuable insights into effects of space radiation on the brain, but they too have sizeable uncertainties. The mouse, for one, has considerable limitations as a laboratory model for the human brain. Further, the dose delivery in experimental studies is brief and of the order of minutes to possibly weeks compared with the years it will take for a mission to Mars. In addition, while the space environment is relatively well-characterized as described above, the experimental simulations of space radiations are far from perfect, i.e. per above, ideally the rodent should be exposed to high energy low-LET protons, high energy alpha particles, a menagerie of HZE particles (from atomic number 3–28), neutrons, pions, other fragmented particles from shielding GCR shielding impacts, and so on. It's an impossible task, so a wide-range of brilliant experiments with, hopefully, equally brilliant assumptions, are brought to bear to provide optimum answers to health effects in mice that are relevant to health effects in women and men (Cucinotta et al. 2013; Norbury et al. 2016).

The current research within the MPS (Table 1) will add to the collection of evidence, albeit with large uncertainties, that are needed to inform radiation protection guidance against the possible health effects of galactic cosmic radiation on brain function (NCRP 2016).

High-level cognitive performance is critically important during space missions, which can involve environmental, physiological and psychological stressors (Basner et al. 2015; Moore et al. 2017). Radiation exposure to GCRs is a possible stressor that may affect brain function (Cucinotta et al. 2014; NCRP 2016). Interestingly, a recent experimental study hints at possible sex-based differences in cosmic ray-induced cognitive decline (Krukowski et al. 2018; Wenz 2018). NASA has supported the development of tests that assess multiple domains of neurocognitive functions: Cognition Test Battery and WinSCAT (Moore et al. 2017) which have been administered to astronauts. Investigators within the MPS are considering the value of evaluating cognition among workers who have had continuous brain tissue exposure to alpha particles for up to 40 years after the intake of plutonium or other radionuclides. Some workers today come back to national laboratories voluntarily and provide urine samples, suggesting a willingness to participate in health research. Over 1000 workers could be identified with known radiation doses to brain who might be contacted and asked to complete a modified cognitive test/s similar to ones taken by astronauts on the International Space Station or elsewhere. A cognition test could be administered at home over the internet and take <30 min. In addition, for workers who had reached the age of 65 years in or after 1996, Medicare and other nursing home records might be searched for any diagnoses of cognitive decline, dementia, Alzheimer's disease or other motor neuron conditions as well as for any quantitative scores from neuropsychological testing (Rector et al. 2004; CMS 2018a, 2018b). Then the diagnoses (and/or cognitive scores) could be correlated with estimated alpha-particle dose to brain tissue, providing another line of evidence for the possibility that high-LET radiation might lead to cognitive and behavioral changes (Table 1).

Risk estimates for ischemic heart disease for prolonged exposures

Love hurts, love scars, love wounds, and mars – Roy Orbison

There appears to be little convincing or consistent evidence that low doses of ionizing radiation below about 0.5 Gy cause cardiovascular disease, particularly ischemic heart disease (NCRP 2018b; Shore et al. 2019). Nonetheless, there is concern whether cumulative exposures to chronic low-dose radiation and to high-LET galactic cosmic rays might be a detriment of space travel (Cucinotta et al. 2015; NASA 2015). The MPS has substantial statistical power to evaluate any risk of ischemic heart disease (IHD) related to radiation doses below 300–500 mGy. Further, radionuclides intakes among DOE workers also contribute to dose to all soft tissue, including the heart, and can be evaluated. Ongoing evaluations, including dose reconstructions, would contribute to understanding the possibility of IHD risks following low-LET radiation and high-LET radiation to heart tissue. Thus far, MPS studies have examined the risk of IHD following exposures to nuclear weapons test participants (Boice 2017b), nuclear power plant workers (NCRP 2018b; Golden et al. 2018), industrial radiographers (NCRP 2018b) and workers at the Mound (Boice et al. 2014), Rocketdyne (Boice et al. 2011), and Mallinckrodt (Golden et al. 2019) facilities. Combining these and ongoing studies, and refining the radiation dose estimates to heart tissue, will make a substantial contribution to knowledge on the possible risk of IHD following prolonged exposures to ionizing radiation accumulating to relatively high levels (Table 1).

Summary

The future belongs to those who believe in the beauty of their dreams – Eleanor Roosevelt

The Million Person Study will address scientific and public health issues on radiation health effects. Important research gaps will be filled to benefit scientific understanding, regulatory agencies, radiation protection organizations, and society (GAO 2017; Boice et al. 2019a). The scientific information will benefit decision makers and policy setters whose responsibility is to provide protection guidance. NASA has the challenging mission of going beyond Earth orbit and potentially to Mars. Informed judgments have to be made today amid a recognized sea of uncertainties (Cucinotta et al. 2013, 2015b; Scoles 2017; Chancellor et al. 2018). Radiation is a limiting factor for long-term missions so that as much knowledge as can be obtained now will provide guidance for the protection of flight crews (Shay et al. 2011; NASA 2015). Ways to reduce the level of uncertainties are a hallmark of the MPS. The MPS of healthy men and women exposed to radiation over years are more representative of astronauts and their exposure circumstances than is the Japanese atomic bomb survivor study of acute exposures in 1945. The large size of the MPS will provide more precise estimates of lifetime cancer risk that would reduce statistical variation and thus allow more time in space, all things being equal. Differences in sex-specific lung cancer risks from chronic radiation exposures is being evaluated that would affect female astronauts and their time in space. Whether high-LET radiation to the brain affects the late development of Alzheimer's disease is being evaluated; ways to evaluate cognition, behavior and cognitive impairment are being developed. As the United States continues to have unbounded vision and imagination to travel to other planets, it is critical that the best science available be used in the judgments to be made today on the protection of space travelers. If we can perceive it, we can achieve it. But the achievement can be based only on a strong scientific foundation (Table 1).

It's kind of fun to do the impossible – Walt Disney

Disclosure statement

The author reports no conflicts of interest. The author alone is responsible for the content and writing of the paper.

Funding

This work was supported in part by a grant from the U.S. Nuclear Regulatory Commission [NRC-HQ-60-14-G-0011], grants from the Centers for Disease Control and Prevention [5UE1EH000989, 5NUE1EH001315], grants from the National Aeronautics and Space Administration [NNX15AU88G, 80NSSC17M0016], and grants from the U.S. Department of Energy (Grant No. DE-SC0008944 awarded to the National Council on Radiation Protection and Measurements, which included interagency support from the U.S. Nuclear Regulatory Commission, the U.S. Environmental Protection Agency, and the National Aeronautics and Space Administration; and more recent Grants No. DE-AU0000042 and DE-AU0000046). Contract support also was recently received from the Naval Sea Systems Command [N00024-17-C-4322]. Further, contract support was received by Oak Ridge National Laboratory from the Office of Radiation and Indoor Air, U.S. Environmental Protection Agency, under Interagency Agreement DOE No. 1824 S581-A1, under contract no. DE-AC05-00OR22725 with UT-Battelle; and contract support was received by Oak Ridge Associated Universities from the U.S. Department of Energy under contract no. DE-SC0014664.

ORCID

John D. Boice http://orcid.org/0000-0002-8755-1299

References

Avtandilashvili M, Puncher M, McComish SL, Tolmachev SY. 2015. U.S. Transuranium and Uranium Registries case study on accidental exposure to uranium hexafluoride. J Radiol Prot. 35:129–151.

Basner M, Savitt A, Moore TM, Port AM, McGuire S, Ecker AJ, Nasrini J, Mollicone DJ, Mott CM, McCann T, et al. 2015. Development and validation of the cognition test battery for spaceflight. Aerosp Med Hum Perform. 86:942–952.

Boice JD Jr. 2014a. NASA – the final frontier in radiation protection. Health Phys News. XLII:20–22. [accessed 2018 October 15]. https://ncrponline.org/wp-content/themes/ncrp/PDFs/BOICE-HPnews/22%20NASA_FINAL_Mar%202014_white.pdf

Boice JD Jr. 2014b. Men and women on Mars. Health Phys News. XLII:16–17. [accessed 2018 October 15]. https://ncrponline.org/wp-content/themes/ncrp/PDFs/BOICE-HPnews/27_NASA-to-Mars_Aug2014.pdf

Boice JD Jr. 2015a. Going to Mars is more than curiosity. Health Phys News. XLIII:23–25. [accessed 2018 October 15]. https://ncrponline.org/wp-content/themes/ncrp/PDFs/BOICE-HPnews/34_Space_Radiation_Mars_Submariners_Mar2015.pdf

Boice JD Jr. 2015b. The million worker study – from whence it came. Health Phys News. XLIII:22–23. [accessed 2018 September 9]. http://ncrponline.org/wp-content/themes/ncrp/PDFs/BOICE-HPnews/37_MWS_June2015.pdf

Boice JD Jr. 2016. Mars matters! Health Phys News. XLIV:32–33. [accessed 2018 October 15]. http://ncrponline.org/wp-content/themes/ncrp/PDFs/BOICE-HPnews/46_Mars_Matters_Mar2016.pdf

Boice JD Jr 2017a. The linear nonthreshold (LNT) model as used in radiation protection: an NCRP update. Int J Radiat Biol. 93:1079–1092.

Boice JD Jr 2017b. Space: the final Frontier-Research Relevant to Mars. Health Phys. 112:392–397.

Boice JD Jr, Cohen SS, Mumma MT, Ellis ED, Eckerman KF, Leggett RW, Boecker BB, Brill AB, Henderson BE. 2011. Updated mortality analysis of radiation workers at Rocketdyne (Atomics International), 1948-2008. Radiat Res. 176:244–258.

Boice JD Jr, Cohen SS, Mumma MT, Ellis ED, Cragle DL, Eckerman KF, Wallace PW, Chadda B, Sonderman JS, Wiggs LD. 2014. Mortality among mound workers exposed to polonium-210 and other sources of radiation, 1944-1979. Radiat Res. 181:208–228.

Boice JD Jr, Cohen SS, Mumma MT, Ellis ED. 2019a. The million person study, from whence it came and why. Int J Radiat Biol. doi:10.1080/09553002.2019.1589015

Boice JD Jr, Ellis ED, Golden AP, Girardi DJ, Cohen SS, Chen H, Mumma MT, Shore RE, Leggett RW. 2018a. The past informs the future: an overview of the million worker study and the Mallinckrodt Chemical Works cohort. Health Phys. 114:381–385.

Boice JD Jr, Ellis ED, Golden AP, Zablotska LB, Mumma MT, Cohen SS. 2019b. Sex-specific lung cancer risk among radiation workers in the Million Person Study and among TB-fluoroscopy patients. Int J Radiat Biol. doi:10.1080/09553002.2018.1547441

Borak TB, Heilbronn LH, Townsend LW, McBeth RA, de Wet W. 2014. Quality factors for space radiation: a new approach. Life Sci Space Res (Amst). 1:96–102.

Boice JD Jr, Leggett RW, Eckerman KF, Tolmachev SY, Woloschak GE, Golden AP, Ellis ED. 2018b. Response to Mortazavi et al. On detecting bone-seeking radionuclides in brain tissue. Health Phys. 115:389–390.

Boice JD Jr, Leggett RW, Ellis ED, Wallace PW, Mumma M, Cohen SS, Brill AB, Chadda B, Boecker BB, Yoder RC. 2006. A comprehensive dose reconstruction methodology for former Rocketdyne/Atomics International radiation workers. Health Phys. 90:409–430.

Bouville A, Toohey RE, Boice JD Jr, Beck HL, Dauer LT, Eckerman KF, Hagemeyer D, Leggett RW, Mumma MT, Napier B, et al. 2015. Dose reconstruction for the million worker study: status and guidelines. Health Phys. 108:206–220.

Cahoon EK, Preston DL, Pierce DA, Grant E, Brenner AV, Mabuchi K, Utada M, Ozasa K. 2017. Lung, laryngeal and other respiratory cancer incidence among Japanese atomic bomb survivors: an updated analysis from 1958 through 2009. Radiat Res. 187:538–548.

[CMS] Centers for Medicare and Medicaid Serves. 2018a. Minimum Data Set 3.0 Public Reports. [accessed 2019 February 2]. https://www.cms.gov/Research-Statistics-Data-and-Systems/Computer-Data-and-Systems/Minimum-Data-Set-3-0-Public-Reports/index.html

[CMS] Centers for Medicare and Medicaid Services. 2018b. Long-Term Care Facility Resident Assessment Instrument 3.0 User's Manual. Version 1.16. [accessed 2019 February 2]. https://downloads.cms.gov/files/1-MDS-30-RAI-Manual-v1-16-October-1-2018.pdf

Chancellor JC, Blue RS, Cengel KA, Auñón-Chancellor SM, Rubins KH, Katzgraber HG, Kennedy AR. 2018. Limitations in predicting the space radiation health risk for exploration astronauts. NPJ Microgravity. 4:8.

Chen S, Paunesku T, Yuan Y, Jin Q, Hornberger B, Flachenecker C, Lai B, Brister K, Jacobsen C, Woloschak G, et al. 2015. The bionanoprobe: synchrotron-based hard x-ray fluorescence microscopy for 2D/3D trace element mapping. Micros Today. 23:26–29.

Cherry JD, Liu B, Frost JL, Lemere CA, Williams JP, Olschowka JA, O'Banion MK. 2012. Galactic cosmic radiation leads to cognitive impairment and increased aβ plaque accumulation in a mouse model of Alzheimer's disease. PLoS One. 7:e53275.

Cucinotta FA. 2015. Review of NASA approach to space radiation risk assessments for Mars exploration. Health Phys. 108:131–142.

Cucinotta FA, Alp M, Rowedder B, Kim MH. 2015. Safe days in space with acceptable uncertainty from space radiation exposure. Life Sci Space Res (Amst). 5:31–38.

Cucinotta FA, Alp M, Sulzman FM, Wang M. 2014. Space radiation risks to the central nervous system. Life Sci Space Res. 2:54–69.

Cucinotta FA, Hamada N, Little MP. 2016. No evidence for an increase in circulatory disease mortality in astronauts following space radiation exposures. Life Sci Space Res (Amst). 10:53–56.

Cucinotta FA, Kim M-H, Chappell LJ. 2013. Space radiation cancer risk projections and uncertainties—2012. NASA/TP-2013-217375. Houston (TX): NASA Johnson Space Center; [accessed 2018 October 19]. https://spaceradiation.jsc.nasa.gov/irModels/TP-2013-217375.pdf

Dauer LT, Bouville A, Toohey RE, Boice JD Jr, Beck HL, Eckerman KF, Hagemeyer D, Leggett RW, Mumma MT, Napier B, et al. 2018. Dosimetry and uncertainty approaches for the million-worker study of radiation workers and veterans: overview of the recommendations in NCRP Report No. 178. Review. Int J Rad Biol. DOI: 10.1080/09553002.

Davis J, Dewji SA, Abelquist E, Hertel N. 2019. Synopsis of the oak ridge radiation protection research needs workshop. Health Phys. 116:69–80.

Davies P, Schulze-Makuch D, editors. 2011. The human mission to Mars: colonizing the red planet. Cambridge (MA): Cosmology Scientific Publishers..

Delp MD, Charvat JM, Limoli CL, Globus RK, Ghosh P. 2016. Apollo lunar astronauts show higher cardiovascular disease mortality: possible deep space radiation effects on the vascular endothelium. Sci Rep. 6:29901.

Elgart SR, Little MP, Chappell LJ, Milder CM, Shavers MR, Huff JL, Patel ZS. 2018. Radiation exposure and mortality from cardiovascular disease and cancer in early NASA astronauts. Sci Rep. 8:8480.

Ellis ED, Boice JD Jr, Golden AP, Girardi DJ, Cohen SS, Mumma MT, Shore RE, Leggett RW, Kerr GD. 2018. Dosimetry is key to good epidemiology: workers at Mallinckrodt Chemical Works had seven different source exposures. Health Phys. 114:386–397.

Filipy RE, Russell JJ. 2003. The United States Transuranium and Uranium Registries as sources for actinide dosimetry and bioeffects. Radiat Prot Dosimetry. 105:185–187.

Golden AP, Cohen SS, Chen H, Ellis ED, Boice JD Jr. 2018. Evaluation of statistical modeling approaches for epidemiologic studies of low-dose radiation health effects. Int J Radiat Biol. doi:10.1080/09553002.2018.1554924

Golden AP, Ellis ED, Cohen SS, Mumma MT, Leggett RW, Wallace PW, Girardi DJ, Watkins JP, Shore RE, Boice JD Jr. 2019. Updated mortality analysis of the Mallinckrodt uranium processing workers, 1942-2012. Int J Radiat Biol. doi:10.1080/09553002.2019.1569773

Grant EJ, Brenner A, Sugiyama H, Sakata R, Sadakane A, Utada M, Cahoon EK, Milder CM, Soda M, Cullings HM, et al. 2017. Solid cancer incidence among the life span study of atomic bomb survivors: 1958-2009. Radiat Res. 187:513–537.

[ICRP] International Commission on Radiological Protection. 2007. The 2007 Recommendations of the International Commission on Radiological Protection. ICRP Publication 103. Ann ICRP. 37(2–4): 1–332.

James AC, Filipy RE, Russell JJ, McInroy JF. 2003. USTUR case 0259 whole body donation: a comprehensive test of the current ICRP models for the behavior of inhaled $^{238}PuO_2$ ceramic particles. U.S. Transuranium and Uranium Registries. Health Phys. 84:2–33.

James AC, Sasser LB, Stuit DB, Wood TG, Glover SE, Lynch TP, Dagle GE. 2007. USTUR whole body case 0262: 33-y follow-up of PuO2 in a skin wound and associated axillary node. Radiat Prot Dosimetry. 127:114–119.

Kathren RL, Tolmachev SY. 2015. Natural uranium tissue content of three Caucasian males. Health Phys. 109:187–197.

Kiernan D. 2013. The girls of atomic city: the untold story of the women who helped win World War II. New York: Touchstone.

Krukowski K, Grue K, Frias ES, Pietrykowski J, Jones T, Nelson G, Rosi S. 2018. Female mice are protected from space radiation-induced maladaptive responses. Brain Behav Immun. 74:106–120.

Leggett RW, Eckerman KF, Bellamy M. 2018b. MPS dose reconstruction for internal emitters: some site-specific issues and approaches. Int J Radiat Biol. doi:10.1080/09553002.2018.1558302

Leggett RW, Tolmachev SY, Boice JD Jr. 2018a. Potential improvements in brain dose estimates for internal emitters. Int J Rad Biol. doi:10.1080/09553002.2018.1554923

Limoli C. 2015. Your brain on Mars. Radiat Res. 18:1–2.

McInroy JF, Boyd HA, Eutsler BC, Romero D. 1985. The U.S. Transuranium Registry report of the 241Am content of a whole body. Part IV: preparation and analysis of the tissues and bones. Health Phys. 49:587–621.

McInroy JF, Gonzales ER, Miglio JJ. 1992. Measurement of thorium isotopes and 228Ra in soft tissues and bones of a deceased Thorotrast patient. Health Phys. 63:54–71. Erratum: Health Phys. 63:604, 1992.

McInroy JF, Kathren RL, Voelz GL, Swint MJ. 1991. U.S. Transuranium Registry report on the 239Pu distribution in a human body. Health Phys. 60:307–333.

Moore TM, Basner M, Nasrini J, Hermosillo E, Kabadi S, Roalf DR, McGuire S, Ecker AJ, Ruparel K, Port AM, et al. 2017. Validation of the cognition test battery for spaceflight in a sample of highly educated adults. Aerosp Med Hum Perform. 88:937–946.

Nathwani AC, Down JF, Goldstone J, Yassin J, Dargan PI, Virchis A, Gent N, Lloyd D, Harrison JD. 2016. Polonium-210 poisoning: a first-hand account. Lancet. 388:1075–1080.

[NA/NRC] National Academies/National Research Council. 2012. Technical evaluation of the NASA model for cancer risk to astronauts due to space radiation. Washington (DC): National Academies Press; [accessed 2019 February 2]. https://www.nap.edu/catalog/

13343/technical-evaluation-of-the-nasa-model-for-cancer-risk-to-astronauts-due-to-space-radiation

[NASA] National Aeronautics and Space Administration. 2015. NASA space flight human-system standard, volume 1, revision A: crew health. NASA STD-3001. [accessed 2018 September 16]. https://standards.nasa.gov/standard/nasa/nasa-std-3001-vol-1

[NCRP] National Council on Radiation Protection and Measurements. 2000. Radiation protection guidance for activities in low-earth orbit. NCRP Report 132. Bethesda (MD): NCRP.

[NCRP] National Council on Radiation Protection and Measurements. 2006. Information needed to make radiation protection recommendations for space missions beyond low-earth orbit. NCRP Report 153. Bethesda (MD): NCRP.

[NCRP] National Council on Radiation Protection and Measurements. 2014. Radiation protection for space activities: supplement to previous recommendations. NCRP Commentary 23. Bethesda (MD): NCRP.

[NCRP] National Council on Radiation Protection and Measurements. 2016. Potential for central nervous system effects from radiation exposure during space activities. Phase 1: overview. NCRP Commentary 25. Bethesda (MD): NCRP.

[NCRP] National Council on Radiation Protection and Measurements. 2018a. Deriving organ doses and their uncertainty for epidemiologic studies (with a focus on the one million U.S. workers and veterans study of low-dose radiation health effects. NCRP Report 178. Bethesda (MD): NCRP.

[NCRP] National Council on Radiation Protection and Measurements. 2018b. Implications of recent epidemiologic studies for the linear nonthreshold model and radiation protection. NCRP Commentary 27. Bethesda (MD): NCRP.

Norbury JW, Schimmerling W, Slaba TC, Azzam EI, Badavi FF, Baiocco G, Benton E, Bindi V, Blakely EA, Blattnig SR, et al. 2016. Galactic cosmic ray simulation at the NASA Space Radiation Laboratory. Life Sci Space Res (Amst). 8:38–51.

Ozasa K, Grant EJ, Kodama K. 2018. Japanese legacy cohorts: the life span study atomic bomb survivor cohort and survivors' offspring. J Epidemiol. 28:162–169.

Ozasa K, Shimizu Y, Suyama A, Kasagi F, Soda M, Grant EJ, Sakata R, Sugiyama H, Kodama K. 2012. Studies of the mortality of atomic bomb survivors, report 14, 1950-2003: an overview of cancer and noncancer diseases. Radiat Res. 177:229–243. Erratum: Radiat Res. 179:e40–1, 2013.

Paunesku T, Wanzer MB, Kirillova EN, Muksinova KN, Revina VS, Lyubchansky ER, Grosche B, Birschwilks M, Vogt S, Finney L, et al. 2012. GE. X-ray fluorescence microscopy for investigation of archival tissues. Health Phys. 103:181–186.

Preston RJ. 2015. Integrating basic radiobiological science and epidemiological studies: why and how. Health Phys. 108:125–130.

Rector TS, Wickstrom SL, Shah M, Thomas Greeenlee N, Rheault P, Rogowski J, Freedman V, Adams J, Escarce JJ. 2004. Specificity and sensitivity of claims-based algorithms for identifying members of Medicare + Choice health plans that have chronic medical conditions. Health Serv Res. 39:1839–1857.

Rühm W, Eidemüller M, Kaiser JC. 2017. Biologically-based mechanistic models of radiation-related carcinogenesis applied to epidemiological data. Int J Radiat Biol. 93:1093–1117.

Schlenker RA, Keane AT, Holtzman RB. 1982. The retention of ^{226}Ra in human soft tissue and bone; implications for the ICRP 20 alkaline earth model. Health Phys. 42:671–693.

Scoles S. 2017. NASA likely to break radiation rules to go to Mars – NOVA Next – PBS. [accessed 2018 September 16]. http://www.pbs.org/wgbh/nova/next/space/nasa-mars-radiation-rule/

Shay JW, Cucinotta FA, Sulzman FM, Coleman CN, Minna JD. 2011. From mice and men to earth and space: joint NASA-NCI workshop on lung cancer risk resulting from space and terrestrial radiation. Cancer Res. 71:6926–6929.

Shore RE, Beck HL, Boice JD, Caffrey EA, Davis S, Grogan HA, Mettler FA, Preston RJ, Till JE, Wakeford R, et al. 2018. Implications of recent epidemiologic studies for the linear nonthreshold model and radiation protection. J Radiol Prot. 38:1217–1233.

Shore RE, Beck HL, Boice JD Jr, Caffrey EA, Davis S, Grogan HA, Mettler FA Jr, Preston RJ, Till JE, Wakeford R, et al. 2019. Recent epidemiologic studies and the linear no-threshold model for radiation protection - considerations regarding NCRP Commentary 27. Health Phys. 116:235–246.

Shore R, Walsh L, Azizova T, Rühm W. 2017. Risk of solid cancer in low dose-rate radiation epidemiological studies and the dose-rate effectiveness factor. Int J Radiat Biol. 10:1064–1078.

Simonsen LC, Zeitlin C. 2017. Mars radiation environment – what have we learned? Briefing to NAC HEO/SMD Joint Committee Meeting, July 25, 2017. [accessed 2018 October 15]. https://www.nasa.gov/sites/default/files/atoms/files/mars_radiation_environment_nac_july_2017_final.pdf

[USTUR] The United States Transuranium and Uranium Registries. 2018. [accessed 2018 October 17]. www.ustur.wsu.edu

Tolmachev SY, Ketterer ME, Hare D, Doble P, James AC. 2011. The US Transuranium and Uranium Registries: forty years' experience and new directions in the analysis of actinides in human tissues. Proc Radiochim Acta. 1:173–181.

Tolmachev SH, McComish SL. 2016. United States Transuranium and Uranium Registries annual report: April 1, 2015 - March 31, 2016. USTUR-0403-16. Richland (WA): Washington State University; [accessed 2018 September 3]. https://s3.wp.wsu.edu/uploads/sites/1058/2017/07/USTUR-0403-16.pdf

[UNSCEAR] United Nations Scientific Committee on the Effects of Atomic Radiation. 2008. Effects of ionizing radiation. UNSCEAR 2006 report. Report to the general assembly. Volume 1. Scientific Annex A. Epidemiological studies of radiation and cancer. Publication E.08.IX.6. New York (NY): United Nations.

[GAO] U.S. Government Accountability Office. 2017. Low-dose radiation: interagency collaboration on planning research could improve information on health effects. GAO-17-546. Washington (DC): GAO; [accessed 2018 September 8]. https://www.gao.gov/products/GAO-17-546A

Wenz J. 2018. Forget "manned" missions – females may be more mentally resilient in deep space. Scientific American, August 24. [accessed 2018 September 16]. https://www.scientificamerican.com/article/forget-manned-missions-females-may-be-more-mentally-resilient-in-deep-space/?print=true

Yamada M, Kasagi F, Mimori Y, Miyachi T, Ohshita T, Sasaki H. 2009. Incidence of dementia among atomic-bomb survivors-Radiation Effects Research Foundation Adult Health Study. J Neurol Sci. 281:11–14.

Yoder RC, Dauer L, Balter S, Boice JD Jr, Grogan H, Mumma M, Passmore CN, Rothenberg LN, Vetter RJ. 2018. Dosimetry for the study of medical radiation workers with a focus on the mean absorbed dose to the lung, brain and other organs. Int J Radiat Biol. doi:10.1080/09553002.2018.1549756

Zeitlin C, Hassler DM, Cucinotta FA, Ehresmann B, Wimmer-Schweingruber RF, Brinza DE, Kang S, Weigle G, Böttcher S, Böhm E, et al. 2013. Measurements of energetic particle radiation in transit to Mars on the Mars Science Laboratory. Science. 340:1080–1084.

Historical perspective on the department of energy mortality studies: focus on the collection and storage of individual worker data

Elizabeth D. Ellis, David Girardi, Ashley P. Golden, Phillip W. Wallace, Joyce Phillips, and Donna L. Cragle

ABSTRACT

Purpose: To present how the Department of Energy's (DOE) Comprehensive Epidemiologic Data Resource (CEDR) is integrated into the Million Person Study of Low Dose Health Effects (MPS). The history of DOE's worker health surveillance and its epidemiology program are described.

Methods and materials: A standard protocol is used to extract data from CEDR for use in the MPS. The data files are pulled from CEDR to form the basis for the cohort analyzed in the MPS. The previous study data are reviewed to formulate the protocol for the MPS study cohort. The activities needed to update the data to construct the new analytic files are carried out in parallel. The primary efforts relate to updating the vital status, retrieving cause of death information and calculating annual radiation doses for the specific organs of interest. Working data files containing the updated data are produced for construction of analytic data files used in the biostatistical analysis. At study completion the working and analytic data files are placed into CEDR for use by other researchers.

Results: The use of CEDR to study the scientific and maintenance workers at the Los Alamos National Laboratory which is currently underway is used to demonstrate the process.

Conclusions: There is a pressing need to answer the question of the health risk of exposure to chronic low-level exposure to ionizing radiation. Using CEDR as the starting point to identify new cohorts to include in the MPS is a cost-effective and a time efficient way to expedite answering this question.

Introduction

From its roots in the Manhattan Project (MP), the Department of Energy (DOE) and its predecessors have had a concern about the possible health effects among workers in its nuclear facilities and have taken a proactive approach in surveillance of exposure and health. Beginning in 1943, MP workers in the uranium processing operations were receiving annual medical examinations, being monitored for radiation exposure by wearing film badges, pocket chambers and finger rings, providing breath radon samples, and radium and uranium urinalysis samples (Mears and Engel 1945; USAEC 1949). Area dust monitoring began at the same time. With the growth of the DOE complex to include about 80 facilities, the total number of individuals ever employed in a DOE facility is estimated to be about 600,000 (Lushbaugh et al. 1983).

The ability to include the DOE nuclear workers in the Million Person Study of Low Dose Health Effects (MPS) hinges on the existence of the various types and completeness of the records required to carry out a retrospective occupational epidemiologic mortality study. The MPS is an epidemiologic study of US workers who were exposed to radiation at low doses and dose rates gradually over time. The goal is to provide relevant and needed data on radiation risks at low doses, particularly below 100 mGy (Boice 2012). DOE's early and continued focus on the effects of radiation exposure on the health of its workforce makes the DOE workers a significant component of the MPS. Their inclusion in the MPS is important because of the size of the workforce, the unique radiation exposures in these facilities, and the extended length of follow up of these workers. The DOE Comprehensive Epidemiologic Data Resource (CEDR) is an invaluable resource to the MPS (https://apps.orau.gov/cedr/default.aspx accessed 6 Sep 2018). CEDR is an electronic database made up of the analytic (AF) and working (WF) data files used in retrospective cohort mortality studies of DOE workers. Its stated purpose is to 'provide independent researchers and educators with access to .... data collected since the Department's early production years'. CEDR will be used to prioritize studies added to the MPS and to expedite the updates to past studies by leveraging the effort already expended by DOE to study health effects in its nuclear workforce.

The early history of the DOE epidemiology program (1960–1979)

In the early 1960s the Atomic Energy Commission (AEC), the successor to the MP, began to evaluate the feasibility of using the existing employee and other facility records to

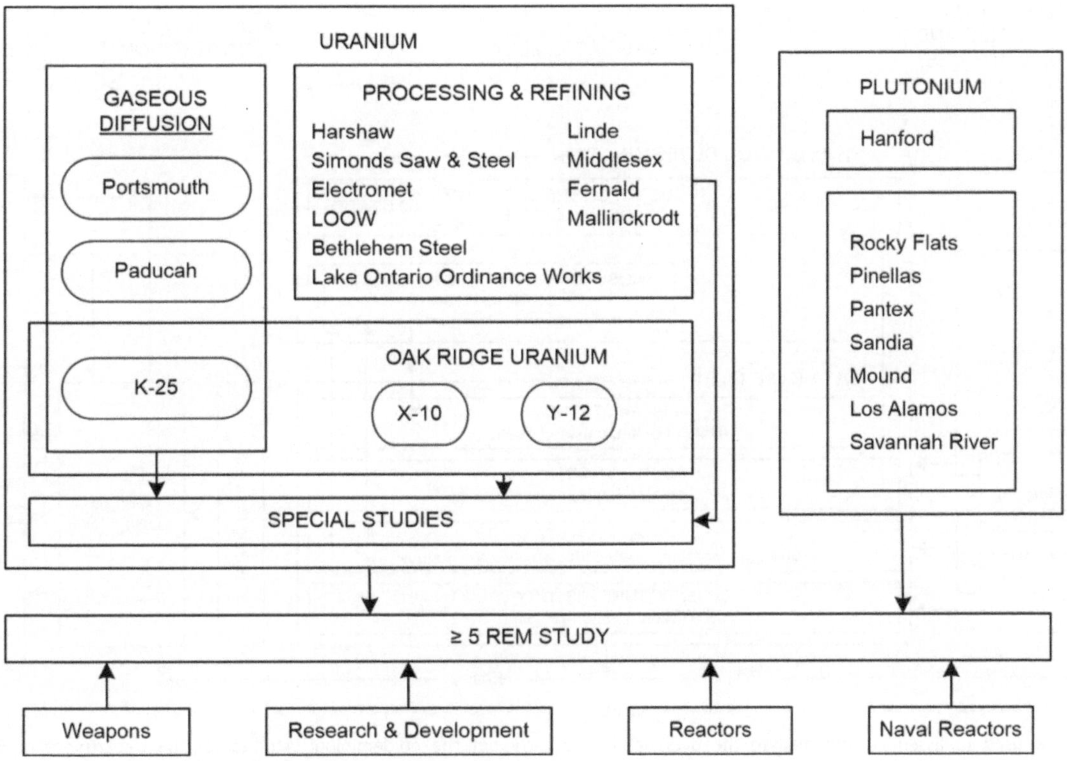

Figure 1. Facilities included in the Department of Energy Health and Mortality Studies.

conduct epidemiologic mortality studies to investigate the relationship between occupational radiation exposure and mortality. Carried out by the University of Colorado, looking at records of two uranium processing sites that began operations in the 1940s and early 1950s, the study demonstrated the feasibility of using plant records to carry out such studies (Bell et al. 1965; 1966; Quigley 1967).

From 1964–1969 the AEC funded a more broadly based five-year pilot study which confirmed the suitability of the existing records in conducting retrospective epidemiology mortality studies that could provide an upper bound of cancer risk associated with exposure to chronic low levels of ionizing radiation. This study conducted by the University of Pittsburgh added several more facilities and concentrated on identifying relevant original records located at the facilities and at off-site federal and other records repositories. Protocols and procedures for records preservation, storage and retrieval, and computer entry were developed. Records that might have been destroyed over time were identified and preserved for future studies. Few records were in electronic machine-readable form. The most relevant data elements for constructing a worker roster and ascertaining the vital status of the workers were abstracted and processed into machine-readable form. The University of Pittsburgh established a process of vital status determination of the workers. The researchers established interfaces with the Social Security Administration (SSA) to determine vital status and with the 50 states to retrieve death certificates (DC) to establish the cause of death. From 1970–1977 focus shifted to completing a mortality analysis of the workers at the Hanford site which was published in 1977 (Mancuso et al. 1977).

At the end of 1977 the Energy Research and Development Administration (ERDA), the predecessor to the DOE, transitioned the project to three epidemiology groups: Oak Ridge Associated Universities (ORAU), Los Alamos National Laboratory (LANL), and Hanford Environmental Health Foundation/Pacific Northwest Laboratory (HEHF/PNL). Each site was assigned specific programmatic responsibilities. HEHF/PNL was responsible for Hanford, LANL for selected sites that processed plutonium, and ORAU for the selected sites that processed uranium (Figure 1). ORAU was also given the responsibility to manage the interface between ERDA and the SSA obtaining the vital status for the worker rosters constructed by all three epidemiology groups as well as DC retrieval as requested by each epidemiology group. In 1979 the DOE accepted ORAU's proposal for a 'Comprehensive Epidemiologic Study of Atomic Workers' which expanded the study to all inactive and active DOE facilities. It was understood that the 80 identified sites with an estimated total workforce of 600,000 would be studied as funding allowed. Except for the interaction required with ORAU to interface with SSA, the three DOE epidemiology programs worked independently to establish the framework for studying their assigned portion of the DOE workforce.

ORAU's epidemiology program (1979–1990)

The epidemiology program set up by ORAU which is referred to as the Health and Mortality Study of DOE Workers (HMS) is germane to this paper because the data structure developed for this program formed the backbone of DOE's Comprehensive Epidemiologic Data Resource

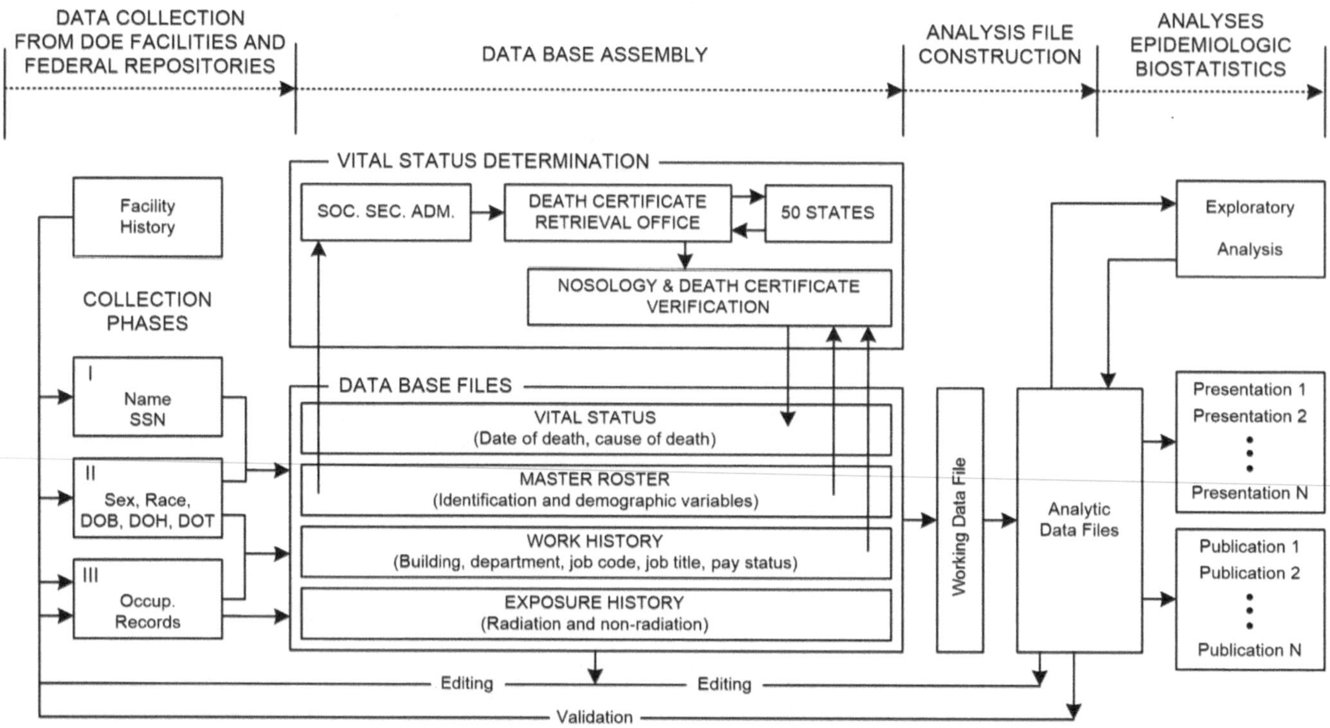

Figure 2. The process used to assemble and analyze the data needed to carry out the epidemiologic studies for the Department of Energy Health and Mortality Studies.

(CEDR) which is a major source of data used to define worker cohorts for the MPS.

The principal purpose of the HMS was to evaluate the effects of chronic occupational exposure to low dose ionizing radiation from external and internal radiation sources on the health of the DOE nuclear workers (Lushbaugh et al. 1983). Since multiple DOE sites were included in the HMS, a standardized study process was developed to ensure consistency in the way data were collected, compiled and processed for analysis (Figure 2). Major components of this process were data collection, data management, study file construction, statistical analysis and presentation. The intent was to ensure consistency in the methods used when working concurrently with data from multiple facilities and to minimize the introduction of systematic bias. An integrated relational database was developed to facilitate the incorporation of information from multiple facilities and the linkage of all data for a given worker.

The study process began with a collection of occupational records from the facility and records storage areas. The focus of the data collection was on records containing information useful for construction of worker rosters, personal identifying information (PII), demographic data, job history and radiation exposure history. With the nuclear operations at these facilities dating back to World War II much of the data collected was in the form of hard copy records. Pertinent information on the hard copy records was computerized. The data in the new electronic files were added to the appropriate table in the relational database. At that time, each worker was assigned a unique pseudo-ID. Since seven percent of the individuals worked at more than one DOE facility, this identifier was useful to link all the records for a given worker together across multiple facilities and record types in the database, and allowed the removal of PII (i.e. name and Social Security number) for sequestration in two tables, a 'Key' table which contains the SSN and the 'Name' table which contains full names. The pseudo-ID protected the privacy of the workers by restricting access to PII to only those tasks for which it was required. Data went through a series of logic checks prior to and during loading to the relational database. Errors were identified and flagged for resolution.

Personal radiation monitoring data were not added to the relational database. The pseudo-ID was linked to the exposure data for each worker. The sheer volume of these data files exceeded the capability of what the relational databases could accommodate at the time. Because of the differing radiation risks across the DOE facilities and the autonomy of each site to set up its health physics program, the variability in the type of monitoring and in the formatting both within and between sites precluded the standardization of the exposure record formats across the DOE facilities. These files were maintained separately for each facility.

Since vital status and cause of death were not widely available from facility records, the HMS built upon the system set up by the University of Pittsburgh. Briefly, the Social Security Administration (SSA) was the primary source for determining the current vital status of the workers. For workers identified as deceased, the state and date of death identified by SSA were used as the starting point for requesting the DC from a state. If the DC was not found in the SSA state, additional states were queried using an algorithm based on the state in which the facility was located. Once the DC was located, it was verified as belonging to the worker, and underlying cause of death and any cancer cause not selected as the underlying cause were coded according

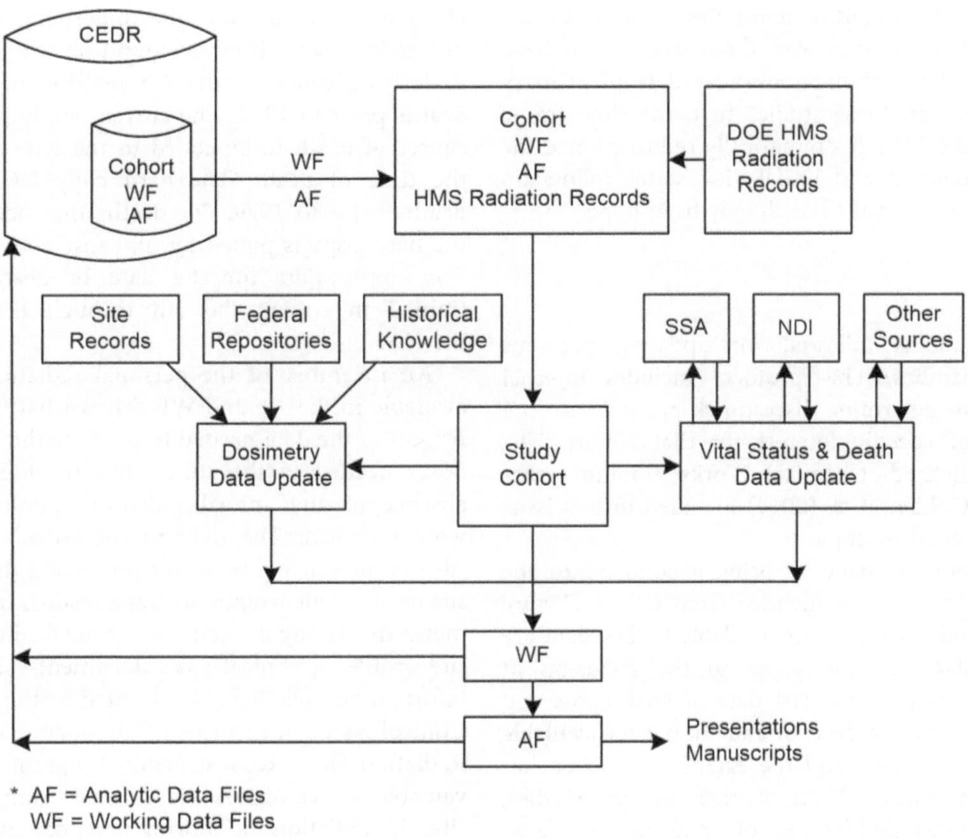

Figure 3. The interaction between of the Department of Energy Comprehensive Epidemiologic Research Database and the Million Person Study of Low Dose Health Effects.

to the rules of the Eighth Revision of the International Classification of Diseases, Adapted (ICDA-8; WHO 1968) by a trained nosologist. After its inception in the early 1980s, the National Death Index (NDI) was used to ascertain causes of death for deaths occurring in or after 1979, the first year of data in the NDI. The date of death and causes of death were entered into the Death table. If the worker was not found to be alive or deceased by SSA, other records systems used to determine vital status included state's Department of Motor Vehicles, Civil Service Retirement Board, Centers for Medicare and Medicaid Services and credit bureaus.

When a researcher was ready to put together files for analysis, the worker data specific for the study were assembled into Working files (WF) to facilitate the creation of the Analysis files (AF) used for the statistical analysis that resulted in a presentation or publication. The WF contained data that would be needed for the variables in the AF. However, the data that went into the WF could still contain errors which were corrected as they were detected. The radiation monitoring data which were kept in separate files were never cleaned prior to use in a study. The AF could contain variables that were derived from manipulation of data in the WF.

Comprehensive epidemiologic resource (CEDR) (1990-present)

In 1989 the DOE commissioned two independent but complimentary committees to review the epidemiologic research efforts within the Department (NRC 1990; SPEERA 1990). The recommendation to make the data collected under the DOE epidemiology research programs available to the broader research community to permit independent analysis and to provide a resource for training in research resulted in CEDR (NRC 1990; SPEERA 1990). The data from the three DOE epidemiology programs were consolidated into CEDR. To preserve worker privacy and confidentiality, the full name and SSN of each worker were replaced with the pseudo-ID, dates of birth and death were truncated to year and all other dates truncated to month and year. A unique pseud-ID was assigned to each worker. The AF and WF for each completed study were compiled and placed in CEDR. The structure and content of the data files varied widely based on the nature of the study. In order to be placed into CEDR, each file had to be accompanied by a data dictionary providing the name, definition, and all values for each variable in the file (https://apps.orau.gov/cedr/data-file-sets.aspx accessed 6 Sep 2018).

Usefulness of CEDR to the MPS

The overall goal of the MPS is to include all DOE nuclear workers. The near-term goal is to update the studies that have already been published from these cohorts. This approach furthers the investment that DOE has already made in studying the possible health effects of working in its nuclear facilities. Many of these studies are ripe for an update with the potential to add up to 25 additional years of follow up to the study. CEDR contains the AF and WF for most of these mortality studies with those studies that have internal and external radiation monitoring data of primary interest. By focusing on these studies first, the time frame for completion of the MPS is dramatically reduced since the cohort is already defined and initial vital status follow up and death certificate retrieval have already been done.

Methods

A standard protocol is followed for updating previous cohort mortality studies. This protocol includes internal radiation analyses to determine disease risk per unit of dose for organ/tissues of specific interest for that cohort. The study of the Mallinckrodt Chemical Works uranium processing workers by Golden et al. (2019) included in this issue of this journal is a good example.

When the decision is made to bring a cohort into the MPS, the AF and WF are downloaded from CEDR (Figure 3). An inventory and review of the available CEDR data are done with particular attention given to the inclusion of women in previous studies; the last date of vital status and DC ascertainment; and the type of radiation data available for the site, its completeness and the extent to which it has been analyzed. In earlier DOE worker cohort studies, women were rarely included because of small numbers, generally lower radiation exposure levels and the difficulty in determining vital status and retrieving DCs. With greater emphasis on including women in the occupational epidemiologic studies, a review is done to decide whether to expand the cohort to include women. The decision is made based on the availability of data on women in the CEDR files and the number employed at the site. For most cohort studies that are in CEDR, radiation exposure was included as no more than annual film badge readings and annual response of 'yes/no' to 'ever monitored for internal radiation exposure'. In addition, the amount of time that has elapsed since the most recent publication of these studies is at least 10 years and the dosimetry methodology to calculate annual organ/tissue radiation doses from internal and external radiation monitoring results has advanced significantly in that time period. Therefore, the vital status and causes of death have to be updated, and the dosimetry calculations have to be done to produce the annual organ-specific doses for each worker. Since these three activities tend to take the longest time to complete, starting each of them as early as possible in the process expedites completion of the study. To accomplish the updates to vital status and dosimetry, the personal identifying information for each cohort member is required. With permission from DOE to use the identifiable information to update the study and approval of the study protocol by the Institutional Review Board, the pseudo-ID is replaced with the personal identifying information and complete dates.

Updating vital status and retrieving DCs for workers identified as deceased follow the procedure outlined in the article by Mumma et al. (2019) in this issue of the journal. Before the inception of the NDI, the practice of the DOE HMS was to code only the underlying cause of death and one cancer cause if it was found on the DC but was not the underlying cause, to the 8th revision of the ICD-A for all deaths prior to 1979. The current study protocol requires all causes of death to be coded to the ICD revision in effect at the time of death. The hard copy DCs are available for deaths prior to 1979. For deaths that occurred before 1969, the hard copy is pulled for all-cause coding to the ICD revision appropriate for the date of death by a nosologist trained in coding the 5th through the 7th revisions of the ICD.

An inventory of the personal radiation monitoring data available in the AF and WF drives what has to be done with respect to the data needed to produce the annual organ/tissue doses needed to the study. The most extreme case is the absence of any annual radiation monitoring data for the worker cohort. The methods described using this extreme case as the starting point for producing the annual organ/tissue doses. Concurrently with the records inventory, any documents describing the activities at the facility are collected. The site profiles, technical basis documents (TBD) and technical information bulletins (TIB) created by the Centers for Disease Control National Institute of Occupational Safety and Health Radiation Dose Reconstruction Program (NIOSH-DR) are a valuable source of information concerning the history of the site, the radiation monitoring practices and the availability of radiation monitoring data (https://www.cdc.gov/niosh/ocas/ocastbds.html accessed 6 Sep 2018). While the purpose of the MPS is different from the NIOSH-DR, particularly with respect to assumptions made to reconstruct doses, the content of these documents is a valuable source of historical information. Current and former employees knowledgeable about the radiation monitoring at the site, are identified. The results of the data inventory of CEDR data combined with the potential contacts and information collected about the facility are used to develop a specific plan to acquire the data needed to produce the annual organ-specific radiation doses for the study. The approach for the dosimetry is detailed in the forthcoming National Council on Radiation Protection and Measurements (NCRP) Report 178 (Bouville et al. 2015).

When the vital status follow-up and DC retrieval efforts are completed and the annual organ doses have been calculated, WF are produced containing the information needed to populate the variables in the AF which is used for the statistical analysis, manuscript and presentation development. Within a year of the first publication, the WF and AF will be added back to CEDR for use in future studies.

An example of the use of CEDR can be demonstrated with the worker cohort at the Los Alamos National Laboratory (LANL) which is the most recent study to be added to the MPS. Within CEDR there are 2 worker cohorts at LANL, the employees of LANL and the employees of the Zia Company (Zia) which is a principal subcontractor for LANL. Within CEDR, there are a number of paths to identify the site of interest. The approach that is the most inclusive is the following starting at the CEDR homepage: select Browse Data Sets > Analytic Data File Sets > Select a Site. Since this study is of LANL and Zia workers, a separate selection was made for each site. Another selection is LANL (including Zia), but the results returned are for a case-control study which is not of interest for the MPS since its focus is on updating cohort mortality studies. The AF files for both the cohort study of females and of males at LANL and the AF for the 1992 cohort study for Zia were downloaded. The option of 'Working Data File Sets' was made to identify any files associated with these two sites. Brief descriptions of the contents of each data file are listed on the 'Data File Set Results' page. By clicking on the data file, the first 10 records of the file are displayed. Clicking on any variable name links to the page containing a description of the variable, its unit of measure and a link to the allowable codes for the variable (as applicable). No WF are available for Zia. The nine data files listed by selecting the 'Combined LANL and Zia Working files' were downloaded. These files were reviewed with special attention given to the end of follow-up, the number of women, the completeness of the data for each worker, and the type of personal radiation exposure data available. In parallel, the NIOSH-DR website was queried for site profiles, TBDs, and TIBs available for LANL and Zia. No entry was listed for Zia. A site profile containing six TBDs and two TIBs was downloaded for LANL. Each was reviewed for information pertinent to the occupational cohort study.

This study is currently in progress with the following decisions made and subsequent activities being carried out in parallel. Based on the evaluation of the collective material, the decision was made to combine the LANL and Zia populations into one study since all employees were employed on the LANL site. Women are included in the study. This decision was made based on the fact that a cohort of over 6000 women was already identified in a previous study with vital status follow up through at least 1981 (Wiggs 1987). Once this decision was made, the potential study cohort was created from the previous AFs for the LANL Female and Male cohort studies and the Zia cohort study (Wiggs et al. 1994). This study cohort file was used to create the file used for vital status follow up and DC retrieval. The vital status file contained anyone without a cause of death in the previous studies. The algorithm described by Mumma et al. (2019) is being executed. Current and former staff at LANL have been contacted to participate as subject matter experts in efforts to identify the location of files needed to update personal radiation exposure and to provide details about the radiation monitoring programs at LANL.

In addition to the use of subject matter experts from the site, autopsy data from the U.S Transuranium and Uranium Registry (USTUR) will be used to refine the brain dosimetry for alpha emitters through model validation. There is increased interest in the potential adverse effects of alpha emitters on the brain as surrogates for the exposures that astronauts will receive during space travel to Mars. The most relevant brain-specific data available for modeling brain kinetics from alpha emitters are USTUR autopsy data from Pu workers, many of whom worked at LANL. Postmortem measurements of ^{239}Pu activity in the brain are being used to improve brain dosimetry by explicitly modeling brain kinetics to replace the current implicit approach of including brain as a fraction of 'other tissue'. These improved models will be used to calculate brain dose in the LANL study and is an example of how different DOE assets (like CEDR and USTUR) can be leveraged to improve study quality.

Discussion

The DOE has a long history of concern of health issues among its nuclear workers resulting from exposures received in its facilities. Its focus on epidemiologic studies of this workforce began in 1960 with the initial pilot studies and culminated with the assembling of the data files used in these epidemiologic studies into the on-line electronic data repository called CEDR. The inclusion of the DOE worker cohorts in the MPS would be impossible today without the data in CEDR. The efforts at the University of Pittsburgh prevented the destruction of essential occupational records essential for an epidemiologic study. The current difficulties and expense of retrieving DCs for deaths occurring prior to 1979 would increase many-fold the cost and time required to complete the studies. In addition, the success in retrieving these pre-1979 DCs might be low enough to jeopardize the results of the studies.

The data in CEDR provide a valuable resource for educators. Most conventional data sets used in the classroom have been 'cleaned' so that no errors exist and the decisions required to assemble the data for analysis have already been made. This deprives the student of getting practice with the messy data that exists in the 'real world'. Using the actual data from DOE facilities in the classroom provides the opportunity for the student to become familiar with issues that arise, and the decisions and assumptions that have to be made to assemble the data needed to do an epidemiologic study. The CEDR data are also a source of data for a thesis or dissertation. Several of the studies in CEDR were done as dissertations (Wiggs 1987; Dupree 1989).

The data in CEDR are the products of previous epidemiologic studies and contain the limitations of those studies. Given the overall goal of the MPS to investigate the relationship between health outcomes and radiation exposure, limitations related to the disease outcomes of interest and the exposures may affect the dose-response relationships identified. The standard methods used to update vital status and DC retrieval ameliorate issues with the disease outcomes. To date, vital status follow-up and DC retrieval for MPS cohorts exceed 95% which is considered good practice for a cohort study. Addressing issues with the exposure data are more difficult. Standardization of monitoring practices for radiation exposure across the DOE complex was minimal. Gaps in knowledge around these data can result in random and systematic misclassification of exposure. Random misclassification typically results in exposure estimates biased toward the null so that the resulting disease risk is overestimated. The effects of systematic misclassification are more difficult to determine and the effect on the disease risk can be in either direction. Knowledge of chemical exposures is even more limited than radiation monitoring but is important with respect to confounding in the disease-response relationship. The established practice for consideration of chemical exposures is a job-exposure matrix. A limited number of studies in CEDR have included such a matrix (Dupree et al. 1987; Kubale et al. 2008). Because of the level of effort required to develop and implement these matrices is labor intensive and time-consuming, the inclusion of chemical exposures in the MPS studies will be limited.

Despite the numerous limitations described, CEDR serves as an excellent starting point for cohort and data identification so that efficiencies can be gained where resources are limited for extended the follow-up for radiation-exposed cohorts that are part of the MPS. One of the main tenants of the MPS is to build on past studies instead of starting from scratch. For instance, in the example presented in this paper for LANL, significant time and effort were saved by using the CEDR analytic files from the previous studies as the foundation of the cohort. This allows us to focus on updating the vital status, refining the dosimetry, and standardizing the dosimetric and analytic methods with those of other studies in the MPS. Coming full circle, as MPS cohorts are finalized, the updated data will be placed into CEDR to allow for pooling on a larger scale (i.e. international) and for use by researchers in the future for the next follow-up of these important populations. To date, small study populations have contributed to the ambiguous results from previous studies addressing the health effects of chronic low dose radiation exposure on human populations. Pooling data from multiple studies will increase statistical power and the likelihood of answering this important question. The standardization of methods used in the MPS, especially for vital status ascertainment and for calculation of organ doses, increases the successful pooling of the data to better refine risk estimates for the health effects of human exposure to radiation.

Disclosure statement

The authors report no conflicts of interest. This report was prepared as an account of work sponsored by the United States Government. Neither the United States Government nor the U.S. Department of Energy, nor any of their employees, makes any warranty, expressed or implied, or assumes any legal liability or responsibility for the accuracy, completeness, or usefulness of any information, apparatus, product, or process disclosed, or represents that its use would not infringe on privately owned rights. Reference herein to any specific commercial product, process, or service by trade name, mark, manufacturer, or otherwise, does not necessarily constitute or imply its endorsement or recommendation, or favor by the U.S. Government or any agency thereof. The views and opinions of authors expressed herein do not necessarily state or reflect those of the U.S. Government or any agency thereof.

Funding

This document describes activities performed under contract number DE-SC0014664 between the U.S. Department of Energy and Oak Ridge Associated Universities.

ORCID

Elizabeth D. Ellis http://orcid.org/0000-0003-2871-1587
Ashley P. Golden http://orcid.org/0000-0002-9430-7702

References

Bell RF, Gilliland JC, Bell JA, Johnson DM. 1965. A report of an epidemiological feasibility study at three uranium feed material sites. Denver (CO): University of Colorado Medical Center. Vol I and II.

Bell RF, Gilliland JC, Bell JA, Johnson DM, Evans EE. 1966. Epidemiologic feasibility report. Denver (CO): University of Colorado Medical Center.

Boice JD. 2012. A study of one million US radiation workers and veterans, a new NCRP initiative. Health Physics News. November 2012 pp 7–10. https://ncrponline.org/wp-content/themes/ncrp/PDFs/BOICE-HPnews/Nov-2012_Million_Worker.pdf [accessed 2018 Dec 12].

Bouville A, Toohey RE, Boice JD, Jr, Beck HL, Dauer LT, Eckerman KF, Hagemeyer D, Leggett RW, Mumma MT, Napier B. 2015. Dose reconstruction for the million worker study: status and guidelines. Health Phys. 108:206–220.

Dupree EA. 1989. An epidemiologic investigation of mortality among workers occupationally exposed to uranium compounds at a uranium processing plant [dissertation]. Chapel Hill (NC): University of North Carolina.

Dupree EA, Cragle DL, McLain RW, Crawford-Brown DJ, Teta MJ. 1987. Mortality among workers at a uranium processing facility, the

Linde Air Products Company Ceramics Plant, 1943-1949. Scand J Work Environ Health. 13:100–107.

Golden AP, Ellis ED, Cohen SS, Mumma MT, Leggett RW, Wallace PW, Girardi D, Watkins JP, Shore R, Boice JD. Jr. Forthcoming 2019. Updated mortality analysis of the Mallinckrodt uranium processing workers, 1942-2012. Int J Radiat Biol.

Kubale TL, Hiratzka S, Henn S, Markey A, Daniels R, Utterback D, Waters K, Silver S, Robinson C, Macievic G, et al. 2008. A cohort mortality study of chemical workers at Deoartment of Energy nuclear plants. Am J Ind Med. 51:656–667.

Lushbaugh CC, Fry SA, Shy CM, Frome EL. 1983. The DOE Health and Mortality Study at Oak Ridge. In: Proceedings of the Sixteenth Midyear topical symposium; Albuquerque (NM): US Department of Energy Report Conf-830101. p. 105–114.

Mancuso FT, Stewart A, Kneale G. 1977. Radiation exposures of Hanford workers dying from cancer and other causes. Health Phys. 33: 369–385.

Mears BJ, Engel B. 1945. Medical history of Middlesex Warehouse. New Jersey: Middlesex.

Mumma M, Cohen SS, Sirko JL, Ellis ED, Boice JD. 2019. Obtaining vital status and cause of death on a million persons. Int J Radiat Biol. Forthcoming.

[NRC] National Research Council, Committee on DoE research on health effects of ionizing radiation 1990. Providing access to epidemiological data. Washington, D.C.: National Academy Press.

Quigley JA. 1967. Epidemiologic study of uranium workers – feasibility. Cincinnati (OH): National Lead of Ohio.

[SPEERA] Report to the Secretary. 1990. The Secretarial panel for the evaluation of epidemiologic research activities for the U.S Department of Energy. Washington, D.C.: U.S Department of Energy.

Wiggs LD. 1987. Mortality among females employed by the Los Alamos National Laboratory: An epidemiologic investigation [dissertation]. Tulsa (OK): University of Oklahoma.

Wiggs LD, Johnson ER, Cox-DeVore CA, Voelz GL. 1994. Mortality through 1990 among white male workers at the Los Alamos National Laboratory: Considering exposures to plutonium and external ionizing radiation. Health Phys. 67:577–588.

[WHO] World Health Organization 1968. International classification of diseases, adapted 1968. Geneva (Switzerland): World Health Organization.

[USAEC] US Atomic Energy Commission 1949. Health hazards in NYOO facilities producing and processing uranium (A status report – April 1, 1945). New York (NY): NY Operations Office Medical Division.

50 Years of the radiation exposure information and reporting system and importance to the Million Person Study

Derek A. Hagemeyer, Gregory P. Nichols, Michael T. Mumma, John D. Boice, and Terry A. Brock

ABSTRACT

Purpose: As the Radiation Exposure Information and Reporting System (REIRS) celebrates 50 years of existence, this is an appropriate time to reflect on the innovative and novel system and how it has shaped the study of occupational radiation exposure. It is also fitting to appreciate the vision and initiative of the individuals who recognized the future value of the collection and analysis of this information to better inform regulations, policies, and epidemiologic studies, and thus contribute to the protection of workers and the public from the adverse health effects of radiation exposure.

Conclusions: REIRS has evolved and expanded over its 50-year history and has played a central role in providing the radiation exposure monitoring records for the Million Person Study for individuals monitored as NRC licensees and at DOE facilities. REIRS has played two major functions in epidemiologic studies. First, it has provided dosimetry information on individual workers in occupational studies to ensure nearly complete ascertainment of career doses. Second, REIRS was used as the primary data source for large cohorts within the Million Person Study: nuclear power plant workers (n~140,000) and industrial radiographers (n~130,000). The legacy that REIRS continues to uphold is a model for creating and maintaining a successful tool throughout decades of political, technological, and demographic change.

Introduction

In 1968, the U.S. Atomic Energy Commission (AEC) required the reporting of an annual summary regarding occupational radiation exposure information in a central repository. This was the catalyst for the creation of the Radiation Exposure Information and Reporting System (REIRS), which has become the gold standard in occupational radiation dose collection and reporting (REIRS 2018).

Over the past 50 years, REIRS has proven to be a reliable, continuously improving, and evolving system for the management and regulatory oversight of radiation exposure monitoring records and has contributed to numerous epidemiologic studies (Cardis et al. 2005; Schubauer-Berigan et al. 2015; Boice et al. 2019a). As REIRS celebrates 50 years of existence, this seems an appropriate time to reflect on the innovative and novel system and how it has shaped the collection and evaluation of occupational radiation exposure. It is also fitting to appreciate the vision and initiative of the individuals who recognized the future value of the collection and analysis of this information to better inform regulations, policies, and studies in order to protect workers and the public from the potential adverse health effects of radiation exposure.

Overview and history

In November 1968, the AEC amended 10 CFR Part 20 to require the reporting of a statistical summary of occupational radiation exposure information, thus creating REIRS. The records were submitted on paper and manually entered into the database, with a few facilities experimenting with reporting electronically on magnetic tapes as a pilot project. REIRS was initially located at the Oak Ridge National Laboratory (ORNL).

In January 1975, the AEC was separated into two agencies, the Energy Research and Development Administration (ERDA) and the U.S. Nuclear Regulatory Commission (NRC). Each agency assumed responsibility for collecting and maintaining occupational radiation exposure information reported by the facilities under its jurisdiction. A further federal reorganization in 1977 replaced ERDA with the Department of Energy (DOE).

Subsequently, the REIRS database was split into two systems, one located at the Oak Ridge K-25 facility containing the monitoring records for NRC licensees, and one located at the Idaho National Laboratory containing the monitoring records for DOE facilities. The basic systems and functional requirements did not change significantly during this period while advancements in database management software were advancing rapidly. The systems continued to provide information to the NRC and DOE for the compilation of annual reports on occupational exposure; however, the annual publications began falling behind schedule with the maintenance and operation of the systems insufficient to keep up with the increasing number of records and information requests.

NRC staff changed the technical specifications for operating nuclear power plants (NPPs) and required them to submit an annual report that indicated number of individuals exposed and the cumulative doses broken down by personnel, work function, and occupation in accordance with NRC Regulatory Guide 1.16 (AEC 1975). On 4 February 1974, 10 CFR Part 20.407 was amended, requiring licensed NPPs to submit an annual statistical report indicating the distribution of the whole-body doses of all individuals monitored at each facility, allowing an estimate to be made of the total collective dose, and of the number of workers receiving measurable doses (NCRP 2018).

Originally, REIRS only collected information on four types of licensees. However, in 1982–1983, 10 CFR 20.408(a) was amended to include three additional categories of licensees required to submit reports: geological repositories for high-level radioactive waste; independent spent fuel storage installations; and facilities for the land disposal of low-level radioactive waste.

In 1990, the NRC transferred the operation and management of REIRS to a company under contract to the NRC to modernize the system into a relational database management system. The system underwent a complete overhaul including the processes for collecting and entering data into the system and the annual reports. A website (which was a brand-new thing in those days) was created to exchange information on the collection and analysis of occupational exposure, as well as relevant regulations and guidance.

Although DOE had been maintaining a parallel system since the 1975 division, the two agencies (DOE and NRC) signed an interagency agreement in 1994–1996, which resulted in a new DOE system that benefited from the experience of the NRC called the Radiation Exposure Monitoring System or REMS. This agreement also established the share and transfer of technology used by NRC to support DOE programs, including REMS, and assisted in the development of a new DOE annual report of exposure. In some ways, after nearly 20 years had passed, the two systems had a sort of homecoming.

Utility of the system

Because of the volume and type of data contained in REIRS, it has become a key resource to any major study regarding human health related to occupational radiation exposure (NCRP 2018; Boice et al. 2019a). The demographic and dosimetric data in REIRS makes it not only a unique resource, but also an extremely valuable one for use by occupational health and safety professionals and regulators. REIRS data has applications in estimating exposure, as well as evaluating the data with health outcomes for epidemiological studies, and evaluating the relationship used to assess dose-response relationships. It also can be used to evaluate the impact on changes in regulation to occupational exposure and the distribution of radiation dose across the worker population (Anzenberg et al. 2010). In this sense, REIRS has the power to be used as a tool for guiding regulatory frameworks regarding radiation protection.

Other countries have similar databases that function as registries, and many of them have been used for health studies (Lee et al. 2010; Hill 2014; Mayer et al. 2016; Lim 2017; GOC 2018). REIRS is the closest dataset to a U.S. occupational radiation exposure registry in existence, particularly for NPP workers. Initially, NRC licensees were only required to submit radiation-monitoring records upon termination of employment; however, in 1994, as part of the implementation of the 1991 revisions of 10 CFR Part 20 (NRC et al. 1991; NRC 1992), the NRC began requiring annual radiation exposure records for every monitored worker. In addition, NRC Generic Letter 94-04 (NRC 1994) requested that licensed utilities report voluntarily the career doses of current employees for input into the REIRS system. Utilities responded favourably to this request. In the mid-1990s, Dr. John Boice, Branch Chief of the Radiation Epidemiology Branch of the National Cancer Institute (NCI) and his colleague, Dr. Gilbert Beebe, sent a letter to the NRC via the Director of the NCI requesting a change in reporting requirements and the addition of data fields to REIRS in order to create a better foundation for what could become a radiation exposure registry suitable for epidemiologic study (NRC et al. 1991, 1994; Muirhead et al. 1996; Boice 2015; Boice et al. 2019a). The NRC revisited the idea of using REIRS as a radiation work registry in 2011 (ORISE 2011).

The relevance and utility of REIRS has been shown time-and-time again as evidenced by its use in several studies and proposed studies. REIRS has played a supporting role in occupational health studies, including the International Agency for Research on Cancer (IARC) "15-country" study (Cardis et al. 2005) and in U.S. nuclear worker studies (Howe et al. 2004; Schubauer-Berigan et al. 2015). In 2001, the U.S. Department of Labor implemented a compensation program for former DOE workers who may have developed certain types of cancer from their time working at DOE facilities post-World War II. The Energy Employees Occupational Illness Compensation Program Act (EEOICPA) (Neton 2014) provides monetary and healthcare resources to certain individuals and utilizes data from REMS and REIRS.

One of the most important uses of REIRS data is its support of the Million Person Study of Low-Dose Health Effects (MPS), which is one of the largest of its kind (Boice 2015, 2017; NCRP 2018; Boice et al. 2018, 2019a). Over the years, REIRS has provided additional dose information for workers in the MPS to assure that nearly complete career doses are available for analysis. These cohorts include Rocketdyne (Boice et al. 2006a, 2006b, 2011), Mound (Boice et al. 2014), nuclear weapons test participants (Caldwell et al. 2016) and Mallinckrodt (Ellis et al. 2018; Golden et al. 2019). Further, the REIRS database was used for the first time as the primary source of two large cohorts in the

MPS: the study of nuclear power plant workers (NCRP 2018; Boice et al. 2019b) and the study of industrial radiographers (NCRP 2018; Boice et al. 2019c). These two cohorts alone comprise about 270,000 workers or approximately 27% of the MPS study populations.

Information and data architectures

One of the most unique things about REIRS is that it began before the digital revolution and is still in existence. Therefore, the rapid change in technology not only forced the REIRS system to evolve, but REIRS became better because of technology. REIRS initially began as paper records transferred to tapes. In the late 1980s, these tapes were converted to an Oracle relational database. As the Internet evolved, so did the power of REIRS, and in May of 1995, the NRC began disseminating radiation exposure information on the Internet. This was one of the first websites the NRC initiated and maintained.

Although information was reported online in 1995, dose records were submitted on compact discs (CDs) until the Anthrax attacks of 2001. In response to the biological threat, all U.S. mail was irradiated prior to delivery. While this killed anthrax spores, it ruined CDs, so data could no longer be submitted on electronic media. This led to the development of a web portal for the secure submittal of annual radiation exposure records. In 2006, DOE followed suit and issued a policy stating that the only acceptable method for submitting radiation exposure data was through web pages. The requirement for online submissions improved functionality, efficiency, and availability of records. Dose histories are now also requested online, saving time and reducing frustration from individuals and licensees. Responses to dose history reports, including NRC and DOE occupational exposure, are now provided within two days, with the vast majority of responses being the same day or the next business day. The process ensures the necessary protections under the Privacy Act of 1974 and related cyber-security requirements for personally identifiable information.

Summary

It is challenging for a program to survive 1 year, let alone 50. The legacy that REIRS continues to uphold is a model for creating and maintaining a successful tool throughout decades of political, technological, and demographic change. REIRS continues to be on the forefront in supporting epidemiologic studies of health effects, most importantly the ongoing Million Person Study of Low-Dose Health Effects.

Disclosure statement

No potential conflict of interest was reported by the authors.

References

Anzenberg V, Lewis DE, Dickson ED, Bush-Goddard SP. 2010. The U.S. nuclear regulatory commission radiation exposure information reporting system (REIRS). Radiat Res. 173:254–255.

Atomic Energy Commission (AEC 1975). Regulatory Guide 1.16: Reporting of Operating Information - Appendix A, Technical Specifications. Washington DC: U.S. Nuclear Regulatory Commission, Office of Nuclear Regulatory Research. Available at https://www.nrc.gov/docs/ML1230/ML12305A256.pdf (accessed September 13, 2018).

Boice JD. Jr. 2015. The million worker study – from whence it came. Health Phys News 43:22–23. http://ncrponline.org/wp-content/themes/ncrp/PDFs/BOICE-HPnews/37_MWS_June2015.pdf (accessed September 9, 2018).

Boice JD. Jr. 2017. Space: the final frontier-research relevant to mars. Health Phys. 112:392–397.

Boice JD Jr, Cohen SS, Mumma MT, Dupree Ellis E, Eckerman KF, Leggett RW, Boecker BB, Brill AB, Henderson BE. 2006a. Mortality among radiation workers at Rocketdyne (Atomics International), 1948-1999. Radiat Res. 166:98–115. Erratum in: Radiat Res. 166:566.

Boice JD, Jr, Leggett RW, Ellis ED, Wallace PW, Mumma M, Cohen SS, Brill AB, Chadda B, Boecker BB, Yoder RC, et al. 2006. A comprehensive dose reconstruction methodology for former Rocketdyne/Atomics International radiation workers. Health Phys. 90:409–430.

Boice JD, Jr, Cohen SS, Mumma MT, Ellis ED, Eckerman KF, Leggett RW, Boecker BB, Brill AB, Henderson BE. 2011. Updated mortality analysis of radiation workers at Rocketdyne (Atomics International), 1948–2008. Radiat Res. 176:244–258.

Boice JD, Jr, Cohen SS, Mumma MT, Ellis ED, Cragle DL, Eckerman KF, Wallace PW, Chadda B, Sonderman JS, Wiggs LD, et al. 2014. Mortality among Mound workers exposed to polonium-210 and other sources of radiation, 1944-1979. Radiat Res. 181:208–228.

Boice JD, Jr, Ellis ED, Golden AP, Girardi DJ, Cohen SS, Chen H, Mumma MT, Shore RE, Leggett RW. 2018. The past informs the future: an overview of the million worker study and the Mallinckrodt Chemical Works cohort. Health Phys. 114:381–385.

Boice JD, Jr, Cohen SS, Mumma MT, Ellis ED. 2019a. The Million Person Study, from whence it came and why. Int J Radiat Biol. (Submitted).

Boice JD, Jr, Cohen SS, Mumma MT, Hagemeyer D, Chen H, Yoder RC, Dauer LT. 2019b. Leukemia among early nuclear power plant workers employed between 1957 and 1984 in the United States. Int J Radiat Biol. (Submitted).

Boice JD, Jr, Cohen S, Mumma M, Hegemeier D, Golden A, Yoder RC, Dauer L, et al. 2019c. Leukemia among industrial radiographers employed between 1939-2011 in the United States. Int J Radiat Biol. (Submitted).

Caldwell GG, Zack M, Mumma M, Falk H, Heath CW, Till JE, Chen H, Boice JD. Jr. 2016. Mortality among military participants at the 1957 PLUMBBOB nuclear weapons test series and from leukemia among participants at the SMOKY test. J Radiol Prot. 36:474–489.

Cardis E, Vrijheid M, Blettner M, Gilbert E, Hakama M, Hill C, Howe G, Kaldor J, Muirhead CR, Schubauer-Berigan M, et al. 2005. Risk of cancer after low doses of ionising radiation: retrospective cohort study in 15 countries. BMJ. 331:77

Ellis ED, Boice JD, Jr, Golden AP, Girardi DJ, Cohen SS, Mumma MT, Shore RE, Leggett RW, Kerr GD. 2018. Dosimetry is key to good epidemiology: workers at Mallinckrodt Chemical Works had seven different source exposures. Health Phys. 114:386–397.

Golden AP, Ellis ED, Cohen SS, Mumma MT, Leggett RW, Wallace PW, Girardi D, Watkins JP, Shore R, Boice JD. Jr. 2019. Updated mortality analysis of the Mallinckrodt uranium processing workers, 1942-2012. Int J Radiat Biol. (Submitted).

Government of Canada (GOC) 2018. National Dose Registry. https://www.canada.ca/en/health-canada/services/environmental-workplace-health/occupational-health-safety/occupational-radiation/national-dose-registry.html (accessed September 13, 2018).

Hill P. 2014. The German system of monitoring workers dose. VI International conference 'Semipalatinsk Test Site: radiation legacy and development perspectives. Kurchatow, Kazakhstan. https://juser.fz-juelich.de/record/173055?ln=en (accessed September 13, 2018).

Howe GR, Zablotska LB, Fix JJ, Egel J, Buchanan J. 2004. Analysis of the mortality experience amongst U.S. nuclear power industry workers after chronic low-dose exposure to ionizing radiation. Radiat Res. 162:517–526.

Lee BI, Kim SI, Suh DH, Jin YW, Kim JI, Choi H, Lim YK. 2010. Radiation dose distribution for workers in South Korean nuclear power plants. Radiat Prot Dosimetry. 140:202–206.

Lim YK. 2017. Radiation exposure from nuclear power plants in Korea: 2011-2015. J Radiat Prot Res. 42:222–228. http://jrpr.org/journal/view.php?doi=10.14407/jrpr.2017.42.4.222 (accessed September 11, 2018).

Mayer S, Baechler S, Damet J, Elmiger R, Frei D, Giannini S, Leupin A, Sarott F, Schuh R. 2016. Occupational exposure to external radiation in Switzerland. Radiat Prot Dosimetry. 170:433–436.

Muirhead CR, Boice JD, Jr, Raddatz CT, Yoder RC. 1996. Comparison of dose histories for U.S. nuclear power plant workers, based on records held by a major dosimetry service company and on the NRC REIRS database. Health Phys. 70:645–650.

National Council on Radiation Protection and Measurements (NCRP) 2018. Deriving organ doses and their uncertainty for epidemiologic studies (with a focus on the one million U.S. workers and veterans study of low-dose radiation health effects). NCRP Report 178. Bethesda, MD: NCRP.

Neton JW. 2014. Characterization of exposures to workers covered under the U.S. Energy Employees Compensation Act. Health Phys. 106:249–258.

Oak Ridge Institute for Science and Education (ORISE) 2011. Establishing a U.S. nuclear regulatory commission worker registry for epidemiologic investigations. A feasibility study using the Radiation Exposure Information and Reporting System. Submitted to the Nuclear Regulatory Commission. (accessed September 8, 2018).

Radiation Exposure Information and Reporting System (REIRS) 2018. Nuclear Regulatory Commission. Available at: http://www.nrc.gov/reading-rm/doc-collections/gils/rad-exp.html (accessed August 30, 2018).

Schubauer-Berigan MK, Daniels RD, Bertke SJ, Tseng CY, Richardson DB. 2015. Cancer Mortality through 2005 among a Pooled Cohort of U.S. Nuclear Workers Exposed to External Ionizing Radiation. Radiat Res. 183:620–631.

U.S. Nuclear Regulatory Commission (NRC). 1992. Instructions for recording and reporting occupational radiation exposure data. Regulatory guide (RG) 8.37. Washington, DC: NRC. https://www.orau.org/ptp/PTP%20Library/library/NRC/Reguide/08-007.pdf (accessed September 8, 2018).

U.S. Nuclear Regulatory Commission (NRC). 1994. Voluntary reporting of additional occupational radiation of exposure data (generic letter 94-04). https://www.nrc.gov/docs/ML1204/ML12047A243.pdf (accessed September 13, 2018).

U.S. Nuclear Regulatory Commission, Cool DA, Peterson HT. 1991. Standards for protection against radiation, 10 CFR Part 20. Available at https://www.osti.gov/servlets/purl/6025037 (accessed September 13, 2018).

[illegible]

Methods

Evaluation of statistical modeling approaches for epidemiologic studies of low-dose radiation health effects

Ashley P. Golden, Sarah S. Cohen, Heidi Chen, Elizabeth D. Ellis, and John D. Boice

ABSTRACT

Purpose: A substantial body of epidemiologic literature addresses risks associated with occupational radiation exposure but comparing results between studies is often difficult as different statistical models are commonly used. It is unclear whether different methods produce similar results for estimates of radiation risk when applied to the same data. The goal of this study was to compare the radiation risk estimates for leukemia other than chronic lymphocytic leukemia (non-CLL) and ischemic heart disease (IHD) produced by both Cox and Poisson regression models for time-dependent dose-response analyses of occupational exposure.

Materials and methods: For brevity, this methods paper presents the results from one cohort, the Nuclear Power Plant workers (NPP), though the evaluation considered five cohorts of varying size and exposure as part of the Million Worker Study. Cox Proportional Hazards models, with age as the underlying timescale for hazard, were conducted using three computer software programs: SAS, R, and Epicure. Doses lagged 2 years for non-CLL and 10 years for ischemic heart disease were treated as time-dependent exposures at the annual level and were examined both in categories and as a continuous term. Hazard ratios (HR) and 95% confidence intervals (CI) were reported for each model in SAS and R, while the Peanuts program of Epicure was utilized to produce Excess Relative Risk (ERR) estimates and 95% CI. All models were adjusted for gender and year of birth. Four piece-wise exponential Poisson models (log-linear regression for rate) were developed with varying cutpoints for age strata from very fine to broad categories using both R and the Amfit program in Epicure for ERR estimates.

Results: Comparable estimates of risk (both RR and ERR) were observed from Cox and Poisson models, regardless of software utilized, as long as appropriately narrow categories of age were utilized to control the confounding of age in Poisson models. The ERR estimates produced in Epicure tended to agree very closely with the HR or RR estimates, and the statistical software program used had no impact to risk estimates for the same model.

Conclusions: As computational power is no longer the burden today as it has been in the past, the results of this evaluation support the use of the Cox proportional hazards or the ungrouped Poisson approach to analyzing time-dependent dose-response relationships to ensure that maximum control over the confounding of age is achieved in studies of mortality for cohorts occupationally exposed to radiation.

Introduction

While the studies of the atomic bomb survivors (Preston et al. 2007; Yamada et al. 2009; Shimizu et al. 2010; Ozasa et al. 2012; Hsu et al. 2013; Sera et al. 2013; Ozasa et al. 2017) have provided estimates for health risks resulting from exposure to ionizing radiation, the acute exposure received by this population is not characteristic of the chronic exposure to low-dose ionizing radiation encountered more broadly in the occupational setting. The body of epidemiologic literature addressing this workplace risk has grown significantly in the last 10–15 years (Pinkerton et al. 2004; Boice et al. 2006; McGeoghegan and Binks 2006; Boice et al. 2007, 2008, 2011, 2014; Cardis et al. 2007; Hornung et al. 2008; Schubauer-Berigan et al. 2009; Yiin et al. 2009, 2017, 2018 2018; Guseva Canu et al. 2010; Anderson et al. 2012; Eidemuller et al. 2012; Guseva Canu et al. 2012; Silver et al. 2013; Zablotska et al. 2013, 2014, 2018; Kreuzer et al. 2015; Zhivin et al. 2016; Richardson et al. 2018; Shore et al. 2018). In reviewing these studies, there are a number of important methodologic issues that make comparison of results difficult. Two different statistical approaches are commonly used in occupational studies of radiation exposure: Cox proportional hazards modeling (Cox 1972) and Poisson piece-wise exponential regression (Frome 1983, 1985). Cox models produce hazards ratios (HR) while Poisson regression results are presented as relative risk (RR).

Both methods can be utilized to estimate excess relative risk (ERR) or excess absolute risk (EAR).

Poisson regression has been viewed as an extension of an age-standardized rate ratio (such as a Standardized Mortality Ratio, SMR, for mortality studies), where the data are typically cross-classified by levels of exposure and other covariates, and person-years grouped for each resulting combination or cell (Breslow and Day 1987). Thus, changes in exposure or other covariates, including age, over time are taken into account by changing the covariate pattern to which the person-years are assigned, which is referred to as 'stratified', 'grouped', or 'piece-wise exponential' Poisson regression. The Poisson model assumes that the rate ratio, or relative risk, among the exposed and non-exposed groups is constant within exposure-age stratum (Checkoway et al. 2004; Allison 2010). As death rates typically rise with age (Breslow et al. 1983) this assumption may not be met if the intervals of time are wide (Callas et al. 1998; Richardson and Loomis 2004; Allison 2010).

As age is often a strong confounder for mortality studies, it is desirable to obtain maximum possible control over the time-dependent effects by stratifying very finely on age (Checkoway et al. 2004). This is achieved such that every age-stratum would contain at most one death through the specification of Cox proportional hazards model (Cox 1972) as an extension of Poisson regression, rather than as a special case of survival analysis (Pearce et al. 1988). Unlike Poisson regression, the age-strata are not included as covariates in the model; rather, age is the underlying timescale that defines the persons at risk at a given time (age) for each case of death (Allison 2010). Thus, the calculated rate ratio is assumed to be independent of time (age) (Checkoway et al. 2004), such that the Cox model does not require any assumption to be made about the reference population's death rate, or baseline hazard (Breslow 1974; Cox and Oakes 1984; Allison 2010).

While it is known that the Poisson model will converge to the Cox proportional hazards model as the age strata are made infinitely small (Checkoway et al. 2004), and several studies of empirical comparisons have demonstrated comparable results between the two models (Ingram and Kleinman 1989; Callas et al. 1994, 1998; Loomis et al. 2005), it is nevertheless suggested that researchers should assess the sensitivity of study findings on the analytic method chosen by comparing findings from different approaches (Breslow and Day 1987; Ingram and Kleinman 1989). Additionally, Poisson modeling has most commonly been done using Epicure software (Preston et al. 2015) in the radiation epidemiology literature, but multiple software packages have been used for Cox modeling, including: The SAS System, R, and Epicure. While it would be expected that the software packages produce the same results for a given model in a study population, there does not appear to be any published evaluation of whether the capabilities for time-dependent modeling of exposure and calculations of ERR are comparable, though it is recognized that some authors may perform this as a sensitivity analyses without necessarily reporting the results.

Thus, the goal of this study was to compare the radiation risk estimates for leukemia other than chronic lymphocytic leukemia (non-CLL) and ischemic heart disease (IHD) produced by both Cox and Poisson regression models in several occupationally-exposed cohorts of varying size and exposure as part of the study of a *Million U.S. Radiation Workers and Atomic Veterans* (MWS) (Boice et al. 2018). For brevity, this methods paper presents the results of the evaluation of one cohort, the Nuclear Power Plant workers (NPP). As the intention of this evaluation was to focus on the possible differences in regression methods and associated software, simple models were defined with only a basic set of covariates which may not represent the biologic relevance of all covariates. Thus, the results presented here are not necessarily recommended for use in precisely defining the radiation risk and dose response relationship, but rather are for analytical comparison.

Methods

Study cohorts

Five occupational cohorts from within the MWS were evaluated and are described elsewhere: Atomic Veterans (AV) (Boice, Till, et al. In press); Industrial Radiographers (IR) (Boice et al. In press); Mallinckrodt Chemical Works (MCW) (Dupree-Ellis et al. 2000; Ellis et al. 2018; Golden et al. In press); Mound workers (Boice et al. 2014); and Nuclear Power Plant workers (NPP) (Boice et al. In press);

As a first consideration, the type of radiation exposure varied among the cohorts. The NPP and IR cohorts were primarily exposed to external ionizing radiation, while the Mound and MCW workers had potential for exposure to internal as well as external radiation. Unlike the other cohorts, the Atomic Veterans were present at above-ground nuclear tests resulting in an acute relatively low radiation exposure. Second, the cohorts varied in size. The demographic characteristics of the NPP cohort are presented in Table 1, while characteristics of the other cohorts are provided in Supplemental Table 1. With 145,209 workers, NPP is the largest cohort followed by IR (123, 556), AV (113, 807), Mound (4954), and MCW (2514). The IR cohort has the most females (12,946) compared to NPP (4,975), and Mound (971). MCW and AV are all male cohorts. NPP was chosen as the representative cohort for this paper as it is large, with a substantial number of females, and slightly older than IR but younger than the others when comparing year of birth categories and average length of follow-up.

Vital status ascertainment, cause of death determination, and radiation dose reconstruction were done similarly for all five cohorts. The reconstruction of radiation organ doses for each of these cohorts was done following the guidance outlined in the compendium from the National Commission on Radiation Protection and Measurements entitled 'Deriving Organ Doses and Their Uncertainty for Epidemiologic Studies—Guidance for the One Million U.S. Persons Study of Low-Dose Radiation health Effects' (Bouville et al. 2015; Dauer et al. 2018). Deaths were obtained via extensive tracing as described in detail elsewhere (Mumma et al. 2018). Briefly,

Table 1. Characteristics of the nuclear power plant workers cohort.

	Nuclear power plant workers	
	N	%
Sex		
Male	140,234	97%
Female	4975	3%
Year of birth categories		
<1916	1156	1%
1916–1920	2514	2%
1921–1925	4951	3%
1926–1930	6941	5%
1931–1935	9109	6%
1936–1940	12,806	9%
1941–1945	19,391	13%
1946–1950	28,166	19%
1951–1955	30,270	21%
>1955	29,905	21%
Education[a]		
Low	35,996	25%
Medium	72,765	50%
High	36,448	25%
Duration of employment (years)		
<1	33,414	23%
1≤5	26,066	18%
5≤10	19,198	13%
10≤20	27,336	19%
20≤30	25,233	17%
30+	13,962	10%
Length of follow-up (years)		
Mean [Std]	30.1 [6.7]	
Range [Min, Max]	[0.02, 43.0]	

[a]SES was categorized as low, medium, or high based on the highest ever census block-level mean educational attainment.

linkages to the Social Security Administration death master file as well as state mortality records using the Centers for Disease Control and Prevention LinkPlus program, a probabilistic scoring system, were used to obtain fact of death information. Both underlying and contributing causes of death were obtained via linkage with the National Death Index (NDI), state mortality data, or death certificates obtained from state departments of health. Two outcomes were chosen to investigate the effect of breaking the 'rare event' assumption of Poisson regression. The 'rare event' examined was death from leukemia other than chronic lymphocytic leukemia (hereafter called non-CLL), and for comparison, the 'prevalent event' outcome examined was death from ischemic heart disease (IHD). International classification of disease (ICD) codes used to define non-CLL and IHD are shown in Supplemental Table 2. For IHD, contributing causes were not included in the analyses.

Statistical modeling

For modeling purposes, to allow for a possible latent period between radiation exposure and any effect consequent to it, doses were lagged by two years for analyses of non-CLL deaths (i.e. doses were excluded if they occurred within two years before the date of death) and doses were lagged by 10 years for analyses of IHD deaths. Lagging was achieved by constructing an annual-level file for each cohort with each record representing one year of follow-up. Radiation exposure in each year was included in the annual-level file and then lagged two or ten years. For doses achieved prior to

Table 2. Number of workers and deaths by dose categories for ischemic heart disease and leukemia other than chronic lymphocytic leukemia.

Dose categories (mGy)	# Workers	# Deaths
Ischemic heart disease		
<5	30,152	1,876
5–24	53,861	2,179
25–49	27,439	914
50–74	12,369	395
75–99	6,913	180
100–124	4,542	134
125–149	2,880	73
150–199	3,388	91
200–249	1,699	41
250+	1,966	71
TOTAL	**145,209**	**5,954**
Leukemia		
<5	27,803	78
5–49	86,798	192
50–99	18,237	42
100+	12,371	25
TOTAL	**145,209**	**337**

the follow-up window for each cohort, doses were summed to the first year of follow-up, less the lag period. Follow-up time was calculated from the start of follow-up to the first of death, loss to follow-up, age 95, or the end of the study period for each cohort. All cohorts had available measures of gender and year of birth which were included as covariates to all simple statistical models. Year of birth was categorized in five- or 10-year categories for each cohort.

Cox Proportional Hazards models were conducted using SAS/STAT software (SAS Institute 2013), R software (R Core Team 2012), and Epicure (Preston et al. 2015). In the Cox models, age was used as the timescale for the hazard function. Doses were treated as time-dependent exposures at the annual-level and were examined both in categories and as a linear (continuous) function. Hazard ratios (HR) and Wald's 95% confidence intervals (CI) were reported for each model using SAS and R (glm function within the stats package), while the Peanuts program of Epicure was utilized to produce ERR estimates and Wald's 95% CI. Four piece-wise exponential Poisson models using log-linear regression for the rate function were developed with varying cutpoints for age strata from very fine to broad categories using both R and the Amfit program in Epicure to produce ERR estimates. The relative risks (RR) and ERRs with corresponding 95% confidence intervals were compared to those from the Cox models. For the continuous models, the beta parameter (slope), standard error of the slope, and p-value for linear trend were also included for comparison.

Results

The demographic characteristics of the NPP cohort are presented in Table 1. The cohort is primarily male, mostly born after 1940, followed for an average of 30 years, have medium or low education levels, and were employed for less than 20 years. Table 2 presents the distribution of workers and deaths across dose categories. There were 5,954 total deaths due to IHD and 337 deaths due to non-CLL.

Comparison of models for leukemia other than chronic lymphocytic leukemia

Estimates of risk for non-CLL from the Cox proportional hazards model are compared with those from four Poisson models for categorical and continuous dose in Table 3. Between models, estimates tended to agree within each category except for the Poisson model with the fewest age categories, where risk estimates were elevated compared to the other models. However, 95% CI were still overlapping, indicating there was no statistical difference. A similar trend was observed for the continuous estimates for RR and ERR at 100 mGy, such that the Cox and Poisson models with numerous age categories had very close agreement. The Poisson model with only three age categories had the highest risk estimates [RR 1.2 (95% CI 1.02, 1.44) and ERR 0.20 (95% CI 0.02. 0.38)] and also exhibited the only statistically significant result, with all other CIs including the null values of RR = 1 or ERR = 0. All three software systems evaluated (SAS, R, and Epicure) produced the exact same results for the Cox estimates (data not shown). Only R and Epicure were used to evaluate the Poisson models and they produced the exact same estimates for each of the models. Further, the ERR estimates produced in Epicure tended to agree very closely with the HR or RR estimates, following the theoretical relationship ERR = RR-1 (Preedy and Watson 2010). The results are not shown for brevity, but similar trends were observed in the other five cohorts for non-CLL estimates. However, the smaller cohorts (Mound and MCW) tended to exhibit slightly more pronounced differences in model estimates, with the Poisson models with fewer age categories tending to produce higher estimates compared to other models. Additionally, the ERRs did not track as closely to the RR-1, regardless of Cox or Poisson, for the smaller cohorts compared to the larger cohorts.

Comparison of models for ischemic heart disease

Estimates of risk for IHD from the Cox proportional hazards model is compared with those from four Poisson models for categorical and continuous dose in Table 4. As seen for non-CLL, estimates for IHD risk within categories tended to closely agree between the Cox and Poisson models with age cutpoints for every event and with small 5-year age categories. However, agreement tended to decrease as age categories widened. While several statistically significant category estimates were observed, the overall linear trend, both for RR and ERR, was null, indicating no significant dose-response for any model except for the Poisson model with the fewest age categories [RR 1.16 (95% CI 1.11, 1.21) and ERR 0.15 (95% CI 0.11, 0.19)]. Once again, the computer software programs had exact agreement and the ERR and RR values were comparable. Similar to the patterns observed for non-CLL, more variation in risk estimates were observed in the smaller cohorts compared to the large (data not shown).

Discussion

In this comparison of common statistical methods utilized for analysis of time-dependent dose response relationship for radiation in occupational cohorts, the results confirm that comparable estimates of risk (both RR and ERR) can be expected from Cox and Poisson models, regardless of software utilized, as long as appropriately narrow categories of age are utilized to control the confounding of age in Poisson models. Other theoretical and empirical comparisons of Cox and Poisson methods for time-dependent analyses have also reported instances of close agreement in estimates (Pearce et al. 1988; Callas et al. 1994, 1998; Steenland and Deddens 2004). In fact, Cox and Poisson estimates are expected to

Table 3. Comparison of different relative risk models for leukemia other than chronic lymphocytic leukemia for Nuclear Power Plant workers by radiation dose (mGy).

	Cox Proportional Hazards model		Poisson model with extremely fine age categories[a]		Poisson model with 5-year interval age categories[b]		Poisson model with 10-year interval age categories[c]		Poisson model with few age categories[d]	
					Categorical dose model					
Dose category	HR	95% CI	HR	95% CI	HR	95% CI	HR	95% CI	HR	95% CI
<5	1.0	Ref	1.0	Ref	1.0	Ref	1.0	Ref	1.0	Ref
5–49	1.15	0.88, 1.50	1.17	0.90, 1.54	1.15	0.88, 1.51	1.16	0.89, 1.52	1.29	0.99, 1.69
50–99	1.42	0.97, 2.09	1.45	0.99, 2.13	1.43	0.97, 2.09	1.44	0.98, 2.11	1.64	1.12, 2.40
100+	1.27	0.81, 2.01	1.3	0.82, 2.06	1.28	0.81, 2.02	1.29	0.82, 2.04	1.47	0.93, 2.33
					Continuous dose model					
	Beta [std] *p value*	RR 95% CI	Beta [std] *p-value*	RR 95% CI	Beta [std] *p-value*	RR 95% CI	Beta [std] *p-value*	RR 95% CI	Beta [std] *p-value*	RR 95% CI
RR at 100 mGy	0.00146 [0.00092] 0.11	1.16 0.97, 1.39	0.00149 [0.00092] 0.10500	1.16 0.97, 1.39	0.00146 [0.00092] 0.11200	1.16 [0.97,1.39]	0.00151 [0.0009] 0.10200	1.16 [0.97, 1.39]	0.00192 [0.0009] 0.03200	1.21 [1.02, 1.44]
ERR at 100 mGy	0.0015 [0.0015] 0.238	0.15 −0.04, 0.33	0.0016 [0.0014] 0.238	0.16 −0.05, 0.39	0.0019 [0.0015] 0.194	0.16 −0.03, 0.38	0.0022 [0.0015] 0.153	0.16 −0.03, 0.33	0.0026 [0.0008] 0.09	0.20 0.02, 0.38

Leukemia other than chronic lymphocytic leukemia cases based on underlying and contributing causes of death. Radiation dose has 2 year lag applied. All models adjusted for gender and year of birth (categories).

[a] Age cut points at all event points.

[b] Age cut points at 5-year intervals: <30, <35, <40, <45, <50, <55, <60, <65, <70, < 75, <80, <85, <90, 90+ years.

[c] Age cut points at 10-year intervals: <30, <40, <50, <60, <70, <80, <90, 90+ years.

[d] Age cut points at < 30, < 50, and 50+ year based on distribution.

Table 4. Comparison of different relative risk models for ischemic heart disease for Nuclear Power Plant workers by radiation dose (mGy).

	Cox Proportional Hazards model		Poisson model with extremely fine age categories[a]		Poisson model with 5-year interval age categories[b]		Poisson model with 10-year interval age categories[c]		Poisson model with few age categories[d]	
	Categorical dose model									
Dose category	HR	95% CI	HR	95% CI	HR	95% CI	HR	95% CI	HR	95% CI
<5	1.0	Ref	1.0	Ref	1.0	Ref	1.0	Ref	1.0	Ref
5–24	1.23	1.15, 1.31	1.23	1.15, 1.32	1.24	1.16, 1.33	1.30	1.21, 1.38	1.86	1.75, 1.99
25–49	1.16	1.07, 1.26	1.16	1.07, 1.26	1.17	1.08, 1.28	1.23	1.13, 1.34	1.81	1.67, 1.97
50–74	1.20	1.07, 1.34	1.20	1.07, 1.34	1.21	1.08, 1.36	1.27	1.14, 1.42	1.89	1.69, 2.11
75–99	1.04	0.89, 1.21	1.04	0.89, 1.21	1.05	0.90, 1.22	1.10	0.94, 1.29	1.63	1.40, 1.90
100–124	1.27	1.06 1.52	1.27	1.06, 1.52	1.28	1.07, 1.53	1.35	1.13, 1.61	2.01	1.68, 2.39
125–149	1.08	0.86, 1.37	1.08	0.86, 1.37	1.10	0.87, 1.39	1.15	0.91, 1.46	1.70	1.35, 2.15
150–199	1.10	0.89, 1.36	1.10	0.89, 1.36	1.11	0.90 1.37	1.17	0.94, 1.44	1.73	1.40, 2.13
200–249	0.99	0.73, 1.36	0.99	0.73, 1.36	1.00	0.74, 1.37	1.05	0.77, 1.44	1.54	1.13, 2.09
250+	1.05	0.83, 1.33	1.05	0.83, 1.33	1.06	0.84, 1.35	1.12	0.88, 1.42	1.68	1.33, 2.14
	Continuous dose model									
	Beta [std] *p-value*	RR 95% CI	Beta [std] *p-value*	RR 95% CI	Beta [std] *p-value*	RR 95% CI	Beta [std] *p-value*	RR 95% CI	Beta [std] *p-value*	RR 95% CI
RR at 100 mGy	−0.00004 [0.00024] 0.87	1.00 0.95, 1.05	−0.00004 [0.00024] 0.86	1.00 0.95, 1.04	−0.00000 [0.00024] 0.99	1.00 0.95, 1.05	0.00016 [0.0002] 0.50	1.02 0.97, 1.06	0.0015 [0.0002] < 0.001	1.16 1.11, 1.21
ERR at 100 mGy	−0.00004 [0.0003] >0.5	0.00 −0.05, 0.04	−0.00004 [0.0003] >0.5	0.00 −0.05, 0.04	0.00000 [0.00024] >0.5	0.00 −0.05, 0.05	0.0002 [0.0002] 0.442	0.02 −0.03, 0.07	0.0015 [0.0002] <0.001	0.15 0.11, 0.19

Ischemic heart disease based on underlying cause of death only. Radiation dose has 10 year lag applied. All models adjusted for gender and year of birth (categories).
[a]Age cut points at all event points.
[b]Age cut points at 5-year intervals: <30, <35, <40, <45, <50, <55, <60, <65, <70, < 75, <80, <85, <90, 90+ years.
[c]Age cut points at 10-year intervals: <30, <40, <50, <60, <70, <80, <90, 90+ years.
[d]Age cut points at < 30, < 50, and 50+ year based on distribution.

converge as the age strata in a Poisson model are made infinitely small (Checkoway et al. 2004). However, potential issues may arise when age is a strong confounder, which is often the case for occupational studies (Breslow et al. 1983), and strata for Poisson regression cannot be made infinitely small, typically because there are too few deaths to support numerous categories. One study (Callas et al. 1998) evaluated numerous models with varying widths of age and calendar time categories, concluding that when wide age intervals were used, there was a marked change with as much as 8% difference in Poisson estimates compared to those from Cox. The intervals for calendar time had less of a pronounced impact. They concluded that Poisson and Cox proportional hazards yield nearly identical estimates of risk and confidence intervals except when confounding by age could not be closely controlled (Callas et al. 1998). It is important to obtain maximum possible control for age in mortality studies as death rates are expected to rise with increasing age (Breslow et al. 1983; Korn et al. 1997). Since the Cox proportional hazards model produces a stratum for each case, age-strata are not explicitly included as terms in the model. Thus, age is the 'time' variable that defines the risk set at the time (age) for each case such that the rate ratio is independent of age (Breslow et al. 1983; Stayner et al. 1995; Checkoway et al. 2004; Allison 2010). This provides an additional advantage of the Cox proportional hazards model over the Poisson by assuming that the rate is constant within strata (Checkoway et al. 2004). The stricter assumption of the Poisson model of constant hazard relative to the Cox model's proportional hazards one, could lead to bias in estimates if time intervals are wide (Callas et al. 1998; Richardson and Loomis 2004; Allison 2010), as was consistently seen in this evaluation for the Poisson models with fewer age strata.

Nearly the exact same estimates and CIs were observed for the Cox model and the Poisson model with age strata for every event for all outcome scenarios across all cohorts. Others have described and evaluated an 'ungrouped' Poisson methodology to avoid potential biases associated with stratification (Loomis et al. 2005). In the ungrouped analyses, the analytic dataset was constructed in the same manner as for Cox proportional hazards such that there is an unique observation for each unit of person time at risk, resulting in a file with one observation for each year of follow up for each person (in contrast to the tabular form for stratified Poisson analyses). The authors of that study concluded that the ungrouped Poisson analysis can be used to estimate quantitative exposure-response relationships and address concerns regarding potential bias in categorical analyses in the same manner as Cox proportional hazards.

It was reassuring to observe that the choice of statistical software package had no impact on risk estimates for models specified in the same manner. It was expected that the estimates of ERR and HR (Cox) or RR (Poisson) would mostly adhere to the theoretical expectation of ERR = RR-1 (Preedy and Watson 2010), though some deviation was observed for the smaller cohorts, but not so important as to affect the interpretation with regards to significance.

In conclusion, as computational power is no longer the burden today as it has been in the past, the results of this evaluation support the use of the Cox proportional hazards or the ungrouped Poisson approach to analyzing time-dependent dose–response relationships to ensure that maximum control of confounding by age is achieved in studies of mortality for cohorts occupationally exposed to radiation. Stratified Poisson models may be limited in their ability to precisely control for age effects, particularly when the cohort and/or number of deaths are small. As one of the goals of the study of a *Million U.S. Radiation Workers and Atomic Veterans* (MWS) is to promote harmonization of many different cohorts utilizing the most optimal methods for vital status tracing, dosimetry, and statistical analyses, it was important to evaluate methodological inconsistencies and uncertainties observed in the existing literature. This methods paper highlights the importance of careful consideration of the statistical approach and demonstrates both the prudence in performing sensitivity analyses and the need for authors to be explicit in the description of statistical approaches in the literature to allow for valid comparisons between studies. There are several other aspects of statistical modelling approaches that were beyond the scope of this current evaluation, including: incorporation of dosimetric errors, evaluation of fit with regards to dose–response shapes other than linear, model implications for excess absolute risk models, and the inclusion or exclusion of important covariates, both fixed and time-varying (such as duration of employment). Additional efforts are underway within the MWS to optimize the approach to addressing these issues in order to refine and enhance estimates of risk to improve our understanding of low-dose radiation exposures.

Disclosure statement

The authors report no conflicts of interest. The authors alone are responsible for the content and writing of the paper.

Funding

This work was supported by the Oak Ridge Associated Universities Directed Research and Development program, an internally funded competitive grant program for ORAU employees and member universities. Additionally, the data analyzed were provided through support by a grant from the U.S. Nuclear Regulatory Commission (NRC-HQ-60-14-G-0011), a grant from the Centers for Disease Control and Prevention (5UE1EH000989), a grant from the National Aeronautics and Space Administration (Grant No. NNX15AU88G), a grant from the National Cancer Institute (Grant No. U01 CA137026), a Discovery Grant from the Vanderbilt-Ingram Cancer Center (Center no. 404-357-9682), and grants from the U.S. Department of Energy (Grant No. DE-SC0008944 and Grant No. DE-AU0000042 awarded to the National Council on Radiation Protection and Measurements, which included interagency support from the U.S. Nuclear Regulatory Commission, the U.S. Environmental Protection Agency and the National Aeronautics and Space Administration).

ORCID

Ashley P. Golden http://orcid.org/0000-0002-9430-7702
Elizabeth D. Ellis http://orcid.org/0000-0003-2871-1587

References

Allison P. 2010. Survival analysis using the SAS system: a practical guide. 2nd ed. Cary, NC: SAS Institue, Inc.

Anderson JL, Daniels RD, Fleming DA, Tseng CY. 2012. Exposure assessment for a cohort of workers at a former uranium processing facility. J Expo Sci Environ Epidemiol. 22:324–330.

Boice JD, Cohen SS, Mumma MT, Chadda B, Blot WJ. 2007. Mortality among residents of Uravan, Colorado who lived near a uranium mill, 1936–84. J Radiol Prot. 27:299–319.

Boice JD, Cohen SS, Mumma MT, Chadda B, Blot WJ. 2008. A cohort study of uranium millers and miners of Grants, New Mexico, 1979–2005. J Radiol Prot. 28:303–325.

Boice JD, Cohen SS, Mumma MT, Dupree-Ellis E, Eckerman KF, Leggett RW, Boecker B, Brill A, Henderson B. 2006. Mortality among radiation workers at Rocketdyne (Atomics International), 1948–1999. Radiat Res. 166:98–115.

Boice JD, Cohen SS, Mumma MT, Ellis ED, Cragle DL, Eckerman KF, Wallace PW, Chadda B, Sonderman JS, Wiggs LD, et al. 2014. Mortality among mound workers exposed to Polonium-210 and other sources of radiation, 1944–1979. Radiat Res. 181:208–228.

Boice JD, Cohen SS, Mumma MT, Ellis ED, Eckerman KF, Leggett RW, Boecker BB, Brill AB, Henderson BE. 2011. Updated mortality analysis of radiation workers at Rocketdyne (Atomics International), 1948–2008. Radiat Res. 176:244–258.

Boice JD, Cohen SS, Mumma MT, Golden AP, Hagemeyer D, Yoder C, Dauer LT. In press. Leukemia Among Industrial Radiographers. Int J Rad Biol.

Boice JD, Cohen SS, Mumma MT, Hagemeyer D, Chen H, Golden AP, Yoder RC, Dauer LT. In press. Leukemia among Nuclearl Power Plant Workers Monitored 1957-1984 in the United States. Int J Rad Biol.

Boice JD, Ellis ED, Golden AP, Girardi DJ, Cohen SS, Chen H, Mumma MT, Shore RE, Leggett RW. 2018. The past informs the future: an overview of the million worker study and the mallinckrodt chemical works cohort. Health Phys. 114:381–385.

Boice JD, Till J, Mumma MT, Chen H, Cohen SS. In press. Mortality among US military participants at seven aboveground nuclear weapons test series. Int J Rad Biol.

Bouville A, Toohey RE, Boice JD, Beck HL, Dauer LT, Eckerman KF, Hagemeyer D, Leggett RW, Mumma MT, Napier B, et al. 2015. Dose Reconstruction for the Million Worker Study: Status and Guidelines. Health Phys. 108:206–220.

Breslow N. 1974. Covariance analysis of censored survival data. Biometrics. 30:89–89.

Breslow N, Day N. 1987. The Design and Analysis of Cohort Studies. Lyon, France: International Agency fo Research on Cancer.

Breslow NE, Lubin J, Marek P, Langholz D. 1983. Multiplicative models and cohort analysis. J Am Stat Assoc. 78:1–12.

Callas PW, Pastides H, Hosmer DW. 1994. Survey of methods and statistical models used in the analysis of occupational cohort studies. Occup Environ Med. 51:649–655.

Callas PW, Pastides H, Hosmer DW. 1998. Empirical comparisons of proportional hazards, Poisson, and logistic regression modeling of occupational cohort data. Am J Ind Med. 33:33–47.

Cardis E, Vrijheid M, Blettner M, Gilbert E, Hakama M, Hill C, Howe G, Kaldor J, Muirhead CR, Schubauer-Berigan M, et al. 2007. The 15-country collaborative study of cancer risk among radiation workers in the nuclear industry: Estimates of radiation-related cancer risks. Radiat Res. 167:396–416.

Checkoway H, Pearce N, Kreibel D. 2004. Research methods in occupational epidemiology. 2nd ed. New York, New York: Oxford University Press.

Cox DR. 1972. Regression models and life-tables. J R Stat Soc Ser B. 34:187–220.

Cox DR, Oakes D. 1984. Analysis of Survival DAta. New York: Chapman and Hall.

Dauer LT, Bouville A, Toohey RE, Boice JD, Beck H, Eckerman KF, Hagemeyer D, Leggett RW, Mumma MT, Napier B, et al. 2018. Dosimetry and uncertainty approaches for the million person study of low-dose radiation health effects: overview of the recommendations in NCRP Report No. 178. Int J Rad Biol. Nov 19:1–10. Epub 2018 Nov 19.

Dupree-Ellis E, Watkins J, Ingle J, Phillips J. 2000. External radiation exposure and mortality in a cohort of Uranium processing workers. Am J Epidemiol. 152:91–95.

Eidemuller M, Jacob P, Lane R, Frost S, Zablotska L. 2012. Lung cancer mortality (1950-1999) among Eldorado uranium workers: a comparison of models of carcinogenesis and empirical excess risk models. PLoS One. 7:e41431.

Ellis ED, Boice JD, Golden AP, Girardi DJ, Cohen SS, Mumma MT, Shore RE, Leggett RW, Kerr GD. 2018. Dosimetry is key to good epidemiology: workers at mallinckrodt chemical works had seven different source exposures. Health Phys. 114:386–397.

Frome EL. 1983. The analysis of rates using Poisson regression models. Biometrics. 39:665–674.

Frome EL, Checkoway H. 1985. Epidemiologic programs for computers and calculators. Use of Poisson regression models in estimating incidence rates and ratios. Am J Epidemiol. 121:309–323.

Golden AP, Ellis ED, Cohen SS, Mumma MT, Leggett RW, Wallace PW, Watkins J, Shore RE, Boice JD. In press. Updated mortality analysis of the Mallinckrodt uranium processing workers, 1942-2012. Int J Rad Biol.

Guseva Canu I, Cardis E, Metz-Flamant C, Caër-Lorho S, Auriol B, Wild P, Laurier D, Tirmarche M. 2010. French cohort of the uranium processing workers: mortality pattern after 30-year follow-up. Int Arch Occup Environ Health. 83:301–308.

Guseva Canu I, Garsi J, Caër-Lorho S, Jacob S, Collomb P, Acker A, Laurier D. 2012. Does uranium induce circulatory diseases? First results from a French cohort of uranium workers. Occup Environ Med. 69:404–409.

Hornung RW, Pinney SM, Lodwick J, Killough GG, Brewer DE, Nasuta J. 2008. Estimation of radon exposures to workers at the Fernald Feed Materials Production Center 1952-1988. J Expo Sci Environ Epidemiol. 18:512–523.

Hsu W-L, Preston DL, Soda M, Sugiyama H, Funamoto S, Kodama K, Kimura A, Kamada N, Dohy H, Tomonaga M, et al. 2013. The incidence of leukemia, lymphoma and multiple myeloma among atomic bomb survivors: 1950-2001. Radiat Res. 179:361–382.

Ingram DD, Kleinman JC. 1989. Empirical comparisons of proportional hazards and logistic regression models. Stat Med. 8:525–538.

Korn EL, Graubard BI, Midthune D. 1997. Time-to-event analysis of longitudinal follow-up of a survey: Choice of the time-scale. Am J Epidemiol. 145:72–80.

Kreuzer M, Dufey F, Laurier D, Nowak D, Marsh JW, Schnelzer M, Sogl M, Walsh L. 2015. Mortality from internal and external radiation exposure in a cohort of male German uranium millers, 1946–2008. Int Arch Occup Environ Health. 88:431–441.

Loomis D, Richardson DB, Elliott L. 2005. Poisson regression analysis of ungrouped data. Occup Environ Med. 62:325–329.

McGeoghegan D, Binks K. 2006. The mortality and cancer morbidity experience of workers at the Springfields uranium production facility, 1946–1995 (vol 20, pg 111, 2006) [Correction]. J Radiol Prot. 26: 455–455.

Mumma MT, Cohen SS, Ellis ED, Boice JD. 2018. Obtaining vital status and cause of death on a million persons. Int J Rad Biol. doi: 10.1080/09553002.2018.1539884. [Epub ahead of print].

Ozasa K, Shimizu Y, Suyama A, Kasagi F, Soda M, Grant EJ, Sakata R, Sugiyama H, Kodama K. 2012. Studies of the mortality of atomic bomb survivors, report 14, 1950-2003: an overview of cancer and noncancer diseases. Radiat Res. 177:229–243.

Ozasa K, Takahashi I, Grant EJ, Kodama K. 2017. Cardiovascular disease among atomic bomb survivors. Int J Radiat Biol. 93:1145–1150.

Pearce N, Checkoway H, Dement J. 1988. Exponential models for analyses of time-related factors, illustrated with asbestos textile worker mortality data. J Occcup Med. 30:517–522.

Pinkerton LE, Bloom TF, Hein MJ, Ward EM. 2004. Mortality among a cohort of uranium mill workers: an update. Occup Environ Med. 61:57–64.

Preedy VR, Watson RR. 2010. Excess Relative Risk. Handbook of Disease Burdens and Quality of Life Measures. New York, NY: Springer.

Preston DL, Lubin J, Pierce D, McConney M, Shilnikova N. 2015. Epicure Risk Regression and Person-Year Computation Software: Command summary and user guide Ottawa, Canada: Risk Sciences International.

Preston DL, Ron E, Tokuoka S, Funamoto S, Nishi N, Soda M, Mabuchi K, Kodama K. 2007. Solid cancer incidence in atomic bomb survivors: 1958-1998. Radiat Res. 168:1–64.

R Core Team 2012. R: A language and environment for statistical computing. Vienna, Austria: R Foundation for Statistical Computing.

Richardson DB, Cardis E, Daniels RD, Gillies M, Haylock R, Leuraud K, Laurier D, Moissonnier M, Schubauer-Berigan MK, Thierry-Chef I, et al. 2018. Site-specific solid cancer mortality after exposure to ionizing radiation: a cohort study of workers (INWORKS). Epidemiology. 29:31–40.

Richardson DB, Loomis D. 2004. The impact of exposure categorisation for grouped analyses of cohort data. Occup Environ Med. 61: 930–935.

SAS Institute I 2013. SAS/STAT software Version 9.4 for Windows. Cary, NC: The SAS System.

Schubauer-Berigan M, Daniels R, Pinkerton L. 2009. Radon Exposure and Mortality Among White and American Indian Uranium miners: an update of the Colorado Plateau cohort. Am J Epidemiol. 169: 718–730.

Sera N, Hida A, Imaizumi M, Nakashima E, Akahoshi M. 2013. The association between chronic kidney disease and cardiovascular disease risk factors in atomic bomb survivors. Radiat Res. 179:46–52.

Shimizu Y, Kodama K, Nishi N, Kasagi F, Suyama A, Soda M, Grant EJ, Sugiyama H, Sakata R, Moriwaki H, et al. 2010. Radiation exposure and circulatory disease risk: Hiroshima and Nagasaki atomic bomb survivor data, 1950-2003. Br Med J. 340:8.

Shore RE, Beck HL, Caffrey EA, Davis S, Grogan HA, Mettler FA, Preston RJ, Till JE, Wakeford R, Walsh L, et al. 2018. Implications of recent epidemiologic studies for the linear nonthreshold model and radiation protection. J Rad Prot. 38(3). DOI: 10.1088/1361-6498/aad348.

Silver SR, Bertke SJ, Hein MJ, Daniels RD, Fleming DA, Anderson JL, Pinney SM, Hornung RW, Tseng CY. 2013. Mortality and ionising radiation exposures among workers employed at the Fernald Feed Materials Production Center (1951-1985). Occup Environ Med. 70: 453–463.

Stayner L, Smith R, Bailer J, Luebeck EG, Moolgavkar SH. 1995. Modeling Epidemiologic Studies of Occupational Cohorts for the Quantitative Assessment of Carcinogenic Hazards. Am J Ind Med. 27:155–170.

Steenland K, Deddens JA. 2004. A practical guide to dose-response analyses and risk assessment in occupational epidemiology. Epidemiology. 15:63–70.

Yamada M, Kasagi F, Mimori Y, Miyachi T, Ohshita T, Sasaki H. 2009. Incidence of dementia among atomic-bomb survivors – Radiation Effects Research Foundation Adult Health Study. J Neurol Sci. 281: 11–14.

Yiin JH, Anderson JL, Bertke SJ, Tollerud DJ. 2018. Dose-response relationships between internally-deposited uranium and select health outcomes in gaseous diffusion plant workers, 1948-2011. Am J Ind Med. 61:605–614.

Yiin JH, Anderson JL, Daniels RD, Bertke SJ, Fleming DA, Tollerud DJ, Tseng CY, Chen PH, Waters KM. 2017. Mortality in a combined cohort of uranium enrichment workers. Am J Ind Med. 60:96–108.

Yiin JH, Anderson JL, Daniels RD, Seel EA, Fleming DA, Waters KM, Chen PH. 2009. A nested case-control study of multiple myeloma risk and uranium exposure among workers at the oak ridge gaseous diffusion plant. Radiat Res. 171:637–645.

Zablotska LB, Fenske N, Schnelzer M, Zhivin S, Laurier D, Kreuzer M. 2018. Analysis of mortality in a pooled cohort of Canadian and German uranium processing workers with no mining experience. Int Arch Occup Environ Health. 91:91–103.

Zablotska LB, Lane RS, Frost SE. 2013. Mortality (1950–1999) and cancer incidence (1969–1999) of workers in the Port Hope cohort study exposed to a unique combination of radium, uranium and γ-ray doses. BMJ Open. 3:e002159.

Zablotska LB, Lane RS, Thompson PA. 2014. A reanalysis of cancer mortality in Canadian nuclear workers (1956-1994) based on revised exposure and cohort data. Br J Cancer. 110:214–223.

Zhivin S, Canu I, Samson E, Laurent O, Grellier J, Collomb P, Zablotska L, Laurier D. 2016. Mortality (1968-2008) in a french cohort of uranium enrichment workers potentially exposed to rapidly soluble uranium compounds. Occup Environ Med. 73:167–174.

Obtaining vital status and cause of death on a million persons

Michael T. Mumma, Sarah S. Cohen, Jennifer L. Sirko, Elizabeth D. Ellis, and John D. Boice

ABSTRACT

Purpose: To present the methodology used to determine vital status and obtain cause of death (COD) within the Million Person Study of Low-Dose Health Effects (MPS). Data sources and vital status tracing techniques used to obtain vital status and COD for six (n = 424,238 subjects) of the ~20+ cohorts under study are described.

Methods and materials: A multistage approach using multiple sources of vital status information was used to determine vital status (or 'trace') study participants from as early as 1940 to the present. Mortality records from state departments of vital statistics and the Social Security Administration Death Master File (SSA-DMF) were matched to study participants by Social Security Number (SSN), full name, date of birth (DOB), and/or sex using deterministic and probabilistic algorithms. The National Death Index (NDI) and SSA Service for Epidemiological Researchers (SSA-SER) were used to obtain COD (after 1978) and verification of alive status, respectively. Online public records and ancestry services, death certificates, and specialized mortality sources were also utilized.

Results: For the MPS cohorts traced to date (nuclear power plant workers, industrial radiographers, atomic veterans, and workers at Rocketdyne/Atomics International, Mound nuclear facility, and Mallinckrodt Chemical Works), vital status was confirmed for over 90% of all study subjects in all but one cohort (88%). The ascertainment of COD was over 96% for all cohorts.

Conclusions: A hallmark of a high-quality epidemiologic cohort mortality study is a low percentage of subjects with unknown vital status and a low percentage of deaths without a COD. The sources and methods used for vital status tracing and COD determination for the MPS have been successful and should be useful for other investigators tracing large, historic study populations. Some of the approaches would be applicable for use in all cohort studies using regional-specific mortality data or modifications to the approach.

Introduction

In 2005, Buchanich et al. published a two-stage approach to vital status ascertainment, where stage 1 was submission of the study population to the Social Security Administration Service to Epidemiological Researchers (SSA-SER) and stage 2 was submission to the National Death Index (NDI). This approach was developed to obtain vital status in a cost-effective manner given the discovery of the incompleteness of the Social Security Administration Death Master File (SSA-DMF). Since then the incompleteness of the SSA-DMF has increased substantially, particularly since the passage of legislation in 2011 restricting the release of death records provided to the SSA from states.

The tracing process developed for the Million Person Study of Low-Dose Health Effects (MPS) expands upon the Buchanich et al. (2005) approach to efficiently obtain vital status for numerous study populations over nearly 80 years, and the methodology is detailed in this manuscript. Briefly, before going to SSA-SER, the MPS population was matched to a collection of in-house mortality data which included the Social Security Administration Death Master File (SSA-DMF) and mortality files assembled from 33 states. This matching is done using both custom algorithms developed by MPS collaborators with SAS software (SAS Institute Inc., Cary, NC) and the Centers for Disease Control and Prevention LinkPlus program, which does not require exact matches on all variables (Campbell et al. 2008). Unmatched participants are then submitted to the SSA-SER which may provide additional dates of death, but more importantly can provide a verification of alive status. At this stage, study participants known to have died in 1979 or later with cause of death (COD) still unknown or where vital status remains unknown are sent to NDI (which began in 1979). If study follow-up starts before 1979, an additional stage to obtain COD for deaths before 1979 must be completed (if COD was not already obtained from state mortality files). In the MPS, death certificates were obtained from a database of scanned death certificates, state departments of health, or online resources (e.g. Ancestry.com). During this multi-stage and iterative process, information from public record data providers are incorporated to confirm and/or correct key identifying information vital for successful matching and to validate a 'presumed alive' status for the small percentage of participants not verified as alive or deceased. Further details of the approach developed for the MPS are described in the Methods section.

Methods

Mortality data files

For over 40 years, MPS investigators have been assembling and consolidating mortality records, including over 57 million death records from 33 state departments of health and/or vital statistics (SDOH). Files from 21 of these 33 states contained causes of death, and 20 contained date of death only (8 states provided both over different time periods). The availability, years of coverage, cost and permissions required to obtain these files have varied widely among the states, with some states providing death information as early as the 1940s. The data items included in the files also vary. Most contain name, Social Security Number (SSN), sex, and date of birth (DOB), while some state files provide additional information useful in epidemiological investigations such as race and contributing causes of death. In addition, MPS investigators have obtained the SSA Death Master File (SSA DMF), containing over 86 million death records (https://www.ssdmf.com). Jones and Vawdrey (2015) provide an excellent description of the SSA DMF, its limitations, and how mortality data are collected in the United States. MPS investigators combined the state mortality files and SSA DMF into a single integrated mortality database (IMD) to facilitate the matching process.

Department of energy death certificates

In addition to the IMD, a collection of over 80,000 death certificates (DC) for workers employed at facilities in a Department of Energy (DOE) complex (or its predecessor) have been retrieved from SDOH and from Benefits Offices at the DOE facilities. Most of these DCs are for deaths that occurred prior to 1979. States in which DOE facilities were located, states surrounding those of employment, and states where workers were likely to retire were systematically targeted for obtaining DC. Standard algorithms were used to verify that the identifying and demographic data for the worker matched that information on the DC. Death certificates were scanned into electronic format (PDF) and indexed by name and/or SSN.

Governmental vital status services

Social security administration service to epidemiological researchers (SSA-SER)

The Social Security Act directs the SSA to provide epidemiological researchers information on whether study subjects are shown on SSA records to be alive or deceased (https://www.ssa.gov/policy/about/epidemiology.html) (Social Security Administration 2018). Requests to link cohort data to the SSA-SER must be approved through an application process and the study must be judged to 'contribute to a national health interest'. After approval is granted, SSN, full name, fullDOB, and sex of study participants are submitted for linkage.

The return file from SSA-SER places participants into one of three categories: dead, alive, or unknown. Death matches include a date of death and may also provide the state where a SSA benefit claim was filed or state of residence at death. Alive matches indicate the individual is living based on SSA program records such as Internal Revenue System, Medicare, or other benefits data. Unknown status indicates SSA has no record of death, insufficient data to presume the subject is alive, or there is missing or incorrect matching information. In addition, like the SSA DMF, the SSA-SER does not provide information on deaths that were provided from some states. Instead, participants where SSA was provided death information from certain states are categorized as 'unknown' vital status. This increasing proportion of 'unknown' vital status places additional importance on subsequent steps to completely ascertain vital status.

National death index

The NDI (https://www.cdc.gov/nchs/data/factsheets/factsheet_ndi.htm) is a centralized database of death record information compiled from state vital statistics offices and established as a resource to aid epidemiologists and other health and medical investigators with mortality ascertainment (Patterson and Bilgrad 1986). The NDI Plus service provides underlying and contributing causes of death coded according to the International Classification of Diseases (ICD) edition in place at the time of death (National Center for Health Statistics 2018). The NDI is an excellent source for identifying deaths occurring within the United States (Cowper et al. 2002). However, it is only available for deaths occurring in 1979 or later and, depending on the characteristics of the population under study, its completeness ranges between about 90–100% (Acquavella et al. 1986; Boyle and Decouflé 1990). Thus, multiple sources are important for complete ascertainment of deaths even after 1978 (Curb et al. 1985; Schall et al. 2001; Wojcik et al. 2010). The NDI can also be cost-prohibitive (Hermansen et al. 2009), particularly when COD is needed in addition to fact of death. Linking data to the NDI also starts with an application process. Once approved, records are sent to NDI in two groups: known deaths (searching NDI for COD only) and unknown deaths (searching NDI for both fact and COD). The cost for searches varies based on whether the death is known or unknown and the number of years searched. To reduce searching fees, particularly for large cohorts, it is helpful to determine the last date the participant was known to be alive and to search NDI only after this date. For occupational studies such as MPS, this is often the date of last employment or last occupational exposure monitoring. NDI returns linkages with a score indicating the quality of the match. We utilize an approach similar to Fillenbaum et al. (2009) where matches on SSN, last name, and one

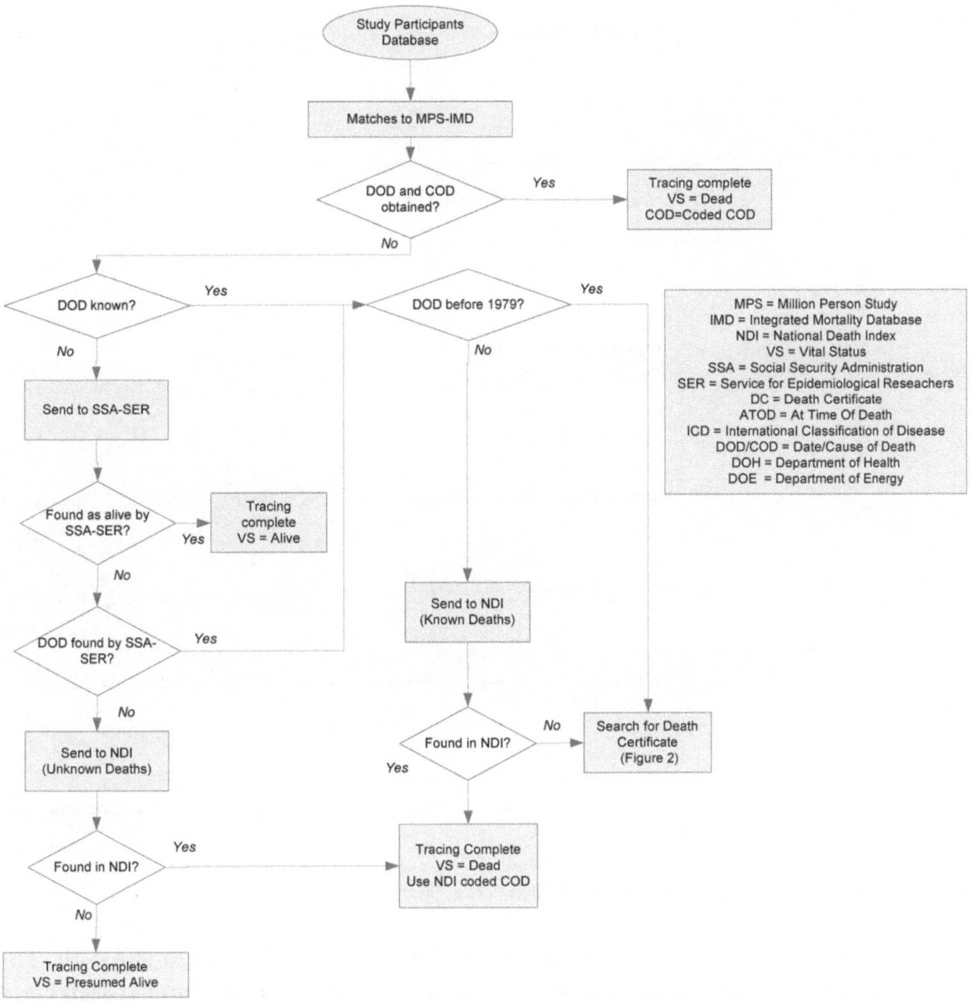

Figure 1. Generalized process to obtain vital status and cause of death utilized in the Million Person Study.

additional unique field (e.g. DOB or middle name) are accepted as matches without further review. The remaining NDI-provided matches are evaluated on a case-by-case basis which can be time consuming.

Special sources

For certain occupational groups, special data sources were available to counter challenges in tracing and the lesser mortality coverage prior to 1979. For example, for military personnel who participated in weapons testing between 1945 and 1962, the Department of Veterans Affairs Beneficiary Identification Records Locator Service (BIRLS) allowed linkages based on the military ID number, particularly helpful as SSN was often not available or widely utilized in these early years (Page et al. 1996; Boyko et al. 2000; Maynard and Chapko 2004).

Vital status tracing process

The flowchart in Figure 1 shows a generalized, multistage approach to the MPS tracing process. The first stage is to link study participants to the combined state mortality and SSDMF mortality files (i.e. the IMD). The first match is conducted using customized SAS software code to obtain 'perfect' or highly likely matches. For example, a match on SSN, first and last name, gender, and DOB would be considered a perfect match. A match on SSN, last name, and one additional piece of information such as DOB or middle name or gender, would be considered as highly likely match. A second match is then conducted using the Centers for Disease Control and Prevention LinkPlus probabilistic matching program (CDC 2018). This freely available software is useful for obtaining matches using a probabilistic scoring system developed by Fellegi and Sunter (1969). Important features of LinkPlus include the ability to link by names phonetically (including nicknames), transposition of digits in SSN, and by individual components of date fields (e.g. month of birth).

During this process, it is not unusual to discover errors within or missing key demographic variables, particularly the SSN. Online resources exist which provide qualified entities with access to public records (such as tax assessments, drivers licenses, etc.). The service we utilized most frequently was LexisNexis® Smartlinx® comprehensive person report. By manually searching these records for participants with known errors in identifying data, errors can be corrected and matches can be improved. Vital status may also be obtained this way, as some states provide mortality information through these online resources. However, as this is a

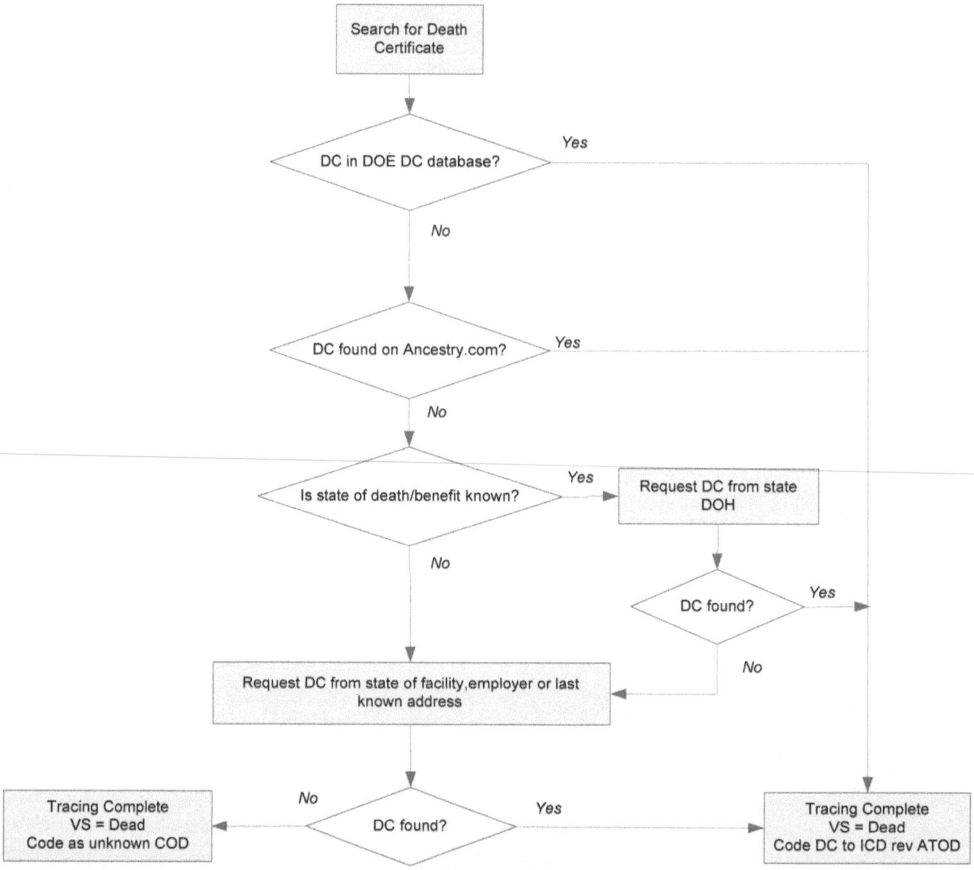

Figure 2. Generalized process to obtain death certificates utilized in the Million Person Study.

time-intensive, manual process, this is only performed after bulk, programmatic tracing options have been exhausted. General web searches for publicly available vital status information, like obituaries, can also be helpful (Sesso et al. 2000).

Corrections or additions to records (either in the participant records or in the IMD record) are entered into dedicated corrections databases. Additional series of matches against the IMD, or submissions to SSA-SER and/or NDI incorporating these corrections are then conducted as needed.

After the matching to the IMD, vital status tracing is considered complete for participants with a matched death date and COD. Remaining participants fall into two groups: (1) Date of death is known but no COD was identified or (2) No vital status determination at all. Those with a date of death available and year of death in 1979 or later are sent to NDI Plus to obtain underlying and contributing causes of death coded to the applicable revision of the ICD. Those with no vital status determination are then sent to SSA-SER to obtain a date of death or verification of alive status, or remain vital status unknown. After the SSA-SER results identify the group of participants with a date of death, those records with death dates 1979 and later are also sent to NDI Plus to obtain causes of death. For those participants verified by SSA-SER as alive after this stage, tracing is complete.

The remaining pool of participants is now those with no match or unknown vital status from the initial linkage to the IMD and the linkage to SSA-SER. This group of workers is sent to NDI Plus for annual searches, beginning with the year last known to be alive or 1979, whichever is later, to obtain both fact and COD.

At this stage, all participants not matched to the IMD, the SSA-SER, or NDI Plus are presumed to be alive. Either a sample or all participants presumed alive (depending on the size of the study population) are individually searched using the LexisNexis® Smartlinx® comprehensive person report. Using alternate methods of searching is often needed when key pieces of identifying information are missing or incorrect. If public records indicate recent activity (e.g. a home sale), and no indication of death is found, it provides additional validation that the participant is alive. If additional data is found (e.g. an alternate name), data fields need corrections, or a date death is provided, these are entered into the corrections databases.

Obtaining cause of death before 1979

Deaths identified prior to 1979 are particularly challenging because the NDI does not include deaths prior to this time. As such, a death certificate must be obtained for all deaths prior to 1979, and the COD must be coded to the ICD revision at the time of death by a trained nosologist. Figure 2 shows the approach MPS investigators used to search for death certificates. The DOE death certificates database was first searched to determine whether a scan of the death certificate is already on hand. Since over 30% of the MPS

Table 1. Summary of vital status tracing activities for selected Million Person Study cohorts.

Study	Total population	Cohort start of follow-up	Vital status year of last follow-up	Death count	Deaths (% of cohort)	COD count	COD (% of deaths)	Confirmed vital status (count)	Confirmed vital status (%)
Atomic veterans (Boice et al.2018, 2019c)	114,270	1945	2010	75,402	66.0	72,803	96.6	110,608	96.8
Nuclear utility workers (Boice et al.2018, 2019a)	145,209	1968/1977	2011	31,952	22.0	31,436	98.4	135,270	93.2
Industrial radiographers (Boice et al.2018, 2019b)	125,867	1968/1977	2011	22,012	17.5	21,598	98.1	110,860	88.1
Mound (Boice et al.2014)	7269	1944	2009	3681	50.6	3629	98.6	6761	93.0
Rocketdyne (Boice et al.2011)	5801	1948	2008	2383	41.1	2372	99.5	5405	93.2
Mallinckrodt chemical works (Golden et al.2019)	2514	1942	2012	1895	75.4	1879	99.2	2493	99.2
Los Alamos National Laboratory (LANL)[a]	23,308	1943	2016+	11,105	47.6[a]	6656	59.9[a]	18,582	79.7[a]

aThe tracing process for LANL is an example of a work in progress. The values reported here reflect the status to date.

covers DOE workers, this unique resource is of special value. If a death certificate is not found within the DOE resource, Ancestry.com, which currently has death certificates from 18 states and Puerto Rico, is searched.

Obtaining approval from SDOH to obtain death certificates can be cost-prohibitive for a number of states, as some states charge based on the number of searches rather than the number of death certificates found and provided. Therefore, a targeted approach was used. The state of death or SSA benefit (if known) is searched first. Next, the state of the facility where the subject was last employed is searched. If a death certificate is obtained, it is then provided to a trained nosologist for coding of the underlying COD and all contributory causes, using the ICD revision in place at the time of death.

Results

Table 1 shows the proportion of subjects successfully traced for vital status in six separate cohorts to date (nuclear power plant workers, industrial radiographers, atomic veterans and workers at Rocketdyne/Atomics International, Mound nuclear facility, and Mallinckrodt chemical works) and one cohort (Los Alamos National Laboratory (LANL workers) in progress. The LANL 'work in progress' indicates the challenges faced to achieve a high location rate after the initial location efforts are made. Within the MPS there are other cohorts in similar stages of tracing, such as the medical radiation workers ($n = 170{,}000$). In the cohorts described in Table 1, the COD ascertainment has been over 96% and verification of vital status has been over 90% of subjects in all but the Industrial Radiographers Cohort, related in part to incomplete identifying information.

Discussion

Other investigators have used a multistage approach for vital status ascertainment in large occupational cohort mortality studies (Buchanich et al. 2005; Hermansen et al. 2009; Wojcik et al. 2010) . We enhanced these approaches in two ways. First, we conducted matches using a collection of state mortality files assembled in-house using both deterministic and probabilistic algorithms. Second, we utilized online service providers such as LexisNexis® Smartlinx® comprehensive person report to access public records to correct and/or confirm key demographic variables used in the linkage process. Using an iterative approach with multiple passes allowed us to incorporate changes and corrections to participant records and to increase the change of obtaining a correct match. In addition, DOE death certificates were extremely valuable to the MPS for obtaining pre-1979 causes of death among facility workers who started working as early as the 1940s.

The completeness of mortality records from the SSA (both in the DMF and SER) varies greatly by age and year of death. Incompleteness may be as high as 40% in recent years (Jones and Vawdrey 2015). However, the SSA data contains the only national source of death information prior to 1979. Therefore, SSA data are very helpful in identifying these early deaths and often provide sufficient information to identify the state of death, from which a death certificate might be requested (Schnorr and Steenland 1997). In the future, investigators may need to rely more heavily on NDI searches, which provide very high-quality matches but can be cost-prohibitive.

While multiple mortality data sources with probabilistic matching algorithms (and manual review) are available to determine fact of death, only one national source (SSA-SER), which uses very strict matching criteria, is currently available to verify alive status. A participant record with common errors such as an alternate, maiden name, or nickname, misspelled last name, a transposed digit within the SSN, or missing month and/or day of birth will likely be probabilistically matched to a SDOH or NDI Plus death record if the participant has died. Because the SSA-SER requires near exact match among SSN, name, and DOB, the participant would not be verified as alive. Therefore, study populations with higher quality or more complete matching variables will have a higher percentage of verified vital status.

The tracing approach presented here for the MPS had the first stage as matching to in-house mortality sources and the second stage as submission to SSA-SER. However, these stages could have been reversed, like Buchanich et al. (2005). Any participants confirmed as alive from SSA-SER could be removed from the tracing process. This approach would reduce the probability of incorrect death matches, and perhaps allow looser matching criteria. Likewise, every mortality resource described herein may not be as useful for cohorts from a different study base. Investigators with occupational cohorts from a single facility might consider obtaining mortality files from one or surrounding states. The utilization of multiple, overlapping sources of mortality information, extensive data cleaning, and identification of specifically targeted resources able to supplement national databases, should assist any study in obtaining as complete mortality follow-up as possible.

Conclusion

The purpose of this article is to generally describe the comprehensive tracing efforts to locate and obtain COD for one million U.S. radiation workers and veterans, which has been ongoing for decades and involves over 20 different cohorts, some of which requiring specialized approaches. Nonetheless, we hope that some of the approaches described might be applicable to all cohorts, large or small, since the goals of achieving a high follow-up rate are the same, but may require cohort-specific modifications.

Disclosure statement

The authors report no conflict of interest. The authors alone are responsible for the content and writing of this article.

Funding

This work was supported in part by a grant from the U.S. Nuclear Regulatory Commission [NRC-HQ-60-14-G-0011], a grant from the Centers for Disease Control and Prevention [5UE1EH000989] a grant from the National Aeronautics and Space Administration [Grant No. NNX15AU88G], and grants from the U.S. Department of Energy [Grant No. DE-SC0008944 and Grant No. DE-AU0000042 awarded to the National Council on Radiation Protection and Measurements, which included interagency support from the U.S. Nuclear Regulatory Commission, the U.S. Environmental Protection Agency, and the National Aeronautics and Space Administration). Further, contract support was received by Oak Ridge National Laboratory from the Office of Radiation and Indoor Air, U.S. Environmental Protection Agency, under Interagency Agreement DOE No. 1824 S581-A1, under contract No. DE-AC05-00OR22725 with UT-Battelle; and contract support was received by Oak Ridge Associated Universities from the U.S. Department of Energy under contract No. DE-SC0014664.

ORCID

Elizabeth D. Ellis http://orcid.org/0000-0003-2871-1587

References

Acquavella JF, Donaleski D, Hanis NM. 1986. An analysis of mortality follow-up through the national death index for a cohort of refinery and petrochemical workers. Am J Ind Med. 9:181–187.

Boice JD, Jr, Cohen SS, Mumma MT, Ellis ED, Eckerman KF, Leggett RW, Boecker BB, Brill AB, Henderson BE. 2011. Updated mortality analysis of radiation workers at Rocketdyne (Atomics International), 1948–2008. Radiat Res. 176:244–258.

Boice JD, Jr, Cohen SS, Mumma MT, Ellis ED, Cragle DL, Eckerman KF, Wallace PW, Chadda B, Sonderman JS, Wiggs LD. 2014. Mortality among mound workers exposed to polonium-210 and other sources of radiation, 1944–1979. Radiat Res. 181:208–228.

Boice JD, Jr, Cohen SS, Mumma MT, Hagemeyer D, Chen H, Yoder RC, Dauer LT. 2019a. Leukemia among early nuclear power plant workers employed between 1957 and 1984 in the United States. Int J Radiat Biol. (Submitted).

Boice JD, Jr, Cohen SS, Mumma MT, Hagemeyer D. 2019b. Leukemia among industrial radiographers employed between 1940 and 2010 in the United States. Int J Radiat Biol. (this issue).

Boice JD, Jr, Ellis ED, Golden AP, Girardi DJ, Cohen SS, Chen H, Mumma MT, Shore RE, Leggett RW. 2018. The past informs the future: an overview of the million worker study and the Mallinckrodt chemical works cohort. Health Phys. 114:381–385.

Boice, JD, Jr, Cohen SS, Mumma MT, Chen H, Golden AP, Beck HL, Till JE. 2019c. Mortality among US military participants at seven aboveground nuclear weapons test series. Int J Radiat Biol. (this issue).

Boyko EJ, Koepsell TD, Gaziano JM, Horner RD, Feussner JR. 2000. U.S. department of veterans affairs medical care system as a resource to epidemiologists. Am J Epidemiol. 151:307–314.

Boyle CA, Decouflé P. 1990. National sources of vital status information: extent of coverage and possible selectivity in reporting. Am J Epidemiol. 131:160–168.

Buchanich JM, Dolan DG, Marsh GM, Madrigano J. 2005. Underascertainment of deaths using social security records: a recommended solution to a little-known problem. Am J Epidemiol. 162: 193–194.

Campbell KM, Deck D, Krupski A. 2008. Record linkage software in the public domain: a comparison of link plus, the link king, and a 'basic' deterministic algorithm. Health Informatics J. 14:5–15.

[CDC] Centers for Disease Control and Prevention. 2018. National program of cancer registries (NPCR). Registry plus™ link plus. National death index. Hyattsville (MD): CDC. (http://www.cdc.gov/cancer/npcr/tools/registryplus/lp.htm)

Cowper DC, Kubal JD, Maynard C, Hynes DM. 2002. A primer and comparative review of major US mortality databases. Ann Epidemiol. 12:462–468.

Curb JD, Ford CE, Pressel S, Palmer M, Babcock C, Hawkins CM. 1985. Ascertainment of vital status through the national death index and the social security administration. Am J Epidemiol. 121: 754–766.

Fellegi IP, Sunter AB. 1969. A theory for record linkage. J Am Stat Assoc. 64:1183–1210.

Fillenbaum GG, Burchett BM, Blazer DG. 2009. Identifying a national death index match. Am J Epidemiol. 170:515–518.

Golden AP, Ellis ED, Cohen SS, Mumma MT, Leggett RW, Wallace PW, Girardi D, Watkins JP, Shore R, Boice JD. Jr. 2019. Updated mortality analysis of the Mallinckrodt uranium processing workers, 1942-2012. Int J Radiat Biol. (this issue).

Hermansen SW, Leitzmann MF, Schatzkin A. 2009. The impact on national death index ascertainment of limiting submission to social security administration death master file matches in epidemiologic studies of mortality. Am J Epidemiol. 169:901–908.

Jones B, Vawdrey D. 2015. Measuring mortality information in clinical data warehouses. AMIA Jt Summits Transl Sci Proc. 2015:450–455.

LexisNexis®. 2018. LexisNexis public records collection, available via Smartlinx® comprehensive person report.

Maynard C, Chapko MK. 2004. Data resources in the department of veterans affairs. Diabetes Care. 27:B22–B26.

National Center for Health Statistics. 2018. Chapter 4, Assessing NDI Output. http://www.cdc.gov/nchs/data/ndi/ndi_users_guide_chapter4.pdf.

Page WF, Mahan CM, Kang HK. 1996. Vital status ascertainment through the files of the department of veterans affairs and the social security administration. Ann Epidemiol. 6:102–109.

Patterson BH, Bilgrad R. 1986. Use of the national death index in cancer studies. J Natl Cancer Inst. 77:877–881.

Schall LC, Buchanich JM, Marsh GM, Bittner GM. 2001. Utilizing multiple vital status tracing services optimizes mortality follow-up in large cohort studies. Ann Epidemiol. 11:292–296.

Schnorr TM, Steenland K. 1997. Identifying deaths before 1979 using the social security administration death master file. Epidemiology. 8: 321–323.

Sesso HD, Paffenbarger RS, Lee IM. 2000. Comparison of national death index and world wide web death searches. Am J Epidemiol. 152:107–111.

Social Security Administration. 2018. Service to epidemiological researchers to provide vital status data on subjects of health research. https:/www.ssa.gov/policy/about/epidemiology.html.

Wojcik NC, Huebner WW, Jorgensen G. 2010. Strategies for using the national death index and the social security administration for death ascertainment in large occupational cohort mortality studies. Am J Epidemiol. 172:469–477.

Validating the use of census data on education as a measure of socioeconomic status

Sarah S. Cohen, Michael T. Mumma, Elizabeth D. Ellis, and John D. Boice

ABSTRACT

Purpose: Adjusting for smoking status or a reliable surrogate [such as socioeconomic status (SES)] is critically important in occupational epidemiology studies when any smoking-related cancer or cardiovascular disease is an outcome of interest. Sometimes, however, data on smoking patterns or individual-level smoking surrogates such as job title, education, pay scale or other measures of SES are not readily available in occupational cohorts.

Methods and materials: To obtain a surrogate measure for missing smoking or individual-level SES data, we demonstrate a method used to obtain and geocode residential address histories which were then linked to area-level SES measures from the United States Census in three test samples and then in a full cohort of workers from the Mound nuclear weapons facility in Dayton, Ohio, USA. The mean educational attainment of the Census Block Group was used to derive a categorical estimate of educational attainment which was compared to self-reported (SR) education available from Mound worker histories using Kappa statistics. Lung cancer mortality patterns between area-derived (AD) and SR education were investigated using Standardized Mortality Ratios (SMR) and Cox Proportional Hazards models with stratification or adjustment by either SR or AD education.

Results: Home address histories were obtained from linkages of individual worker data to online resources. In the test cohorts, mean educational attainment was the Census Block Group measure found to have the largest magnitude association with individual-level SES measures. Among 7251 Mound workers, 5685 (78.4%) had at least one residential address match (mean 4.9 addresses) identified. The SR and AD educational attainment measures were highly correlated (weighted Kappa coefficient 0.10, $p < .0001$). SMR patterns by SR and AD educational attainment were similar, with steadily decreasing mortality with increased educational attainment by either measure. Cox models for lung cancer using AD education produced similar results as those using SR education as an adjustment factor.

Conclusion: When individual-level SES indicators are not available for statistical adjustment, area-level SES measures can serve as a reliable surrogate when investigating outcomes that are affected by lifestyle factors such as smoking.

Introduction

Adjustment for individual-level socioeconomic status (SES) is frequently employed in occupational studies to control for possible confounding factors such as tobacco use and other lifestyle factors that are not available but may influence mortality and morbidity. Often, indicators of SES such as educational attainment, job title, income, or pay-type (hourly versus salaried) are used as surrogates for smoking status, as individuals with lower SES tend to smoke at higher rates than those with higher SES (National Academies 2015). There is a growing body of literature supporting the idea that a person's environment is strongly associated with lifestyle factors, such as smoking and drinking patterns, obesity, risk of accidents and others (e.g. Cohen et al. 2011), and thus while individual-level SES variables are preferred over area-level variables for confounder adjustment, in the absence of individual measures, area-level variables have been shown to be adequate correlates of SES in occupational and other studies of cancer, heart disease and overall (Anderson et al. 1997; Waitzman and Smith 1998; Yen and Kaplan 1999; Diez Roux et al. 2001; Steenland et al. 2004). Thus, an innovative and resourceful approach to determining SES for occupational workers when individual-level measures are not available is to use area-level measures of education obtained from residential histories. In this study, we first demonstrate results from samples from three Million Person Study (MPS) cohorts that evaluated area-level measures derived from census data as surrogates for individual-level measures. Next, using one MPS cohort, we compare various measures of SES [none, individual-level, and area-derived (AD)] as an adjustment factor in models examining the association between radiation dose to the lung and lung cancer mortality.

Methods

Approval for Human Subjects research was received from the Oak Ridge Site-wide Institutional Review Board and the Vanderbilt University Institutional Review Boards.

Test samples

Three MPS cohorts were selected for testing of methodology to establish the best area-level measure from census data to use as a surrogate for individual SES measures. The test samples evaluated include workers from the Mound nuclear weapons facility (hereafter referred to as the Mound cohort), workers at the Mallinckrodt Chemical Works (MCW), and radiation workers at Rocketdyne (Atomics International) facilities in California. Details of each of these MPS cohorts have been published (Boice et al. 2011, 2014, 2018; Ellis et al. 2018; Golden et al. 2019). For the Mound population, a stratified random sample of 100 workers was selected within four categories of individual-level education obtained from work histories. For the Rocketdyne sample, the surrogate measure for SES was hourly or salaried worker status as available from occupational records; a simple random sample of 100 Rocketdyne workers was selected. For MCW workers, hourly or salaried status was the available SES surrogate from the work histories; for this population, the entire cohort of 2514 workers was included in the test sample.

Rosters for each of the three samples were matched via LexisNexis® to all available residential histories (Mumma et al. 2019). All matching residential addresses were geocoded and linked to Census Block Group measures including mean home value, mean educational attainment, and mean income. A Census Block Group contains about 600–3000 persons, or 250–1000 households, and due to this small size, residents are generally similar with regard to education, income, and home value. Within each test sample, a series of AD SES measures using census variables (including the lowest ever, highest ever, and most recent measures) were examined individually in relation to the available individual-level SES measures using ANOVA.

Mound cohort

After using the three test samples to identify the census measure most highly correlated with individual-level SES, we conducted an analysis using the full population of workers from the Mound nuclear weapons facility located near Dayton, Ohio, to compare adjustment with individual-level versus AD SES measures in models examining the association between radiation dose to the lung and lung cancer mortality. Among the Mound workers, vital status as of 31 December 2009, and the underlying cause of death were determined as previously described (Boice et al. 2014; Mumma et al. 2019). Vital status was obtained for 98.7% of the population and 50.7% had died by 2009. Self-reported (SR) education was extracted from employment records as Grade School or less, Some High School, High School Graduate, and Associates degree or higher, and was available for 6859 (94%) of all Mound workers and 4811 (97%) of radiation workers.

Demographic information for all Mound workers was used to conduct residential address history searches, including dates when the address was active. Addresses were then geocoded using SAS Proc Geocode. Addresses not successfully geocoded at the street-level and addresses containing a PO Box or rural delivery route were attempted to be geocoded to the ZIP plus 4 level. Addresses that did not geocode to the ZIP plus 4 were not used to derive educational attainment, as the ZIP code would likely be too large to reflect neighborhood characteristics. Geocoded address locations were then placed into 2000 (addresses active prior to 2005) or 2010 (addresses active after 2004) Census Block Groups. Based on the work done in the test samples described in the Methods above and Results below, the mean educational attainment from the Census Block Group was selected to be the AD measure of SES and was thus merged by Block Group to each address. The highest mean educational attainment for a worker's address history was a continuous measure that was then grouped such that the AD education cut-points were similar to the SR education distribution; cut-points were made at 13.3, 15.1, and 16 years to create a four-level categorical measure.

Standardized mortality ratio (SMR) analyses, using an approach similar to Marsh et al. (1998), compared the numbers of deaths observed among Mound workers with the numbers expected based on general population rates in the United States for persons of the same age, race, and sex over the same time periods using the underlying cause of death. SMR analyses for multiple outcomes were performed on the subset of workers for whom both an SR and AD education measure was available, and SMRs were calculated within categories of each education measure (SR and AD). Person-time began at the date of first hire at Mound and ended either at the date of death; 31 December 2009; age 95, or date of loss to follow-up, whichever came first.

Analyses within cohort were conducted using Cox proportional hazards models to compute risks of lung cancer (Cox 1972). Lung cancer mortality was determined from the underlying cause of death. Radiation doses specifically to the lung from external (photon and neutron) as well as internal intakes of radionuclides (tritium, polonium, and plutonium) were obtained as described previously (Boice et al. 2014; NCRP 2018); for these analyses, the radiation weighting factor is taken to be 1 for internal intakes. Radiation doses were lagged by 10 years and treated as a time-dependent covariate in the model. Age was used as the timescale for the hazard function and all models included adjustment for year of birth (six categories) and sex (male/female). Hazard ratios (HR) and 95% confidence intervals (CI) for lung cancer mortality were produced for categories of radiation exposure in the subset of radiation workers with non-missing values for both SR and AD education. Models were constructed in this subgroup first without any adjustment for SES, then using the SR measure, and finally using the AD measure. Additionally, a model using SR education when available and imputed with AD education when the SR measure was missing was fitted in order to utilize the maximum available number of workers and serve as a demonstration of a hybrid approach that might be used in future studies. All analyses were conducted with SAS/STAT software (SAS/STAT software, Version 9.4 of the SAS System for Windows, SAS Institute Inc., Cary, NC, USA).

Results

Test samples

Address histories were linked to the three test sample rosters using available demographic information. For the Mound sample, 73% of the sample had at least one match to a residence and the average number of linked addresses was 3.7 (range 1–13). For the Rocketdyne sample, 89% of workers had at least one match with an average of 6.3 linked addresses (range 1–14). Among the MCW workers, 1777 (71%) had at least one linked address with an average of 3.9 linked addresses (range 1–16). Addresses for workers who died prior to 1995 were more difficult to obtain. Workers with no matching addresses were disproportionately in the lowest education group in the Mound sample (chi-square $p < .0001$) or were hourly workers as opposed to salaried workers in the Rocketdyne (chi-square $p = .025$) and Mallinckrodt samples (chi-square $p < .0001$).

Census variables including mean home value, mean income, and mean education were generally significantly positively associated with individual-level SES measures across the three test samples (Tables 1–3). Highest mean level of education at any address had the largest magnitude of association with individual-level education in the Mound cohort ($p = .0001$), and was also significantly associated with hourly/salary status in the MCW workers ($p < .0001$) as well as in the Rocketdyne cohort ($p = .007$).

Table 1. Evaluation of individual-level SES (measured as pay type: hourly versus salary) with area-level SES measures associated with home addresses in the Mallinckrodt Chemical Workers cohort ($N = 2514$).

	Individual-level SES Measure: Pay type			
Area-derived SES measure	Hourly Mean	Salary Mean	Missing Mean	*p*-Value*
Median home value ($)				
Lowest at any address	91,838	125,184	114,655	<.0001
Highest at any address	212,396	312,721	316,637	<.0001
Most recent	173,891	254,856	258,678	<.0001
Mean education (years)				
Lowest at any address	12.4	13.0	12.9	<.0001
Highest at any address	13.9	14.9	14.8	<.0001
Most recent	13.4	14.3	14.3	<.0001
Median household income ($)				
Lowest at any address	37,169	43,263	40,122	<.0001
Highest at any address	69,137	88,053	91,566	<.0001
Most recent	59,020	72,140	73,666	<.0001

*p-value from ANOVA. ANOVA model does not include the missing hourly/salary column (i.e. comparison is only hourly/salary).

Mound cohort

After establishing that highest mean level of education at any address was the best census measure to use as a surrogate for individual-level SES, we conducted an analysis to compare effects of adjustment with this AD measure compared to the individual-level measure of education in the full population of Mound workers. The Mound cohort has been previously described (Wiggs et al. 1991; Boice et al. 2014). In brief, 7270 employees first hired between 1944 and 1979 were identified including 4954 who were monitored for any type of radiation exposure. Most of the workers were male (75.2%), white (80.3%), born before 1930 (56.0%), hired before 1960 (51.4%), and followed for more than 30 years (82.3%).

In the Mound workers, the SR measure of education obtained from occupational work histories (four categories) and AD measure of education taken as the Census Block Group highest mean educational attainment in four categories were correlated, with a Kappa coefficient of 0.10 ($p < .0001$). Figure 1 shows the positive linear association between the mean Census Block Group educational attainment and the four SR education categories.

Figure 2 shows that the SMRs in the Mound workers for smoking-related cancers, lung cancer, and heart disease within categories of either SR or AD education were similar with a steady decrease in the SMRs as educational attainment increases. Similar inverse associations were seen over

Table 2. Evaluation of individual-level SES (measured as four levels of education) with area-level SES measures associated with home addresses in the Mound worker test sample ($N = 100$).

	Individual-level SES measure: Education				
Area-derived SES measure	Grade school ($n = 16$) Mean	Some high school ($n = 12$) Mean	High school graduate ($n = 32$) Mean	Associate's degree or more ($n = 13$) Mean	*p*-Value*
Mean home value ($)					
Lowest at any address	110,227	112,531	179,483	188,902	.072
Highest at any address	198,548	161,435	440,524	362,397	.0003
Mean education (years)					
Lowest at any address	12.00	12.08	12.84	13.38	.0055
Highest at any address	13.38	13.58	15.06	15.00	.0001
Mean income ($)					
Lowest at any address	42,277	45,831	45,773	56,209	.42
Highest at any address	64,908	72,191	104,615	99,367	.0023

*p-value from ANOVA.

categories of both SR and AD measures of education for other disease outcomes known to be associated with lifestyle measures such as smoking and alcohol consumption including cirrhosis of the liver, nephritis and nephrosis, and external causes of death (data not shown).

Table 3. Evaluation of individual-level SES (measured as pay type: hourly versus salary) with area-level SES measures associated with home addresses in the Rocketdyne test sample ($N = 100$).

	Individual-level SES measure: Pay type		
Area-derived SES measure	Hourly workers Mean	Salary workers Mean	*p*-Value*
Mean home value			
Lowest at any address	231,060	262,843	.40
Highest at any address	626,256	692,691	.19
Mean education			
Lowest at any address	11.51	12.44	.0005
Highest at any address	14.58	15.38	.007
Mean income			
Lowest at any address	45,617	51,846	.19
Highest at any address	105,689	127,436	.002

**p*-value from ANOVA.

The results of the internal cohort analyses for the Mound workers are shown in Table 4 with separate models being fitted including no adjustment for SES, adjustment for the SR education measure, and adjustment for the AD education measure. The model without adjustment for SES had a lower HR in the dose category ≥500 milligray (mGy) than models with either SR or AD education adjustment. The models including SR and AD adjustment had very similar HRs in all categories of lung dose. The final model was fitted to the full Mound radiation cohort of 4954 workers using SR education when available ($N = 4811$ workers) and imputed with AD education when SR was missing ($N = 143$ workers); the HRs (95% CI) for lung doses of <10, 10–99, 100–499, and 500+ mGy were 1.0 (reference), 1.24 (0.90–1.70), 0.86 (0.56–1.33), and 1.36 (0.79–2.36), respectively compared to HRs (95% CIs) of 1.24 (0.90–1.70), 0.84 (0.54–1.30), and

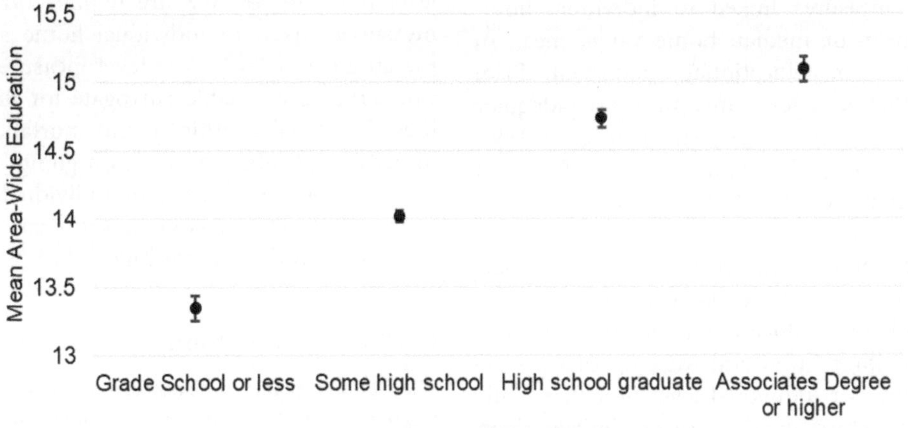

Figure 1. Mean educational attainment at the Census Block Group level by individual-level self-reported educational attainment from worker records among Mound workers. Note: Error bars show two standard errors from the mean

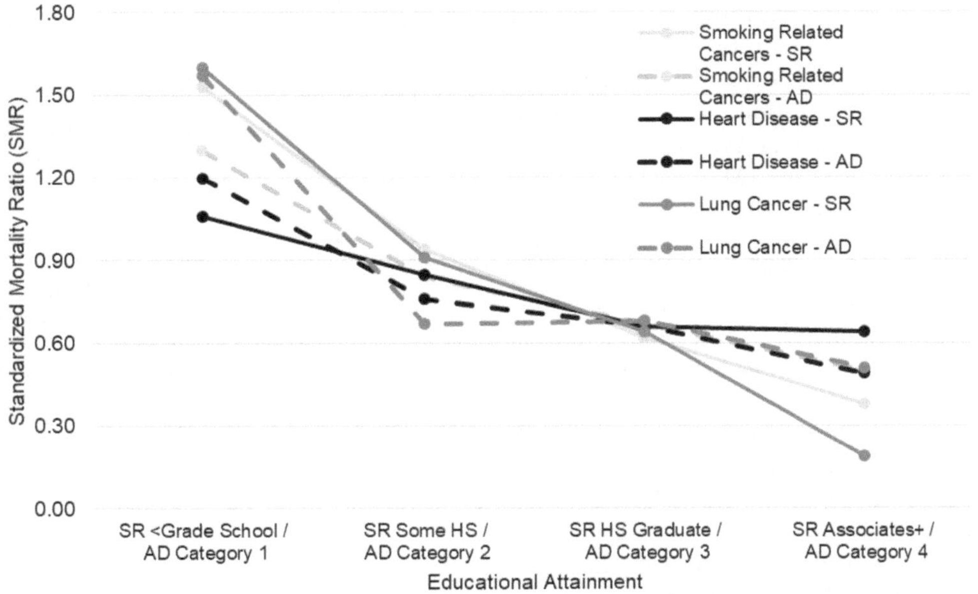

Figure 2. Standardized Mortality Ratio (SMR) for smoking-related cancers, lung cancer, and heart disease by categories of SR and AD education among Mound workers. SR = Self-reported; AD = Area-derived; HS = High school; AD categories of education are based on Census Block Group mean educational attainment using cut-points at 13.3, 15.1, and 16 years.

1.21 (0.70–2.09), respectively, when the model included no SES adjustment.

Discussion

We found that individual-level measures of SES such as SR education or pay type were highly correlated with area-level Census Block Group measures linked to individual home addresses including mean or median home value, mean or median income, and mean educational attainment. Prior work has indicated that area-level measures are adequate correlates of individual-level measures (Anderson et al. 1997; Waitzman and Smith 1998; Yen and Kaplan 1999; Diez Roux et al. 2001; Steenland et al. 2004), and this study supports those findings.

For most of the cohorts in the MPS, SES variables are available in the source files that can be used for statistical adjustment such as job title, education level, military rank, or pay scale. However, three important MPS cohorts do not have consistent measures of individual-level SES that could be used as adjustment factors: nuclear power plant workers, industrial radiographers, and medical radiation workers. The current study was conducted in large part to establish whether the use of area-level measures from census blocks could reasonably serve as surrogates for individual-level SES in these cohorts. The reason why measures of SES are not readily available relates in large part to the data sources used to identify these populations for study which include the Nuclear Regulatory Commission (NRC) Radiation Exposure Information and Reporting System (REIRS) and the Landauer, Inc. dosimetry database (Muirhead et al. 1996; NRC 1994, 2018; Anzenberg et al. 2010; NCRP 2018).

This study has several strengths as well as limitations. The MPS cohorts utilized in both the test samples and the full cohort analysis are occupational cohorts with long follow-up (up to 60 years), high quality capture of occupational radiation doses, a low percentage of workers who were lost to follow-up, and a low percentage of deaths for which a specific cause was not available. For the Mound study, we were able to utilize the entire cohort to compare the statistical effects of adjustment using an individual-level measure of education captured in the work histories to compare with an area-level measure derived from address histories. Limitations include the relatively small number of workers in the Mound population and the lack of individual-level smoking history which precludes a full assessment regarding the ability of AD measures of SES to serve as appropriate surrogates for smoking. Nonetheless, it was informative that the area-level measures of SES correlated so closely with the frequently used individual-level measures derived from occupational histories including job title, salary information, and/or education level.

In conclusion, this study demonstrates that individual-level measures of SES are highly correlated with area-level measures linked to individual home addresses, and that in the absence of individual-level measures, area-level measures can serve as a reliable surrogate for SES when investigating lifestyle-related morbidity and mortality. These findings are directly applicable to work on-going within the MPS, particularly for the cohorts of individuals where the available source data do not contain sufficient information to group individual workers according to SES.

Disclosure statement

The authors report no conflicts of interest. The authors alone are responsible for the content and writing of the paper.

Funding

The Million Person Study of Low-Dose Health effects is supported in part by a grant from the U.S. Nuclear Regulatory Commission [NRC-HQ-60-14-G-0011], grants from the Centers for Disease Control and Prevention [5UE1EH000989, 5NUE1EH001315], grants from the National Aeronautics and Space Administration [NNX15AU88G, 80NSSC17M0016], and grants from the U.S. Department of Energy [Grant No. DE-SC0008944] awarded to the National Council on Radiation Protection and Measurements, which included interagency support from the U.S. Nuclear Regulatory Commission, the U.S. Environmental Protection Agency, and the National Aeronautics and Space Administration; and more recent Grants No. DE-AU0000042 and DE-AU0000046. Contract support was received from the Naval Sea Systems Command [N00024-17-C-4322]. Further, contract support was received by Oak Ridge National Laboratory from the Office of Radiation and Indoor Air, U.S. Environmental Protection Agency, under Interagency Agreement DOE No. 1824 S581-A1, under contract No. DE-AC05-00OR22725 with UT-Battelle; and contract support was received by Oak Ridge Associated Universities from the U.S. Department of Energy under contract No. DE-SC0014664.

ORCID

Sarah S. Cohen http://orcid.org/0000-0003-0421-1983
Michael T. Mumma http://orcid.org/0000-0001-7506-8710
Elizabeth D. Ellis http://orcid.org/0000-0003-2871-1587
John D. Boice http://orcid.org/0000-0002-8755-1299

References

Anderson RT, Sorlie P, Backlund E, Johnson N, Kaplan GA. 1997. Mortality effects of community socioeconomic status. Epidemiology. 8:42–47.

Anzenberg V, Lewis DE, Dickson ED, Bush-Goddard SP. 2010. The U.S. Nuclear Regulatory Commission Radiation Exposure Information Reporting System (REIRS). Radiat Res. 173:254–255.

Boice JD, Jr, Cohen SS, Mumma MT, Ellis ED, Cragle DL, Eckerman KF, Wallace PW, Chadda B, Sonderman JS, Wiggs LD. 2014. Mortality among Mound workers exposed to polonium-210 and other sources of radiation, 1944-1979. Radiat Res. 181:208–228.

Boice JD, Jr, Cohen SS, Mumma MT, Ellis ED, Eckerman KF, Leggett RW, Boecker BB, Brill AB, Henderson BE. 2011. Updated mortality analysis of radiation workers at Rocketdyne (Atomics International), 1948-2008. Radiat Res. 176:244–258.

Boice JD, Jr, Ellis ED, Golden AP, Girardi DJ, Cohen SS, Chen H, Mumma MT, Shore RE, Leggett RW. 2018. The past informs the future: an overview of the million worker study and the Mallinckrodt Chemical Works cohort. Health Phys. 114:381–385.

Cohen SS, Sonderman JS, Mumma MT, Signorello LB, Blot WJ. 2011. Individual and neighborhood-level socioeconomic characteristics in relation to smoking prevalence among black and white adults in the Southeastern United States: a cross-sectional study. BMC Public Health. 11:877.

Cox DR. 1972. Regression models and life-tables. J R Stat Soc B. 34: 187–220.

Diez Roux AV, Merkin SS, Arnett D, Chambless L, Massing M, Nieto FJ, Sorlie P, Szklo M, Tyroler HA, Watson RL. 2001. Neighborhood of residence and incidence of coronary heart disease. N Engl J Med. 345: 99–106.

Ellis ED, Boice JD, Jr, Golden AP, Girardi DJ, Cohen SS, Mumma MT, Shore RE, Leggett RW, Kerr GD. 2018. Dosimetry is key to good epidemiology: workers at Mallinckrodt Chemical Works had seven different source exposures. Health Phys. 114:386–397.

Golden AP, Ellis ED, Cohen SS, Mumma MT, Leggett RW, Wallace PW, Girardi D, Watkins JP, Shore R, Boice JD, Jr. 2019. Updated mortality analysis of the Mallinckrodt uranium processing workers, 1942-2012. Int J Radiat Biol. Submitted.

Marsh GM, Youk AO, Stone RA, Sefcik S, Alcorn C. 1998. OCMAP-PLUS: a program for the comprehensive analysis of occupational cohort data. J Occup Environ Med. 40:351–362.

Muirhead CR, Boice JD, Jr, Raddatz CT, Yoder RC. 1996. Comparison of dose histories for U.S. nuclear power plant workers, based on records held by a major dosimetry service company and on the NRC REIRS database. Health Phys. 70:645–650.

Mumma MT, Cohen SS, Sirko JL, Ellis ED, Boice JD. 2019. Obtaining vital status and cause of death on a million persons. Int J Radiat Biol. Submitted.

National Academies of Sciences, Engineering, and Medicine; Committee on the Long-Run Macroeconomic Effects of the Aging U.S. Population-Phase II; Committee on Population, Division of Behavioral and Social Sciences and Education; Board on Mathematical Sciences and Their Applications; Division on Engineering and Physical Sciences. 2015. The growing gap in life expectancy by income: implications for federal programs and policy responses. Washington (DC): The National Academies Press.

National Council on Radiation Protection and Measurements (NCRP). 2018. Deriving organ doses and their uncertainty for epidemiologic studies (with a focus on the one million U.S. workers and veterans study of low-dose radiation health effects. NCRP Report 178. Bethesda (MD): NCRP.

Nuclear Regulatory Commission (NRC). 1994. Voluntary reporting of additional occupational radiation of exposure data (generic letter 94-04); [accessed 2018 Sep 8]. https://www.nrc.gov/docs/ML1204/ML12047A243.pdf

Nuclear Regulatory Commission (NRC). 2018. Radiation exposure information and reporting system (REIRS) [accessed 2018 Sep 8]. http://www.nrc.gov/reading-rm/doc-collections/gils/rad-exp.html

Steenland K, Henley J, Calle E, Thun M. 2004. Individual- and area-level socioeconomic status variables as predictors of mortality in a cohort of 179,383 persons. Am J Epidemiol. 159:1047–1056.

Waitzman NJ, Smith KR. 1998. Phantom of the area: poverty-area residence and mortality in the United States. Am J Public Health. 88: 973–976.

Wiggs LD, Cox-DeVore CA, Voelz GL. 1991. Mortality among a cohort of workers monitored for 210Po exposure: 1944-1972. Health Phys. 61: 71–76.

Yen IH, Kaplan GA. 1999. Neighborhood social environment and risk of death: multilevel evidence from the Alameda County Study. Am J Epidemiol. 149:898–907.

Cohort profile – MSKCC radiation workers: a pilot sub-cohort of a multicenter medical radiation worker component of the Million Person Study of low-dose radiation health effects

Lawrence T. Dauer, Meghan Woods, Daniel Miodownik, Brian Serencsits, Brian Quinn, Michael Bellamy, R. Craig Yoder, Xiaolin Liang, John D. Boice, and Jonine Bernstein

ABSTRACT

Background: The National Council on Radiation Protection and Measurements (NCRP) is coordinating an expansive epidemiologic effort entitled the Million Person Study of Low-Dose Radiation Health Effects (MPS). Medical workers constitute the largest occupational radiation-exposed group whose doses are typically received gradually over time.

Methods: A unique opportunity exists to establish an Institutional Review Board/Privacy Board (IRB/PB) approved, retrospective feasibility sub-cohort of diseased Memorial Sloan Kettering Cancer Center (MSK) medical radiation workers to reconstruct occupational/work history, estimate organ-specific radiation absorbed doses, and review existing publicly available records for mortality from cancer (including leukemia) and other diseases. Special emphasis will be placed on dose reconstruction approaches as a means to provide valid organ dose estimates that are as accurate and precise as possible based on the available data, and to allow proper evaluation of accompanying uncertainties. Such a study that includes validated dose measurements and information on radiation exposure conditions would significantly reduce dose uncertainties and provided greatly improved information on chronic low-dose risks.

Results: The feasibility sub-cohort will include deceased radiation workers from MSK who worked during the nearly seventy-year timeframe from 1946 through 2010 and were provided individual personal radiation dosimetry monitors. A feasibility assessment focused on obtaining records for about 25–30,000 workers, with over 124,000 annual doses, including personnel/work histories, specific dosimetry data, and appropriate information for epidemiologic mortality tracing will be conducted. MSK radiation dosimetry measurements have followed stringent protocols complying with strict worker protection standards in order to provide accurate dose information for radiation workers that include detailed records of work practices (including specific task exposure conditions, radiation type, energy, geometry, personal protective equipment usage, badge position, and missed doses), as well as recorded measurements. These dose measurements have been ascertained through a variety of techniques that have evolved over the years, from film badges to thermoluminescent dosimetry technology to optically stimulated luminescent methodologies. It is expected that individual total doses for the sub-cohort will have a broad range from <10 mSv to >=1000 mSv.

Conclusions: MSK has pioneered the use of novel radiation diagnostic and therapeutic approaches over time (including initial work with x-rays, radium, and radon), with workplace safety in mind, resulting in a variety of radiation worker exposure scenarios. The results of this feasibility sub-cohort of deceased radiation workers, and associated lessons learned may potentially be applied to an expanded multicenter study of about 170,000 medical radiation worker component of the MPS.

Introduction

Previous attempts to estimate cancer and non-cancer risk from radiation exposures at low doses (<100 mGy) and low dose rates (<5 mGy per hour) have significant uncertainties, but have suggested that risks following low-dose rates may be as harmful as those reported among Japanese atomic-bomb survivors. Even today, 70 years after the exposure, 18,443 atomic-bomb survivors exposed to >100 mSv form the basis of radiation protection standards (NRC 2006; Cardis 2007; ICRP 2007; UNSCEAR 2008; Jacob et al. 2009;

Wakeford 2005; Dauer 2011; NCRP 2018a) and scientific and medical committees grapple over how best to estimate and apply a 'dose and dose rate effectiveness factor' (ICRP 2005; Tubiana et al. 2006; ICRP 2007; Tenforde and Schauer 2008; NCRP 2018a) to scale the risks to situations involving chronic exposures relevant to occupational, environmental, and medical imaging circumstances.

The National Council on Radiation Protection and Measurements (NCRP) is coordinating an expansive epidemiologic effort entitled the Million Person Study of Low-Dose Radiation Health Effects (MPS) (Boice 2012; Bouville et al. 2015; NCRP 2018b). The primary aim of the MPS is to provide scientifically valid information and improve precision on the estimated level of radiation risk when exposures are received gradually over time, as opposed to the acute exposure of the Japanese atomic-bomb survivors. The primary health outcome of interest for the MPS is cancer mortality, but other causes of death such as cardiovascular disease and cerebrovascular disease will also be evaluated. The validity of the MPS is a function of the accuracy of the organ dose estimates (i.e. the absorbed dose averaged over all parts of an organ or tissue) and their accompanying uncertainties. Thus, the focus is on providing the best estimate of organ dose and uncertainty for each individual, and to (when possible) characterize shared uncertainties that affect groups of individuals within the various cohorts considered. The study populations include atomic veterans, Department of Energy workers, nuclear power plant workers, industrial radiographers, and medical radiation workers.

Memorial Sloan Kettering Cancer Center (MSK) consisting of Memorial Hospital (MH) and the Sloan Kettering Institute (SKI) laboratories has a unique history with regard to the use of radiation for the diagnostic and therapeutic treatment of cancer and allied diseases. The initial New York Cancer Hospital (NYCH) was founded in 1884, barely a decade before the seminal burst of discoveries in radiation. Wilhelm Roentgen discovers 'X-Rays' in 1895 and a week later makes his famous first X-ray images of the hand of Mrs. Roentgen (Anna Bertha Ludwig) wearing her wedding ring (Pietzch 2018). Henri Becquerel subsequently discovered 'radioactivity' and radioactive materials in 1896 and this was quickly followed by the discovery of 'polonium' and 'radium' by the Curies (Nobel 2018). Immediate attention is given to the application of these rays and materials to the healing arts. Research and use began almost immediately across the world, even in New York where Thomas Edison demonstrated fluoroscopes in 1896 (King 2012). As early as 1902, the NYCH employed X-Rays and X-ray therapies, practices that continue through the present. Also, as early as 1902, several adverse biological effects began to be identified in some medical radiation workers, both short-term (e.g. reddening of the skin, dermatitis, skin ulceration, epilation, eye irritation) and longer-term (e.g. skin cancers, cataracts, and other cancers) (Linet et al. 2010).

Radium was incredibly scarce in the early years of the 20th century, especially in the United States. In 1913, with help from the US government, Dr. James Douglas, a scientist and mining engineer, formed the National Radium Institute, significantly increasing the amount of radium available. In 1917, Dr. Douglas gave his entire share – worth about $300,000 at the time – to Memorial Hospital (the new name for the NYCH), with a stipulation that radiation would be studied and used to treat cancer (Tontonoz 2016a, 2016b).

At MH, Gioacchino Failla headed up the development of radium to be safely used for treatment, either as a teletherapy (packing a radium source in a shielded lead box for protection along with an opening for treatment), or as implanted needles of radium or later radium 'emanations' (radon gas) as some of the earliest forms of brachytherapy. Failla, who received his Ph.D. with Marie Curie in Paris, developed methods for radium use in treatment and constructed a semi-automatic radon collection system to safely extract and package radon gas in glass tubing or gold needles. By 1926, MH owned and utilized more radium than anywhere else in the world, about one-third of the world's supply at the time (Tontonoz 2016a, 2016b).

One of Failla's major professional accomplishments included those in radiation protection, particularly for the physicians and technologists who handled radium. Over long periods of time, he made careful observations of the changes in their skin and nails and convinced himself that inexperienced and careless individuals exposed themselves the most. Dr. Edith Quimby, Failla's assistant, was concerned with radiation protection of patients and staff. She developed standard dosing tables for safe and effective radiation treatment (Quimby 1934). For staff exposures, although radiologists had previously used either a 'reddening of the skin' or X-ray film to estimate radiation exposure, by 1926, Failla and Quimby were the first to start a full-scale film-badge program using dental film with a filter to distinguish exposure to gamma or beta radiation (Quimby 1926). They employed the use of standard 'dose' films for comparison and later developed the use of densitometers for reading badges.

Although some radioactive isotopes were able to be synthesized prior to the war using cyclotrons (e.g. ^{130}I and ^{131}I in small amounts), after World War II, it became possible to create large amounts of man-made radioactive elements in nuclear reactors. Such development was enhanced beginning in 1953 with the 'Atoms-for-Peace' program (Eisenhower 1953). These new isotopes (e.g. ^{32}P, ^{60}Co, ^{137}Cs, ^{125}I, ^{103}Pd, and many others) were much safer to handle and administer than radium or radon seeds and were employed at MSK for therapeutic purposes. In addition, a revolution in imaging and treating thyroid cancer was enabled by the more general availability of ^{131}I.

Later, with collaboration of the US Department of Energy and National Laboratories, a cyclotron (affectionately known as 'Betsy'), one of the earliest hospital-based cyclotrons in the world and the first in the country, was installed in the SKI laboratory building in order to enable safe research and development and production of other radionuclides for imaging and therapy. Subsequent cyclotrons were installed, including a dual-beam instrument enabling both liquid and solid targets, as well as the most recent Radiochemistry and Cyclotron Facility which routinely produces positron-emitting radionuclides and develops and performs organic synthesis of both single-photon-emitting as well as positron-emitting radiotracers. These facilities were heavily supported by medical health physics engineering, design, and operational assistance to incorporate safety features while facilitating radionuclide production, research, and patient care.

The unique history of radiation use at MSK as well as the associated plethora of occupational exposure scenarios and data collection, suggests that establishing a feasibility sub-cohort of deceased radiation workers from MSK could help inform a larger multicenter medical radiation worker cohort as part of the MPS.

Key considerations in establishing the feasibility for the MSK sub-cohort of deceased radiation workers

Selection of the MSK sub-cohort

Across the nation and all occupations, medical radiation workers represent a large fraction of radiation-exposed workers, with ~2.5 million monitored workers in 2006 (NCRP 2009a). Historically, the average annual occupational effective dose estimates were higher in the past and have trended downward for the medical radiation worker populations (Linet et al. 2010). Most present-day medical radiation workers generally experience very low radiation exposures, almost exclusively from external irradiation. Those individuals who currently perform certain fluoroscopically guided interventional procedures and potentially those who prepare or administer radionuclides for nuclear-medicine procedures are an exception to this generalization (Dauer 2014).

The MPS includes a large medical radiation worker cohort initially assembled using very preliminary estimates of lifetime doses contained in a database supplied by a commercial provider of personal monitoring services (Landauer Inc., Glenwood, IL). Workers were selected to be included in the MPS medical radiation worker cohort so as to ensure complete coverage of the lifetime dose range, and considering the costs associated with epidemiological tracing. To that end, all workers having lifetime doses equal to or greater than 50 mSv (based on personal dose equivalent, $H_P(10)$) were automatically selected. Adding to this group was a random selection of 50% of workers with lifetime doses between 10 and 50 mSv and 2% of those workers with a lifetime dose of less than 10 mSv. The initial identified medical radiation worker cohort study population totals 168,601 individuals selected from an eligible population of 1.71 million medical workers. However, there are large uncertainties regarding the lifetime dose estimates from such national databases and the dose reconstruction process as the derivation of organ doses from such monitoring data poses difficult problems for the medical worker cohort because of, among other factors:

- Difficulties assigning workers to a specific exposure scenario;
- Often extreme inhomogeneity of exposure over the body of personnel for any given procedure type;
- Differing degrees and methods of radiation protection;
- Inconsistent wearing of dosimeters by personnel (i.e. at times choosing not to wear dosimeters in order to avoid investigations) (NCRP 2010), combined with poor information, as well as high variability, on the workloads of physicians and technologists (i.e. the number of procedures of a given type conducted monthly or annually); and
- Changing technology and medical procedure protocols.

Recognizing the need for more specific guidance on these challenges, NCRP recently chartered Scientific Committee 6–11 to provide dosimetry guidance for medical workers with a focus on lung dose reconstruction (Yoder et al. 2019).

Typically, local medical health physics or radiation safety sections at hospitals collect and maintain additional data on medical radiation workers including job titles (i.e. exposure scenario), average radiation energies, use of protective aprons, dosimeter placement, and minimally detectable doses over time. A deceased medical radiation worker feasibility sub-cohort study that includes validated dose measurements and information on radiation exposure conditions would significantly reduce dose uncertainties and provided greatly improved information on chronic low-dose risks.

The feasibility sub-cohort will include deceased radiation workers from MSK who worked during the nearly seventy-year timeframe from 1946 through 2010 and were provided individual personal radiation dosimetry monitors. A feasibility assessment focused on obtaining records for about 25–30,000 workers, with over 124,000 annual doses, including personnel/work histories, specific dosimetry data, and appropriate information for epidemiologic mortality tracing will be conducted.

Initial epidemiology considerations

Although leukemia and/or solid cancers have been observed among pioneering radiologists and early radiation technologists (March 1944; Lewis 1963; Matanoski et al. 1975; Smith and Doll 1981; Wang et al. 1990; Berrington et al. 2001), more recent studies of medical x-ray technologists have been less clear in revealing excess cancers (Jablon and Miller 1978; Doody et al. 1998; Yoshinaga et al. 1999; Mohan et al. 2003; Yoshinaga et al. 2004). However, the absence of reliable dosimetry in these studies has limited risk interpretation (Simon et al. 2006; Bhatti et al. 2007; Simon et al. 2014). In contrast, the MPS, including a multicenter cohort of medical radiation workers, has excellent individualized annual dosimetry measurements, and linkages to worker dosimetry from other occupational exposures received by workers as they may have moved from workplace to workplace over their careers, to ensure as complete as possible information on lifetime doses for individuals in the study. Further, a specific MSK sub-cohort can provide validated dose measurements and information on radiation exposure conditions could significantly reduce dose uncertainties and provide greatly improved information on potential risks.

As a pilot feasibility epidemiologic study, the primary aim is to establish the MSK Medical Radiation Worker Cohort of deceased radiation-exposed workers in order to: assess the feasibility of collecting complete dosimetry and personnel records from about 1946–2010; assess the feasibility of retrieving personnel/work histories, dosimetry data, and personal identifiers suitable for tracing to determine the earliest acceptable date to include; and finally to assess the feasibility of obtaining outcome and overall career doses (e.g. mortality tracing; matching Social Security Number [SSN] for vital status; linking national dosimetry databases for career dose determination). Adverse outcomes will be correlated initially with cumulative whole-body or appropriate organ radiation dose. The secondary aim is to provide dosimetric uncertainty considerations for a larger multicenter medical radiation worker cohort included in the MPS using the more detailed and complete MSK records to ascertain inaccuracies in the MPS dosimetry databases. In addition, appropriate modeling parameters will be suggested for assessing organ doses from dosimetry badge results based on specific radiation worker tasks over time.

Sources of vital status and cause of death will be the Social Security Master Death File and other Social Security Administration (SSA) vital status files, Pension Benefit Information files, Comserv (a computer services firm specializing in mortality searches), state death indexes (e.g. New York) and the National Death Index, LexisNexis search, MicroBilt, credit bureaus and Comserv a computer services firm specializing in locating persons. In a few instances for deaths occurring prior to 1979 when the National Death Index began, death certificates would be requested from states and cause of death coded by a trained nosologist. For linkage with the SSA Master Death File of 83 million deaths, large-scale probabilistic matching programs will be applied to account for the possibility of potential errors in names, dates of birth or SSN.

It is anticipated that the employee data will be more readily available for deceased employees who were employed in recent years (1977–2010) and the employee data collected prior to 1977 will be more difficult to retrieve. We will determine completeness of the available correctly recorded annualized dosimeter information, overall career doses, and cause of death data. As 'proof of principle' we will examine mortality from leukemia associated with cumulative whole-body radiation dose and only generate data in aggregate format.

Source of dosimetry information

The effort and resources required for dose reconstruction depend largely on the availability and form of the dosimetry records. Such records exist at specific institutions under various filing and archival systems. Throughout the study period, MSK personal radiation dosimetry measurements have followed stringent regulatory (e.g. federal, state, and local) requirements and protocols to provide accurate dose information for radiation workers that include detailed records of work practices (including specific task exposure conditions, radiation type, energy, geometry, personal protective gear usage, badge position, and missed doses), as well as specific recorded measurements. Existing data on the workers from about 1946–1976 is captured on paper/microfilm logbook and personnel records, while data on workers from 1976–2010 is primarily captured in electronic dose measurements and personnel records at MSK. The data includes information such as date of birth, full job history at MSK, SSN, and series code (indicating the type of radiation, exposure conditions, protection, and badge position).

Personal dosimetry considerations

The MSK radiation dosimetry measurements have followed stringent protocols to provide accurate dose information for radiation workers that include detailed records of work practices (including specific task exposure conditions, radiation type, energy, geometry, personal protective gear usage, badge position, missed dose), plus recorded measurements.

Personal monitoring and dose measurements have been ascertained through a variety of techniques that have evolved over the years, from film badges to thermoluminescent dosimetry technology to optically stimulated luminescent methodologies. The MSK Department of Medical Physics has maintained several index card and log-book data sets that could be utilized to reconstruct annual worker doses from about 1946–1977. Pilot work with these paper records demonstrated the likelihood of >7000 usable annual entries from 1967–1976 alone, however, the completeness of the data is uncertain. Since 1977, the MSK Department of Medical Physics has maintained an electronic database of >20,000 individual medical radiation workers with >124,000 individual annual entries. The records include available personal identifiers, employment data (specific work function in the hospital, dosimetry inception, termination dates, etc.), and most importantly dose data (annual, cumulative, and previous whole body-deep dose).

A dosimetry service provider managed radiation dosimetry services for MSK since about 1946 and computerized the initial measurement files as of 1977. It is expected that individual total career doses for the sub-cohort will have a broad range from about >10 mSv to ≥1000 mSv. Further modeling can be developed to assess organ doses from dosimeter data by accounting for different average radiation energies, use of protective aprons, dosimeter placement, and minimal detectable doses over time (Gilbert et al. 1996; Simon et al. 2006; Simon et al. 2014; Bouville et al. 2015; Yoder et al. 2019). In addition, an assessment of natural background radiation will be made for the MSK site.

Radiation exposure scenarios for MSK deceased medical radiation workers

Radiation worker exposure scenarios include radiology x-ray, radiology fluoroscopy, nuclear medicine, radiation oncology, radiation protection, radiochemistry, medical physicists, research laboratories, inpatient nursing, brachytherapy, imaging and localization procedures, and others. The source of employee information is dependent on the year in which the data were collected by the MSK Human Resources Department, as various databases have been used over the years to collect employee information, including scanned paper archives, early databases, and present information systems.

MSK workers were assigned each year an occupational code that often reflected a department or specialty within the hospital or laboratory and this code was updated for each worker if their job or area changed. These codes may be useful for understanding the radiation exposure scenarios to which workers in that code were exposed and may be useful for aggregate data analyses.

Dose reconstruction methods

The specifics of practical dose reconstruction and uncertainty analysis discussions for the epidemiologic studies that will be included in the MPS are specific applications of the previous guidance provided in NCRP Reports Nos. 158, 163, 164, and 171 (NCRP 2007, 2009b, 2009c, 2012). More recently, the NCRP has provided focused guidance on the derivation of organ doses and their uncertainty in NCRP Report No. 178, Deriving Organ Doses and their Uncertainty for Epidemiologic Studies with a focus on the One Million U.S. Workers and Veterans Study of Low-Dose Radiation Health Effects (NCRP 2018b). The dose reconstruction and uncertainty methods of these guidance documents will be employed for the MSK deceased sub-cohort.

The conditions of exposure are critical for the dose reconstruction process. In general, the approach for determining annual organ doses for individuals in the MSK deceased sub-cohort will be to normalize available annual dose values (from historical personal monitoring and dosimetry data) to $H_P(10)$, and then to convert these to mean annual organ doses, D_T. With D_T and $H_P(10)$ each being related to common fundamental dosimetry quantities, the NCRP (NCRP 2018b) has generated tables of conversion factors that directly relate $H_P(10)$ to D_T for the various irradiation geometries, as well as methodologies for initially normalizing early exposure data to $H_P(10)$. In addition, more specific guidance on dosimetry (including considerations for the use of protective aprons and/or thyroid shields and monitoring practices) and associated uncertainties are addressed by the NCRP (NCRP 2018b) (e.g. 2D Monte Carlo approaches, probability distribution of estimated doses, etc.) and is being specifically developed for the multicenter medical radiation worker cohort of the MPS (Yoder et al. 2019) and will be applied to this feasibility study.

Conclusions

MSK has been at the forefront of novel radiation diagnostic and therapeutic approaches (including initial work with x-rays, radium, and radon), and radiation worker exposure scenarios include: radiology x-ray, radiology fluoroscopy, nuclear medicine, radiation oncology, radiation protection, radiochemistry, medical physicists, research laboratories, inpatient nursing, brachytherapy, imaging and localization procedures, and others. The results of this feasibility sub-cohort of deceased workers, and associated lessons learned may potentially be applied to an expanded multicenter study of about 170,000 medical radiation worker component of the MPS.

Acknowledgements

We acknowledge the assistance of the MSK Department of Epidemiology& Biostatistics, the Department of Medical Physics, Leslie Ballantyne of Human Resources, William Morgan and Margaret Eliasinski of Information Systems, and Sharon Bayuga-Miller of Epidemiology & Biostatistics. In addition, we appreciate the support of MSK leadership (Roy Cambria, Dr. Roger Wilson, and Dr. Craig Thompson), IRB/PB reviewers, legal staff (Joanna Halperin, Kristen Ahearn, Kristen Kim, Lee Stetson, and Arielle Cireseanu), and Information Systems (William Pustorino). Special thanks to those health physicists who were responsible for maintaining the personal monitoring programs during the nearly seven decades of information covered by this feasibility study.

Disclosure statement

The authors declare no personal conflicts of interest. The views expressed in this paper represent the collective opinions of the authors and are not necessarily those of their professional affiliations.

Funding

This research was funded in part by a grant from the National Institutes of Health (Support Grant No. P30 CA008748).

ORCID

Lawrence T. Dauer https://orcid.org/0000-0002-5629-8462
John D. Boice http://orcid.org/0000-0002-8755-1299

References

Berrington A, Darby SC, Weiss HA, Doll R. 2001. 100 years of observation on British radiologists: mortality from cancer and other causes 1897-1997. Br J Radiol. 74:507–519.

Bhatti P, Preston DL, Doody MM, Hauptmann M, Kampa D, Alexander BH, Petibone D, Simon SL, Weinstock RM, Bouville A. 2007. Retrospective biodosimetry among United States radiologic technologists. Radiat Res. 167:727–734.

Boice JD. Jr. 2012. A study of one million U.S. radiation workers and veterans: a new NCRP initiative. Health Physics News. XL:7–10.

Bouville B, Toohey RE, Boice JD, Jr, Beck HL, Dauer LT, Eckerman KF, Hagemeyer D, Leggett RW, Mumma MT, Napier B, et al. 2015. Dose reconstruction for the million worker study: status and guidelines. Health Phys. 108:206–220.

Cardis E. 2007. Commentary: low dose-rate exposures to ionizing radiation. Int J Epidemiol. 36:1046–1047.

Dauer LT. 2011. Science-informed, justified, and optimized radiation safety policies. Health Phys. 100:332–334.

Dauer LT. 2014. Exposed medical staff: challenges, available tools, and opportunities for improvement. Health Phys. 106:217–224.

Doody M, Mandel J, Lubin J, Boice JD. Jr. 1998. Mortality among United States radiologic technologists, 1926–90. Cancer Causes Control. 9:67–75.

Eisenhower DD. 1953. Atoms for peace. Address by Mr. Dwight D. Eisenhower, President of the United States of America, to the 470th Plenary Meeting of the United Nations General Assembly. Tuesday 9 December 1953, 2:45 pm. [Accessed 2018 Sep 6]. Available from: https://www.iaea.org/about/history/atoms-for-peace-speech

Gilbert ES, Fix JJ, Baumgartner WV. 1996. An approach to evaluating bias and uncertainty in estimates of external dose obtained from personal dosimeters. Health Phys. 70:336–345.

International Commission on Radiological Protection (ICRP). 2005. Low-dose extrapolation of radiation-related cancer risk. ICRP Publication No. 99. Ann ICRP 35(4).

International Commission on Radiological Protection (ICRP). 2007. Recommendations of the International Commission on Radiological Protection: International Commission on Radiological Protection. ICRP Publication No. 103. Ann ICRP 37(2-4).

Jablon S, Miller R. 1978. Army technologists: 29-year followup for cause of death. Radiology. 126:677–679.

Jacob P, Ruhm W, Walsh L, Blettner M, Hammer G, Zeeb H. 2009. Is cancer risk of radiation workers larger than expected? Occup Environ Med. 66:789–796.

King G. 2012. Clarence Dally - the man who gave Thomas Edison X-ray vision. Smithsonian.com. [Accessed 2018 Sep 6]. Available from: https://www.smithsonianmag.com/history/clarence-dally-the-man-who-gave-thomas-edison-x-ray-vision-123713565/

Lewis EB. 1963. Leukemia, multiple myeloma, and aplastic anemia in American radiologists. Science. 142:1492–1494.

Linet MS, Kim KP, Miller DL, Kleinerman RA, Simon SL, de Gonzalez AB. 2010. Historical review of occupational exposures and cancer risks in medical radiation workers. Radiat Res. 174: 793–808.

March H. 1944. Leukemia in radiologists. Radiology. 43:275–278.

Matanoski GM, Seltser R, Sartwell PE, Diamond EL, Elliott EA. 1975. The current mortality rates of radiologists and other physician specialists: deaths from all causes and from cancer. Am J Epidemiol. 101:188–198.

Mohan AK, Hauptmann M, Freedman DM, Ron E, Matanoski GM, Lubin JH, Alexander BH, Boice JD, Doody MM, Linet MS. 2003. Cancer and other causes of mortality among radiologic technologists in the United States. Int J Cancer. 103:259–267.

National Council on Radiation Protection and Measurements (NCRP). 2007. Uncertainties in the measurement and dosimetry of external radiation. NCRP Report No. 158. Bethesda (MD): NCRP.

National Council on Radiation Protection and Measurements (NCRP). 2009a. Ionizing radiation exposure of the population of the United States. NCRP Report No. 160. Bethesda (MD): NCRP.

National Council on Radiation Protection and Measurements (NCRP). 2009b. Radiation dose reconstruction: principles and practice. NCRP Report No. 163. Bethesda (MD): NCRP.

National Council on Radiation Protection and Measurements (NCRP). 2009c. Uncertainties in internal radiation dose assessment. NCRP Report No. 164 Bethesda (MD): NCRP.

National Council on Radiation Protection and Measurements (NCRP). 2010. Radiation dose management for fluoroscopically-guided interventional medical procedures. NCRP Report No. 168. Bethesda (MD): NCRP.

National Council on Radiation Protection and Measurements (NCRP). 2012. Uncertainties in the estimation of radiation risks and probability of disease causation. NCRP Report No. 171. Bethesda (MD): NCRP.

National Council on Radiation Protection and Measurements (NCRP). 2018a. Implications of recent epidemiologic studies for the linear-nonthreshold model and radiation protection. NCRP Commentary No. 27. Bethesda (MD): NCRP.

National Council on Radiation Protection and Measurements (NCRP). 2018b. Deriving organ doses and their uncertainty for epidemiologic studies (with a focus on the one million U.S. workers and veterans study of low-dose radiation health effects). NCRP Report No. 178. Bethesda (MD): NCRP.

Nobel Prize (Nobel). 2018. The Nobel Prize in physics 1903. NobelPrize.org. Nobel Media AB 2018. [Accessed 2018 Sep 6]. Available from: https://www.nobelprize.org/prizes/physics/1903/summary/

National Research Council (NRC). 2006. BEIR VII: health risks from exposure to low levels of ionizing radiation. Washington, DC: National Research Council.

Pietzch J. 2018. A Helping Hand from the Media. NobelPrize.org. Nobel Media AB 2018. [Accessed 2018 Sep 6]. Available from: https://www.nobelprize.org/prizes/physics/1901/perspectives/

Quimby EH. 1926. A method for the study of scattered and secondary radiation in x-ray and radium laboratories. Radiology. 7:211.

Quimby EH. 1934. Determination of dosage for long radium or radon needles. Am J Roentgenol. 31:74–91.

Simon SL, Weinstock RM, Doody MM, Neton J, Wenzl T, Stewart P, Mohan AK, Yoder RC, Hauptmann M, Freedman DM, et al. 2006. Estimating historical radiation doses to a cohort of U.S. radiologic technologists. Radiat Res. 166:174–192.

Simon SL, Preston DL, Linet MS, Miller JS, Sigurdson AJ, Alexander BH, Kwon D, Yoder RC, Bhatti P, Little MP, et al. 2014. Radiation organ doses received in a nationwide cohort of U.S. radiologic technologists: methods and findings. Radiat Res. 182:507–528.

Smith PG, Doll R. 1981. Mortality from cancer and all causes among British radiologists. Br J Radiol. 54:187–194.

Tenforde T, Schauer D. 2008. Low dose and low dose-rate radiation effects and models. NCRP Forty-Fourth Annual Meeting; 2008 April 14-15. Bethesda, MD: Health Physics News 2008.

Tontonoz M. 2016a. Hot times in "Radium Hospital": the history of radium therapy at MSK. New York, NY: Memorial Sloan Kettering Cancer Center.

Tontonoz M. 2016b. Radiation as weapon in cancer cure: MSK doctor unearths surprising history. New York, NY: Memorial Sloan Kettering Cancer Center.

Tubiana M, Aurengo A, Averbeck D, Masse R. 2006. Recent reports on the effect of low doses of ionizing radiation and its dose-effect relationship. Radiat Environ Biophys. 44:245–251.

United Nations Scientific Committee on the Effects of Atomic Radiation (UNSCEAR). 2008. United Nations scientific committee on the effects of atomic radiation 2006 report to the general assembly with scientific annexes, effects of ionizing radiation. New York, NY: United Nations.

Wakeford R. 2005. Cancer risk among nuclear workers. J Radiol Prot. 25: 225–228.

Wang J-X, Inskip PD, Boic JD, Li B-X, Zhang J-Y, Fraumeni JF. Jr. 1990. Cancer incidence among medical diagnostic X-ray workers in China, 1950 to 1985. Int J Cancer. 45:889–895.

Yoder RC, Dauer LT, Balter S, Boice JD, Jr, Grogan HA, Mumma M, Passmore CN, Rothenberg LN, Vetter RJ. 2019. Dosimetry for the study of medical radiation workers with a focus on the mean absorbed dose to the lung, brain and other organs. Int J Radiat Biol. 1–36. DOI: 10.1080/09553002.2018.1549756

Yoshinaga S, Aoyama T, Yoshimoto Y, Sugahara T. 1999. Cancer mortality among radiological technologists in Japan: updated analysis of follow-up data from 1969 to 1993. J Epidemiol Japan Epidemiol Assoc. 9:61–72.

Yoshinaga S, Mabuchi K, Sigurdson AJ, Doody MM, Ron E. 2004. Cancer risks among radiologists and radiologic technologists: review of epidemiologic studies. Radiology. 233:313–321.

Dosimetry

Dosimetry and uncertainty approaches for the million-worker study of radiation workers and veterans: overview of the recommendations in NCRP Report No. 178

Lawrence T. Dauer, André Bouville, Richard E. Toohey, John D. Boice, Harold L. Beck, Keith F. Eckerman, Derek A. Hagemeyer, Richard W. Leggett, Michael T. Mumma, Bruce Napier, Kathy H. Pryor, Marvin Rosenstein, David A. Schauer, Sami Sherbini, Daniel O. Stram, James L. Thompson, John E. Till, R. Craig Yoder, and Cary Zeitlin

ABSTRACT

Purpose: Scientific Committee 6–9 was established by the National Council on Radiation Protection and Measurements (NCRP), charged to provide guidance in the derivation of organ doses and their uncertainty, and produced a report, NCRP Report No. 178, Deriving Organ Doses and their Uncertainty for Epidemiologic Studies with a focus on the Million Person Study of Low-Dose Radiation Health Effects (MPS). This review summarizes the conclusions and recommendations of NCRP Report No. 178, with a concentration on and overview of the dosimetry and uncertainty approaches for the cohorts in the MPS, along with guidelines regarding the essential approaches used to estimate organ doses and their uncertainties (from external and internal sources) within the framework of an epidemiologic study.

Conclusions: The success of the MPS is tied to the validity of the dose reconstruction approaches to provide realistic estimates of organ-specific radiation absorbed doses that are as accurate and precise as possible and to properly evaluate their accompanying uncertainties. The dosimetry aspects for the MPS are challenging in that they address diverse exposure scenarios for diverse occupational groups being studied over a period of up to 70 y. Specific dosimetric reconstruction issues differ among the varied exposed populations that are considered: atomic veterans, U.S. Department of Energy workers exposed to both penetrating radiation and intakes of radionuclides, nuclear power plant workers, medical radiation workers, and industrial radiographers. While a major source of radiation exposure to the study population comes from external gamma- or x-ray sources, for some of the study groups, there is also a meaningful component of radionuclide intakes that requires internal radiation dosimetry assessments.

Introduction

The National Council on Radiation Protection and Measurements (NCRP) is coordinating an expansive epidemiologic effort entitled the One Million U.S. Workers and Veterans Study of Low- Dose Radiation Health Effects [One Million U.S. Workers and Veterans Study (MPS[1])] (Boice 2012a; Bouville et al. 2015; Boice et al. 2018). The primary aim of the MPS is to provide scientifically valid information and improve precision on the level of radiation risk when exposures are received gradually over time, and not acutely as was the case for Japanese atomic-bomb survivors. The major health outcome of interest for the MPS is cancer mortality, but other causes of death such as cardiovascular disease and cerebrovascular disease will be evaluated. The validity of the MPS is tied to the validity of the dose reconstruction approaches to provide accurate estimates of organ doses (i.e. the absorbed dose averaged over all parts of an organ or tissue) and their accompanying uncertainties. The focus is on providing the best estimate of organ dose and uncertainty for each individual, and to (when possible) characterize shared uncertainties that affect groups of individuals within the various cohorts considered.

NCRP established Scientific Committee 6-9 (SC 6-9) and tasked the committee members with developing a report to provide guidance in the derivation of organ doses and their associated uncertainty for epidemiologic studies in general, with a focus on the workers and atomic veterans that make up the MPS coordinated by NCRP (Boice 2012a, 2014a; Bouville et al. 2015). The study populations include atomic veterans[2], Department of Energy (DOE) workers, nuclear power plant (NPP) workers, medical radiation workers, and industrial radiographers. Organ doses from exposure to all the relevant external and internal sources for a given population are being derived. The Report covers the specifics of practical dose reconstruction for the epidemiologic studies included in the MPS with uncertainty analysis discussions and is a specific application of the previous guidance provided in NCRP Reports Nos. 158, 163, 164, and 171 (NCRP 2007, 2009b, 2009c, 2012). An interim status of the work of SC 6-9 was initially presented at the fiftieth annual meeting of the NCRP, 'NCRP: achievements of the past 50 years and addressing the needs of the future' (NCRP 2014), and a synopsis of that update was previously published (Bouville et al. 2015). The final NCRP Report No. 178 (NCRP 2018b) (the Report) was developed by multi-disciplinary experts based on a comprehensive review of available data, developing methods, and reviewed by the council.

This paper summarizes description of the cohorts in the MPS, several guidelines, and key points of the Report. Readers are encouraged to review the complete report for the discussion of specific background material, case studies, as well as cohort-specific considerations and recommendations. While the report was primarily aimed at providing guidance for the MPS, much of the material is applicable to any high-quality radiation epidemiologic study and should be considered for both future and ongoing research.

Overview of cohorts in the MPS

The MPS seeks to combine the radiation dose and mortality experiences from several occupational groups. The dosimetry aspects for the MPS are challenging in that they address diverse exposure scenarios for the various occupational groups being followed-up over a period of up to 70 y (Boice 2012a, 2012b, 2013a, 2013b, 2013c). These sub-cohorts have unique aspects with regard to exposure scenarios and organ dose and accompanying uncertainty estimation. The dosimetric issues differ among the exposed populations: atomic veterans, DOE workers, NPP workers, medical radiation workers, and industrial radiographers.

For the MPS, an attempt will be made to derive annual organ doses and their associated uncertainties for organs typically addressed in an epidemiological study, with particular emphasis on the most relevant organs for which cancer or other disease mortality data are available, especially for the radiosensitive sites [for example, active bone marrow, female breasts, thyroid (for young persons), lungs, heart, brain (dementia and associated conditions)]. To date, the derivation of annual organ doses and their associated uncertainties from the available recorded dose quantities and other information has been performed for part of the U.S. DOE workers, is well under way for the atomic veterans, has been undertaken for the NPP workers and the industrial radiographers, and is at the planning stage for the medical radiation workers. For most of the MPS cohorts, external sources were the predominant mode of exposure. However, preliminary estimates indicate that about half of the DOE workers also were exposed to internal irradiation from intakes of radionuclides.

For the MPS, all career occupational doses are sought. This requires linkages with the Radiation Exposure Information and Reporting System (REIRS), DOE Radiation Exposure Monitoring System (REMS) (DOE 2015), DOE historical databases, military databases, and the Landauer, Inc. (Landauer) database. These linkages are discussed in Boice et al. (2006a). One of the challenges for career dosimetry for early workers is that many of them, up to 25%, worked at facilities other than the index facility and dosimetry for these other facilities have to be accessed, interpreted, and then incorporated, often with lower quality assurances than for the prime facility.

It is recognized that the MPS and its approaches to dosimetry are a work in progress and there will be flexibility and changes in direction as new information is obtained, both with regard to dosimetry and with regard to the epidemiologic features of the study components.

U.S. Department of energy workers

Scores of facilities involved in the production of nuclear weapons, fuel rods, heat sources, or other devices containing large quantities of radioactive material have been operated in the USA since the 1940s. Workers at these facilities have the potential for elevated intake of radionuclides as well as exposure to external irradiation. Most of the populations that constitute the 290,000 DOE workers have been previously studied, but over 20 y ago. The populations recently under investigation include the workers at Rocketdyne (Leggett et al. 2005; Boice et al. 2006a, 2006b, 2011), Mound (Boice et al. 2014), and Mallinckrodt (Dupree-Ellis et al. 2000; Ellis et al. 2018; Golden et al. 2018). Populations that will soon be initiated for study include workers at Los Alamos (Wiggs et al. 1994), Rocky Flats (Gilbert et al. 1993), and Fernald (Silver et al. 2013) and those in the U.S. nuclear weapons complex who received annual occupational recorded doses of 50 mSv or more (Fry et al. 1996). Two case studies, Rocketdyne and Mound, are used in the NCRP Report No. 178 to illustrate the typical issues encountered in a dose reconstruction for radiation workers at a production facility and to describe the common sources of uncertainty in the reconstructed doses, which were mainly due to intakes of radionuclides. In addition, a detailed description of the issues of complex dose reconstruction at the Mallinckrodt Chemical Works, an early uranium processing facility, is presented in an appendix (Ellis et al. 2018).

The first case study (Boice et al. 2006a, 2006b, 2011) is a dose reconstruction for the 5801 workers who were monitored for radiation exposure and employed between 1948 and 1999 at the Rocketdyne (formerly Atomics International) Site near Los Angeles, California. Rocketdyne workers were involved in a wide range of radiological activities, including uranium and plutonium fuel fabrication, spent-fuel evaluation, radiochemistry, and storage of nuclear material. Workers were exposed to a number of radionuclides including isotopes of cerium, cesium, hydrogen (i.e. tritium), plutonium, promethium, strontium, thorium, and uranium.

The second case study (Boice et al. 2014) is a dose reconstruction for the 4977 workers at the Mound Site in Miamisburg, Ohio, who were monitored for radiation exposure and were first hired between 1944 and 1979. The Mound analysis was performed as part of a pilot study of the feasibility of conducting a time- and cost-efficient epidemiologic study of radiation workers at production facilities across the USA. The dose reconstruction for Mound generally followed the scheme laid out earlier for the Rocketdyne Site but differed in some ways from the Rocketdyne analysis, primarily due to differences in the dominant internal emitters at the two sites and in the main sources of uncertainty in organ dose estimates for the internal emitters.

Atomic veterans

The cohort of atomic veterans consists of military personnel included in the Eight Series Study (a cohort of 114,270 military personnel who participated in eight nuclear weapon test series) and were mainly exposed at either:

- the Nevada Test Site (NTS, currently named the Nevada National Security Site) where they participated in military maneuvers, observed nuclear weapons tests, or provided support during related operations that occurred from 1952 through 1957;
- the Pacific Proving Grounds where personnel were aboard ships or stationed on islands in the area during and following the nuclear weapons tests from 1946 through 1962; or
- the first nuclear weapons test, TRINITY, which took place in 1945 in New Mexico, where they participated in security, various test preparation activities, and post-shot monitoring.

Dose or dose-related information is available in the U.S. Department of Defense (DOD) Nuclear Test Personnel Review (NTPR) records for each veteran who participated in the Eight Series Study. The NTPR assembled a wealth of information from classified and unclassified historical records collected during the nuclear atmospheric tests and also records of military personnel that participated in the testing. However, the purpose of the NTPR Program was to estimate organ doses to be used for compensation and these organ doses were often deliberately high-sided in accordance with regulations NTPR was required to follow. The mean cumulative absorbed dose in the whole body calculated by NTPR for the 113,580 veterans included in the MPS is <10 mGy with 2% of these veterans receiving a cumulative whole-body dose >50 mGy.

Efforts have been undertaken to refine and modify as necessary the information available in the NTPR records to produce an organ dose estimate as unbiased as possible and suitable for epidemiology with a corresponding uncertainty estimate. Details about the dose assessment methods currently applied in the study of atomic veterans are reported in Beck et al. (2017) and Till et al. (2014). The initial focus of the atomic veterans study is on leukemia mortality. Thus, the organ of interest is active bone marrow. Because of cost and time, it was found impractical to conduct historical dose reconstructions on all of the 114,270 cohort members in the Eight Series Study. Therefore, organ doses are estimated for ~2,000 individuals consisting of cases based on diagnosis of leukemia and male breast cancer and a 1% random sample of the entire cohort, used for comparison in a case-cohort design. Primary emphasis is given to external exposure since almost all the exposure to active bone marrow is known to be from penetrating external radiation. Future follow-up investigations will include other cancers such as bone, liver, salivary gland, and thyroid.

Nuclear power plant workers

About 600,000 workers have been employed in NPPs in the USA since the first commercial production of electricity in 1957. Because the average annual recorded dose of workers has decreased over the years down to ~2 mSv on average, only the workers at NPPs first employed from 1957 through 1984 are considered in the MPS. About 330,000 cohort members were selected from databases available from the REIRS, which is maintained by the U.S. Nuclear Regulatory Commission (NRC 2011), and Landauer, Inc. (Landauer) which is a dosimetry service provider. Most radiation exposures were due to penetrating external gamma rays with only a few neutron exposures or internal contamination exposures. A 10% sample of workers with cumulative recorded dose <10 mSv (i.e. from personal monitors) was studied for purposes of cost efficiency and provided a strong low-dose category of over 20% of the studied population. The mean recorded cumulative dose for the remaining 145,209 workers is ~50 mSv, with 25% of them receiving a cumulative recorded dose >50 mSv.

Medical workers

Medical radiation workers represent a large fraction of radiation-exposed workers, with ~2.5 million monitored workers in 2006 (NCRP 2009a). Historically, the average annual occupational effective dose estimates have trended downward for the medical radiation worker populations (Linet et al. 2010). Most present day medical radiation workers generally experience very low radiation exposures, essentially from external irradiation. Those individuals who perform certain fluoroscopically guided interventional procedures and potentially those who prepare or administer radionuclides for nuclear-medicine procedures are an exception to this generalization (Dauer 2014). Dosimetric information for 243,448 workers with cumulative recorded dose from personal monitors was obtained from Landauer. The mean cumulative recorded dose for these medical workers is ~100 mSv, with 36% of them receiving a cumulative recorded dose >50 mSv.

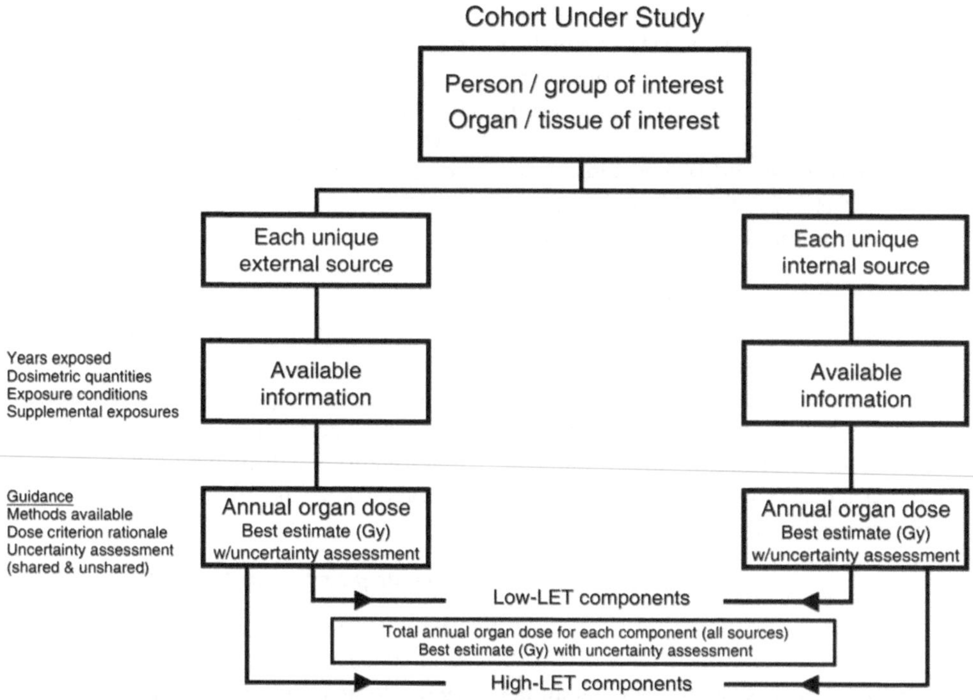

Figure 1. Generic synopsis for the presentation of annual organ dose and its uncertainty.

The derivation of organ doses from monitoring data poses difficult problems and reflects some of the sources of uncertainty for estimating organ doses for the medical worker cohort because of, among other factors:

- often extreme inhomogeneity of exposure over the body of personnel for any given procedure type;
- differing degrees and methods of radiation protection;
- inconsistent wearing of dosimeters by personnel (i.e. at times choosing not to wear dosimeters to avoid investigations) (NCRP 2010), combined with poor information, as well as high variability, on the workloads of physicians and technologists (i.e. the number of procedures of a given type conducted monthly or annually); and
- changing technology and medical procedure protocols.

Recognizing the need for more specific guidance on these challenges, NCRP recently chartered Scientific Committee 6-11 to provide dosimetry guidance for medical workers with a focus on lung dose reconstruction (Yoder et al. 2018). The approaches will rely on various exposure scenarios for medical workers classified as radiology, nuclear medicine, fluoroscopy-guided interventional, and radiation oncology.

Industrial radiographers

The sources of choice for industrial radiographic nondestructive testing are ^{192}Ir and ^{60}Co (NDT 2014). Industrial radiographers are only exposed via external irradiation, generally in an anterior-posterior (AP) geometry. To date, information on annual recorded dose has been collected by the MPS for 127,910 industrial radiographers. The main sources of information are the REIRS and the Landauer databases. The average cumulative recorded dose for these industrial radiographers is ~20 mSv (i.e. from personal monitors), with 10% of them receiving a cumulative recorded dose >50 mSv.

Supplemental populations of interest

In addition to the extensive discussion of the dose assessment of the populations that are part of the MPS, the NCRP Report No. 178 includes a description of supplemental population groups, which are of interest either because of uncommon conditions of exposure (astronauts), exemplary efforts to process and analyze dose records and information from the distant past (Mayak workers, Gilbert et al. 2004), or inclusion into exposure registries [U.S. Transuranium and Uranium Registries (USTUR, Tolmachev et al. 2011) and Radiation Emergency Assistance Center/Training Site (REAC/TS) accident registry].

Conclusions from NCRP Report No. 178

The Report provides guidelines regarding the estimation of organ doses for an epidemiologic study, estimation of organ doses from external irradiation, estimation of organ doses from internal irradiation, and assessment of uncertainties. In addition, several key points are made and discussed with regard to the overall dosimetry and uncertainty approaches.

Within the framework of an epidemiologic study evaluating the health effect attributed to a radiation dose, the quantity of interest is the annual absorbed dose to the organ or tissue (organ dose) that is assumed to be the origin of the radiation-induced cancer. For example, active bone-marrow dose is of interest if leukemia is the disease being considered in the epidemiologic study. Organ doses in most of the organs and tissues of the body can be calculated using

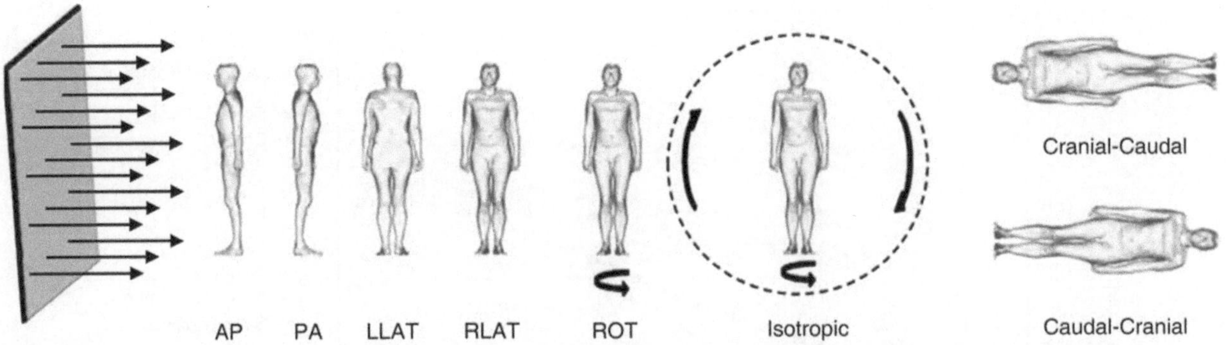

Figure 2. Schematic representation of the idealized geometries considered in NCRP Report No. 178 (2018), including: AP, PA, LLAT (left lateral), RLAT (right lateral), ROT (rotational), isotropic, cranial-caudal (top right) and caudal-cranial (bottom right) geometries.

relationships with measured or estimated personal dose equivalent [$H_p(10)$] values (in the case of external irradiation) or with unit of activity intakes of radionuclides (in the case of internal irradiation) that can be derived from ICRP publications. However, these relationships are not available for all organs and tissues of the body, so that a surrogate organ or tissue is sometimes used to estimate the organ dose to the organ or tissue that is assumed to be the origin of the radiation-induced cancer.

To facilitate time-dependent analyses that account for the accrual of radiation doses over time (Cox 1972; Preston et al. 2017), the organ dose is calculated on an annual basis. The annual organ dose that is considered is estimated for each year of the entire period of time over which the organ dose is delivered, beginning with the date of first exposure and ending with the date of last exposure, the date of cancer diagnosis or death (whichever is earlier). Although it is recognized that the absorbed dose distribution may not be uniform over all parts of the organ or tissue, it is assumed in the NCRP Report No. 178 that the absorbed dose averaged over all parts of the organ or tissue is the quantity of interest in the epidemiologic analysis.

Figure 1 provides a generic synopsis for presenting annual organ doses and their uncertainty for members of the populations included in the MPS. The synopsis is a summary of the exposure situation for each individual or group of individuals similarly exposed. The purpose is to display for that individual or group the characteristics of the exposure situation and either the results of a completed dose assessment or the elements of a dose assessment yet to be conducted. It addresses the total annual organ dose of interest (from all applicable external and internal radiation sources), with that total annual organ dose partitioned into low- and high-LET components (all expressed in Gy) without any weighting for the biological effectiveness of the type of radiation. For each partitioned component, the associated uncertainty is also addressed.

Estimation of organ doses from external irradiation

Most of the workers considered in the Report received radiation doses from external irradiation. Individual monitoring data, in terms of the quantities exposure or personal dose equivalent for strongly penetrating radiation in soft tissue at 10 mm [$H_p(10)$], are often available for these workers. The main effort then consists in defining the exposure scenarios for the various tasks carried out by the workers to determine the irradiation geometry and the energy spectrum of the incident photons on the body. The Report includes guidance on methodologies to estimate the dose coefficients relating the quantities exposure and $H_p(10)$ to organ dose, given the photon-energy spectrum and the irradiation geometry. It is important to note that the relative geometrical relationship between the dosimeter's placement on the worker, the incident source direction, and the organ or tissue is needed to estimate an organ dose.

Most records or personal dosimeter results reflect a single value of exposure or $H_p(10)$ for a monitoring period. The exposure scenario and pathways should assess the applicability of this single value as being appropriate for all areas of the body. Many exposure conditions result in different body areas being irradiated at much different levels as happens when protective aprons are worn during fluoroscopic procedures or when source shielding configurations cause large spatial variations in dose rate. Regulatory guidance requires personal monitoring devices to be worn at the location of highest dose; therefore, the recorded dosimeter results may overstate the $H_p(10)$ for some areas of the body. The derivation of organ doses from external irradiation assumes that an estimate exists of the quantity exposure or personal dose equivalent [$H_p(10)$]. For photons, the quantities assessed from personal monitoring dosimeters generally fall into two categories: early dosimeter data for which the quantity measured was exposure, generally measured at the surface of the body [expressed in roentgen (R)], and later data for which the quantity measured was $H_p(10)$, initially expressed in rem and now expressed in sievert (Sv). Therefore, early dosimeter data should be converted from exposure to $H_p(10)$.

For external exposures, the measured $H_p(10)$ or that value converted from a measurement of exposure provides an initial input to the derivation of the resultant organ doses of interest. Such derivations are aided by the results of computer models that have established relationships for the fundamental quantities of fluence and air kerma with both $H_p(10)$ and organ dose for various standard but idealized exposure conditions (Figure 2). Having a common relationship to fluence or air kerma, $H_p(10)$ and organ dose can be

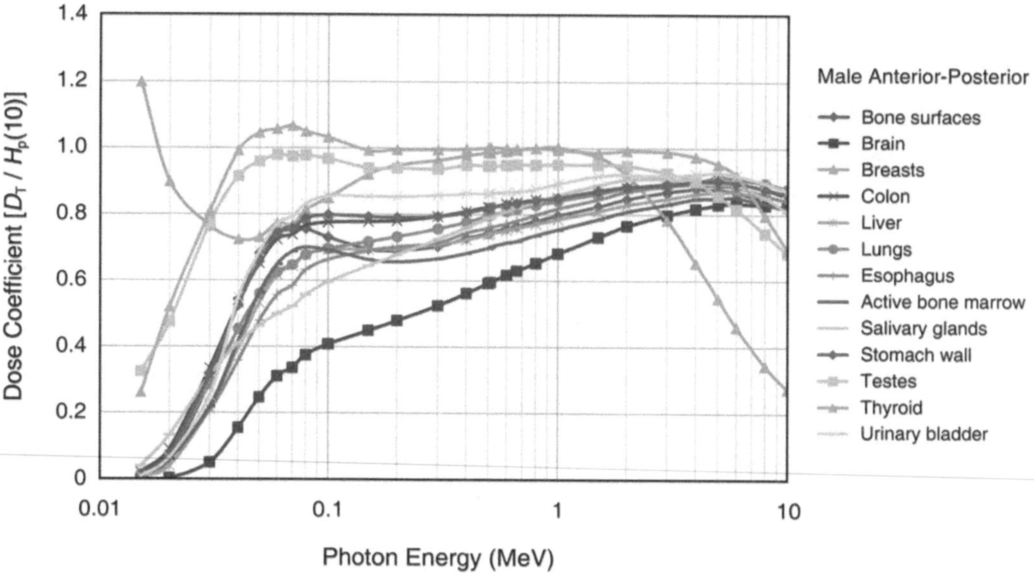

Figure 3. A plot of the relationship (dose coefficient) for an AP irradiation condition between the absorbed dose to the primary organs (DT)(mGy) in the male body and personal dose equivalent [$H_p(10)$] (mSv), for photons.

generally related to each other when the exposure conditions are comparable.

ICRP (1995) recommended conversion coefficients that relate air kerma (free-in-air) for monoenergetic photons to $H_p(10)$ for various photon energies and angles of incidence. These conversion coefficients have remained constant since their publication and continue to form the basis for dosimeter calibrations and performance testing in the USA. This stability is largely due to the unchanging specification of the slab phantom in which $H_p(10)$ is defined.

Factors that relate air kerma (free-in-air) or fluence to organ dose have changed over the years as improvements have been made in computer models, radiation transport equations, and the mathematical constructs used to represent the human body, particularly the size of the body and location of the internal organs. The most recent set of conversion coefficients for photons was published by ICRP (2010); that set is based on voxel phantoms of the human body using medical imaging data, a significant change from the abstract body constructs used in past reports. These latest conversion coefficients relate air kerma and fluence to organ dose for 14 different organs in each of the male and female phantoms for various photon energies and exposure geometries.

The Report provides several dose coefficients [relating $H_p(10)$ and organ dose] that could be applicable to various external exposure scenarios and geometries for the MPS populations where personal dosimeter data were available. Figure 3 provides an example of dose coefficients for AP exposure geometry. The use of dose coefficients typically assumes uniform irradiation of the body from a designated angle of incidence, particularly for the torso where most of the radiosensitive organs reside. Some exposure scenarios and radiation environments do not result in relative uniform irradiation of the torso and head. In these situations, the estimate of $H_p(10)$ may need to be modified to reflect the nonuniform irradiation of different regions of the body. In the absence of detailed information on the irradiation geometry related to work activities, it is recommended to assume typical or representative geometries such as 100% AP or 50% AP plus 50% ROT (NCRP, 2009b), or 50% AP plus 50% isotropic (Thierry-Chef et al. 2007). In some facilities, multiple personal monitoring dosimeters may have been used by a single person during conditions of very non-uniform irradiation of the body. In such cases, the estimate of $H_p(10)$ from the dosimeter located nearest the organ of interest should be used. ANSI/HPS (2011) prescribes a procedure in which the body is divided into compartments for which a separate compartment-weighted $H_p(10)$ value is assessed from a dosimeter located nearest that compartment.

Other considerations are undetected dose (commonly referred to as missed dose) and unmonitored dose. Undetected dose is defined as the dose received that was not measured by the dosimeter, because it fell below the minimum detectable response of the dosimeter. Since the undetected dose may in reality range from zero to the minimum detectable, it is customary to assign some fraction of the minimum detectable dose (or some other clearly stated value) for each monitoring period in which the dosimeter reads zero (i.e. less than the minimum detectable). Unmonitored dose is that assumed to be received when a personal dosimeter was not worn, and often may be reconstructed from knowledge of workplace activities or from coworker data when others in the same location did wear dosimeters. This latter approach is typically applied to atomic veterans, when one or two dosimeters were issued to a unit that may have totaled 40 or more individuals.

Additional complications are:

- recorded doses may include a mixture of radiation types (e.g. photons and neutrons);
- organs or tissues may be only partially irradiated, for example when medical personnel wear lead aprons; and
- recorded doses from all facilities where the worker was exposed during a career are processed, with possible

differences regarding the photon energy spectra, the irradiation geometries, and the measured dose quantities.

Estimation of organ doses from internal irradiation

In the case of occupational exposure from internal irradiation, the recorded dose, which is expressed in terms of committed effective dose or committed effective dose equivalent (CEDE), is not directly related to the dosimetric quantity of interest, which is the annual absorbed dose, averaged over all parts of the organ or tissue under consideration. Whenever possible, it is recommended to reconstruct the annual doses from internal irradiation on the basis of the available records of concentration of workplace air and surface contamination monitoring, and of individual bioassay data. The available information is processed to: estimate the radionuclide intake and its characteristics; and calculate the annual organ dose for each year of the entire period of time over which the organ dose is delivered, beginning with the date of first exposure and ending with the date of last exposure, the date of cancer diagnosis or death (whichever is earlier) (Boice et al. 2006a; Ellis et al. 2018).

Guidelines for estimating organ doses from internal emitters encountered in a large-scale production facility are described in detail in the Report. The Report notes that detailed dose reconstructions for internal emitters often require considerable time and effort for purposes of a given epidemiologic study and may not be justified below some level of intake of a given radionuclide. If comprehensive dose reconstructions for internal emitters would require prohibitive resources, it is useful to devise rapid screening techniques that yield conservative organ dose estimates based on available monitoring data. Dose reconstructions for internal emitters can then be limited to workers whose initial dose-related information is above a criterion level that represents insignificant organ doses for purposes of the epidemiologic study.

The same observation applies to organ doses that need to be reconstructed on the basis of environmental measurements for an external source, as is the case for some of the atomic veterans. Organ dose estimates for internal emitters should be radionuclide-specific (i.e. specific to each parent radionuclide taken into the body) and each parent radionuclide should be partitioned into organ dose from low-LET components and from high-LET components [all expressed in absorbed dose (in Gy)]. Organ dose estimates for external irradiation and for all addressed internal emitters should be summed to get the total annual organ doses (separately for the various low- and high-LET components) for individual workers or similar groups of workers.

Assessment of uncertainties

Estimates of organ dose obtained in a dose reconstruction have limitations and are uncertain. Limitations and lack of certainty in organ doses can result from factors such as:

- lack of complete knowledge of an exposure scenario;
- uncertainty in relevant measurements;
- lack of relevant data at locations and times of exposure;
- uncertainty in internal dosimetry; and
- conversion of externally measured quantities to organ doses.

All uncertainties, including uncertainties in exposure scenarios and uncertainties in data and models used to estimate organ dose, should be considered and taken into account in an appropriate manner in a dose reconstruction. As discussed in previous NCRP reports (NCRP 2007, 2009b, 2009c, 2012), the uncertainties can be classified as classical or Berkson, random or systematic, aleatory or epistemic, Type A or Type B, and shared or unshared. In the NCRP Report No. 178, the uncertainties are usually classified as shared or unshared. In the context of epidemiologic studies, knowledge of the existence and likely magnitude of correlations arising from shared parameters is essential to proper interpretation and use of the organ dose estimates (e.g. Stram and Kopecky 2003; Li et al. 2007; NCRP 2009c; Kwon et al. 2016). Depending on the relative importance of shared and unshared uncertainties, several approaches may be used to represent uncertainty in dosimetry numerically.

The first approach consists in attaching to each dose estimate an uncertainty estimate applicable to that specific person's dose. This approach may be used if almost all uncertainty is due to unshared uncertainty [due for example to imperfections in individual input data (e.g. location, shielding) that can be treated as independent from individual to individual]. This uncertainty estimate, which is based on an analysis of the uncertainties associated with each parameter used in the dose-estimation process, can be calculated either analytically or by means of a single Monte–Carlo procedure. It can also be an expert guess based on subjective judgment (NCRP 2009c).

In a second approach, which can be used if shared uncertainty is expected to be significant, a single uncertainty estimate can be applied to all subjects in a portion of a cohort affected by the same source of uncertainty, resulting in a covariance matrix of dose uncertainties. The idea of using both organ dose estimates for the study participants and a covariance matrix (representing both shared and unshared uncertainty) was described, for example, by Stram et al. (2015).

A third approach to representing uncertainty for complex exposure situations in which the shared uncertainty is expected to be significant has been to represent the shared/unshared uncertainties as repeated draws from a two-dimensional Monte–Carlo procedure that provides many as opposed to just one estimate of organ dose for each subject (e.g. Simon et al. 2015).

At the onset of the epidemiologic study, it is highly recommended for the dosimetrists to work closely with the statisticians/epidemiologists to decide how to characterize and present the uncertainties in the dose estimates.

Key points of NCRP Report No. 178

Several key points are made and discussed in the Report as general guidance and for assessment of organ doses from external and internal sources, as well as evaluation of uncertainties.

General guidance:

- The goal of the dosimetry is to estimate annual absorbed doses to the organ or tissue (annual organ dose) that is assumed to be the origin of the radiation-induced cancer of interest.
- When performing dose reconstructions for estimating annual organ doses, it is important to recognize that each sub-cohort may require a different methodology.
- Although each radiation sub-cohort is unique, common principles for dose assessment will facilitate the combination of sub-cohorts in the MPS.
- Where possible, annual organ doses obtained at other facilities where the individual may have worked are important to consider.
- Applying a decision level below which detailed dose reconstruction need not to be done is appropriate and can result in a considerable reduction in dosimetry effort without affecting the epidemiologic results.
- The coordination and close interaction of the dosimetric and epidemiologic teams is critical to the success of an epidemiologic study.

Assessment of organ doses from external sources:

- The basic procedure for estimating organ doses from external irradiation is similar for all the sub-cohorts and starts with using personal or environmental measurement data that can be applied to a scenario of exposure.
- The process of deriving organ doses from dosimeter results is relatively straightforward for photons. For many of the sub-cohorts in the MPS, the sensitivity of the derivations is not very dependent on the photon energies and on the geometry of irradiation. However, exceptions are found in those sub-cohorts exposed to low photon energies, when the organ of interest is small and asymmetrically located in the body. In this case, the geometry of irradiation causes large spatial differences across the body.

Assessment of organ doses from internal sources:

- Although the major source of radiation exposure for many sub-cohorts of the MPS comes from external penetrating radiation, some sub-cohorts may have a meaningful component of radionuclide intakes that requires addressing specific assessment techniques. However, detailed dose reconstructions for such intakes may not be justified (based on considerable time and effort) below some screening level of intake of a given radionuclide.
- Organ doses from internal sources can be estimated using available records on concentration of radionuclides in air or on surfaces, from bioassays or other biological techniques and combined with organ doses from external sources for epidemiologic studies of cancer risk.

Evaluation of uncertainties:

- All organ dose estimates obtained in a dose reconstruction have limitations and uncertainties that should be identified, considered, and taken into account in an appropriate manner in a dose reconstruction.

Notes

1. The MPS has had a number of name changes over the years. The most recent is the Million Person Study of Low-Dose Radiation Health Effects (MPS). Both Million Worker Study (MWS) and MPS refer to the same program of studies.
2. Atomic veterans are military personnel who participated in the nuclear weapons testing program.

Disclosure statement

The authors declare no personal conflicts of interest. The views expressed in this paper represent collective opinions of the authors and are not necessarily those of their professional affiliations.

Funding

This work was supported in part by grants from the U.S. Department of Energy [grant numbers: DE-AU0000042 and DE-SC0008944] awarded to the National Council on Radiation Protection and Measurements, which included interagency support from the U.S. Nuclear Regulatory Commission, the U.S. Environmental Protection Agency, and the National Aeronautics and Space Administration; a grant from the U.S. Nuclear Regulatory Commission [grant number: NRC-HQ-60-14-G-0011], grants from the Centers for Disease Control and Prevention [grant numbers: 5UE1EH000989 and 5NUE1EH001315], and grants from the National Aeronautics and Space Administration [grant numbers: NNX15AU88G and 80NSSC17M0016]. We acknowledge the support provided by the U.S. Department of Defense/Defense Threat Reduction Agency and the U.S. Department of Veterans Affairs. We acknowledge the leadership by the Chair of SC 6-9 Andre Bouville, the major support of the NCRP Secretariat, including, Managing Editor Cindy L. O'Brien, Office Manager Laura J. Atwell, and Executive Directors James Cassata (2012–2014), David Smith (2014–2016), and Kathy Held (2016–2018). We express our gratitude to the Council members for the time and effort devoted to the review of NCRP Report No. 178.

ORCID

John E. Till http://orcid.org/0000-0003-3666-8806

References

American National Standards Institute/Health Physics Society (ANSI/HPS). 2011. Criteria for performing multiple dosimetry, ANSI N13.41-2011. McLean (VA): Health Physics Society.

Beck HL, Till JE, Grogan HA, AAnenson JW, Mohler HJ, Mohler SS, Voilleque PG. 2017. Red bone marrow and male breast doses for a cohort of atomic veterans. Radiat Res. 187:221–228.

Boice JD, Jr. 2012a. A study of one million U.S. radiation workers and veterans: a new NCRP initiative. Health Phys. News XL. (11): 7–10.

Boice JD, Jr. 2012b. Study of atomic veterans who participated at U.S. aboveground atmospheric nuclear weapons tests, 1945–1962. Health Phys. News XL. (12): 6–9.

Boice JD, Jr. 2013a. Study of U.S. nuclear power plant workers. Health Phys. News XLI. (1):11–12.

Boice JD, Jr. 2013b. DOE worker studies. Health Phys. News XLI. 22–23.

Boice JD, Jr. 2013c. NCRP dosimetry committee for the Million Worker Study: Oak Ridge, Tennessee, April 2013. Health Phys. News XLI. 15–17.

Boice JD, Jr. 2014. Editorial: the importance of radiation worker studies. J Radiol Prot. 34:E7–E12.

Boice JD, Jr. 2018. The Million Person Study, Whence it came and why. Int J Radiat Biol. (submitted).

Boice JD, Jr, Cohen SS, Mumma MT, Dupree-Ellis E, Cragle DL, Eckerman KF, Wallace PW, Chadda B, Sonderman JS, Wiggs LD, et al. 2014. Mortality among Mound workers exposed to polonium-210 and other sources of radiation, 1944–1979. Radiat Res. 181:208–228.

Boice JD, Jr, Cohen SS, Mumma MT, Dupree-Ellis E, Eckerman KF, Leggett RW, Boecker BB, Brill AB, Henderson BE. 2006. Mortality among radiation workers at Rocketdyne (Atomics International), 1948–1999. Radiat Res. 166:98–115.

Boice JD, Jr, Cohen SS, Mumma MT, Dupree-Ellis E, Eckerman KF, Leggett RW, Boecker BB, Brill AB, Henderson BE. 2011. Updated mortality analysis of radiation workers at Rocketdyne (Atomics International), 1948–2008. Radiat Res. 176:244–258.

Boice JD, Jr, Leggett RW, Dupree-Ellis E, Wallace PW, Mumma M, Cohen SS, Brill AB, Chadda B, Boecker BB, Yoder RC, et al. 2006. A comprehensive dose reconstruction methodology for former Rocketdyne/Atomics International radiation workers. Health Phys. 90:409–430.

Bouville B, Toohey RE, Boice JD, Jr, Beck HL, Dauer LT, Eckerman KF, Hagemeyer D, Leggett RW, Mumma MT, Napier B, et al. 2015. Dose reconstruction for the million worker study: status and guidelines. Health Phys. 108:206–220.

Cox DR. 1972. Regression models and life-tables. J R Stat Soc B. 34: 187–220.

Dauer LT. 2014. Exposed medical staff: challenges, available tools, and opportunities for improvement. Health Phys. 106:217–224.

Dupree-Ellis E, Watkins J, Ingle JN, Phillips J. 2000. External radiation exposure and mortality in a cohort of uranium processing workers. Am J Epidemiol. 152:91–95.

Ellis ED, Boice JD, Jr, Golden AP, Girardi DJ, Cohen SS, Mumma MT, Shore RE, Leggett RW, Kerr GD. 2018. Dosimetry is key to good epidemiology: workers at Mallinckrodt Chemical Works had seven different source exposures. Health Phys. 114:386–397.

Fry SA, Dupree EA, Sipe AH, Seiler DL, Wallace PW. 1996. A study of mortality and morbidity among persons occupationally exposed to </=50 mSv in a year: Phase I, mortality through 1984. Appl Occup Environ Hyg. 11:334–343.

Gilbert ES, Cragle DL, Wiggs LD. 1993. Updated analyses of combined mortality data for workers at the Hanford Site, Oak Ridge National Laboratory, and Rocky Flats Weapons Plant. Radiat Res. 136: 408–421.

Gilbert ES, Koshurnikova NA, Sokolnikov ME, Shilnikova NS, Preston DL, Ron E, Okatenko PV, Khokhryakov VF, Vasilenko EK, Miller S, et al. 2004. Lung cancer in Mayak workers. Radiat Res. 162:505–516.

Golden AP, Ellis ED, Cohen SS, Mumma MT, Leggett RW, Wallace PW, Girardi D, Watkins JP, Shore R, Boice JD. Jr. 2018. Updated mortality analysis of the Mallinckrodt uranium processing workers, 1942–2012. Int J Radiat Biol. (submitted).

International Commission on Radiological Protection (ICRP). 1995. Conversion coefficients for use in radiological protection against external radiation. ICRP Publication 74. Ann ICRP. 26(3–4):1–205.

International Commission on Radiological Protection (ICRP). 2010. Conversion coefficients for radiological protection quantities for external radiation. ICRP Publication 116. Ann ICRP. 40(2–5):1–257.

Kwon D, Hoffman FO, Moroz BE, Simon SL. 2016. Bayesian dose-response analysis for epidemiological studies with complex uncertainty in dose estimation. Statist Med. 35:339–423.

Leggett RW, Eckerman KF, Boice JD, Jr. 2005. A respiratory model for uranium aluminide based on occupational data. J Radiol Prot. 25: 405–416.

Li Y, Guolo A, Hoffman FO, Carroll RJ. 2007. Shared uncertainty in measurement error problems, with application to Nevada Test Site fallout data. Biometrics. 63:1226–1236.

Linet MS, Kim KP, Miller DL, Kleinerman RA, Simon SL, de Gonzalez AB. 2010. Historical review of occupational exposures and cancer risks in medical radiation workers. Radiat Res. 174:793–808.

National Council on Radiation Protection and Measurements (NCRP). 2007. Uncertainties in the measurement and dosimetry of external radiation. Bethesda (MD): NCRP. NCRP Report No.: 158.

National Council on Radiation Protection and Measurements (NCRP). 2009a. Ionizing radiation exposure of the population of the United States. Bethesda (MD): NCRP. NCRP Report No.: 160.

National Council on Radiation Protection and Measurements (NCRP). 2009b. Radiation dose reconstruction: principles and practice. Bethesda (MD): NCRP. NCRP Report No.: 163.

National Council on Radiation Protection and Measurements (NCRP). 2009c. Uncertainties in internal radiation dose assessment. Bethesda (MD): NCRP. NCRP Report No.: 164.

National Council on Radiation Protection and Measurements (NCRP). 2010. Radiation dose management for fluoroscopically-guided interventional medical procedures. Bethesda (MD): NCRP. NCRP Report No.: 168.

National Council on Radiation Protection and Measurements (NCRP). 2012. Uncertainties in the estimation of radiation risks and probability of disease causation. Bethesda (MD): NCRP. NCRP Report No.: 171.

National Council on Radiation Protection and Measurements (NCRP). 2014. Fiftieth annual meeting program. NCRP: Achievements of the past 50 years and addressing the needs of the future. Bethesda (MD): NCRP (cited 2018 Sep 5). Available from: https://ncrponline.org/wp-content/uploads/2018/08/Annual/pp50.pdf.

National Council on Radiation Protection and Measurements (NCRP). 2018. Deriving organ doses and their uncertainty for epidemiologic studies (with a focus on the one million U.S. workers and veterans study of low-dose radiation health effects). Bethesda (MD): NCRP. NCRP Report No.: 178.

NDT Education Resource Center (NDT). 2014. Nondestructive testing. Ames (IA): Iowa State University (cited 2018 Sep 5). Available from: http://www.ndt-ed.org [accessed 2018 Sep 7].

Preston DL, Lubin JH, Pierce DA, McConney ME, Shilnikova NS. 2017. Epicure risk regression and person-year computation software: command summary and user guide. Version 2.00.03. Risk Sciences International, Ottawa, Canada. [Software]. http://epicurehelp.risksciences.com [accessed 2018 September 7].

Simon SL, Hoffman FO, Hofer E. 2015. The two-dimensional Monte Carlo: a new methodologic paradigm for dose reconstruction for epidemiological studies. Radiat Res. 183:27–41.

Silver SR, Bertke SJ, Hein MJ, Daniels RD, Fleming DA, Anderson JL, Pinney SM, Hornung RW, Tseng CY. 2013. Mortality and ionising radiation exposures among workers employed at the Fernald Feed Materials Production Center (1951–1985). Occup Environ Med. 70: 453–463.

Stram DO, Kopecky KJ. 2003. Power and uncertainty analysis of epidemiological studies of radiation-related disease risk in which dose estimates are based on a complex dosimetry system: some observations. Radiat Res. 160:408–417.

Stram DO, Preston DL, Sokolnikov M, Napier B, Kopecky KJ, Boice J, Beck H, Till J, Bouville A. 2015. Shared dosimetry error in epidemiological dose-response analyses. PLoS One. 10:e0119418.

Thierry-Chef I, Marshall M, Fix JJ, Bermann F, Gilbert ES, Hacker C, Heinmiller B, Murray W, Pearce MS, Utterback D, et al. 2007. The 15-Country Collaborative Study of Cancer Risk Among Radiation Workers in the Nuclear Industry: study of errors in dosimetry. Radiat Res. 167:380–395.

Tolmachev SY, Ketterer ME, Hare D, Doble P, James AC. 2011. The US Transuranium and Uranium Registries: forty years' experience and new directions in the analysis of actinides in human tissues. Proc Radiochim Acta. 1:173–181.

Till JE, Beck HL, Aanenson JW, Grogan HA, Mohler HJ, Mohler SS, Voillequé PG. 2014. Military participants at the U.S. atmospheric nuclear weapons testing—methodology for estimating dose and uncertainty. Radiat Res. 181:471–484.

U.S. Department of Energy (DOE). 2015. Occupational Radiation Exposure. Washington (DC). Department of Energy, Washington (cited 2018 Sep 5). Available from: http://energy.gov/ehss/policy-guidance-reports/databases/occupational-radiation-exposure.

U.S. Nuclear Regulatory Commission (NRC). 2011. Radiation Exposure Information and Reporting System (REIRS) for radiation workers. Washington (DC): U.S. Nuclear Regulatory Commission (cited 2018 Sep 5). Available from: http://www.nrc.gov/site-help/templates/reirs/riers.com-index.html.

Wiggs LD, Johnson ER, Cox-Devore CA, Voelz GL. 1994. Mortality through 1990 among white male workers at the Los Alamos National Laboratory: considering exposures to plutonium and external ionizing radiation. Health Phys. 67(6):577–588.

Yoder RC, Dauer LT, Balter S, Boice JD, Jr, Grogan HA, Mumma M, Passmore CN, Rothenberg LN, Vetter RJ. 2018. Dosimetry for the study of medical radiation workers with a focus on the mean absorbed dose to the lung, brain and other organs. Int J Radiat Biol. (submitted).

Dosimetry associated with veterans who participated in nuclear weapons testing

John E. Till, Harold L. Beck, Jill W. Aanenson, Helen A. Grogan, H. Justin Mohler, S. Shawn Mohler, and Paul G. Voillequé

ABSTRACT

Background: This article summarizes the methodology, results, and challenges of the reconstruction of red bone marrow and male breast doses for a 1982-person sub-cohort of ~114,270 U.S. military veterans who participated in eight atmospheric nuclear weapons tests between 1945 and 1962. These doses are being used in an epidemiological investigation of leukemia and male breast cancer as part of a study of one million U.S. persons to investigate risk from chronic low-dose radiation exposure.

Methods: Previous doses to these veterans had been estimated for compensation and tended to be biased high but newly available documentation made calculating individual doses and uncertainties using detailed exposure scenarios for each veteran possible. The techniques outlined in this report detail the methodology for developing individual scenarios and accounting for bias and uncertainty in dose based on the assumptions made about exposure.

Results: Doses to the atomic veterans in this sub-cohort were relatively low, with about two-thirds receiving red bone marrow doses <5 mGy and only four individuals receiving a red bone marrow dose >50 mGy. The average red bone marrow dose for members of the sub-cohort was 5.9 mGy. Doses to male breast were approximately 20% higher than red bone marrow doses.

Discussion and challenges: Relatively low uncertainty was achieved as a result of our methodology for reconstructing exposures based on knowledge of the individual veterans' locations and activities from military records. Challenges did arise from use of military records to determine probability of participation in specific activities but accounted for in estimates of uncertainty.

Background

This paper summarizes the methodology used to reconstruct radiation doses and associated uncertainties in an epidemiological analysis of military personnel who participated in the U.S. nuclear testing program. This cohort of atomic veterans is one component in a broader-scope study of about one million U.S. persons designed to investigate risk from chronic low-dose radiation exposure (Bouville et al. 2015; Boice et al. 2019). Military veterans participated in atmospheric nuclear weapons testing conducted by the United States between 1945 and 1963 and the reconstructed doses described here have been used in a case-cohort design to estimate the risk of leukemia and male breast cancer mortality resulting from chronic vs. acute radiation exposure (Boice et al. 2019). We also discuss some of the challenges and limitations of the methodology as well as opportunities for further research. Details of the methodology are reported by Till et al. (2014) and detailed doses are reported in Beck et al. (2017).

The U.S atmospheric weapons tests (DOE 2000) took place at:

- The Nevada Test Site (NTS) from 1951 through 1957 (Test series RANGER, BUSTER-JANGLE, TUMBLER-SNAPPER, UPSHOT-KNOTHOLE, TEAPOT, PLUMBBOB, and HARDTACK II). Military personnel participated in military maneuvers immediately following a nuclear test, observed tests, or provided support during test-related operations.
- The Pacific Proving Grounds[1] (PPG) from 1946 through 1962 (Test series CROSSROADS, SANDSTONE, GREENHOUSE, IVY, CASTLE, WIGWAM, REDWING, HARDTACK I, and DOMINIC). Military personnel were aboard ships or stationed on islands in the area during and following the tests.
- Alamogordo, NM, where the first nuclear test, TRINITY, took place on 16 July 1945.

The atomic veterans were primarily exposed to gamma and beta radiation from fission products resulting from nuclear weapons detonations. Many of the participants at the PPG were stationed on islands or ships impacted by fallout and thus received radiation exposure from fallout deposited on the ground or on ship surfaces as well as from descending fallout and re-suspended fallout. Other examples of activities that may have resulted in exposure include boarding contaminated target ships,[2] operating small boats in contaminated sea water, shore leave on fallout-contaminated islands, troop maneuvers at NTS during and following tests and observing tests at the NTS. Previous attempts to investigate disease among nuclear test participants in the U.S. have yielded mixed results, and these studies lacked the detailed dosimetry on individuals studied (Till et al. 2014).

Methods

The fundamental approach to dose estimation is to first determine the scenario of exposure for each veteran and then to assign a dose accounting for sources of bias and uncertainty. Because calculation of doses for all ~115,000 members of the cohort would have been prohibitive both from a cost and time perspective, an efficient epidemiologic study design, the case-cohort approach, was used that required detailed dose estimates for a smaller sub-cohort (Wacholder 1991). The sub-cohort in the current study included 1982 individuals, including all leukemia and male breast cancer cases and a 1% random comparison sample of the entire cohort.

Doses to U.S. military veterans were estimated previously under the Nuclear Test Personnel Review (NTPR) program directed by the Defense Threat Reduction Agency (DTRA)[3] within the Department of Defense (DOD). In 1988, legislation was passed that authorized a compensation program for veterans based on doses estimated by DOD personnel and their contractors. To estimate doses to veterans, the DOD undertook a comprehensive review of historical records and developed a dose reconstruction methodology that was designed to be high sided and favor veterans seeking compensation. The goal for this study was to estimate individual doses with uncertainties without the high-sided bias that was often introduced in the methodology developed during the veterans' compensation program.

Our approach built upon available film badge dosimetry and other measurement data recorded at the time of the tests and incorporated detailed scenarios of exposure for each veteran based on personal, unit, and other available historical records. Approximately, 25% of the military personnel had film badge records that accounted for at least 80% of their dose. These records were used to estimate specific organ doses to individuals wearing a badge and for groups of veterans performing similar duties at the same time and location. The distributions of film badge readings for individuals in specific military units with likely similar exposures were also used for estimating uncertainties.

One or more members of a unit performing similar duties were often issued a "cohort" film badge, which often could be used in estimating dose to the entire unit. The availability of film badge data varied over time. A relatively large percentage of individuals had film badge data for the later test series (e.g. PLUMBBOB-1957 and HARDTACK I-1958), while a relatively small percentage had film badge data for the earlier series (e.g. CROSSROADS-1946, GREENHOUSE-1951, and UPSHOT-KNOTHOLE-1953). In addition, while half of the CASTLE-1954 participants had reported badge exposures, the majority were based on a cohort badge wearer whose dose was assigned as a badge dose to other members of the cohort. For CASTLE in particular, the cohort badge assignments were determined to not necessarily reflect the activities an individual was involved in. Likewise, for REDWING, although many individuals wore badges, many of the badges were damaged from high heat and humidity and did not reflect the actual exposures received (NRC 1989).

Contributing to the total body of knowledge about an individual's potential for radiation exposure are documents related to military units, ships' logs, photographs of facilities, and activities at the NTS and the PPG; the indicator of military rate or rating (indicating an individual's job); and numerous other historical records. All available sources of information were incorporated into our methodology (Till et al. 2014).

Cancers studied and pathways of exposure

Leukemia and male breast cancer were selected as the focus of the study. Leukemia was selected because previous U.S. studies suggest a general pattern of increased leukemia risks (Seltzer and Jablon 1974; 1977; Caldwell et al. 1980; 1983; Robinette et al. 1985; Thaul et al. 2000), but these studies did not include individual dosimetry or uncertainty. Male breast cancer was chosen because a previous study of atomic veterans had reported a non-statistically significant increased risk of 39% (Thaul et al. 2000). The study of atomic bomb survivors (Ron et al. 2005) reported a statistical association of male breast cancer with radiation dose but based on small numbers, and the number of male breast cancers in our study, about 30, would be the largest study to date.

Veterans could have been exposed to both external and internal radiation. Based on conservative NTPR analyses as well as estimated doses to residents of the Marshall Islands, internal exposure to red bone marrow (organ of interest for leukemia) and male breast would almost always be no more than a few percent of the dose from external exposure (Goetz 1986; Phillips et al. 1985; Weitz et al. 2009). Other researchers have also concluded that when external and internal exposure exists from radioactive fallout of nuclear weapons, the primary exposure pathway for most organs other than thyroid is external (Bouville et al. 2010; Simon et al. 2010).

Criterion for detailed dose reconstruction

In reviewing the historical records such as film badge data, previous dose reconstructions, and other information describing a veteran's military unit and personal activities, if it could clearly be established that the dose was below a given level that did not merit a comprehensive dose reconstruction, then the dose reported by the NTPR program was used. This approach has been taken in similar studies where a decision level (also called "cutoff level") is chosen for the

Table 1. Examples of estimated bias or error[a] in NTPR dose assignments (Taken from Till et al. 2014; Table 4).

Source of exposure	NTPR source of bias or error	~ Range of bias or error[b]
Cohort badges at CASTLE	NTPR assigned doses based on random cohort assignments that resulted in errors due to both overestimating and underestimating doses to particular individuals	0.2–3.0
Film badge interpretation	NTPR assumed film badge reading equaled the whole body dose that resulted in all NTPR doses being biased high and did not correct for other known biases	1.2–2.0
Engine room duty	NTPR assumed high-sided dose (bias) and placed unlikely rates and ratings in the engine room (error).	1.2–10
Hull and seawater contamination	NTPR assumed all personnel worked and berthed below the water level where exposure would be highest resulting in errors in individual exposures	0.4–9.0
Bias correction for exposure to fallout on residence islands and ships	NTPR used available exposure rate measurements and model estimates of shielding that did not accurately reflect the true mean exposure. This generally biased high the mean dose to all those in a particular unit	0.5–6.0

[a]'Bias' refers to the tendency to systematically either overestimate or underestimate doses for a given exposure scenario. 'Error' refers to random errors in individual doses such as errors in assumed individual exposure scenario or errors in NTPR records. Random errors for a particular exposure scenario can be in either direction, while bias is always in the same direction.
[b]Range of ratios of NTPR to AVS doses.

practical reasons of cost and feasibility (Boice et al. 2006). Veterans with doses below the decision level were not excluded from the study and their estimated dose was still used in the analysis.

We selected a decision level of 5 mGy. We assigned an uncertainty to all NTPR doses clearly less than this level characterized by a geometric standard deviation (GSD) of 1.4, consistent with NTPR estimates of uncertainty in external exposure at those levels.

Process for dose estimation

In the first step of our methodology, we evaluated all information available in the Nuclear Test Review and Information Systems database (NuTRIS)[4] and NTPR literature on the veteran's potential sources of exposure along with the NTPR-estimated doses for each source. If the individual's exposure scenario was well established and the sum of all the NTPR doses was clearly less than the decision level of 5 mSv, the NTPR dose was converted to organ dose using appropriate organ dose coefficients and recorded in a database. If the total dose is not clearly less than the decision level, the doses recorded in NuTRIS were adjusted to provide a more realistic dose. These adjustments included

- Making use of more current information.
- Correcting obvious errors in the NuTRIS entries.
- Replacing suspect film badge data with reconstructed doses.
- Accounting for sources and time periods of exposure not included in NuTRIS.

More detailed custom dose reconstructions were performed when the probability of occurrence of a potential scenario needed to be estimated to calculate a dose, when there was no NTPR-estimated dose for a potential exposure, or when there was no generic NTPR dose reconstruction that applied to the particular source of exposure. About two-thirds of the cases with total dose deemed not clearly less than the decision level required some custom dose reconstruction.

Next, the uncertainty in our final dose estimate for each source of exposure was estimated. The final step was to total the exposures from all sources, convert to annual organ dose, and calculate a total uncertainty for each annual dose.

Scenarios of exposure

The scenario of exposure accounts for the time, duration, location, duties, and other factors that resulted in a veteran's exposure to radiation. Developing realistic scenarios of exposure was a key step toward estimating dose. Although individual scenario development was carried out during the NTPR program in support of claims for compensation, NTPR dose estimates for most of our cohort did not account for possible individual-specific activities and instead assumed the same generic exposure for all members of a unit.

Scenarios of exposure were divided by location, series, and type allowing for efficient consideration of the commonalities across the following broad categories:

- Ship-based scenarios at the PPG.
- Land-based scenarios at the PPG.
- Maneuvers, observers, and other activities at the NTS.

Broken down in this fashion, scenario similarities allowed for transfer of knowledge gained in one situation to be applied to similar situations. This categorization was particularly important for unique pathways that were appropriate for more than one type of individual.

Duties for most military personnel could be generally inferred from the branch of service, the unit, and the person's rate or pay grade (enlisted) or rank (officers).

Adjusting for bias and error in the NTPR dose

There are instances where the NTPR dose for a particular activity was known or believed to be systematically high-sided (biased) or where errors were made related to assumptions about exposures or model parameters. These instances were identified and adjusted to remove the effect of bias or error in the NTPR dose. Table 1 lists examples of several key instances of NTPR bias or error that required an adjustment to the NTPR dose and the approximate range of the adjustment. Detailed information about the methodology used to correct for errors and deliberate bias is provided in Till et al. (2014).

Accounting for uncertainty

A key component of dose estimation is the uncertainty. Our methodology separated uncertainty into two fundamental areas: uncertainty associated with defining the scenario of exposure and uncertainty in the dose for the given scenario.

Scenario uncertainty

Scenario uncertainty represents the degree to which an individual's location, responsibilities, and activities at a specific time are known. This uncertainty was first addressed separately in a qualitative manner for each potential source of exposure using the following three categories:

- A = individual's activities and duties are well known, or potential dose for the individual clearly falls below the dose reference level for the study, described previously.
- B = individual's activities and duties are less well known, but some aspects can be inferred from other information to estimate the probability of actually being exposed.
- C = individual could possibly have been exposed, but very little to nothing is known about the individual's activities and duties and, thus, very little can be inferred about the probability of actually being exposed.

These three categories of uncertainty were quantified by assigning estimated probabilities to the likelihood that each exposure occurred (Beck et al. 2017).

Individual variability in exposures

The uncertainty in an individual's dose depends both on the uncertainty in the mean exposure corrected for NTPR bias and on uncertainty due to individual variability. Any individual's dose will vary about the assigned mean dose, for example, due to differences between an individual's time spent shielded while indoors or below deck, spatial differences in exposure rates, and differences in specific duties within the unit. In some cases, we removed some of this uncertainty by applying corrections based on a crew member's rating and rank. For most cases, all members of a unit were assigned the same mean dose for a given activity by NTPR. In these cases, because we had no precise information on the actual location and specific duties of an individual crew member on an island or ship, it was important to include this variability as part of the total uncertainty in the veteran's exposure. The estimated coefficient of variance based on film badge readings on ships and islands varied from as little as 20% to as much as 70–80%.

Thus, for each source of exposure, the uncertainty in the exposure, assuming such exposure actually occurred, was estimated either from the mean and the dispersion in available film data if sufficient film data were available, from NTPR model calculations, or both. Of particular help was an initiative by NTPR to revise its previous high-sided uncertainty estimates for many scenarios to develop unbiased stochastic estimates of total uncertainty (Weitz et al. 2009). These stochastic estimates of uncertainty generally confirmed that most NTPR point estimates of upper-bound doses from external radiation exposure (95% CI) were less than about a factor of 3 above the estimated "unbiased" mean dose.

Probability of participating in specific activities

The NTPR generally assigned all individuals who could have participated in a particular activity a dose based on having participated, even if the probability of their participation was low. Examples are observing a particular shot at NTS, operating a small boat in contaminated water, boarding a target ship, or participating in rest and recreation on an island. This often resulted in a significant overestimate in the dose to some cohort members.

At the PPG, an activity that could have resulted in significant exposure was the re-boarding of target ships, usually by their original crews, for repairs, decontamination, or the retrieval of salvageable items (Till et al. 2014). In some cases, the ship's log notes which crew members participated in these boarding parties, along with the time a boarding party embarked and when it disembarked the target ship. In other cases, the ship's log notes a boarding party embarked but does not provide details about who participated. Knowledge about which crew members were most likely to re-board target ships was key to estimating uncertainty.

The probability of participating in any particular re-boarding was estimated by knowing the rank and rating of veterans who participated and the total number of these rank and ratings for the target ship crew. By combining available data from several types of ships, a general pattern emerged about which crew members were most likely to have participated in boarding parties and the probability of their boarding. This information allowed us to develop probabilities of boarding a target ship by rank and rating.

A similar approach was taken for calculating the probability of other activities. Based on all available data, the probability, p, of participation in a given activity was estimated, along with an uncertainty in this probability estimate. Often, the uncertainty in p was only crudely estimated due to lack of actual data. This estimated probability of participation, p, was combined with the uncertainty in the exposure, if the exposure actually occurred, to estimate the unconditional uncertainty in exposure. For example, if the probability of boarding a target ship on a particular day was $p = .5$ and the uncertainty in the exposure, E, assuming it occurred, had a coefficient of variation of ~0.35–0.40, the coefficient of variation in the assigned exposure (0.5 * E) would be ~1.1.

Shared and unshared uncertainties

As for any estimate of risk based on dose for an epidemiological study, both the total uncertainty and the type of uncertainty are important. In particular, it is important to understand how much uncertainty is shared among individuals (Stram and Kopecky 2003). An example of shared uncertainty would be the uncertainty that arises if a single-dose estimate is assigned to all members of a unit participating in a maneuver after a detonation, since any error in the models and measurements used to calculate the estimate would be the same for all exposed. If a model overestimates the average dose by 50%, everyone's dose would be overestimated by an average of 50%. In contrast, the uncertainty in dose to two individuals exposed to fallout at different test series is unshared (uncorrelated) since the doses are based on completely different sets of measurements. Note that shared uncertainty, when present, can coexist with unshared uncertainties. Continuing the example, each member of a military unit had a certain probability of being absent from his unit's maneuver following a test detonation, and it is plausible that one person's absence was unrelated to the absence of anyone else in his unit. Similarly, each person's actual dose coefficient for a given organ differs from the mean value by an unknown amount, which is independent of everyone else's dose coefficient.

Consideration of classical (measurement) error as opposed to Berkson-type error is also important since classical-type errors can result in a reduction of the slope of the dose-response curve while unshared Berkson-type errors will not (Stram and Kopecky 2003; NCRP 2009). Our review of shared and unshared uncertainty in this study suggests there is, in the aggregate, relatively little shared uncertainty among members of the study group. This finding is due largely to the small number of study participants who served in the same unit at the same time and who were assigned a dose based on the same measurements. Most of the unshared uncertainty for members of the same unit is Berkson-type error. For example, although the same dose due to fallout is assigned to all unbadged personnel present on Enewetak island for a given time period, based on the mean of available film badge data for those badged, the uncertainty in any individual's "true" dose is due primarily to individual variations about this mean from variations in shielding (time indoors), location, and specific duties. The uncertainty in the mean itself, while shared classical type uncertainty, is usually small compared to the variability about the mean.

Total organ dose

The total radiation exposures from all sources are combined, and the total associated uncertainty was computed by combining the unconditional variances for all the individual's sources of exposure. The sum of all free-in-air exposures was then multiplied by a dose coefficient that converts the estimate of total exposure to organ dose, in this case to dose to red bone marrow (6.6 mGy R^{-1}) or male breast (7.9 mGy R^{-1}). Although a rotational incidence is generally more representative than an isotropic or other angular distribution (NCRP 2009) because the angular incidence will differ depending on activity and location (ship versus land), values between rotational and isotropic were adopted with a bias more toward the rotational (Beck et al. 2017). The conversion factors for converting from exposure to organ dose were assumed to have an uncertainty (GSD =1.2) which is due mainly to individual variations in actual angular incidence, in body and organ size, and in energy spectra of the incident gamma radiation.

Results

The doses to the atomic veterans in the sub-cohort were relatively low with about two-thirds of the atomic veterans receiving red bone marrow (RBM) doses <5 mGy and only a small fraction receiving doses >50 mGy. The average RBM dose for the cohort subset was 5.9 mGy. Each male breast dose is ~20% greater than the corresponding RBM dose. Detailed breakdowns of estimated doses and corresponding uncertainty by test series, test site, military service, and military rank are given in Beck et al. (2017).

The doses varied somewhat by test series with the highest doses to participants in the PPG CASTLE test series in 1954 due in a large part to the heavy fallout encountered by some naval vessels after the 1 March 1954, BRAVO test. Although 34% of the sub-cohort participated only in the 1946 CROSSROADS test series, a far greater percentage than for any other test series, their doses were low in comparison to those participating in most of the other later test series.

The mean uncertainty in the estimated doses, expressed in terms of the CV, was much lower for the REDWING, PLUMBBOB, and HARDTACK I test series than for earlier test series, primarily because almost all veterans participating in these test series were issued film badges.[5] Previously, only cohort badges were issued to one or two members of a unit performing similar duties or to individuals expected to receive a meaningful exposure. Although the mean CV for all test series was <0.5, the range of CV varied considerably with a few as high as CV =3.4.[6]

There was little variability in doses between test sites although the environmental characteristics and troop activities at the NTS and the PPG varied greatly. We also found no significant variation in dose between the branches of service (Air Force,[7] Army, Marines, and Navy) at each site.

Table 2. Distribution of total red bone marrow (RBM) and male breast (MB) doses.

	This study – RBM		This study – male breast		NuTRIS[a]	
Dose range (mGy)	# of individuals	% of 1982 calculated doses	# of individuals	% of 1982 calculated doses	% of entire ~115,000 NuTRIS doses	% of entire ~115,000 adjusted NuTRIS doses
<5	1,312	66.2	1,257	63.4	55.7	68.4
5–10	304	15.3	304	15.3	17.5	15.0
10–20	214	10.8	216	10.9	12.6	9.8
20–50	148	7.5	195	9.8	12.5	6.3
50–100	3	0.15	8	0.4	1.5	0.36
[b]1–300	1	0	2	0.1	0.21	0.09
[b]>300	0	0	0	0	0.035[c]	0.02

[a]NuTRIS doses were reported as mean whole-body doses and thus would be comparable to red bone marrow doses and about 20% smaller than male breast doses.
[b]The maximum calculated RBM dose was 108 mGy. The maximum sub-cohort NuTRIS dose is 199 mGy.
[c]The maximum NuTRIS dose (entire cohort) is 908 mGy.

Although the mean dose was similar for those participating only in NTS tests or only in PPG tests, the maximum dose was greater at the PPG. Far more atomic veterans served at the PPG than at the NTS, again due to the large participation, mostly Navy, at the 1946 CROSSROADS test series. Approximately, 80% of the Army participation was at the NTS while almost all the Navy participation (~98%) was at the PPG. Air Force participation was greater at the PPG than at the NTS, and the doses to Air Force personnel at the NTS were much smaller than at the PPG. The higher Air Force doses at the PPG were because Air Force personnel (as well as Army and some Navy and Marine) at the PPG were housed on islands that frequently were contaminated by fallout. Nevertheless, the average uncertainty (CV) is comparable for our reconstructions of NTS and PPG participants even though, as discussed in Till et al. (2014), the exposure scenarios at the NTS were often very different from the exposure scenarios at the PPG. The small number of individuals who participated in tests at both the NTS and PPG received, on average, much higher doses.

Although there were many more enlisted personnel than officers who participated in the tests, there was generally little difference in officer versus enlisted doses. Navy officers received slightly higher doses than enlisted personnel, due in part to the fact that officers almost always led the boarding parties that inspected and decontaminated target ships.

We also compared the calculated RBM doses from the current study with the corresponding NuTRIS doses (Beck et al. 2017). The RBM doses were generally 30–40% lower than those estimated by NTPR from their generic dose assessments. The overall correlation was quite good (RBM dose =0.64 * NuTRIS dose, 95% C.I. = 0.61–0.67), although there were some comparisons that deviated significantly from the general trend. The correlation was somewhat better for higher doses due primarily to the fact that many of those doses were based on film badge dosimetry and thus were more precise. The dose distributions of the calculated RBM and male breast doses are given in Table 2. The distribution of male breast doses differs slightly from the RBM distribution since each male breast dose is ~14% higher than the corresponding RBM dose. The distribution of NuTRIS doses for the entire ~115,000-person cohort is also shown, as well as the distribution after multiplying the cohort NuTRIS doses by the 0.64 ratio of NuTRIS to reconstructed doses described above. Because most of the calculated doses were for the randomly selected 1% of the entire cohort, the distributions of NuTRIS doses for the sub-cohort are very close to that of the NuTRIS doses for the entire 115,000-person cohort. However, while the maximum calculated RBM NuTRIS dose for this sub-cohort was 199 mGy (127 mGy after adjustment) and the maximum calculated RBM dose for this study was 108 mGy, the maximum NuTRIS dose for the entire cohort is ~900 mGy (580 mGy after adjustment). About 0.05% (~60) of the ~115,000 NuTRIS doses are >200 mGy.

Discussion

The average NuTRIS dose to the entire ~115,000-person cohort is 9.0 mGy, which if adjusted by the 0.64 ratio of NUTRIS to reconstructed RBM doses, results in an estimated cohort mean RBM dose of 5.8 mGy, very close to the mean dose for the sub-cohort calculated from the current study.

The uncertainty in most of the dose estimates was also small, averaging about CV =0.4–0.5.[8] The CV ranged from as low as 0.2, when most of the exposure was based on film data, to CV >3 when the exposure scenario was very uncertain. The CV was generally slightly less for the higher doses (mean CV=~0.3 for doses >30 mGy) due primarily to the fact that persons likely to receive significant exposure were more likely to have been issued a film badge.

The magnitude and the range of our uncertainties in organ dose are comparable to the estimated uncertainty in the reconstructed dose to the members of the Japanese Life Span Study (Young and Kerr 2005) and relatively small compared to those typically observed in most historical dose reconstructions of exposure from radionuclides in the environment, particularly studies involving significant internal dose (Beck et al. 2017).

The reasons for the relatively low and narrow range of uncertainty in our study include the availability of some film-badge data, detailed environmental monitoring data in fallout fields at test sites, and comparatively well-known exposure conditions based on knowledge of the individual veterans' locations and activities from military records. Furthermore, reconstructing doses from external exposure generally involves much lower uncertainty than when

Table 3. Organ doses (mGy) from external radiation exposure expressed as a ratio to the RBM dose (mGy).

Organ	Ratio to RBM[a]
Bone surface	1.10
Brain	1.11
Breast (male)	1.14
Heart	0.97
Liver	0.95
Prostate	0.92
Salivary gland	1.20
Testes	1.04
Thyroid	1.13

[a]RBM dose =1.0 (0.65 rad/R or 6.5 mGy/R).

internal exposure needs to be considered since internal dose cannot generally be measured directly.

Limitations of methodology and opportunities for further study

One challenge in this study was to accurately estimate the exposure scenario for individuals who could have participated in a particular activity but it was not known if they did. This is, however, reflected in the estimated uncertainty. Furthermore, estimating the probability of exposure to a given source along with the uncertainty in that exposure, if it occurred, was also a unique feature of this study compared with some of the other case–control studies discussed in this report where individual dose records were available for most participants.

Another limitation was the unavailability of original measurement data to validate various NTPR generic dose assessments. This limitation, coupled with limited availability of film badge data for some units, often made it difficult to estimate the bias in NTPR generic unit dose assessments for the individual unit members.

Assessing the validity of some of the reported film badge data also limited this study, particularly for test series where it was known some of the badge data was suspect.

An unexpected outcome of this study was the relatively good correlation between our reconstructed doses and the corresponding NTPR dose estimates. This suggests that it is possible to use the NuTRIS doses (which are available for the entire cohort), multiplied by an appropriate scaling factor (slope of the dose-reconstructed doses for the sub-cohort versus NuTRIS doses as determined in this study) to obtain estimates of risk from external radiation exposure for additional organs. The NuTRIS adjustment factor of 0.64 identified for RBM doses can be scaled by the respective ratios of exposure-to-dose coefficients for gamma rays for testes, brain, and heart to RBM (ICRP 2010) (Table 3) to estimate the corresponding organ-absorbed doses from external exposure. This assumption is supported by the fact that when the risk estimates for leukemia and male breast cancer based on the case-cohort study design are compared with corresponding risk estimates using the entire 115,000-person cohort with adjusted NuTRIS doses, nearly identical results are obtained (Boice et al. 2019). Comparison of our atomic veteran doses with NuTRIS doses for individual test series suggest the method may even be improved by using separate adjustment factors for each test series.

The dosimetry developed for the Atomic Veterans Study was designed to meet key criteria recommended for environmental dose reconstruction as described in NCRP Commentary 27 (NCRP 2018) and Till et al. (2017). These criteria include consideration of shared and unshared dosimetry, dose validation, and dose confounders among others. Therefore, with the large study size and generally low doses, the Atomic Veterans Study makes one of the largest cohorts with high quality dosimetry and low uncertainties available to researchers in the Million Person Study.

Only minimal additional effort would be required to estimate external dose for additional cancer cases for a more accurate case-cohort analysis. This is because external exposure in air for the 1% random selection component of the case-cohort has already been calculated. Adrenals, brain, kidneys, pancreas, spleen, testes, heart, and bladder are all organs where internal exposure to atomic veterans is also believed to have been negligible compared to external exposure (Weitz et al. 2009; Case et al. 2011). Internal dose to some organs for some veterans, although still generally less than external dose, could have been meaningful for the stomach, intestines, liver, lung, bone, and in particular, thyroid, and would require additional effort and resources to assess.

Although the doses to atomic veterans were found to be generally low, a key contribution of this project has been the development of a methodology to characterize exposure scenarios for any specific veteran in the absence of individual-specific dosimetry or individual interviews. The exposure scenario methodology, which comprises an essential component of the dosimetry, has provided a significant contribution to our understanding of the unique exposure circumstances encountered by these veterans and the likelihood that any individual participated in activities that resulted in his exposure. Previously, such detailed evaluations had only been completed for a small number of veterans filing a compensation claim.

Understanding the extent to which doses may have been overestimated is an important finding, not only for this epidemiologic study and to the veterans themselves but also to policy makers who may design and implement compensation programs in comparable situations.

Disclosure statement

No potential conflict of interest was reported by the authors.

Funding

This research was supported in part by contracts and grants from the National Cancer Institute [Grant no. U01 CA137026], the U.S. Department of Energy [Grant no. DE-SC0008944 awarded to the National Council on Radiation Protection and Measurements], and a Discovery Grant from the Vanderbilt-Ingram Cancer Center [Center no. 404-357-9682]. Dr John D. Boice, Jr is the principal investigator for the Million Person Study. This support is acknowledged and appreciated. The authors also acknowledge Dr Paul Blake, Director, Chief, Nuclear Test Personnel Review, Defense Threat Reduction Agency, Department of Defense and his staff for their technical support of the project.

Notes

1. The Pacific Proving Grounds was renamed Enewetak Proving Ground prior to the HARDTACK I test series in 1958.
2. A large number of unmanned "target" ships were anchored close to surface zero during the CROSSROADS test BAKER. Some of these ships were later re-boarded to assess damage and to decontaminate if possible.
3. The predecessor of DTRA was the Defense Nuclear Agency (DNA) whose legacy goes back to the Manhattan Project, which directed the development of the first atomic weapons. The DTRA program has been ongoing since the early 1980s.
4. NuTRIS (Nuclear Test Review Information System) is a database developed and populated by NTPR that contains all of the pertinent information about each veteran including the dose estimated by NTPR.
5. Although, as discussed in Till et al. (2014), some of the film badges for REDWING suffered environmental damage, those doses were replaced by reconstructed doses and thus the uncertainties in total dose were not significantly impacted.
6. A CV of 0.5 would correspond to a geometric standard deviation (GSD) of 1.6 while a CV of 3.4 would correspond to a GSD of ~5, assuming the uncertainty distributions are approximately lognormal.
7. The Air Force was not a separate branch of service until 1947; thus, "Air Force" participants at TRINITY and CROSSROADS were actually in the Army Air Force.
8. Corresponding to GSD ~1.5–1.6, assuming an approximate lognormal distribution.

ORCID

John E. Till http://orcid.org/0000-0003-3666-8806
Helen A. Grogan http://orcid.org/0000-0003-3532-5166

References

Beck HB, Till JE, Grogan HA, Aanenson JW, Mohler HJ, Mohler SS, Voillequé PG. 2017. Red bone marrow and male breast doses for a cohort of atomic veterans. Radiat Res. 187:221–228.

Boice JD Jr, Leggett RW, Ellis ED, Wallace PW, Mumma M, Cohen SS, Brill AB, Chadda B, Boecker BB, Yoder RC, et al. 2006. A comprehensive dose reconstruction methodology for former Rocketdyne/Atomics International radiation workers. Health Phys. 90:409–430.

Boice JD Jr, Till JE, Mumma MT, Chen H. 2019. Mortality among US military participants at seven aboveground nuclear weapons test series. Int J Radiat Biol. [Epub ahead of print].

Bouville A, Beck HL, Simon SL. 2010. Doses from external irradiation to Marshall Islanders from Bikini and Enewetak nuclear weapons tests. Health Phys. 99:143–156.

Bouville A, Toohey RE, Boice JD Jr, Beck HL, Dauer LT, Eckerman KF, Hagemeyer D, Leggett RQ, Mumma MT, Napier B, et al. 2015. Dose reconstruction for the Million Worker Study: status and guidelines. Health Phys. 108:206–220.

Caldwell GG, Kelley D, Zack M, Falk H, Heath CW Jr. 1983. Mortality and cancer frequency among military nuclear test (SMOKY) participants, 1957 through 1979. JAMA. 250:620–624.

Caldwell GG, Kelley D, Heath CW Jr. 1980. Leukemia among participants in military maneuvers at a nuclear bomb test: a preliminary report. JAMA. 244:1575–1578.

Case DR, Chehata M, Dunavant JD, Egbert SD, Mason CL, McKenzie-Carter MA, Singer HA, Weitz RL. 2011. A technical approach to expedited processing of NTPR radiation dose assessments. Fort Belvoir (VA): Defense Threat Reduction Agency; Report No: DTRA-TR-10-29.

Department of Energy (DOE). 2000. United States Nuclear Tests July 1945 through September 1992. DOE/NV—209-REV 15. U.S. Department of Energy Nevada Operations Office.

Goetz J. 1986. Ingestion Doses at CROSSROADS. Memorandum to Dr. Auton. DNA-STBE/NTPR, 21 February.

International Commission on Radiological Protection (ICRP). 2010. Conversion coefficients for radiological protection quantities for external radiation exposures. ICRP Publication 116, Ann. ICRP 40(2–5). Thousand Oaks (CA): Sage Publications.

National Council on Radiation Protection and Measurements (NCRP). 2009. Radiation dose reconstruction: principles and practices. NCRP Report 163, p. 2084–3095. Bethesda, MD.

National Council on Radiation Protection and Measurements (NCRP). 2018. Implications of Recent Epidemiologic Studies for the Linear-Nonthreshold Model and Radiation Protection. NCRP Commentary No. 27, Bethesda, MD.

National Research Council (NRC). 1989. Film badge dosimetry in atmospheric nuclear tests. Energy and Engineering Board. Washington, DC: National Academies Press.

Phillips J, Klemm J, Goetz J. 1985. Internal dose assessment-operation CROSSROADS. McLean, VA: Science Applications International Corporation. Report DNA-TR-84-119.

Robinette CD, Jablon S, Preston TL. 1985. Mortality of nuclear weapons test participants. Washington, DC: National Academy Press.

Ron E, Ikeda T, Preston DL, Tokuoka S. 2005. Male breast cancer incidence among atomic bomb survivors. J Natl Cancer Inst. 97: 603–605.

Seltzer CC, Jablon S. 1974. Effects of selection on mortality. Am J Epidemiol. 100:367–372.

Seltzer CC, Jablon S. 1977. Army rank and subsequent mortality by cause: 23-year follow-up. Am J Epidemiol. 105:559–566.

Simon SL, Bouville A, Melo D, Beck HL, Weinstock RM. 2010. Acute and chronic intakes of fallout radionuclides by Marshallese from nuclear weapons testing at Bikini and Enewetak and related internal radiation doses. Health Phys. 99:157–200.

Stram DO, Kopecky KJ. 2003. Power and uncertainty analysis of epidemiological studies of radiation-related disease risk in which dose estimates are based on a complex dosimetry system: some observations. Radiat Res. 160:408–417.

Thaul S, Page WF, Crawford H, O'Maonaigh H. 2000. Institute of Medicine, the five series study: mortality of military participants in U.S. nuclear weapons tests. Washington (DC): National Academy Press.

Till JE, Beck HB, Aanenson JW, Grogan HA, Mohler HJ, Mohler SS, Voillequé PG. 2014. Military participants at the U.S. atmospheric nuclear weapons testing-methodology for estimating dose and uncertainty. Radiat Res. 181:471–484.

Till JE, Beck HB, Aanenson JW, Grogan HA, Mohler HJ, Mohler SS, Voillequé PG. 2017. A review of dosimetry used in epidemiological studies considered to evaluate the linear no-threshold (LNT) dose-response model for radiation protection. Int J Radiat Biol. 93: 112–1144.

Wacholder S. 1991. Practical considerations in choosing between the case-cohort and nested case-control designs. Epidemiology. 2: 155–158.

Weitz RL, Case DR, Chehata M, Egbert SD, Mason CL, Singer HA, Martinez DG, McKenzie-Carter MA, Shaw RS, Stiver JS. 2009. A probabilistic approach to uncertainty analysis in NTPR radiation dose assessments. McLean (VA): Science Applications International Corporation. Report DTRA-TR-09-13.

Young RW, Kerr GD, editors. 2005. Reassessment of the atomic bomb radiation dosimetry for Hiroshima and Nagasaki-Dosimetry System-DS-02. Hiroshima, Japan: Radiation Effects Research Foundation. Report of the Joint U.S.–Japan Working Group.

Dosimetry for the study of medical radiation workers with a focus on the mean absorbed dose to the lung, brain and other organs

R. Craig Yoder, Lawrence T. Dauer, Stephen Balter, John D. Boice, Helen A. Grogan, Michael T. Mumma, Christopher N. Passmore, Lawrence N. Rothenberg, and Richard J. Vetter

ABSTRACT

Background: The reconstruction of lifetime radiation doses for medical workers presents special challenges not commonly encountered for the other worker cohorts comprising the Million Worker Study.

Methods: The selection of approximately 175,000 medical radiation workers relies on using estimates of lifetime and annual personal monitoring results collected since 1977. Approaches have been created to adjust the monitoring results so that mean organ absorbed doses can be estimated.

Results: Changes in medical technology and practices have altered the radiation exposure environments to which a worker may have been exposed during their career. Other temporal factors include shifts in regulatory requirements that influenced the conduct of radiation monitoring and the changes in the measured dose quantities.

Conclusions: The use of leaded aprons during exposure to lower energy X rays encountered in fluoroscopically based radiology adds complexity to account for the shielding of the organs located in the torso when dosimeters were worn over leaded aprons. Estimating doses to unshielded tissues such as the brain and lens of the eye become less challenging when dosimeters are worn at the collar above the apron. The absence of leaded aprons in the higher energy photon settings lead to a more straightforward process of relating dosimeter results to mean organ doses.

Introduction

The Million Worker Study (MWS) seeks to combine the radiation dose and mortality experiences from several occupational groups in an effort to better understand the incidence of cancer induction to people occupationally exposed to low levels of radiation for many years. This study is being conducted under the direction of the National Council on Radiation Protection. Medical radiation workers represent one of the occupational groups that includes large numbers of women from which sex-related radiation effects can be potentially identified. Of particular concern is the unexplained difference in the radiation-related incidence of lung cancer between the male and female survivors of the Japanese atomic bombings with females having an approximate three times greater risk of contracting the disease. This difference has caught the attention of the National Aeronautical Space Administration (NASA) as it addresses the risks from space radiations to male and female astronauts making space flights to Mars (Boice 2017, 2019; Boice et al. 2018).

The National Council on Radiation Protection and Measurements (NCRP) expects to publish general guidance directed at the MWS for estimating the mean absorbed dose for the primary and secondary organs considered by the International Commission on Radiological Protection (ICRP) to present the risks of stochastic diseases (ICRP 2010). A summary of the contents of this report was prepared by Bouville et al (2015). In 2018, the NCRP created Scientific Committee 6-11 to draft specific guidance for reconstructing the mean absorbed dose to the lung and selected other organs using personal monitoring results for the medical radiation worker cohort of the MWS. The Committee expects to complete its work in late 2019. The following describes the major issues and their intended resolutions identified by the Committee. These issues include the methods to assimilate decades of dosimetry data pertaining to approximately 175,000 workers who were collectively employed by thousands of hospitals, clinics, and private practices; and for whom little direct knowledge exists about the specific conditions to which a particular person might have been exposed. The conditions of exposure are critical for the dose reconstruction process in which dosimeter results must be converted into estimates of the personal dose equivalent, $H_P(10)$, followed by the conversion of $H_P(10)$ estimates into estimates of mean absorbed dose to an organ or tissue.

The effort and resources required for dose reconstruction depend largely on the availability and form of the dosimetry records. For certain groups of workers, dose information can be retrieved from national databases maintained by the US Nuclear Regulatory Commission and the US Department of Energy created as part of their national regulatory responsibilities. The Federal databases aid the effort to collect doses a particular person may have incurred at different institutions regulated by these government entities. The uses of X rays are regulated individually by each State, and thus this regulatory structure has precluded the formation of a nationwide database of doses received by medical workers. In lieu of a nationally oriented database, the MWS has had to rely on an electronic records database developed by Landauer, Inc., a supplier of personal monitoring services. The firm's history and its sizeable market share pertaining to hospitals and medical clinics have enabled it to create a nationally scoped dosimetry records repository.

Methods and discussion of key topics under consideration for dose reconstruction

Selection of the medical radiation worker cohort

An initial medical radiation worker cohort is to be assembled using very preliminary estimates of lifetime doses (quasi-lifetime doses) contained in the Landauer database. All workers having lifetime doses equal to or greater than 50 mSv are to be automatically selected. Adding to this group will be random selections of 50% of those having lifetime doses between 10 and 50 mSv and 2% of those workers having a lifetime dose less than 10 mSv. As noted below, there are large uncertainties regarding these lifetime dose estimates. The cohort is expected to total approximately 175,000 people selected from an eligible population of 1.71 million medical workers.

The source of dosimetry information

The basic information used to select the medical radiation worker cohort and reconstruct both annual and lifetime mean organ absorbed doses originates from a special digital database created by Landauer, Inc. in the mid-1980s following an informal agreement to assist with an epidemiology study of registered radiation technologists being conducted by the Radiation Epidemiology Branch of the National Cancer Institute (Boice et al. 1992; Simon et al. 2006). The database contains annual doses for 1977 and later years; although annual doses for the years 1960 through 1976 can be obtained through manual examination of microfilm copies of dosimetry reports for many workers. Missing from the epidemiological database are the individual dosimeter results that when summed create the stored annual and quasi-lifetime doses. Individual dosimeter results communicated to the client via a Dosimetry Report indicate the monitoring period (e.g. weekly, monthly, quarterly, etc.), the quantitative dose values with indications of the radiation type and energies detected, and the quarterly, annual and lifetime to date dose totals. These reports have been archived as images, not as digital data. The archived report images remain available for quality control checks and to verify assignments to workers to radiation exposure scenarios as the dose reconstruction process dictates.

Through an enrollment process, each institution assigns their radiation workers to one or more account numbers, hereinafter called accounts. Within a specific account, each worker is given a numerical identifier called a participant number. The participant number is only unique to a worker within one account number. Dosimeter results and their totals are attached to each participant at the account level. Each combination of account number and participant number represents a distinct entity for which annual and quasi-lifetime dose values are stored. The term quasi-lifetime dose refers to the fact that the lifetime dose in the database only refers to the total dose received by a worker as an enrollee in an account and does not necessarily include dose history from other accounts. Linking a worker across accounts requires the use of personal identifiers (name, birthdate, and government provided identification numbers) provided during the enrollment process. However, the provision of personal identifiers is and was not mandatory. Workers with no personal identifier were ineligible for selection into the study cohort. Those workers that could be tracked across multiple accounts may have some lifetime dose histories from prior employers counted multiple times thus artificially increasing their apparent lifetime dose. Fortunately, there exist within the database information that can help resolve these errant conditions during dose reconstruction. Those medical workers routinely receiving dose with long careers at one institution will more likely have more complete dose histories because few instances will exist of having to combine dose from multiple accounts.

Workers could be assigned to a client-defined group called a series or subaccount. The series code often reflects a department or specialty within the hospital and is useful for understanding the radiations to which workers in that series were exposed. Frequently the codes were abbreviations of departments such as RAD for radiology, NUC for nuclear medicine and CAR for cardiology.

Personal dosimetry considerations

The annual and lifetime doses in the database represent the summation of a series of personal monitoring measurements acquired over various monitoring periods. Depending on the year and dosimeter technology; employers issued personal radiation monitors to workers primarily on a weekly, monthly, or quarterly frequency; although some institutions used biweekly and bimonthly exchange periods as well. The monitoring frequency dictates the number of measurements summed to calculate the annual dose totals and thereby affects the overall uncertainty in the annual dose value; however, the medical worker dosimetry database does not indicate monitoring frequency. Monitoring frequency must be inferred from the year of the annual dose value and the predominant monitoring frequencies selected by Landauer's clients at that time. Dosimetry results from the 1950s were predominantly from weekly measurements as then recommended dose limits were expressed in terms of weekly exposures. Later years witnessed a shift from more frequent to less frequent dosimeter exchange periods as the expression of dose limits in terms of 13-week periods enabled monthly and quarterly monitoring. In addition to quarterly dose limits, a lifetime dose limit was also adopted by regulators. The 1960s contained a mix of weekly and monthly monitoring with weekly being more prevalent in the early part of the decade and monthly becoming more widely used by the end of the decade. Landauer's adoption of the more stable thermoluminescence dosimetry (TLD) technology in 1974 allowed the introduction of quarterly monitoring but the use of this frequency never overtook monthly monitoring. Quarterly monitoring did not provide radiation safety staff timely information to prevent highly exposed workers from receiving doses that exceeded the quarterly regulatory limits in force during the period from the early 1960s through the mid-1990s. Consequently, quarterly monitoring was mostly selected for the lesser exposed worker and is expected to be less prevalent among the selected medical radiation worker cohort. As the cohort selection method preferentially includes high dose workers, the Committee has assumed that 100% of annual doses received after 1970 to be from monthly monitoring.

Monitoring frequency also influences the amount of dose that could not be measured due to the dosimetry system's lower limit of detection or minimum reportable dose. Often referred to as 'missed dose', doses that do not exceed the detection limit or minimum reportable dose are given a value of zero for reporting purposes. Epidemiologists have attempted to account for the missed dose by adding an estimate of the dose that could have been received but not detected based on a presumed dose value intermediate between zero and the minimum reportable value. Prior to 1998, the minimum reportable value was 0.1 mSv (10 mrem as reported to the client) irrespective of the quantity being reported, for example, exposure, restricted dose equivalent index, personal dose equivalent. The minimum reportable dose became 0.01 mSv with the introduction of optically stimulated luminescent dosimetry (OSL) in 1998. By 2002, Landauer had converted all of its film and TLD users to OSL thus reducing the amount of dose possibly undetected. For the highly exposed members of the medical worker cohort, the missed dose does not likely contribute a significant amount of unrecorded dose because most dosimeters yielded measurements above the minimum reportable value. However, those workers having lifetime doses of 10 mSv or less probably had many instances of dosimeter results being less than the reportable threshold so that over a career of 35 years or more, the missed dose could represent a meaningful percentage of the worker's total dose. The Committee has decided that for the lowest exposed segment of the cohort, an annual value of 0.4 mSv be assumed for each annual dose record having no measurements above the minimum reportable value. This is based on an estimated median monthly dose of approximately 0.035 mSv for those receiving less than 0.1 mSv per month as assessed from recent OSL results.

An advantage of the Landauer database is the consistent approach to dosimetry practiced over time. While multiple dosimeter technologies (e.g. film, thermoluminescent dosimeters, and optically stimulated dosimeters) and dosimeter models were employed by Landauer at various times from the period 1954 through the present, the results all link back to common calibration and data analysis philosophies that kept dosimeter measurement performance relatively consistent and lessens the uncertainty arising from combining results originating from different dosimeter models compared to combining dosimeter results originating from different dosimetry services or laboratories. A discussion of measurement uncertainties can be found in NCRP Report 158 (NCRP 2007).

The quantity of each annual dose value is the one defined for regulatory compliance at that time: *exposure* (X) for years before 1985; restricted dose equivalent index (DEI) for years 1985 through 1994; personal dose equivalent ($H_P(10)$ also called deep dose equivalent) for the years after 1994 and the effective dose equivalent (H_E) for years after about 1998. Since 1998, the annual doses can be either $H_P(10)$ or H_E depending on whether the worker was involved with fluoroscopic procedures. With the exception of the personal dose equivalent (and to some extent the restricted dose equivalent index), values of *exposure* and effective dose equivalent must be converted to the personal dose equivalent in order to use the conversion factors presented in Figures 2–8. Prior to 1985, all Landauer dosimeter results reflect the quantity, *exposure*, at the surface of the body and this was considered equivalent to the dose equivalent to tissue. Using basic radiation mass energy absorption coefficients it is a relatively simple process of converting values of *exposure* to $H_P(10)$. The roentgen value measured from the dosimeter approximates $H_p(10)$, expressed here in rem for the purpose of the comparison, to within 5% for energies above 150 keV and within a maximum of 15% for energies between 50 and 150 keV.

The commencement in 1985 of dosimeter performance testing in accordance with ANSI N13.11-1983 (ANSI 1983) and the regulatory inducement for dosimetry services to become accredited by the National Voluntary Laboratory Accreditation Program caused Landauer to modify its dosimeter analysis formulas to assess the *restricted* dose equivalent index. Based on the dose equivalent index defined by the International Commission on Radiation Units and Measurements (ICRU 1971), the restricted dose equivalent index is the dose at 10 mm depth in a spherical phantom of tissue equivalent material along the central axis of the radiation beam while the dose equivalent index is the maximum dose at any depth in the sphere along the central axis. The

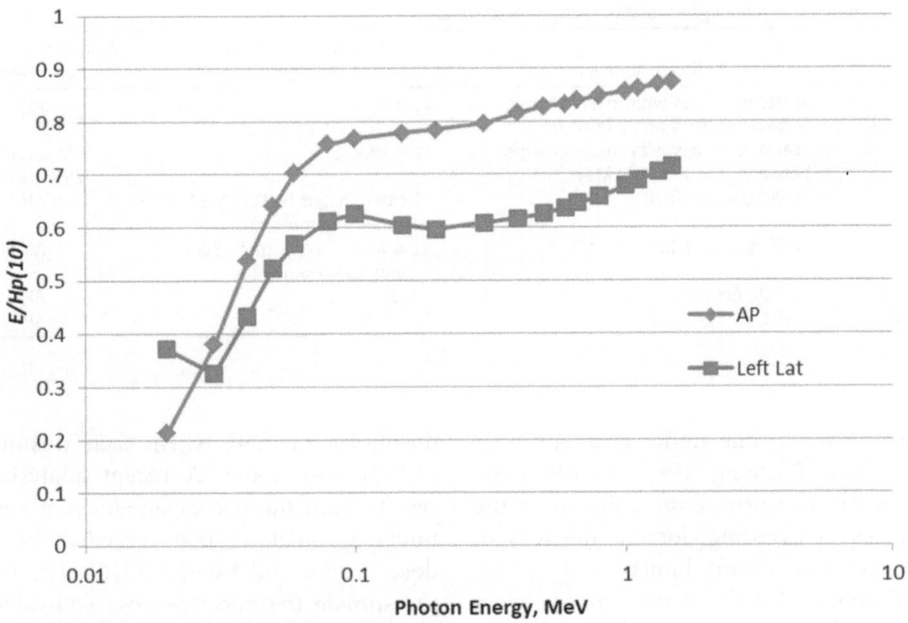

Figure 1. The ratio of Effective Dose, *E*, to the Personal Dose Equivalent, $H_P(10)$ for two irradiation geometries in the male anthropomorphic phantom. Data derived from information presented in ICRP Publications 74 and 116 (ICRP 1996, 2010).

latter quantity suffered from practical problems such as non-additivity when different radiation conditions caused maximum doses at different depths (Johns 1980). A 1995 revision to dosimeter performance testing altered the dose quantity to the deep dose equivalent or the dose at 10 mm depth along the central axis in a slab phantom of tissue equivalent material. The deep dose equivalent is functionally identical to the personal dose equivalent, $H_P(10)$ described by the ICRP (1996). The formal definition of the personal dose equivalent is the dose at a point in the body at a depth in soft tissue, not in a slab phantom; however the dose in the slab phantom composed of ICRU four-element, tissue-equivalent material (ICRU 1993) has generally been accepted as an operational surrogate. The slab phantom features more scattered radiation at the 10 mm depth and therefore, the dose in the slab for a given photon fluence or air kerma exceeds that in the sphere by about 10–15% for X rays with energies commonly encountered in diagnostic radiology and only a few percent for the higher energy gamma rays found in nuclear medicine.

In 1977, the ICRP introduced a new radiation protection quantity for use in setting dose limits, evolving in terminology from the effective dose equivalent (H_E) based on one set of tissue weighting factors to the effective dose (*E*) based on newer tissue weighting factors, (ICRP 1977, 1990, 2007). The quantity in any of its definitions is immeasurable and defined as a formulaic summation of mean organ and tissue absorbed doses adjusted for radiation quality and weighted by the relative risk an organ presents for the development of stochastic health effects. The effective dose equivalent and effective dose are not quantities that exist at a point in the body because organs and tissues are dispersed throughout and a mean absorbed dose in an organ reflects the overall deposition of energy in the entire mass of the organ; meaning that parts of the organ receive more and some parts less than the mean. Personal monitoring has been based on measuring point quantities, one of which is the personal dose equivalent, $H_P(10)$. Federal and State regulatory changes that became effective in the mid-1990s philosophically adopted the effective dose equivalent but equated it to $H_P(10)$ for external irradiation of the body for most exposure situations. Generally, $H_P(10)$ overestimates the effective dose and effective dose equivalent, particularly for areas using X rays for imaging because significant attenuation occurs in the larger organs that contribute most to the stochastic risk and are located at depths in the body much greater than 10 mm (Figure 1). This conservatism becomes excessive when $H_P(10)$ values assessed from dosimeters worn over leaded aprons are assumed to be equal to the effective dose equivalent because most of the organs contributing to stochastic disease risk are located in the torso and thus shielded. To some extent, this overly conservative approach to estimating effective dose has been practically rectified by various proposals to adjust the measured values of $H_P(10)$ as discussed below.

Regulations and regulatory guidance dictate the conduct of personal monitoring. When personal dosimeters are properly used, monitoring relies on the assumption that the same radiation field is incident on the dosimeter and worker at the same time for the same duration. This permits the dose value measured by the dosimeter to be considered the same as that received by the person. Regulations have required dosimeters be located at the highest point of dose on the whole body (Extremities are separately monitored). In most situations when the radiation source is in front of the body and no protective apron is used, it has been acceptable to wear dosimeters on the chest or abdomen. When a protective apron is used, the torso is shielded so that the head becomes the point of highest dose. For practical reasons, dosimeters came to be worn over the apron at the collar; although this has not been a universal practice with some radiation safety officials preferring to have the dosimeters worn under the apron and some electing to

Table 1. Characteristics and assumptions pertaining to radiation exposure scenarios.

Scenario	Radiation sources	Assumed weighted energy	Assumed irradiation geometry	Assumed protection
General radiology	Scattered X rays with mean energies between 0.04 and 0.06 MeV	0.05 MeV	AP	Barriers/leaded aprons
Fluoroscopically guided interventions	Scattered X rays with mean energies between 0.04 and 0.06 MeV	0.05 MeV	75% AP 25% Lateral	Leaded aprons
Nuclear medicine before 2000	Tc-99m, I-131, Tl-201	Mix dominated by 0.14 MeV Tc-99m gamma ray	AP	None
Nuclear medicine after 2000	Tc-99m, F-18, I-131	75% associated with Tc-99m and 25% from F-18	AP	None
Radiation oncology before 1970	Ra-226, Au-198, Cs-137, Co-60	1 MeV	AP	None
Radiation oncology after 1970	Ir-192, Cs-137, I-125	0.35 MeV	AP	None

require two dosimeters be worn, one under and one over the apron (Brateman 1989; Bushong 1989; NCRP 1989; Simon et al. 2006). Dosimeters worn at the collar over the apron also provide means of assessing dose to the lens of the eye, a tissue with separate regulatory limits.

As noted earlier, a dosimeter located at the collar above a leaded apron provides an inappropriate estimate of the effective dose equivalent. Numerous medical and health physicists have suggested formulas that translate a dosimeter measurement made over the apron or a combination of dosimeter measurements, one made under and one made over the apron into an estimate of effective dose equivalent (Webster 1989; Niklason et al. 1993, NCRP 1995). There has been a continued debate both nationally and internationally as to the preferred approach to monitor medical staff in the fluoroscopy suite; one dosimeter over the apron or two dosimeters, one under and one over the apron; with the ICRP favoring the latter (ICRP 2018). In the late 1990s, State regulatory agencies began to allow doses for medical staff involved with fluoroscopically guided procedures to be reported as the effective dose using the NCRP recommended formulas (NCRP 1995). Equation 1 is the formula recommended for a single dosimeter worn over the apron at the collar and Equation 2 is the formula for two dosimeters, one worn over the apron at the collar and one worn under the apron on the chest.

$$\text{Effective dose equivalent} = 0.3 \times H_P(10)_{\text{over}} \quad (1)$$

$$\text{Effective dose equivalent} = 0.04 \times H_P(10)_{\text{over}} + 1.5 \times H_P(10)_{\text{under}} \quad (2)$$

$H_P(10)_{\text{over}}$ is from the dosimeter located over the apron while $H_P(10)_{\text{under}}$ is from the dosimeter worn under the apron.

Landauer enabled its clients to select either monitoring method for those identified by the client as working in fluoroscopic situations. Radiation safety officers have demonstrated a preference to monitor fluoroscopic staff using the one dosimeter method. A recent review of Landauer monitoring results for 2012 found that 70% of fluoroscopic staff were monitored with the one dosimeter method (Yoder and Salasky 2016). It should be noted that when no dose is measured under the apron, the effective dose equivalent for the two dosimeter monitoring protocol reduces to 0.04 $H_P(10)_{\text{over}}$ or about 13% of the dose estimated for the one dosimeter method when each dosimeter yields the same $H_P(10)$ dose value. A recent analysis of 2012 monitoring results from the two dosimeter method found only 11% had under apron doses that exceeded the minimum measurable dose (Yoder and Salasky 2016). The multiple accepted ways to estimate the effective dose equivalent introduces complications for the dose reconstruction process because the dosimetry database does not distinguish which method of computing effective dose equivalent was selected for a person. Prior to 2000, the predominant method for monitoring staff wearing leaded aprons was with one dosimeter located over the apron at the collar with no adjustment of the measured $H_P(10)$ value. After 2000, the database contains a mix of effective dose equivalent analysis results. The difficulty from the two dosimeter approach is that neither value of $H_P(10)$ can be predicted from only the effective dose equivalent value because numerous combinations of $H_P(10)_{\text{over}}$ and $H_P(10)_{\text{under}}$ can result in the same dose value.

In summary, the annual dose values represent different dosimetry quantities depending on the year and regulations in force at that time. Lifetime dose values were not recomputed as the quantities changed; therefore, the lifetime whole body dose estimates in the database equal the sum of the annual doses without respect to the measured quantity. This further limits the utility of the lifetime dose values in the database other than giving a relative ranking to how much dose an individual might have received during their employment history.

Radiation exposure scenarios for medical radiation workers

Many medical procedures involve the use of radiation but these can be distilled into six general exposure scenarios: general diagnostic radiology, fluoroscopically guided procedures, nuclear medicine up through 1999, nuclear medicine from 2000 to the present, radiation oncology up through 1969, and radiation oncology from 1970 to the present (Table 1). These can be described as either low energy photon exposure situations (predominant energies less than 0.1 MeV) or high energy photon energy situations (predominant energies above 0.1 MeV). The demarcation on energy corresponds to the likelihood of the use of leaded aprons and the transition zone at which the conversion factors that relate $H_P(10)$ to mean organ absorbed dose become less energy dependent as discussed later. The time separation for nuclear medicine arises from the introduction of positron emission tomography that mostly relies on fluorine-18 and the decay of the emitted positron into 0.511 MeV photons. Prior to 2000, nuclear medicine procedures were mostly conducted with technicium-99m and to a lesser extent iodine-131 and thallium-204.

The time division for radiation oncology is based on changes in the sources used in brachytherapy and the shift from cobalt-60 to high energy linear accelerators and cyclotrons for teletherapy. The time distinction has minimal impact on dose reconstruction as all radiation oncology that presents an opportunity for significant worker dose would involve high energy photons for which dose conversion factors exhibit little change with increasing energy.

With the exception of fluoroscopically guided procedures, all scenarios assume that 100% of the measured doses were predominantly received with the worker facing the source so that the radiations traversed the body in an anterior to posterior (AP) direction. For fluoroscopic exposures, 75% of the dose is assumed to be received in the AP direction with the remainder from the sides (left lateral, LLAT and right lateral, RLAT geometries). The left side is expected to be preferentially exposed for physicians performing many fluoroscopically guided interventional procedures based on their normal position adjacent to the right side of the patient near the thigh with the physician's body oriented toward the video image displays; although this is not universal. The fluoroscopic and nuclear medicine scenarios are expected to be the most common for the workers having the highest doses. The general radiology scenario presents little risk of high dose because operators of the X ray unit usually stand in shielded control areas or at some distance from the X ray unit during image acquisition. Many radiation workers fall into this category and it is expected that a majority of the randomly selected workers for the MWS cohort with lifetime doses less than 50 mSv will fall into this exposure scenario. A similar situation exists for radiation oncology as radiation therapy staff are not present in the treatment room during irradiation of the patient. An exception is possible prior to the adoption of high dose rate brachytherapy when staff would be present during the insertion and retraction of the radioactive seeds.

Medical staff exposed during nuclear medicine and radiation therapy procedures are assumed not to have worn leaded aprons. The energy of radiations is such that leaded aprons provide little dose reduction. General radiology workers are also assumed not to have worn leaded aprons.

Assigning workers to radiation exposure scenarios

The impracticality of contacting over a thousand institutions to gain information about exposures to workers acquired over many years dictates the need to use account and series code information in the Landauer database to infer the exposure scenario for each member of the cohort for each year of their work. Selected examination of archived dosimetry reports will provide a means to verify the computer assignment routines; in particular, the energies of radiation detected by the dosimeters and whether one or two dosimeters were issued to those using leaded aprons. In addition, database information will be compared with the stored records maintained by a few large hospitals as an alternate, more precise approach to verify the correctness of the assignments and dates of service.

A customer category code assigned by Landauer enables hospitals, small clinics, private practices, and other medical facilities to be separately identified from non-medical facilities. Hospitals and larger clinics are most likely to have most if not all of the exposure scenarios. Fluoroscopically guided interventions are assumed to be constrained to hospitals only. Small clinics will typically be limited to general radiology, nuclear medicine and/or radiation oncology. The institution's name frequently provides clues as to the type of medical procedures practiced. The series code to which workers are frequently associated in an account can provide information about the type of radiation procedures performed by the department or location when multiple environments may exist. When a code does not indicate a medical specialty, dosimetry reports along with the energy detected by dosimeters in the series can be used to identify the scenario most likely to be found.

The process to normalize annual dose values to $H_P(10)$

The ICRP has published the results of recent calculations (particularly Monte Carlo simulations) to generate absorbed dose distributions in anthropomorphic phantoms for monoenergetic photons incident on the body in various geometries and compiled the results to create tables of conversion coefficients for use in estimating the mean organ absorbed dose (D_T) in various organs per unit of radiation fluence and kerma in air (ICRP 2010). The ICRP has published similar conversion factors that relate fluence and kerma in air to $H_P(10)$ (ICRP 1996). With D_T and $H_P(10)$ each being related to common fundamental dosimetry quantities, it becomes a straightforward process to compute conversion factors that directly relate $H_P(10)$ to D_T for the various irradiation geometries.

The Committee advising the MWS epidemiologists has specified the following adjustments to the annual dose values in the Landauer database so that all values are expressed as $H_P(10)$. Values of mR should be multiplied by 0.9 for radiation scenarios involving lower energy photons (mean energy of 0.05 MeV) and 0.95 for higher energy photons (energies above 0.1 Mev). Values of the restricted dose equivalent index should be multiplied by 1.1 for the scenarios involving lower energy photons and 1.03 for higher energy photon scenarios based on comparisons of the data published in ANSI N13.11-1983 (1983) for the restricted dose equivalent index and in a revised version of the ANSI standard (ANSI 2001) for the slab phantom. With the exception of annual doses for fluoroscopy workers that are stored as the effective dose equivalent, all other annual dose values should be considered as the quantity, $H_P(10)$;

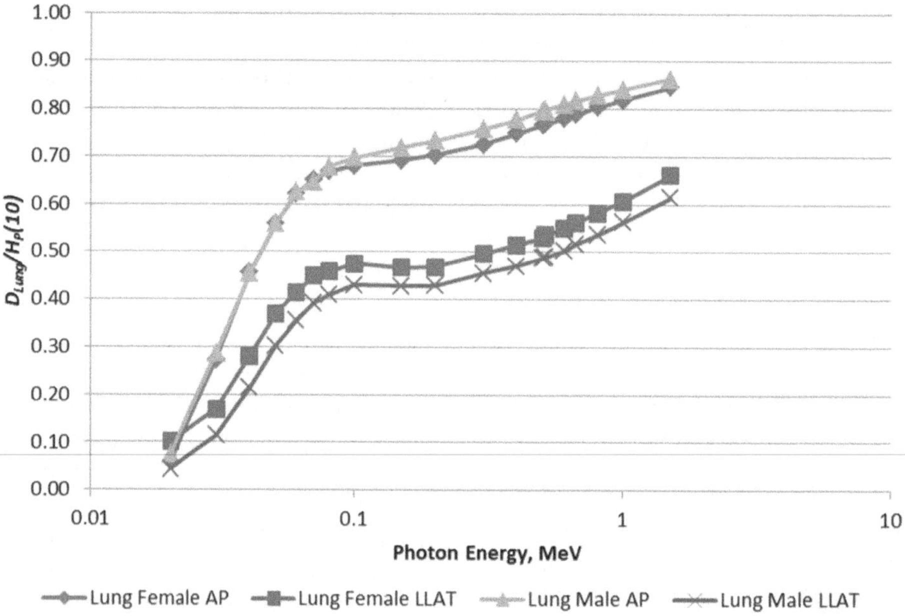

Figure 2. The mean absorbed dose to the lung for males and females for two irradiation geometries. Data derived from information presented in ICRP Publications 74 and 116 (ICRP 1996, 2010).

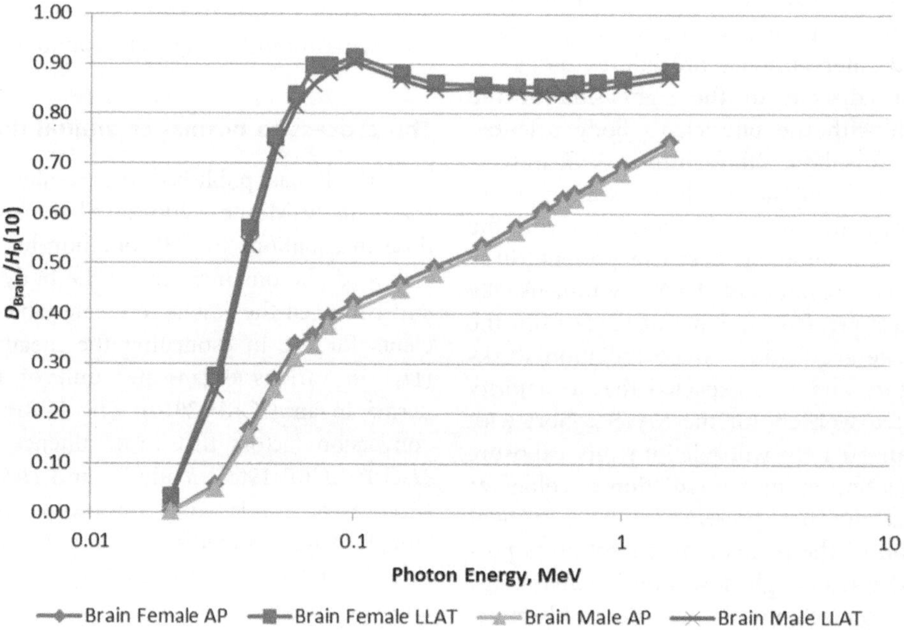

Figure 3. The mean absorbed dose to the brain per $H_P(10)$ for males and females for two irradiation geometries. Data derived from information presented in ICRP Publications 74 and 116 (ICRP 1996, 2010).

although adjustments for the use of a leaded apron must be employed.

The use of leaded aprons introduces the greatest challenges for dose reconstruction. After adjusting for the measured quantity, the dose reconstruction process aims to establish an annual value for $H_P(10)$ that would have been measured on the body (e.g. the chest) as if no apron had been present. Before 2000 when the effective dose equivalent formulas were sparsely used, the annual dose values represent a dose measured above the apron. The measured value of $H_P(10)$ is appropriate for the head region but must be adjusted for the torso organs.

The adjustment for the effects of the leaded apron can be inferred from an assessment of ratios of dosimeters worn over the apron to those worn under the apron. This ratio largely depends on the attenuation characteristics of the apron but also any spatial or geometrical differences in the dose rates at the locations of the dosimeters, differences in dosimeter performance when the dosimeters are not directed at the same angle toward the source and from the change in dose per unit fluence due to shifts in the energy spectra that occur from photon absorption and scattering in the apron. Apron thicknesses may vary from 0.25 mm of lead equivalent to 0.5 mm and the placement of a dosimeter over the

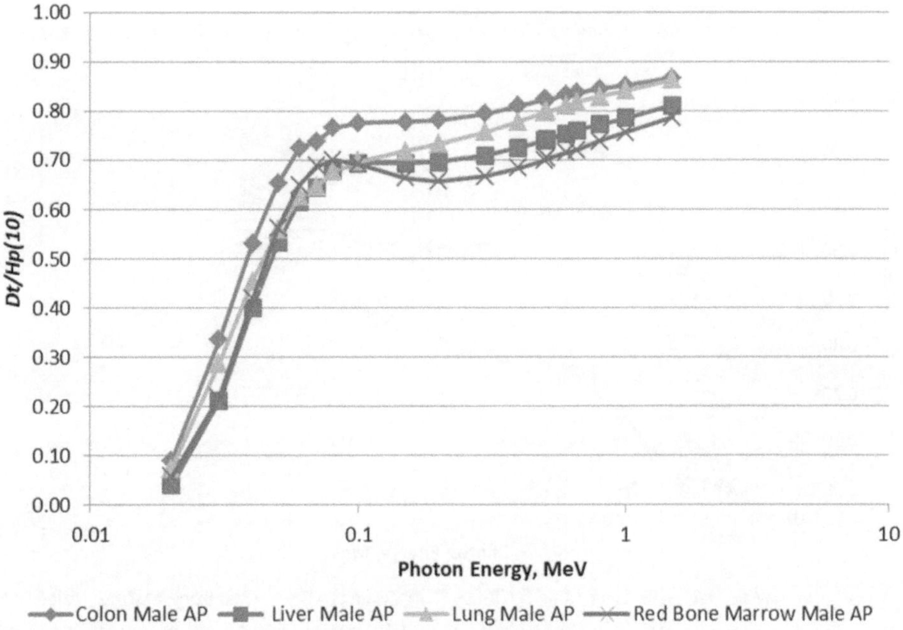

Figure 4. The mean absorbed dose to four tissues per $H_P(10)$ for the male body in the AP geometry. Data derived from information presented in ICRP Publications 74 and 116 (ICRP 1996, 2010).

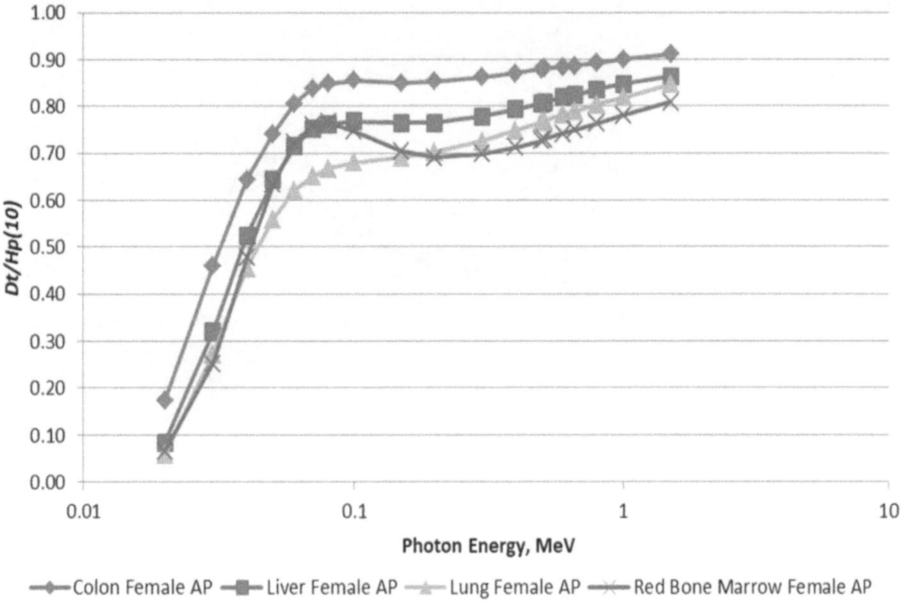

Figure 5. The mean absorbed dose to four tissues per $H_P(10)$ for the female body in the AP geometry. Data derived from information presented in ICRP Publications 74 and 116 (ICRP 1996, 2010).

apron is not likely to be in the same place as that worn under the apron. Over apron dosimeters tend to be worn high on the body at the collar level and under apron dosimeters worn on the chest or even waist. Yoder and Salasky (2016) observed over to under dosimeter dose ratios that ranged from 4 to 200 for fluoroscopically-based procedures. The range was influenced by the very low doses measured under the apron but when only under apron dose values above 0.1 mSv were considered, the median ratio was 7.6. These ratios shift when annual summations of over and under apron measurements are examined because many under apron dosimeter values observed during the year failed to exceed the 0.02 mSv detection limit for the dosimeter. The median ratio for annual doses was 17 when the annual under apron dose exceeded 0.1 mSv. Combining this information with calculated attenuation factors for 0.5 mm of lead for 50 keV photons, the Committee has tentatively decided to use a 95% dose reduction factor created by the apron so that all over apron dose values will be multiplied by 0.05.

Relating $H_P(10)$ to mean organ absorbed dose for the lung, brain and other select organs

ICRP Publication 116 presents a series of tables of conversion factors that relate air kerma or photon fluence to

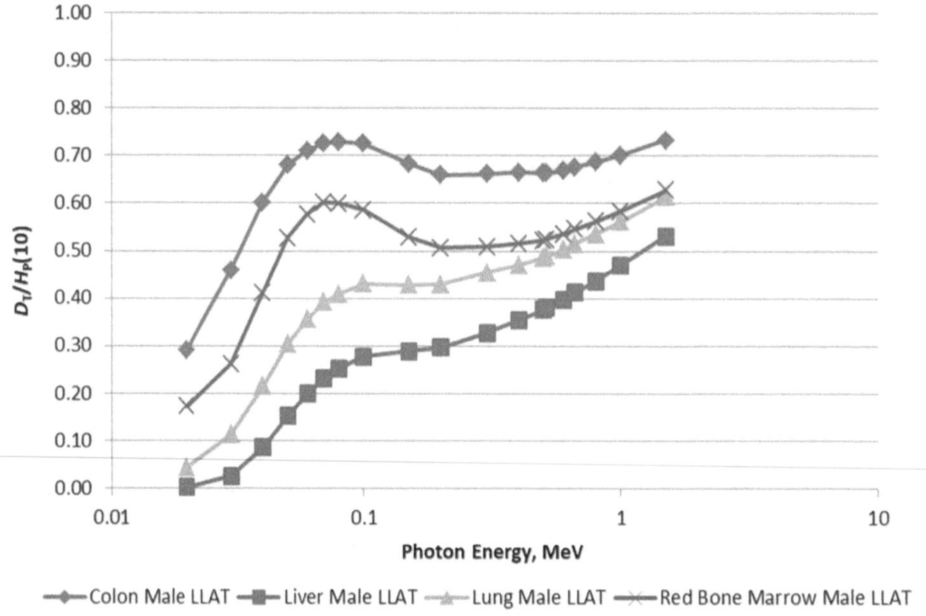

Figure 6. The mean absorbed dose to four tissues per $H_P(10)$ for the male body in the Left Lateral geometry. Data derived from information presented in ICRP Publications 74 and 116 (ICRP 1996, 2010).

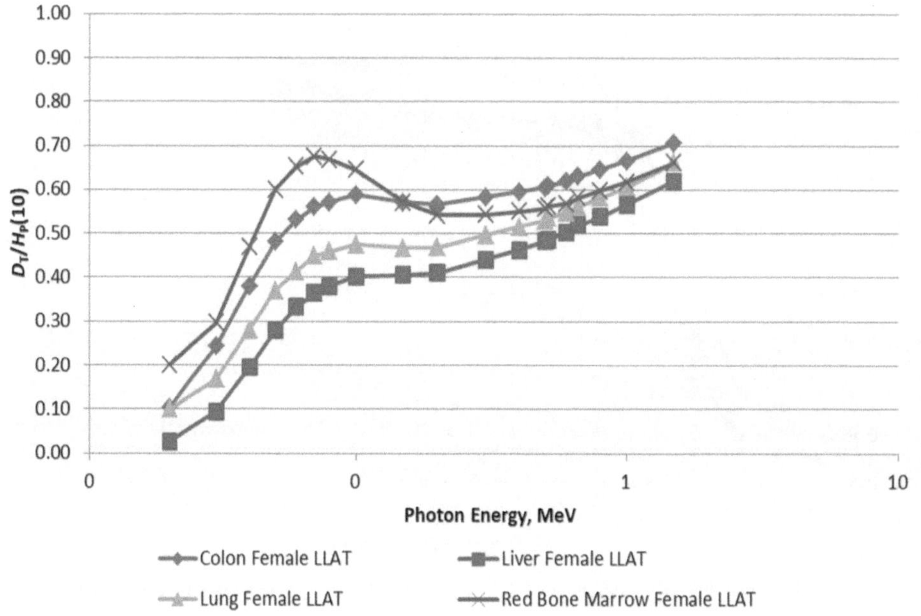

Figure 7. The mean absorbed dose to four tissues per $H_P(10)$ for the female body in the Left Lateral geometry. Data derived from information presented in ICRP Publications 74 and 116 (ICRP 1996, 2010).

effective dose and the mean absorbed dose in 13 primary and 12 secondary or remainder organs and tissues in the male and female body for several irradiation geometries (ICRP 2010). Similarly, ICRP Publication 74 presents conversion factors that relate air kerma and photon fluence to $H_P(10)$ (ICRP 1996). Using both sets of conversion factors enables the derivation of conversion factors that directly relate $H_P(10)$ to effective dose and mean organ absorbed dose. Examples of such conversion factors are plotted in Figures 1–8.

Studies of the survivors of the Hiroshima and Nagasaki atomic bombings have indicated a difference between males and females in the risk of developing lung cancer from exposure to radiation. The radiobiological basis for this risk difference is not clear and NASA has expressed interest to the NCRP in determining whether lung cancer incidence differs between men and women occupationally exposed to radiation as part of the risk assessment for astronauts making an expedition to Mars (Boice 2017). In addition, some reports have examined the incidence of brain cancers in physicians performing fluoroscopically guided interventional procedures (Roguin et al. 2013) and in radiological technologists (Kitahara et al. 2017). The size of the medical worker cohort may shed information about the incidence of brain cancer and other effects on the central nervous system. Leukemia has been of historical interest stemming from

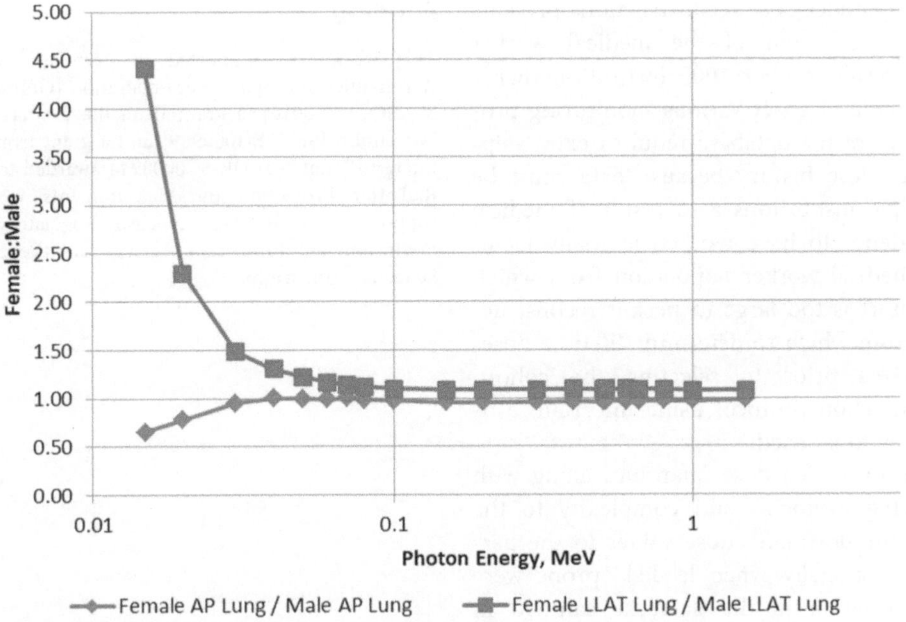

Figure 8. The ratio of the female to male mean absorbed dose to the lung in the AP and Left Lateral geometries. Data derived from information presented in ICRP Publications 74 and 116 (ICRP 1996, 2010).

Table 2. Conversion factors presently considered to relate $H_P(10)$ to mean absorbed dose to the male and female lung and brain.

Scenario	Male lung	Female lung	Male brain	Female brain
General radiology	0.56	0.56	0.56	0.56
Fluoroscopically guided interventions	0.50	0.51	0.5	0.51
Nuclear medicine before 2000	0.72	0.69	0.45	0.46
Nuclear medicine after 2000	0.78	0.75	0.57	0.58
Radiation oncology before 1970	0.84	0.82	0.68	0.69
Radiation oncology after 1970	0.77	0.74	0.54	0.55

early regulations that limited exposure of the red bone marrow. The dose–response for cancer incidence in other organs is an ongoing subject behind the linear no threshold hypothesis (LNT) (NCRP 2018). Figures 2 and 3 present the conversion factors for the lung and brain, respectively as these have been stated to be of most interest to the MWS. Figures 4–7 present information for three organs found in the torso and the red bone marrow that is more widely distributed within the body. These figures demonstrate the efficacy of assessing mean organ doses to other organs once values for $H_P(10)$ have been established. All figures present conversion factors for both sexes. Figure 8 plots the ratio of the female to male conversion factors for the lung in the AP geometry.

Generally, the AP geometry creates the largest mean organ absorbed dose of the torso organs considered in this review; however, the lateral irradiation geometry leads to larger mean absorbed doses to the brain. Organ and body morphology as well as the structure and composition of tissues intervening between the surface of the body and the organ of interest influence the conversion factors as demonstrated by the lung and brain plots. The conversion factors relating $H_P(10)$ to D_T vary significantly for photons with energies below 0.1 MeV; however, the range of effective energies encountered with the use of X rays that lead to higher worker doses ranges between 0.04 and 0.06 MeV with 0.05 MeV being selected as most typical. Above 0.1 MeV, the conversion factors become less influenced by energy allowing less precision in defining the specific photon energies involved with the higher energy exposure scenarios.

The conversion factors presented in the figures pertain to mono-energetic photons and weighting is required to derive conversion factors applicable to the work environment. Based on the radiation scenario information presented in Table 1, starting values of the energy-weighted conversion factors for the six scenarios are presented in Table 2. The conversion factor for the fluoroscopically guided interventions has been weighted by the relative proportions of the dose received from the AP (75%) and LAT (25%) geometries. The energy-weighted conversion factors presented in Table 2 indicate the relative sensitivity of defining the effective energies. For the lung, the conversion factors increase by about 50% from the X ray to the nuclear medicine environments and for the brain, the change is even less.

The conversion factors represent upper bounds on the organ absorbed dose because the models on which the ICRP data were based cause the body to be uniformly irradiated by a parallel and aligned radiation field. This is an idealized condition not encountered in medical work environments. Most work environments will have spatial variations in dose rate causing the body to be unevenly irradiated. Such uneven irradiation may alter the absorbed energy distribution in an organ. Spatial variations combined with dosimeters being worn at the higher dose points on the body suggest that the conversion factors derived from the ICRP data could over-estimate by an unknown amount the true mean absorbed dose to an organ.

Conclusions

The use of the Landauer database of annual and lifetime doses as the primary tool to select the cohort of medical radiation workers and conduct dose reconstructions presents special challenges. The selection of the medical worker cohort is expected to result in over 1000 institutions being represented with each able to enact various monitoring protocols. The file structure of the database requires extra steps to develop a complete dose history because data must be assembled from multiple institutions as a result of medical workers having a tendency to have worked at many facilities. The size of the medical worker population from which to select the study cohort is too large to permit reconstruction of annual doses from which to determine lifetime doses for all medical workers prior to selecting the cohort. Therefore, an initial selection protocol using uncertain estimates of lifetime dose is to be used.

The changes in the measured dose quantities along with variations in monitoring protocols add complexity to the process of converting the dosimeter dose values to the personal dose equivalent; especially when leaded aprons were used. Descriptive information in the database can aid in placing workers into a radiation exposure scenario but additional information from other databases and dosimetry reports will likely be necessary to separate those workers who used a leaded apron from those that did not. The effect of photon energy is secondary to the effect of the leaded apron for estimating the mean absorbed dose to the lung.

Once a worker has been associated with a radiation exposure scenario, the process of converting annual values of $H_P(10)$ to mean organ absorbed doses will be relatively straightforward. There is extensive knowledge about the predominant radiations and their energies that workers normally encounter in a medical facility. The regulatory influence to wear dosimeters at the point of highest dose imparts an unknown bias toward over-stating organ doses.

The Landauer database enables an efficient means of studying the medical radiation worker population. An element of consistency is achieved because the dosimetry calibration approaches used by Landauer have been constant over many years and the data is in a standard electronic format that avoids manual data entry from archived files. The Landauer database contains annual dose information back to 1977 and earlier information is available from microfilmed images of dosimetry reports that go back to the early 1960s. The cost to manually retrieve and digitize dose records has been a significant cost for large epidemiological studies such as the MWS. The use of the Landauer database is one approach being used to mitigate these costs.

Disclosure statement

The author's report no conflicts of interest. RCY was employed by Landauer Inc. who provided the database being used by the NCRP for the Million Person Study. CNP is currently employed by Landauer, Inc. The work in preparing this paper is entirely voluntary, entails no financial implications and only involves the scientific interpretation and application of the radiation measurement data acquired many years ago by Landauer.

Funding

This work was supported in part by grants from the National Aeronautics and Space Administration [Grants No. NNX15AU88G and 80NSCC17M0016], a grant from the Centers for Disease Control and Prevention [5EU1EH000989] and a grant from the U.S. Department of Energy [Grant No. DE-SC0008944 awarded to the National Council on Radiation Protection and Measurements, which included interagency support from the U.S. Nuclear Regulatory Commission, the U.S. Environmental Protection Agency and the National Aeronautics and Space Administration].

References

[ANSI] American National Standards Institute. 1983. American National Standard for dosimetry – personnel dosimetry performance – criteria for testing. ANSI N13.11-1983. New York: American National Standards Institute.

[ANSI] American National Standards Institute. 2001. American National Standard for dosimetry – personnel dosimetry performance – criteria for testing. ANSI N13.11-2001. New York: American National Standards Institute.

Boice JD. Jr. 2017. Space: the final frontier-research relevant to Mars. Health Phys. 112:392–397.

Boice JD. Jr. 2019. Relevance of the Million Person Study to research needs for NASA and space exploration. Int J Radiat Biol. Submitted.

Boice JD, Jr, Ellis ED, Golden AP, Girardi DJ, Cohen SS, Chen H, Mumma MT, Shore RE, Leggett RW. 2018. The past informs the future: an overview of the Million Worker Study and The Mallinckrodt Chemical Works Cohort. Health Phys. 114:381–385.

Boice JD, Jr, Mandel JS, Doody MM, Yoder RC, McGowan R. 1992. A health survey of radiologic technologists. Cancer. 69:586–598.

Bouville A, Toohey RE, Boice JD, Jr., Beck HL, Dauer LT, Eckerman KF, Hagemeyer D, Leggett RW, Mumma MT, Napier B, et al. 2015. "Dose reconstruction for the Million Worker Study: Status and guidelines". Health Phys. 108:206–220.

Brateman L. 1989. A plea for sensible film badge placement in diagnostic radiology. Health Phys. 56:567–568.

Bushong SC. 1989. Personnel monitoring in diagnostic radiology: revisited – again! Health Phys. 56:565–566.

[ICRP] International Commission on Radiological Protection. 1977. Recommendations of he International Commission on Radiological Protection. ICRP Publication 26. Ann. ICRP 1(3). ICRP.

[ICRP] International Commission on Radiological Protection. 1990. 1990 recommendations of the International Commission on Radiological Protection. ICRP Publication 60. Ann. ICRP 21(1–3). ICRP.

[ICRP] International Commission on Radiological Protection. 1996. Conversion coefficients for use in radiation protection against external radiation. ICRP Publication 74. Ann. ICRP 26(3-4). ICRP.

[ICRP] International Commission on Radiological Protection. 2007. The 2007 recommendations of the International Commission on Radiological Protection. ICRP Publication 103. Ann ICRP 37(2-4).

[ICRP] International Commission on Radiological Protection. 2010. Conversion coefficients for radiological protection quantities for external radiation exposures. ICRP Publication 116. Ann. ICRP 40(2-5). ICRP.

[ICRP] International Commission on Radiological Protection. 2018. Occupational radiological protection in interventional procedures. ICRP Publication 139. Ann. ICRP 47(2). ICRP.

[ICRU] International Commission on Radiation Units and Measurements. 1971. Radiation Quantities and Units. ICRU Report 19. Bethesda (MD): ICRU.

[ICRU] International Commission on Radiation Units and Measurements. 1993. Quantities and units in radiation protection dosimetry. ICRU Report 51. Bethesda (MD): ICRU.

Johns TF. 1980. Dose equivalent index and personal dosimetry. Health Phys. 39:92–95.

Kitahara CM, Linet MS, Balter S, Miller DL, Rajaraman P, Cahoon EK, Velazquez-Konen R, Simon SL, Little MP, Doody MM, et al. 2017. Occupational radiation exposure and deaths from malignant intracranial neoplasms of the brain and CNS in R.S. radiologic technologists, 1983-2012. AJR. 208:1278–1284.

[NCRP] National Council on Radiation Protection and Measurements. 1989. Radiation protection for medical and allied health personnel. NCRP Report 105. Bethesda (MD): NCRP.

[NCRP] National Council on Radiation Protection and Measurements. 1995. Use of personal monitors to estimate effective dose equivalent and effective dose to workers for external exposure to low-LET radiation. NCRP Report 122. Bethesda (MD): NCRP.

[NCRP] National Council on Radiation Protection and Measurements. 2007. Uncertainties in the measurement and dosimetry of external radiation. NCRP Report 158. Bethesda (MD): NCRP.

[NCRP] National Council on Radiation Protection and Measurements. 2018. Implications of recent epidemiologic studies for the linear non-threshold model and radiation protection. NCRP Commentary 27. Bethesda (MD): NCRP.

Niklason LT, Marx MV, Chan H-P. 1993. Interventional radiologists: occupational doses and risks. Radiology. 187:729–733.

Roguin A, Goldstein J, Bar O, Goldstein JA. 2013. Brain and neck tumors among physicians performing interventional procedures. Am J Card. 111:1368–1372.

Simon SL, Weinstock RM, Doody MM, Neton J, Wenzl T, Stewart P, Mohan AK, Yoder RC, Hauptmann M, Freedman DM, et al. 2006. Estimating historical radiation doses to a cohort of U.S. radiologic technologists. Rad Res. 166:174–192.

Webster EW. 1989. EDE exposure with protective aprons. Health Phys. 56:568–569.

Yoder RC, Salasky M. 2016. A review of two methods used in the USA to assess H_E during fluoroscopic-based radiology. Radiat Prot Dosimetry. 170:307–310.

MPS dose reconstruction for internal emitters: some site-specific issues and approaches

Richard W. Leggett, Keith F. Eckerman, and Michael Bellamy

ABSTRACT

Background: As part of the Million Person Study (MPS), dose reconstructions for internal emitters have been performed for several U.S. facilities where large quantities of radionuclides were handled. The main challenges and dominant sources of potential error in retrospective dose estimates for internally exposed workers have been found to vary from site to site. This article discusses some important issues encountered in dose reconstructions performed for selected MPS sites and the approaches used to address those issues. The focus is on some foundational components of retrospective dose assessments that have received little attention in the literature.

Methods: The discussion is built around illustrative exposure data and dose reconstructions for workers at selected facilities addressed in the MPS. Related findings at some non-MPS sites are also discussed.

Results: Each of the following items has been found to be a major source of potential error in reconstructed tissue doses for some MPS sites: identification of all dosimetrically important internal emitters; the time pattern of intake; the mode(s) of intake; reliability of bioassay measurements; application of surrogate (coworker) information in lieu of, or in conjunction with, worker-specific monitoring data; the chemical and physical forms of inhaled radionuclides; and the relation of air monitoring data to actual intake.

Conclusions: (1) Much of the dose reconstruction effort for internal emitters should be devoted to development of best feasible exposure scenarios. (2) Coworker data should be used to assign exposure scenarios or dose estimates to workers with missing exposure data only if there is compelling evidence of similar coworker exposure. (3) Bioassay data for some radionuclides and periods of operation at MPS sites are of questionable reliability due to sizable uncertainties associated with contamination, recovery, or background issues. (4) Dose estimates derived solely from air monitoring data should be treated as highly uncertain values in the absence of site-specific information demonstrating that the data are reasonably predictive of intake. (5) For intakes known or assumed to be via inhalation, the uncertainty in lung dose typically is much greater than the uncertainty in dose to systemic tissues, when dose estimates are based on urinary excretion data. (6) The lung dose estimate often can be improved through development of site-specific respiratory absorption parameter values. (7) There is generally insufficient site-specific information to justify development of site-specific systemic models.

Introduction

As part of the Million Person Study (MPS), dose reconstructions have been performed for several U.S. facilities where large quantities of radionuclides were handled. Operations at some of these facilities began in the 1940s and continued for several decades, and some of the facilities are still in operation. Thousands of workers at these facilities were exposed to elevated levels of external radiation, and a generally smaller but considerable number of workers had elevated intake of radionuclides, particularly in the early years of operation. The MPS dose reconstructions have provided yearly dose estimates for workers exposed from external or internal sources, as well as broad assessments of the sources and extent of uncertainties in the reconstructed doses.

MPS estimates of external dose to a worker typically are based on personal dosimeter readings recorded over the worker's career. A personal dosimeter reading generally represents an operational dose quantity needed to demonstrate compliance with regulatory limits and involves nontrivial uncertainties when extrapolated to tissue doses (NCRP 2018). Nevertheless, dosimeter readings are expected to provide meaningful and reasonably complete records of external exposure to most radiation workers.

Monitoring data from which to reconstruct doses from internal emitters at MPS sites generally are more uneven and difficult to interpret than data for external radiation sources. Bioassay data are the preferred monitoring data for dose reconstructions, but often are sparse or absent for workers with high potential for intake of radionuclides. Also, interpretation of bioassay data often relies on incomplete information on the nature of the exposure such as the time pattern of intake, mode(s) of intake, and the chemical and physical form of the internally deposited radionuclide(s). Air monitoring data are often available for specific work areas, but are crude and often unreliable predictors of the level of intake of airborne radionuclides. In some cases, no monitoring data are available for workers with the potential for elevated intake of radionuclides.

For such reasons, a reasonably complete and meaningful set of retrospective dose estimates for internal emitters for an MPS site requires a carefully designed study built on a wider set of site-specific information than bioassay or air monitoring records. The study should begin with a detailed review of the site history including specific operations, materials handled, incident reports, worker-specific records, and, if feasible, interviews of former workers. This type of information is needed to construct reasonably realistic exposure scenarios for individual workers or groups of workers, to assess the need for site-specific or worker-specific biokinetic models, and to gauge the validity of applying coworker data to potentially exposed workers with sketchy or missing records. A critical review of bioassay measurement techniques over time is also needed to assess the reliability of these data. These early steps in the dose reconstruction process provide the foundation for interpreting monitoring data, selection of appropriate biokinetic models for exposures to individuals or groups, and the feasibility of characterizing site-specific or worker-specific respiratory retention and absorption for forms of radionuclides whose bioassay data are not adequately reproduced by standard models.

This article discusses and illustrates the importance of these foundational steps in the dose reconstruction process in the MPS, some differences between sites in the nature of the information available to develop exposure scenarios or derive dose estimates, and some site- or worker-specific models and methods that have been applied in MPS dose reconstructions. The following topics are addressed:

- Methods of developing exposure scenarios and potential errors arising from different components of an exposure scenario
- Improved dose estimates through development of site-specific biokinetic models
- Reliability of bioassay data
- Reliability of air monitoring data

This article does not address the following topics concerning retrospective dose assessments for internal emitters that have been addressed frequently in the literature: methods of back-calculation of radionuclide intakes based on model fits to bioassay data, uncertainties in predictions of biokinetic models used in radiation protection or dose reconstruction, and uncertainties in dose estimates derived from bioassay data. The reader is referred to the following articles and reports for discussions and bibliographies on those topics (Leggett et al. 1998, Harrison et al. 2001, 2002; Leggett 2001; 2007; Boice et al. 2006; Bess et al. 2007; Pawel et al. 2007; NCRP 2009, 2018; Poudel et al. 2018).

Most of the illustrations in this article come from dose reconstructions performed for four MPS sites referred to as Rocketdyne (Boice et al. 2006), Mound (Boice et al. 2014), Mallinckrodt (Ellis et al. 2018), and Los Alamos National Laboratory (LANL) (Wiggs et al. 1994). Some illustrations from related sites studied by the authors or addressed in the literature are also provided.

Methods

This article consists largely of illustrations of the types, quantity, and quality of data available to derive retrospective dose estimates for internal emitters and some of the methods and models used to derive those estimates. The illustrations are taken mainly from monitoring data or other information collected for selected MPS sites. Four MPS sites are featured:

- Rocketdyne nuclear facility, near Los Angeles, California,
- Mound nuclear facility, Miamisburg, Ohio,
- Mallinckrodt Chemical Works, St. Louis, Missouri,
- LANL, Los Alamos, New Mexico.

The dose reconstructions for Rocketdyne, Mound, and Mallinckrodt have been completed and are summarized in the open literature (Boice et al. 2006, 2014, 2018; Ellis et al. 2018) and in a report of the National Council on Radiation Protection and Measurements (NCRP) (2018). Dose reconstructions for LANL workers are in progress at the time of preparation of this article; these will update and expand results of an earlier study (Wiggs et al. 1994). The general methods of dose reconstruction and evaluation of uncertainties in dose estimates in the MPS are described in detail in the NCRP report (NCRP 2018; Dauer et al. 2019).

Rocketdyne was involved in a wide range of radiological projects from 1948 to 1999, including uranium fuel fabrication, sodium-cooled breeder reactor technology, spent fuel evaluation, radiography, radiochemistry, plutonium (Pu) fuel fabrication, and storage of nuclear material. Most nuclear operations were ended by the early 1980s, the main exceptions being decladding of irradiated reactor fuel, decontamination and decommissioning of the nuclear facilities, storage of radioactive material, and applied physics experiments. Over 2200 workers were monitored for intake of radionuclides. Monitoring data are available for many different radionuclides including ^{234}U, ^{235}U, ^{238}U, ^{232}Th, ^{239}Pu, ^{241}Am, ^{210}Po, ^{137}Cs, and ^{90}Sr. The most important internal emitters at Rocketdyne were ^{234}U, ^{235}U, and ^{238}U (Leggett et al. 2005; Boice et al. 2006).

The worker cohort in the Mound study includes workers involved in a ^{210}Po project conducted at a Laboratory in Dayton, Ohio, from 1943 to 1948, when the project was transferred to the Mound site. The Dayton facility separated ^{210}Po from natural materials for use in polonium-beryllium neutron generators. This project was continued at Mound until 1959. Several other projects involving various radionuclides were conducted at Mound between 1949 and 1995, including production of ^{210}Po heat sources for radioisotope thermoelectric generators from 1953 to 1969 and ^{238}Pu heat sources from 1959 to 1995. Over 2300 workers had biossays for polonium and about 1500 had bioassays for Pu (Boice et al. 2014).

The Mallinckrodt Chemical Works site was used from 1942 to 1957 for production of forms of uranium needed for the Manhattan Project. Purified uranium was produced by processing pitchblende from the Belgian Congo. Mallinckrodt operations included extraction of ^{226}Ra, because the company that provided the uranium ore retained ownership of its radium content. Other processes with high potential for intake of ^{238}U, ^{235}U, and their chain members included recovery of uranium from scrap uranium metal, machining of natural uranium metal rods to make reactor fuel slugs, reverting UF_4 to UO_2 or U_3O_8, production of UO_2F_2, and extraction and concentration of ^{230}Th from pitchblende raffinate. The most important internal emitters at Mallinckrodt appear to be the natural uranium isotopes and ^{226}Ra. Over 1900 workers had bioassays for uranium and about 500 had bioassays for ^{226}Ra (Ellis et al. 2018).

LANL has handled most natural or man-made radionuclides at one time or another, but mostly in small quantities. The most important source of dose to LANL workers from internal emitters is Pu, which has been handled in relatively large quantities at LANL since 1943. Early roles of LANL included reduction of Pu to metallic form, determination of its physical and metallurgical properties, and development of the technology required to build a nuclear bomb using Pu. Later programs have included peacetime applications of Pu such as development or testing of ^{238}Pu heat sources for space electric power generation or powering artificial organs. Over 1700 workers had positive bioassays for ^{239}Pu and nearly 600 had positive bioassays for ^{238}Pu.

For the most part, the biokinetic models applied in the dose reconstructions conducted to this point in the MPS are the models applied in ICRP Publication 68 (1994a). For the LANL study and for some special cases at Rocketdyne and Mound described in a later section of this article, the models of Publication 68 have been replaced with updated or site-specific biokinetic models. For most inhalation exposures at Rocketdyne, Mound, and Mallinckrodt, the applied respiratory model was the Human respiratory tract model (HRTM) introduced in ICRP Publication 66 (1994b) and applied in Publication 68. The ICRP originally provided three sets of parameter values for the HRTM describing different levels of solubility ('absorption types') of inhaled particles in the respiratory tract: Type F, representing fast dissolution and a high rate of absorption to blood; Type M, representing a moderate rate of dissolution and a moderate rate of absorption to blood; and Type S, representing slow dissolution and slow absorption to blood. These absorption types are referred to frequently in this article and, unless otherwise indicated, refer to their original definitions.

An updated version of the HRTM described in ICRP Publication 130 (2015) was applied to ^{239}Pu and ^{238}Pu inhalation cases in the LANL study. As discussed later, worker-specific parameter values were applied to Pu intakes in some cases.

Results

Importance of the exposure scenario

Definition

An exposure scenario used for a dose assessment of a known or suspected intake of a radionuclide is essentially a model describing the details of the exposure required to develop a meaningful dose estimate for a worker or a group of workers thought to have similar exposures. An exposure scenario for internally deposited activity includes the mode of intake (e.g. inhalation, wound, ingestion, or absorption through intact skin); the internally deposited radionuclide(s); the date(s) and pattern of intake (e.g. acute intake at a specific time, or chronic intake over a specified period); and the physical and chemical form(s) of the internally deposited radionuclide(s). Identification of the actual exposure conditions where feasible, or assignment of the most likely exposure conditions where information is incomplete, is a critical step toward derivation of meaningful retrospective dose estimates for an internal emitter.

Mode of intake

Bioassay records may provide clear evidence that a worker had an acute or extended intake of a radionuclide, or health physics records other than bioassay data may indicate that a worker was present in areas of high exposure to a radionuclide. The primary mode of the known or assumed intake often cannot be determined with much confidence from health physics records or available bioassay data. In the absence of specific information, it is generally assumed that intake was by inhalation because experience indicates that this is the most common mode of exposure to radionuclides in occupational settings. The second most common mode of intake at MPS sites studied thus far appears to be via wounds from contaminated objects. In a compilation of incidents at LANL with potential intake of ^{238}Pu or ^{239}Pu, roughly one-fifth of the cases involved wounds. Available incident reports for the Mound site suggest that some of the highest intakes of ^{210}Po may have been through wounds or absorption through intact skin. As illustrated in Figure 1, it is generally not possible to distinguish between an inhalation intake and a wound intake based on patterns of excretion of a radionuclide. The default assumption that internally deposited radionuclides entered the body in inhaled air can result in a large overestimate of the lung dose if this was not the actual intake mode or primary intake mode.

For dose estimates based on reasonably detailed urinary excretion data, the intake route tends to be a much less important source of uncertainty for estimated doses to systemic tissues than for estimated doses to tissues along the route of entry into the body, particularly the respiratory

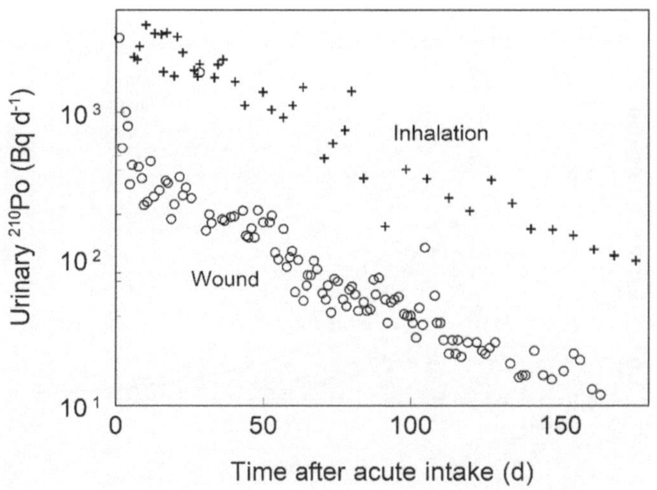

Figure 1. Patterns of time-dependent urinary excretion of ^{210}Po by two acutely exposed Mound workers, one via inhalation and the other via a contaminated wound.

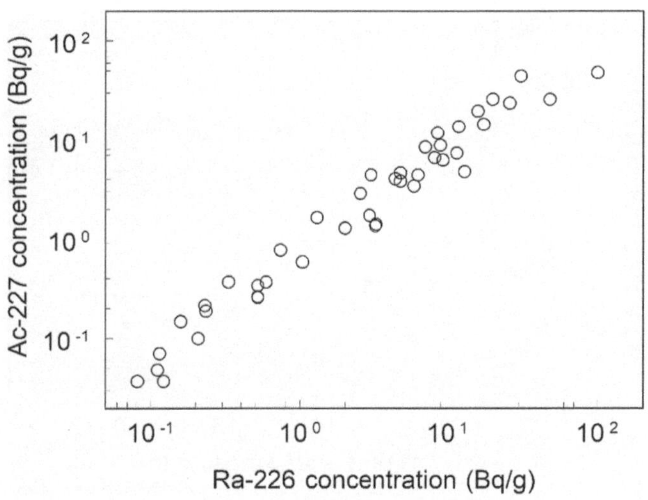

Figure 2. Relation of ^{226}Ra and ^{227}Ac concentrations in radioactive waste material originating at the Mallinckrodt site.

tract. This is because the rate of urinary excretion of a radionuclide does not definitively reveal the mode of intake or, for presumably inhaled activity, the distribution and retention time in the respiratory tract. Regardless of the route of intake, urinary activity is expected to reflect the level of activity that reached the systemic circulation and the characteristic systemic kinetics of the radionuclide, with subject-specific variations due to biological variability.

Identification of internal emitters

As a practical matter, the internal emitters addressed in exposure scenarios and dose reconstructions for MPS sites usually have been only those that site health physicists considered sufficiently important to include in their internal dosimetry monitoring programs. As illustrated below, however, assessments of the uncertainty in estimated total doses for MPS sites should address radionuclides that were not monitored but perhaps represent important missed doses for some workers.

Much of the work performed at the Mallinckrodt site involved extraction of U from pitchblende ore and concentration of the extract (Ellis et al. 2018). The operations also included extraction of ^{226}Ra, which was required because the company that provided the ore retained ownership of its radium content. Bioassay programs at Mallinckrodt focused on intake of natural U isotopes, ^{226}Ra, and ^{222}Rn. Urinalyses for thorium, presumably for monitoring ^{230}Th intake, were also performed for a brief period. The air monitoring program focused on measurement of U isotopes, total alpha activity, and ^{222}Rn. The MPS dose reconstruction for Malinckrodt was limited to intake of U isotopes, ^{226}Ra, and ^{222}Rn and its short-lived progeny. There is also suggestive evidence of exposure of Mallinckrodt workers to the alpha emitter ^{227}Ac, which was not addressed in the dose reconstruction due to lack of monitoring data. Records indicate that ^{227}Ac may have been precipitated in high concentration in a process used to extract ^{226}Ra. Radiological surveys of a site that received residues from radionuclide separation processes at Mallinckrodt revealed similarly high concentrations of ^{226}Ra and ^{227}Ac in soil (Figure 2). A post-operation survey at Mallinckrodt also revealed hot spots in soil with high concentrations of both ^{226}Ra and ^{227}Ac. Such findings provide suggestive evidence that Mallinckrodt workers may have sometimes inhaled similar levels of ^{227}Ac and ^{226}Ra, but not sufficient evidence to warrant the assumption of intake of ^{227}Ac in a Mallinckrodt dose reconstruction performed for epidemiological purposes. Nevertheless, potentially missed doses from exposure to ^{227}Ac, which can yield far higher dose per inhaled Bq than ^{226}Ra, should be considered when addressing uncertainties in retrospective dose estimates for Mallinckrodt workers.

Dose reconstruction based on bioassay data often is complicated by a lack of specificity in the measurement technique. For example, bioassay records for Rocketdyne often describe measured activity in urine as 'mixed fission products' or 'total alpha.' In such cases, there may be no compelling reason to assume any specific mixture of fission products or alpha emitters, so that dose estimates require subjective choices of radionuclide mixtures or dosimetrically dominant radionuclides. In dose reconstructions for the Rocketdyne site, mixed fission products were often assumed to consist entirely of ^{90}Sr. The rationale was that exposures to fission products at Rocketdyne generally occurred when workers were handling equipment contaminated with aged fission products; the longest-lived fission products expected to be present were ^{90}Sr and ^{137}Cs; and in cases of acute exposure the urinary excretion curves often were reasonably consistent with a ^{90}Sr intake (Figure 3) but not ^{137}Cs intake. This approach may have substantially underestimated doses to soft tissues, or ^{90}Sr may have indeed been the dosimetrically dominant radionuclide. In any case, lack of identification of specific internal emitters resulted in non-trivial uncertainties in dose estimates for urinary activity identified in such broad terms as mixed fission products or total alpha.

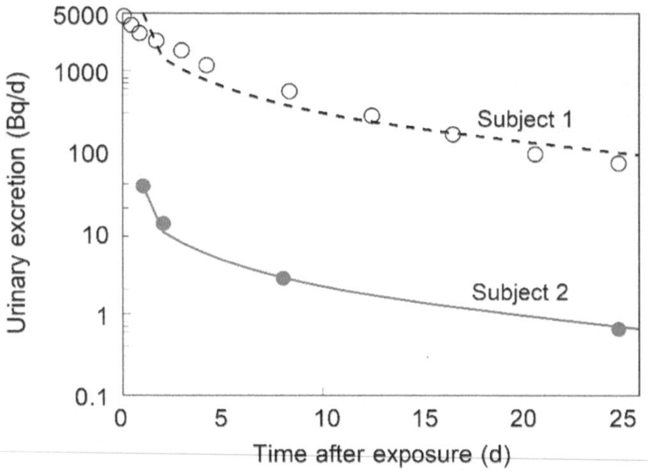

Figure 3. Urinary activity for two Rocketdyne workers, acutely exposed to mixed fission products at different times and locations. The indicated curve fits are based on intake of a relatively soluble form (Type F) of ^{90}Sr.

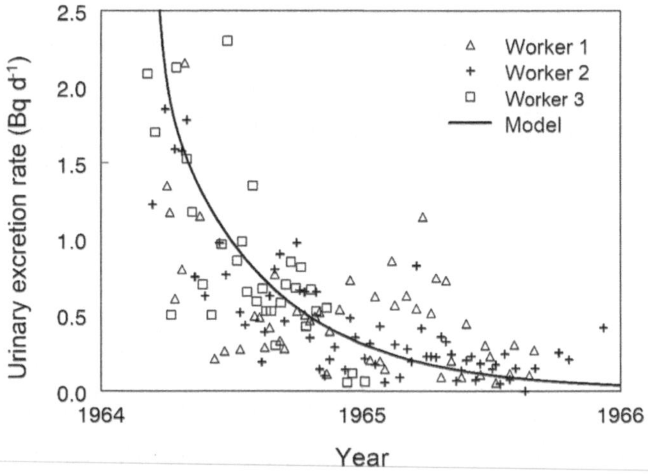

Figure 4. Urinary U data for three Rocketdyne workers acutely exposed to airborne U in the same incident. The model curve is based on acute inhalation of 10,000 Bq U, Type M, 5 μm AMAD.

Time pattern of intake

In many cases of accidental intake of radionuclides, the time of intake was recorded in an incident report and appeared to have been determined closely enough to introduce little or no uncertainty into the dose reconstruction. If an intake was not recognized at the time of occurrence but was revealed later by a routine bioassay measurement, the assumed time pattern of intake sometimes represented a significant or even dominant source of uncertainty in the dose estimate. For example, bioassay records for a Rocketdyne worker indicated intake of ^{239}Pu sometime within a 2.5-month period. Calculations based on different times of intake and a default form of inhaled Pu (Type M, particle size 5 μm AMAD) yielded estimates of ^{239}Pu intake ranging from 1 to 1200 Bq. Taking all sources of uncertainty into account including the form of inhaled material and errors in measurement of urinary ^{239}Pu, the uncertainty band on intake ranged from zero to several thousand Bq, with the time of intake being the dominant uncertainty in the dose estimate.

When bioassay data have indicated consistently low but nonzero intake of a radionuclide over an extended period, the standard assumption has been that the worker was chronically exposed at a constant rate. A time-dependent rate of intake has been applied to MPS sites when there is a sound basis for such a scenario, such as the existence of air monitoring data for uranium in the work area of interest.

LANL maintains a machine-readable database of incidents involving potential intake of ^{238}Pu or ^{239}Pu. For each incident and each worker involved in that incident, the database contains the date, worker identifier, and an abbreviation representing the nature of the exposure or reason for suspecting an intake (for example, w = wound, e = wound with excision, h = high nose count, r = high room count, u = unspecified accident). This database is a valuable aid in matching an increase in urinary ^{238}Pu or ^{239}Pu in a LANL worker with a likely time of intake. An incident database is particularly useful for dose reconstruction for Pu workers because intake of Pu appears to be associated more with incidents than with chronic exposure.

Form of an inhaled radionuclide

For inhalation exposure cases addressed in the MPS, the form of the inhaled material often can be reasonably well identified or inferred from health physics reports, information on the type of work performed, or other information. For example, intake of U at Rocketdyne was frequently associated with spontaneous uranium fires. For such cases, the intakes were generally assumed, and sometimes specified in reports, to be U oxide. For a sizable number of Rocketdyne workers, chronic exposure to relatively high levels of U occurred in a work area known as the powder room, where U aluminide (UAl_x) was produced. The urinary excretion curves, fecal to urinary excretion ratios, and external lungs measurements for workers exposed to U in the powder room generally followed a characteristic pattern for UAl_x (Leggett et al. 2005). Similarly, elevated intakes of ^{238}Pu at Mound or LANL often occurred during production or testing of ^{238}Pu heat sources, and the excretion curves for workers exposed to ^{238}Pu in those operations often followed a pattern characteristic of a form of ^{238}Pu used in those heat sources. In the above situations, information on the formation of the material together with bioassay data provided sufficient support for assigning a specific form of the radionuclide to the exposure scenario.

For most intakes known or suspected to be via inhalation, however, the choice of a respiratory model has been based on less compelling information. For the Rocketdyne site, for example, the ICRP's HRTM for Type M material was used as the default respiratory model for inhaled U based on patterns of urinary excretion of U by workers with known acute intakes, which in most cases were broadly consistent with inhalation of Type M material as illustrated in Figures 4–6. Similarly, bioassay data for Rocketdyne workers identified as exposure to ^{90}Sr, and sometimes as mixed fission products as indicated earlier, often were suggestive of inhalation of relatively soluble material (Figure 7), in which case Type F was used as the default absorption type for such exposures. Excretion data for inhaled ^{210}Po at the Mound site were not consistent with any of the ICRP's standard absorption types. A default absorption type for ^{210}Po inhalation at Mound was

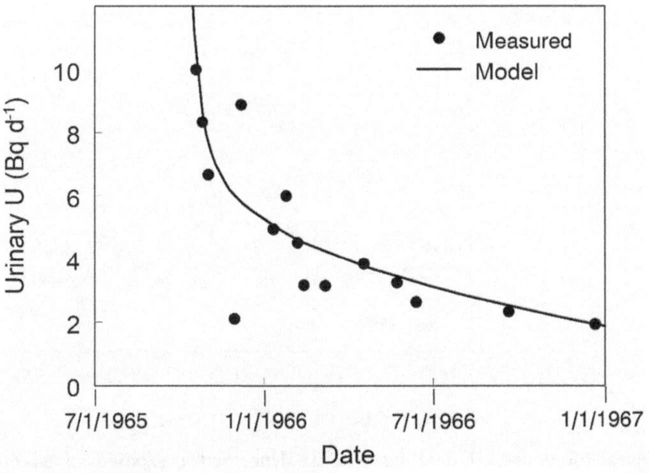

Figure 5. Observed and modeled urinary excretion rate for a worker acutely exposed to of U oxide in late 1965 and chronically exposed to lower quantities of U in 1966. The model predictions are based on inhalation of Type M material.

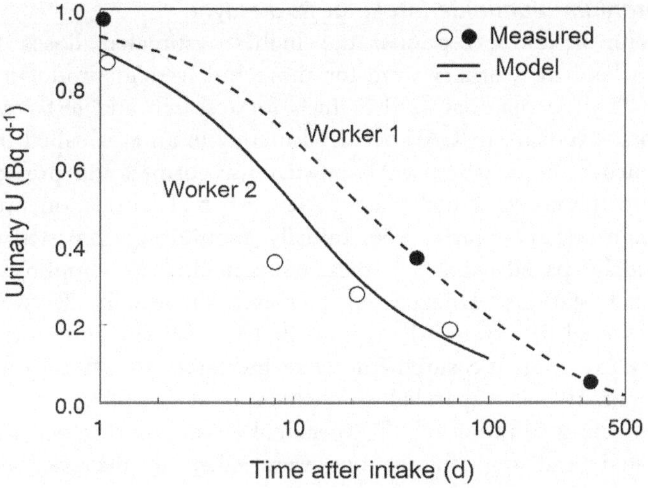

Figure 6. Observed and modeled urinary U for two workers acutely exposed to U at different times. Model predictions are based on inhalation of Type M material.

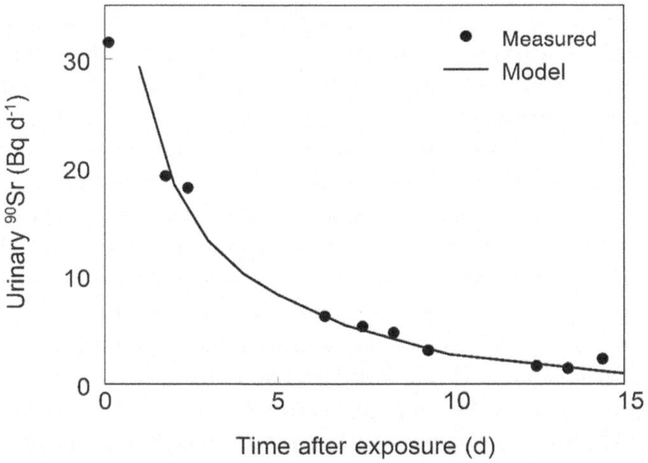

Figure 7. Observed and modeled urinary excretion rate for a worker presumed to have an acute intake of ^{90}Sr. The model predictions are based on inhalation of Type F material.

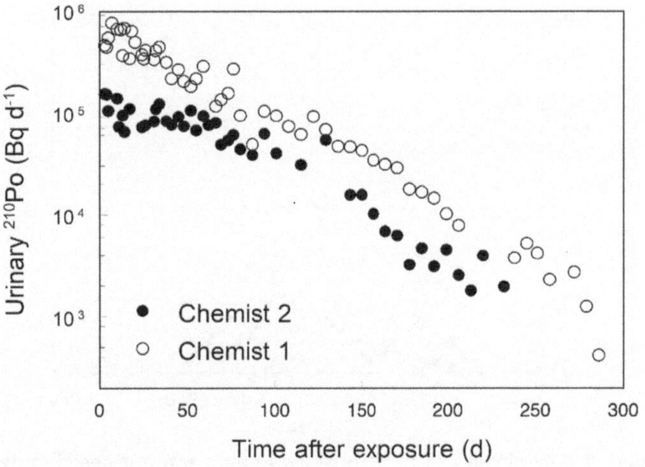

Figure 8. Urinary ^{210}Po data for two Mound chemists exposed to airborne ^{210}Po in the same incident.

developed from site-specific data together with published data on the behavior of inhaled ^{210}Po in human subjects and laboratory animals.

A later section discusses the site-specific respiratory models applied to cases of inhalation of ^{210}Po at Mound, UAl_x at Rockedyne, and ^{238}Pu at LANL and Mound.

Application of data for coworkers (surrogate data)

In dose reconstructions for the sites addressed here, the situation sometimes arose that follow-up urinary excretion measurements were available for some but not all workers listed in a report as being involved in an incident. In such cases, the same exposure scenario, but not necessarily the same level of intake, was assigned to each of the listed workers. In some cases, the radionuclide intake during the incident could be scaled from measurements of activity in nasal smears taken soon after the incident. If no data for nasal smears were available, a judgment had to be made as to whether to assign a zero intake to workers with no follow-up bioassay, or some positive intake extrapolated from data for workers with follow-up data. For example, if the incident occurred in a small, enclosed area, it seemed reasonable to assign an intake based on coworker data. If the incident occurred in a large open area, it seemed much less likely that all workers identified in the incident report had similar intake.

In any case, there is always considerable uncertainty in assigning intake data for a worker to an unmonitored coworker, as indicated by MPS case studies of follow-up monitoring of multiple workers involved in the same incident. For example, a Rocketdyne report lists 14 workers with potential intake of U resulting from a fire involving U carbide. As illustrated in Figure 4, three of the workers showed similar patterns and levels of urinary U over the following 1.5 years. The urinary excretion rate in a fourth worker was about a factor of 2 lower. The other 10 workers apparently had little intake of U.

In another case study, cumulative urinary ^{210}Po differed by a factor of 3 for two chemists at the Mound site who

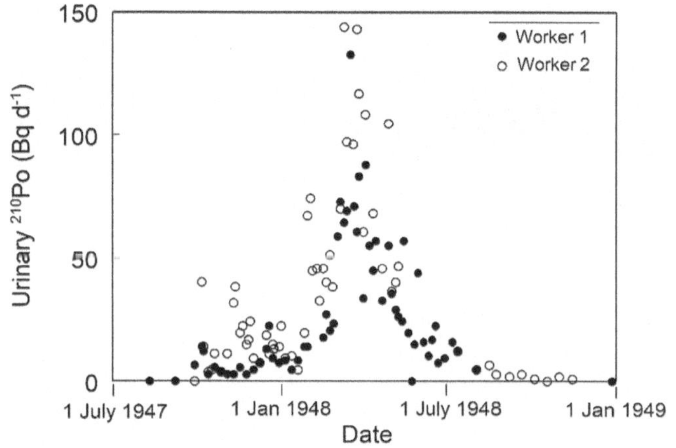

Figure 9. Similarity in urinary ^{210}Po patterns over a two-year period for two Mound workers.

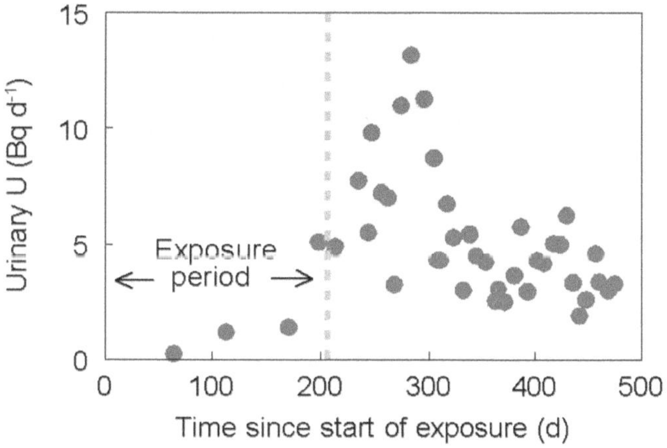

Figure 10. Urinary U data for a Rocketdyne worker exposed to airborne U aluminide.

were acutely exposed in the same incident to high levels of airborne ^{210}Po (Figure 8).

In MPS dose reconstructions, common exposure scenarios sometimes have been applied to groups of workers based on simultaneous changes in their bioassay data. Common exposure scenarios have been applied to as few as two workers, such as the two Mound workers whose urinary ^{210}Po data are shown in Figure 9, and to relatively large groups of workers as illustrated by the following example. Bioassay records for the Rocketdyne site revealed an unusually large cluster of positive urinary U measurements for the first few weeks of 1963, with over 50 workers showing elevated urinary U during that period. Similar patterns of decline in urinary U over the following months suggested an unrecognized short-term elevation of airborne U in a work area in early January 1963. In their dose reconstructions, these 50+ workers were all assumed to have been exposed acutely to a moderately soluble form of airborne U (Type M) on 2 January 1963. The level of intake of U for each worker was derived from the level of that worker's urinary excretion data.

As an aid in assigning exposure scenarios to groups of workers with little direct subject-specific information on exposure conditions, a history of known or suspected internal exposures to the Rocketdyne/AI workers was developed in the form of a time line. This is analogous to LANL's machine-readable database of incidents involving potential intake of ^{238}Pu or ^{239}Pu (described earlier) but contains verbal descriptions of exposures and includes some prolonged elevated intakes in addition to intakes resulting from incidents. This exposure time line was based on incident reports and bioassay data extracted from exposure histories of individual workers. Additionally, interviews of former radiation workers were conducted to learn directly from Rocketdyne workers about incidents, exposure circumstances, and work conditions. Incidents or prolonged elevated intakes involving a few hundred workers were identified. The largest single incident involved 35 workers exposed to a low level of airborne mixed fission products.

Illustrations of site-specific biokinetic models used in the MPS

Uranium aluminide intake at Rocketdyne

With a few exceptions, the highest estimated doses to Rocketdyne workers were for those involved in production of UAl_x, a material used in fuels for research and test reactors. Exposure to UAl_x occurred mainly in an area called the powder room, where UAl_x powder was formed and pressed into the cores of fuel plates. The UAl_x fuel fabrication program began in early 1966. Initially, monitoring consisted of widely spaced urinary U measurements and air sampling at fixed stations. Unexpected increases in urinary U were observed in some UAl_x workers in 1967. Subsequently, urinary excretion measurements were increased, the monitoring program was expanded to include fecal sampling and in vivo lung counting of ^{235}U, personal air monitoring was initiated, and remedial actions were taken to decrease airborne U.

The unexpected increase over time in the rate of urinary excretion of U presumably reflected an unusual pattern of behavior of UAl_x in the lungs. Initially, UAl_x deposited in the lungs dissolved at an extremely slow rate, but over time the inhaled particles apparently began to dissolve more rapidly, leading to a rapidly increasing rate of urinary excretion of U. The pattern of excretion relative to the time of exposure is illustrated in Figure 10, which shows data for a worker who was removed from exposure after a routine urine measurement revealed a large increase in urinary U since the previous measurement. His urinary U levels continued to increase for a few months after the end of exposure and then declined sharply. Because U is typically excreted at a high rate after absorption to blood and the rate of absorption from lungs to blood depends on the rate of dissolution in the lungs, the data indicate a substantial increase over time in the dissolution rate in the lungs.

Because none of the ICRP's default absorption types were applicable to UAl_x, special parameter values of the ICRP's HRTM were developed to describe the respiratory behavior of this unusual form of U (Leggett et al. 2005). The parameter values were chosen to reproduce as closely as feasible the urinary and fecal excretion patterns of U indicated by

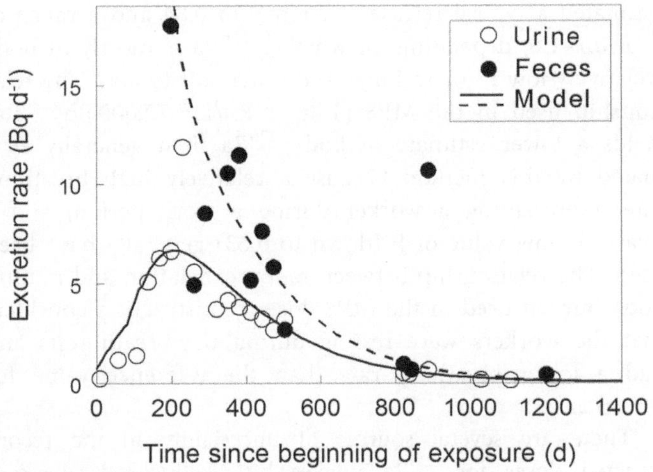

Figure 11. Model fit to urinary and fecal excretion data for a worker exposed to UAl_x.

extensive bioassay data for the relatively large set of workers exposed to UAl_x and the clearance rate from the lungs as indicated by in vivo lung counts for UAl_x workers. A particle size of 1 μm AMAD rather than the default size of 5 μm AMAD was assumed for inhaled UAl_x based on information obtained from interviews of workers who had been involved in its production.

Figure 11 illustrates the fit of the UAl_x model predictions to urinary and fecal excretion data for a worker who was exposed to UAl_x for a few months. The exposure scenario is based on the relative change with time in the concentration of U in air in the powder room. That is, the rate of intake of UAl_x was assumed to be proportional to the time-dependent concentration of UAl_x in air throughout the exposure period. The absolute air concentrations of UAl_x were not used in the dose reconstruction for powder room workers.

Polonium-210 intake at Mound

At the time of the MPS dose reconstruction for Mound, the ICRP's systemic model for internally deposited polonium was a model adopted in ICRP Publication 67 (1993), and the ICRP's default absorption type for inhaled polonium was Type M. Different models were applied in MPS dose reconstructions for ^{210}Po intakes at Mound. A systemic model of Leggett and Eckerman (2001) was applied. The ICRP's parameter values for Type M material were modified for consistency with patterns of excretion of ^{210}Po for Mound workers as well as consistency with published data on the behavior of inhaled ^{210}Po in the respiratory tract. Data for laboratory animals (Figure 12) and accidentally exposed Mound workers (Figure 13) suggested that Type M parameter values substantially overestimated the retention time of Po in the lungs for the form of Po likely to be encountered at Mound. Also, data for a worker exposed to ^{210}Po oxide, presumably one of the less soluble forms of polonium, at another site (Scott and West 1975) indicate much faster removal from the lungs than depicted by Type M parameter values. For derivation of doses from inhaled ^{210}Po, the long-term biological half-time of ~140 d for polonium in the deep lungs specified in ICRP Publication 66 for Type M

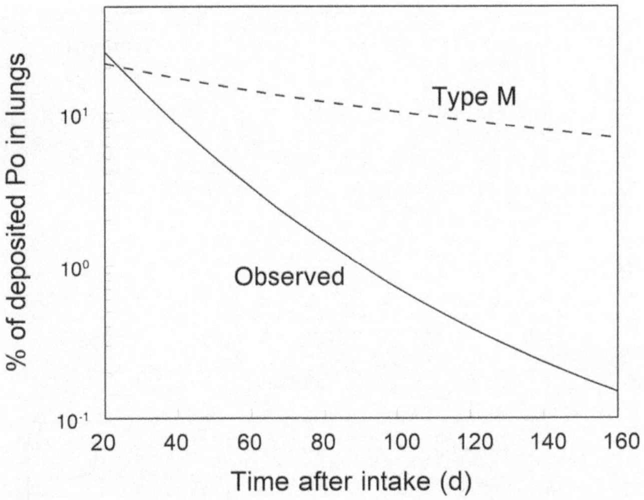

Figure 12. Lung retention of an inhaled ^{210}Po aerosol in dogs (curve labeled "observed" based on data of Smith et al. 1961), compared with predictions based on absorption Type M.

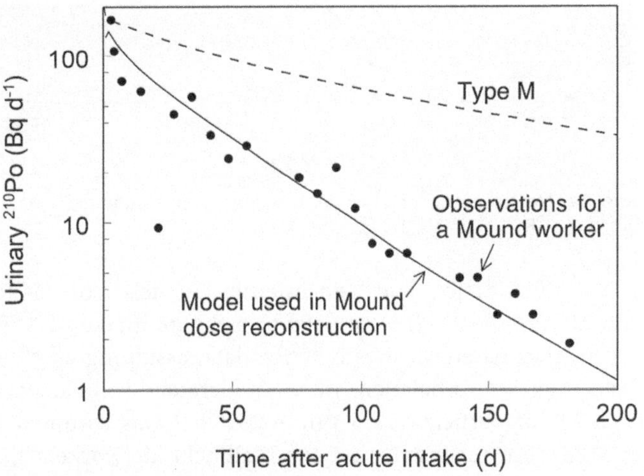

Figure 13. Model predictions of rate of decline of urinary ^{210}Po following inhalation, based on alternate respiratory models, compared with data (close circles) representative of the typical pattern of excretion of inhaled ^{210}Po in Mound workers.

material was replaced with a half-time of 30 d. A particle size of 1 μm AMAD was assumed for inhaled polonium rather than the default size of 5 μm AMAD because airborne ^{210}Po seemed likely to exist as fine aerosols in view of the types of operations conducted at Mound.

Radium-226 intake at Mallinckrodt

A non-standard approach to modeling intake of ^{226}Ra at Mallinckrodt arose from the type of bioassay data used to monitor ^{226}Ra exposure, rather than from specific information on the chemical or physical form of inhaled ^{226}Ra.

Exposure to ^{226}Ra was monitored at Mallinckrodt through measurement of exhaled ^{222}Rn rather than its rate of loss in urine. Measurement of ^{222}Rn in breath was made at Mallinckrodt during the period 1948–1959. Figure 14 shows ^{222}Rn breath measurements for two Mallinckrodt workers.

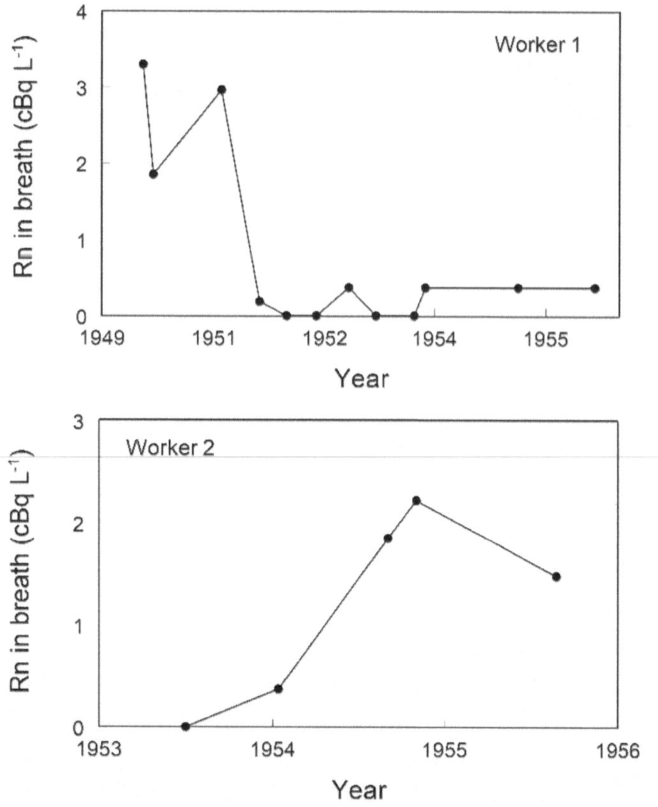

Figure 14. Measurements of ^{222}Rn in breath in two Mallinckrodt workers exposed to ^{226}Ra.

The biokinetic and dosimetric models of ICRP Publication 68 (1994a) were used to estimate intake of ^{226}Ra for a worker based on breath ^{222}Rn data, assuming all ^{226}Ra intake was via inhalation of a moderately soluble form (Type M) of particle size 5 μm AMAD. It was assumed in the MPS dose reconstruction for Mallinckrodt workers that the integrated ^{222}Rn activity in breath over the entire monitoring period for a radium worker was proportional to the integrated ^{226}Ra in the body over that period, and the ^{226}Ra intake rate was constant over that period. The constant ^{226}Ra intake rate for an individual worker could then be derived using the biokinetic models of ICRP Publication 68, the worker's continuous breath ^{222}Rn curve, and a conversion factor that relates breath ^{222}Rn to a typical body content of ^{226}Ra over an extended period.

To convert from ^{222}Rn in breath to ^{226}Ra in the body it was assumed that 1 Bq ^{222}Rn per liter of air exhaled during a breath test corresponds to a body ^{226}Ra content of 72,000 Bq. This relation is derived from a formula proposed by Srivastava et al. (1986) and is reasonably consistent with predictions of the biokinetic models used in this study assuming extended exposure to airborne ^{226}Ra. Srivastava and coworkers concluded that the ^{226}Ra body burden could be estimated from the rate of exhalation of ^{222}Rn by the relationship

$$^{226}\text{Ra in the body (Bq)} = (\text{C BR})/(\lambda \text{F}) \qquad (1)$$

where BR is the breathing rate in liters (h^{-1}), λ is the decay constant of ^{222}Rn (h^{-1}), C is the concentration of ^{222}Rn in the breath sample (Bq L^{-1}), and F is the fraction of ^{222}Rn produced in the body that is exhaled. Srivastava et al. estimated a typical release fraction F of 0.84 and a range of F of 0.7–1.0, depending on whether ^{226}Ra is mostly in bone (relatively low F) or in lungs and other soft tissues. The relationship used in the MPS (1 Bq ^{222}Rn/L = 72,000 Bq ^{226}Ra) yields a lower estimate of body ^{226}Ra than generally estimated by this method because a relatively high breathing rate, representing a worker during a work period, and a relatively low value of F (down to 0.63) generally have been used. The relationship between radon exhalation and radium body burden used in the MPS dose reconstruction considers that the workers were resting during the breath tests and had a lower breathing rate than the reference value for a worker.

There are several sources of uncertainty in the reconstructed doses for ^{226}Ra intakes by Mallinckrodt workers, including the cumulative ^{222}Rn exhalation, the relation of ^{222}Rn in breath to the content of ^{226}Ra in the body, the solubility of inhaled ^{226}Ra from the lungs, and the possibility of ^{226}Ra intake via other pathways including wounds and ingestion. Overall, the largest uncertainties may be associated with the estimate of cumulative ^{222}Rn in breath, in view of the high variability in the ^{222}Rn exhalation rate and the fact that a worker typically had only 1–4 breath tests per year.

Plutonium-238 intake at LANL

The production of ^{238}Pu heat sources was started at Mound in 1961. Determination of properties of the heat sources and some production support began at LANL in the late 1960s. Production of ^{238}Pu heat sources was largely transferred to LANL by the late 1970s. Published data and internal incident reports, together with bioassay data for the workers involved, indicate a sizable number of acute LANL to the high-fired ^{238}Pu dioxide used in ^{238}Pu heat sources. Generally, the observed pattern of urinary excretion of activity following inhalation of ^{238}Pu dioxide is broadly similar to the pattern described above for inhaled UAl_x. That is, there is initially little urinary excretion of activity after exposure to relatively high concentrations of airborne ^{238}Pu, but the inhaled material begins to dissolve more rapidly in the lungs over an extended period and subsequently is excreted at a substantially increased rate. Daily loss of ^{238}Pu in urine typically peaks at least a few months, and up to a few years in some cases, after acute intake and then declines gradually thereafter.

Dose estimates for inhaled ^{238}Pu dioxide at LANL are based on worker-specific variations of respiratory parameter values for ^{238}Pu dioxide that have been adopted by the ICRP for use in an upcoming report (Part 4 of an ICRP series on occupational intake of radionuclides, in press). A similar but not identical model with worker-specific parameter values was applied to ^{238}Pu dioxide intakes at Mound, prior to the completion of the ICRP's ^{238}Pu dioxide model. The worker-specific parameter values are set to reproduce the time and level of peak urinary excretion of ^{238}Pu following acute inhalation of ^{238}Pu dioxide. Figure 15 shows model fits to ^{238}Pu excretion data for two LANL workers and two Mound workers acutely exposed to airborne ^{238}Pu-dioxide.

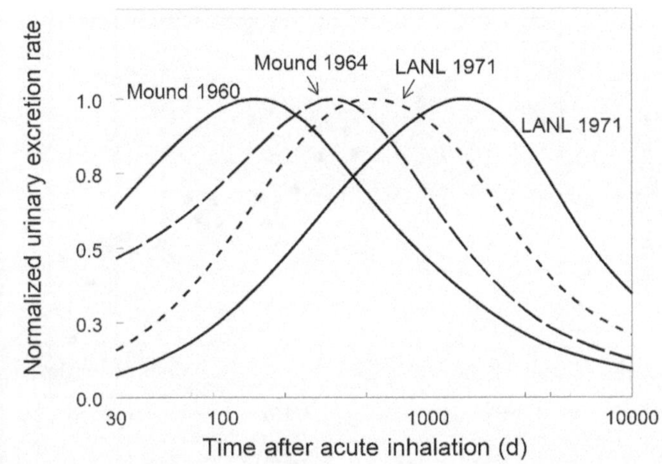

Figure 15. Differences in peak urinary excretion times for workers acutely exposed to ^{238}Pu dioxide. Each curve is a model fit to urinary ^{238}Pu data for an individual worker.

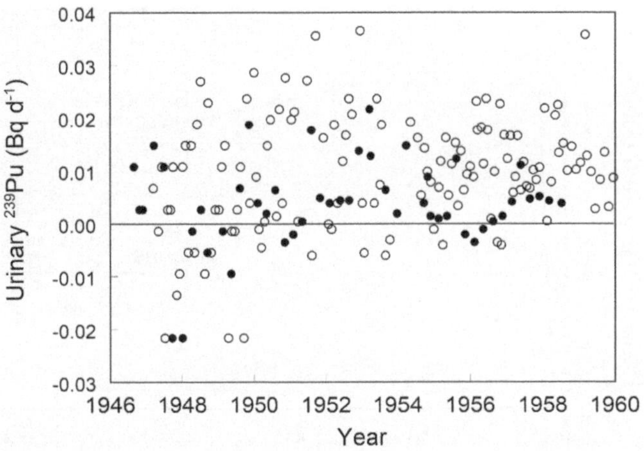

Figure 16. Urinary ^{239}Pu measurements for two LANL workers, illustrating a reduction over time in the scatter of measurements and the magnitude of negative results.

Illustrations of problematic bioassay data

Two types of uncertainty associated with bioassay data

The components of uncertainty in a bioassay measurement may be divided into two main categories commonly referred to as Type A and Type B uncertainties (ICRP 2015). Type A uncertainties are those arising from counting statistics, and Type B components are those associated with all other sources of uncertainty. Type A uncertainties tend to decrease with increasing activity or counting time, while uncertainties associated with Type B may be largely independent of activity or counting time. Type B uncertainties associated with measurement of activity in a urine sample, for example, would include uncertainties in sample volume, calibration, recovery, background, contamination, and conversion of results for a spot sample to a 24-h excretion rate.

The illustrations in this section address some important uncertainties (largely Type B) that have been encountered in MPS dose reconstructions in interpretation of urinary excretion data, because this is the type of bioassay data most commonly available for MPS sites.

Uncertainties in urinary Pu data for LANL workers

Urinary Pu measurements frequently involve important Type A uncertainties due to the low rate of excretion of Pu following its uptake to blood. The following brief history of the urinary Pu measurement program at LANL during the 1940s and 1950s (Campbell et al. 1972; Hempelmann et al. 1973; Lawrence 1978; Miller et al. 2008) describes some important Type B uncertainties such as high background (blank) counts and contamination issues that, together with common counting issues for urinary Pu, appear to have resulted in unreliable urinary Pu measurements in the early years of the LANL bioassay program.

A program for measurement of urinary activity in Pu workers was instituted at LANL in 1944. Twenty-four-hour urine samples were collected, with all urine voided by a worker during the day collected in a single bottle in presumably clean areas after decontamination of the worker and a change of clothes. The bottle was taken home at night and returned the next morning to complete the 24-h collection. Occasional extremely high activities suggested contamination problems, perhaps ^{210}Po contamination in some cases as Pu and Po were handled in the same work area. New methods were started in 1945 to reduce contamination issues. Workers were given two paid days to be spent away from work and away from Los Alamos, after which the workers reported to a hospital where they provided urine samples in an environment that was kept as clean as feasible.

From 1944 to 1957, LANL used the cupferron method of urinary Pu measurement in which Pu with an iron carrier was extracted by cupferron, a precipitant for iron and copper. The original counting system had high background counts that made it difficult to achieve a meaningful determination of the concentration of Pu in urine because of the low rate of excretion of absorbed Pu. The counting system was replaced in 1945 with a system that reduced background counts by as much as a factor of 30, and later improvements further reduced background counts by another order of magnitude. Figure 16 shows urinary ^{239}Pu measurements for two LANL workers, both involved in multiple incidents in the 1940s with apparent intake of ^{239}Pu. The figure illustrates a decrease over time in data scatter and in the magnitude of negative results for early LANL worker, presumably reflecting improved measurements due to reduced background counts.

In 1948, the two-day vacation system was abandoned due to the expense. The only time off work for purposes of limiting contamination of urine samples was the 1 d during which the sample was collected in the hospital. Urine collection methods were further simplified in 1952 to allow relatively frequent sampling of a larger group of workers and again in 1958 to solicit more favorable employee cooperation. The method for determining the concentration of Pu in urine was updated in 1957, when the cupferron method was replaced by the nuclear-tract method of alpha counting (NTA method) in an effort to increase the sensitivity of the bioassay procedure. Further improvements were made frequently thereafter.

To summarize, it appears that urinary Pu measurements during the first year or so of the bioassay program at LANL involve large uncertainties due to high background counts

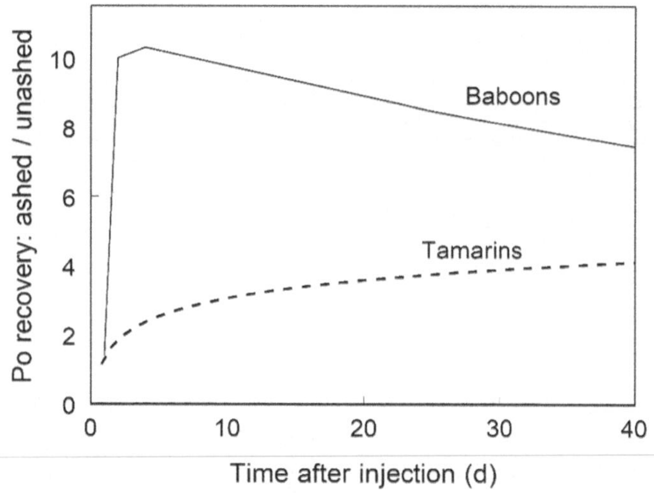

Figure 17. For different primates, the time-dependent increase in recovery of ^{210}Po from urine resulting from wet ashing of urine before measurement of ^{210}Po by the technique used at Mound.

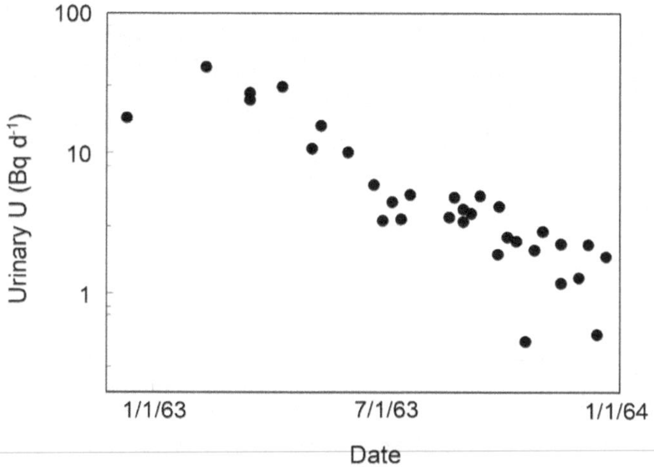

Figure 18. Urinary U levels in a worker at a U fuel production plant that had relied solely on air monitoring for the previous two years.

and high potential for contamination of samples. Considerable improvements were made in 1945 regarding reduction of background counts and contamination, but the reliability of the improved method is difficult to quantify. The potential for contamination of urine samples may have increased again in the late 1940s and early 1950s due to efforts to monitor more workers more frequently and encourage workers to cooperate with the bioassay program. A considerable improvement in the Pu assay methods apparently was made in 1957, when a more sensitive measurement technique was introduced.

Dominant uncertainty in urinary ^{210}Po data for Mound workers

An important uncertainty in reconstruction of doses from intake of ^{210}Po by Mound workers was the level of recovery of ^{210}Po from urine (a Type B uncertainty). The technique used at Mound to measure the ^{210}Po concentration in urine samples involved spontaneous deposition of ^{210}Po onto a metal disc from which its decays were counted. An issue apparently not recognized until after completion of the ^{210}Po program at Mound was that Po excreted in urine may not be recovered to the same extent as tracer Po added to urine to determine the fraction recovered, unless there is wet ashing (acid digestion) of the urine prior to deposition on the disc. The Mound technique did not involve wet ashing of the urine. Results of human and animal studies of the percentage of excreted polonium recovered from unashed urine are highly variable (Black 1956; Sedlet and Robinson 1971; Bale et al. 1975; Fellman et al. 1989, 1994). The available data suggest that the increase in recovery of Po from urine due to wet ashing may vary with time after intake and with animal species. For example, with the measurement technique applied at Mound, wet ashing increased Po recovery from urine by a factor of 7–10 in baboons (excluding day 1), depending on time after intake, and a factor of about 2–4 in tamarins, depending on time after intake (Figure 17). In the MPS dose reconstruction for Mound, it was assumed that that recovery was 20%, a best estimate based on the collective data. This applied value might have overestimated actual recovery by a factor of 2 or underestimated recovery by a factor of 3 or more.

Limitations of air monitoring data

Air monitoring can provide a useful warning system for release of radionuclides into a work area, but measured air concentrations of radionuclides often are crude and unreliable predictors of intake. This conclusion is based on: (1) comparisons of intake derived from air monitoring data with intake based on bioassay data at MPS sites (e.g. in the Rocketdyne 'powder room' where UAl_x was produced) as well as other facilities (Britcher and Strong 1994; Britcher et al. 1998; Eckerman and Kerr 1999) and (2) the fact that different methods of air monitoring often yield considerably different results (Marshall and Stevens 1980; Britcher and Strong 1994).

Marshall and Stevens (1980) found that concentrations of airborne radionuclides based on personal air samplers (PAS) can be up to 50 times greater than values based on static air samplers (SAS). Britcher and Strong (1994) concluded from a review of monitoring data for a group of workers in the UK that intakes assessed from PAS data were about an order of magnitude greater than those implied by SAS data. In a review of the use of PAS as part of internal dosimetry monitoring programs at two facilities in the UK, Britcher and Strong (1994) concluded that the correlation of radionuclide intake between assessments using PAS and biological samples was poor. Both SAS and PAS data appear to be particularly unreliable for reconstructing acute intakes of radionuclides. Estimation of radionuclide intake based on air sampling is further complicated by inconsistent use of masks, which can reduce intake by as much as a factor of 50. For these reasons, air monitoring data generally have not been used in the MPS to derive intake of radionuclides except for radon measurement or as an indicator of relative intakes of a radionuclide over an extended period.

The following example describes a failure in an air monitoring system at a U.S. facility (not an MPS site) that produced U metals and compounds from natural and enriched uranium feed stocks for use as fuel in nuclear reactors. Routine urinary U measurements were started in 1957 to supplement an air monitoring program. This bioassay program was discontinued in early 1961, but reinstated in late 1962 at the direction of the U.S. Atomic Energy Commission. The first set of urinary U measurements after the restart of the bioassay program revealed far higher intakes of U than had been indicated by air sampling, in an area where enriched U was handled. The highest intakes were associated with milling of UO_2. Urinary U data for one of the highly exposed workers are shown in Figure 18. Based on the assumption that urinary excretion data represent inhaled U, the estimated lung dose was at least 0.1 Sv assuming intake of Type M material. This could be a substantial underestimate if much of the intake occurred long before the restart of the bioassay program or if the inhaled material was less soluble than assumed.

Conclusions

1. Much of the dose reconstruction effort for internal emitters should be devoted to development of best feasible exposure scenarios, including the time pattern and mode(s) of intake, the internally deposited radionuclide(s), and the form(s) of the internally deposited radionuclide(s).
2. Coworker data should be used to assign exposure scenarios or dose estimates to workers with missing exposure data if, and only if, there is compelling evidence of similar coworker exposure.
3. Bioassay data for some radionuclides and periods of operation at MPS sites are of questionable reliability due to sizable uncertainties associated with contamination, recovery, or background issues.
4. Dose estimates derived solely from air monitoring data should be treated as highly uncertain values in the absence of information demonstrating that the data are reasonably predictive of intake.
5. For intakes known or assumed to be via inhalation, the uncertainty in lung dose typically is much greater than the uncertainty in dose to systemic tissues, when dose estimates are based on urinary excretion data.
6. The lung dose estimate often can be improved through development of site-specific respiratory absorption parameter values.
7. There is generally insufficient site-specific information to justify development of site-specific systemic models. However, it may be justified to replace outdated ICRP models with more recently developed models.

Acknowledgements

The United States Government retains and the publisher, by accepting the article for publication, acknowledges that the United States Government retains a non-exclusive, paid-up, irrevocable, world-wide license to publish or reproduce the published form of this manuscript, or allow others to do so, for United States Government purposes. The U.S. Department of Energy will provide public access to these results of federally sponsored research in accordance with the DOE Public Access Plan (http://energy.gov/downloads/doe-public-access-plan). The authors also are grateful for the financial support received by the NCRP from the U.S. Department of Energy.

Disclosure statement

The authors declare no conflict of interest. The authors alone are responsible for the content and writing of the article.

Funding

This manuscript and research were supported by UT-Battelle, LLC under Contract No. DE-AC05-00OR22725 with the U.S. Department of Energy. The work described in this manuscript was sponsored by the Centers for Disease Control and Prevention (CDC) Office of Noncommunicable Diseases, Injury and Environmental Health, National Center for Environmental Health, under Interagency Agreement DOE No. 2220-Z051-16, under Contract No. DE-AC05-00OR22725 with UT-Battelle, under grant No. 5NUE1EH001315 with the National Council on Radiation Protection and Measurements (NCRP), and by the U.S. Department of Energy under WAS Project No. 2018-AU-2000MPS. The authors also are grateful for the financial support received by the NCRP from the U.S. Department of Energy (Grant No. DE-AU0000042) and from the National Aeronautics and Space Administration (Grant No. 80NSSC17M0016) and US National Council on Radiation Protection and Measurements.

References

Bale WF, Helmkamp RW, Hrynyszyn V, Contreras MA. 1975. The determination of ^{210}Po in urine. Health Phys. 29:663–671.

Bess JD, Krahenbuhl MP, Miller SC, Slaughter DM, Khokhryakov VV, Khokhryakov VF, Suslova KG, Vostrotin VV. 2007. Uncertainties analysis for the plutonium dosimetry model, DOSES-2005, using Mayak bioassay data. Health Phys. 93:207–219.

Black SC. 1956. Low-level polonium determination of tissue and urine. Rochester (NY): University of Rochester. University of Rochester Atomic Energy Commission Report UR-463. [accessed 2018 Sep 20]. https://www.osti.gov/servlets/purl/4338659.

Boice JD, Jr, Cohen SS, Mumma MT, Ellis ED, Cragle DL, Eckerman KF, Wallace PW, Chadda B, Sonderman JS, Wiggs LD. 2014. Mortality among Mound workers exposed to polonium-210 and other sources of radiation, 1944–1979. Radiat Res. 181:208–228.

Boice JD, Ellis ED, Golden AP, Girardi DJ, Cohen SS, Chen H, Mumma MT, Shore RE, Leggett RW. 2018. The past informs the future: an overview of the Million Worker Study and the Mallinckrodt Chemical Works Cohort. Health Phys. 114:381–385.

Boice JD, Jr, Leggett RW, Ellis ED, Wallace PW, Mumma M, Cohen SS, Brill AB, Chadda B, Boecker BB, Yoder RC, et al. 2006. A comprehensive dose reconstruction methodology for former Rocketdyne/Atomics International radiation workers. Health Phys. 90:409–430.

Britcher AR, Strong R. 1994. Personal air sampling - a technique for the assessment of chronic low-level exposure? Radiat Prot Dosim. 53:59–62.

Britcher AR, Battersby WP, Peace MS. 1998. The practical application of models for assessing intakes of radionuclides by workers. Radiat Prot Dosim. 79:71–74.

Campbell EE, Milligan MF, Moss WD, Schulte HF. 1972. History of the plutonium bioassay program at the Los Alamos Scientific Laboratory, 1944–1972. LA-5008, UC-41. October 1972. [accessed 2018 Sep 20]. https://www.osti.gov/servlets/purl/4587709.

Dauer LT, Bouville A, Toohey RE, Boice JD, Jr, Beck HL, Eckerman KF, Hagemeyer D, Leggett RW, Mumma MT, Napier B, et al. 2019. Dosimetry and uncertainty approaches for the million-worker study of radiation workers and veterans: overview of the recommendations in NCRP Report No. 178. Review. Int J Rad Biol. 1-10 (Submitted).

Eckerman KF, Kerr GD. 1999. Y12 uranium exposure study. Oak Ridge (TN): Oak Ridge National Laboratory ORNL/TM-1999-114. [accessed 2018 Sep 20]. https://www.ornl.gov/sites/default/files/TM1999-114.pdf.

Ellis ED, Boice JD, Jr, Golden AP, Girardi DJ, Cohen SS, Mumma MT, Shore RE, Leggett RW, Kerr GD. 2018. Dosimetry is key to good epidemiology: workers at Mallinckrodt Chemical Works had seven different source exposures. Health Phys. 114:386–397.

Fellman A, Ralston L, Hickman D, Ayres L, Cohen N. 1994. Polonium metabolism in adult female baboons. Radiat Res. 137:238–250.

Fellman A, Ralston L, Hickman D, Ayres L, Cohen N, Spitz H, Robinson B. 1989. The importance of acid digestion of urine prior to spontaneous deposition of 210Po. Health Phys. 57:615–6221.

Harrison JD, Leggett RW, Nosske D, Paquet F, Phipps AW, Taylor DM, Metivier H. 2001. Reliability of the ICRP's dose coefficients for members of the public, II. Uncertainties in the absorption of ingested radionuclides and the effect on dose estimates. Radiat Prot Dosim. 95:295–308.

Harrison JD, Khursheed A, Lambert BE. 2002. Uncertainties in dose coefficients for intakes of tritiated water and organically bound forms of tritium by members of the public. Radiat Prot Dosimetry. 98:299–311.

Hempelmann LH, Langham WH, Richmond CR, Voelz GL. 1973. Manhattan project plutonium workers: a twenty-seven year follow-up study of selected cases. Health Phys. 25:461–479.

International Commission on Radiological Protection (ICRP). 1993. ICRP Publication 67. Age-dependent doses to members of the public from intake of radionuclides by workers. Part 2. Ingestion dose coefficients. Ann ICRP. 23(3–4):1–167.

International Commission on Radiological Protection (ICRP). 1994a. ICRP Publication 68. Dose coefficients for intake of radionuclides by workers. Ann ICRP. 24(4):1–83.

International Commission on Radiological Protection (ICRP). 1994b. ICRP Publication 66. Human Respiratory Tract Model for Radiological Protection. Ann ICRP. 24(1–3):1–482.

International Commission on Radiological Protection (ICRP). 2015. ICRP Publication 130. Occupational intake of radionuclides, Part 1. Ann ICRP. 44(2):5–188.

Lawrence JNP. 1978. A history of PUQFUA – Plutonium body burden (Q) from urine assays. Los Alamos, New Mexico USA: Los Alamos Scientific Laboratory. LA-7403-H, History, UC-41. October 1978.

Leggett RW. 2001. Reliability of the ICRP's dose coefficients for members of the public. 1. Sources of uncertainty in the biokinetic models. Radiat Prot Dosimetry. 95:199–213.

Leggett RW, Eckerman KF. 2001. A systemic biokinetic model for polonium. Sci Total Environ. 275:109–125.

Leggett RW, Bouville A, Eckerman KF. 1998. Reliability of the ICRP's systemic biokinetic models. Radiat Prot Dosim. 79:335–342.

Leggett RW, Eckerman KF, Boice JD. Jr. 2005. A respiratory model for uranium aluminide based on occupational data. J Radiol Prot. 25: 405–416.

Leggett RW, Harrison J, Phipps A. 2007. Reliability of the ICRP's dose coefficients for members of the public: IV. Basis of the human alimentary tract model and uncertainties in model predictions. Radiat Prot Dosim. 123:156–170.

Marshall M, Stevens DC. 1980. The purposes, methods and accuracy of sampling for airborne particulate radioactive materials. Health Phys. 39:409–423.

Miller G, Bertelli L, Guilmette R, McNaughton MW, Eisele WF. 2008. A study of early Los Alamos internal exposures to plutonium. Radiat Prot Dosimetry. 130:503–509.

NCRP. 2009. National council on radiation protection and measurements. Uncertainties in internal radiation dose assessment. Bethesda (MD): NCRP. NCRP Report No. 164.

National Council on Radiation Protection and Measurements (NCRP). 2018. Deriving organ doses and their uncertainty for epidemiologic studies (with a focus on the one million U.S. workers and veterans study of low-dose radiation health effects. Bethesda (MD): NCRP. NCRP Report 178.

Pawel DJ, Leggett RW, Eckerman KF, Nelson CD. 2007. Uncertainties in cancer risk coefficients for environmental exposure to radionuclides: an Uncertainty Analysis for Risk Coefficients Reported in Federal Guidance Report No. 13. Vienna, Austria: International Atomic Energy Agency (IAEA). ORNL/TM-2006/583.

Poudel D, Miller G, Klumpp JA, Bertelli L, Waters TL. 2018. Bayesian analysis of plutonium bioassay data at Los Alamos National Laboratory. Health Phys. 115:712–726. [accessed 2018 Sep 20]. https://www.epa.gov/sites/production/files/2015-05/documents/ornl-tm-2006-583.pdf.

Scott LM, West CM. 1975. Excretion of 210-Po oxide following accidental inhalation. Health Phys. 28:563–565.

Sedlet J, Robinson JJ. 1971. Elimination of polonium-210 following accidental inhalation of neutron-irradiated bismuth (abstract P;121], Health Phys. 21:62.

Smith FA, Morrow PE, Gibb FR, Della Rosa RJ, Casarett LJ, Scott JK, Morken DA, Stannard JN. 1961. Distribution and excretion studies in dogs exposed to an aerosol containing polonium-210. Am Ind Hyg Assoc J. 22:201–208.

Srivastava GK, Raghavayya M, Kotrappa P, Somasundaram S. 1986. Radium-226 body burden in U miners by measurement of Rn in exhaled breath. Health Phys. 50:217–221.

Wiggs LD, Johnson ER, Cox-DeVore CA, Voelz GL. 1994. Mortality through 1990 among white male workers at the Los Alamos National Laboratory: considering exposures to plutonium and external ionizing radiation. Health Phys. 67:577–588.

Potential improvements in brain dose estimates for internal emitters

Richard W. Leggett, Sergei Y. Tolmachev, and John D. Boice

ABSTRACT

Background: Element-specific biokinetic models are used to reconstruct doses to systemic tissues from internal emitters. Typically, a systemic model for a radionuclide explicitly depicts only its dominant repositories. Remaining tissues and fluids are aggregated into a pool called *Other tissue* in which the radionuclide is assumed to be uniformly distributed. In the systemic biokinetic models used in radiation protection, the brain usually is addressed as an implicit mass fraction of *Other tissue* rather than an explicitly depicted repository. Due to increasing interest in radiation effects on the brain, efforts are underway to improve brain dosimetry for internal radiation sources.

Methods: We assessed potential improvements in brain dosimetry for internal emitters by explicitly modeling brain kinetics rather than treating the brain as a mass fraction of *Other tissue*. We selected 10 elements for which brain kinetics can be modeled using published biokinetic data. Injection dose coefficients were calculated for a relatively long-lived radioisotope of each element using each of two versions of the ICRP's latest systemic biokinetic model for the element, the original version and a modified version differing only in the treatment of brain. If the ICRP model contained an explicit brain pool, the modified version depicted brain instead as a mass fraction of *Other tissue*. If the ICRP model included brain in *Other tissue*, the modified version included an explicit brain pool with kinetics based on best available brain-specific data.

Results: The result for a given radionuclide is expressed as a ratio A:B, where A and B are the dose coefficients based on the versions of the model with and without an explicit brain pool, respectively. The following ratios A:B were obtained for the 10 radionuclides addressed here: ^{241}Am, 0.13; ^{207}Bi, 0.57; ^{234}U, 0.81; ^{239}Pu, 0.96; ^{203}Hg (vapor), 1.4; ^{134}Cs, 1.5; ^{54}Mn, 1.7; ^{210}Po, 1.7; ^{226}Ra, 1.9; ^{210}Pb, 3.3. These ratios indicate that a dose estimate for brain based on a biokinetic model with brain implicitly contained in *Other tissue* may substantially underestimate or substantially overestimate a dose estimate that reflects best available brain-specific biokinetic data. Of course, the reliability of the latter estimate depends on the quality of the underlying biokinetic data.

Conclusions: Where feasible, the brain should be depicted explicitly in biokinetic models used in epidemiological studies addressing adverse effects of ionizing radiation.

Introduction

There is an increasing interest in potential adverse effects of ionizing radiation on the brain. For example, the Million Person Study (MPS) is evaluating dementia as well as more specific brain diseases such as Alzheimer's, Parkinson's, and motor neuron disease as possible adverse outcomes of combined high- and low-LET exposures to brain tissue (Boice 2017; Boice et al. 2018). The National Aeronautics and Space Administration (NASA) is evaluating the level of protection needed against in-flight behavioral and cognitive impairments from galactic cosmic rays (GCRs). NASA is interested in epidemiological findings on alpha-particle dose to brain as a limited analogy that might inform their decisions regarding protection against the high-energy charged particles encountered in space (NCRP 2016; Boice 2017).

Element-specific biokinetic models are used to reconstruct doses to systemic tissues from internal emitters. These models typically depict explicitly only those tissues that tend to dominate the systemic behavior of the element over time. The remaining tissues are aggregated into a pool called *Other tissue* in which activity is assumed to be uniformly distributed. Explicitly identified tissues usually consist of some subset of liver, kidneys, bone, bone marrow, gonads, thyroid, spleen, and skin. The brain is included explicitly in systemic biokinetic models for a few elements but typically is addressed as an implicit mass fraction of *Other tissue*.

This paper assesses potential improvements in brain dosimetry for internal emitters from explicit modelling of brain kinetics rather than treating the brain as an implicit mass fraction of *Other tissue*. The analysis serves more generally as a case study of potential errors in radiation dose estimates resulting from assuming common biokinetics and uniform activity concentration across a large collection of tissues with different compositions and physiological functions. Comparisons are made of dose coefficients for selected radionuclides based on alternate versions of the systemic biokinetic model for each radionuclide, differing only with respect to explicit versus implicit treatment of brain tissue.

Methods

Biokinetic data for elements are considerably more limited for brain than for tissues, such as bone, liver, and kidneys, that are commonly addressed explicitly in the biokinetic models used in radiation protection and dose reconstruction for radionuclides. Typically, quantitative data on uptake and retention of a given element by the brain come primarily from a small number of studies on laboratory animals. Thus, the explicit models for brain considered in the present study generally rely heavily on data for laboratory animals, mixed with available relevant information for humans such as autopsy measurements indicating the relative contents of an element in the brain and other tissues. The reader is referred to an article by Leggett (2001) for a critical examination of the basis for interspecies extrapolation of biokinetic data, the sources and extent of uncertainties inherent in such extrapolation, and the animal species that have proven to be the most reliable laboratory models for human biokinetics of different families of elements.

We selected 10 radionuclides for which informative element-specific data are available on relative levels of accumulation by the brain and other systemic repositories. Half of these radionuclides are included because they are important internal emitters encountered in the MPS: ^{210}Po (Boice et al. 2006, 2014), ^{226}Ra (Silver et al. 2013; Ellis et al. 2018), ^{234}U (Boice et al. 2006; Ellis et al. 2018), ^{239}Pu (Gilbert et al. 1993; Wiggs et al. 1994; Boice et al. 2014), and ^{241}Am (Boice et al. 2006). Four other radionuclides, ^{54}Mn, ^{134}Cs, ^{203}Hg (inhaled as a vapor), and ^{210}Pb, are included as convenient case studies for exploring the value of an explicit brain model for estimating doses to brain because a well-founded brain model already existed for each of these elements. The remaining radionuclide, ^{207}Bi, illustrates a situation that may frequently arise in assessing the usefulness of an explicit brain model, that is, data on brain kinetics of the element are available but of questionable relevance for modeling biokinetics in humans.

For each of the 10 radionuclides, we compared alternate dose coefficients for brain, one based on the ICRP's latest systemic model for the element and the other based on a modification of that model with a different treatment of brain as described below. Two different variations of the ICRP model were considered for ^{210}Po. It was assumed that the radionuclide is injected into blood of a male worker. The ICRP's updated dosimetry system for workers (ICRP 2016a) was applied.

If the ICRP model for the element already contained an explicit brain pool, a modified version of the ICRP model was constructed by removing the brain pool so that brain was included implicitly in *Other tissue*. The deposition fractions (fractional outflows from a central blood compartment to specified destinations) for compartments of *Other tissue* were increased proportionally so that the sum of the new deposition fractions for compartments of *Other tissue* equaled the sum of all deposition fractions for the brain and *Other tissue* in the ICRP model. Suppose, for example, that in the ICRP model the deposition fraction for the brain is 0.006 and that *Other tissue* contains two compartments with deposition fractions 0.1 and 0.2. In the alternate model, the deposition fractions for the two compartments of *Other tissue* would be increased to $0.1 + 0.002 = 0.102$ and $0.2 + 0.004 = 0.204$.

If the brain was implicitly contained in *Other tissue* in the ICRP model for the element, then a modified version of the ICRP model was constructed by adding an explicit brain pool that exchanged activity with blood and assigning transfer coefficients between brain and blood based on brain-specific biokinetic data. If needed, the deposition fractions for compartments of *Other tissue* were decreased proportionally so that the sum of all deposition fractions for *Other tissue* and the brain equaled the sum of the deposition fractions for compartments of *Other tissue* in the ICRP model. Such an adjustment was not needed if the blood compartment feeding the added brain compartment(s) was not directly connected to *Other tissue*.

Results

Polonium

The analysis for ^{210}Po is described in some detail to illustrate the general approach used here and the considerations that went into selection of biokinetic data.

The ICRP's latest model for systemic polonium (Po) is described in Publication 137 (2017), the third part of a series of reports that provide models, dose coefficients, and

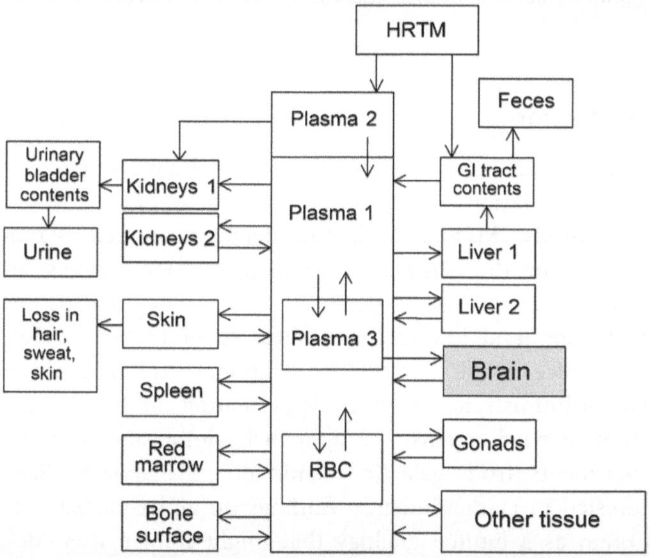

Figure 1. Model structure for systemic Po. The unshaded boxes and associated arrows represent the ICRP's latest model (ICRP 2017). The modified version of the ICRP model considered here adds the brain compartment and associated arrows. RBC: red blood cells; HRTM: Human Respiratory Tract Model (ICRP, 2015).

Table 1. Transfer coefficients in the ICRP's model for systemic polonium.

From	To	Transfer coefficient (d^{-1})
Plasma 2	Plasma 1	800
Plasma 2	Kidneys 1	200
Plasma 1	Plasma 3	4
Plasma 1	RBC	6
Plasma 1	Liver 1	17.5
Plasma 1	Liver 2	17.5
Plasma 1	Kidneys 1	5
Plasma 1	Kidneys 2	5
Plasma 1	Skin	5
Plasma 1	Red Marrow	4
Plasma 1	Bone Surface	1.5
Plasma 1	Spleen	2
Plasma 1	Testes	0.1
Plasma 1	Ovaries	0.05
Plasma 1	Other	32.35
Plasma 3	Plasma 1	0.099
RBC	Plasma 1	0.099
Liver 1	GI Tract	0.139
Liver 2	Plasma 1	0.099
Kidneys 1	Urinary bladder content	0.173
Kidneys 2	Plasma 1	0.099
Skin	Plasma 1	0.00693
Skin	Excreta	0.00693
Red Marrow	Plasma 1	0.099
Bone Surface	Plasma 1	0.0231
Spleen	Plasma 1	0.099
Testes or Ovaries	Plasma 1	0.0139
Other	Plasma 1	0.099

RBC: Red blood cells.

reference bioassay data for assessment of occupational exposure to radionuclides (the 'OIR' series). A more detailed description of the basis for the model is given in an article by Leggett and Eckerman (2001). The structure of the model is shown in Figure 1. Transfer coefficients for a reference worker are listed in Table 1. The brain is not depicted explicitly in the model and hence is contained implicitly in the single compartment representing *Other tissue*.

The ICRP's model for Po was based to a large extent on data for human subjects exposed to ^{210}Po, but data for laboratory animals were used to fill sizable gaps in data for humans. Data for baboons and dogs were given higher weight than findings for other studied animals due to the relatively detailed information available for baboons and dogs; quantitative similarities in the distribution and retention of Po in humans, baboons, and dogs; and the general consideration that primates and dogs have proven to be reasonably good laboratory models for human biokinetics of many elements (Leggett 2001). For the same reasons, data for baboons were used here to develop a modified version of the ICRP's model for Po that contains an explicit brain model. The brain was not addressed in the dog studies used in development of the ICRP's model for Po.

Autopsy data on the distribution of ^{210}Po in tissues are available for uranium miners (Blanchard and Moore 1971) but are not useful for modeling brain kinetics of ^{210}Po because much of the ^{210}Po in tissues of these subjects probably arose from radionuclides that sequentially decayed to ^{210}Po *in vivo*. Polonium-210 was measured at autopsy in an adult male who died 23 days after being poisoned with ^{210}Po (Nathwani et al. 2016; Harrison et al. 2017). The kidneys showed the highest activity concentration. The activity concentration in brain (5500 Bq g^{-1}) was about 10% of that

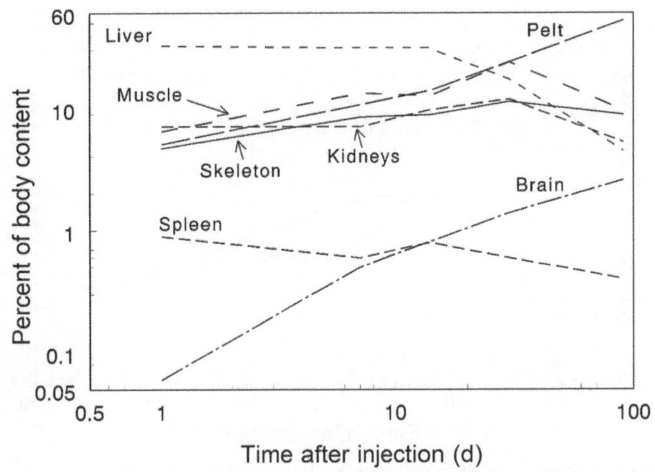

Figure 2. Relative ^{210}Po content in tissues of baboons as a function of time after intravenous injection (data from Cohen et al. 1989; Fellman et al. 1994).

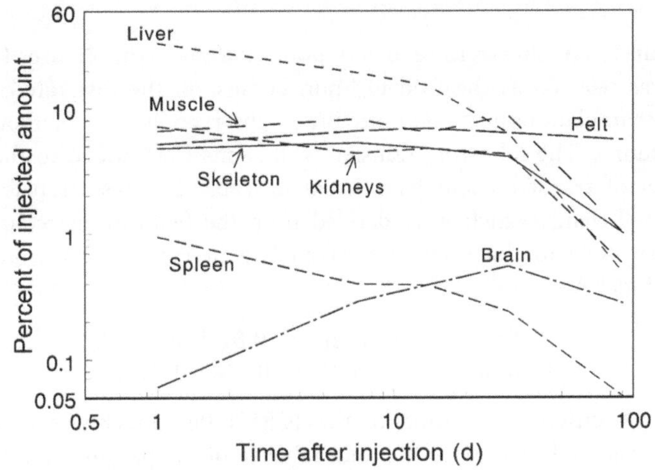

Figure 3. Retention on Po in baboon tissues as a function of time after intravenous injection of ^{210}Po (data of Cohen et al. 1989; Fellman et al. 1994).

in the kidneys, 20% of that in the liver, and 50% of that in the spleen. The concentration in brain was considerably higher than in muscle. Data for this subject were not used in the derivation of a biokinetic model for brain due to the distinct possibility of a substantially increased level of entry of ^{210}Po into brain resulting from alpha-particle breakdown of the blood-brain barrier.

Following intravenous administration of ^{210}Po to baboons, the activity concentration was substantially lower in the brain than in other studied tissues over the first week (Cohen et al. 1989, Fellman et al. 1994). The brain content as a fraction of total-body content of ^{210}Po increased monotonically over the 3-month study period (Figure 2). The data indicate that the residence time for Po in the brain was substantially greater than in other tissues except pelt (Figures 2 and 3).

A variation of the structure of the ICRP's systemic model for Po in workers was developed by addition of a single compartment representing the brain (shaded box in Figure 1). Transfer coefficients describing exchange of Po between brain and compartments of plasma were set to approximate the time-dependent retention of ^{210}Po in brain determined in the study on baboons summarized above (Figure 4, solid

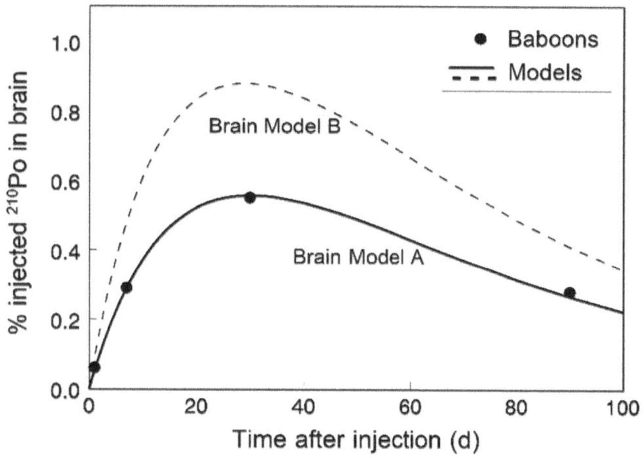

Figure 4. Predictions of brain retention of Po based on two variations of the ICRP's systemic model with different brain kinetics (Brain Model A and Brain Model B defined in text).

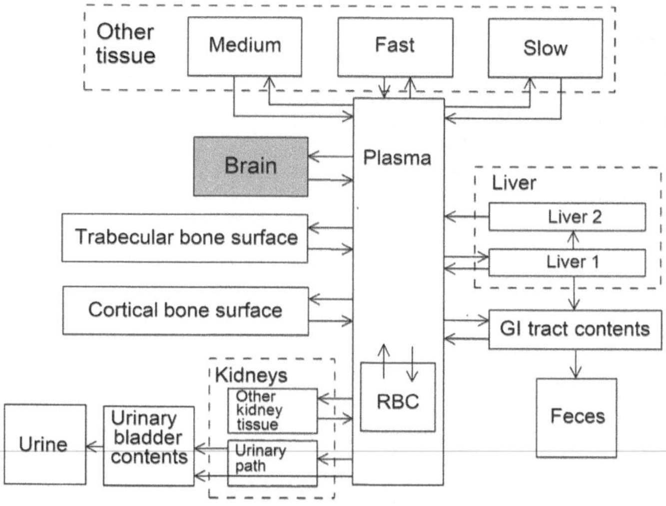

Figure 5. Model structure for systemic Bi. The unshaded boxes and associated arrows represent the ICRP's latest model (ICRP2017). The modified version of the ICRP model considered here adds the brain compartment and associated arrows. RBC: red blood cells; GI: gastrointestinal.

curve). A slow-release blood pool (Plasma 3 in Figure 1) was selected as the feed to brain because of the low rate of accumulation of ^{210}Po into brain observed in the baboon studies. The following transfer coefficients were added to the set of transfer coefficients listed in Table 1. These transfer coefficients, which were derived from the fit to brain retention data for baboons, are referred to in Figure 4 as Brain Model A:

$$\text{Plasma 3 to Brain} = 0.0135\ d^{-1},$$
$$\text{Brain to Plasma 1} = 0.0385\ d^{-1}.$$

A different variation of the ICRP's Po model was also considered in view of the total reliance of the parameter values for the brain compartment on data for a nonhuman species. The mass of the brain relative to total-body mass appears to be roughly 70% higher for adult humans than adult baboons, based on comparison of data of Cohen et al. (1989) for baboons with data of ICRP Publication 89 (2002) for reference adult humans. It was considered that extrapolation of brain kinetics of Po from baboons to humans should perhaps reflect that difference through assignment of a larger deposition fraction to brain than derived from the baboon data. An alternate modification of the ICRP model was performed by assigning a deposition fraction to brain that is 1.7 times the deposition fraction in Brain Model A. The removal rate from brain to blood was left at the value assigned in Brain Model A. This second set of parameter values for brain is referred to in Figure 4 as Brain Model B:

$$\text{Plasma 3 to Brain} = 0.02295\ d^{-1},$$
$$\text{Brain to Plasma 1} = 0.0385\ d^{-1}.$$

Dose coefficients for brain were derived for injection of ^{210}Po into blood of a male worker, using the current ICRP model for Po and each of the two modifications of the ICRP model described above, i.e., using Brain Model A or Brain Model B. The dose coefficient based on the modification of the ICRP model using Brain Model A is 1.3 times the value based on the ICRP model. The dose coefficient based on the modification of the ICRP model using Brain Model B is 2.0 times the value based on the ICRP model.

Bismuth

The ICRP's current systemic model for bismuth (Bi) is described in ICRP Publication 137 (2017). The model structure is shown in Figure 5. Parameter values for an adult worker are listed in the section on bismuth in Publication 137. In the model, the brain is implicitly contained in *Other tissue*, which consists of three compartments representing fast, medium, and slow phases of removal of Bi to blood.

Autopsy measurements on patients who had received intramuscular Bi treatments indicate the following central estimates of the relative Bi concentrations in tissues: kidney, 1.0; liver, 0.2; spleen, 0.048; colon, 0.039; lung, 0.027; brain, 0.018; blood, 0.015 (Sollmann and Seifter 1942; Sollmann 1957; Fowler and Vouk 1986). The patients apparently lived for extended periods following the end of Bi treatment. The findings suggest relatively low accumulation of Bi by the human brain compared with the other studied tissues.

Biokinetic data for Bi in rats indicate a lower rate of uptake but a longer retention time for Bi in brain than in other soft tissues. In the first few days after parenteral injection of Bi compounds into rats, the brain showed considerably lower uptake of Bi than other tissues (Gregus and Klaassen 1986; Zidenberg-Cherr et al. 1987). Following oral administration of Bi to rats over a 14-month period, the Bi concentration in brain was of the same order as that of most other studied tissues with the main exception of kidneys, the primary repository for Bi (Lee et al. 1980).

A variation of the structure of the ICRP's systemic model for Bi in workers was developed by addition of a compartment representing the brain that exchanges Bi with blood plasma (Figure 5). Parameter values for brain were set for reasonable consistency with the semi-quantitative data on brain kinetics of Bi in human subjects (the times between Bi treatments and death were not found in reports of these studies) and the quantitative data for rats. It was assumed that the brain receives 0.01% of outflow from plasma and loses Bi to plasma with a biological half-time of 100 d. The implied transfer coefficients describing exchange of Bi between plasma and brain are

$$\text{Plasma to Brain} = 0.04\ d^{-1},$$
$$\text{Brain to Plasma} = 0.00693\ d^{-1}.$$

The total outflow rate from plasma to all destinations combined was left unchanged by reducing the transfer coefficients from plasma to each of the compartments of *Other tissue* proportionally to their values in the ICRP model.

Alternate dose coefficients for brain were derived for the case of injection of the long-lived bismuth radioisotope ^{207}Bi into blood of a male worker, using the current ICRP model for Po and the modification of the ICRP model described above. The dose coefficient based on the modified ICRP model with an explicit brain compartment is 0.57 times the value based on the ICRP model.

Lead

The ICRP's current systemic model for lead (Pb) is described in ICRP Publication 137 (2017). In this model, the brain is implicitly contained in *Other tissue*, which consists of three compartments representing fast, medium, and slow phases of removal of Pb to blood.

The ICRP's model for Pb is a modified version of a model developed by Leggett (1993) for application to Pb as a chemical hazard as well as a radiation source. The brain was addressed explicitly in the original model because of the brain's sensitivity to Pb as a chemical toxin.

The distribution of Pb in the brain is uneven and varies with the level of exposure. For low-level exposure, the Pb concentration in different regions of the human brain appears to be associated with the potassium concentration, suggesting accumulation of Pb in cell-rich parts of the brain such as the hippocampus (Grandjean 1978; Bryce-Smith and Stephens 1983; Petit et al. 1983). For high exposure, Pb may gain substantial access to neural tissue following breakdown of the blood brain barrier (Petit et al. 1983).

Due to limitations in available region-specific biokinetic data, the brain was represented in the original model (Leggett 1993) as a single compartment that exchanges Pb slowly with a compartment representing diffusible Pb in blood plasma. Parameter values describing the kinetics of Pb in brain were selected for consistency of model predictions with short- to intermediate-term data for laboratory animals including baboons and dogs (Cohen et al. 1970; Lloyd et al. 1975) and autopsy data for Pb in tissues of occupationally or environmentally exposed persons (Leggett 1993).

A modified version of the ICRP's model for systemic Pb was developed by adding a compartment representing brain and assigning flow rates between plasma and brain consistent with predictions of the more detailed model of Leggett (1993). That is, the parameter values describing uptake and outflow of Brain Pb were set for consistency with the brain retention curve predicted by the original model. The following transfer coefficients between plasma and brain were assigned:

$$\text{Plasma to Brain} = 0.018\ d^{-1},$$
$$\text{Brain to Plasma} = 0.00095\ d^{-1}.$$

The total outflow rate from plasma to all destinations combined in the ICRP model was left unchanged by reducing the transfer coefficients from plasma to each of the compartments of *Other tissue* proportionally to their values in the ICRP model.

Alternate dose coefficients were derived for injection of ^{210}Pb into blood of an adult male worker based on the systemic model for Pb applied in ICRP Publication 137 and the modified model with an explicitly modeled brain compartment described above. The dose coefficients for ^{210}Pb included the contribution to dose from its radioactive progeny ^{210}Bi and ^{210}Po produced in the body after intake of ^{210}Pb. Biokinetic models described in ICRP Publication 137 for radioisotopes of Bi and Po produced in systemic compartments by decay of a radioisotope of Pb were used together with the ICRP's model for Pb to derive one of the dose coefficients for brain. For calculation of the alternate dose coefficient based on explicit modeling of brain, the model for Bi as progeny of Pb applied in Publication 137 was modified by addition of a brain compartment with parameter values described above in the discussion of Bi. For ^{210}Po produced in brain after intake of ^{210}Pb, we applied the model for Po as Bi progeny applied in Publication 137, modified by addition of a brain pool as described above for Po. We examined the sensitivity of the dose coefficient for ^{210}Pb to the alternate brain kinetics of Po considered earlier by deriving dose coefficients for ^{210}Pb based on each set of parameter values (Brain Model A and Brain Model B described above).

The dose coefficient based on the modified model with an explicit brain compartment is 3.0 times the value based on the ICRP model if Brain Model A is applied to the progeny ^{210}Po, and 3.5 times the value based on the ICRP models if Brain Model B is applied to ^{210}Po.

Plutonium

The ICRP's systemic model for plutonium (Pu) is updated in Part 4 of the series on occupational intake of radionuclides (to appear). See Leggett et al. (2005) for a description of that model including a list of transfer coefficients. As in previous ICRP biokinetic models for systemic Pu, the brain is included implicitly in *Other tissue*.

The U.S. Transuranium and Uranium Registries (USTUR) has studied the distribution of Pu in tissues of former nuclear workers (Kathren and Tolmachev 2019). The activity of ^{239}Pu in liver and skeleton has been measured routinely, and activity in brain has been measured in some subjects. As a central estimate (mean, median, or geometric mean) for these individuals, the brain contains ~0.2% as much ^{239}Pu as liver and skeleton combined (Figure 6).

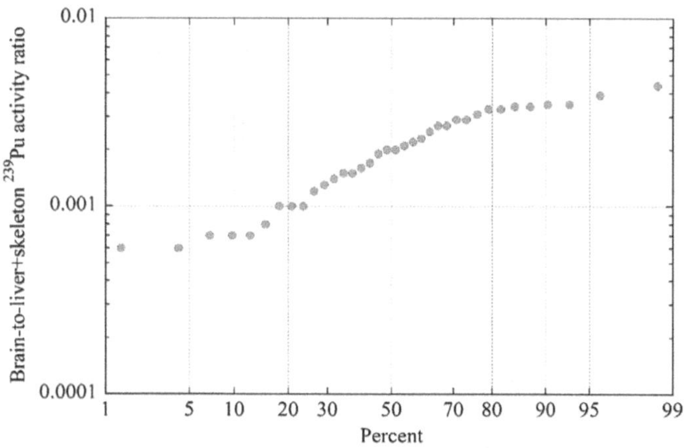

Figure 6. Brain activity of ^{239}Pu relative to activity in liver plus skeleton in Pu workers.

Uptake and retention of Pu by the brain has been measured in laboratory animals following its parenteral administration. In dogs, the ratio of activity in brain to that in the liver plus skeleton varied from about 0.00004 to 0.0023 at 13–30 d after injection, with a central value of about 0.0013 (mean, 0.0012; median, 0.0014) (Lloyd et al. 1976).

A modified version of the ICRP's model for systemic Pu was developed by adding a compartment representing brain and assigning flow rates between blood and brain consistent with the relative contents of Pu in brain and in liver + skeleton in Pu workers at late times and dogs at early times after exposure. The following transfer coefficients were assigned:

$$\begin{aligned}\text{Blood 1 to Brain} &= 0.0009\ \text{d}^{-1},\\ \text{Brain to Blood 2} &= 0.0002\ \text{d}^{-1},\end{aligned}$$

where Blood 1 and Blood 2 are blood compartments that receive absorbed Pu and Pu returning from tissues to blood, respectively.

Dose coefficients were derived for brain for the case of injection of ^{239}Pu into blood of a male worker, using the ICRP's systemic model for Pu and the modified version of that model with an explicit brain compartment. The dose coefficient based on the modified ICRP model with an explicit brain compartment is 0.96 times the value based on the ICRP model.

Americium

The ICRP's systemic model for americium (Am) is updated in Part 4 of the series on occupational intake of radionuclides (to appear). The updated model is a modified version of the ICRP's model for systemic Am for adult members of the public adopted in Publication 67 (1993). The main changes from the earlier model are an added liver compartment and depiction of redeposition onto local cortical bone surface of a portion of the Am removed from cortical bone surface by bone restructuring processes. As in previous ICRP models for systemic Am, the brain is included implicitly in *Other tissue* in the updated model.

The USTUR whole-body analyses include a subject who apparently was exposed to pure ^{241}Am about 25 y before his death (Breitenstein and Newton 1985; McInroy et al. 1985). The tissue of his left hand contained about 2% of the total-body content, suggesting that a puncture wound to his hand may have been the mode of exposure. The brain showed the lowest ^{241}Am concentration of all systemic tissues. The concentration in brain was roughly 9% of the average concentration in soft tissues and 2% of the average concentration in the total body. The ^{241}Am content in the brain represented about 0.037% of systemic ^{241}Am. This is slightly below the lower end of the range of corresponding values for Pu indicated by USTUR data (Figure 6), assuming that liver + skeleton represents at least 90% of systemic Pu at times remote from exposure.

Filipy and Kathren (1996) compared the concentrations of $^{239+240}$Pu and ^{241}Am in brain and liver in each of eight USTUR donors with occupational exposures to these radionuclides. Much of the ^{241}Am measured in these subjects may have arisen from production of ^{241}Am in the body following intake of its parent ^{241}Pu, which usually is found together with ^{239}Pu and ^{240}Pu. The concentration ratio $^{239+240}$Pu: ^{241}Am in liver ranged from 1.7 to 21, with mean and median values of 10.6 ± 6.4 (SD) and 8.6, respectively. The mean and median concentration ratios of $^{239+240}$Pu in liver to that in brain were 173 ± 180 (SD) and 105, respectively. The mean and median concentration ratios of ^{241}Am in liver to that in brain were 94 ± 104 (SD) and 55, respectively. The results suggest that the brain accumulates and/or retains Am to a lesser extent than Pu, particularly taking into account that the liver typically contains a much smaller portion of systemic Am than Pu at times remote from exposure, either in Pu workers or in subjects exposed directly to ^{241}Am.

In baboons, the brain contained ~0.019% of intravenously injected ^{241}Am at 1-86 d after intravenous injection, 0.0081% at 206 d, and 0.0024% at 817 d (Guilmette et al. 1980). In dogs, the brain contained <0.001% of intravenously administered ^{241}Am at 1–22 d and 0.008% at 401–4448 d (Lloyd et al. 1970).

A modified version of the ICRP's model for systemic Am was developed by adding a compartment representing brain and assigning flow rates between blood and brain consistent with the relative contents of Am in brain and liver plus skeleton in human subjects at late times and in laboratory animals in the early months after exposure. The following transfer coefficients were assigned:

$$\begin{aligned}\text{Blood to Brain} &= 0.0024\ \text{d}^{-1},\\ \text{Brain to Blood} &= 0.000077\ \text{d}^{-1}.\end{aligned}$$

Alternate dose coefficients were derived for brain for the case of injection of ^{214}Am into blood of a male worker, using the ICRP's systemic model for Am and the modified version of that model with an explicit brain compartment. The dose coefficient based on the modified ICRP model with an explicit brain pool is 0.13 times the value based on the ICRP model.

Cesium

A close physiological relationship of the alkali metals potassium (K), rubidium (Rb), and cesium (Cs) is well established.

These elements compete for transport into and out of cells of the body with transport rates across cell membranes generally decreasing in the order K > Rb > Cs (Leggett et al. 2003).

The rate of transfer of K, Rb, or Cs from plasma into a tissue or organ can be estimated as the product of the blood flow rate to that tissue (plasma volumes per day) and an element- and tissue-specific extraction fraction, denoted by $E_{T,X}$ for tissue T and element X, defined as the fraction of atoms extracted by that tissue in a single passage from arterial to venous plasma. The rate of return from a tissue to plasma can be estimated from the relative contents of the element in plasma and the tissue at equilibrium as estimated from reported autopsy measurements for the element.

Experimentally determined tissue-specific extraction fractions typically are in the range 0.6–0.9 for K, 0.4–0.8 for Rb, and 0.05–0.2 for Cs for most tissues, but much lower extraction fractions are seen for the brain. Estimated extraction fractions for the brain are on the order of 0.015 for K, 0.01 for Rb, and 0.002 for Cs. For a reference adult male, the transfer coefficient describing flow of Cs, for example, from plasma into brain can be calculated as $0.002 \times 0.12 \times 1766$ d^{-1} = 0.424 d^{-1}, where 0.002 is the extraction fraction for brain for Cs, 0.12 is the reference fraction of cardiac output received by brain, and 1766 d^{-1} is the reference cardiac output in terms of the volume of plasma pumped through the heart each day. The transfer coefficient describing total outflow of Cs from brain is estimated as 0.002×0.424 d^{-1}/ $0.01 = 0.0848$ d^{-1} (corresponding to a half-time of ~8 d, a relatively long half-time for escape of Cs from tissues), where 0.002 and 0.01 are reference fractions of total-body Cs in plasma and brain, respectively, at equilibrium, as estimated from reported autopsy measurements of concentrations of Cs in tissues of the adult human body. Based on these concepts, K, Rb, or Cs is estimated to be taken up much more slowly by the brain than by other tissues but also to be retained much longer by the brain than by other tissues.

These concepts were used to develop the ICRP's updated model for Cs described in ICRP Publication 137 (2017). The model was originally constructed around a framework representing the paths of blood flow in the human body (Leggett et al. 2003). The original model structure was later replaced by a more conventional model structure built around a central blood pool (Leggett, 2014). The latter structure was used in ICRP Publication 137.

A modified version of the model for systemic Cs used in ICRP Publication 137 was developed for this study by eliminating the brain compartment and increasing the deposition fractions in the compartments of *Other tissue* so that deposition fractions for all tissues and excretion pathways still add to 1.0.

Alternate dose coefficients were derived for the case of injection of ^{134}Cs into blood of a male worker, using the ICRP's systemic model for Cs and the modified version of that model with brain implicitly contained in *Other tissue.* The dose coefficient based on the ICRP model with an explicit brain compartment is 1.5 times the value based on the modified model.

Manganese

Manganese (Mn) is an essential trace element required for metabolism of amino acids, proteins, carbohydrates, and lipids in mammals. Ingested Mn is absorbed from the gut via the portal vein and is removed by the liver as required to achieve homeostasis. Manganese entering blood by other routes, for example, by absorption from the lungs after inhalation, initially bypasses the control processes in the liver and ultimately becomes largely bound to transferrin, which may be transferred to the brain or other tissues by transferrin receptors. Thus, intake of Mn by routes other than ingestion affect its biokinetics and homeostatic balance and result in elevated accumulation in the brain. Excessive inhalation of Mn can result in adverse health effects including progressive neurodegenerative damage with an associated motor dysfunction syndrome similar to that seen in Parkinson's disease. Most reported cases of Mn intoxication have been linked to chronic occupational exposure to airborne Mn, particularly among manganese miners, welders, smelters, and workers in dry cell battery factories (Leggett 2011)

Isotopic studies on laboratory animals show that absorbed or intravenously injected manganese leaves blood rapidly and initially concentrates mainly in organs rich in mitochondria, particularly the liver, pancreas, and kidneys. Over time, other organs, especially brain and bone, contain increasingly larger portions of the retained activity. Studies on rats indicate that for most tissues the concentration as a function of time roughly paralleled whole-body retention, but bone and brain had slower loss than other tissues (Leggett 2011). The adult human brain has been estimated to contain about 5% of the total body content of stable manganese.

The ICRP's latest biokinetic model for Mn is described briefly in ICRP Publication 134 (2016b), where the model is used to describe the behavior of Mn as a progeny of iron (Fe). A more detailed description including a figure showing the model structure and a table of transfer coefficients can be found in an article by Leggett (2011). The model was developed from data on the time-dependent behavior of Mn in human subjects and laboratory animals, together with autopsy studies of the distribution of Mn in the human body. The model contains a compartment explicitly representing the brain. As depicted in the model, the brain takes up only 0.1% of the outflow of Mn from blood plasma following its inhalation and absorption to blood, but the retention time is much longer in the brain than in other soft tissues, resulting in a gradual increase in the Mn concentration in brain relative to other soft tissues.

A modified version of the ICRP's model for systemic Mn was developed by eliminating the brain compartment and increasing the deposition fractions in the compartments of *Other tissue* so that deposition fractions for all tissues and excretion pathways still add to 1.

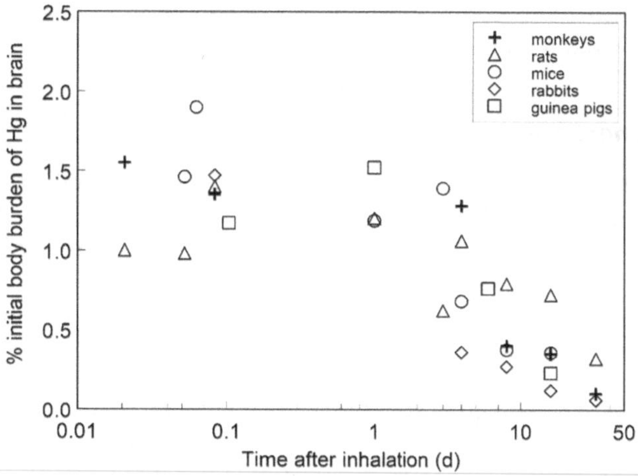

Figure 7. Observations of time-dependent content of Hg in the brain in laboratory animals after acute inhalation of Hg vapor. Data from Khayat and Dencker 1983, 1984; Magos 1967; Berlin et al. 1969; Nordberg and Serenius 1969; Hursh et al. 1976.

Alternate dose coefficients were derived for the case of injection of ^{54}Mn into blood of a male worker, using the ICRP's systemic model for Mn and the modified version of that model with brain implicitly contained in *Other tissue*. The dose coefficient based on the ICRP model with an explicit brain pool is 1.7 times the value based on the modified model.

Mercury vapor

Inhaled mercury (Hg) vapor is carried in plasma to tissues including the brain, where it readily crosses the blood–brain barrier. After entering brain tissue, Hg vapor is oxidized to the divalent form of Hg, which is trapped in the brain because it is difficult for the divalent form to cross the blood–brain barrier. Studies on mice, rats, rabbits, and monkeys indicate that fractional uptake of mercury by the brain is an order of magnitude greater after inhalation of Hg vapor than after intravenous injection of Hg salts (Berlin et al. 1966, 1969).

After acute inhalation of Hg vapor by squirrel monkeys, rats, mice, rabbits, and guinea pigs, the peak mercury content in the brain typically was 1–2% of the initial body burden (Figure 7). The pattern of uptake and retention was broadly consistent across species, despite the large variation in brain size as a fraction of total body weight in the studied species. The data for laboratory animals indicate a biological half-time on the order of 10 d for the preponderance of inorganic Hg deposited in the brain. Estimates of the removal half-time of Hg from the brain in human subjects exposed to Hg vapor or divalent Hg are in the range 14–29 d (Hursh et al. 1976; Newton and Fry 1978).

A long-term component of retention of Hg in the brain is suggested by some studies on laboratory animals and by autopsy data for human subjects who were occupationally exposed to Hg vapor. Studies involving acute exposure to laboratory animals suggest that long-term retention (weeks) of Hg in the brain following inhalation of Hg vapor represents at most 2% of the inhaled amount.

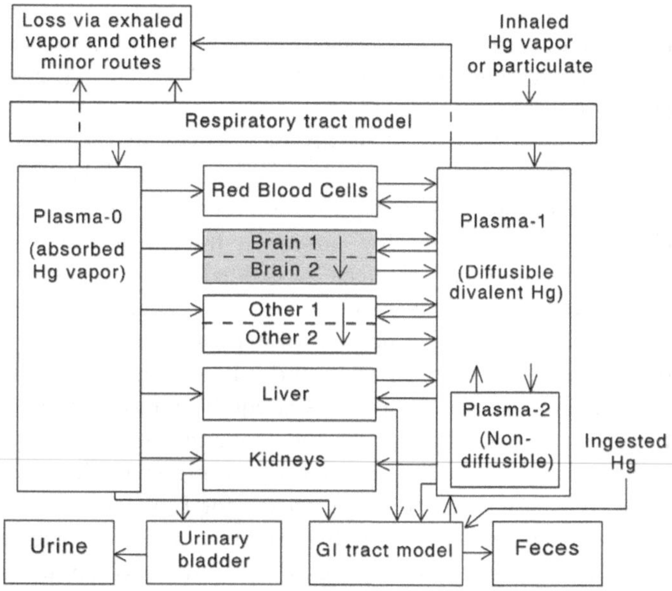

Figure 8. Structure of the ICRP's provisional model for mercury taken into the body as Hg vapor or divalent Hg.

Table 2. Transfer coefficients in the ICRP's provisional model for vapor and divalent forms of inorganic mercury.

Path	Transfer coefficient (d^{-1})
Plasma-0 (Hg vapor) to:	
Red blood cells	100
Brain 1	20
Kidneys	120
Liver	50
Other 1	560
SI content	80
Minor excretion pathways	70
Plasma 1 (Diffusible divalent Hg) to:	
Red Blood Cells	0.665
Plasma 2 (Nondiffusible divalent Hg)	1.996
Brain 1	0.0333
Kidneys	5.823
Liver	3.322
Other 1	2.296
Small intestine content	1.664
Minor excretion pathways	0.832
Red Blood Cells to Plasma 1	0.231
Plasma 2 to Plasma 1	0.693
Brain 1 to Plasma 1	0.0329
Brain 1 to Brain 2	0.00173
Brain 2 to Plasma 1	0.000380
Kidneys to Urinary bladder content	0.0198
Liver to Small intestine content	0.0347
Liver to Plasma 1	0.0347
Other 1 to Plasma 1	0.0260
Other 1 to Other 2	0.00866
Other 2 to Plasma 1	0.00693

A provisional biokinetic model for systemic Hg applicable to different forms of inorganic Hg including Hg vapor has been developed for the ICRP's OIR series. The model structure is shown in Figure 8. Parameter values are not yet available in ICRP reports or in the open literature and hence are provided here (Table 2).

A modified version of the ICRP's provisional model for systemic Hg vapor was developed by eliminating the two brain compartments and increasing the deposition fraction in the compartment of *Other tissue* named Other 1, so that deposition fractions for all tissues and excretion pathways still add to 1.

Alternate dose coefficients were derived for the case of injection of ^{203}Hg vapor into blood of a male worker, using the provisional ICRP model for Hg and the modified version of that model with brain implicitly contained in *Other tissue*. The dose coefficient based on the ICRP model with an explicit brain pool is 1.4 times the value based on the modified model.

Radium

The ICRP's systemic model for radium was updated in ICRP Publication 137 (2017). The updated model is a modified version of the ICRP's model for systemic Ra for adult members of the public adopted in Publication 67 (1993). The main changes from the earlier model are an added liver compartment and compartments explicitly representing the kidneys. As in previous ICRP models for systemic Ra, the brain is included implicitly in *Other tissue*.

Schlenker et al. (1982) reviewed autopsy data on the ^{226}Ra content of soft tissues in 17 subjects who received ^{226}Ra by injection or ingestion at times from 5 d to 53 y before death. Schlenker and coworkers normalized ^{226}Ra concentrations to total body activity. That is, ^{226}Ra concentrations in tissues were tabulated as specific activity in individual tissues divided by total body activity. Data for brain were given for four times or time periods post injection: 230–1800 d; 4200 d; 17,000 d; and 17,700 d. Comparison of the normalized activity in brain to corresponding values for other sampled tissues contained in *Other tissue* in the ICRP's model for Ra suggest that the normalized concentrations in brain may be about 1.5–3.0 times higher than the median normalized concentration in *Other tissue*, depending on the time after intake and the method of deriving the median concentration in *Other tissue* from the uneven data set. The median value was used here as the measure of central tendency in view of the wide scatter in normalized concentrations for soft tissues.

Following intravenous administration of ^{224}Ra chloride to dogs, the activity in brain expressed as a percentage of total activity in soft-tissues excluding liver and kidneys (i.e. in *Other tissue* in the ICRP's model for systemic Ra) increased continually with time over the 7-d observation period: 0.07% at 1 h, 0.13% at 8 h, 0.28% at 1 d, 0.61% at 3 d, and 1.43% at 7 d.

A modified version of the ICRP's model for systemic Ra was developed by adding a compartment representing brain and assigning transfer rates between plasma and the brain compartment that yield model predictions reasonably consistent with the data for human subjects at times remote from exposure and the data for dogs at early times after injection. The following transfer coefficients between blood and brain were assigned:

$$\text{Blood to Brain} = 0.0035\ d^{-1},$$
$$\text{Brain to Blood} = 0.00038\ d^{-1}.$$

The transfer coefficient from brain to blood is the value applied to the slow turnover compartment of *Other tissue* in

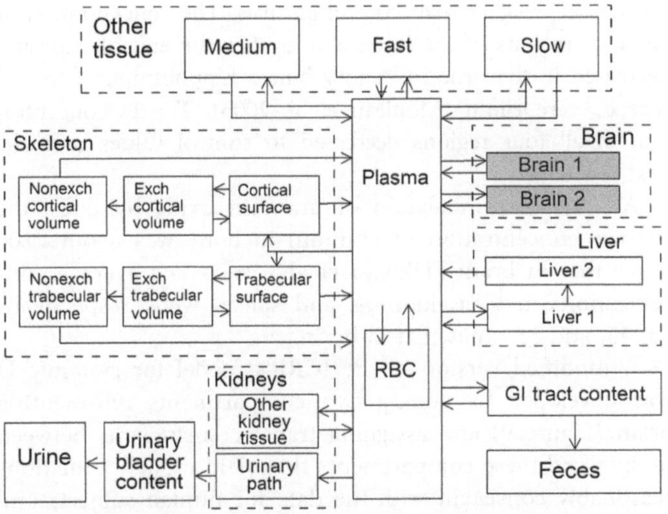

Figure 9. Model structure for systemic U. The unshaded boxes and associated arrows represent the ICRP's latest model (ICRP2017). The modified version of the ICRP model considered here adds two brain compartments and associated arrows. RBC: red blood cells; Exch: exchangeable; Nonexch: nonexchangeable; GI: gastrointestinal.

the ICRP model. The total outflow rate from plasma to all destinations combined was left unchanged by reducing the transfer coefficient from the slow turnover compartment of *Other tissue* by the transfer coefficient from blood to brain.

Alternate dose coefficients were derived for brain for the case of injection of ^{226}Ra into blood of a male worker, using the ICRP's systemic model for Ra and the modified version of that model with an explicit brain compartment. For simplicity, attention was restricted to dose from the parent radionuclide ^{226}Ra only, which contributes more than 90% of the dose to brain in the ICRP's model. The dose coefficient based on the modified ICRP model with an added explicit brain pool is 1.9 times the value based on the ICRP's unmodified model.

Uranium

The ICRP's systemic model for uranium is described in ICRP Publication 137 (2017). The model structure is shown in Figure 9. In this model, the brain is included implicitly in *Other tissue*.

The tissue distribution of uranium was determined in three USTUR donors (Registrants) thought to have only environmental exposure to U (Kathren and Tolmachev 2015) and in two Registrants with occupational exposure to U (Russell and Kathren 2004; Avtandilashvili et al. 2015). The U content of brain as a percentage of systemic U was 0.4–1.0% in the environmentally exposed subjects and 0.4–1.2% in the occupationally exposed subject. The average for the five subjects was 0.7%.

Six months after surgical implant of depleted uranium pellets into rats, an unexpectedly high, dose-dependent accumulation of U was observed in the rat brain (Pellmar et al. 1999). Uranium was found to be heterogeneously distributed in the brain.

Rats exposed repeatedly over 3 weeks to depleted uranium dioxide via inhalation showed U concentrations in brain substantially higher than control animals. The concentration in different regions of the brain at 1 d after the end of exposure decreased in the order: olfactory bulb > hippocampus > frontal cortex > cerebellum (Monleau et al. 2005). The U concentration in all four regions decreased to control values over the next few days.

At 5 y after inhalation of uranium oxide by dogs, the average concentration of uranium in bone was about 2200 times that in brain (Tikhaya et al. 1965). The average concentrations in liver, kidneys, and spleen were, respectively, 73, 45, and 35 times that in brain.

A modified version of the ICRP's model for systemic U was developed by adding two compartments representing brain (Figure 9) and assigning transfer coefficients between plasma and these compartments that yield model predictions reasonably consistent with the data for human subjects and laboratory animals. The following transfer coefficients between plasma and brain compartments were assigned:

$$\begin{aligned} \text{Plasma to Brain 1} &= 0.0326\ \text{d}^{-1}, \\ \text{Plasma to Brain 2} &= 0.00147\ \text{d}^{-1}, \\ \text{Brain 1 to Plasma} &= 0.0347\ \text{d}^{-1}, \\ \text{Brain 2 to Plasma} &= 0.000019\ \text{d}^{-1}. \end{aligned}$$

The outflow rates from Brain 1 and Brain 2 to plasma were assumed to be the same as the outflow rates from the medium and slow exchange compartments of *Other tissue* in the ICRP's systemic model for U. The total outflow rate from plasma to all destinations combined was left unchanged by reducing the transfer coefficients from plasma to the medium and slow exchange compartments of *Other tissue* proportionally to their values in the ICRP model.

Alternate dose coefficients were derived for brain for the case of injection of ^{234}U into blood of a male worker, using the ICRP's systemic model for U and the modified version of that model with explicit brain compartments. The dose coefficient based on the modified ICRP model with an explicit brain pool is 0.81 times the value based on the ICRP model.

Summary and conclusions

Typically, an element-specific systemic biokinetic model used in dose reconstruction for a radionuclide depicts explicitly only the dominant repositories of the element. Remaining tissues are aggregated into a pool called *Other tissue* in which the element is assumed to be uniformly distributed.

In the systemic biokinetic models currently used in dose reconstructions, the brain usually is addressed as an implicit mass fraction of *Other tissue* rather than an explicitly depicted repository. Due to increasing interest in potential adverse effects of radiation on the brain, efforts are underway within the Million Person Study to improve brain dosimetry for both internal and external radiation sources (Boice et al. 2006, 2014; Ellis et al. 2018). As part of this effort, we are assessing potential improvements in brain dosimetry for internal emitters resulting from explicit modelling of brain kinetics rather than treating the brain as an implicit mass fraction of *Other tissue*. This paper summarizes results of 10 case studies in which we compared dose coefficients for the brain based on two versions of the ICRP's latest model for the radionuclide of interest. If the ICRP model contained an explicit brain pool, we built an alternate version of the model depicting brain as a mass fraction of *Other tissue*. If the ICRP model included brain in *Other tissue*, we built an alternate version of the model that included an explicit brain pool with kinetics based on best available brain-specific data.

Table 3. Summary of ratios of injection dose coefficients for brain in a reference male worker, based on systemic biokinetic models with or without an explicitly modeled brain.

Radionuclide	Ratio of dose coefficients for brain Explicit Brain : Implicit Brain
Americium-241	0.13
Bismuth-207	0.57
Uranium-234	0.81
Plutonium-239	0.96
Mercury-203 (vapor)	1.4
Cesium-134	1.5
Manganese-54	1.7
Polonium-210	1.7[a]
Radium-226	1.9[b]
Lead-210	3.3[c]

[a]Average ratio. Alternate explicit brain models were addressed (see text).
[b]Based on dose from ^{226}Ra only; contribution from progeny not considered.
[c]Average ratio. Alternate explicit brain models were addressed for ^{210}Po produced in the body after intake of ^{210}Pb (see text).

The results of the study are summarized in Table 3. The values in the right column of Table 3 are ratios A:B, where A is the dose coefficient based on the version of the model with an explicit brain pool, and B is the dose coefficient based on the version of the model in which the brain is implicitly contained in *Other tissue*. It is assumed in each case that the radionuclide is injected into blood of an adult male worker.

Based on reviews of biokinetic data for brain for the present case studies and for additional elements not addressed here, it appears that the brain typically has a lower rate of uptake per gram of tissue but a longer residence time than do most other studied soft tissues. Thus, an initially low uptake of a radionuclide by brain should not be interpreted as indicating that the dose to brain is smaller than that of most other tissues.

For the 10 radionuclides addressed in this study, the ratio A:B defined above ranged from 0.13 to 3.3 and were in the range 1.4–3.3 in six of the 10 case studies. This indicates that a dose estimate for brain based on a biokinetic model with brain implicitly contained in *Other tissue* may substantially underestimate or substantially overestimate a dose estimate that reflects best available brain-specific biokinetic data.

Regarding those radionuclides addressed here that are frequently encountered in the MPS, the results of this study indicate that addition of an explicit brain model in the current systemic models for U and Pu does not appear to be needed for epidemiological purposes because this would result in little change in dose estimates for brain for important isotopes of these elements. Addition of an explicit brain model to current systemic models for Po, Ra, and Am does appear to be warranted at least for epidemiological purposes because best available data suggest that the current models may substantially over- or underestimate the dose to brain from important isotopes of these elements.

In interpreting the results of the study as summarized in Table 3, it should be kept in mind that the 10 elements were not selected entirely at random. Four of the elements (Mn, Cs, Hg, and Pb) were chosen for convenience, in that the latest ICRP models for these elements (or a more detailed, published version of the latest ICRP model, in the case of Pb) already included an explicit brain pool. This helped to minimize biokinetic modeling efforts in the study. Of the four elements with existing brain models, three (Mn, Hg as vapor, and Pb) are known to have relatively high uptake by the brain. Thus, regarding the ratio A:B of dose coefficients for brain based on biokinetic models with or without explicit brain pools, the overall results of this study are biased on the high side.

For a given radionuclide, the potential for improvement of dose estimates for brain through explicit depiction of the brain in the systemic model depends on the quantity and quality of available element-specific biokinetic data for brain and the associated uncertainties in a predictive brain model based on those data. Ideally, only data for human subjects would be used in the development of a biokinetic model for humans. In practice, it is often necessary to supplement data for human subjects with findings for laboratory animals and in some cases to rely solely on animal data. The reader is referred to an article by Leggett (2001) for a critical examination of the basis for interspecies extrapolation of biokinetic data, the sources and extent of uncertainties inherent in such extrapolation, and the preferred animal species to use as laboratory models for different families of elements.

The preponderance of published data on accumulation of elements in the brain come from studies on laboratory animals. Best available data for the human brain generally come from autopsy studies of occupationally or environmentally exposed subjects. Such autopsy data were found for seven of the elements addressed here: Cs, Mn, Pb, Ra, U, Pu, Am. The brain models applied here to these elements were required to be consistent with the element-specific brain contents (as a percentage of the systemic content) determined in the autopsy studies. The brain model for Cs is a well-supported model based on data on the blood perfusion rate in the human brain and measurements of the fraction of blood-borne Cs extracted by the brain during passage of blood through the brain, in addition to autopsy measurements of environmental ^{137}Cs in human tissues including brain. For Mn, Pb, Ra, U, Pu, and Am, data for laboratory animals were used to model the rate of accumulation in the brain starting at the beginning of exposure. For radioisotopes of these six elements addressed here (^{54}Mn, ^{210}Pb, ^{226}Ra, ^{234}U, ^{239}Pu, and ^{241}Am), analyses were performed to determine the sensitivity of the dose coefficients for brain, based on explicitly depicted brain models, to uncertainties in the interspecies extrapolation of the data. It was found that the dose estimates for brain were largely determined by the requirement that predictions of the brain model be consistent with the results of the autopsy data. Stated differently, the dose coefficients for brain for ^{54}Mn, ^{210}Pb, ^{226}Ra, ^{234}U, ^{239}Pu, and ^{241}Am based on explicit brain models were found to be only moderately sensitive to uncertainties (within reasonable uncertainty bounds) in the use of animal data to predict brain kinetics of the element at early times after intake. For example, if the rate of uptake of ^{239}Pu by the brain was varied within a factor of 2 of the value indicated by data for dogs and the biological half-time in the brain was covaried so that model predictions remained consistent with USTUR autopsy data, the derived committed dose to the human brain varied within 20% of the value based on the explicit brain model applied here to ^{239}Pu.

The most problematic data set underlying any of the brain models applied in this analysis is the collection of information for bismuth, consisting of data for rats and autopsy data for patients who had received Bi treatments. The value of the autopsy data is limited by a lack of information on the time frame between Bi intake and death, the state of health of the subjects, and the possibility of a mass effect on the biokinetics of Bi due to the relatively high masses of administered Bi. The rat is not the preferred laboratory model for human biokinetics of elements and may be a particularly poor model for human brain kinetics in view of a variety of differences between rats and humans including brain size compared with total body size.

The behavior of inhaled Hg vapor has been studied extensively in laboratory animals, and some data are available from controlled studies on human subjects. The collective data provided a relatively strong foundation for development of the model for Hg vapor applied in this paper.

The construction of the brain model for Po differed from those of the other elements addressed here in that data for humans were not used. This was due to the difficulty in interpreting available autopsy data for ^{210}Po as the subjects were also exposed not only to ^{210}Po but also to radionuclides that sequentially decayed to ^{210}Po *in vivo* (Blanchard and Moore 1971) and concern that accumulation of ^{210}Po in brain of a subject who died from ^{210}Po poisoning did not reflect typical brain kinetics of Po (Harrison et al. 2017). Data on ^{210}Po kinetics in baboons were judged to be the most relevant data for modeling brain kinetics of Po in humans. Alternate methods of extrapolation of Po uptake data by brain were considered. It was assumed that the residence time of Po in the brain is the same in baboons and humans, which is another source of uncertainty in the application of baboon data to humans. The baboon data seem more likely to underestimate than overestimate the residence time of Po in the brain in humans in view of a moderately shorter residence time of Po in baboons (Leggett and Eckerman 2001). In any case, the present analysis for ^{210}Po suggests that inclusion of brain in *Other tissue* is likely to result in an underestimate of the dose to the brain from internally deposited ^{210}Po.

Disclosure statement

The authors declare no conflict of interest. The authors alone are responsible for the content and writing of the paper.

Funding

This manuscript and research were supported by UT-Battelle, LLC under Contract No. DE-AC05-00OR22725 with the U.S. Department of Energy. The United States Government retains and the publisher, by accepting the article for publication, acknowledges that the United States Government retains a non-exclusive, paid-up, irrevocable, world-wide license to publish or reproduce the published form of this manuscript, or allow others to do so, for United States Government purposes. The U.S. Department of Energy will provide public access to these results of federally sponsored research in accordance with the DOE Public Access Plan (http://energy.gov/downloads/doe-public-access-plan). The work described in this manuscript was sponsored by the U.S. Department of Energy under WAS Project No. 2018-AU-2000MPS, the U.S. Department of Energy, Office of Domestic and International Health Studies (AU-13), under grant award number DE-HS0000073 (USTUR funding), the Centers for Disease Control and Prevention (CDC) Office of Noncommunicable Diseases, Injury and Environmental Health, National Center for Environmental Health, under Interagency Agreement DOE No. 2220-Z051-16, under contract No. DE-AC05-00OR22725 with UT-Battelle, nd under grant No. 5NUE1EH001315 with the National Council on Radiation Protection and Measurements (NCRP). The authors also are grateful for the financial support received by the NCRP from the U.S. Department of Energy (Grant No. DE-AU0000042) and from the National Aeronautics and Space Administration (Grant No. 80NSSC17M0016).

References

Avtandilashvili M, Puncher M, McComish SL, Tolmachev SY. 2015. US Transuranium and Uranium Registries case study on accidental exposure to uranium hexafluoride. J Radiol Prot. 35:129–151.

Berlin M, Fazackerley J, Nordberg G. 1969. The uptake of mercury in the brains of mammals exposed to mercury vapor and to mercuric salts. Arch Environ Health. 18:719–729.

Berlin M, Jerksell LG, von Ubisch H. 1966. Uptake and retention of mercury in the mouse brain. A comparison of exposure to mercury vapor and intravenous injection of mercuric salt. Arch Environ Health. 12:33–42.

Blanchard RL, Moore JB. 1971. Body burden, distribution and internal dose of ^{210}Pb and ^{210}Po in a uranium miner population. Health Phys. 21:499–518.

Boice JD. 2017. Space: the final Frontier-Research Relevant to Mars. Health Phys. 112:392–397.

Boice JD, Jr, Cohen SS, Mumma MT, Ellis ED, Cragle DL, Eckerman KF, Wallace PW, Chadda B, Sonderman JS, Wiggs LD. 2014. Mortality among Mound workers exposed to polonium-210 and other sources of radiation, 1944-1979. Radiat Res. 181:208–228.

Boice JD, Ellis ED, Golden AP, Girardi DJ, Cohen SS, Chen H, Mumma MT, Shore RE, Leggett RW. 2018. The past informs the future: an overview of the Million Worker Study and the Mallinckrodt Chemical Works Cohort. Health Phys. 114:381–385.

Boice JD, Jr, Leggett RW, Ellis ED, Wallace PW, Mumma M, Cohen SS, Brill AB, Chadda B, Boecker BB, Yoder RC, et al. 2006. A comprehensive dose reconstruction methodology for former Rocketdyne/Atomics International radiation workers. Health Phys. 90:409–430.

Breitenstein BD, Newton CE. 1985. Part I: Introduction and history of the case. In: Health Phys. 49, Special issue on The U.S. Transuranium Registry report on the Am-241 content of a whole body; 565–567.

Bryce-Smith D, Stephens R. 1983. Sources and effects of environmental lead. In: Trace elements in health (Rose J, ed). London: Butterworths, 83–131.

Cohen N, Eisenbud M, Wrenn ME. 1970. The retention and distribution of lead-210 in the adult baboon. Progress report, radioactivity studies. New York: NYU Medical Center.

Cohen N, Fellman AL, Hickman DP, Ralston LG, Ayres LS. 1989. Primate polonium metabolic models and their use in estimation of systemic radiation doses from bioassay data. Miamisburg, OH: Mound Laboratory.

Ellis ED, Boice JD, Jr, Golden AP, Girardi DJ, Cohen SS, Mumma MT, Shore RE, Leggett RW, Kerr GD. 2018. Dosimetry is key to good epidemiology: workers at Mallinckrodt Chemical Works had seven different source exposures. Health Phys. 114:386–397.

Fellman A, Ralston L, Hickman D, Ayres L, Cohen N. 1994. Polonium metabolism in adult female baboons. Radiat Res. 137:238–250.

Filipy RE, Kathren RL. 1996. Changes in soft tissue concentrations of plutonium and americium with time after human occupational exposure. Health Phys. 70:153–159.

Fowler BA, Vouk VB. 1986. Chapter 6, Bismuth. In Handbook on the toxicology of metals. Vol. II. Friberg L, Nordberg GF, Vouk VB, eds. New York: Elsevier, 117–129.

Gilbert ES, Cragle DL, Wiggs LD. 1993. Updated analyses of combined mortality data for workers at the Hanford Site, Oak Ridge National Laboratory, and Rocky Flats Weapons Plant. Radiat Res. 136: 408–421.

Grandjean P. 1978. Regional distribution of lead in human brains. Toxicol Lett. 2:65–69.

Gregus Z, Klaassen CD. 1986. Disposition of metals in rats: a comparative study of fecal, urinary, and biliary excretion and tissue distribution of eighteen metals. Toxicol Appl Pharmacol. 85:24–38.

Guilmette RA, Cohen N, Wrenn ME. 1980. Distribution and retention of ^{241}Am in the baboon. Radiat Res. 81:100–119.

Harrison J, Fell T, Leggett R, Lloyd D, Puncher M, Youngman M. 2017. The polonium-210 poisoning of Mr. Alexander Litvinenko. J Radiol Prot. 37:266–278.

Hursh JB, Clarkson TW, Cherian MG, Vostal JJ, Mallie RV. 1976. Clearance of mercury (Hg-197, Hg-203) vapor inhaled by human subjects. Arch Environ Health. 31:302–309.

International Commission on Radiological Protection (ICRP). 1993. ICRP Publication 67. Age-Dependent Doses to Members of the

Public from Intake of Radionuclides Part 2 Ingestion Dose Coefficients. Ann ICRP. 23 (3/4). Oxford: Pergamon Press.

International Commission on Radiological Protection (ICRP). 2002. ICRP Publication 89. Basic anatomical and physiological data for use in radiological protection: reference values. Ann ICRP. 32 (3/4). Oxford: Pergamon Press.

International Commission on Radiological Protection (ICRP). 2016a. ICRP Publication 133. The ICRP computational framework for internal dose assessment for reference adults - specific absorbed fractions. Ann ICRP. 45(2). London: Sage.

International Commission on Radiological Protection (ICRP). 2016b. ICRP Publication 134. Occupational intake of radionuclides, Part 2. Ann ICRP. 45(3/4). London: Sage.

International Commission on Radiological Protection (ICRP). 2017. ICRP Publication 137. Occupational intake of radionuclides, part 3. Ann ICRP. 46(3/4). London: Sage.

Kathren RL, Tolmachev SY. 2015. Natural uranium tissue content of three Caucasian males. Health Phys. 109:187–197.

Kathren RL, Tolmachev SY. 2019. The United States Transuranium and Uranium Registries (USTUR): A five-decade follow-up of plutonium and uranium workers. Health Phys. 116: 000-000.

Khayat A, Dencker L. 1984. Organ and cellular distribution of inhaled metallic mercury in the rat and marmoset monkey (Callitrix jacchus): influence of ethyl alcohol pretreatment. Acta Pharmacol et Toxicol. 55:145–152.

Khayat A, Dencker L. 1983. Whole body and liver distribution of inhaled mercury vapor in the mouse: Influence of ethanol and aminotriazole pretreatment. J Appl Toxicol. 3:66–74.

Lee SP, Lim TH, Pybus J, Clarke AC. 1980. Tissue distribution of orally administered bismuth in the rat. Clin Exp Pharmacol Physiol. 7: 319–324.

Leggett RW. 1993. An age-specific kinetic model of lead metabolism in humans. Environ Health Perspect. 101:598–616.

Leggett RW. 2001. Reliability of the ICRP's dose coefficients for members of the public. 1. Sources of uncertainty in the biokinetic models. Radiat Prot Dosimetry. 95:199–213.

Leggett RW. 2011. A biokinetic model for manganese. Sci Total Environ. 409:4179–4186.

Leggett RW. 2013. Biokinetic models for radiocaesium and its progeny. J Radiol Prot. 33:123–140.

Leggett RW, Eckerman KF. 2001. A systemic biokinetic model for polonium. Sci Total Environ. 275:109–125.

Leggett RW, Eckerman KF, Khokhryakov VF, Suslova KG, Krahenbuhl MP, Miller SC. 2005. Mayak worker study: an improved biokinetic model for reconstructing doses from internally deposited plutonium. Radiat Res. 164:111–122.

Leggett RW, Williams LR, Melo DR, Lipsztein JL. 2003. A physiologically based biokinetic model for cesium in the human body. Sci Total Environ. 317:235–255.

Lemercier V, Millot X, Ansoborlo E, Ménétrier F, Flüry-Hérard A, Rousselle C, Scherrmann JM. 2003. Study of uranium transfer across the blood-brain barrier. Radiat Prot Dosimetry. 105:243–245.

Lloyd RD, Atherton DR, McFarland SS, Mays CW, Stevens W, Williams JL, Taylor GN. 1976. Studies of injected ^{237}Pu(IV) citrate in beagles. Health Phys. 30:47–52.

Lloyd RD, Mays CW, Atherton DR, Bruenger FW. 1975. 210Pb studies in beagles. Health Phys. 28:575–583.

Lloyd RD, Mays CW, Taylor GN, Atherton DR. 1970. Americium-241 studies in beagles. Health Phys. 18:149–156.

Lloyd RD, Mays CW, Taylor GN, Atherton DR, Bruenger FW, Jones CW. 1982. Radium-224 retention, distribution, and dosimetry in beagles. Radiat Res. 92:280–295.

Magos L. 1967. Mercury-blood interaction and mercury uptake by the brain after vapor exposure. Environ Res. 1:323–337.

McInroy JF, Boyd HA, Eutsler BC, Romero D. 1985. The U.S. Transuranium Registry report of the ^{241}Am content of a whole body. Part IV: Preparation and analysis of the tissues and bones. Health Phys. 49:587–621.

Monleau M, Bussy C, Lestaevel P, Houpert P, Paquet F, Chazel V. 2005. Bioaccumulation and behavioural effects of depleted uranium in rats exposed to repeated inhalations. Neurosci Lett. 390:31–36.

Nathwani AC, Down JF, Goldstone J, Yassin J, Dargan PI, Virchis A, Gent N, Lloyd D, Harrison JD. 2016. Polonium-210 poisoning: a first-hand account. Lancet. 388:1075–1080.

National Council on Radiation Protection and Measurements (NCRP). 2016. Potential for central nervous system effects from radiation exposure during space activities. Phase 1: overview. Commentary No. 25. Bethesda, MD: NCRP.

Newton D, Fry FA. 1978. The retention and distribution of radioactive mercuric oxide following accidental inhalation. Ann Occup Hyg. 21: 21–32.

Nordberg GF, Serenius F. 1969. Distribution of inorganic mercury in the guinea pig brain. Acta Pharmacol Toxicol (Copenh). 27:269–283.

Pellmar TC, Fuciarelli AF, Ejnik JW, Hamilton M, Hogan J, Strocko S, Emond C, Mottaz HM, Landauer MR. 1999. Distribution of uranium in rats implanted with depleted uranium pellets. Toxicol Sci. 49:29–39.

Petit TL, Alfano DP, LeBoutillier JC. 1983. Early lead exposure and the hippocampus: a review and recent advances. NeuroToxicology. 4: 79–94.

Russell JJ, Kathren RL. 2004. Uranium deposition and retention in a USTUR whole body case. Health Phys. 86:273–284.

Schlenker RA, Keane AT, Holtzman RB. 1982. The retention of ^{226}Ra in human soft tissue and bone; implications for the ICRP 20 alkaline earth model. Health Phys. 42:671–693.

Silver SR, Bertke SJ, Hein MJ, Daniels RD, Fleming DA, Anderson JL, Pinney SM, Hornung RW, Tseng CY. 2013. Mortality and ionising radiation exposures among workers employed at the Fernald Feed Materials Production Center (1951-1985). Occup Environ Med. 70: 453–463.

Sollmann TA. 1957. Manual of pharmacology and its applications to therapeutics and toxicology. Philadelphia: W.B. Saunders Co.

Sollmann T, Seifter J. 1942. Intravenous injections of soluble bismuth compounds: Their toxicity, and their sojourn in the blood and organs. J Pharmacol Exp Ther. 74:134–154.

Tikhaya MG, Novikova AP, Parfenov YuD. 1965. Body distribution of uranium in dogs long periods after the inhalation of uranium oxide (in Russian). Med. Radiol. 10(10): 50–54.

Wiggs LD, Johnson ER, Cox-DeVore CA, Voelz GL. 1994. Mortality through 1990 among white male workers at the Los Alamos National Laboratory: considering exposures to plutonium and external ionizing radiation. Health Phys. 67:577–588.

Zidenberg-Cherr S, Parks NJ, Keen CL. 1987. Tissue and subcellular distribution of bismuth radiotracer in the rat: considerations of cytotoxicity and microdosimetry for bismuth radiopharmaceuticals. Radiat Res. 111:119–129.

[illegible]

Epidemiology

Mortality from leukemia, cancer and heart disease among U.S. nuclear power plant workers, 1957–2011

John D. Boice, Sarah S. Cohen, Michael T. Mumma, Derek A. Hagemeyer, Heidi Chen, Ashley P. Golden, R. Craig Yoder, and Lawrence T. Dauer

ABSTRACT

Background: The aim of the Million Person Study of Low-Dose Health Effects (MPS) is to examine the level of radiation risk for chronic exposures received gradually over time and not acutely as was the case for the Japanese atomic bomb survivors. Nuclear power plant (NPP) workers comprise nearly 15 percent of the MPS. Leukemia, selected cancers, Parkinson's disease, ischemic heart disease (IHD) and other causes of death are evaluated.

Methods and material: The U.S. Nuclear Regulatory Commission's Radiation Exposure Information and Reporting System (REIRS) and the Landauer, Inc. dosimetry databases identified 135,193 NPP workers first monitored 1957–1984. Annual personal dose equivalents [*Hp(10)*] were available for each worker. Radiation records from all places of employment were sought. Vital status was determined through 2011. Mean absorbed doses to red bone marrow (RBM), esophagus, lung, colon, brain and heart were estimated by adjusting the recorded *Hp(10)* for each worker by scaling factors, accounting for exposure geometry and energy of the incident gamma radiation. Standardized mortality ratios (SMR) were calculated. Radiation risks were estimated using Cox proportional hazards models.

Results: Nearly 50% of workers were employed for more than 20 years. The mean duration of follow-up was 30.2 y. Overall, 29,124 total deaths occurred, 296 from leukemia other than chronic lymphocytic leukemia (CLL), 3382 from lung cancer, 140 from Parkinson's disease and 5410 from IHD. The mean dose to RBM was 37.9 mGy (maximum 1.0 Gy; percent >100 mGy was 9.2%), 43.2 mGy to lung, 43.7 mGy to colon, 33.2 mGy to brain, and 43.9 mGy to heart. The SMRs (95% CI) were 1.06 (0.94; 1.19) for leukemia other than CLL, 1.10 (1.07; 1.14) for lung cancer, 0.90 (0.76; 1.06) for Parkinson's disease, and 0.80 (0.78; 0.82) for IHD. The excess relative risk (ERR) per 100 mGy for leukemia other than CLL was 0.15 (90% CI −0.001; 0.31). For all solid cancers the ERR per 100 mGy (95% CI) was 0.01 (−0.03; 0.05), for lung cancer −0.04 (−0.11; 0.02), for Parkinson's disease 0.24 (−0.02; 0.50), and for IHD −0.01 (−0.06; 0.04).

Conclusion: Prolonged exposure to radiation increased the risk of leukemia other than CLL among NPP workers. There was little evidence for a radiation association for all solid cancers, lung cancer or ischemic heart disease. Increased precision will be forthcoming as the different cohorts within the MPS are combined, such as industrial radiographers and medical radiation workers who were assembled and evaluated in like manner.

Introduction

The study of a *Million U.S. Radiation Workers and Veterans* (also known as the Million Person Study of Low-Dose Health Effects (MPS)) is designed to examine the level of radiation risk for chronic exposures that are received gradually over time (Boice, Cohen et al. 2019) and not acutely, as occurred for the Japanese atomic bomb survivors (Ozasa et al. 2012; Grant et al. 2015; McLean et al. 2017). Individualized and annualized dosimetry is an essential aspect of the MPS (Boice et al. 2006; Till et al. 2014; Bouville et al. 2015; Beck 2017; Dauer et al. 2018; Leggett et al. 2018; Ellis, Boice et al. 2018; NCRP 2018a, 2020; Yoder et al. 2021). Previous studies have been published on workers at U.S. Department of Energy (DOE) facilities (Boice et al. 2011, 2014; Boice, Cohen, Mumma, Golden et al. 2021; Boice 2017a; Boice 2017b; Golden et al. 2019) and veterans who participated in above-ground nuclear weapons tests (Boice 2014; Till et al. 2014; Caldwell et al. 2016; Boice et al. 2020).

Workers in the nuclear power plant industry comprise nearly 15% of the MPS population. The consistent reporting of annual worker doses required by the U.S. Nuclear Regulatory Commission (NRC) for their licensees provided a high-quality dosimetry database that was redesigned in 1994 to facilitate epidemiologic study (Hagemeyer et al. 2018). The number of nuclear power plant workers under study ($n = 135{,}193$) is nearly three times the number of adults over age 20 years at exposure ($n = 51{,}215$) in the study of Japanese atomic bomb survivors (Ozasa et al. 2012), and over seven times the number of adult males ($n = 18{,}860$). Herein we report cause-specific analyses for over 40 causes of death and dose-response analyses for leukemia, myelodysplastic disease, all solid cancers, esophageal cancer, lung cancer, Parkinson's disease, and ischemic heart disease (IHD) among early workers in the nuclear power industry first monitored between 1957 and 1984 in the United States and followed through 2011.

Methods

Human subjects research approval for the study was received from Vanderbilt University.

Cohort definition

A cohort of 1,065,703 workers employed at nuclear power plants in the United States in 1957 through 2011 was available for study: 1,001,958 workers within the Radiation Exposure Information and Reporting System (REIRS) maintained by the NRC (Anzenberg et al. 2010; Hagemeyer et al. 2018; USNRC 2021) and 63,745 workers within the Landauer, Inc. dosimetry database (NCRP 2018a, 2020; Yoder et al. 2021) (Figure 1). Workers were excluded if the date of first monitoring was after 1984 ($n = 592{,}935$) or if personal identifiers needed for tracing, specifically a Social Security Number ($n = 5738$) were absent. Workers first monitored before 1985 were selected on the basis of cost and information value. Workers employed after 1984 were less likely to receive moderate to high cumulative doses compared with workers employed earlier, and were younger and less likely to have died. The choice of 1985 rather than 1980 was because the Three Mile Island reactor accident in 1979 led to new NRC safety requirements involving reactor modifications that increased workers' exposures until about 1985 (Anzenberg et al. 2010; Blevins and Andersen 2011; NCRP 2018a). There were 307,553 workers with cumulative

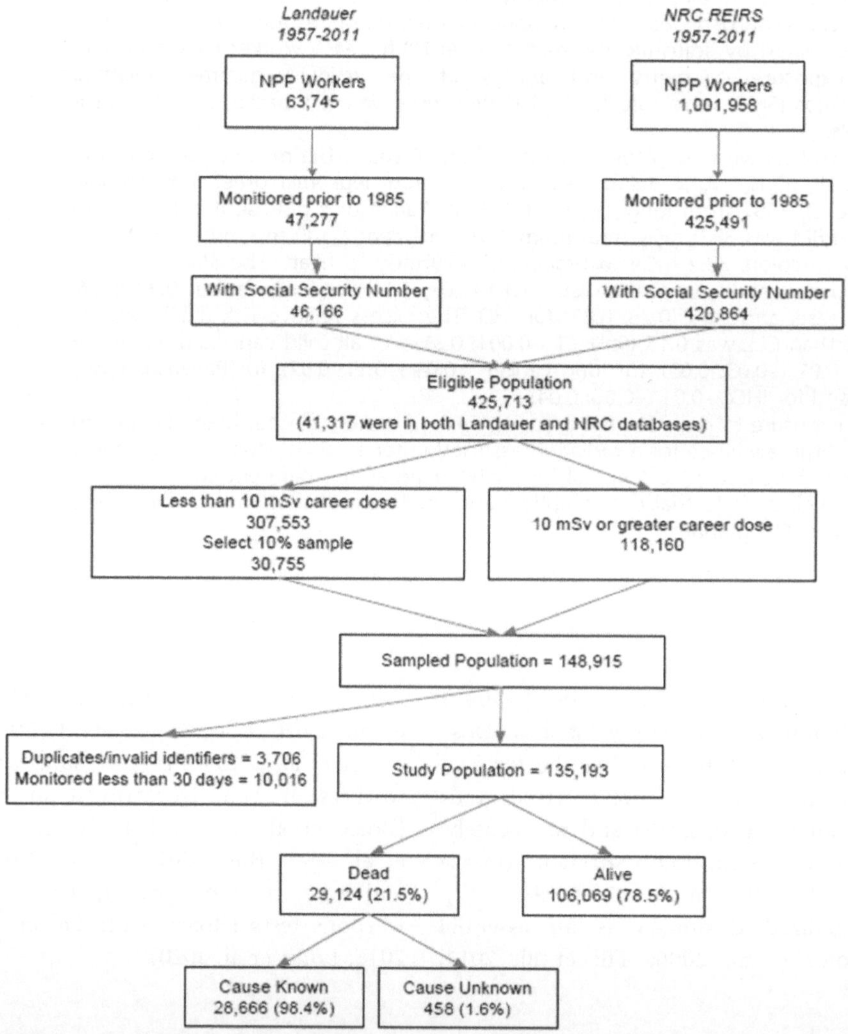

Figure 1. Schematic of the selection and vital status of the 135,193 nuclear power plant workers first monitored between 1957 and 1984 in the United States and followed through 2011.

doses <10 mSv, and it would have been prohibitively expensive to trace and collect death information on them all. A 10% sampling of these very low dose workers was for cost savings without a noticeable diminution of statistical power for dose-response analyses. Short-term workers monitored for less than 30 days ($n = 10{,}016$), duplicate records and those with invalid identifiers ($n = 3706$) were removed. The final study population of 135,193 consisted of all workers with a cumulative dose ≥10 mSv and a 10% sample of workers with a cumulative dose <10 mSv (Table 1). The short-term workers monitored for less than 30 days were traced and the cause of death determined in the same manner as the final study population and evaluated separately.

Just over 15% of the cohort had dosimetry records in both REIRS and Landauer files. The final cohort was comprised almost entirely of workers identified within REIRS (99.2%) with supplemental information from Landauer files (Table 1). In the past, the REIRS and Landauer databases have been used in the planning and conduct of epidemiologic studies (Goldsmith et al. 1989; Jablon and Boice 1993; Muirhead et al. 1996; Boice et al. 2006) but not in defining a cohort for study.

Although first monitoring at a nuclear power plant could begin as early as 1957 (Supplement Table 2), follow-up started in 1969 for the 14% of workers who were identified within the REIRS database as being monitored before 1970. The Atomic Energy Act of 1969 provided the legal requirements for NRC licensees to record and report worker doses. However, these reports to the NRC were not optimal for epidemiologic study (Goldsmith et al. 1989; Jablon and Boice 1993; Muirhead et al. 1996) until 1994 when reporting requirements changed and near complete voluntary reporting back to 1969 essentially created a system of records with personal identifiers and annual doses that could be used for health studies (USNRC 1994; Oak Ridge Institute for Science and Education (ORISE) 2011; Hagemeyer et al. 2018). Follow-up started in 1977 for the 0.8% of workers who were identified only from the Landauer files. Although Landauer has microfiche records dating back to the 1950s, they are not easily accessible. Electronic files suitable for epidemiologic studies were first available in 1977 and included cumulative exposures before 1977 if the worker had been employed as a Landauer client. After removing 3706 workers with invalid identifiers or who were duplicates and 10,016 workers who were monitored for less than 30 d, the cohort was reduced from the sampled population of 148,915 workers to the final study cohort of 135,193 workers (Figure 1).

Landauer developed a special database of annual and lifetime personal monitoring results in response to a mid-1980s request from the Radiation Epidemiology Branch of the National Institutes of Health as an ongoing tool to support their study of registered radiologic technologists (Villoing et al. 2021). The expanded and extensive nature of the database has allowed its use as a confirmatory and supplemental source of dose information residing in other governmental dose registers or worker records (Ellis, Girardi et al. 2018; Yoder et al. 2021; Boice, Cohen, Mumma, Howard et al. 2021).

Table 1. Demographic and occupational characteristics of 135,193 nuclear power plant workers first monitored for radiation between 1957 and 1984 and followed through 2011 in the United States.

Characteristic	*N*	%
Source of the population		
NRC REIRS only	111,681	82.6
Landauer only	1123	0.8
Both REIRS and Landauer	22,389	16.6
Sex		
Female	4420	3.3
Male	130,773	96.7
Year of birth (YOB)		
1902–	2529	1.9
1920–	9906	7.3
1930–	18,941	14.0
1940–	42,026	31.1
1950–	52,990	39.2
1960–1966	8801	6.5
Mean YOB = 1946		
Year first monitored for occupational radiation at an NPP		
1957–	1823	1.4
1970–	17,649	13.1
1975–	46,821	34.6
1980–1984	68,900	51.0
Mean year of first NPP monitoring = 1978		
Number of nuclear power plants at which a worker was monitored		
1	64,241	47.4
2	27,351	20.2
3	13,256	9.8
4	7417	5.5
5–9	13,773	10.2
10 or more facilities	9155	6.8
Years monitored for radiation at a nuclear power plant		
<1	20,569	15.2
1–	28,881	21.4
5–	19,201	14.2
10–	27,331	20.2
20–	25,241	18.7
30–47	13,970	10.3
Mean number of years monitored at a NPP = 12.8		
Years of follow-up		
<1	118	0.1
1–	3092	2.3
10–	7303	5.4
20–	46,473	34.4
30–	74,711	55.3
40–42	3496	2.6
Mean duration of follow-up (years) = 30.2		
Mean age at start of follow-up (years) = 32.0		
Career cumulative radiation dose, personal dose equivalent		
< 5 mSv	21,048	15.6
5–	72,085	53.3
50–	13,786	10.2
75–	8293	6.1
100–	13,659	10.1
200–	4024	3.0
300 mSv or greater	2298	1.7
Mean career dose = 52.6 mSv		
Vital status as of 12/31/2011		
Alive/assumed alive	106,069	78.5
Dead	29,124[a]	21.5
Mean age at end of follow-up (years) = 62.2		

[a]29,076 deaths occurred in the US and were eligible for the SMR analyses. 13 deaths occurred outside the US, and 35 deaths occurred after age 94 y.

NRC: U.S. Nuclear Regulatory Commission; REIRS: Radiation Exposure Information and Reporting System; NPP: Nuclear Power Plant; SMR: Standardized Mortality Ratio.

Vital status and outcome determination

Vital status was determined through 2011. The Social Security Administration (SSA) Service for Epidemiological Researchers (https://www.ssa.gov/policy/about/epidemiology.html) was relied upon to determine alive status. This service

Table 2. Standardized mortality ratios (SMR) and 95% confidence intervals (CI) for 135,193 Nuclear Power Plant Workers, first monitored for radiation 1957–1984 with follow-up through 2011 in the United States.

	Males		Females		Total		
Persons-at-risk	130,773		4420		135,193		
Person-years-at-risk	3,947,966		131,654		4,079,620		
Cause of death (ICD9)[a]	Obs	SMR	Obs	SMR	Obs	SMR	95% CI
All causes of death (001–999)	28,646	0.88*	430	1.00	29,076	0.89*	0.88–0.90
All malignant neoplasms (140–208)	9157	1.03*	172	1.12	9329	1.03*	1.01–1.05
All solid cancers (140–199)	8287	1.04*	158	1.12	8445	1.04*	1.01–1.06
Buccal cavity & pharynx (140–149)	163	0.82*	3	2.11	166	0.83*	0.71–0.96
Esophagus (150)	351	1.07	4	3.13	355	1.07	0.96–1.19
Stomach (151)	216	1.01	0	0.00	216	1.00	0.87–1.14
Colon (153)	629	0.94	16	1.64	645	0.95	0.88–1.03
Rectum (154)	131	0.96	1	0.54	132	0.95	0.80–1.13
Biliary passages & liver (155,156)	253	0.81*	4	1.32	257	0.81*	0.72–0.92
Pancreas (157)	498	1.01	10	1.40	508	1.01	0.92–1.10
Pleura & peritoneum (158.8, 158.9, 163) & mesothelioma (ICD10 C45)[b]	251	5.69*	0	0.00	251	5.66*	4.98–6.40
Larynx (161)	97	0.96	0	0.00	97	0.96	0.78–1.17
Bronchus, trachea & lung (162)	3334	1.10*	48	1.27	3382	1.10*	1.07–1.14
Bone (170)	8	0.40*	0	0.00	8	0.39*	0.17–0.77
Connective & other soft tissue (171)	63	1.00	2	1.44	65	1.01	0.78–1.29
Female breast (174)	n/a	n/a	37	1.15	37	1.15	0.81–1.58
Male breast (175)	4	0.35*	n/a	n/a	4	0.35*	0.09–0.89
All uterine (females) (179–182)	n/a	n/a	3	0.37	3	0.37	0.07–1.08
Cervix uteri (180)	n/a	n/a	1	0.23	1	0.23	0.00–1.26
Ovary and other female genital organs (183–184)	n/a	n/a	7	0.67	7	0.67	0.27–1.38
Prostate (185)	527	0.99	n/a	n/a	527	0.99	0.91–1.08
Testes & other male genital organs (186, 187)	19	0.72	n/a	n/a	19	0.72	0.43–1.12
Kidney (189.0–189.2)	229	0.86*	3	1.21	232	0.87*	0.76–0.99
Bladder & other urinary (188, 189.3–189.9)	209	0.92	0	0.00	209	0.91	0.79–1.05
Melanoma of skin (172)	217	1.01	3	1.08	220	1.01	0.88–1.15
Other malignant neoplasm of skin (173)	41	0.73*	0	0.00	41	0.73*	0.52–0.99
Eye (190)	6	1.25	0	0.00	6	1.23	0.45–2.67
Brain & central nervous system (191–192)	267	0.90	7	1.41	274	0.90	0.80–1.02
Thyroid & other endocrine glands (193–194)	31	1.04	1	1.57	32	1.06	0.72–1.49
All lymphatic, hematopoietic tissue (200–208)	870	0.96	14	1.15	884	0.96	0.90–1.03
Non-Hodgkin lymphoma (200, 202)	316	0.86*	3	0.64	319	0.86*	0.77–0.96
Hodgkin lymphoma (201)	24	0.65*	0	0.00	24	0.63*	0.41–0.94
Multiple myeloma (203)	166	1.09	4	2.01	170	1.11	0.95–1.29
Leukemia (202.4, 204–208)	360	1.03	6	1.26	366	1.04	0.93–1.15
Acute lymphocytic leukemia (204.0)	21	0.99	0	0.00	21	0.97	0.60–1.49
Acute myeloid leukemia (205.0, 205.3, 206.0, 207.0, 207.2)	159	1.12	2	0.88	161	1.12	0.95–1.30
Chronic myeloid leukemia (205.1, 206.1)	37	1.08	0	0.00	37	1.06	0.75–1.47
Chronic lymphocytic leukemia (CLL) (202.4, 204.1, 204.9)	67	0.95	0	0.00	67	0.95	0.73–1.20
Leukemia other than CLL	290	1.05	6	1.41	296	1.06	0.94–1.19
Smoking-related cancers (140–150, 157, 161–162, 188–189)	4881	1.05*	68	1.31*	4949	1.05*	1.02–1.08
Nonsmoking-related cancers	4276	1.00	104	1.02	4380	1.00	0.97–1.03
Polycythemia vera (238.4)	1	0.27	0	0.00	1	0.27	0.00–1.49
Myelodysplastic syndrome (238.7)	48	1.09	0	0.00	48	1.08	0.80–1.43
Diabetes (250)	578	0.68*	13	1.00	591	0.68*	0.63–0.74
Mental and behavioral disorders (290–319)	419	0.77*	6	0.84	425	0.77*	0.70–0.85
Dementia, Alzheimer's, Parkinson's & motor neuron disease (290.0–290.4, 331.0, 332, 335.2)	651	0.93	6	0.55	657	0.93	0.86–1.00
Dementia and Alzheimer's (290.0–290.4, 331.0)	405	0.93	6	0.74	411	0.92	0.84–1.02
Parkinson's disease (332)	140	0.91	0	0.00	140	0.90	0.76–1.06
Diseases of the nervous system (320–389)	664	0.83*	9	0.63	673	0.82*	0.76–0.89
All heart disease (390–398, 404, 410–429)	7497	0.82*	64	0.81	7561	0.82*	0.80–0.84
Ischemic heart disease (410–414)	5372	0.80*	38	0.77	5410	0.80*	0.78–0.82
Cerebrovascular disease (430–438)	1056	0.88*	22	1.12	1,078	0.88*	0.83–0.94
Nonmalignant Respiratory Disease (460–519)	2155	0.89*	25	0.72	2,180	0.89*	0.85–0.92
Bronchitis, emphysema, asthma (490–493)	1080	0.97	15	0.80	1,095	0.96	0.91–1.02
Asbestosis (501)	87	9.15*	0	0.00	87	9.15*	7.33–11.3
Cirrhosis of liver (571)	714	0.81*	4	0.42	718	0.81*	0.75–0.87
AIDS (042–044, 795.8)	144	0.29*	1	0.44	145	0.29*	0.24–0.34
Nephritis & nephrosis (580–589)	281	0.78*	5	0.96	286	0.78*	0.69–0.88
All external causes of death (800–999)	3076	0.84*	53	1.25	3,129	0.85*	0.82–0.88
Accidents (850–949)	1960	0.91*	37	1.42	1,997	0.92*	0.88–0.96
Suicides (950–959)	817	0.79*	10	1.00	827	0.79*	0.74–0.85
Unknown causes of death	446	–	12	–	458	–	

[a]ICD = International classification of disease. Revision 9 codes are shown. n/a denotes not applicable.
[b]Mesothelioma did not have an explicit code in ICD9 but did in ICD10 as denoted.
*Indicates SMR is significantly different from 1.0.

accesses the Internal Revenue Service records of persons who recently filed a tax return, Medicare beneficiaries for those 65 y of age or older, and other sources of vital status. Death notifications relied upon the National Death Index (NDI) which began in 1979, the SSA Death Master File which began in 1960, and state mortality files available from 33 states (Mumma et al. 2018). A matching algorithm that uses a probabilistic scoring system for all matching variables, e.g. date of birth, was used to confirm the fact of death (Campbell et al. 2008). The NDI, state mortality data files, and death certificates obtained from state departments of health identified both the underlying and contributing causes of death. Information from LexisNexis (www.lexis-nexis.com) and/or credit bureaus was employed to validate or correct key matching variables.

Overall, 106,069 workers (78.5%) were confirmed or assumed to be alive, and 29,124 workers (21.5%) were found to have died. Cause of death was obtained for all but 458 (1.6%) of those who died. Of the 106,069 workers confirmed or assumed alive at the end of follow-up (Table 1), 90.0% were confirmed alive by the SSA Service for Epidemiological Researchers. The remaining workers were presumed alive based on the absence of a match against the NDI, the SSA Death Master File, and 33 state mortality files. The over 11,000 workers known to be alive after 1978, based on dates recorded on dosimetry records, were submitted to the NDI for a full search of death records from 1979 through 2011. Further, a 1% sample of workers presumed to be alive was intensely searched through credit bureaus and LexisNexis and 94% were confirmed alive as of the end of follow-up and no deaths were found.

Of the 29,124 deaths overall, only 370 (1.3%) occurred before 1979 when the NDI began. Of these 370 deaths, a cause of death was obtained for 281 (76.0%) from state mortality files or from a death certificate. For 89 pre-NDI deaths (0.3% of all deaths), a death certificate or cause of death could not be determined.

Radiation dose estimation

Most radiation exposures received by nuclear power plant workers were from penetrating external gamma radiation with only a negligible contribution coming from neutrons or intakes of radioactive material (Bouville et al. 2015; Dauer et al. 2018; NCRP 2018a). Annual personal dose equivalents, *Hp(10)*, were available from the REIRS and Landauer files. Radiation doses from all places of employment were sought by linking the study roster to the Radiation Exposure Monitoring System (REMS) maintained by the U.S. Department of Energy (DOE) (DOE 2021), other DOE dosimetry databases, and U.S. military service dosimetry databases. All doses were combined to create annual dose estimates for each worker following methods previously described (Boice et al. 2006; Ellis, Boice et al. 2018). Overall, the mean cumulative *Hp(10)* was estimated as 19.5 mSv for all 425,713 workers eligible for the study, 52.6 mSv for the study cohort of 135,193 workers, and 61.3 mSv for the 115,288 workers with cumulative doses >10 mSv (maximum 1.32 Sv).

Practically all recorded doses were from external exposures to high-energy low-LET photons, predominantly from ^{58}Co, ^{60}Co and ^{137}Cs. As in previous MPS studies (Boice et al. 2006), a workshop was held with former nuclear power radiation workers, dosimetrists, and health physicists to learn firsthand about exposure circumstances, monitoring practices, and work conditions (Boice 2016). Most exposures were received during maintenance, modification, and repair work during outages where the external exposure photon fields were generally anterior to posterior (AP) and with degraded energy through piping, valves or tank steel. The methods employed to estimate organ doses for the NPP cohort are described in NCRP Report 178 (NCRP 2018a). Mean absorbed doses (mGy) to red bone marrow (RBM), lung, esophagus, colon, brain and heart were estimated by adjusting the recorded *Hp(10)* for each worker by scaling factors (conversion coefficients), accounting for exposure geometry and energy of the incident gamma radiation (ICRP 2010). The conversion coefficients were assumed to be stable and not variable over time or with exposure circumstances following the recommendation in NCRP Report 178 (NCRP 2018a) when only annual dosimetry results are available. The activity-weighted average photon energy was typically around 0.7 MeV, ranging from about 0.6 to 1.5 MeV (NCRP 2018a). The associated scaling factors were 0.72 for RBM, 0.81 for lung, 0.79 for esophagus, 0.83 for colon, 0.63 for brain, and 0.83 for heart. The mean cumulative absorbed dose to RBM in the study cohort was 37.9 mGy (maximum 1.0 Gy; percent >100 mGy was 9.2%); 43.2 mGy to lung (maximum 1.1 Gy; percent >100 mGy was 11.3%); 41.6 mGy to esophagus (maximum 1.05 Gy; percent > 100 mGy was 10.7%); 43.7 mGy to colon (maximum 1.1 Gy; percent > 100 mGy was 11.5%); 33.2 mGy to brain (maximum 0.83 Gy; percent >100 mGy was 7.4%); and 43.9 mGy to heart (maximum 1.1 Gy; percent >100 mGy was 11.6%).

Outcome classification for leukemias

The classifications for the leukemia subtypes used in the analyses are based on the International Classification of Diseases (ICD) codes for leukemia over three ICD Revisions: ICD-8, ICD-9 and ICD-10 (Supplement Table 1). The specific categories were selected to be consistent with the incident study of Japanese atomic bomb survivors (Hsu et al. 2013). The Japanese investigators reclassified early leukemia diagnoses using the French–American–British (FAB) classification system and subsequently coded leukemia diagnoses using the ICD for Oncology (ICD-O) codes that were current at the time of diagnosis. The specific analyses conducted are 'leukemia other than CLL', CLL, acute myelogenous leukemia (AML), acute lymphocytic leukemia (ALL), chronic myelogenous leukemia (CML) and myelodysplastic syndrome (MDS). While up to a third of the cases of MDS may progress to AML, it appears that MDS is biologically and clinically different from AML and should be considered separately (Albitar et al 2002; Komrokji et al. 2019; Zeidan et al. 2019), noting also that its association with radiation exposure appears uncertain (UNSCEAR 2020).

Analytic methods

Standardized mortality ratio (SMR) analyses were employed to compare the observed number of deaths from specific causes with the number expected based on general population rates in the United States for persons of the same age and sex over the same calendar years. Rates for white males and females were used in the SMR analysis since the most common ethnicity among nuclear power plant workers over the years of study was white. Follow-up started on the date of first radiation monitoring at a nuclear power facility or 1 January 1969 (if the worker was identified in REIRS) or 1 January 1977 (if the worker was identified only from Landauer records), whichever was earlier. The end of the follow-up was the date of death, 95th birthday, or 31 December 2011, whichever came first. Of the 29,124 deaths overall, 29,076 occurred in the US and were eligible for the SMR analyses which are based on mortality rates for the US. Excluded from the SMR analyses were 13 deaths that occurred outside the US and 35 deaths that occurred after age 94 y.

Internal (within-cohort) analyses were conducted to account in part for the healthy worker effect that is often present in occupational studies when comparisons are made with the general population (Monson 1986; Howe et al. 1988; Checkoway et al. 1989; Buckley et al. 2015). Internal analyses were conducted using Cox proportional hazards models (Cox 1972) to compute the risks of leukemia, all solid cancers, lung cancer, esophageal cancer, non-Hodgkin lymphoma, multiple myeloma, MDS, Parkinson's disease, and IHD across categories of radiation dose. The Cox models included adjustment for sex, year of birth, and socioeconomic status (SES) (categorized as a low, medium, high based on census block group mean education levels) (Cohen et al. 2018). Age was used as the timescale for the hazard function. The primary analysis used categories of radiation exposure as the primary exposure. Dose-response functions for continuous measurement of radiation exposure were additionally modeled as linear functions. Cox analyses were conducted using SAS/STAT software (version 9.4 of the SAS System for Windows, SAS Institute Inc., Cary, NC).

Cox and Poisson excess relative risk (ERR) models were constructed using EPICURE software (Preston et al. 2015). The PEANUTS model for ERRs included the same covariates as the final Cox models (sex, year of birth, and SES). Akaike Information Criterion (AIC) values were used during some analyses to compare different regression models (Akaike 1974). Poisson regression models were also used. For each individual worker, the personal dose equivalents in mSv were all converted to organ-specific mGy doses. The dose-response models were all based on the estimated organ-specific absorbed doses in mGy. The dose-response graphics presented are based on a linear model with 95% confidence bands about the regression line representing the range of *true* regression lines at 95% confidence.

Adjustment for SES is frequently done in occupational studies to control for possible confounding factors that are not available on an individual basis, such as tobacco use and other lifestyle factors that may influence mortality and disease occurrence. Estimates of SES for each nuclear power plant worker were obtained by geocoding residential histories and acquiring area-wide estimates of education based on census-block group data (Cohen et al. 2018; Mumma et al. 2018). LexisNexis provided residential histories for the over 130,000 workers in our cohort. Then, based on the geocoded address, each worker was placed into a census-block group with an estimate of mean educational attainment characterized as low, medium or high.

To account for the possibility that workers who frequently worked at different utilities each year, i.e. so called transient workers, differed from workers who remained most of their career at a single facility, an adjustment for mobility was made. Worker mobility was classified into two categories: high mobility (workers who were monitored at 2 to >10 facilities) and low mobility (workers monitored at a single facility) (Table 1). The adjustment for mobility did not significantly change the estimates of risk so was not retained in the final models (Supplement Table 4). Analyses with and without an adjustment for the duration of radiation monitoring were performed. To adjust for the duration of radiation monitoring, workers were classified into one of six categories of years monitored (<1, 1–4, 5–10, 10–19, 20–29, >30). Because time must elapse between exposure and the appearance of a consequent health event, radiation doses were lagged in the analysis. For leukemia other than CLL and for MDS, the dose lagging was 2 years; for CLL all solid cancers, lung cancer, esophageal cancer, non-Hodgkin lymphoma, multiple myeloma, Parkinson's disease and IHD, doses were lagged 10 years. A 5-year lag was also evaluated for lung cancer.

For the leukemia internal analyses, the underlying cause of death and most contributing causes of death were included to increase statistical precision. Leukemia as a contributing cause of death, however, was not included if the underlying cause was cancer. This was done to reduce the likelihood that the leukemia was secondary to any treatment for the initial primary cancer. Including contributing causes of death in studies of occupational cohorts has been recognized as providing additional information on risk assessment (Steenland et al. 1992; Rushton 1994).

All exposure estimates were based on recorded personal dosimeter measurements. There was no imputation of missing dosimetric data which were judged to be minimal among employees at nuclear power facilities (Bouville et al. 2015; NCRP 2018a). The classical (Berkson) error structure of the individual worker data dose estimates indicates little sharing of errors, and, accordingly, accounting for this type of dosimetric uncertainty in the dose-response analyses does not have a meaningful effect (Stram et al. 2015). To address the influence of unshared dosimetric errors (NCRP 2008), the procedures outlined in NCRP Report 178 (NCRP 2018a)

Table 3. Observed and expected deaths due to leukemia other than chronic lymphocytic leukemia and risk estimates by radiation dose to red bone marrow (RBM) among 135,193 nuclear power plant workers, first monitored for radiation 1957–1984 and followed through 2011 in the United States.

	Radiation dose to RBM (mGy)							
	<5	5–	50–	75–	100–	200–	≥300	Total
No. of workers	22,968	81,739	11,710	6466	9519	2078	713	135,193
Mean dose (mGy)	0.88	21.0	61.1	86.6	136.7	238.5	377.8	37.6
Leukemia deaths[a]								
Observed	58	176	27	13	13	6	3	296
Expected	60.5	166.3	20.6	11.0	15.4	3.8	1.7	279.5
O/E (SMR)	0.96	1.06	1.31	1.18	0.84	1.58	1.73	1.06
Leukemia cases[b]	62	182	28	14	14	7	4	311
Hazards Ratio[c]	1.0 (R)	1.20	1.51	1.43	1.01	2.05	2.52	1.16[d]
90% Confidence intervals	–	0.93; 1.53	1.03; 2.21	0.87; 2.34	0.62; 1.66	1.06; 3.97	1.08; 5.91	0.99; 1.36

[a]Number of leukemia deaths other than chronic lymphocytic leukemia recorded as the underlying cause of death on the death certificate

[b]Number of leukemia deaths other than chronic lymphocytic leukemia recorded on the death certificate as either the underlying cause of death or the contributing cause of death, excluding contributing causes where the underlying cause was recorded as a cancer

[c]Doses lagged by 2 years. Cox proportional hazards models adjusted for sex, year of birth, and area-level education. Age used as underlying timescale. HR denotes hazards ratio.

[d]Model estimate of HR (90% CI) at 100 mGy = 1.16 (90% CI 0.999; 1.36), one-sided p for trend = 0.051 (+). The ERR per 100 mGy = 0.15 (90% CI −0.001; 0.31).

Table 4. Observed and expected deaths due to chronic lymphocytic leukemia (CLL) and risk estimates by radiation dose to red bone marrow (RBM) among 135,193 nuclear power plant workers first monitored for radiation 1957–1984 and followed through 2011 in the United States.

	Radiation dose to RBM (mGy)				
	<5	5–	50–	≥100	Total
No. of workers	26,027	80,273	17,604	11,289	135,193
Mean dose (mGy)	0.83	20.9	70.2	165.9	35.6
CLL deaths					
Observed[a]	12	49	4	2	67
Expected	15.5	42.2	7.8	5.3	70.9
O/E (SMR)	0.84	1.12	0.39	0.38	0.92
CLL cases[b]	17	52	3	2	74
Hazards ratio[c]	1.0 (R)	1.68	0.53	0.54	0.46[d]
95% Confidence intervals	–	0.94; 3.00	0.15; 1.86	0.12; 2.36	0.21; 1.05

[a]Number of CLL deaths recorded as the underlying cause of death on the death certificate

[b]Number of CLL deaths recorded on the death certificate as either the underlying cause of death or the contributing cause of death, excluding contributing causes where the underlying cause was recorded as a cancer

[c]Doses lagged by 10 years. Cox proportional hazards models adjusted for year of birth and area-level education. Sex was not included as an adjustment factor due to the small numbers of female cases. Age used as underlying timescale. HR denotes hazards ratio.

[d]Model estimate of HR at 100 mGy = 0.46 (95% CI 0.21; 1.05), two-sided p for trend = 0.066 (-). ERR per 100 mGy = −0.77 (95% CI −1.59; 0.05).

are being followed as are additional analyses explicitly addressing issues of undetected dose-related in part to the minimum detectable response of the dosimeters in use during the study period. Compared with other occupational dose reconstruction circumstances (Gilbert 1998), the NPP industry is required to provide precise dosimetry data that are collected under strict criteria in compliance with legal regulations (USNRC 2021).

Results

Most of the 135,193 NPP workers were male (96.7%), born before 1950 (54.3%), and followed for more than 30 years (57.9%). The date of first radiation monitoring at an NPP was before 1975 for 14.5% of the workers, and 25.3% of workers were under the age of 25 years at first monitoring. The mean number of years monitored was 12.8 years. Because of the completeness of the personal identifiers used in the matching procedures, vital status was obtained essentially for all workers. Overall, 29,076 workers (21.5%) had died before 2012 (SMR 0.89; 95% CI 0.88; 0.90). The average follow-up was 30.2 years. The mean cumulative personal dose equivalent from external radiation was 52.6 mSv (maximum 1.32 Sv; percent workers >100 mSv, 14.8%). There were 27,518 workers (20.4%) who received occupational exposures at other facilities before ($n = 10{,}592$) or after leaving ($n = 16{,}926$) employment at a nuclear power facility. The mean personal dose equivalent received elsewhere was 20.4 mSv.

The observed and expected deaths for over 50 specific and grouped causes of death are shown in Table 2 by sex. For all causes of death, the SMR was 0.89 (95% CI 0.88; 0.90) and significantly low based on 29,076 deaths, related in part to the deficit observed for ischemic heart disease (SMR 0.80; 95% CI 0.78; 0.82; $n = 5{,}410$). In contrast, all cancer deaths were significantly increased (SMR 1.03; 95% CI 1.01; 1.05; $n = 9{,}329$), related in part to the excess observed for lung cancer (SMR 1.10; 95% CI 1.07; 1.14; $n = 3{,}382$). Cancers of the pleura, peritoneum and mesothelioma were significantly increased (SMR 5.66; 95% CI 4.98; 6.40; $n = 251$) as were deaths due to asbestosis (SMR 9.15; 95% CI 7.33; 11.3; $n = 87$). There were 296 deaths from leukemia other than CLL (SMR 1.06; 95% CI 0.94; 1.19), and 67 deaths from CLL (SMR 0.95; 95% CI 0.73;

1.20). Hodgkin lymphoma (SMR 0.63; 95% CI 0.41; 0.94; $n = 24$) and non-Hodgkin lymphoma (SMR 0.86; 95% CI 0.77; 0.96; $n = 319$) were significantly below expectation. Multiple myeloma (SMR 1.11; 95% CI 0.95; 1.29; $n = 170$) and MDS (SMR 1.08; 95% CI 0.80; 1.43; $n = 48$) were elevated but the increases were not statistically meaningful. The SMR for Parkinson's disease was 0.90 (95% CI 0.76; 1.09; $n = 140$), and the SMR for prostate cancer was 0.99 (95% CI 0.91; 1.08; $n = 527$).

For females, the only statistically significant SMR was for all smoking-related cancers SMR 1.31 (95% CI 1.02;1.66; $n = 68$). Cancers of the female breast (SMR 1.15; 95% CI 0.81; 1.59; $n = 37$) and lung (SMR 1.27; 95%CI 0.94; 1.68; $n = 48$), and leukemia other than CLL (SMR 1.41; 95%CI 0.51; 3.07; $n = 6$) were increased but not significantly. There were no deaths due to mesothelioma or asbestosis among female workers.

The 10,043 workers monitored for <30 d had 298,458 person-years of observation (29.7 y on average). For all causes of death, the SMR was 0.98 (95% CI 0.95; 1.02) based on 2820 deaths. The site-specific SMRs for most causes of death were compatible with workers who were monitored for longer times, including all cancers (SMR 1.05; 95% CI 0.98; 1.13; $n = 832$), lung cancer (SMR 1.14; 95% CI 1.02; 1.28; $n = 305$), smoking-related cancers (SMR 1.13; 95% CI 1.02; 1.123; $n = 455$), ischemic heart disease (SMR 0.85; 95% CI 0.78; 0.93; $n = 531$), and all external causes of death (SMR 1.15; 95% CI 1.02; 1.28; $n = 303$). There were 23 deaths from leukemia other than CLL (SMR 0.97; 95% CI 0.67; 1.45).

The internal analyses included an additional 15 cases of non-CLL and an additional 7 cases of CLL that were recorded on the death certificate as a contributing, but not underlying, cause of death. The internal analyses were thus based on a total of 311 cases of non-CLL (using a 2-year dose lag) and 74 cases of CLL (using a 10-year dose lag). For leukemia other than CLL (Table 3, Figure 2), the ERR at 100 mGy was 0.15 (90% CI −0.001; 0.31). Both linear and linear-quadratic ERR models fit the data equally well, i.e. the Akaike Information Criteria (AIC) values of 600.19 and 601.85, respectively, were very similar.

The SMR for CLL was not elevated and there was no evidence of a dose-response (Table 4). Based on a linear model, the ERR at 100 mGy was −0.77 (95% CI −1.59; 0.05).

The SMRs for subtypes of leukemia were: 0.97 (95% CI 0.60; 1.49) for ALL, 1.12 (95% CI 0.95; 1.30) for AML, and 1.06 (95% CI 0.75; 1.47) for CML (Table 2). The patterns of risk over categories of dose showed increasing trends for each subtype (Table 5). For ALL the estimate of HR at 100 mGy was 1.59 (90% CI 1.10; 2.30; p for trend <0.01); for AML it was 1.27 (90% CI 1.06; 2.53; p for trend 0.046); and for CML it was 1.21 (90% CI 0.79; 1.85; p for trend 0.45). Standard Cox proportional hazards models (HR-1) and Cox proportional hazards regression models to estimate ERRs (Preston et al. 2015) provided equivalent results (Table 5). There was no significant trend over categories of RBM dose for MDS alone (ERR per 100 mGy −0.34 (95% CI −1.11; 0.42) $n = 48$) or in combination with leukemia other than CLL (data not shown). For multiple myeloma and non-Hodgkin lymphoma the comparable ERRs (95% CI) at 100 mGy were 0.01 (−0.29; 0.31; $n = 170$) and −0.09 (−0.34; 0.16; $n = 319$), respectively.

The SMR for all solid cancers, excluding leukemias and lymphomas, was significantly increased [SMR 1.04 (95%CI 1.01; 1.06; $n = 8,445$)] and this increase was due almost entirely to increases in smoking-related cancers. Internal dose-response analyses for all solid cancers found no evidence for an increase in risk over categories of absorbed dose to the colon, taken as an estimate of whole-body dose [ERR (95% CI) at 100 mGy was 0.01 (-0.03; 0.05); $n = 8,445$].

The SMR for lung cancer was significantly increased [SMR 1.10 (95% CI 1.07; 1.14; $n = 3,382$)] as was the SMR for cancers of the pleura and mesothelioma [SMR 5.66 (95% CI 4.98; 6.40; $n = 251$)] (Table 2). The highest SMR for any cause of death was for asbestosis [SMR 9.15 (95% CI 7.33; 11.3; $n = 87$)]. Internal dose-response analyses for lung cancer found no evidence for an increase in risk over categories of absorbed dose to the lung: ERR (95% CI) at 100 mGy was −0.04 (−0.11; 0.02) (Table 6, Figure 3). The sex-specific ERRs (95% CI) at 100 mGy were −0.06 (−0.11; 0.01; $n = 3334$) among the 130,773 male workers and 0.63 (−0.91; 2.17; $n = 48$) among the 4420 female workers; a difference that was not statistically significance (p-value for heterogeneity = 0.37). The mean lung dose was 44.0 mGy (max 1.08 Gy) for males and 19.3 mGy (max 0.47 Gy) for females.

Cancer of the esophagus was elevated in comparison with the general population although the increase was not statistically significant (SMR 1.07, 95% CI 0.96; 1.19, $n = 355$).

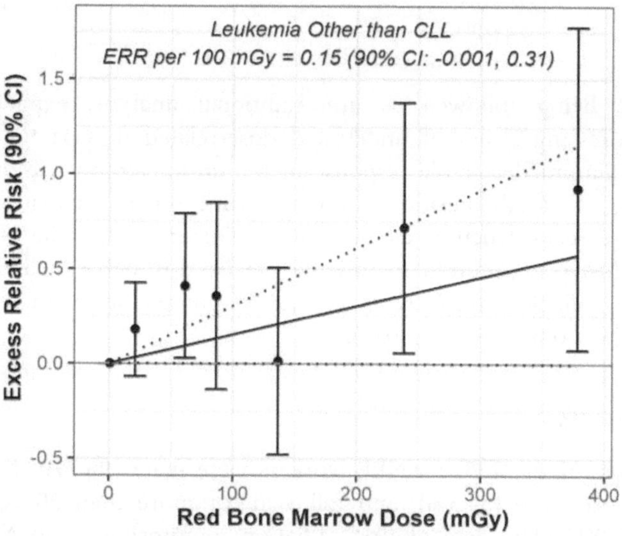

Figure 2. Excess Relative Risk (ERR) of leukemia other than chronic lymphocytic leukemia (CLL) by radiation dose to red bone marrow among 135,193 workers at U.S. nuclear power plants first monitored between 1957–1984 and followed through 2011. Linear model presented (solid line) based on a continuous measure of dose, and 95% confidence bands (dotted lines) presented about the regression line. Doses were lagged by 2 years and adjustments were made for sex, year of birth, and socioeconomic status. ERRs and 90% confidence limits are calculated for selected categories of dose to the red bone marrow and displayed at the mean dose in each category (categories <5 (referent), 5-, 50-, 75-, 100-, 200-, ≥300 with respective mean doses of 0.88, 21.0, 61.1, 86.6, 136.6, 238.5, and 377.8 mGy (see Table 3)).

Table 5. Excess relative risk (ERR) estimates at 100 mGy for leukemia subtypes by analytical method among 135,193 nuclear power plant workers, first monitored for radiation 1957–1984 and followed through 2011 in the United States.

		Standard cox model		EPICURE[d]	
Leukemia subtypes[a]	Number of cases	HR-1 at 100 mGy	90% CI	ERR per 100 mGy	90% CI
Acute lymphocytic leukemia (ALL)	25	0.59	0.10; 1.30	0.44	0.07; 0.81
Acute myeloid leukemia (AML)	165	0.27	0.06; 1.53	0.24	0.06; 0.42
Chronic myeloid leukemia (CML)	40	0.21	−0.21; 0.85	0.18	−0.25; 0.61
Chronic lymphocytic leukemia (CLL)	74	−0.54	−0.79; 0.05[b]	−0.77	−1.59; 0.05[b]
Leukemia other than CLL[c]	311	0.16	−0.001; 0.36	0.15	−0.001; 0.31

[a]ICD definitions and numbers found in Supplement Table 1. Counts include underlying and contributing causes of death. Doses are lagged two years in models for all subtypes except CLL. CLL model uses doses with 10-year lagging. All models adjusted for sex, year of birth, and area-level education except CLL which was not adjusted for sex due to small numbers of cases among female workers. Age used as underlying timescale in Cox models. HR denotes hazards ratio.
[b]95% Confidence intervals are provided for CLL because it is not seen to be consistently increased following radiation (UNSCEAR 2008).
[c]The counts for leukemia other than CLL include acute leukemias of unspecified cell type as listed in Supplement Table 1.
[d]Cox proportional hazards regression model results obtained using the PEANUTS program within EPICURE (Preston et al. 2015).

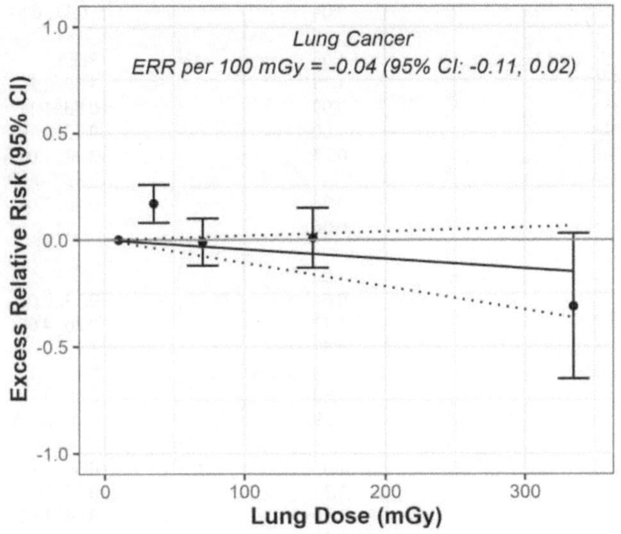

Figure 3. Excess Relative Risk (ERR) of lung cancer by radiation dose to lung among 135,193 workers at U.S. nuclear power plants first monitored between 1957–1984 and followed through 2011. Linear model presented (solid line) based on a continuous measure of dose, and 95% confidence bands (dotted lines) presented about the regression line. Doses were lagged by 10 years and adjustments were made for sex, year of birth, and socioeconomic status. ERRs and 95% confidence limits are calculated for selected categories of dose to the lung and displayed at the mean dose in each category (categories <25 (referent), 25-, 50-, 100-, and >250 with respective mean doses of 10.1, 35.5, 70.3, 148.5, and 335.0 mGy).

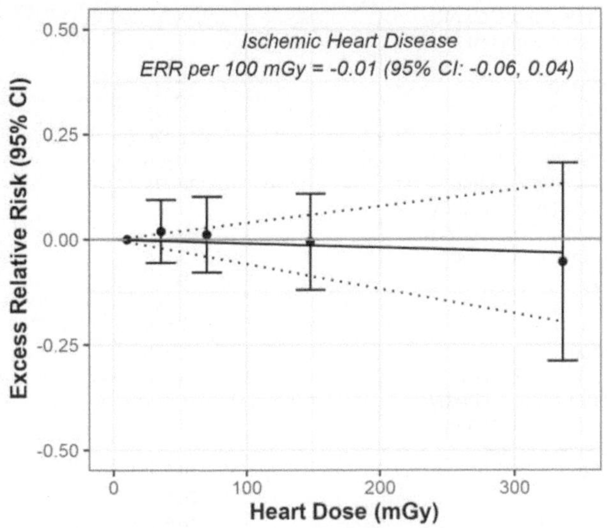

Figure 4. Excess Relative Risk (ERR) of ischemic heart disease by radiation dose to heart among 135,193 workers at U.S. nuclear power plants first monitored between 1957–1984 and followed through 2011. Linear model presented (solid line) based on a continuous measure of dose, and 95% confidence bands (dotted lines) presented about the regression line. Doses were lagged by 10 years and adjustments were made for sex, year of birth, and socioeconomic status. ERRs and 95% confidence limits are calculated for selected categories of dose to the heart and displayed at the mean dose in each category (categories <25 (referent), 25-, 50-, 100- and >250 with respective mean doses of 10.1, 35.5, 70.3, 148.7, and 335.9 mGy).

Internal dose-response analyses for esophageal cancer showed the ERR (95% CI) at 100 mGy to be 0.11 (-0.05; 0.28) (Table 6). The SMR for all heart disease was significantly low at 0.82 (95% CI 0.80; 0.84, $n=7{,}561$). Similarly, death from IHD occurred significantly below expectation [SMR 0.80 (95% CI 0.78; 0.82, $n=5{,}410$)]. Internal dose-response analyses found no evidence for an increase in IHD over categories of absorbed dose to the heart. Over 14,000 workers had heart doses in excess of 100 mGy and nearly 2000 had doses in excess of 250 mGy. The ERR (95% CI) at 100 mGy was −0.01 (-0.06; 0.04) (Table 6, Figure 4).

The SMRs for non-Hodgkin lymphoma (NHL), multiple myeloma, and myelodysplastic syndrome (MDS) were 0.86 (95% CI 0.77, 0.96; n = 319), 1.11 (95% CI 0.95; 1.29; n = 170), and 1.08 (95% CI 0.80, 1.43; n = 48), respectively (Table 2). The ERRs (95% CI at 100 mGy) for NHL, multiple myeloma, and MDS were −0.09 (−0.34; 0.16), 0.01 (−0.29; 0.31), and −0.34 (−1.11; 0.42), respectively (analyses not presented).

The adjustment for duration of monitoring in our study of NPP workers was found to increase both the estimate of risk and the significance of the dose-response for leukemia other than CLL (Supplement Tables 3 and 4). Without an adjustment for the duration of monitoring, the ERR at 100 mGy was 0.15 (90% CI −0.001, 0.31), whereas with an adjustment it was increased to 0.32 (90% CI 0.17, 0.46). Although the risk estimates are statistically indistinguishable, the risk estimate with adjustment for the duration of monitoring was double the estimate without such an adjustment. The choice of the number of duration categories used for adjustment did not make a difference. If two categories were used, less than or more than 10 y duration, the estimate of

Table 6. Internal cohort dose-response analyses for all solid cancers, cancers of the lung and esophagus, Parkinson's disease, and ischemic heart disease by organ-specific dose among 135,193 nuclear power plant workers first monitored for radiation 1957–1984 and followed through 2011 in the United States[a].

Dose (mGy)	Workers	Cases	HR	95% CI
All solid cancers excluding leukemias and lymphomas (ICD-9 140–199)				
<25[b]	74,707	5275	1.0	REF
25–<50	27,084	1,591	1.16	1.09; 1.23
50–<100	19,145	904	1.00	0.93; 1.08
100–<250	12,335	569	1.06	0.97; 1.15
≥250	1,922	106	0.90	0.74; 1.09
Totals and *p* for trend	135,193	8445		*0.60 (+)*
HR (95% CI) at 100 mGy			1.01	0.97; 1.05
ERR (95% CI) at 100 mGy			0.01	−0.03; 0.05
Lung cancer (total)				
<25[c]	75,217	2131	1.0	REF
25–<50	26,985	648	1.18	1.08; 1.29
50–<100	18,990	353	0.99	0.88; 1.11
100–<250	12,161	216	1.01	0.88; 1.17
≥250	1840	34	0.73	0.52; 1.03
Totals and *p* for trend	135,193	3382		0.17 (–)
HR (95% CI) at 100 mGy			0.96	0.90; 1.02
ERR (95% CI) at 100 mGy			−0.04	-0.11; 0.02
Lung cancer (males)				
<25[c]	71,784	2091	1.0	REF
25–<50	26,442	645	1.19	1.09; 1.30
50–<100	18,675	351	0.99	0.88; 1.11
100–<250	12,036	213	1.00	0.87; 1.16
≥250	1836	34	0.73	0.52; 1.03
Totals and *p* for trend	130,773	3334		0.15 (–)
HR (95% CI) at 100 mGy			0.96	0.90; 1.02
ERR (95% CI) at 100 mGy			-0.06	-0.11; 0.01
Lung cancer (females)				
<25[c]	3433	40	1.0	REF
25–<50	543	3	0.64	0.19; 2.08
50–<100	315	2	1.11	0.26; 4.66
≥100	129	3	3.59	1.09; 11.8
Totals and *p* for trend	4420	48		0.12 (+)
HR (95% CI) at 100 mGy			1.80	0.86; 3.78
ERR (95% CI) at 100 mGy			0.59	−0.16; 1.33
Esophageal cancer				
<25[d]	76,799	205	1.0	REF
25–<50	26,681	71	1.31	0.99; 1.72
50–<100	18,491	46	1.30	0.94; 1.80
100–<250	11,564	29	1.39	0.94; 2.06
≥250	1658	4	0.97	0.36; 2.61
Totals and *p* for trend	135,193	355		0.17 (+)
HR (95% CI) at 100 mGy			1.12	0.95; 1.32
ERR (95% CI) at 100 mGy			0.11	-0.05; 0.28
Parkinson's disease				
<25[e]	86,335	97	1.0	REF
25–<50	24,048	21	1.03	0.64; 1.65
50–<100	15,909	8	0.71	0.34; 1.46
100–<250	8175	11	1.74	0.93; 3.25
≥250	726	3	2.35	0.75; 7.44
Totals and *p* for trend	135,193	140		0.07 (+)
HR (95% CI) at 100 mGy			1.27	0.98; 1.65
ERR (95% CI) at 100 mGy			0.24	-0.02; 0.50
Ischemic heart disease				
<25[f]	74,427	3535	1.0	REF
25–<50	27,149	896	1.02	0.95; 1.10
50–<100	19,215	574	1.01	0.93; 1.11
100–<250	12,448	334	1.00	0.89; 1.12
≥250	1954	71	0.95	0.75; 1.20
Totals and *p* for trend	135,193	5410		0.70 (–)
HR (95% CI) at 100 mGy			0.99	0.94; 1.04
ERR (95% CI) at 100 mGy			−0.01	−0.06; 0.04

[a]Models utilize radiation doses lagged by 10 years. Outcomes include deaths from underlying causes of death only. Cox proportional hazards models adjusted for sex, year of birth, and area-level education. Age used as underlying timescale. HR denotes hazards ratio. CI denotes confidence interval. ERR denotes excess relative risk.

[b]Model utilizes dose to the colon.

[c]Model utilizes dose to the lungs. The p-value for heterogeneity was 0.37 for the difference in lung cancer sex-specific risk estimates.

[d]Model utilizes dose to the esophagus.

[e]Model utilizes dose to the brain.

[f]Model utilizes dose to the heart.

Table 7. Excess relative risk (ERR) estimates per 100 mGy for leukemia other than chronic lymphocytic leukemia for selected studies of radiation-exposed populations.

Study[a]	Persons	Cases	ERR per 100 mGy	Confidence intervals	Reference
U.S. nuclear power plant workers	135,193	311	0.15	0.00; 0.31[b]	Current study
U.S. industrial radiographers	123,556	173	0.17	−0.02; 0.35[b]	NCRP (2018b)
U.S. medical radiation workers	109,019	126	0.10	−0.34; 0.54[c]	Boice, Cohen, Mumma, Howard et al. (2021)
Mallinckrodt uranium processing	2514	18	0.14	−0.60; 0.33[c]	Golden, Ellis, et al. (2019)
Mound	4954	21	0.04	−0.37; 0.71[b]	Boice et al. (2014)
Atomic veterans	113,807	710	−0.37	−1.08; 0.33[b]	Boice et al. (2020)
Atomic bomb survivors (adult males)[d]	24,845	98	0.09	−0.16; 0.37[b]	Preston[d]
INWORKS	308,297	531	0.30	0.12; 0.52[b]	Leuraud (2015), Hamra et al. (2016)
IARC 15-country	407,391	198	0.19	<0; 0.85[c]	Cardis et al. (2005, 2007)
IARC 3-country	95,673	119	0.22	0.01; 0.57[b]	Cardis et al. (1995)
UK National Registry of Radiation Workers	174,541	198	0.17	0.01; 0.43[c]	Muirhead et al. (2009)[e]
NIOSH U.S. case-control	NR	369	0.09	−0.17; 0.65[c]	Daniels et al. (2013)
French nuclear workers	59,021	60	0.40	<0; 1.68[b]	Metz-Flamant et al. (2013)
U.S. 15-utilities	53,698	26	0.57	−0.26; 3.04[c]	Howe (2004)
U.S. radiologic technologists	110,297	155	0.05	< −0.09; 0.24[c]	Linet et al. (2020)
Mayak Production Association	22,373	56	0.36	0.16; 0.82[c]	Kuznetsova et al. (2016)

[a]Studies are not all independent. The INWORKS study, for example, contains workers in the IARC 3-country, IARC 15-country, UK NRRW and NIOSH U.S. case-control studies.
[b]90% CI; also cases include underlying and contributing causes.
[c]95% CI.
[d]Dale Preston (Personal Communication, 31 May 2016).
[e]A recent update with 269 deaths from leukemia other than CLL is generally consistent with the earlier reports [ERR per 100 mGy 0.10 (90%CI −0.23 to 2.90)] (Gillies et al. 2019).
NR denotes not reported.

Table 8. Excess relative risk (ERR) estimates per 100 mGy for subtypes of leukemia for selected studies of radiation-exposed populations.

Study[a,b] (reference)	ALL (*n*) (90% CI)	AML (*n*) (90% CI)	CML (*n*) (90% CI)	CLL (*n*) (90% CI)
Current study – U.S. nuclear power plant workers	0.44 (25) (0.07; 0.81)	0.24 (165) (0.06; 0.42)	0.18 (40) (−0.25; 0.61)	−0.77 (74) (−1.59; 0.05)
INWORKS (Leuraud et al. 2015; Hamra et al. 2016)	0.58 (30) (<0; 3.16)	0.13 (254) (−0.08; 0.43)	1.05 (100) (0.45; 2.00)	−0.11 (138) (<0; 0.18)
IARC 15-country (Cardis et al. 2007)	<0.0 (19) NR	−0.41 (81) (−0.76; 0.17)	1.0 (45) (−0.09; 4.02)	−0.17 (47) (−0.89; 0.40)
IARC 3-country (Cardis et al. 1995)	−0.89 (11) (<0; 7.3)	0.34 (32) (<0; 1.49)	1.1 (28) (0.29; 3.09)	−0.09 (27) (<0; 0.73)
UK NRRW (Muirhead et al. 2009)	0.78 (15) (<0.19; 8.95)	0.12 (102) −0.12; 0.57	0.33 (44) (0.04; 0.93)	<−0.19 (69)[d] (<−0.I9; 0.12)
NIOSH U.S. case-control (Daniels et al. 2013)	NR	0.22 (150) (−0.I9; 1.2)	0.29 (52) (<0; 2.6)	−0.032 (74) (−0.16; 0.74)
French nuclear workers (Metz-Flamant et al. 2013)	NR	1.58 (35) (0.15; 4.61)	NR[c]	−0.14 (NR) (<0; 1.49)
U.S. radiologic technologists (Linet et al. 2020)	0.058 (15) (<−0.02; 1.03)	0.00 (85) (<−0.02; 0.24)	−0.02 (9) (<−0.02; >10)	<−0.02 (23) (<−0.03; 0.18)
Mayak Production Association (Kuznetsova et al. 2016)	NR	1.32 (24) (0.43; 4.9)	0.14 (13) (−0.02; 0.73)	−0.02 (21) NR

[a]Acute lymphocytic leukemia (ALL), acute myeloid leukemia (AML), chronic myeloid leukemia (CML), chronic lymphocytic leukemia (CLL), not reported (NR). Current study includes underlying and contributing causes of death. Note that Linet and colleagues (Linet et al. 2020, Table 4) prepared an informative and detailed comprehensive summary of leukemia subtype ERRs for radiation-exposed populations.
[b]Studies are not independent. The INWORKS study, for example, contains workers in the IARC 3-country, IARC 15-country, UK NRRW and the NIOSH case-control studies.
[c]Estimates are presented only for myeloid leukemia and not separately for AML or CML. ALL not presented ($n \leq 7$).
[d]Using a 10-year lag, the ERR per 100 mSv was −0.03 (90% CI −0.17; 0.31).

risk was similar to the estimate using six categories: the HR at 100 mGy was 1.28 (95% 1.10; 1.48) based on 2 category adjustment whereas it was 1.37 (90% CI 1.19;1.59) based on a finer 6 category adjustment.

Sensitivity analyses were conducted to evaluate the choice of referent group, adjustment factors, and the effect of including contributing causes of death. Excluding all workers with zero recorded dose and using the <10 mGy category as referent had little effect on the dose-response analysis for leukemia other than CLL with the HR at 100 mGy estimated to be 1.17 (95% CI 1.02; 1.34; trend $p = 0.03$). Adjusting for mobility and year of hire increased the estimates of risk slightly but the increase was not statistically significant (Supplement Table 4). Removing adjustments for sex and SES also had minimal effect on the estimates of radiation risk. Using a 2-year lag for CLL rather than a 10-year lag, did not affect the dose-response which remained significantly negative. Using a 5-year lag for lung cancer rather than a 10-year lag, also did not affect the dose response.

Discussion

Previous studies of nuclear power plant workers in the US were limited by low statistical power. An early study of nearly 9000 workers at a single nuclear power facility demonstrated the feasibility of linking together company dosimetry records with NRC REIRS and Landauer data to obtain complete dosimetry information on workers (Goldsmith et al. 1989; Jablon and Boice 1993; Muirhead et al. 1996). A larger study of 53,698 workers was carefully conducted but limited in design (Howe et al. 2004). Workers had to be alive in 1979 and employed at a nuclear power plant operated by one of 15 utilities in the United States. A radiation association for leukemia was not found but statistical power was limited due to small numbers. The 1979 restrictions were for ease of tracing since 1979 is the year when the NDI began, but this restriction resulted in few leukemia deaths, only 26, low cumulative doses (mean 25.7 mSv), and few career doses >100 mSv (about 5%). The current study of 135,193 early NPP workers monitored as early as 1957 is larger, with longer follow-up and a broader range of radiation doses than previous investigations. Consistent with current understanding (Daniels and Schubauer-Berigan 2011; McLean et al. 2017; Gilbert et al. 2020; Hauptmann et al. 2020) (Table 7), prolonged exposure to occupational radiation was seen to increase the mortality risk of leukemia other than CLL among NPP workers.

International occupational studies

Large international occupational studies have been coordinated by the International Agency for Research on Cancer (IARC) (Cardis et al. 1995, 2005, 2007). Results compared with other studies are not independent because of overlapping facilities included in each analysis, e.g. the British Nuclear Fuels plc (BNFL) facility at Sellafield contributed many workers and leukemia cases among those who received doses >200 mSv. A significant risk of leukemia other than CLL was reported in the IARC 3-country study (Cardis et al. 1995), primarily due to relatively high occupational doses (>200 mSv) seen at the Sellafield plant (Douglas et al. 1994). A significant leukemia risk was not observed in the subsequent IARC 15-country study which excluded workers who were monitored for internal intakes of radioactive material and other factors (Cardis et al. 2007). Among the over 49,346 U.S. nuclear power plant (NPP) workers in the 15-country study, leukemia excluding CLL also was not significantly increased (ERR per 100 mSv 0.13; 90% CI <0.00, 0.18; n = 65). It is not entirely clear, but this U.S. NPP cohort of 15 facilities (Vrijheid et al. 2008) apparently derived from the larger study conducted by Howe et al. (2004).

The latest and largest international investigation is the International Nuclear Workers Study (INWORKS), which includes 308,297 radiation workers from France, the United Kingdom, and the United States (Leuraud et al. 2015; Richardson et al. 2015, 2018; Thierry-Chef et al. 2015; Hamra et al. 2016; Daniels et al. 2017; Laurier et al. 2017; Haylock et al. 2018). In contrast to the 15-country study, U.S. NPPs were not included. INWORKS reported an association between the cumulative external photon dose to the RBM and mortality from leukemia other than CLL: ERR per 100 mGy = 0.30 (90% CI: 0.17; 0.52). These data are not independent of other studies and include radiation workers from the UK National Registry for Radiation Workers (NRRW) (Muirhead et al. 2009), nuclear industry workers in France (Metz-Flamant et al. 2013), and five nuclear facilities in the United States (Schubauer-Berigan et al. 2015). Despite the many strengths of the INWORKS investigation, possible confounding tempers interpretations (NCRP 2018b). Statistically significant risks were reported for cancers of the testis, rectum and peritoneum, sites not convincingly linked to ionizing radiation (UNSCEAR 2008), and positive dose responses were reported for cancers of the pleura and mesothelioma which are caused by asbestos and not radiation (Mumma et al. 2019). Removing France in a sensitivity analysis resulted in statistical non-significance, even though the French data were the smallest component of INWORKS, and there were rather puzzling results regarding neutron exposures (NCRP 2018b). That is, neutron doses were not computed but adjustment based upon a neutron monitoring flag was made. If no adjustment for neutron monitoring is made the ERR per Gy estimate for solid cancers was reduced from 0.48 (95% CI 0.20; 0.79) to 0.20 (90% CI −0.03; 0.46) (Richardson et al. 2015). The reasons behind this large drop in the ERR estimate, and from statistical significance to nonsignificance, are unclear, 'but may relate to some confounding, especially since one might expect missing neutron doses in the risk modeling to cause overestimation, rather than underestimation' (NCRP 2018b).

The UK National Registry for Radiation Workers (NRRW) study of 174,541 workers found a significant risk of leukemia other than CLL, based on 198 deaths, mean dose 25.5 mSv and with 6.2% of workers with occupational doses >100 mSv (excess relative risk (ERR) per Sv = 1.38; 90% CI 0.08; 3.24) (Muirhead et al. 2009; Gillies and Haylock 2014; Gillies et al. 2019). More than half of the workers with high cumulative doses came from the Sellafield facility. A study of 64,956 workers at four BNFL facilities, again with nearly 50% of workers coming from Sellafield, reported a significant risk of leukemia other than CLL (ERR per Sv = 2.60; 90% CI 0.28; 7.01) based on 85 deaths and mean dose of 53.0 mSv among the 42,431 radiation workers (Gillies and Haylock 2014). The mortality analysis based on 269 deaths from leukemia other than CLL is consistent with earlier reports [ERR per 100 mGy 0.10 (90%CI −0.23; 2.90)] (Gillies et al. 2019).

A study of 45,468 Canadian nuclear power industry workers and workers at Atomic Energy of Canada Limited (a research facility) failed to observe a significant association between occupational dose and leukemia but analyses were based on only 17 leukemia deaths (Zablotska et al. 2014). A study of 59,021 French nuclear workers reported a nonsignificant association between occupational exposure and leukemia other than CLL based on 60 deaths and a mean dose of 22.5 mSv (Metz-Flamant et al. 2013). A study of 200,583 Japanese nuclear workers reported a nonsignificant negative association with cumulative dose and leukemia based on 80 deaths and mean dose of 12.2 mSv (with 2.6% of the workers > 100 mSv) (Akiba and Mizuno 2012; Akiba 2018).

Combined U.S. occupational studies

One of the first combined mortality analyses of U.S. worker studies was conducted by Gilbert and colleagues in 1993 of 44,943 workers at three weapons facilities (Gilbert et al. 1993). There were 67 deaths due to leukemia other than CLL, the mean dose was <50 mSv and the correlation between radiation was negative (ERR per Sv = −1.0; 90% CI <0.0; 2.2). A later combined case-control study of leukemia among workers in the United States involved six facilities (including the Portsmouth Naval Shipyard) and 105,245 workers (Daniels et al. 2013). There were 264 deaths due to leukemia other than CLL, the mean dose was 26.5 mSv, and the association between radiation was positive but not statistically significant (ERR per 100 mSv = 0.09 (95% CI −0.17; 0.65). There were 74 deaths due to CLL and similar to the current study, there was a negative association with radiation (ERR per 100 mSv = −0.032 (95% CI −0.16; 0.74)). A recent cohort evaluation of workers at five U.S. facilities ($n = 119{,}195$), excluded the Los Alamos National Laboratory, changed the eligibility criteria for inclusion, included contributing causes of CLL (total cases 128) and confirmed the previous findings based on 369 deaths from leukemia other than CLL and a mean external dose of 20.2 mS (Schubauer-Berigan et al. 2015).

A feasibility study at the Calvert Cliffs Nuclear Power Plant published in 1993 envisioned the creation of a radiation registry of workers which might form the basis of a comprehensive study of over 100,000 NPP workers in the United States (Jablon and Boice 1993). This vision became a reality (Hagemeyer et al. 2018) as demonstrated by the current study. The pilot study at the Calvert Cliffs facility included 9132 workers, of whom 5093 were plant employees and 4093 were contract workers. The contract workers experienced the higher cumulative doses (Goldsmith et al. 1989). No significant findings were reported. However, in the current follow-up of 6264 plant employees at Calvert Cliffs (Supplement Table 2), a significant SMR of mesothelioma was observed based on 5 deaths (SMR 3.19; 95% CI 1.03; 7.44), and similar to the asbestos-related risk reported for all workers in the current study.

Leukemia other than CLL

Most occupational studies of combined facilities discussed above are characterized as having well-defined populations, good occupational records, comprehensive follow-up procedures, nearly complete ascertainment of deaths and causes of death, individual personal dosimetry measurements to estimate cumulative occupational dose, and careful and complete statistical analyses (NCRP 2018b; UNSCEAR 2018). There is a consistency in the presence of a positive correlation between radiation and death due to leukemia other than CLL (Table 7), and the absence of a correlation between radiation and death due to CLL (Table 8).

Leukemia other than CLL is found to be increased in studies of Japanese atomic bomb survivors (Hsu et al. 2013), patients receiving high-dose radiation procedures (Boice et al. 1987; Curtis et al. 1994; Little et al. 1999), pioneering radiologists (Matanoski et al 1975; NA/NRC 2006; UNSCEAR 2008), populations exposed to environmental contamination (Krestinina et al. 2010; Preston et al. 2017; NCRP 2018b) and more recently in large occupational studies (NCRP 2018b; Linet et al. 2020) (Table 7). The current study of nuclear power plant workers in the United States adds to this evidence. The numbers of cases of leukemia subtypes, ALL, AML and CML, also were sufficient to compute estimates of mortality risk that were generally consistent with those reported in other studies although the confidence intervals about the risk estimates were broad (Table 8).

Among Japanese atomic bomb survivors, linear-quadratic and purely quadratic dose-response models provided better fits to the non-CLL data than a linear model (UNSCEAR 2008; Hsu et al. 2013). Among bomb survivors over the age 20 years when exposed in 1945, there were 121 deaths due to leukemia other than CLL; the ERR per Sv was approximately 3.1 (95% CI 1.7; 5.0) (UNSCEAR 2008) or about twice the estimate of the current study. In the most recent incidence study of atomic bomb survivors (Hsu et al. 2013), there were 98 incident leukemia cases other than CLL or adult T-cell leukemia (ATL) among 24,845 adult males, the mean dose was about 10 mGy, and the estimate of ERR per 100 mGy was 0.09 (95% CI −0.16; 0.37) (Table 7). Although the confidence intervals overlap for the risk estimates in the current study, the acute exposure of atomic bomb survivors may carry a somewhat higher risk of radiation-induced leukemia than the study of NPP workers who received prolonged exposures over many years. This lower risk following protracted exposures is consistent with chromosome aberration data comparing atomic bomb survivors with Sellafield workers (Tucker et al. 1997). A more statistically powerful evaluation of the risk of leukemia following prolonged radiation exposure will come within the MPS when large cohorts are combined, including industrial radiographers (NCRP 2018b), medical radiation workers (Boice, Cohen, Mumma, Howard et al. 2021), above-ground nuclear weapons test participants (Boice et al. 2020) and DOE radiation workers (Boice, Cohen et al. 2019; Boice, Cohen, Mumma, Golden 2021). A study of 112,726 nuclear submariners, including the 1954 Nautilus crew, has begun that increases the study size of previous investigations by including officers and by extending the years of follow-up by 18 years, from 1982 to 2020 (Charpentier et al. 1993; Mueller et al. 2020; NDC 2021; Friedman-Jimenez et al. 2021). The study of nuclear submariners will provide additional statistical power for future pooled analyses of comparable populations within the MPS.

CLL

For NPP workers in the U.S., the dose-response for CLL was negative based on either a 10-year dose lag or a 2-year dose lag. CLL is not generally considered to be caused by ionizing radiation (UNSCEAR 2008). Practically all studies fail to find positive correlations, including patients treated with radiation (Boice et al. 1987; Curtis et al. 1994) and occupationally exposed workers (Cardis et al. 2007; Schubauer-Berigan et al. 2007, 2015; Vrijheid et al. 2008; Muirhead et al. 2009; Daniels et al. 2013; Leuraud et al. 2015; Kuznetsova et al. 2016; Linet et al. 2020). Negative dose-response relationships for CLL are seen for most occupational studies (Table 8).

A significant correlation between radiation and CLL was reported in the study of Japanese atomic bomb survivors (Hsu et al. 2013) based on only 12 cases. Results are uncertain because 2 of the 12 cases (16.7%) were hairy cell leukemias, which should be classified as non-Hodgkin lymphoma (NHL) and not CLL (Linet et al. 2020). In comparison with the current study, 2 of the 74 cases (2.7%) of CLL among NPP workers were hairy cell leukemias; the population expected value was 2.7 and excluding the 2 cases did not affect the negative dose-response. Further, NHL is not seen to be a consistent consequence following radiation exposure (UNSCEAR 2008), and we found no evidence for a dose response (ERR per 100 mGy −0.09; 95% CI −0.34; 0.16; $n = 319$). Some reports of increases in CLL (Zablotska et al. 2013) are not consistent with previous studies and may reflect peculiarities in the dose reconstruction (based on worker interviews), case ascertainment and diagnostic criteria.

MDS

Radiation has been associated with MDS among long-term atomic bomb survivors at Nagasaki (Iwanaga et al. 2011). Neither MDS nor MDS in combination with AML were found to be correlated with radiation dose in the current study, similar to three other MPS studies, i.e. for medical radiation workers, workers at the Los Alamos National Laboratory, and military participants at above-ground nuclear weapons tests (Boice et al. 2020; Boice, Cohen, Mumma, Golden 2021; Boice, Cohen, Mumma, Howard 2021). Studies of U.S. radiologic technologists (Linet et al. 2020) also report no association between MDS and radiation exposures. High dose radiotherapy, as well as chemotherapy, are major risk factors for MDS (Komrokji et al. 2019). MDS appears to be distinct from AML (Albitar et al. 2002; Zeidan et al. 2019), and it has been questioned recently whether MDS should be included in the assessment of leukemia risk following radiation (UNSCEAR 2020). MDS is a grouping of heterogeneous diseases, and its classification has evolved over the years (Komrokji et al. 2019). It was not uncommon for MDS cases to be misclassified as AML which might affect mortality evaluations. Whether low doses of radiation can cause MDS is an area of scientific interest, and will be addressed further in combined studies within the MPS.

Lung cancer

The SMR for lung cancer was significantly elevated in comparison with the general population, reflecting perhaps residual confounding due to cigarette smoking or to intakes of asbestos from thermal insulation received by many of the workers. The NPP workers had previously been found to be at high risk of asbestos-related diseases including mesothelioma and asbestosis (Mumma et al. 2019). Further, it has been estimated that up to 12% of lung cancers might be related to occupational levels of exposure to asbestos and that up to 20–25% of workers with heavy exposure to asbestos might develop lung cancer (Agency for Toxic Substances and Disease Registry (ATSDR) 2013). Given the substantial excesses of deaths from mesothelioma and asbestosis among these NPP workers, it is possible that some lung cancer deaths were similarly related to asbestos exposure. Unfortunately, job classification was not available to separate NPP workers with likely exposure to asbestos as was possible in several other MPS investigations, i.e. industrial radiographers, shipyard workers and atomic veterans (Till et al. 2018; Mumma et al. 2019).

There was little evidence for a radiation dose response for lung cancer nor was there any discernible difference between men and women and radiation risk, (p for heterogeneity 0.37), consistent with eight other MPS cohorts (Boice 2019; Boice, Ellis et al. 2019; Boice et al. 2020; Boice, Cohen, Mumma, Golden et al. 2021; Boice, Cohen, Mumma, Howard et al. 2021; Boice, Quinn et al. 2021; NASEM 2021a). These findings of healthy populations who experienced chronic exposures during employment are relevant to understanding and assessing the radiation risks faced by astronauts on long-term missions beyond Earth orbit. The lung cancer risk estimates and standards currently used by NASA to protect flight crew are based on the Japanese atomic bomb survivor study that indicates that women are at 2-3 times higher lifetime risk of lung cancer than are men on a relative scale (NCRP 2014; Boice 2019). The current radiation standard for astronauts limits the time women are allowed in space compared with men, although these standards are being reviewed (NASEM 2021a).

Other studies of occupationally exposed workers that failed to find a significant radiation association for lung cancer include the Hanford workers (Petersen et al. 1990), medical radiologic technicians (Velazquez-Kronen et al. 2020), and the UK National Registry of Radiation Workers (Muirhead et al. 2009; Haylock et al. 2018). Increases in lung cancer have been reported among atomic bomb survivors acutely exposed to radiation in 1945 (Ozasa et al. 2012); workers in the 15-country study, although a negative dose response was seen for the subcohort of NPP workers (ERR per 100 mSv −0.12; 90% CI <0.00, 0.05; $n = 316$) (Cardis et al. 2007); Mayak workers in Russia (Gillies et al. 2017), and the INWORKS occupationally-exposed workers (Richardson et al. 2018). Differences in results might be

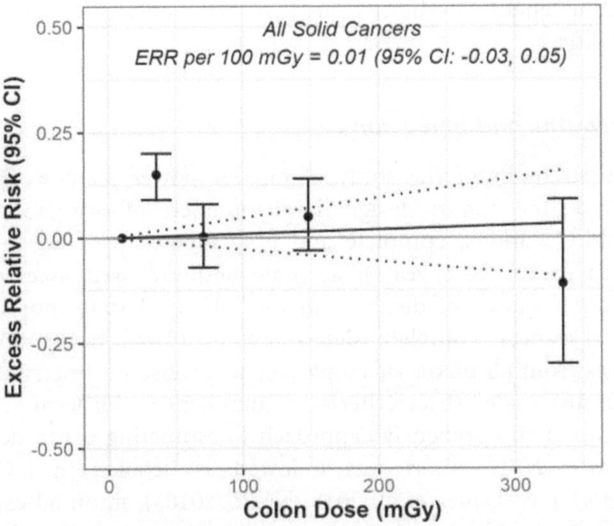

Figure 5. Excess Relative Risk (ERR) of all solid cancers by radiation dose to colon among 135,193 workers at U.S. nuclear power plants first monitored between 1957–1984 and followed through 2011. Linear model presented (solid line) based on a continuous measure of dose, and 95% confidence bands (dotted lines) presented about the regression line. Doses were lagged by 10 years and adjustments were made for sex, year of birth, and socioeconomic status. ERRs and 95% confidence limits are calculated for selected categories of dose to the colon and displayed at the mean dose in each category (categories <25 (referent), 25-, 50-, 100-, and >250 with respective mean doses to the colon of 10.1, 35.5, 70.3, 148.6, and 335.3 mGy).

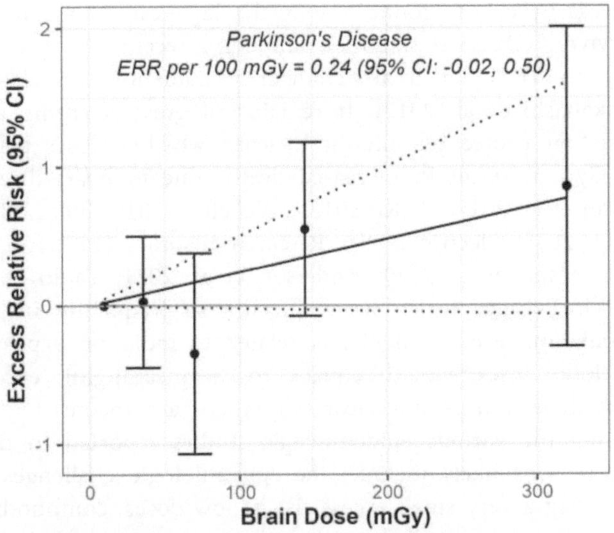

Figure 6. Excess Relative Risk (ERR) of Parkinson's disease by radiation dose to brain among 135,193 workers at U.S. nuclear power plants first monitored between 1957–1984 and followed through 2011. Linear model presented (solid line) based on a continuous measure of dose, and 95% confidence bands (dotted lines) presented about the regression line. Doses were lagged by 10 years and adjustments were made for sex, year of birth, and socioeconomic status. ERRs and 95% confidence limits are calculated for selected categories of dose to the brain and displayed at the mean dose in each category (categories <25 (referent), 25-, 50-, 100-, and >250 with respective mean doses to the brain of 9.6, 35.4, 69.9, 144.8, and 319.7 mGy).

related to characteristics of exposure such as brief compared with prolonged; the quality of the radiation such as low-LET compared with high-LET; uncontrolled confounding related to tobacco use or asbestos exposure; or possible biases in dosimetry, including the manner in which intakes of radionuclides and exposure to neutrons were handled.

All solid cancers

For completion, an analysis of all solid cancers was conducted and found little evidence for a dose-response relationship (Figure 5). Sampling variability was such, however, that ERRs per 100 mGy higher than 0.05 could be excluded with 95% confidence. Thus, the data are not inconsistent, in a statistical sense, with the Japanese atomic bomb survivor ERR per 100 mGy estimate of 0.042 (95% CI 0.032; 0.053) for all solid cancer based on a linear model (Ozasa et al. 2012). Estimation of radiation associations for all solid cancers is frequently done in epidemiologic studies to increase statistical precision, has value for purposes of radiation protection, but lacks biological plausibility (NCRP 2012). The value of combining all solid cancers as a single outcome has been recently questioned given the heterogeneity of background rates for individual cancers and the differences that arise when selected cancers are removed from the combined cancer dose-response relationship (Cologne et al. 2019). NCRP (2012) recommended caution in interpreting radiation associations, either positive or negative, based on combining all cancers: 'Clearly, there are differences in the etiologies and the radiation responses between lung cancer (caused by smoking), breast cancer (influenced by hormonal factors), cervical cancer (caused by human papillomavirus), rectal cancer (seen in excess only following radiation therapy), bone cancer (a high-dose effect), and brain cancer (nondividing neural cells, where the association is seen primarily among children)…' The different shapes of the dose-response curves for individual cancer sites, e.g. linearity for breast and thyroid cancer, nonlinearity for nonmelanoma skin and lung cancer, and a flat response (no risk) for cervical cancer and other cancers, tempers a biologically meaningful interpretation of the shape of the dose-response relationship when all cancers are taken together.

Ischemic heart disease

Nuclear power plant workers were significantly less likely to die than comparable persons in the general population. This is not unexpected because there are selection factors related to health that influence a person's ability to work, the so-called 'healthy worker effect' (Monson 1986; Checkoway et al. 1989). Internal dose-response comparisons are conducted to minimize this bias. Based on 5410 deaths from IHD, there was no evidence for a radiation association. The ERR at 100 mGy was slightly negative, −0.01; and was influenced by the lower risk apparent among the over 14,000 workers who received >100 mGy dose to the heart. The absence of a radiation association for IHD has been seen in all MPS cohorts reported to date (Boice, Quinn et al. 2021), including Rocketdyne (Boice et al. 2006, 2011), Mound (Boice et al. 2014), Mallinckrodt (Ellis, Boice et al. 2018; Golden et al. 2019), Los Alamos National Laboratory (Boice et al. 2021a), medical radiation workers (Boice, Cohen, Mumma, Howard et al. 2021), and industrial radiographers (NCRP 2018b). Significant radiation associations for heart disease have been found among the Japanese atomic bomb survivors (Ozasa et al. 2012), although a recent report focusing on IHD failed to confirm a radiation association (Takahashi et al. 2017). It remains of great scientific and radiation protection interest to learn whether doses below 0.5 Gy can result in increased deaths due to heart disease (Tran et al. 2017; NCRP 2018b; Wakeford 2019, Boice, Held et al. 2019; Electric Power Research Institute (EPRI) 2020; Richardson et al. 2020; Anderson et al. 2021; Tapio et al. 2021). Because heart disease occurs so frequently in the population, even a small risk related to radiation exposure at low doses would equate to a meaningful excess. Nonetheless, it is not clear why there are inconsistencies among the various epidemiologic studies reported to date but reasons likely include: the epidemiologic challenges in detecting a very small excess risk at low doses; confounding effects of uncontrolled cigarette smoking or other health conditions related to heart disease such as high blood pressure, diabetes, and obesity; radiation effects on the heart related to any concomitant kidney exposure; and changes in the coding of death certificates over time or misclassification of causes of death (Ozasa et al. 2017; de Vocht et al. 2020).

Other sites

Cancer of the esophagus was elevated and there was a suggestion of a dose-response consistent with several of the other studies within the MPS including Mound (Boice et al. 2014), atomic veterans (Boice et al. 2020), and Los Alamos National Laboratory (Boice, Cohen, Mumma, Golden et al. 2021). Cancer of the prostate was not elevated: the SMR based on 527 deaths was 0.99 (95% CI 0.91, 1.08). Previous MPS studies have not revealed significant dose- response relationships for prostate cancer (Boice et al. 2020; Boice, Cohen, Mumma, Golden et al. 2021; Boice, Cohen, Mumma, Howard 2021), in contrast to a recent publication on the Japanese atomic bomb survivors (Mabuchi et al. 2021). Dementia, Alzheimer's, Parkinson's, and motor neuron diseases were not increased. However, a nonsignificant dose-response for Parkinson's was seen (ERR per 100 mGy 0.24; 95% CI −0.02, 0.50; $n = 140$) (Figure 6) similar to those reported in other large MPS studies, notably for medical radiation workers (ERR per 100 mGy 0.07; 95% CI −0.20, 0.54; $n = 87$), industrial radiographers (ERR per 100 mGy 0.24; 95% CI −0.02, 0.50; $n = 128$), and workers at the Los Alamos National Laboratory (ERR per 100 mGy 0.16; 95% CI −0.17, 0.40; $n = 193$) (Boice and Dauer 2021; Boice, Cohen, Mumma, Golden et al. 2021; Boice, Cohen, Mumma, Howard 2021; Boice, Quinn et al. 2021). These estimates are broadly consistent with the association recently reported among Mayak workers for a cumulative gamma-ray dose to the brain (ERR per Gy 1.02; 95% CI 0.59, 1.63; $n = 300$ diagnoses) (Azizova et al. 2020). Improved estimates of radiation risk following prolonged exposures are anticipated from the future pooling of MPS cohorts for Parkinson's disease, leukemia and cancers of the lung, stomach, bladder, brain, thyroid, kidney and other causes of death, including those from nonmalignant conditions such as IHD and cerebrovascular disease.

Strengths and limitations

The strengths of the study of nuclear power plant workers include the cohort design; large numbers of over 135,000 healthy workers; complete and long follow-up of up to 60 years (mean 30.2 years); accurate and unbiased ascertainment of causes of death from the NDI and state mortality indices; near complete determination of all occupational doses from all places of employment; precise dosimetry data collected under strict criteria in compliance with legal regulations; a comprehensive approach to estimating organ doses for the NPP cohort was followed as detailed in NCRP Report 178 (Dauer et al. 2018; NCRP 2018a); minimal exposures to neutrons or intakes of radioactive materials a large number of cause-specific deaths available for analysis (e.g. 311 deaths from leukemia other than CLL compared with 121 among atomic bomb survivors exposed over age 19 (UNSCEAR 2008)); a large number of workers exposed to over 100 mSv (e.g. 19,981 compared with 4097 adult male atomic bomb survivors over age 19 years); sufficient numbers of the four subtypes of leukemia for statistical analyses of risk; and substantial statistical power to evaluate a radiation effect following chronic exposures over many years. Limitations include the small number of female workers such that sex-specific differences could not be thoroughly evaluated; the relatively small number of excess leukemia cases that precluded clear differentiation between various dose-response models; reliance on death certificate information; the possibility of missed doses, although the legal requirements to report worker doses at nuclear power plants might mitigate somewhat this concern as does the multiple independent sources of dosimetry information available; and limited evaluation of uncertainty in the analyses to date (Ron and Hoffman 1999; Schafer and Gilbert 2006; NCRP 2008, 2018a; UNSCEAR 2014). Most of these limitations will be met in the larger context of the MPS and the pooling of all cohorts, i.e. up to a million radiation workers and veterans.

Cigarette smoking causes lung cancer, heart disease, leukemia and many other diseases (USDHSS 2010). In the absence of individual information on tobacco use or other lifestyle behaviors that may influence mortality, adjustment for SES is frequently done as a way to account for such potentially confounding factors. For the over 130,000 NPP workers, individual SES adjustment was based on area-wide educational levels obtained from small-scale census-block group residential histories (Cohen et al. 2018; Mumma et al. 2018). While individual-level SES variables are preferred over area-level variables (Steenland et al. 2004), area-level variables have been shown to be adequate correlates of SES in occupational and other studies of cancer, heart disease and overall mortality (Anderson 1997; Waitzman and Smith 1998; Yen and Kaplan 1999; Diez Roux 2001). Further, it is of note that education level has been strongly correlated with tobacco use among early DOE radiation workers in interview studies (Mahoney and Wilkinson 1987). Nonetheless, despite our best efforts to adjust for correlates of smoking, the possibility of residual confounding is recognized as a limitation.

Other uncertainties include the choice of analytical strategies such as whether adjustments for the duration of monitoring should be made. In addition to conducting internal cohort analyses as a way to minimize the selection bias associated with having to be healthy in order to work, i.e. the 'healthy worker effect' (Monson 1986; Howe et al. 1988; Checkoway et al. 1989), adjusting for duration of employment is often done to account for any 'healthy worker survivor effect' (Buckley et al. 2015). Duration of employment can be correlated, however, with cumulative exposures, length of follow-up and other factors that might affect risk estimates (Boice 2010; Golden et al. 2018). In some radiation studies, the adjustment led to sizeable increases in the estimates of risk (Cardis et al. 2007), while in others, the effect of adjustment tended to reduce estimates of risk (Muirhead et al. 2009). It is also of interest that early departure from employment is not always an indicator of poor health, e.g. for early radiation workers it was not uncommon for many healthy workers to leave employment at a radiation facility for employment elsewhere (Boice et al. 2006). The over 10,000 NPP workers monitored for <30 d, evaluated separately, appeared as healthy as the longer-termed worker. Although the length of follow-up was similar, the workers monitored for <30 d may have been visitors or managers and not comparable to other workers. Nonetheless, they do not provide strong support for the notion that those who continued to work were especially healthy compared with shorter-term NPP workers. Further, the NPP workers who moved from plant to plant, presumably because of unique skills or the desire for improved positions, had slightly higher radiation risks (related to higher exposures) than those who remained at a single plant, although the differences were not statistically meaningful.

A recent Government Accountability Office report endorsed the MPS and stressed the need for enhanced scientific understanding in order to reduce uncertainties of the health consequences of low-dose radiation received by the public and workers (USGAO 2017). There is broad consensus support by national and international radiation research and radiation protection communities for the renewal and reinvigoration of low-dose radiation research (NASEM 2019, 2021b). Acts passed by the U.S. Congress and appropriation bills signed by the President of the U.S. have supported the MPS and confirmed the need for scientific research that will inform decision-makers and improve risk-management strategies (Boice, Held et al. 2019).

Conclusions

The study of nuclear power plant workers employed 1957–1984 provides convincing confirmatory evidence that low-dose and low-dose-rate exposures over time significantly increase the risk of leukemia other than CLL. A negative association between radiation and CLL is consistent with most radiation studies and indicates that this type of leukemia should not be considered highly sensitive to radiation exposures. Consistent with MPS cohort studies completed to date, there was little evidence for a radiation association with lung cancer or ischemic heart disease. In contrast, associations between radiation and Parkinson's disease are seen in most MPS cohort studies. More statistically precise estimates will be forthcoming when other cohorts within the MPS are combined, in particular the studies of 123,000 industrial radiographers, 114,000 nuclear weapons test participants, 109,000 medical radiation workers, 112,000 nuclear submariners, and the over 150,000 workers at the Department of Energy nuclear facilities.

Acknowledgments

We are indebted to the DOE (Nimi Rao), the NRC (Doris Lewis), the U.S. Army Dosimetry Center (William S. Harris, Jr., CHP), the U.S. Air Force Radiation Dosimetry Laboratory (Gerald M. Achenbach), and the Naval Dosimetry Center (CAPT Thad Sharp and LCDR Nakima McCormack) for facilitating linkages with their respective military dosimetry files. Special thanks to Dale Preston who provided leukemia analysis with the atomic bomb data consistent with our parameters (adult males, age range, follow-up time), and who provided insights into the power of Epicure in conducting a wide range of analyses. We are most grateful to Sara C. Howard at ORAU who provided Epicure analyses and created the graphics for Figures 2–7 in this article. The results presented herein represent the conclusions and opinions solely of the authors. Its publication does not imply endorsement by the National Council on Radiation Protection and Measurements, Vanderbilt University or any of the acknowledged agencies.

Disclosure statement

No potential conflict of interest was reported by the author(s).

Funding

The study of U.S. nuclear power plant workers, a component of the Million Person Study, was supported in part by a research grant from the U.S. Department of Energy [Grant No. DE-SC0008944] awarded to the National Council on Radiation Protection and Measurements, which included interagency support from the U.S. Nuclear Regulatory Commission (NRC), the U.S. Environmental Protection Agency and the National Aeronautics and Space Administration (NASA)); and more recent Grants from the Department of Energy [No. DE-AU0000042 and DE-AU0000046]. Additional support included grants from the NRC [NRC-HQ-60-14-G-0011]; the Centers for Disease Control and Prevention [5UE1EH000989, 5NUE1EH001315]; NASA [NNX15AU88G, 80NSSC17M0016], and a Discovery Grant from the Vanderbilt-Ingram Cancer Center. [Center no. 404-357-9682]. Contract support also was received from the Naval Sea Systems Command [N00024-17-C-4322] for dosimetry linkage.

ORCID

John D. Boice http://orcid.org/0000-0002-8755-1299
Sarah S. Cohen http://orcid.org/0000-0003-0421-1983
Michael T. Mumma http://orcid.org/0000-0001-7506-8710
Ashley P. Golden http://orcid.org/0000-0002-9430-7702
R. Craig Yoder http://orcid.org/0000-0002-5132-967X
Lawrence T. Dauer http://orcid.org/0000-0002-5629-8462

References

Agency for Toxic Substances and Disease Registry (ATSDR). 2013. Case studies in environmental medicine: asbestos toxicity. Atlanta (GA): ATSDR, Public Health Service, US Department of Health and Human Services. [accessed 2021 January 16]. https://www.atsdr.cdc.gov/csem/asbestos_2014/docs/asbestos.pdf

Akaike H. 1974. A new look at the statistical model identification. IEEE Trans Automat Contr. 19(6):716–723.

Akiba S. 2018. The latest study of Japanese nuclear workers. J Radiol Prot. 38(2):E26–E28.

Akiba S, Mizuno S. 2012. The third analysis of cancer mortality among Japanese nuclear workers, 1991–2002: estimation of excess relative risk per radiation dose. J Radiol Prot. 32(1):73–83.

Albitar M, Manshouri T, Shen Y, Liu D, Beran M, Kantarjian HM, Rogers A, Jilani I, Lin CW, Pierce S, et al. 2002. Myelodysplastic syndrome is not merely "preleukemia." Blood. 100(3):791–798.

Anderson JL, Bertke SJ, Yiin J, Kelly-Reif K, Daniels RD. 2021. Ischaemic heart and cerebrovascular disease mortality in uranium enrichment workers. Occup Environ Med. 78(2):105–111.

Anderson RT, Sorlie P, Backlund E, Johnson N, Kaplan GA. 1997. Mortality effects of community socioeconomic status. Epidemiology. 8(1):42–47.

Anzenberg V, Lewis DE, Dickson ED, Bush-Goddard SP. 2010. The U.S. Nuclear Regulatory Commission Radiation Exposure Information Reporting System (REIRS). Radiat Res. 173(2):254–255.

Azizova TV, Bannikova MV, Grigoryeva ES, Rybkina VL, Hamada N. 2020. Occupational exposure to chronic ionizing radiation increases risk of Parkinson's disease incidence in Russian Mayak workers. Int J Epidemiol. 49(2):435–447.

Beck HL, Till JE, Grogan HA, Aanenson JW, Mohler HJ, Mohler SS, Voillequé PG. 2017. Red bone marrow and male breast doses for a cohort of atomic veterans. Radiat Res. 187(2):221–228.

Blevins MR, Andersen RL. 2011. Radiation protection at U.S. nuclear power plants - today and tomorrow. Health Phys. 100(1):35–38.

Boice JD. 2010. Uncertainties in studies of low statistical power (editorial). J Radiol Prot. 30(2):115–120.

Boice JD, Jr. 2014. The importance of radiation worker studies (editorial). J Radiol Prot. 34(3):E7–E12.

Boice JD, Jr. 2016. From Oak Ridge to Indian Point Studying workers at nuclear power plants. Health Physics News. 44:23–24. [accessed 2021 November 12]. https://ncrponline.org/wp-content/themes/ncrp/PDFs/BOICE-HPnews/45-From-Oak-Ridge-to-Indian-Point-Feb-2016.pdf

Boice JD, Jr. 2017a. The linear nonthreshold (LNT) model as used in radiation protection: an NCRP update. Int J Radiat Biol. 93(10): 1079–1092.

Boice JD, Jr. 2017b. Space: the final frontier–research relevant to Mars. Health Phys. 112(4):392–397.

Boice JD, Jr. 2019. The Million Person Study relevance to space exploration and Mars. Int J Radiat Biol. DOI:10.1080/09553002.2019.1589020

Boice JD, Jr, Blettner M, Kleinerman RA, Stovall M, Moloney WC, Engholm G, Austin DF, Bosch A, Cookfair DL, Krementz ET, et al. 1987. Radiation dose and leukemia risk in patients treated for cancer of the cervix. J Natl Cancer Inst. 79(6):1295–1311.

Boice JD, Jr, Cohen SS, Mumma MT, Chen H, Golden AP, Beck HL, Till JE. 2020. Mortality among US military participants at eight above ground nuclear weapons test series. Int J Radiat Biol. DOI:10.1080/09553002.2020.1787543

Boice JD, Jr, Cohen SS, Mumma MT, Ellis ED, Cragle DL, Eckerman KF, Wallace PW, Chadda B, Sonderman JS, Wiggs LD, et al. 2014. Mortality among Mound workers exposed to polonium-210 and other sources of radiation, 1944–1979. Radiat Res. 181(2):208–228.

Boice JD, Jr, Cohen SS, Mumma MT, Ellis ED, Eckerman KF, Leggett RW, Boecker BB, Brill AB, Henderson BE. 2011. Updated mortality analysis of radiation workers at Rocketdyne (Atomics International), 1948–2008. Radiat Res. 176(2):244–258.

Boice JD, Jr, Cohen SS, Mumma MT, Ellis ED. 2019. The Million Person Study, whence it came and why. Int J Radiat Biol. DOI:10.1080/09553002.2019.1589015

Boice JD, Jr, Cohen SS, Mumma MT, Golden AP, Howard S, Girardi DJ, Ellis ED, Bellamy M, Dauer LT, Samuels C, et al. 2021. Mortality among workers at the Los Alamos National Laboratory, 1943–2017. Int J Radiat Biol. DOI:10.1080/09553002.2021.1917784

Boice JD, Jr, Cohen SS, Mumma MT, Howard SC, Yoder C, Dauer LT. 2021. Mortality among medical radiation workers in the United States, 1965–2016. Int J Radiat Biol. DOI:10.1080/09553002.2021.1967508.

Boice JD, Jr, Dauer LT. 2021. Million Person study of low-dose radiation health effects. Trans Am Nucl Soc. (in press)

Boice JD, Jr, Ellis ED, Golden AP, Zablotska LB, Mumma MT, Cohen SS. 2019. Sex-specific lung cancer risk among radiation workers in the Million-Person Study and among TB-fluoroscopy patients. Int J Radiat Biol. DOI:10.1080/09553002.2018.1547441

Boice JD, Held KD, Shore RE. 2019. Radiation epidemiology and health effects following low-level radiation exposure. J Radiol Prot. 39(4): S14–S27.

Boice JD, Jr, Leggett RW, Ellis ED, Wallace PW, Mumma M, Cohen SS, Brill AB, Chadda B, Boecker BB, Yoder RC, et al. 2006. A comprehensive dose reconstruction methodology for former Rocketdyne/Atomics International radiation workers. Health Phys. 90(5):409–430.

Boice JD, Jr, Quinn B, Al-Nabulsi I, Ansari A, Blake PK, Blattnig SR, Caffrey EA, Cohen SS, Golden AP, Held KD, et al. 2021. A million persons, a million dreams: a vision for a National Center of Radiation Epidemiology and Biology. Int J Radiat Biol. DOI:10.1080/09553002.2021.1988183.

Bouville A, Toohey RE, Boice JD, Jr, Beck HL, Dauer LT, Eckerman KF, Hagemeyer D, Leggett RW, Mumma MT, Napier B, et al. 2015. Dose reconstruction for the million worker study: status and guidelines. Health Phys. 108(2):206–220.

Buckley JP, Keil AP, McGrath LJ, Edwards JK. 2015. Evolving methods for inference in the presence of healthy worker survivor bias. Epidemiology. 26(2):204–212.

Caldwell GG, Zack MM, Mumma MT, Falk H, Heath CW, Till JE, Chen H, Boice JD. 2016. Mortality among military participants at the 1957 PLUMBBOB nuclear weapons test series and from leukemia among participants at the SMOKY test. J Radiol Prot. 36(3): 474–489.

Campbell KM, Deck D, Krupski A. 2008. Record linkage software in the public domain: a comparison of Link Plus, The Link King, and a 'basic' deterministic algorithm. Health Informatics J. 14(1):5–15.

Cardis E, Gilbert ES, Carpenter L, Howe G, Kato I, Armstrong BK, Beral V, Cowper G, Douglas A, Fix J. 1995. Effects of low doses and low dose rates of external ionizing radiation: cancer mortality among nuclear industry workers in three countries. Radiat Res. 142(2):117–132.

Cardis E, Vrijheid M, Blettner M, Gilbert E, Hakama M, Hill C, Howe G, Kaldor J, Muirhead CR, Schubauer-Berigan M, et al. 2005. Risk of cancer after low doses of ionising radiation: retrospective cohort study in 15 countries. BMJ. 331(7508):77. DOI:10.1136/bmj.38499.599861.E0

Cardis E, Vrijheid M, Blettner M, Gilbert E, Hakama M, Hill C, Howe G, Kaldor J, Muirhead CR, Schubauer-Berigan M, et al. 2007. The 15-country collaborative study of cancer risk among radiation workers in the nuclear industry: estimates of radiation-related cancer risks. Radiat Res. 167(4):396–416.

Charpentier P, Ostfeld AM, Hadjimichael OC, Hester R. 1993. The mortality of US nuclear submariners, 1969–1982. J Occup Med. 35(5):501–509.

Checkoway H, Pearce N, Crawford-Brown DJ. 1989. Research Methods in Occupational Epidemiology. New York (NY): Oxford University Press.

Cohen SS, Mumma MT, Ellis ED, Boice JD, Jr. 2018. Validating the use of census data on education as a measure of socioeconomic status in an occupational cohort. Int J Radiat Biol. DOI:10.1080/09553002.2018.1549758

Cologne J, Kim J, Sugiyama H, French B, Cullings HM, Preston DL, Mabuchi K, Ozasa K. 2019. Effect of heterogeneity in background incidence on inference about the solid-cancer radiation dose response in atomic bomb survivors. Radiat Res. 192(4):388–398.

Cox DR. 1972. Regression models and life-tables. J R Stat Soc B. 34(2): 187–220.

Curtis RE, Boice JD, Jr, Stovall M, Bernstein L, Holowaty E, Karjalainen S, Langmark F, Nasca PC, Schwartz AG, Schymura MJ. 1994. Relationship of leukemia risk to radiation dose following cancer of the uterine corpus. J Natl Cancer Inst. 86(17):1315–1324.

Daniels RD, Bertke S, Waters KM, Schubauer-Berigan MK. 2013. Risk of leukaemia mortality from exposure to ionising radiation in US nuclear workers: a pooled case-control study. Occup Environ Med. 70(1):41–48.

Daniels RD, Bertke SJ, Richardson DB, Cardis E, Gillies M, O'Hagan JA, Haylock R, Laurier D, Leuraud K, Moissonnier M, et al. 2017. Examining temporal effects on cancer risk in the international nuclear workers' study. Int J Cancer. 140(6):1260–1269.

Daniels RD, Schubauer-Berigan MK. 2011. A meta-analysis of leukaemia risk from protracted exposure to low-dose gamma radiation. Occup Environ Med. 68(6):457–464.

Dauer LT, Bouville A, Toohey RE, Boice JD, Jr, Beck HL, Eckerman KF, Hagemeyer D, Leggett RW, Mumma MT, Napier B, et al. 2018. Dosimetry and uncertainty approaches for the million-person study of low-dose radiation health effects: overview of the recommendations in NCRP Report No. 178. Int J Radiat Biol. DOI:10.1080/09553002.2018.1536299

de Vocht F, Hidajat M, Martin RM, Agius R, Wakeford R. 2020. Ischemic heart disease mortality and occupational radiation exposure in a nested matched case-control study of British nuclear fuel cycle workers: investigation of confounding by lifestyle, physiological traits and occupational exposures. Radiat Res. 194(4):431–444.

Diez Roux AV, Merkin SS, Arnett D, Chambless L, Massing M, Nieto FJ, Sorlie P, Szklo M, Tyroler HA, Watson RL. 2001. Neighborhood of residence and incidence of coronary heart disease. N Engl J Med. 345(2):99–106.

Douglas AJ, Omar RZ, Smith PG. 1994. Cancer mortality and morbidity among workers at the Sellafield plant of British Nuclear Fuels. Br J Cancer. 70(6):1232–1243.

Electric Power Research Institute (EPRI). 2020. Cardiovascular risks from low dose radiation exposure: review and scientific appraisal of the literature. 3002018408. Palo Alto (CA): EPRI. [accessed 2021 January 12]. https://www.epri.com/research/products/000000003002018408

Ellis ED, Boice JD, Golden AP, Girardi DJ, Cohen SS, Mumma MT, Shore RE, Leggett RW, Kerr GD. 2018. Dosimetry is key to good epidemiology: workers at Mallinckrodt Chemical Works had seven different source exposures. Health Phys. 114(4):386–397.

Ellis ED, Girardi D, Golden AP, Wallace PW, Phillips J, Cragle DL. 2018. Historical perspective on the Department of Energy mortality studies: focus on the collection and storage of individual worker data. Int J Radiat Biol. DOI:10.1080/09553002.2018.1547851

Friedman-Jimenez G, Kato I, Factor-Litvak P, Shore R. 2021. Mortality of enlisted men who served on nuclear-powered submarines in the United States Navy. J Occup Environ Med. DOI: 10.1097/JOM.0000000000002364.

Gilbert ES. 1998. Accounting for errors in dose estimates used in studies of workers exposed to external radiation. Health Phys. 74(1): 22–29.

Gilbert ES, Cragle DL, Wiggs LD. 1993. Updated analyses of combined mortality data for workers at the Hanford Site, Oak Ridge National Laboratory, and Rocky Flats Weapons Plant. Radiat Res. 136(3): 408–421.

Gilbert ES, Little MP, Preston DL, Stram DO. 2020. Issues in interpreting epidemiologic studies of populations exposed to low-dose, high-energy photon radiation. J Natl Cancer Inst Monogr. 2020(56):176-187.

Gillies M, Haylock R, Hunter N, Zhang W. 2019. Risk of leukemia associated with protracted low-dose radiation exposure: updated results from the National Registry for Radiation Workers Study. Radiat Res. 192(5):527–537.

Gillies M, Haylock R. 2014. The cancer mortality and incidence experience of workers at British Nuclear Fuels plc, 1946–2005. J Radiol Prot. 34(3):595–623.

Gillies M, Kuznetsova I, Sokolnikov M, Haylock R, O'Hagan J, Tsareva Y, Labutina E. 2017. Lung cancer risk from plutonium: a pooled analysis of the Mayak and Sellafield worker cohorts. Radiat Res. 188(6):645–660.

Golden AP, Cohen SS, Chen H, Ellis ED, Boice JD, Jr. 2018. Evaluation of statistical modeling approaches for epidemiologic studies of low dose radiation health effects. Int J Radiat Biol. DOI: 10.1080/09553002.2018.1554924

Golden AP, Ellis ED, Cohen SS, Mumma MT, Leggett RW, Wallace PW, Girardi DJ, Watkins JP, Shore RE, Boice JD, Jr. 2019. Updated mortality analysis of the Mallinckrodt uranium processing workers, 1942–2012. Int J Radiat Biol. DOI:10.1080/09553002.2019.1569773

Goldsmith R, Boice JD, Jr, Hrubec Z, Hurwitz PE, Goff TE, Wilson J. 1989. Mortality and career radiation doses for workers at a commercial nuclear power plant: feasibility study. Health Phys. 56(2): 139–150.

Grant EJ, Ozasa K, Ban N, de González AB, Cologne J, Cullings HM, Doi K, Furukawa K, Imaoka T, Kodama K, et al. 2015. A report from the 2013 international symposium: the evaluation of the effects of low-dose radiation exposure in the life span study of atomic bomb survivors and other similar studies. Health Phys. 108(5): 551–556.

Hagemeyer D, Nichols G, Mumma MT, Boice JD, Brock TA. 2018. 50 Years of the Radiation exposure information and reporting system. Int J Radiat Biol. DOI:10.1080/09553002.2018.1540896

Hamra GB, Richardson DB, Cardis E, Daniels RD, Gillies M, O'Hagan JA, Haylock R, Laurier D, Leuraud K, Moissonnier M, et al. 2016. Cohort profile: the International Nuclear Workers Study (INWORKS). Int J Epidemiol. 45(3):693–699.

Hauptmann M, Daniels RD, Cardis E, Cullings HM, Kendall G, Laurier D, Linet MS, Little MP, Lubin JH, Preston DL, et al. 2020. Epidemiological studies of low-dose ionizing radiation and cancer: summary bias assessment and meta-analysis. J Natl Cancer Inst Monogr. 2020(56):188–200.

Haylock RGE, Gillies M, Hunter N, Zhang W, Phillipson M. 2018. Cancer mortality and incidence following external occupational radiation exposure: an update of the 3rd analysis of the UK National Registry for Radiation Workers. Br J Cancer. 119(5):631–637.

Howe GR, Chiarelli AM, Lindsay JP. 1988. Components and modifiers of the healthy worker effect: evidence from three occupational cohorts and implications for industrial compensation. Am J Epidemiol. 128(6):1364–1375.

Howe GR, Zablotska LB, Fix JJ, Egel J, Buchanan J. 2004. Analysis of the mortality experience amongst U.S. nuclear power industry workers after chronic low-dose exposure to ionizing radiation. Radiat Res. 162(5):517–526.

Hsu WL, Preston DL, Soda M, Sugiyama H, Funamoto S, Kodama K, Kimura A, Kamada N, Dohy H, Tomonaga M, et al. 2013. The incidence of leukemia, lymphoma and multiple myeloma among atomic bomb survivors: 1950–2001. Radiat Res. 179(3):361–382.

International Commission on Radiological Protection (ICRP). 2010. Conversion coefficients for radiological protection quantities for external radiation exposures. ICRP Publication 116. Ann ICRP. 40(2–5):1–257.

Iwanaga M, Hsu WL, Soda M, Takasaki Y, Tawara M, Joh T, Amenomori T, Yamamura M, Yoshida Y, Koba T, et al. 2011. Risk of myelodysplastic syndromes in people exposed to ionizing radiation: a retrospective cohort study of Nagasaki atomic bomb survivors. J Clin Oncol. 29(4):428–434.

Jablon S, Boice JD, Jr. 1993. Mortality among workers at a nuclear power plant in the United States. Cancer Causes Control. 4(5): 427–430.

Komrokji RS, Padron E, List AF. 2019. Chapter 106: Myelodysplastic syndromes. In: DeVita VT, Lawrence TS, Rosenberg SA, editors. DeVita, Hellman, and Rosenberg's Cancer: Principles and Practice of Oncology. 11th ed. Philadelphia (PA): Wolters Kluwer; p. 1802–1817.

Krestinina L, Preston DL, Davis FG, Epifanova S, Ostroumova E, Ron E, Akleyev A. 2010. Leukemia incidence among people exposed to chronic radiation from the contaminated Techa River, 1953–2005. Radiat Environ Biophys. 49(2):195–201.

Kuznetsova IS, Labutina EV, Hunter N. 2016. Radiation risks of leukemia, lymphoma and multiple myeloma incidence in the Mayak cohort: 1948–2004. PLoS One. 11(9):e0162710.

Laurier D, Richardson DB, Cardis E, Daniels RD, Gillies M, O'Hagan J, Hamra GB, Haylock R, Leuraud K, Moissonnier M, et al. 2017. The International Nuclear Workers Study (Inworks): a collaborative epidemiological study to improve knowledge about health effects of protracted low-dose exposure. Radiat Prot Dosimetry. 173(1–3): 21–25.

Leggett RW, Eckerman KF, Bellamy M. 2018. MPS dose reconstruction for internal emitters: some site-specific issues and approaches. Int J Radiat Biol. DOI:10.1080/09553002.2018.1558302

Leuraud K, Richardson DB, Cardis E, Daniels RD, Gillies M, O'Hagan JA, Hamra GB, Haylock R, Laurier D, Moissonnier M, et al. 2015. Ionising radiation and risk of death from leukaemia and lymphoma in radiation-monitored workers (INWORKS): an international cohort study. Lancet Haematol. 2(7):e276–e281.

Linet MS, Little MP, Kitahara CM, Cahoon EK, Doody MM, Simon SL, Alexander BH, Preston DL. 2020. Occupational radiation and haematopoietic malignancy mortality in the retrospective cohort study of US radiologic technologists, 1983–2012. Occup Environ Med. 77(12):822–831.

Little MP, Weiss HA, Boice JD, Jr, Darby SC, Day NE, Muirhead CR. 1999. Risks of leukemia in Japanese atomic bomb survivors, in women treated for cervical cancer, and in patients treated for ankylosing spondylitis. Radiat Res. 152(3):280–292.

Mabuchi K, Preston DL, Brenner AV, Sugiyama H, Utada M, Sakata R, Sadakane A, Grant EJ, French B, Cahoon EK, et al. 2021. Risk of prostate cancer incidence among atomic bomb survivors: 1958–2009. Radiat Res. 195(1):60–65.

Mahoney MC, Wilkinson GS. 1987. Smoking patterns among Los Alamos National Laboratory employees. Report LA-10650. Los Alamos (NM): Los Alamos National Laboratory. [accessed 2021 July 16]. https://www.osti.gov/biblio/6403953-smoking-patterns-among-los-alamos-national-laboratory-employees

Matanoski GM, Seltser R, Sartwell PE, Diamond EL, Elliott EA. 1975. The current mortality rates of radiologists and other physician specialists: specific causes of death. Am J Epidemiol. 101(3):199–210.

McLean AR, Adlen EK, Cardis E, Elliott A, Goodhead DT, Harms-Ringdahl M, Hendry JH, Hoskin P, Jeggo PA, Mackay DJC, et al. 2017. A restatement of the natural science evidence base concerning the health effects of low-level ionizing radiation. Proc Biol Sci. 284(1862):20171070.

Metz-Flamant C, Laurent O, Samson E, Caër-Lorho S, Acker A, Hubert E, Richardson DB, Laurier D. 2013. Mortality associated with chronic external radiation exposure in the French combined cohort of nuclear workers. Occup Environ Med. 70(9):630–638.

Monson RR. 1986. Observations on the healthy worker effect. J Occup Med. 28(6):425–433.

Mueller TJ, Weishar TM, Hallworth JM, Draper DA. 2020. Occupational radiation exposure from U.S. Naval nuclear plants and their support facilities. NT-20-2. Washington (DC): Naval Nuclear Propulsion Program. [accessed 2021 January 18]. https://www.energy.gov/sites/prod/files/2020/07/f77/NT-20-2.pdf

Muirhead CR, Boice JD, Jr, Raddatz CT, Yoder RC. 1996. Comparison of dose histories for U.S. nuclear power plant workers, based on records held by a major dosimetry service company and on the NRC REIRS database. Health Phys. 70(5):645–650.

Muirhead CR, O'Hagan JA, Haylock RG, Phillipson MA, Willcock T, Berridge GL, Zhang W. 2009. Mortality and cancer incidence following occupational radiation exposure: third analysis of the National Registry for Radiation Workers. Br J Cancer. 100(1):206–212.

Mumma MT, Cohen SS, Sirko SS, Ellis ED, Boice JD, Jr. 2018. Obtaining vital status and cause of death on a million persons. Int J Radiat Biol. DOI:10.1080/09553002.2018.1539884

Mumma MT, Sirko JL, Boice JD, Blot WJ. 2019. Mesothelioma mortality within two radiation monitored occupational cohorts. Int J Radiat Biol. DOI:10.1080/09553002.2019.1642540

National Academies of Sciences, Engineering, and Medicine (NASEM). 2019. The future of low dose radiation research in the United States. Proceedings of a Symposium. Washington (DC): National Academies Press. https://doi.org/10.17226/25578

National Academies of Sciences, Engineering, and Medicine (NASEM). 2021a. Space radiation and astronaut health: managing and communicating cancer risks. Washington (DC): The National Academies Press. https://doi.org/10.17226/26155

National Academies of Sciences, Engineering, and Medicine (NASEM). 2021b. Developing a long-term strategy for low-dose radiation research in the United States. https://www.nationalacademies.org/our-work/developing-a-long-term-strategy-for-low-dose-radiation-research-in-the-united-states

National Academies/National Research Council (NA/NRC). 2006. Health risks from exposure to low levels of ionizing radiation (BEIR VII phase 2). Washington (DC): National Academies Press.

National Council on Radiation Protection and Measurements (NCRP). 2008. Uncertainties in the measurement and dosimetry of external radiation. NCRP Report No. 158. Bethesda (MD): NCRP.

National Council on Radiation Protection and Measurements (NCRP). 2012. Uncertainties in the estimation of radiation risks and probability of disease causation. NCRP Report No. 171. Bethesda (MD): NCRP.

National Council on Radiation Protection and Measurements (NCRP). 2014. Radiation protection for space activities: supplement to previous recommendations. NCRP Commentary No. 23. Bethesda (MD): NCRP.

National Council on Radiation Protection and Measurements (NCRP). 2018a. Deriving organ doses and their uncertainty for epidemiologic studies (with a focus on the one million U.S. workers and veterans study of low-dose radiation health effects). NCRP Report 178. Bethesda (MD): NCRP.

National Council on Radiation Protection and Measurements (NCRP). 2018b. Implications of recent epidemiologic studies for the linear-nonthreshold model and radiation protection. NCRP Commentary 27. Bethesda (MD): NCRP.

National Council on Radiation Protection and Measurements (NCRP). 2020. Dosimetry guidance for medical radiation workers with a focus on lung dose reconstruction. NCRP Commentary 30. Bethesda (MD): NCRP.

Naval Dosimetry Center (NDC). 2021. Lifetime exposure history repository for the U.S. Navy's nuclear-powered fleet personnel. [accessed 2021 January 12]. https://www.cnic.navy.mil/regions/ndw/installations/nsa_bethesda/about/tenant_orgs/ndc.html

Oak Ridge Institute for Science and Education (ORISE). 2011. Establishing a U.S. Nuclear Regulatory Commission worker registry for epidemiologic investigations. A feasibility study using the Radiation Exposure Information and Reporting System. Submitted to the Nuclear Regulatory Commission. [accessed 2021 January 12].

Ozasa K, Shimizu Y, Suyama A, Kasagi F, Soda M, Grant EJ, Sakata R, Sugiyama H, Kodama K. 2012. Studies of the mortality of atomic bomb survivors, Report 14, 1950–2003: an overview of cancer and noncancer diseases. Radiat Res. 177(3):229–243.

Ozasa K, Takahashi I, Grant EJ, Kodama K. 2017. Cardiovascular disease among atomic bomb survivors. Int J Radiat Biol. 93(10): 1145–1150.

Petersen GR, Gilbert ES, Buchanan JA, Stevens RG. 1990. A case-cohort study of lung cancer, ionizing radiation, and tobacco smoking among males at the Hanford Site. Health Phys. 58(1):3–11.

Preston DL, Lubin J, Pierce D, McConney M, Shilnikova N. 2015. Epicure risk regression and person-year computation software: command summary and user guide. Ottawa (CA): Risk Sciences International.

Preston DL, Sokolnikov ME, Krestinina LY, Stram DO. 2017. Estimates of radiation effects on cancer risks in the Mayak Worker, Techa River and atomic bomb survivor studies. Radiat Prot Dosimetry. 173(1–3):26–31.

Richardson DB, Abalo K, Bernier MO, Rage E, Leuraud K, Laurier D, Keil AP, Little MP. 2020. Meta-analysis of published excess relative risk estimates. Radiat Environ Biophys. 59(4):631–641.

Richardson DB, Cardis E, Daniels RD, Gillies M, Haylock R, Leuraud K, Laurier D, Moissonnier M, Schubauer-Berigan MK, Thierry-Chef I, et al. 2018. Site-specific solid cancer mortality after exposure to ionizing radiation: a cohort study of workers (INWORKS). Epidemiology. 29(1):31–40.

Richardson DB, Cardis E, Daniels RD, Gillies M, O'Hagan JA, Hamra GB, Haylock R, Laurier D, Leuraud K, Moissonnier M, et al. 2015. Risk of cancer from occupational exposure to ionising radiation: retrospective cohort study of workers in France, the United Kingdom, and the United States (INWORKS). BMJ. 351:h5359.

Ron E, Hoffman O. 1999. Uncertainties in radiation dosimetry and their impact on dose-response analyses. Publication No. 99-4541. Bethesda (MD): National Cancer Institute.

Rushton L. 1994. Use of multiple causes of death in the analysis of occupational cohorts-an example from the oil industry. Occup Environ Med. 51(11):722–729.

Schafer DW, Gilbert ES. 2006. Some statistical implications of dose uncertainty in radiation dose-response analyses. Radiat Res. 166(1 Pt 2):303–312.

Schubauer-Berigan MK, Daniels RD, Bertke SJ, Tseng CY, Richardson DB. 2015. Cancer mortality through 2005 among a pooled cohort of U.S. nuclear workers exposed to external ionizing radiation. Radiat Res. 183(6):620–631.

Schubauer-Berigan MK, Daniels RD, Fleming DA, Markey AM, Couch JR, Ahrenholz SH, Burphy JS, Anderson JL, Tseng CY. 2007.

Chronic lymphocytic leukaemia and radiation: findings among workers at five US nuclear facilities and a review of the recent literature. Br J Haematol. 139(5):799–808.

Steenland K, Henley J, Calle E, Thun M. 2004. Individual- and area-level socioeconomic status variables as predictors of mortality in a cohort of 179,383 persons. Am J Epidemiol. 159(11):1047–1056.

Steenland K, Nowlin S, Ryan B, Adams S. 1992. Use of multiple-cause mortality data in epidemiologic analyses: US rate and proportion files developed by the National Institute for Occupational Safety and Health and the National Cancer Institute. Am J Epidemiol. 136(7): 855–862.

Stram DO, Preston DL, Sokolnikov M, Napier B, Kopecky KJ, Boice J, Beck H, Till J, Bouville A. 2015. Shared dosimetry error in epidemiological dose-response analyses. PLOS One. 10(3):e0119418.

Takahashi I, Shimizu Y, Grant EJ, Cologne J, Ozasa K, Kodama K. 2017. Heart disease mortality in the life span study, 1950–2008. Radiat Res. 187(3):319–332.

Tapio S, Little MP, Kaiser JC, Impens N, Hamada N, Georgakilas AG, Simar D, Salomaa S. 2021. Ionizing radiation-induced circulatory and metabolic diseases. Environ Int. 146:106235.

Thierry-Chef I, Richardson DB, Daniels RD, Gillies M, Hamra GB, Haylock R, Kesminiene A, Laurier D, Leuraud K, Moissonnier M, et al. 2015. Dose estimation for a study of nuclear workers in France, the United Kingdom and the United States of America: methods for the International Nuclear Workers Study (INWORKS). Radiat Res. 183(6):632–642.

Till JE, Beck HL, Aanenson JW, Grogan HA, Mohler HJ, Mohler SS, Voillequé PG. 2014. Military participants at U.S. atmospheric nuclear weapons testing-methodology for estimating dose and uncertainty. Radiat Res. 181(5):471–484.

Till JE, Beck HL, Boice JD, Jr, Mohler HJ, Mumma MT, Aanenson JW, Grogan HA. 2018. Asbestos exposure and mesothelioma mortality among atomic veterans. Int J Rad Biol. DOI:10.1080/09553002.2018.1551641.

Tran V, Zablotska LB, Brenner AV, Little MP. 2017. Radiation-associated circulatory disease mortality in a pooled analysis of 77,275 patients from the Massachusetts and Canadian tuberculosis fluoroscopy cohorts. Sci Rep. 7(1):44147.

Tucker JD, Tawn EJ, Holdsworth D, Morris S, Langlois R, Ramsey MJ, Kato P, Boice JD, Jr, Tarone RE, Jensen RH. 1997. Biological dosimetry of radiation workers at the Sellafield nuclear facility. Radiat Res. 148(3):216–226.

U.S. Department of Energy (DOE). 2021. DOE 2019 Occupational radiation exposure. Washington (DC): US Department of Energy, Office of Environment, Health, Safety and Security, The Radiation Exposure Monitoring System (REMS). [accessed 2021 April 29]. http://energy.gov/ehss/occupational-radiation-exposure

U.S. Department of Health and Human Services (USDHHS). 2010. How tobacco smoke causes disease: the biology and behavioral basis for smoking-attributable disease: a report of the Surgeon General. Atlanta (GA): U.S. Department of Health and Human Services.

U.S. Government Accountability Office (USGAO). 2017. Low-dose radiation: interagency collaboration on planning research could improve information on health effects. GAO-17-546. Washington (DC): GAO. [accessed 2021 January 12]. https://www.gao.gov/products/GAO-17-546

U.S. Nuclear Regulatory Commission (USNRC). 1994. Voluntary reporting of additional occupational radiation of exposure data (generic letter 94-04). [accessed 2021 January 12]. https://www.nrc.gov/docs/ML1204/ML12047A243.pdf

U.S. Nuclear Regulatory Commission (USNRC). 2021. Radiation exposure information and reporting system (REIRS). [accessed 2021 January 19]. https://www.reirs.com/ and http://www.nrc.gov/reading-rm/doc-collections/gils/rad-exp.html

United Nations Scientific Committee on the Effects of Atomic Radiation (UNSCEAR). 2008. Effects of ionizing radiation. UNSCEAR 2006 report. Report to the general assembly. Volume 1. Scientific annex A. Epidemiological studies of radiation and cancer. Publication E.08.IX.6. New York (NY): United Nations.

United Nations Scientific Committee on the Effects of Atomic Radiation (UNSCEAR). 2014. UNSCEAR 2012 Report to the General Assembly with Scientific Annexes. Scientific annex B. Uncertainties in risk estimates for radiation-induced cancer. New York (NY): United Nations.

United Nations Scientific Committee on the Effects of Atomic Radiation (UNSCEAR). 2018. UNSCEAR 2017 report. Principles and criteria for ensuring the quality of the committee's reviews of epidemiological studies of radiation exposure. Scientific annex A. New York (NY): United Nations. [accessed 2021 January 15]. https://www.unscear.org/docs/publications/2017/UNSCEAR_2017_Annex-A.pdf

United Nations Scientific Committee on the Effects of Atomic Radiation (UNSCEAR). 2020. Sources, effects and risks of ionizing radiation. UNSCEAR 2019 Report to the General Assembly. Scientific Annex A: evaluation of selected health effects and inference of risk due to radiation exposure. New York (NY): UNSCEAR. [accessed 2021 April 29]. https://www.unscear.org/docs/publications/2019/UNSCEAR_2019_Annex-A.pdf

Velazquez-Kronen R, Gilbert ES, Linet MS, Moysich KB, Freudenheim JL, Wactawski-Wende J, Simon SL, Cahoon EK, Alexander BH, Doody MM, et al. 2020. Lung cancer mortality associated with protracted low-dose occupational radiation exposures and smoking behaviors in U.S. radiologic technologists, 1983–2012. Int J Cancer. 147(11):3130–3138.

Villoing D, Borrego D, Preston DL, Alexander BH, Rose A, Salasky M, Linet MS, Lee C, Kitahara CM. 2021. Trends in occupational radiation doses for U.S. radiologic technologists performing general radiologic and nuclear medicine procedures, 1980-2015. Radiology. 300(3):605–612.

Vrijheid M, Cardis E, Ashmore P, Auvinen A, Gilbert E, Habib RR, Malker H, Muirhead CR, Richardson DB, Rogel A, et al. 2008. Ionizing radiation and risk of chronic lymphocytic leukemia in the 15-country study of nuclear industry workers. Radiat Res. 170(5): 661–665.

Waitzman NJ, Smith KR. 1998. Phantom of the area: poverty-area residence and mortality in the United States. Am J Public Health. 88(6): 973–976.

Wakeford R. 2019. Does low-level exposure to ionizing radiation increase the risk of cardiovascular disease? Hypertension. 73(6): 1170–1171.

Yen IH, Kaplan GA. 1999. Neighborhood social environment and risk of death: multilevel evidence from the Alameda County Study. Am J Epidemiol. 149(10):898–907.

Yoder RC, Balter S, Boice JD, Jr, Grogan H, Mumma M, Rothenberg L, Passmore C, Vetter R, Dauer L. 2021. Using personal monitoring data to derive organ doses for medical radiation workers in the Million Person Study – considerations regarding NCRP Commentary No. 30. J Radiol Prot. DOI:10.1088/1361-6498/abcfcb

Zablotska LB, Bazyka D, Lubin JH, Gudzenko N, Little MP, Hatch M, Finch S, Dyagil I, Reiss RF, Chumak VV, et al. 2013. Radiation and the risk of chronic lymphocytic and other leukemias among Chornobyl cleanup workers. Environ Health Perspect. 121(1):59–65.

Zablotska LB, Lane RS, Thompson PA. 2014. A reanalysis of cancer mortality in Canadian nuclear workers (1956–1994) based on revised exposure and cohort data. Br J Cancer. 110(1):214–223.

Zeidan AM, Shallis RM, Wang R, Davidoff A, Ma X. 2019. Epidemiology of myelodysplastic syndromes: why characterizing the beast is a prerequisite to taming it. Blood Rev. 34:1–15.

Mortality among U.S. military participants at eight aboveground nuclear weapons test series

John D. Boice, Sarah S. Cohen, Michael T. Mumma, Heidi Chen, Ashley P. Golden, and John E. Till

ABSTRACT

Background: Approximately 235,000 military personnel participated at one of 230 U.S. atmospheric nuclear weapons tests from 1945 through 1962. At the Nevada Test Site (NTS), the atomic veterans participated in military maneuvers, observed nuclear weapons tests, or provided technical support. At the Pacific Proving Ground (PPG), they served aboard ships or were stationed on islands during or after nuclear weapons tests.

Material and methods: Participants at seven test series, previously studied with high-quality dosimetry and personnel records, and the first test at TRINITY formed the cohort of 114,270 male military participants traced for vital status from 1945 through 2010. Dose reconstructions were based on Nuclear Test Personnel Review records, Department of Defense. Standardized mortality ratios (SMR) and Cox and Poisson regression models were used in the analysis.

Results: Most atomic veterans were enlisted men, served in the Navy at the PPG, and were born before 1930. Vital status was determined for 96.8% of the veterans; 60% had died. Enlisted men had significantly high all-causes mortality SMR (1.06); officers had significantly low all-causes mortality SMR (0.71). The pattern of risk over time showed a diminution of the 'healthy soldier effect': the all-causes mortality SMR after 50 years of follow-up was 1.00. The healthy soldier effect for all cancers also diminished over time. The all-cancer SMR was significantly high after 50 years (SMR 1.10) primarily from smoking-related cancers, attributed in part to the availability of cigarettes in military rations. The highest SMR was for mesothelioma (SMR 1.56) which was correlated with asbestos exposure in naval ships. Prostate cancer was significantly high (SMR 1.13). Ischemic heart disease was significantly low (SMR 0.84). Estimated mean doses varied by organ were low; e.g., the mean red bone marrow dose was 6 mGy (maximum 108 mGy). Internal cohort dose–response analyses provided no evidence for increasing trends with radiation dose for leukemia (excluding chronic lymphocytic leukemia (CLL)) [ERR (95% CI) per 100 mGy −0.37 (−1.08, 0.33); $n = 710$], CLL, myelodysplastic syndrome, multiple myeloma, ischemic heart disease, or cancers of the lung, prostate, breast, and brain.

Conclusion: No statistically significant radiation associations were observed among 114,270 nuclear weapons test participants followed for up to 65 years. The 95% confidence limits were narrow and excluded mortality risks per unit dose that are two to four times higher than those reported in other investigations. Significantly elevated SMRs were seen for mesothelioma and asbestosis, attributed to asbestos exposure aboard ships.

Introduction

The study of nuclear weapons test participants is a component of the Million Person Study (MPS) of low-dose health effects being conducted to evaluate radiation risks among healthy U.S. workers and veterans (Boice et al. 2019a). In contrast to the study of Japanese atomic bomb survivors exposed briefly to moderate-dose radiation in 1945, the atomic veteran exposure was at a lower dose rate (Till et al. 2014, 2018a; Bouville et al. 2015; NCRP 2018a).

Approximately 235,000 military and civilian personnel participated in at least one of the 19 atmospheric U.S. nuclear weapons operations (test series) from 1945 (the TRINTY test) until the Limited Nuclear Test Ban Treaty in 1963 (Gladeck and Johnson 1996; IOM 2000; DTRA 2019). More than 230 aboveground detonations (i.e., tests, explosions, blasts, shots) were conducted primarily at the Nevada Test Site (NTS) and the Pacific Proving Ground (PPG). NTS was called the Nevada Proving Ground from 1952 to 1955; PPG was the primary oceanic site and sometimes termed the

Eniwetok Proving Ground or Bikini Proving Ground. A particular test series could have involved the detonation of up to 35 nuclear weapons over several weeks. Military code names were applied to both test series and individual detonations; e.g., ABLE and BAKER were the two detonations within the OPERATION CROSSROADS test series in 1946. Personnel at the NTS could have been exposed to gamma radiation from ground contamination and fission products when they either witnessed the explosions or entered areas following the blasts for tactical maneuvers and scientific studies. Fallout from detonations or from re-suspended radioactive debris might have been inhaled. Personnel in the Pacific also were exposed to gamma and beta radiation from fission fragments when boarding target ships placed at varying distances from the blast, entering contaminated lagoons or going ashore on the atolls after detonations, or when unexpectedly caught in the main fallout of detonations (Till et al. 2018a; DTRA 2019).

The Eight Series Study includes 114,270 military participants at seven test series with high-quality dosimetry and personnel records that were previously studied by the Medical Follow-up Agency, Institute of Medicine: the OPERATION CROSSROADS series (IOM 1996; Johnson et al. 1997), the Five Series cohorts (Robinette et al. 1985; IOM 2000), and the OPERATION HARDTACK I series (Watanabe et al. 1995). The eighth series was Project TRINITY, the first weapons test in 1945. The Five Series Study included two operations at NTS (UPSHOT-KNOTHOLE and PLUMBBOB) and three at the Bikini and Enewetak Atolls, Marshall Islands, PPG (GREENHOUSE, CASTLE, and REDWING). The 68,000 participants in the Five Series Study had been followed for mortality through 1996 (IOM 2000). OPERATION CROSSROADS included approximately 42,000 participants who were followed for mortality through 1992 (IOM 1996; Johnson et al. 1997). OPERATION HARDTACK I included over 13,000 participants of whom 8500 sailors had been followed through 1991 (Watanabe et al. 1995). There were 35 atmospheric nuclear detonations within the OPERATION HARDTACK I test series in 1958; all but two were at the PPG. The two other detonations in OPERATION HARDTACK I were above Johnson Island, which lies about 800 miles west-southwest of the Hawaiian Islands. Project TRINITY, the world's first nuclear test in 1945, included 396 participants who were studied for the first time (DTRA 2019).

The current investigation adds nearly 20 years of follow-up to the previous studies. Unique aspects include the large number of atomic veterans, the near lifetime follow-up, and the comprehensive dose reconstruction program to estimate radiation doses to specific organs. Previous studies were not able to benefit from the advances in dosimetric techniques, access to DTRA historical records (Till et al. 2014, 2018a), and tracing capabilities (Mumma et al. 2018) available today.

Previous studies of atomic veterans within the MPS included a second follow-up of the 3020 participants at the 1979 SMOKY shot within the OPERATION PLUMBBOB test series, which included 12,219 participants overall. The first follow-up in 1983 revealed an excess of leukemia (Caldwell et al. 1983), which led to further studies and compensation programs (Robinette et al. 1985; IOM 2000; NA/NRC 2003). The recent follow-up of SMOKY participants confirmed the leukemia excess, but the increased risk was not related to radiation dose to red bone marrow [ERR mGy^{-1} (95% CI) −0.05 (−0.14, 0.04)] (Caldwell et al. 2016). A high risk of mesothelioma and asbestosis among sailors at the PPG was linked to work in boiler rooms and other areas in ships with high likelihood of asbestos exposure (Till et al. 2018b). Preliminary results for ischemic heart disease found no association with low-dose radiation exposure (Boice 2017; NCRP 2018b).

Another study of cancer mortality (Dalager et al. 2000) focused on 1010 U.S. atmospheric nuclear weapons test participants who exceeded the federal occupation guideline for a gamma radiation dose of 50 mSv (5 rem, assuming the film badge gamma radiation dose in the Nuclear Test Review and Information System (NuTRIS) database was equivalent to the effective dose guideline, Gladeck and Johnson 1996) in a 12-month period. Follow-up was through 1996. Mortality from all causes and from all lymphopoietic cancers was significantly elevated compared with participants at OPERATION HARDTACK I ($n = 2870$) who received a minimal radiation dose. Leukemia and respiratory cancer were not significantly increased. These atomic veterans are included in the current study with longer follow-up and individual organ dose estimations.

The HARRY detonation within the OPERATION UPSHOT-KNOTHOLE test series in 1953 was notable in that heavy fallout was deposited in populated areas near and around St. George, Utah, when an unexpected change in wind direction occurred (Fehner and Gosling 2006). The larger than expected fallout raised concern about possible health risks, several epidemiologic studies were conducted (Stevens et al. 1990; Kerber et al. 1993), and subsequent compensation programs for 'downwinders' in Utah, Nevada, and Arizona were enacted (NA/NRC 2005).

The most comprehensive study of nuclear weapons test participants from other countries was conducted in the United Kingdom on 21,357 veterans (Doll et al. 1998; Muirhead and Kendall 2003; Muirhead et al. 2003a, 2003b; Kendall et al. 2004). Muirhead et al. (2004) provide an excellent review of studies of test participants from the United Kingdom, the United States, New Zealand, and Australia. Subsequent studies of nuclear test participants have been conducted in New Zealand (Pearce et al. 1997), Australia (Gun et al. 2008), and France (Pédrono et al. 2011). While there is no compelling evidence that test participation influenced the induction of cancer, a small absolute risk of leukemia other than chronic lymphocytic leukemia (CLL) could not be ruled out (Muirhead et al. 2003a). One aim of the current investigation of 114,270 nuclear weapons test participants (U.S. atomic veterans) is to provide quantitative organ-specific estimates of health effects following low level and low rate exposures.

Methods

Human subject approval was obtained from the Vanderbilt Institutional Review Board.

Population identification cohort definition

There were approximately 235,000 U.S. military personnel who participated at one of the 230 atmospheric nuclear weapons tests from 1945 through 1962 (Gladeck and Johnson 1996; IOM 2000; DTRA 2019). The detonation of a nuclear device was called a test, and each test was grouped into one of 19 test series or OPERATIONS. The eight test series selected for study were: CASTLE, CROSSROADS, GREENHOUSE, HARDTACK I, PLUMBBOB, REDWING, Project TRINITY, and UPSHOT-KNOTHOLE and consisted of 114,270 veterans. The study population of atmospheric weapons test participants was obtained from databases within the Department of Defense (DoD) Nuclear Test Personnel Review (NTPR) program coupled with epidemiologic research investigations conducted by the Medical Follow-up Agency (IOM 1996, 2000; Johnson et al. 1997; DTRA 2019) and the Department of Veterans Affairs (VA) (Watanabe et al. 1995). The test series were selected because of the availability of high quality personnel and dosimetry records, and indications that radiation exposures were higher than at other tests series (Gladeck and Johnson 1996; Dalager et al. 2000). There were 98 weapon detonations (or shots) within the eight test series studied, 34 at the NTS and 64 at the PPG. Available information on the atomic veterans included: name, military service number, date of birth, sex, pay grade (enlisted/officer), rank (for officers) or rating (for enlisted men), unit membership and dates during participation, permanent unit, and dates of entry to and separation from the military unit. Table 1 shows the number of participants at each of the eight nuclear weapons test series, the number from each military service, and other military and demographic characteristics.

The Eight Series Study cohort is comprised only of military servicemen. Female military participants ($n=83$) were excluded as were civilian and nonmilitary observers. The civilian participants and observers, perhaps as many as 11,900, were difficult to enumerate and to track (Gladeck and Johnson 1996). They included civilian employees of DoD, DOE, and other federal agencies as well as their contractors; U.S. observers such as members of Congress and corporate executives; and foreign observers both military and civilian. Just over 10,600 civilian participants at the atmospheric tests are identified in the NuTRIS data system with nearly 6000 employed by the U.S. Navy and the others primarily by the U.S. Army, U.S. Air Force, and DoD (DTRA 2020). Of historical note, both Robert Oppenheimer and Richard Feynman were at the TRINITY test in 1945. Oppenheimer is included in the study as a badged military participant. He had been given a military commission while working at the Los Alamos Scientific Laboratory and was in the NuTRIS military database. Feynman, on the other hand, did not have a military commission and is not included in the study population of military participants.

Table 1. Demographics and military service characteristics of the 114,270 nuclear weapons test participants in the Eight Series Study, 1945–2010.

	Number	Percent
Year of birth		
1874–1919	16,878	14.8%
1920–1924	17,796	15.6%
1925–1929	42,893	37.5%
1930–1934	22,610	19.8%
1934–1941	14,091	12.3%
Missing	2	<0.1%
Service		
Air Force	12,848	11.2%
Army	26,091	22.8%
Marines	5003	4.4%
Navy	70,309	61.5%
Other	19	<0.1%
Pay grade		
Enlisted	96,071	84.1%
Officer	18,199	15.9%
Test series theater		
Pacific proving ground	79,313	69.4%
Nevada test site	28,589	25.0%
Multiple/other	6368	5.6%
NuTRIS[a] gamma radiation film badge dose (mSv)		
<5	63,192	55.3%
5–19	34,216	22.9%
20–99	15,903	13.9%
100–299	236	0.21%
300–908	33	0.03%
Missing	690	0.6%
Age at first nuclear weapons test (years)		
16–19	30,012	26.3%
20–24	47,217	41.3%
25–29	14,969	13.1%
30–71	22,072	19.3%
Age at end of follow-up (December 31, 2010)		
<40 years	3385	3.0%
40–49 years	5162	4.5%
50–59 years	10,711	9.4%
60–69 years	19,226	16.8%
70–79 years	43,865	38.4%
80–89 years	28,953	25.3%
90–95 years	2966	2.6%
Year of first test participation		
1945–1947	38,733	33.9%
1948–1952	9477	8.3%
1953–1955	34,077	29.8%
1956–1958	31,983	30.0%
Eight test series		
CASTLE	16,223	14.2%
CROSSROADS	38,376	33.6%
GREENHOUSE	9606	8.4%
HARDTACK I	10,329	9.0%
PLUMBBOB	12,219	10.7%
REDWING	13,623	11.9%
TRINITY	396	0.3%
UPSHOT-KNOTHOLE	18,558	16.2%
Number of test series		
1 series	99,628	87.2%
2 series	11,633	10.1%
3 series	2072	1.8%
4 series	533	0.5%
5 series	209	0.2%
6 series	92	<0.1%
7 series	44	<0.1%
8 series	26	<0.1%
9 series	17	<0.1%
10–13 series	16	<0.1%
Vital status as of December 31, 2010		
Confirmed dead	75,402	66.0%
Confirmed alive	35,206	30.8%
Assumed alive	3662	3.2%

[a]NuTRIS denotes Nuclear Test Review and Information System maintained by the U.S. Defense Threat Reduction Agency, Department of Defense.

Population vital status and outcome determination

The determination of vital status was challenging because the social security number was not used as the military identification number until 1968 and was not otherwise readily available (Mumma et al. 2018). Thus, we relied heavily on mortality linkages based on military identification number, name, and date of birth. The VA BIRLS system (Beneficiary Identification Records Locator Subsystem) was a primary source of vital status information, supplemented with standard location sources such as the Social Security Administration (SSA) records and the National Death Index (NDI).

The VA BIRLS system has been used extensively for epidemiologic research on veterans (Watanabe et al. 1991, 1995; Watanabe and Kang 1995; Thomas et al. 1991; Kang and Bullman 1996; Kang et al. 2000, 2002, 2006, 2015; Bullman and Kang 2000; Bullman et al. 2005; Smith et al. 2004). BIRLS contains identifying information on individuals who have submitted claims for veterans' benefits, including educational benefits and home loans (Fisher et al. 1995; Page et al. 1996; Boyko et al. 2000; Maynard and Chapko 2004). BIRLS data files can be searched on name and military service number and were used in previous studies to verify and update information for atomic veterans. For veterans who have died, death certificates can be retrieved from the VA claims record folder when a death benefit is awarded.

The Eight Series Study is comprised of military participants in the Five Series and OPERATION CROSSROADS studies conducted by the Medical Follow-up Agency of the IOM, the OPERATION HARDTACK I study conducted by the VA, and Project TRINITY (which is newly studied). NDI started in 1979 but a substantial number of deaths occurred before then. Accordingly, cause of death was first obtained from the available information in the previous studies. Next, mortality linkages with state mortality files and the NDI were conducted. If cause of death were not obtained from these sources, the VA BIRLS system and individual state departments of vital statistics were contacted. Approximately 25% of the causes of death were obtained from previous studies, 25% from state mortality electronic records, 44% from NDI, and 1% from death certificates obtained from state departments of health or VA BIRLS records. VA BIRLS was also a source to confirm alive status. This approach is described in Mumma et al. (2018).

The vital status for the atomic veterans cohort through December 31, 2010, is shown in Figure 1. Overall, 27.9% of the veterans were confirmed alive, 68.2% had died, and 2.9% were likely to be alive since they had been identified in BIRLS and no indication of death was found after extensive searching. Of those known to have died, cause of death was obtained for 96.2%. There were 286 veterans who died outside the United States (0.3%), including 55 deaths related to war, for whom the cause of death was not readily available. Of the 3.4% of deaths without a known cause of death, 65.1% occurred between the 1940s and 1978 and thus were most difficult to identify, in part because the National Death Index did not begin until 1979 and deaths before then had to be obtained by contacting individual state departments of health.

Radiation dose estimation

Most radiation doses were received from exposures to radioactive fallout and not from direct or acute exposures to the nuclear-weapons detonations. External gamma radiation exposure from fission and activation products, radioactive debris, or other irradiated materials was the primary source of exposure. Internal radiation exposure and organ-specific doses were judged to be minimal with only a few exceptions. Therefore, the reported associations are based on estimated doses from external radiation exposure. The estimation of dose was facilitated by record systems maintained by the U.S. Defense Threat Reduction Agency (DTRA) which included film badge readings (NA/NRC 1989; Till et al. 2014, 2018a).

Dose reconstruction and the handling of uncertainty has been described (Till et al. 2014, 2018a; Stram et al. 2015; Beck et al. 2017; NCRP 2018a; Dauer et al. 2018). Uncertainty and shared error was addressed in some detail, particularly in Stram et al. (2015), and was determined to be so small that no adjustment was required. Two important resources for individual dose estimation were available: the Nuclear Test Personnel Review (NTPR) program and NuTRIS (Gladeck and Johnson 1996). The NuTRIS database included detailed dosimetric information on each of the 230 atmospheric detonations as well as on individual and representative veterans present at each of the nuclear tests (Till et al. 2014, 2018a). General information concerning issuance and processing of film badges and about bias and uncertainty in the results have been described (NA/NRC 1989). Film badge results and NTPR generic dose reconstructions for each military unit were available as a starting point for estimating both external and internal exposure and organ doses. In addition to NuTRIS documents, data and other reference material were available on military units, ship logs, photographs of facilities, and maneuvers at the NTS and the PPG. The NuTRIS indicator of military rate (for officers) or rating (for enlisted men), the individual's job category, and numerous historical records were available. Much of the detailed dosimetry information available had been compiled to support estimation of dose for compensation programs for exposed veterans (NA/NRC 2003; DVA 2011a, 2011b; Blake and Komp 2014).

High-yield weapons, with a fusion component, were tested in the Pacific at the PPG. Only relatively low-yield, fission-only devices were tested at the NTS. The NTS tests involved mostly detonations conducted on steel towers or from tethered balloons at heights of 100–200 m over land. Forward observers were exposed to gamma and neutron radiation from the blasts. Soldiers involved in post-shot maneuvers were exposed to external gamma radiation from fallout fission and neutron-activation products, and may have received internal exposures from any inhaled resuspended fission and activation products or unfissioned plutonium or uranium. Resuspension of radioactive fallout from previous tests also was possible as a result of the blast's shock wave or when troop and equipment movement disturbed those previous deposits during maneuvers.

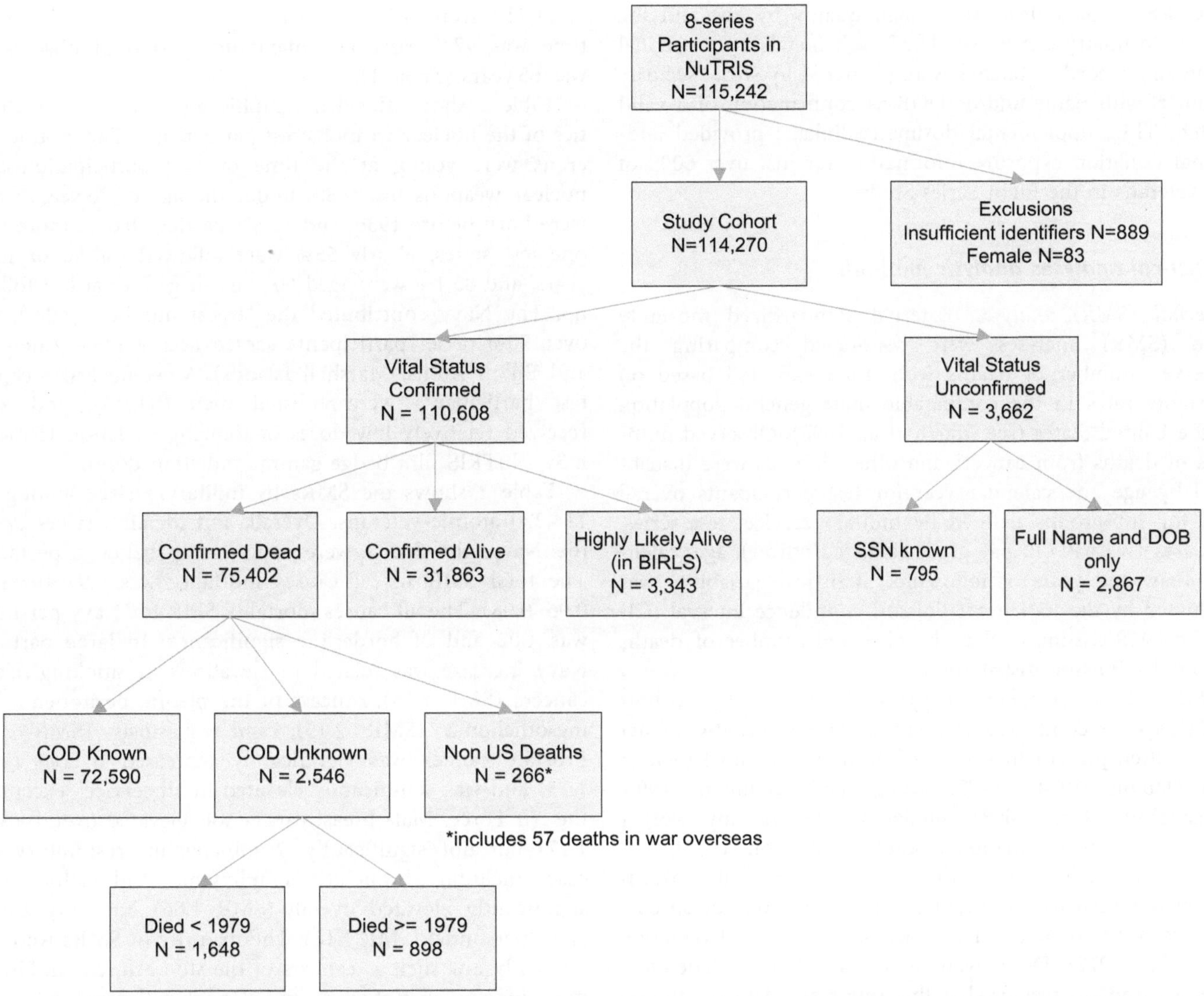

Figure 1. Schematic of the selection and vital status tracing results as of December 31, 2010, for the study population of 114,270 United States military participants at atmospheric nuclear weapons tests between 1945 and 1962. COD indicates Cause of Death; DOB, Date of Birth; and BIRLS, Beneficiary Identification Records Locator Subsystem (U.S. Department of Veterans Affairs).

For many of the PPG surface-level tests, the devices were detonated on barges anchored in lagoons. A few tests included target ships that were heavily contaminated and then boarded after the explosion to collect scientific apparatus and other data. The test participants had external exposures from particulate and gaseous fission and activation products that were airborne, deposited on the ground and on target and support ships, and in sea water. They also had internal exposures from inhalation and ingestion of fission products, activation products, and plutonium or uranium. However, the number of individuals potentially exposed to internal exposure is expected to be small relative to the total cohort.

The comprehensive radiation dose reconstructions were initially for all leukemia and breast cancer deaths and a 1% random sample of all participants within the Eight Series Study. This assessment included all exposures at any of the 19 test series in which a veteran might have participated. The sub-cohort of 1106 randomly selected participants formed a comparison group for which comprehensive dose reconstruction was performed for a case-cohort analyses (Till et al. 2014). During the dose reconstruction (Beck et al. 2017), it was found that the NuTRIS doses (NA/NRC 1989, 2003), available for all veterans, could be adjusted with an organ-specific scaling factor to bring them into close alignment with the comprehensive dose estimates for red bone marrow and breast tissue. Adjusted NuTRIS organ-specific dose estimates are used in the internal within-cohort analyses.

To obtain as complete an assessment of all occupational radiation exposure as possible, both before and after participation as a nuclear weapons test participant, the rosters of atomic veterans were matched against dosimetry registries maintained by the military services (Navy, Army, and Air Force), the U.S. Nuclear Regulatory Commission's Radiation Exposure Information Reporting System (REIRS), the U.S. Department of Energy's Radiation Exposure Monitoring System (REMS), and a private dosimetry service (Landauer, Inc.) (Boice et al. 2006; Hagemeyer et al. 2018; Dauer et al. 2018). Linkage variables included the military service number, social security number, and/or name and date of birth. The dosimetry data bases were comprehensive; e.g., the

NRC REIRS data files are of high quality by law and the Naval Dosimetry Center validated each match from original dosimetry records. Matches were primarily by social security number, with name and/or DOB as confirmation of a valid match. This supplemental dosimetry linkage provided additional radiation exposure information for just over 6000 of the veterans in the Eight Series Study.

Statistical analyses analytic methods

External (SMR) analyses. External standardized mortality ratio (SMR) analyses were conducted comparing the observed number of deaths with those expected based on mortality rates in the comparable male general population of the United States (see Marsh et al. 1998). Observed numbers of deaths from cancers and other diseases were distributed by age and calendar year for test participants overall and for sub-groups defined by military service, test series, test area (NTS/PPG), pay grade (enlisted/officer), age, calendar year, and years of follow-up. Statistical variability was evaluated by the 95% exact Poisson confidence interval (CI) of the SMR assuming that the observed number of deaths followed a Poisson distribution.

Internal within-cohort analyses. Internal within-cohort analyses were conducted to account for the 'healthy soldier effect' often present in studies of military personnel (Seltzer and Jablon 1974, 1977; Kang and Bullman 1996; McLaughlin et al. 2008), similar to the 'healthy worker effect' seen in occupational studies (Monson 1986). For multivariate adjustment, the Cox proportional hazards model was used to compute hazards ratios (HR) for specific cancers and non-cancer deaths across categories of radiation dose (Cox 1972). Doses were analyzed in a time-dependent manner, and age was used as the time scale. All Cox models were conducted using SAS/STAT software (SAS/STAT software, Version 9.4 of the SAS System for Windows, SAS Institute Inc., Cary, NC). Excess Relative Risks (ERR) per 100 mGy were estimated using Cox regression within the Peanuts module of Epicure (Preston et al. 2015). All models were adjusted for year of birth, test area and military pay grade (enlisted/officer). Pay grade was considered a surrogate measure of socioeconomic status (SES) to indirectly account for smoking and other life choices. Radiation doses were lagged by 2 years for leukemia other than CLL and myelodysplastic syndrome (MDS) and lagged by 10 years for all other outcomes. Outcomes were deaths coded as the underlying causes of death, unless otherwise noted when outcome deaths were sparse and contributing causes of death were included. Dose–response functions for continuous measurement of radiation exposure were modeled as linear and quadratic functions (Golden et al. 2018).

Results

The Eight Series Study consisted of 114,270 male veterans of whom 75,404 were found to have died as of December 31, 2010 (Figure 1). Vital status was determined for 96.8% of the test participants. The person-years of observation totaled 5,371,110 from 1945 to 2010, and the average follow-up time was 47.0 years. The maximum period of observation was 65 years (Table 1).

Table 1 shows the demographic and service characteristics of the nuclear weapons test participants. The atomic veterans were young at the time of first participation at a nuclear weapons test (68% under the age of 25 years); 68% were born before 1930, and 12.8% participated in more than one test series. Nearly 55% were followed for 50 or more years, and 83.1% were aged 60 years or greater at last follow-up. The Navy contributed the largest numbers (62%), and over 70% of all participants served at the PPG (Enewetak and Bikini Atolls, Marshall Islands). Most nuclear weapons test participants were enlisted men (84.1%), and most received relatively low doses of ionizing radiation (55% <5 mSv, NuTRIS film badge gamma radiation dose).

Table 2 shows the SMRs by military service among the 114,270 atomic veterans. Overall, and for all services except the Navy, the SMRs were significantly below expectation. The total SMR for all-causes mortality was 0.97 (95% CI 0.96, 0.98). The all-causes mortality SMR for Navy personnel was 1.02 and of borderline significance. In large part, the Navy increase was related to elevations in smoking-related cancers (SMR 1.15), cancers of the pleura, peritoneum and mesothelioma (SMR 2.15), and asbestosis (SMR 3.22) Prostate cancer was significantly increased overall (SMR 1.13) and was significantly elevated in all services except for the Air Force. Male breast cancer was elevated overall (SMR 1.17) but not significantly. Nonmalignant respiratory diseases including bronchitis, emphysema, and asthma were significantly elevated overall (SMR 1.06) but only among Navy personnel (SMR 1.16). The majority of SMRs were significantly low such as cancers of the stomach, colon, kidney, non-Hodgkin lymphoma, and Hodgkin lymphoma. Also, significantly low values were seen for diabetes, mental and behavioral disorders, diseases of the nervous system, ischemic heart disease, cerebrovascular disease, nonmalignant kidney disease, accidents, and suicides.

Table 3 shows SMRs among the atomic veterans by pay grade (enlisted/officer). The majority of veterans were enlisted men (96,071); there were 18,199 officers. Enlisted men were at a significantly higher risk of dying than the general population (SMR 1.06; 95% CI 1.05, 1.07), whereas officers were at a significantly lower risk (SMR 0.71; 95% CI 0.70, 0.72). The only significantly high risk observed in officers was death from prostate cancer (SMR 1.12), which was also significantly elevated among enlisted men (SMR 1.13). Enlisted men had significantly high rates for all cancers, esophageal cancer, liver cancer, pleura, peritoneum and mesothelioma, lung cancer, smoking-related cancers, and nonmalignant respiratory disease.

Table 4 presents SMRs by test series. The first nuclear weapons test was Project TRINITY in New Mexico in 1945. The tests at the PPG and Johnson Island were OPERATIONS CROSSROADS (1946), GREENHOUSE (1951), CASTLE (1954), REDWING (1954), and HARDTACK I (1958). The tests at the NTS were OPERATIONS UPSHOT-KNOTHOLE (1953) and

Table 2. Standardized mortality ratios (SMR) by military service among the 114,270 nuclear weapons test participants in the Eight Series Study, 1945–2010.

	Navy		Army		Air Force		Marines		Total		
Persons-at-risk	70,309		26,091		12,848		5003		114,270		
Person-years	3,394,674		1,162,652		579,465		232,768		5,370,306		
Cause of death (ICD9[a])	Obs	SMR	Obs	SMR	Obs	SMR	Obs	SMR	Obs	SMR	95% CI
All causes of death (0011-9–999)	46,902	1.02*	16,860	0.92*	8373	0.86*	2772	0.99	74,921	0.97*	0.96–0.98
All malignant neoplasms (140–208)	13,809	1.10*	4658	0.96*	2319	0.91*	842	1.06	21,633	1.04*	1.03–1.06
Buccal cavity and pharynx (140–149)	297	1.18*	109	1.11	52	1.02	20	1.24	478	1.14*	1.04–1.25
Esophagus (150)	468	1.27*	151	1.08	93	1.28*	31	1.25	743	1.23*	1.14–1.32
Stomach (151)	329	1.00	101	0.77*	44	0.65*	21	1.05	495	0.90*	0.83–0.99
Colon (153)	939	0.94	320	0.81*	168	0.80*	54	0.88	1483	0.89*	0.85–0.94
Rectum (154)	178	0.91	71	0.90	36	0.89	10	0.83	295	0.90	0.80–1.01
Biliary passages and liver (155,156)	319	1.02	139	1.15	56	0.88	20	0.96	534	1.03	0.95–1.12
Pancreas (157)	633	1.00	217	0.89	108	0.84	44	1.08	1002	0.96	0.90–1.02
Pleura and peritoneum (158.8,158.9,163) and mesothelioma (ICD10 C45)[b]	130	2.15*	10	0.45*	10	0.85	3	0.75	153	1.56*	1.32–1.82
Larynx (161)	159	1.14	57	1.05	20	0.71	10	1.13	246	1.07	0.94–1.21
Bronchus, trachea, and lung (162)	5278	1.18*	1619	0.95*	793	0.89*	287	1.00	7978	1.08*	1.06–1.11
Bone (170)	21	0.77	9	0.88	4	0.77	3	1.79	37	0.83	0.59–1.15
Connective and other soft tissue (171)	57	0.87	23	0.93	13	1.00	5	1.14	98	0.91	0.74–1.11
Male breast (175)	15	1.00	8	1.37	4	1.30	2	2.10	29	1.17	0.78–1.67
Prostate (185)	1163	1.13*	496	1.16*	239	1.01	77	1.32*	1977	1.13*	1.08–1.18
Testes and other male genital organs (186, 187)	26	0.64*	11	0.81	3	0.44	1	0.39	41	0.65*	0.46–0.88
Kidney (189.0–189.2)	322	1.00	103	0.83	49	0.76*	28	1.32	502	0.94	0.86–1.03
Bladder and other urinary (188, 189.3–189.9)	369	1.03	123	0.85	62	0.80	17	0.78	571	0.95	0.87–1.03
Melanoma of skin (172)	226	1.03	72	0.89	42	0.99	17	1.14	357	1.00	0.90–1.11
Other malignant neoplasm of skin (173)	71	1.04	27	1.01	20	1.41	9	2.06	127	1.12	0.93–1.33
Eye (190)	7	1.06	3	1.16	1	0.75	0	0.00	11	1.01	0.50–1.80
Brain and central nervous system (191–192)	310	0.96	106	0.88	61	0.98	18	0.84	495	0.94	0.86–1.03
Thyroid and other endocrine glands (193–194)	36	0.98	15	1.07	2	0.28*	1	0.41	54	0.90	0.67–1.17
All lymphatic, hematopoietic tissue (200–208)	1208	0.94*	455	0.93	236	0.91	93	1.14	1992	0.94*	0.90–0.99
Non-Hodgkin lymphoma (200, 202)	446	0.92	152	0.82*	95	0.96	34	1.09	727	0.91*	0.84–0.98
Hodgkin lymphoma (201)	52	0.75*	22	0.91	5	0.41*	5	1.17	84	0.77*	0.61–0.95
Leukemia (204–208)	504	1.00	189	0.98	94	0.92	35	1.11	822	0.99	0.93–1.06
Chronic lymphocytic leukemia (CLL) (204.1)	98	0.89	42	0.97	13	0.56*	6	0.89	159	0.87	0.74–1.01
Leukemia other than CLL	406	1.04	147	0.99	81	1.03	29	1.16	663	1.03	0.95–1.11
Multiple myeloma (203)	201	0.93	89	1.07	42	0.95	18	1.31	350	0.98	0.88–1.09
Smoking-related cancers (140–150, 157, 161–162, 188–189)	7526	1.15*	2379	0.95*	1177	0.89*	437	1.04	11,520	1.07*	1.05–1.09
Nonsmoking-related cancers	6283	1.05*	2279	0.97	1142	0.92*	405	1.09	10,113	1.02	1.00–1.04
Polycythemia vera (238.4)	3	0.38	2	0.60	0	0.00	0	0.00	5	0.37*	0.12–0.87
Myelodysplastic syndrome (238.7)	72	1.04	25	0.97	20	1.33	2	0.47	119	1.04	0.86–1.25
Diabetes (250)	931	0.89*	363	0.89*	140	0.64*	51	0.75*	1486	0.85*	0.81–0.90
Mental and behavioral disorders (290–319)	602	0.92*	269	1.04	113	0.74*	37	0.96	1021	0.92*	0.87–0.98
Dementia and Alzheimer's disease (290.0–290.4, 331.0)	804	0.93*	308	0.90	169	0.77*	49	1.04	1330	0.90*	0.86–0.95
Diseases of the nervous system (320–389)	1,101	0.83*	433	0.83*	262	0.89	74	0.97	1871	0.84*	0.81–0.88
All heart disease (390–398, 404, 410–429)	14,103	0.92*	5134	0.82*	2559	0.77*	791	0.88*	22,592	0.88*	0.86–0.89
Ischemic heart disease (410–414)	10,495	0.90*	3733	0.78*	1891	0.75*	581	0.86*	16,704	0.85*	0.84–0.86
Cerebrovascular disease (430–438)	1936	0.90*	756	0.82*	366	0.75*	103	0.84	3,161	0.86*	0.83–0.89
Nonmalignant respiratory disease (460–519)	4289	1.04*	1468	0.87*	761	0.83*	240	0.98	6758	0.97*	0.95–0.99
Bronchitis, emphysema, asthma (490–493)	1916	1.16*	583	0.91*	329	0.94	99	0.93	2927	1.06*	1.02–1.10
Asbestosis (501)	60	3.22*	4	0.58	1	0.25	0	0.00	65	2.13*	1.64–2.71
Cirrhosis of liver (571)	891	0.98	325	0.95	161	0.92	55	0.93	1432	0.97	0.92–1.02
Nephritis and nephrosis (580–589)	499	0.86*	221	0.98	84	0.66*	31	0.90	835	0.86*	0.80–0.92
AIDS (042–044, 795.8)	37	0.72*	12	0.63	6	0.68	0	0.00	55	0.65*	0.49–0.85
All external causes of death (800–999)	3238	0.95*	1039	0.87*	557	0.92*	232	1.03	5067	0.93*	0.90–0.96
Accidents (850–949)	2099	0.90*	684	0.85*	387	0.95	164	1.10	3335	0.90*	0.87–0.93
Suicides (950–959)	745	0.94	228	0.79*	124	0.84	45	0.84	1142	0.89*	0.84–0.94
Unknown causes of death	1534	—	635	—	293	—	81	—	2543	-	-

[a]ICD = International Classification of Disease. Revision 9 codes are shown.
[b]Mesothelioma did not have an explicit code in ICD9 but did in ICD10 as denoted.
*Indicates SMR is significantly different from 1.0.

PLUMBBOB (1957). Each series had a different number of bomb detonations; overall, there were 98 detonations. In order of increasing number of detonations at each operation: TRINITY (1), CROSSROADS (2), GREENHOUSE (4), CASTLE (6), UPSHOT-KNOTHOLE (11), REDWING (17), PLUMBBOB (23), HARDTACK I (35) (DTRA 2019). Two-thirds of the bomb detonations (64) occurred at the NTS. Participants at the OPERATION CROSSROADS test had the only significantly high SMR for all-cause mortality (SMR 1.03, 95% CI 1.02, 1.04; $n = 29{,}306$). This increase was related primarily to smoking-related cancers including the lung, esophagus, and nonmalignant respiratory disease. Statistically significant excesses of mesothelioma deaths ($n = 153$) were seen only among Navy personnel participating in tests at the PPG with over half ($n = 78$) occurring among participants at OPERATION CROSSROADS.

Table 3. Standardized mortality ratios (SMR) by pay grade (Enlisted v. Officer) among the 114,270 nuclear weapons test participants in the Eight Series Study, 1945–2010.

		Enlisted			Officers	
Persons-at-risk		96,071			18,199	
Person-years		4,588,339			781,967	
Cause of death (ICD9[a])	Obs	SMR	95% CI	Obs	SMR	95% CI
All causes of death (001–999)	61,440	1.06*	1.05–1.07	13,481	0.71*	0.70–0.72
All malignant neoplasms (140–208)	18,009	1.11*	1.10–1.13	3624	0.79*	0.77–0.82
Buccal cavity and pharynx (140–149)	387	1.18*	1.07–1.31	91	1.00	0.81–1.23
Esophagus (150)	630	1.29*	1.19–1.40	113	0.96	0.79–1.16
Stomach (151)	422	1.02	0.93–1.12	73	0.54*	0.43–0.68
Colon (153)	1187	0.94*	0.88–0.99	296	0.74*	0.66–0.83
Rectum (154)	250	1.01	0.89–1.14	45	0.57*	0.41–0.76
Biliary passages and liver (155,156)	457	1.11*	1.01–1.22	77	0.72*	0.57–0.90
Pancreas (157)	808	0.99	0.92–1.06	194	0.86*	0.74–0.99
Pleura and peritoneum (158.8, 158.9, 163) and mesothelioma (ICD10 C45)[b]	145	1.81*	1.53–2.13	8	0.44*	0.19–0.87
Larynx (161)	221	1.23*	1.07–1.40	25	0.50*	0.32–0.74
Bronchus, trachea, and lung (162)	7019	1.20*	1.17–1.23	959	0.63*	0.59–0.67
Bone (170)	28	0.80	0.53–1.16	9	0.93	0.43–1.77
Connective and other soft tissue (171)	76	0.88	0.69–1.10	22	1.04	0.65–1.58
Male breast (175)	24	1.24	0.80–1.85	5	0.90	0.29–2.09
Prostate (185)	1404	1.13*	1.08–1.19	573	1.12*	1.03–1.22
Testes and other male genital organs (186, 187)	37	0.69*	0.49–0.95	4	0.40	0.11–1.03
Kidney (189.0–189.2)	415	0.98	0.89–1.08	87	0.79*	0.64–0.98
Bladder and other urinary (188, 189.3–189.9)	461	1.03	0.94–1.13	110	0.72*	0.59–0.86
Melanoma of skin (172)	283	0.97	0.86–1.09	74	1.14	0.90–1.43
Other malignant neoplasm of skin (173)	102	1.16	0.95–1.41	25	0.98	0.63–1.45
Eye (190)	9	1.06	0.49–2.02	2	0.81	0.09–2.93
Brain and central nervous system (191–192)	387	0.91	0.82–1.00	108	1.10	0.90–1.33
Thyroid and other endocrine glands (193–194)	47	0.98	0.72–1.30	7	0.58	0.23–1.19
All lymphatic, hematopoietic tissue (200–208)	1581	0.95	0.91–1.00	411	0.91*	0.82–1.00
Non-Hodgkin lymphoma (200, 202)	598	0.95	0.88–1.03	129	0.76*	0.64–0.90
Hodgkin lymphoma (201)	69	0.77*	0.60–0.97	15	0.76	0.43–1.25
Leukemia (204–208)	635	0.98	0.91–1.06	187	1.03	0.89–1.19
Chronic lymphocytic leukemia (CLL) (204.1)	117	0.84	0.70–1.01	42	0.95	0.68–1.28
Leukemia other than CLL	518	1.02	0.93–1.11	145	1.06	0.89–1.24
Multiple myeloma (203)	273	0.98	0.86–1.10	77	0.99	0.78–1.23
Smoking-related cancers (140–150, 157, 161–162, 188–189)	9941	1.17*	1.14–1.19	1579	0.69*	0.66–0.73
Nonsmoking-related cancers	8068	1.06*	1.03–1.08	2045	0.89*	0.85–0.93
Polycythemia vera (238.4)	5	0.52	0.17–1.22	0	0.00	0.00–0.94
Myelodysplastic syndrome (238.7)	90	1.02	0.82–1.25	29	1.11	0.74–1.59
Diabetes (250)	1321	0.97	0.92–1.02	165	0.43*	0.37–0.50
Mental and behavioral disorders (290–319)	759	0.95	0.88–1.02	262	0.86*	0.76–0.97
Dementia and Alzheimer's disease (290.0–290.4, 331.0)	884	0.87*	0.81–0.93	446	0.98	0.89–1.08
Diseases of the nervous system (320–389)	1312	0.81*	0.77–0.85	559	0.94	0.86–1.02
All heart disease (390–398, 404, 410–429)	18,364	0.98*	0.96–0.99	4228	0.61*	0.59–0.63
Ischemic heart disease (410–414)	13,634	0.95*	0.94–0.97	3070	0.57*	0.55–0.59
Cerebrovascular disease (430–438)	2451	0.95*	0.91–0.99	710	0.65*	0.60–0.70
Nonmalignant respiratory disease (460–519)	5626	1.10*	1.07–1.13	1132	0.60*	0.57–0.64
Bronchitis, emphysema, asthma (490–493)	2568	1.20*	1.15–1.25	359	0.58*	0.52–0.65
Asbestosis (501)	63	2.67*	2.05–3.41	2	0.29	0.03–1.04
Cirrhosis of liver (571)	1211	1.01	0.96–1.07	221	0.78*	0.68–0.89
Nephritis and nephrosis (580–589)	678	0.94	0.87–1.01	157	0.63*	0.54–0.74
AIDS (042–044, 795.8)	47	0.60*	0.44–0.80	8	1.24	0.53–2.44
All external causes of death (800–999)	4380	0.97*	0.94–1.00	687	0.74*	0.69–0.80
Accidents (850–949)	2848	0.93*	0.90–0.97	487	0.77*	0.70–0.84
Suicides (950–959)	978	0.92*	0.87–0.98	164	0.72*	0.61–0.84
Unknown causes of death	2015	–	–	528	–	–

[a]ICD: International Classification of Disease. Revision 9 codes are shown.
[b]Mesothelioma did not have an explicit code in ICD9 but did in ICD10 as denoted.
*Indicates SMR is significantly different from 1.0.

Prostate cancer was elevated among participants at all eight test series and significantly high at OPERATIONS CROSSROADS and PLUMBBOB. Ischemic heart disease was significantly low among participants at all of the eight series. Leukemia other than CLL was not significantly elevated in any series nor was MDS. Male breast cancer was significantly elevated among participants at OPERATION GREENHOUSE (SMR 3.15; $n=7$), and connective and other soft tissues were significantly elevated at Project TRINITY (SMR 8.64; $n=4$).

Table 5 shows the SMRs by time since first participation at a nuclear weapons test through the end of follow-up 50–65 years later. With the passage of time the 'healthy soldier effect' diminished and rates of death approached, and for some causes of death exceeded, the comparable rates seen in the general population (Seltzer and Jablon 1974, 1977; Kang and Bullman 1996; McLaughlin et al. 2008). For

Table 4. Standardized mortality ratios (SMR) by test series among the 114,270 nuclear weapons test participants in the Eight Series Study, 1945–2010.

	CASTLE		CROSSROADS		GREENHOUSE		HARDTACK I		PLUMBBOB		REDWING		TRINITY		UPSHOT-KNOTHOLE	
Persons-at-risk	16,223		38,376		9606		10,329		12,219		13,623		396		18,558	
Person-years	767,077		1,914,522		444,392		466,031		524,504		622,211		18,884		843,064	
Cause of death (ICD9[a])	Obs	SMR	Obs	SMR	Obs	SMR	Obs	SMR	Obs	SMR	Obs	SMR	Obs	SMR	Obs	SMR
All causes of death (001-999)	9413	0.95*	29,306	1.03*	6899	0.98	5249	1.02	7446	0.87*	7495	0.98*	319	0.71*	12,340	0.89*
All malignant neoplasms (140-208)	2874	1.03	8403	1.12*	1878	1.01	1585	1.08*	2114	0.94*	2277	1.06*	97	0.95	3377	0.93*
Buccal cavity and pharynx (140-149)	64	1.14	178	1.19*	42	1.12	36	1.19	43	0.95	57	1.30	2	0.98	77	1.05
Esophagus (150)	104	1.21	286	1.36*	61	1.16	61	1.30	82	1.26*	88	1.31*	1	0.40	109	1.05
Stomach (151)	68	0.96	215	1.07	28	0.56*	29	0.79	44	0.73*	53	0.97	3	0.95	85	0.85
Colon (153)	196	0.90	564	0.93	142	0.94	108	0.96	144	0.78*	154	0.91	9	0.99	241	0.81*
Rectum (154)	37	0.88	104	0.87	29	0.98	21	0.94	30	0.83	36	1.09	1	0.54	47	0.80
Biliary passages and liver (155,156)	63	0.87	193	1.07	53	1.17	35	0.90	55	0.98	45	0.80	1	0.44	102	1.13
Pancreas (157)	125	0.88	377	1.01	87	0.94	76	1.00	115	1.01	108	0.97	1	0.20	159	0.87
Pleura and peritoneum (158.8,158.9,163) and mesothelioma (ICD10 C45)[b]	31	2.19*	76	2.18*	7	0.82	18	2.46*	5	0.50	9	0.85	0	0.00	9	0.54
Larynx (161)	30	0.96	95	1.15	37	1.80*	11	0.66	20	0.80	21	0.87	4	3.59	37	0.91
Bronchus, trachea, and lung (162)	1,062	1.05	3,237	1.22*	663	1.01	615	1.17*	730	0.93*	888	1.15*	21	0.64*	1,125	0.88*
Bone (170)	9	1.56	10	0.60	1	0.26	5	1.57	5	1.07	2	0.44	0	0.00	6	0.79
Connective and other soft tissue (171)	12	0.79	33	0.88	3	0.32*	13	1.55	10	0.86	16	1.34	4	8.64*	14	0.76
Male breast (175)	4	1.20	8	0.90	7	3.15*	0	0.00	3	1.10	3	1.15	0	0.00	4	0.91
Prostate (185)	224	1.07	756	1.14*	187	1.13	112	1.11	235	1.15*	171	1.07	16	1.30	384	1.16*
Testes and other male genital organs (186, 187)	8	0.92	14	0.58*	3	0.55	3	0.62	2	0.34	6	0.89	0	0.00	8	0.79
Kidney (189.0-189.2)	74	1.00	193	1.03	46	0.98	44	1.09	49	0.85	47	0.82	0	0.00	76	0.82
Bladder and other urinary (188, 189.3-189.9)	76	0.98	238	1.07	38	0.69*	40	1.03	54	0.8	48	0.81	4	1.13	92	0.84
Melanoma of skin (172)	58	1.13	127	1.03	28	0.91	12	0.42*	35	0.93	43	1.07	1	0.72	60	1.00
Other malignant neoplasm of skin (173)	14	0.92	45	1.11	19	1.87*	4	0.50	13	1.05	13	1.10	0	0.00	25	1.24
Eye (190)	3	2.07	4	1.01	1	1.02	1	1.28	0	0.00	3	2.65	0	0.00	1	0.52
Brain and central nervous system (191-192)	66	0.89	188	1.02	42	0.92	38	0.92	56	1.01	55	0.94	1	0.46	73	0.82
Thyroid and other endocrine glands (193-194)	5	0.59	23	1.08	4	0.75	4	0.88	5	0.78	7	1.07	1	3.73	6	0.58
All lymphatic, hematopoietic tissue (200-208)	274	0.96	704	0.92*	177	0.94	139	0.93	215	0.95	210	0.96	11	1.08	333	0.90
Non-Hodgkin lymphoma (200, 202)	108	0.99	258	0.90	79	1.11	57	0.99	70	0.81	66	0.79	6	1.59	111	0.80*
Hodgkin lymphoma (201)	9	0.61	26	0.62*	7	0.73	9	1.13	8	0.76	11	0.97	0	0.00	15	0.83
Leukemia (204-208)	106	0.95	296	0.98	65	0.88	51	0.88	90	1.01	93	1.09	5	1.22	141	0.97
Chronic lymphocytic leukemia (CLL) (204.1)	14	0.58*	69	1.03	13	0.78	8	0.67	16	0.79	15	0.82	2	1.97	31	0.94
Leukemia other than CLL	92	1.05	227	0.97	52	0.91	43	0.94	74	1.08	78	1.16	3	0.97	110	0.98
Multiple myeloma (203)	50	1.03	120	0.93	26	0.81	21	0.84	45	1.16	38	1.02	0	0.00	66	1.05
Smoking-related cancers (140-150, 157, 161-162, 188-189)	1535	1.04	4604	1.19*	974	1.01	883	1.14*	1093	0.94*	1257	1.11*	33	0.67*	1,675	0.89*
Nonsmoking-related cancers	1339	1.02	3799	1.05*	904	1.01	702	1.02	1021	0.94*	1020	1.01	64	1.22	1,702	0.96
Polycythemia vera (238.4)	0	0.00	3	0.59	–	0.00	0	0.00	1	0.65	0	0.00	0	0.00	1	0.39
Myelodysplastic syndrome (238.7)	21	1.36	43	1.02	13	1.25	6	0.81	11	0.90	8	0.71	1	1.71	23	1.14
Diabetes (250)	167	0.70*	587	0.95	133	0.86	114	0.90	163	0.85*	147	0.80*	8	0.94	240	0.78*
Mental and behavioral disorders (290-319)	124	0.91	369	0.89*	94	0.91	78	1.12	110	0.86	109	1.03	6	0.80	188	0.93
Dementia and Alzheimer's disease (290.0-290.4, 331.0)	159	0.94	529	0.92	108	0.75*	74	0.95	154	0.90	111	0.87	10	0.87	266	0.97
Diseases of the nervous system (320-389)	219	0.80*	700	0.83*	154	0.74*	127	0.95	224	0.91	180	0.87	10	0.69	366	0.9
All heart disease (390-398, 404, 410-429)	2733	0.86*	9044	0.94*	2175	0.91*	1494	0.92*	2218	0.75*	2156	0.87*	79	0.47*	3,770	0.79*
Ischemic heart disease (410-414)	2045	0.85*	6719	0.92*	1590	0.87*	1095	0.89*	1590	0.71*	1622	0.87*	64	0.50*	2,742	0.75*
Cerebrovascular disease (430-438)	405	0.92	1238	0.90*	314	0.91	206	0.94	324	0.75*	257	0.76*	16	0.60*	561	0.79*
Nonmalignant respiratory disease (460-519)	780	0.89*	2760	1.07*	675	1.04	495	1.15*	635	0.80*	684	1.02	31	0.69*	1,044	0.80*
Bronchitis, emphysema, asthma (490-493)	363	0.96	1194	1.21*	286	1.15*	239	1.27*	253	0.85*	326	1.15*	9	0.66	406	0.83*
Asbestosis (501)	7	1.74	42	3.58*	7	2.49*	5	2.69	1	0.31	2	0.69	0	0.00	3	0.56
Cirrhosis of liver (571)	214	1.05	509	0.97	121	0.93	100	0.88	130	0.83*	159	0.98	3	0.47	250	1.00
Nephritis and nephrosis (580-589)	121	0.98	294	0.81*	80	0.89	53	0.87	92	0.86	69	0.74*	6	1.00	158	0.90
AIDS (042-044, 795.8)	10	0.70	14	0.86	4	0.78	10	0.69	4	0.36*	7	0.46*	0	0.00	6	0.51

(continued)

Table 4. Continued.

	CASTLE		CROSSROADS		GREENHOUSE		HARDTACK I		PLUMBBOB		REDWING		TRINITY		UPSHOT-KNOTHOLE	
Persons-at-risk	16,223		38,376		9606		10,329		12,219		13,623		396		18,558	
Person-years	767,077		1,914,522		444,392		466,031		524,504		622,211		18,884		843,064	
Cause of death (ICD9[a])	Obs	SMR	Obs	SMR	Obs	SMR	Obs	SMR	Obs	SMR	Obs	SMR	Obs	SMR	Obs	SMR
All external causes of death (800-999)	710	0.95	1879	0.94*	438	0.95	404	0.91	524	0.97	568	0.94	10	0.43*	742	0.84*
Accidents (850-949)	495	0.99	1172	0.85*	296	0.94	266	0.92	368	1.02	401	1.01	4	0.25*	487	0.82*
Suicides (950-959)	156	0.87	444	0.98	112	1.02	83	0.78*	108	0.81*	113	0.78*	6	1.12	159	0.75*
Unknown causes of death	222	–	1065	–	238	–	131	–	229	–	212	–	30	–	509	–

[a]ICD: International Classification of Disease. Revision 9 codes are shown.
[b]Mesothelioma did not have an explicit code in ICD9 but did in ICD10 as denoted.
*Indicates SMR is significantly different from 1.0

all-causes mortality, statistically significant low SMRs were observed for all follow-up periods up to 40 years after first participation and then remained essentially flat at 1.00 from 40 to 65 years. For all-cancer mortality, a similar pattern was seen with significantly low cancer rates up to 30 years after first participation, but then significant elevations continued to rise through 50–65 years of observation. Lung cancer reflected this general pattern, starting with SMRs significantly low, increasing with every decade of follow-up, and then reaching statistically significant elevated rates 40–65 years after first test series participation. Prostate cancer was not significantly elevated until 40 years had passed. Mesothelioma was significantly elevated only 40–65 years after follow-up began. Ischemic heart disease was significantly low over all periods of follow-up but less so with each passing decade. Deaths from suicide were significantly below expectation compared with the male general population for 30 years after first participation at a nuclear weapons test, but the SMRs steadily rose until significance was reached among those followed for more than 50 years (SMR 1.26).

Table 6 presents the within-cohort internal dose–response analysis for selected causes of death over categories of organ-specific radiation doses among 113,804 atomic veterans. These analyses compare atomic veterans with atomic veterans over different categories of radiation absorbed dose, adjusting for important risk indicators for health events such as age and pay grade (enlisted/officer). The numbers differ slightly from those used in the SMR analyses ($n = 114{,}270$) because the within-cohort analyses exclude veterans with unknown exposure estimates ($n = 466$) within the NuTRIS system.

Leukemia other than CLL is one of the most frequent malignancies identified in excess following exposures to radiation. Among atomic veterans, there were 671 deaths due to leukemia other than CLL and 36 contributing causes of death. There was no evidence for an increased radiation risk, and no evidence for a dose response over the range of estimated doses to active bone marrow [ERR (95% CI) per 100 mGy −0.37 (−1.08, 0.33) $n = 710$]. Similarly, for CLL there was no evidence for an increasing risk with radiation dose [ERR (95% CI) per 100 mGy −0.66 (−0.98, 0.85) $n = 190$].

Lung cancer was not associated with radiation dose to lung and there was no evidence for a dose–response trend [ERR (95% CI) per 100 mGy 0.04 (−0.11, 0.19) $n = 7980$].

Mesothelioma ($n = 153$) showed a significantly high SMR (1.56), and the elevated risk was seen only among Navy personnel who served on ships in the PPG (SMR 2.15; $n = 130$). The internal analyses for mesothelioma indicated a radiation association [ERR (95% CI) per 100 mGy 0.45 (0.04, 0.86)]. The hazard ratios (HR) for the dose categories presented were high but significant only at the lowest dose categories. Because mesothelioma SMR risks were concentrated among Navy personnel with high potential for asbestos exposure aboard ships (Till et al. 2018b), an internal dose–response analysis was conducted of the 43,574 non-Navy military personnel. This analysis of veterans with

Table 5. Standardized mortality ratios (SMR) by time since exposure (years) among the 114,270 nuclear weapons test participants in the Eight Series Study, 1945–2010.

	Less than 10 years		10–19 years		20–29 years		30–39 years		40–49 years		50 or more years	
Persons-at-risk	114,270		111,911		107,737		99,270		84,517		62,445	
Person-years	1,130,577		1,100,867		1,039,433		925,010		740,383		434,036	
Cause of death (ICD9[a])	Obs	SMR	Obs	SMR	Obs	SMR	Obs	SMR	Obs	SMR	Obs	SMR
All causes of death (001-999)	2098	0.80*	4169	0.90*	8467	0.95*	14736	0.96*	21933	0.99	23518	1.00
All malignant neoplasms (140-208)	178	0.57*	613	0.76*	2087	0.99	4714	1.04*	7254	1.07*	6787	1.10*
Buccal cavity and pharynx (140-149)	2	0.24*	18	0.69	85	1.26*	126	1.09	145	1.21*	102	1.28*
Esophagus (150)	–	0.00	17	1.10	74	1.42*	148	1.14	266	1.25*	238	1.25*
Stomach (151)	4	0.27*	26	0.73	63	0.83	131	0.98	140	0.87	131	1.04
Colon (153)	6	0.31*	41	0.69*	140	0.81*	361	0.94	539	0.97	396	0.83*
Rectum (154)	5	0.56	18	0.82	31	0.69*	59	0.75*	107	1.07	75	1.04
Biliary passages and liver (155,156)	3	0.85	15	1.36	37	1.03	98	0.97	184	0.99	197	1.09
Pancreas (157)	6	0.53	37	0.91	90	0.82	206	0.93	351	1.04	312	0.96
Pleura and peritoneum (158.8,158.9,163) and mesothelioma (ICD10 C45)[b]	–	0.00	3	1.20	4	0.93	8	0.91	44	1.48*	94	1.84*
Larynx (161)	–	0.00	6	0.57	37	1.21	58	0.95	80	1.08	65	1.27
Bronchus, trachea, and lung (162)	31	0.57*	161	0.66*	794	0.99	1852	1.04	2771	1.11*	2369	1.19*
Bone (170)	3	0.47	6	1.15	9	1.31	7	0.88	7	0.74	5	0.58
Connective and other soft tissue (171)	3	0.78	5	0.83	15	1.28	23	1.01	29	0.88	23	0.75
Male breast (175)	1	2.46	–	0.00	3	1.09	5	0.99	12	1.54	8	1.06
Prostate (185)	3	1.02	12	0.73	98	1.22	320	1.10	723	1.17*	821	1.11*
Testes and other male genital organs (186, 187)	8	0.43*	10	0.63	3	0.33*	8	1.19	8	1.13	4	0.64
Kidney (189.0-189.2)	3	0.45	16	0.74	54	0.90	122	1.00	160	0.93	147	0.98
Bladder and other urinary (188, 189.3-189.9)	–	0.00	7	0.52	43	1.03	78	0.79*	171	0.89	272	1.08
Melanoma of skin (172)	11	0.96	30	1.22	49	1.08	64	0.86	88	0.87	115	1.14
Other malignant neoplasm of skin (173)	–	0.00	1	0.28	8	0.83	33	1.40	31	0.88	54	1.36*
Eye (190)	–	0.00	1	1.09	1	0.64	1	0.40	6	1.97	2	0.80
Brain and central nervous system (191-192)	24	0.93	22	0.47*	79	1.00	118	0.96	142	0.98	110	1.05
Thyroid and other endocrine glands (193-194)	–	0.00	2	0.43	7	0.91	14	1.18	19	1.10	12	0.74
All lymphatic, hematopoietic tissue (200-208)	48	0.61*	85	0.72*	180	0.89	367	0.95	647	1.00	665	0.98
Non-Hodgkin lymphoma (200, 202)	9	0.49*	26	0.72	71	0.96	136	0.88	249	0.97	236	0.91
Hodgkin lymphoma (201)	15	0.57*	18	0.66	13	0.66	10	0.71	16	1.31	12	1.20
Leukemia (204-208)	23	0.72	37	0.84	69	0.89	147	1.00	251	1.01	295	1.06
Chronic lymphocytic leukemia (CLL) (204.1)	3	1.32	2	0.44	11	0.92	25	0.76	52	0.85	66	0.93
Leukemia other than CLL	20	0.69	35	0.89	58	0.89	122	1.07	199	1.06	229	1.10
Multiple myeloma (203)	1	0.43	4	0.46	26	0.93	73	1.05	127	1.04	119	0.94
Smoking-related cancers (140-150, 157, 161-162, 188-189)	42	0.45*	262	0.70*	1177	1.01	2590	1.02	3944	1.09*	3505	1.16*
Nonsmoking-related cancers	136	0.62*	351	0.82*	910	0.96	2124	1.05*	3310	1.04*	3282	1.04*
Polycythemia vera (238.4)	–	0.00	1	0.65	–	0.00	1	0.43	2	0.61	1	0.28
Myelodysplastic syndrome (238.7)	–	0.00	–	0.00	1	1.00	2	0.43	25	0.84	91	1.16
Diabetes (250)	6	0.18*	23	0.34*	80	0.65*	219	0.80*	527	0.91*	631	0.95
Mental and behavioral disorders (290-319)	7	0.38*	32	0.74	72	0.90	132	1.05	235	0.94	543	0.92
Dementia and Alzheimers disease (290.0-290.4, 331.0)	–	0.00	–	0.00	9	1.10	66	1.09	300	0.92	955	0.89*
Diseases of the nervous system (320-389)	5	0.05*	40	0.29*	83	0.75*	217	1.08	505	0.94	1021	0.90*
All heart disease (390-398, 404, 410-429)	254	0.47*	1136	0.72*	3027	0.86*	5102	0.88*	6768	0.92*	6305	0.90*
Ischemic heart disease (410-414)	200	0.49*	968	0.70*	2396	0.80*	3655	0.84*	4865	0.90*	4620	0.90*
Cerebrovascular disease (430-438)	39	0.52*	119	0.64*	298	0.79*	527	0.82*	992	0.92*	1186	0.90*
Nonmalignant respiratory disease (460-519)	34	0.41*	130	0.68*	359	0.79*	1040	0.93*	2174	0.96	3021	1.05*
Bronchitis, emphysema, asthma (490-493)	6	0.35*	34	0.57*	89	0.78*	188	0.91	809	1.01	1801	1.16*
Asbestosis (501)	0	0.00	0	0.00	0	0.00	4	1.20	27	2.63*	34	2.10*
Cirrhosis of liver (571)	23	0.43*	169	0.88	383	1.04	412	1.05	302	0.98	143	0.86
Nephritis and nephrosis (580-589)	18	0.39*	28	0.76	38	0.90	97	0.96	219	0.83*	435	0.91*
AIDS (042-044, 795.8)	–	0.00	–	0.00	2	0.37	25	0.54*	23	0.84	5	0.91
All external causes of death (800-999)	848	0.75*	1017	0.96	926	0.93*	817	0.99	720	0.98	739	1.06
Accidents (850-949)	679	0.75*	680	0.95	560	0.93	471	0.97	438	0.93	507	1.00
Suicides (950-959)	86	0.53*	182	0.77*	214	0.80*	246	0.95	229	1.08	185	1.26*
Unknown causes of death	588	–	580	–	489	–	234	–	283	–	369	–

[a]ICD: International Classification of Disease. Revision 9 codes are shown.
[b]Mesothelioma did not have an explicit code in ICD9 but did in ICD10 as denoted.
*Indicates SMR is significantly different from 1.0.

minimal to no asbestos exposure found no evidence for a radiation trend [ERR (95% CI) per 100 mGy −0.34 (−3.51, 2.81) $n = 23$]. A significantly high SMR (2.13) also was seen for death due to asbestosis ($n = 65$), and the elevated risk occurred only among Navy personnel serving in the PPG (SMR 3.22; $n = 60$). There was no evidence for a radiation trend for asbestosis [ERR (95% CI) per 100 mGy −0.18 (−2.15, 1.80) $n = 65$]. Similar to the mesothelioma analysis, the highest HR estimates for asbestosis were for personnel with the lowest doses.

Prostate cancer normally would not have been selected for internal analyses since it's not been convincingly linked to ionizing radiation, but the SMR analyses in Tables 2–5 showed statistically significant risks for prostate cancer across practically all military service characteristics. The internal dose response analysis in Table 6, however, provides no evidence that radiation exposure was related to prostate cancer [ERR (95% CI) per 100 mGy 0.03 (−0.27, 0.33)].

Male breast cancer was not significantly associated with radiation dose [ERR (95% CI) per 100 mGy 0.23 (−1.21, 1.67) $n = 34$]; neither was testicular cancer [ERR (95% CI) per 100 mGy 0.17 (−1.63, 1.97) $n = 49$] nor multiple myeloma [ERR (95% CI) per 100 mGy −0.16 (−1.03, 0.72) $n = 351$]. A radiation dose response was not seen for MDS [ERR (95% CI) per 100 mGy −0.48 (−2.18, 1.23) $n = 120$], and a significant negative trend was seen for brain cancer [ERR (95% CI) per 100 mGy −1.40 (−2.50, −0.32) $n = 495$].

Ischemic heart disease among the atomic veterans was not associated with estimated radiation dose to the heart. The radiation dose response was flat over the dose categories [ERR (95% CI) per 100 mGy −0.01 (−0.12, 0.11) $n = 16{,}709$].

Discussion

No statistically significant associations with radiation dose were observed among 114,270 nuclear weapons test participants at eight atmospheric weapons test series. The vital status follow-up was complete and the organ dose reconstructions comprehensive. In comparison with the general population, significant SMRs were observed for mesothelioma and asbestosis, which was attributed to asbestosis exposure, and for prostate cancer but was not related to radiation dose. Within-cohort internal analyses revealed no significant associations with organ-specific radiation doses for leukemia, MDS, multiple myeloma, ischemic heart disease, and cancers of the lung, prostate, testes, male breast cancer, and brain. The large population size and comprehensive dosimetry, however, provide rather narrow 95% confidence limits so that levels of risk higher than some value (ERR per 100 mGy) might be excluded with some confidence. Such levels of possible radiation risk from low dose and low-dose rate exposure data, coupled with the other studies within the MPS (Boice et al. 2019a), should enhance understanding of the possible range of health effects from prolonged exposures to radiation among healthy workers and veterans. Such knowledge may be helpful in setting or revising guidelines to protect workers and the public (NCRP 2018b, 2018c) and, conceivably, in improving compensation schemes currently in place for nuclear weapons test participants (NA/NRC 2003; DVA 2011a, 2011b; VBDR 2016; Caldwell et al. 2016), radiation workers (Neton 2014; Boice et al. 2019a), and down-winders who lived in proximity to weapons testing areas (NA/NRC 2005).

All cancer mortality

Compared with the general male population a slight 4% increase in all cancer mortality was observed among all nuclear weapons test participants, reflecting a significant elevation among Navy personal and significant decreases among Army and Air Force personnel. The increases among Navy personnel were primarily due to malignant respiratory disease, other smoking-related cancers, prostate cancer and mesothelioma. The all-cancer pattern of risk over time showed a diminution of the 'healthy soldier effect': the all-cancer SMR was significantly low after 20 years of follow-up and then increased to be significantly high after 30 years, primarily due to smoking-related cancers, attributed in part to the availability of cigarettes in military rations, and to asbestos exposure. Estimated radiation dose to specific organs was not found to be associated with increased risk for any cancer.

Leukemia, CLL, and MDS

Of all cancers, the malignancy with the highest risk coefficient per dose of radiation (ERR per Gy) is leukemia other than CLL. However, leukemia other than CLL ($n = 710$) was not significantly elevated overall [ERR (95% CI) per 100 mGy −0.37 (−1.08, 0.33); $n = 710$] or in any test series, nor was MDS ($n = 120$). For CLL ($n = 190$), a malignancy not convincingly linked to ionizing radiation (UNSCEAR 2008), there was no evidence for a radiation dose response. For leukemia other than CLL, CLL, and MDS, the dose–response trends were all negative and the excess relative risk (ERR) estimates per 100 mGy were also negative but significantly so only for CLL. Thus, there was no evidence for a radiation-related increase in leukemia or MDS at the low doses received by the atomic veterans (mean dose to red bone marrow, 6 mGy with a maximum of 108 mGy). Nonetheless, the upper 95% CI about the ERR per 100 mGy of 0.33 indicates that mortality risks higher than this level can be excluded with 95% confidence. Thus, the sampling variability is such that risks for leukemia other than CLL are not inconsistent in a statistical sense with those reported in studies of male atomic bomb survivors (Hsu et al. 2013) and in large occupational studies (NCRP 2018b; Leuraud et al. 2015; Gillies et al. 2019). The 95% confidence limits, however, exclude mortality risks per unit dose that are two to four times those reported in these investigations.

Lung cancer

There was no evidence for an increase in lung cancer related to radiation dose to lung. The ERR (95% CI) per 100 mGy was 0.04 (−0.11, 0.19). The absorbed dose to the lung from external exposure was low (mean 6.2 mGy with a maximum of 972 mGy). The mean and maximum absorbed doses were 6.3 and 972 mGy for enlisted men and 5.7 and 507 mGy for officers. The SMR of 1.20 among enlisted men was significantly high and could be attributed to cigarette smoking. There was no evidence for a lung cancer radiation dose–response among enlisted men. The SMR of 0.63 among officers was significantly low. These data are consistent with recent studies of workers at the Mallinckrodt Chemical Works, the Mound facility and the Los Alamos National Laboratory, nuclear power plant workers, industrial radiographers, and Canadian tuberculosis patients monitored with frequent chest fluoroscopies (Boice et al. 2019b). Together they provided little to no evidence that chronic or fractionated radiation exposures increase the risk of lung cancer, in contrast with the acute exposure received by Japanese atomic bomb

Table 6. Internal cohort dose–response analyses for select causes of death over categories of organ-specific radiation doses among 113,806 Nuclear Weapons Test Participants for whom dose estimates could be made.

Dose (mGy)	Number of veterans	Number of cases	HR	95% CI
Leukemia other than CLL[a,*]				
<2.5	58,203	373	1.0	Ref
2.5–<5	19,092	115	1.00	0.81, 1.23
5–<10	17,050	115	1.10	0.89, 1.36
10–<25	14,195	76	0.86	0.67, 1.11
≥25	5266	28	0.86	0.59, 1.27
Total	113,806	707	*p for trend*	*0.30 (-)*
HR (95% CI) at 100 mGy				0.69 (0.34, 1.39)
ERR (95% CI) at 100 mGy				−0.37 (−1.08, 0.33)
CLL (ICD9 204.1)				
<2.5	59,092	108	1.0	Ref
2.5–<5	18,797	28	0.90	0.59, 1.36
5–<10	16,763	29	1.03	0.68, 1.56
10–<25	13,974	20	0.85	0.53, 1.38
≥25	5180	5	0.57	0.23, 1.41
Total	113,806	190	*p for trend*	*0.20 (-)*
HR (95% CI) at 100 mGy				0.31 (0.05, 1.85)
ERR (95% CI) at 100 mGy				−0.66 (−0.98, 0.85)
Myelodysplastic syndrome (ICD9 238.7)				
<2.5	58,203	68	1.0	Ref
2.5–<5	19,092	20	1.01	0.61, 1.66
5–<10	17,050	16	0.86	0.50, 1.49
10–<25	14,195	12	0.75	0.41, 1.40
≥25	5266	3	0.50	0.16, 1.60
Total	113,806	119	*p for trend*	*0.50 (-)*
HR (95% CI) at 100 mGy				0.52 (0.08, 3.45)
ERR (95% CI) at 100 mGy				−0.48 (−2.18, 1.23)
Lung cancer (ICD9 162)				
<2.5	58,905	4108	1.0	Ref
2.5–<5	18,597	1313	1.01	0.95, 1.07
5–<10	16,915	1229	1.04	0.98, 1.11
10–<25	14,013	940	0.98	0.91, 1.05
≥25	5376	369	1.03	0.92, 1.14
Total	113,806	7959	*p for trend*	*0.63 (+)*
HR (95% CI) at 100 mGy				1.04 (0.90, 1.21)
ERR (95% CI) at 100 mGy				0.04 (−0.11, 0.19)
Prostate cancer (ICD9 185)				
<2.5	60,721	1090	1.0	Ref
2.5–<5	19,104	330	1.06	0.94, 1.20
5–<10	15,792	257	1.00	0.87, 1.15
10–<25	13,714	230	1.06	0.92, 1.22
≥25	4475	65	0.90	0.70, 1.16
Total	113,806	1972	*p for trend*	*0.55 (-)*
HR (95% CI) at 100 mGy				0.89 (0.61, 1.30)
ERR (95% CI) at 100 mGy				−0.12 (−0.49, 0.26)
Male breast cancer [*] (ICD9 175)				
<2.5	56,569	14	1.0	Ref
2.5–<5	17,091	4	0.97	0.32, 2.95
5–<10	19,012	6	1.33	0.51, 3.47
10–<25	14,131	8	2.41	1.00, 5.79
≥25	7003	2	1.20	0.27, 5.31
Total	113,806	34	*p for trend*	*0.76 (+)*
HR (95% CI) at 100 mGy				1.25 (0.29, 5.39)
ERR (95% CI) at 100 mGy				0.23 (−1.21, 1.67)
Testicular cancer [*] (ICD9 186, 187)				
<2.5	58,560	30	1.0	Ref
2.5–<5	18,207	8	1.13	0.50, 2.52
5–<10	17,292	2	0.29	0.07, 1.24
10–<25	14,048	6	1.08	0.44, 2.67
≥25	5699	3	1.36	0.41, 4.58
Total	113,806	49	*p for trend*	*0.84 (+)*
HR (95% CI) at 100 mGy				1.20 (0.20, 7.09)
ERR (95% CI) at 100 mGy				0.17 (−1.63, 1.97)
Brain cancer (ICD9 191, 192)				
<2.5	57,326	270	1.0	Ref
2.5–<5	17,332	86	1.12	0.88, 1.43
5–<10	18,349	67	0.83	0.63, 1.09
10–<25	14,079	48	0.76	0.56, 1.04
≥25	6720	23	0.77	0.51, 1.19
Total	113,806	494	*p for trend*	*0.012 (-)*
HR (95% CI) at 100 mGy				0.25 (0.08, 0.73)
ERR (95% CI) at 100 mGy				−1.40 (−2.49, −0.31)

(continued)

Table 6. Continued.

Dose (mGy)	Number of veterans	Number of cases	HR	95% CI
Multiple myeloma (ICD9 203)				
<2.5	59,092	171	1.0	Ref
2.5–<5	18,797	78	1.48	1.13, 1.94
5–<10	16,763	39	0.83	0.59, 1.18
10–<25	13,974	47	1.20	0.87, 1.67
≥25	5180	15	1.04	0.61, 1.76
Total	113,806	350	*p for trend*	*0.72 (-)*
HR (95% CI) at 100 mGy				0.85 (0.36, 2.04)
ERR (95% CI) at 100 mGy				−0.16 (−1.03, 0.72)
Pleura and peritoneum (158.8,158.9,163) and Mesothelioma (ICD10 C45) (All Veterans)				
<2.5	58,905	65	1.0	Ref
2.5–<5	18,597	32	1.55	1.02, 2.37
5–<10	16,915	32	1.65	1.08, 2.52
10–<25	14,013	16	0.99	0.57, 1.72
≥25	5376	8	1.35	0.65, 2.83
Total	113,806	153	*p for trend*	*0.032 (+)*
HR (95% CI) at 100 mGy				1.56 (1.04, 2.35)
ERR (95% CI) at 100 mGy				0.45 (0.04, 0.86)
Pleura and peritoneum (158.8,158.9,163) and Mesothelioma (ICD10 C45) (Non-Navy Veterans)				
<2.5	17,931	7	1.0	Ref
2.5–<5	5263	6	2.67	0.89, 8.04
5–<10	6687	3	0.98	0.25, 3.80
10–<25	9718	6	1.32	0.44, 3.96
≥25	3975	1	0.56	0.07, 4.54
Total	43,574	23	*p for trend*	*0.88 (-)*
HR (95% CI) at 100 mGy				0.79 (0.04, 15.9)
ERR (95% CI) at 100 mGy				−0.35 (−3.51, 2.81)
Ischemic heart disease (ICD9 410-414)				
<2.5	59,092	8911	1.0	Ref
2.5–<5	18,797	2672	1.00	0.96, 1.05
5–<10	16,763	2395	1.02	0.97, 1.06
10–<25	13,974	1947	1.01	0.96, 1.06
≥25	5180	700	0.99	0.92, 1.07
Total	113,806	16,625	*p for trend*	*0.94 (-)*
HR (95% CI) at 100 mGy				1.00 (0.89, 1.12)
ERR (95% CI) at 100 mGy				−0.001 (−0.12, 0.11)
Asbestosis (ICD9 501)				
<2.5	58,905	27	1.0	Ref
2.5–<5	18,597	20	2.38	1.33, 4.24
5–<10	16,915	7	0.93	0.41, 2.14
10–<25	14,013	8	1.34	0.61, 2.97
≥25	5376	3	1.37	0.42, 4.54
Total	113,806	65	*p for trend*	*0.86 (-)*
HR (95% CI) at 100 mGy				0.84 (0.12, 6.02)
ERR (95% CI) at 100 mGy				−0.18 (−2.15, 1.80)

[a]Leukemia other than CLL includes ICD9 codes 204.0, 204.2, 204.3-204.9, 205-208.

ICD: International Classification of Disease. Revision 9 codes are shown.

Veterans with missing exposure information ($n = 466$) were excluded from these analyses. Models include radiation doses lagged by 2 years for non-CLL leukemia and MDS, and lagged by 10 years for all other outcomes. Doses were analyzed as a time-dependent measure. Outcomes include deaths from underlying causes of death unless noted with *, in which case contributing causes were also included. All models adjusted for military pay grade (officer v. enlisted) and year of birth (in six categories except for male breast cancer, testicular cancer, asbestosis and MDS which used three categories due to sparse data). *P*-values for test for linear trend in the hazards ratio treated organ dose as a continuous measure. (+) indicates a positive beta coefficient for the organ dose and (−) indicates a negative beta coefficient for the organ dose in the continuous model. *P*-values are two-sided. CI: confidence interval; CLL: chronic lymphocytic leukemia; ERR: excess relative risk; HR: hazard ratio; REF: referent category.

survivors (Cahoon et al. 2017) and reports from other epidemiologic studies (Boice et al. 2019b). The upper 95% CI about the ERR per 100 mGy of 0.19 indicates that risks higher than this level can be excluded with 95% confidence. An ongoing study of medical radiation workers within the MPS of approximately 109,000 men and women was designed to evaluate the effectiveness of gradual exposures over time to cause lung cancer and whether such exposure results in different radiation lung cancer risks between men and women (Boice et al. 2019a; Yoder et al. 2018).

Mesothelioma and asbestosis

Mesothelioma mortality was significantly increased in comparison with the general population. The increase, however, was seen only among the Navy personnel who served on ships in the PPG, and not among the other military services where all SMRs were less than 1.00. Significantly high mesothelioma risks were seen among enlisted men whereas significantly low mesothelioma risks were seen among officers. The excess mesothelioma among enlisted men was attributed to asbestos exposure among Navy personnel with specific job classifications such as work in boiler rooms (Till et al. 2018b). The internal analysis showed a positive association between radiation dose and mesothelioma but is attributed to confounding by asbestos. A negative radiation association was seen for non-Navy personnel with little to no potential for asbestos exposure.

Confounding by asbestos is also indicated by the high risk of asbestosis seen only among Navy personnel serving in the PPG and with job classifications indicating high potential for asbestos exposure (Till et al. 2018b). The internal analysis showed no evidence for an association between asbestosis and radiation lung dose. Further evidence for confounding by asbestos in occupational studies of radiation comes from studies of nuclear power plant workers, industrial radiographers, and shipyard workers (Mumma et al. 2019), workers at the Idaho National Laboratory (Schubauer-Berigan et al. 2005) and at other nuclear facilities (Schubauer-Berigan et al. 2015; Mumma et al. 2019).

Male breast cancer

Male breast cancer was of special interest because the Five Series Study of atomic veterans had reported an increased risk based on 10 cases, although not statistically significant [HR (95% CL) 1.39 (0.53, 3.66)] (IOM 2000), and preliminary SMR analyses within the Eight Series Study suggested a radiation dose–response based on 34 cases. The comprehensive dose-reconstructed doses for all male breast cancer deaths were compared with the corresponding NuTRIS doses (NA/NRC 2003) and it was determined that NuTRIS doses, available for all atomic veterans, could be used with appropriate scaling factors for internal within-cohort analyses for practically all organs of interest (Beck et al. 2017; Till et al. 2018a; NCRP 2018a). Finally, there are few studies that provide information on radiation-induced male breast cancer (Thomas et al. 1994; Ron et al. 2005; Little and McElvenny 2017).

There was a slight elevation in the breast cancer SMR that was not significant (SMR 1.17), and the within-cohort internal analyses diminished the evidence for a radiation dose response suggested in early SMR analyses [ERR (95% CI) per 100 mGy 0.23 (−1.21, 1.67), $n = 34$]. A significant radiation dose response based on 10 incident cases was reported among the Japanese atomic bomb survivors exposed to much higher doses and at a higher dose rate [ERR (95% CI) per 100 mSv 0.80 (0.08, 4.8)] The risk estimate is extremely imprecise, however, because of the small number of cases (Ron et al. 2005). Recently, Little and McElvenny (2017) analyzed the atomic bomb survivor data on seven incident cases of male breast cancer and six male breast cancer deaths [ERR (95% CI) per 100 mSv 0.89 (0.06, 9.2), $n = 6$] and concluded that there was evidence of a much higher relative risk for radiation-associated male than for female breast cancer. However, they were cautious in this interpretation because of the very small number of male breast cancer cases and deaths.

Prostate cancer

The prostate cancer SMRs were significantly elevated in practically all categories of atomic veterans: enlisted men and officers, all eight test series, and particularly for those followed for the longest time. However, there was no evidence for a radiation dose response, which was essentially flat over all dose categories [ERR (95% CI) per 100 mGy was 0.01 (−0.29, 0.31)]. There is no clear explanation for the increase in prostate cancer, a disease of elderly men and particularly susceptible to screening, in comparison with the male general population of the U.S.. It is only speculative, but perhaps military veterans have an increased likelihood compared with the general population for medical care and screening activities, which would lead to a better diagnosis of prostate cancer during the calendar years these veterans were followed. Prostate cancer mortality is rarely found to be associated with radiation exposure in epidemiologic studies (UNSCEAR 2008), such as the Japanese atomic bomb survivor study [ERR (95% CI) per 100 mGy 0.03 (<0.0, 0.12), $n = 130$] (Ozasa and Mabuchi 2012[AQ]) or the recent INWORKS investigation ERR (90% CI) per 100 mGy −0.01 (−0.07, 0.07), $n = 1685$] (Richardson et al. 2018).

Testicular cancer

Few analyses of radiation-related testicular cancer have been informative largely because testicular cancer is relatively rare; e.g., there were only 18 incident cases of testicular cancer among Japanese atomic bomb survivors (Grant et al. 2017), nine deaths in the three-country worker study (Cardis et al. 1995), 27 deaths in the 15-country worker study (Cardis et al. 2007), and 48 deaths in the INWORKS study (Richardson et al. 2018). Only the INWORKS study indicated a positive association, but interpretation is not straightforward because different analytical approaches gave different results: significance was reported for maximum-likelihood regression analyses but not for hierarchical Poisson regression analyses. Based on 49 deaths among atomic veterans in the Eight Series Study, there was no significant association with radiation dose to testes [ERR (95% CI) per 100 mGy was 0.17 (−1.63, 1.97)]. The upper 95% CI about the ERR per 100 mGy of 1.97 indicates that a doubling of risk can be excluded with 95% CI.

Brain cancer

Ionizing radiation can induce several types of brain cancers following exposures in childhood but less so following exposures in adulthood (UNSCEAR 2013). Susceptibility to radiation-induced brain cancer appears to decrease as brain development nears completion. The study of Japanese atomic bomb survivors exposed to acute high-dose radiation found no significant associations with brain cancer among adults (Brenner et al. 2020); an increase was seen for schwannoma, a type of benign tumor (Preston et al. 2002). Occupational worker studies provide no evidence for an association between radiation and brain cancer (Cardis et al.

2007; Muirhead et al. 2009; Richardson et al. 2018; Golden et al. 2018). The study of nuclear weapons test participants is consistent with previous epidemiologic study of low-dose radiation exposures in that no association between brain cancer and radiation was found. In fact, the dose–response based on 495 deaths was negative and statistically significant [ERR (95% CI) per 100 mGy −1.40 (−2.50, −0.32)].

Multiple myeloma

There is little evidence to support an association between low-LET radiation exposure and multiple myeloma (UNSCEAR 2008). The study of Japanese atomic bomb survivors finds no evidence for a statistically significant radiation-associated risk of incident multiple myeloma based on 136 cases (Hsu et al. 2013). The ERRs per Gy for multiple myeloma were elevated but were not significantly different from zero in the 15-country study based on 83 deaths (Cardis et al. 2007) and in INWORKS based on 283 deaths (Leuraud et al. 2015). The most recent follow-up of the U.K. weapons test participants found no evidence of an increased risk for multiple myeloma, and the suggestion of a raised risk in the first follow-up was judged most likely to be a chance finding (Muirhead et al. 2003a, 2003b). The Eight Series Study of U.S. atomic veterans shows a negative dose response [ERR (95% CI) per 100 mGy −0.16 (−1.03, 0.72) based on 351 deaths, and thus consistent with previous investigations providing little support for an association between radiation and multiple myeloma.

Ischemic heart disease

Consistent with preliminary analyses of ischemic heart disease (IHD) among atomic veterans in the Eight Series Study (Boice 2017; NCRP 2018b), there was no evidence that low-dose radiation increased the risk of IHD. There were nearly 17,000 deaths from IHD and the dose–response trend over the dose categories was essentially flat. The ERR (95% CI) per 100 mGy was −0.01 (−0.12, 0.11). These data are also consistent with MPS studies showing no increase in IHD over categories of radiation dose to the heart (Mallinckrodt, Rocketdyne, nuclear power plant workers, and industrial radiographers) (NCRP 2018b). Whether radiation doses to the heart below 0.5 Gy can result in increased deaths due to heart disease is an area of intense interest and study. The many studies conducted to date are inconsistent with limited ability to control for potential confounding factors such as cigarette smoking and other lifestyle behaviors (by good measures of SES), diabetes, high blood pressure, and others. In this study, the risk of IHD was significantly low for both enlisted men and officers. The SMR for enlisted men (SMR 0.95) was significantly higher than the SMR for officers (SMR 0.57) indicating the importance of adjusting for risk factors for heart disease analyses as possible, but for SES at a minimum.

Suicide

Suicide was significantly low overall in comparison with the male general population (SMR 0.89). Navy personnel had a higher risk for suicide (SMR 0.94) than military personnel in the other services (SMR range from 0.79 to 0.84). Suicide among enlisted men (SMR 0.92) was higher than among officers (SMR 0.72). The SMRs due to suicide increased progressively and significantly over time: from a significantly low SMR of 0.53 at <10 years after first participation at a nuclear weapons test to a significantly high SMR of 1.26 at >50 years of follow-up, i.e., among veterans over the age of 70.

Veterans are known to be at increased risk for suicide in comparison with the general population of the United States (DVA 2019). In 2017, the suicide rate for veterans was 1.5 times the rate for non-veterans. Suicide is the tenth leading cause of death in the United States. Of the 45,390 American adults who died from suicide in 2017, 6,139 (13.5%) were veterans. Firearms were the most frequent method used to commit suicide (70.7%). Each day in 2017, 16.8 veterans committed suicide, an increase from 2015 when it was 15.9 per day. The rate of death among young veterans is higher than among older veterans, but the absolute numbers are much higher among veterans over age 60 years (Fanning and Pietrzak 2013; DVA 2019). There is no single cause for suicide. Suicide rates among veterans have been associated with suicidal ideation (i.e., among those who think about suicide), serving in combat, and not being enrolled in Veteran Health Administration (VHA) services for health care (Sher et al. 2012; DVA 2019). The many interconnecting risk factors for suicide depend not only on specific individual characteristics such as age, sex, ethnicity, residence, and marital status, but also on circumstances such as combat service, deployment, post-traumatic stress disorder (PTSD), other mental health conditions and related medication use, economic disparities, homelessness, unemployment, level of military service connected disability status, community connection, personal health and well-being, and suicide ideation (McCarthy et al. 2015; DVA 2019). Predictive models are being refined to identify veterans at high-risk for suicide who were not so identified on clinical grounds (Kessler et al. 2017). The Department of Veterans Affairs, Environmental Epidemiology Service has conducted recent studies of suicide among military personnel, but none have addressed atomic veterans or servicemen with exposures to ionizing radiation (Kang and Bullman 2009; Kang et al. 2015; Bullman and Schneiderman 2019; Bullman et al. 2015, 2018, 2019; Reger et al. 2015, 2017, 2018).

Because of the uniqueness of the atomic veteran cohort and because suicide prevention is a national and VA priority (DVA 2019), additional studies are planned to examine the determinants of suicide in this veteran population followed since 1945 and essentially for life. This approach is similar in concept to our study attributing the high risk of mesothelioma to asbestos exposure on ships in the PPG (Till et al. 2018b). Questions to address would include: was risk notably related to age at exposure and/or age at follow-up, job category, rank or rate, pay grade, participation at more than one nuclear weapons tests (some veterans participated in 13 test series), to radiation exposure (as an indicator perhaps of a stressful job category (or occupation), such as 'cloud samplers' flying through radioactive clouds or radiation safety specialists), to length of service or service during World War II (which is likely for many since the OPERATION CROSSROADS test series was in 1946, and four other test series were conducted in the early 1950s), to service-connected disability, and perhaps other factors that could be identified within BIRLS (Fisher et al. 1995; Kang et al. 2015) and within the Patient Treatment Files and other databases within the VA medical health care system (Boyko et al. 2000). Further, previously conducted psychological studies of troop reactions to atomic explosions at test sites could be evaluated in light of subsequent risk of suicide (Bordes et al. 1953).

An additional 10 years of follow-up for mortality will be conducted through 2020. Contributing causes of death will be evaluated (such as mental and behavioral conditions), as will suicide as a possible 'other significant condition,' i.e., something that contributed to death but did not cause it such as a history of depression or a potentially fatal disease such as metastatic pancreatic cancer (Hanzlick 2006). For the over 50,000 atomic veterans who reached the age of 65 in or after 1999, Medicare and nursing home records would be searched for any clinically-based diagnoses of depression, PTSD, other medical health conditions and psychiatric disorders, dementia, motor neuron conditions, advanced-stage cancer, treatment for possible fatal diseases, as well as any evidence of non-fatal suicide attempts, suicide ideation, substance abuse, alcoholism, cognitive decline, quantitative scores from neuropsychological testing, and other factors or conditions related to health and well-being (Rector et al. 2004; CMS 2018a, 2018b; DVA 2019). The aim would be to identify new or confirm known factors associated with suicide among a veteran population followed for life, with particular focus on veterans over the age of 70.

Personal medical X-ray procedures

A unique feature of the Eight Series Study was the ability to address personal medical X-ray procedures such as CT examinations (Boice 2019a). Medical imaging procedures are frequent today, but much less so in the years covered by this study which began follow-up in 1945. The VA has an elaborate system of electronic records, including patient treatment files, that facilitated the evaluation of radiological procedures for all veterans with leukemia and lung cancer and a sample of the cohort (Fisher et al. 1995; Boyko et al. 2000). The frequency of CT examinations was the same for the leukemia cases and the cohort sample and did not differ by year of death. On the other hand, veterans who died of lung cancer had more CT examinations than the cohort sample. However, the increase was entirely due to procedures performed within 1–2 years of death, indicating that they were conducted for the purposes of diagnosis and treatment planning.

Strengths and limitations

The strengths of the study of nuclear weapons test participants include the large numbers of veterans (114,270), the nearly complete follow-up for vital status for up to 65 years after first participation at a weapons test (96.8%), and the comprehensive dose reconstructions (Till et al. 2014, 2018a; Beck et al. 2017). The dose reconstruction was based on substantial radiation exposure information available for all 230 individual aboveground nuclear weapons tests and for all military veterans. Access to the records within the NTPR program and the NuTRIS raised the quality of the dose estimation procedure to a high level. The follow-up was of exceptional high quality, given that the veteran participation at weapons testing spanned 1945–1962. The social security number was not used as the military identification number until 1968 so access to BIRLS (the VA Beneficiary Identification Records Locator Subsystem) was important to trace the military veterans. Access to the veteran medical record system was a strength in evaluating medical radiographic procedures.

It is also noteworthy that the complex and comprehensive dose reconstruction scheme was validated by biodosimetry (Simon et al. 2019; McKenna et al. 2019). A biodosimetry dose reconstruction study was conducted of servicemen who received relatively high-doses to active bone marrow (>200 mGy) from fallout from the BRAVO detonation (the first of six shots during the OPERATION CASTLE series in 1954) and at several other PPG and NTS detonations. Of the 59 atomic veterans with >200 mGy, 16 were alive and 12 agreed to be interviewed and have a blood sample drawn. Six veterans were weather observers stationed on Rongerik in 1954 who were exposed to direct fallout after the BRAVO detonation (DTRA 2019), two were close-in test observers, two were 'cloud samplers' (men who flew through radioactive clouds to obtain air and debris samples), and two provided technical or research support. To evaluate biological signatures of exposures that occurred over 60 years ago, three forms of dose estimation were compared: historical personnel monitoring data; dose reconstruction based on detailed exposure scenarios; and two chromosome aberration assays and a telomere length assay.

Dose to red bone marrow was estimated (Beck et al. 2017), and directional genomic hybridization (dGH) was used to detect chromosomal translocations and inversions (McKenna et al. 2019). Inversions were used for the first time in retrospective biodosimetry evaluations. Significant correlations were found between chromosomal translocations and inversions, taken together, with the doses estimated from film badge readings and from comprehensive scenario-specific dose reconstructions (Simon et al. 2019; McKenna et al. 2019). These results indicated that biological signatures of exposure were still present 60 years after radiation exposures to fallout, and that chromosomal translocations and inversions, taken together, were reliable retrospective biodosimeters. Perhaps equally important, the chromosomal indicators of dose were consistent and validated the dose reconstruction approach developed for organ dose estimation. One intriguing observation was that telomere length was inversely correlated with the estimates of radiation dose. The willingness of veterans to participate in blood draws for biodosimetry studies was a unique integration of epidemiology and biology.

A limitation of the study is that the mean organ doses were relatively small (the mean NuTRIS film badge gamma radiation dose was 6 mSv, ranged up to 908 mSv, and only a few veterans received organ-specific doses over 300 mGy), and thus the statistical power to identify any underlying relationships if they existed were accordingly limited. The large numbers and the exceptional dosimetry, however, provide rather narrow 95% confidence limits so that levels of risk higher than some value (ERR per 100 mGy) might be excluded with some confidence. For example, an ERR per 100 mGy higher than 0.33 could be excluded with 95% confidence for leukemia other than CLL. The corresponding ERRs per 100 mGy that could be excluded with 95% confidence for lung cancer is 0.19, for MDS 1.23, for multiple myeloma 0.72, for prostate cancer 0.27 and for IHD 0.11. The data from the Eight Series Study also will provide a strong 'low-dose' component to the future combination of the cohorts within the MPS (see Boice et al. 2019b).

Pay grade (enlisted/officer) was the indicator of SES used as a proxy measure of lifestyle and behavior characteristics, such as tobacco use. Pay grade did predict higher rates of lung cancer and other causes of death among enlisted men compared with officers, indicating the importance of adjustment. Nonetheless, a limitation, as in practically all large scale occupational studies, is the inability to adjust for known (and unknown) determinants of disease for which pay grade as a measure of SES may be inadequate.

Another limitation is uncertainty associated with organ doses from intakes of radionuclides. While the number of veterans with 'potential' for internal exposure appears small relative to the entire cohort, for some organs, such as the lung, the organ dose might be of the order of the external dose for some veterans. It is apparent that for many organs, however, that do not concentrate radionuclides, there is a minimal contribution to organ dose, such as bone marrow, brain, kidneys, pancreas, spleen, testes, breast, heart, and bladder (Till et al. 2018a). Nonetheless, an evaluation is ongoing to evaluate the internal organ dose contribution to total dose considering knowledge gained from our initial evaluations (Till et al. 2014, 2018a; Beck et al. 2017), from previous evaluations of urine samples taken among nuclear weapons test participants over the years and from other sources indicating the potential for internal exposures (Phillips et al. 1985; Schaeffer et al. 2007; Weitz et al. 2009; Harris et al. 2010; Case et al. 2011; DTRA 2015; Till et al. 2018a). Although radioactive iodine (^{131}I) is a potential source of internal exposure to the thyroid gland, because the adult thyroid gland has a low susceptibility to radiation-related cancer (Dickman et al. 2003; Furukawa et al. 2013), evaluations were not informative. There were only 54 deaths from thyroid cancer (SMR 0.90), risk was not seen to increase over time, radiation doses were low, and mortality is not a good indicator of radiation associations for thyroid cancer because of favorable survival characteristics.

Summary

In summary, there was no evidence that the low doses of radiation experienced by atomic veterans increased the risk of cancer or ischemic heart disease. Mesothelioma and prostate cancer, however, were significantly elevated but not related to radiation dose. The mesothelioma excess could be explained by asbestos exposure aboard ships at the PPG. The emergence of increased mortality from suicide 40 or more years after participation at a nuclear weapons test deserves further study. The study of atomic veterans exposed to radioactive fallout should be relevant to understanding the consequences of possible terrorist attacks with dirty bombs or improvised nuclear devices. The study results may help inform ongoing compensation programs for atomic veterans, radiation workers, and 'down-winders.' The study of nuclear weapons test participants provides sound information to the veteran and his family on the possible health risks related to service to our country.

Disclosure statement

The authors report no conflicts of interest. The authors alone are responsible for the content and writing of the paper.

Funding

This research was supported in part by contracts and grants from the National Cancer Institute [Grant No. U01 CA137026]; the U.S. Department of Energy [Grant No. DE-SC0008944 awarded to the National Council on Radiation Protection and Measurements], which included interagency support from the U.S. Nuclear Regulatory Commission, the U.S. Environmental Protection Agency and the National Aeronautics and Space Administration; and a Discovery Grant from the Vanderbilt-Ingram Cancer Center (Center No. 404-357-9682). The biodosimetry work was supported by the NCI Intramural Research Program through an interagency agreement with the Radiation Nuclear Countermeasures Program of the National Institute of Allergy and Infectious Diseases (NIAID) under NIAID agreement #Y2-Al-5077 and NCI agreement #Y3-CO-5117. We acknowledge Dr. Paul Blake, Chief, Nuclear Test Personnel Review, Defense Threat Reduction Agency, U.S. Department of Defense and his staff for their technical support of the project, and Hanson Gaugler at L-3 Services for his critical contributions to our accessing and interpreting the comprehensive data sources necessary for study conduct. Similarly, Han Kang and Tim Bullman, Environmental Epidemiology Service, U.S. Department of Veterans Affairs, were instrumental in providing support and assistance throughout the conduct of the Eight Series Study of atomic veterans. We also are indebted to: the Department of Energy (Nimi Rao), the Nuclear Regulatory Commission (Doris Lewis), Oak Ridge Associated Universities (Derek A. Hagemeyer), Landauer, Inc. (Craig Yoder, PhD), the U.S. Army Dosimetry Center (William S. Harris, Jr., CHP), the U.S. Air Force Radiation Dosimetry Laboratory (Ms. Linda Wilson), and the Naval Dosimetry Center (CDR Anthony Williams and Lt Selena Hayes) for facilitating linkages with their respective dosimetry files. The findings and conclusions in this paper are those of the authors. Its publication does not necessarily represent the official position of or imply endorsement by the National Council on Radiation Protection and Measurements, Oak Ridge Associated Universities, Vanderbilt University, or any of the acknowledged agencies.

ORCID

John D. Boice http://orcid.org/0000-0002-8755-1299
Sarah S. Cohen http://orcid.org/0000-0003-0421-1983
Michael T. Mumma http://orcid.org/0000-0001-7506-8710
Ashley P. Golden http://orcid.org/0000-0002-9430-7702
Harold L. Beck http://orcid.org/0000-0003-4355-0507
John E. Till http://orcid.org/0000-0003-3666-8806

References

Beck HL, Till JE, Grogan HA, Aanenson JW, Mohler HJ, Mohler SS, Voillequé PG. 2017. Red bone marrow and male breast doses for a cohort of atomic veterans. Radiat Res. 187(2):221–228.

Blake PK, Komp GR. 2014. Radiation exposure of U.S. military individuals. Health Phys. 106(2):272–278.

Boice JD, Jr, Cohen SS, Mumma MT, Ellis ED. 2019a. The Million Person Study, whence it came and why. Int J Radiat Biol. 1–14. DOI:10.1080/09553002.2019.1589015

Boice JD, Jr, Ellis ED, Golden AP, Zablotska LB, Mumma MT, Cohen SS. 2019b. Sex-specific lung cancer risk among radiation workers in the Million Person Study and among TB-fluoroscopy patients. Int J Radiat Biol. 1–12. DOI:10.1080/09553002.2018.1547441

Boice JD, Jr, Leggett RW, Ellis ED, Wallace PW, Mumma M, Cohen SS, Brill AB, Chadda B, Boecker BB, Yoder RC, et al. 2006. A comprehensive dose reconstruction methodology for former Rocketdyne/Atomics International radiation workers. Health Phys. 90(5):409–430.

Boice JD. Jr. 2017. Space: the final frontier-research relevant to mars. Health Phys. 112(4):392–397.

Bordes PA, Finan JL, Hochstim JR, McFann HH, Schwartz SG. 1953. DESERT ROCK I: a psychological study of troop reactions to an atomic explosion. The Human Resources Research Organization (HumRRO) -TR-1. ACHRE No. CORP-111694-A. Available as Document Number 6252 from Nuclear Testing Archive, Mission Support & Test Services (MSTS), Las Vegas, NV, p. 89193–88521.

Bouville A, Toohey RE, Boice JD, Jr, Beck HL, Dauer LT, Eckerman KF, Hagemeyer D, Leggett RW, Mumma MT, Napier B, et al. 2015. Dose reconstruction for the million worker study: status and guidelines. Health Phys. 108(2):206–220.

Boyko EJ, Koepsell TD, Gaziano JM, Horner RD, Feussner JR. 2000. U.S. Department of Veterans Affairs medical care system as a resource to epidemiologists. Am J Epidemiol. 151(3):307–314.

Brenner AV, Sugiyama H, Preston DL, Sakata R, French B, Sadakane A, Cahoon EK, Utada M, Mabuchi K, Ozasa K. 2020. Radiation risk of central nervous system tumors in the Life Span Study of atomic bomb survivors, 1958–2009. Eur J Epidemiol. 35(6):591–600.

Bullman T, Hoffmire C, Schneiderman A, Bossarte R. 2015. Time dependent gender differences in suicide risk among operation enduring freedom and operation Iraqi freedom veterans. Ann Epidemiol. 25(12):964–965.

Bullman T, Schneiderman A, Bossarte R. 2018. Suicide risk by unit component among veterans who served in Iraq or Afghanistan. Arch Suicide Res. 22(1):1–10.

Bullman T, Schneiderman A, Gradus JL. 2019. Relative importance of posttraumatic stress disorder and depression in predicting risk of suicide among a cohort of Vietnam veterans. Suicide Life Threat Behav. 49(3):838–845.

Bullman T, Schneiderman A. 2019. Suicide risk among US peacekeepers serving in the Bosnia and Kosovo Theater, 1996-2002. Am J Epidemiol. 188(10):1768–1773.

Bullman TA, Kang HK. 2000. A 50 year mortality follow-up of veterans exposed to low level chemical warfare agent, mustard gas. Ann Epidemiol. 10(5):333–338.

Bullman TA, Mahan CM, Kang HK, Page WF. 2005. Mortality in US Army Gulf War veterans exposed to 1991 Khamisiyah chemical munitions destruction. Am J Public Health. 95(8):1382–1388.

Cahoon EK, Preston DL, Pierce DA, Grant E, Brenner AV, Mabuchi K, Utada M, Ozasa K. 2017. Lung, laryngeal and other respiratory cancer incidence among Japanese atomic bomb survivors: an updated analysis from 1958 through 2009. Radiat Res. 187(5): 538–548.

Caldwell GG, Kelley D, Zack M, Falk H, Heath CW. Jr. 1983. Mortality and cancer frequency among military nuclear test (SMOKY) participants, 1957 through 1979. JAMA. 250(5):620–624.

Caldwell GG, Zack M, Mumma M, Falk H, Heath CW, Till JE, Chen H, Boice JD. Jr. 2016. Mortality among military participants at the 1957 PLUMBBOB nuclear weapons test series and from leukemia among participants at the SMOKY test. J Radiol Prot. 36(3): 474–489.

Cardis E, Gilbert E, Carpenter L, Howe G, Kato I, Armstrong B, Beral V, Cowper G, Douglas A, Fix J. 1995. Effects of low doses and low dose rates of external ionizing radiation: cancer mortality among nuclear industry workers in three countries. Radiat Res. 142(2): 117–132.

Cardis E, Vrijheid M, Blettner M, Gilbert E, Hakama M, Hill C, Howe G, Kaldor J, Muirhead CR, Schubauer–Berigan M, et al. 2007. The 15-country collaborative study of cancer risk among radiation workers in the nuclear industry: estimates of radiation-related cancer risks. Radiat Res. 167(4):396–416.

Case DR, Chehata M, Dunavant JD, Egbert SD, Mason CL, McKenzie-Carter MA, Singer HA, Weitz RL. 2011. A technical approach to expedited processing of NTPR radiation dose assessments. Fort Belvoir (VA): Defense Threat Reduction Agency; Report No: DTRA-TR-10-29. [accessed 2019 Nov 15]. https://www.dtra.mil/Portals/61/Documents/NTPR/4-Rad_Exp_Rpts/27_DTRA-TR-10-29_Technical_Basis_for_NTPR_Expeditied_RDA.pdf.

[CMS] Centers for Medicare and Medicaid Serves. 2018a. Minimum Data Set 3.0 Public Reports. [accessed 2019 Oct 12]. https://www.cms.gov/Research-Statistics-Data-and-Systems/Computer-Data-and-Systems/Minimum-Data-Set-3-0-Public-Reports/index.html

[CMS] Centers for Medicare and Medicaid Services. 2018b. Long-Term Care Facility Resident Assessment Instrument 3.0 User's Manual. Version 1.16. October 2018. [accessed 2019 Oct 12]. https://downloads.cms.gov/files/1-MDS-30-RAI-Manual-v1-16-October-1-2018.pdf

Cox DR. 1972. Regression models and life-tables. J R Stat Soc B. 34(2): 187–220.

Dalager NA, Kang HK, Mahan CM. 2000. Cancer mortality among the highest exposed US atmospheric nuclear test participants. J Occup Environ Med. 42(8):798–805.

Dauer LT, Bouville A, Toohey RE, Boice JD, Jr, Beck HL, Eckerman KF, Hagemeyer D, Leggett RW, Mumma MT, Napier B, et al. 2018. Dosimetry and uncertainty approaches for the million-worker study of radiation workers and veterans: overview of the recommendations in NCRP Report No. 178. Int J Rad Biol. 1–10. DOI:10.1080/09553002.2018.1536299

[DTRA] Defense Threat Reduction Agency. 2015. Plutonium Bioassay Testing of U.S. Atmospheric Nuclear Test Participants and U.S. Occupation Forces of Hiroshima and Nagasaki, Japan. DTRA-TR-15-058. [accessed 2019 Nov 15]. https://www.dtra.mil/Portals/61/Documents/NTPR/4-Rad_Exp_Rpts/42_DTRA-TR-15-058_Pu_Bioassay_Rpt.pdf

[DTRA] Defense Threat Reduction Agency. 2019. Nuclear Test Personnel Review (NTPR) and Fact Sheets: Operation CASTLE, Operation CROSSROADS, Operation GREENHOUSE, Operation PLUMBBOB, Operation REDWING, Operation UPSHOT-KNOTHOLE, Operation HARDTACK I, Project TRINITY. [accessed 2019 Oct 07]. https://www.dtra.mil/Mission/Nuclear-Test-Personnel-Review/NTPR-Fact-Sheets/.

[DTRA] Defense Threat Reduction Agency. 2020. Design Aspects of DTRA's NuTRIS-Web Software Application. Fort Belvoir, VA: DTRA-TR-19-039.

Dickman PW, Holm LE, Lundell G, Boice JD, Jr, Hall P. 2003. Thyroid cancer risk after thyroid examination with 131I: a population-based cohort study in Sweden. Int J Cancer. 106(4):580–587.

Doll R, Sharp C, Muirhead C, Darby S. 1998. Study of UK men who had participated in the UK nuclear weapons test programme. J Radiol Prot. 18(3):209–210.

Fanning JR, Pietrzak RH. 2013. Suicidality among older male veterans in the United States: results from the National Health and Resilience in Veterans Study. J Psychiatr Res. 47(11):1766–1775.

Fehner TR, Gosling FG. 2006. Battlefield of the cold war: the Nevada test site. Atmospheric Nuclear Weapons Testing, 1951-1963 (DOE/MA-0003). Washington, DC: Office of History and Heritage Resources, Department of Energy.

Fisher SB, Weber L, Goldberg J, Davis F. 1995. Mortality ascertainment in the veteran population: alternatives to the National Death Index. Am J Epidemiol. 141(3):242–250.

Furukawa K, Preston D, Funamoto S, Yonehara S, Ito M, Tokuoka S, Sugiyama H, Soda M, Ozasa K, Mabuchi K. 2013. Long-term trend of thyroid cancer risk among Japanese atomic-bomb survivors: 60 years after exposure. Int J Cancer. 132(5):1222–1226.

Gillies M, Haylock R, Hunter N, Zhang W. 2019. Risk of leukemia associated with protracted low-dose radiation exposure: updated results from the National Registry for Radiation Workers study. Radiat Res. 192(5):527–537.

Gladeck F, Johnson A. 1996. For the record-a history of the nuclear test personnel review program, 1978-1993. Washington, DC: Defense Nuclear Agency. DNA-604IF. [accessed 6 April 2020] https://www.dtra.mil/Portals/61/Documents/NTPR/4-Rad_Exp_Rpts/38_DNA-6041F_History_of_the_NTPR_Program_1978-1993.pdf

Golden AP, Cohen SS, Chen H, Ellis ED, Boice JD. Jr. 2018. Evaluation of statistical modeling approaches for epidemiologic studies of low-dose radiation health effects. Int J Radiat Biol. :1–8.

Grant EJ, Brenner A, Sugiyama H, Sakata R, Sadakane A, Utada M, Cahoon EK, Milder CM, Soda M, Cullings HM, et al. 2017. Solid cancer incidence among the life span study of atomic bomb survivors: 1958-2009. Radiat Res. 187(5):513–537.

Gun RT, Parsons J, Crouch P, Ryan P, Hiller JE. 2008. Mortality and cancer incidence of Australian participants in the British nuclear tests in Australia. Occup Environ Med. 65(12):843–848.

Hagemeyer D, Nichols G, Mumma MT, Boice JD, Jr, Brock TA. 2018. 50 years of the radiation exposure information and reporting system and importance to the Million Person Study. Int J Radiat Biol. 1–4. DOI:10.1080/09553002.2018.1540896

Hanzlick R (ed). 2006. Cause of death and the death certificate - Important information for physicians, coroners, medical examiners, and the public. Northfield, IL: College of American Pathologists.

Harris PS, Simon SL, Ibrahim SA. 2010. Urinary excretion of radionuclides from Marshallese exposed to fallout from the 1954 Bravo nuclear test. Health Phys. 99(2):217–232.

Hsu WL, Preston DL, Soda M, Sugiyama H, Funamoto S, Kodama K, Kimura A, Kamada N, Dohy H, Tomonaga M, et al. 2013. The incidence of leukemia, lymphoma and multiple myeloma among atomic bomb survivors: 1950-2001. Radiat Res. 179(3):361–382.

[IOM] Institute of Medicine. 1996. Mortality of veteran participants in the crossroads nuclear test. Washington, DC: National Academy

Press. [accessed 2019 Oct 13] https://www.nap.edu/catalog/5428/mortality-of-veteran-participants-in-the-crossroads-nuclear-test

[IOM] Institute of Medicine. 2000. The five series study: mortality of military participants in U.S. nuclear weapons tests. Washington, DC: National Academy Press. [accessed 2019 Oct 12] https://www.nap.edu/catalog/9697/the-five-series-study-mortality-of-military-participants-in-us

Johnson JC, Thaul S, Page WF, Crawford H. 1997. Mortality of veteran participants in the CROSSROADS nuclear test. Health Phys. 73:187–189.

Kang HK, Bullman TA, Macfarlane GJ, Gray GC. 2002. Mortality among US and UK veterans of the Persian Gulf War: a review. Occup Environ Med. 59(12):794–799.

Kang HK, Bullman TA, Mahan CM. 2000. A mortality follow-up study of WWII submariners who received nasopharyngeal radium irradiation treatment. Am J Ind Med. 38(4):441–446.

Kang HK, Bullman TA, Smolenski DJ, Skopp NA, Gahm GA, Reger MA. 2015. Suicide risk among 1.3 million veterans who were on active duty during the Iraq and Afghanistan wars. Ann Epidemiol. 25(2):96–100.

Kang HK, Bullman TA, Taylor JW. 2006. Risk of selected cardiovascular diseases and posttraumatic stress disorder among former World War II prisoners of war. Ann Epidemiol. 16(5):381–386.

Kang HK, Bullman TA. 1996. Mortality among U.S. veterans of the Persian Gulf War. N Engl J Med. 335(20):1498–1504.

Kang HK, Bullman TA. 2009. Is there an epidemic of suicides among current and former U.S. military personnel? Ann Epidemiol. 19(10):757–760.

Kendall GM, Muirhead CR, Darby SC, Doll R, Arnold L, O'Hagan JA. 2004. Epidemiological studies of UK test veterans: I. General description. J Radiol Prot. 24(3):199–217.

Kerber RA, Till JE, Simon SL, Lyon JL, Thomas DC, Preston-Martin S, Rallison ML, Lloyd RD, Stevens W. 1993. A cohort study of thyroid disease in relation to fallout from nuclear weapons testing. JAMA. 270(17):2076–2082.

Kessler RC, Hwang I, Hoffmire CA, McCarthy JF, Petukhova MV, Rosellini AJ, Sampson NA, Schneider AL, Bradley PA, Katz IR, et al. 2017. Developing a practical suicide risk prediction model for targeting high-risk patients in the Veterans health Administration. Int J Methods Psychiatr Res. 26(3):28675617. [PMID

Leuraud K, Richardson DB, Cardis E, Daniels RD, Gillies M, O'Hagan JA, Hamra GB, Haylock R, Laurier D, Moissonnier M, et al. 2015. Ionising radiation and risk of death from leukaemia and lymphoma in radiation-monitored workers (INWORKS): an international cohort study. Lancet Haematol. 2(7):e276-281–e281.

Little MP, McElvenny DM. 2017. Male breast cancer incidence and mortality risk in the Japanese atomic bomb survivors - Differences in excess relative and absolute risk from female breast cancer. Environ Health Perspect. 125(2):223–229.

Marsh GM, Youk AO, Stone RA, Sefcik S, Alcorn C. 1998. OCMAP-PLUS: a program for the comprehensive analysis of occupational cohort data. J Occup Environ Med. 40(4):351–362.

Maynard C, Chapko MK. 2004. Data resources in the department of veterans affairs. Diabetes Care. 27(Supplement 2):B22–B26.

McCarthy JF, Bossarte R, Katz IR, Thompson C, Kemp J, Hannemann C, Nielson C, Schoenbaum M. 2015. Predictive modeling and concentration of the risk of suicide: implications for preventive interventions in the US Department of Veterans Affairs. Am J Public Health. 105(9):1935–1942.

McKenna MJ, Robinson E, Taylor L, Tompkins C, Cornforth MN, Simon SL, Bailey SM. 2019. Chromosome translocations, inversions and telomere length for retrospective biodosimetry on exposed U.S. Atomic Veterans. Radiat Res. 191(4):311–322.

McLaughlin R, Nielsen L, Waller M. 2008. An evaluation of the effect of military service on mortality: quantifying the healthy soldier effect. Ann Epidemiol. 18(12):928–936.

Monson RR. 1986. Observations on the healthy worker effect. J Occup Med. 28(6):425–433.

Muirhead CR, Bingham D, Haylock RGE, O'Hagan JA, Goodill AA, Berridge GLC, English MA, Hunter N, Kendall GM. 2003b. Mortality and cancer incidence 1952-1998 in UK participants in the UK atmospheric nuclear weapons tests and experimental programmes NRPB-W27. Oxon, UK: National Radiological Protection Board, p. 2003b.

Muirhead CR, Bingham D, Haylock RGE, O'Hagan JA, Goodill AA, Berridge GLC, English MA, Hunter N, Kendall GM. 2003a. Follow up of mortality and incidence of cancer 1952-98 in men from the UK who participated in the UK's atmospheric nuclear weapon tests and experimental programmes. Occup Environ Med. 60(3):165–172.

Muirhead CR, Kendall GM, Darby SC, Doll R, Haylock RG, O'Hagan JA, Berridge GL, Phillipson MA, Hunter N. 2004. Epidemiological studies of UK test veterans: II. Mortality and cancer incidence. J Radiol Prot. 24(3):219–241.

Muirhead CR, Kendall GM. 2003. UK nuclear-test veterans. Lancet. 362(9380):331–332.

Muirhead CR, O'Hagan JA, Haylock RG, Phillipson MA, Willcock T, Berridge GL, Zhang W. 2009. Mortality and cancer incidence following occupational radiation exposure: third analysis of the National Registry for Radiation Workers. Br J Cancer. 100(1):206–212.

Mumma MT, Cohen SS, Sirko JL, Ellis ED, Boice JD. Jr. 2018. Obtaining vital status and cause of death on a million persons. Int J Radiat Biol. 9:1–21.

Mumma MT, Sirko JL, Boice JD, Jr., Blot WJ. 2019. Mesothelioma mortality within two radiation monitored occupational cohorts. Int J Radiat Biol. 1–9. DOI:10.1080/09553002.2019.1642540

[NA/NRC] National Academies/National Research Council. 1989. Film badge dosimetry in atmospheric nuclear tests. Washington, DC: National Academies Press. [accessed 2019 Oct 13] https://www.nap.edu/catalog/1404/film-badge-dosimetry-in-atmospheric-nuclear-tests

[NA/NRC] National Academies/National Research Council. 2003. A review of the dose reconstruction program of the Defense Threat Reduction Agency. Washington, DC: National Academies Press.

[NA/NRC] National Academies/National Research Council. 2005. Assessment of the Scientific Information for the Radiation Exposure Screening and Education Program. Washington, DC: The National Academies Press. [accessed 2019 Oct 08] https://www.nap.edu/download/11279

[NCRP] National Council for Radiation Protection and Measurements. 2018a. Deriving organ doses and their uncertainty for epidemiologic studies (With a focus on the One Million U.S. Workers and Veterans Study of Low-Dose Radiation Health Effects). Bethesda, MD: NCRP.

[NCRP] National Council on Radiation Protection and Measurements. 2018b. Implications of recent epidemiologic studies for the linear nonthreshold model and radiation protection. Bethesda, MD:NCRP.

[NCRP] National Council on Radiation Protection and Measurements. 2018c. Management of exposure to ionizing radiation: radiation protection guidance for the United States. Bethesda, MD: NCRP.

Neton JW. 2014. Characterization of exposures to workers covered under the U.S. energy employees compensation act. Health Phys. 106(2):249–258.

Page WF, Mahan CM, Kang HK. 1996. Vital status ascertainment through the files of the Department of Veterans Affairs and the Social Security Administration. Ann Epidemiol. 6(2):102–109.

Pearce N, Winkelmann R, Kennedy J, Lewis S, Purdie G, Slater T, Prior I, Fraser J. 1997. Further follow-up of New Zealand participants in United Kingdom atmospheric nuclear weapons tests in the Pacific. Cancer Causes Control. 8(2):139–145.

Pédrono G, Pachol H, Ségala C. 2011. Mortality study among veterans with dosimeter monitoring during the French nuclear tests in the Pacific. Rev Epidemiol Sante Publique. 59(3):187–196.

Phillips J, Klemm J, Goetz J. 1985. Internal dose assessment—operation CROSSROADS. McLean, VA: Science Applications International Corporation. [assessed 2019 15 Nov] https://www.dtra.mil/Portals/61/Documents/NTPR/4-Rad_Exp_Rpts/25_DNA-TR-84-119_Internal_Dose_Assessment-Operation_CROSSROADS.pdf

Preston DL, Lubin J, Pierce D, McConney M, Shilnikova N. 2015. Epicure Risk Regression and Person-Year Computation Software: Command summary and user guide. Ottawa, Canada: Risk Sciences International.

Preston DL, Ron E, Yonehara S, Kobuke T, Fujii H, Kishikawa M, Tokunaga M, Tokuoka S, Mabuchi K. 2002. Tumors of the nervous system and pituitary gland associated with atomic bomb radiation exposure. J Natl Cancer Inst. 94(20):1555–1563.

Rector TS, Wickstrom SL, Shah M, Greeenlee NT, Rheault P, Rogowski J, Freedman V, Adams J, Escarce JJ. 2004. Specificity and sensitivity of claims-based algorithms for identifying members of Medicare + Choice health plans that have chronic medical conditions. Health Serv Res. 39(6 Pt 1):1839–1857.

Reger MA, Smolenski DJ, Skopp NA, Metzger-Abamukang MJ, Kang HK, Bullman TA, Perdue S, Gahm GA. 2015. Risk of suicide among US military service members following operation enduring freedom or operation Iraqi freedom deployment and separation from the US military. JAMA Psychiatry. 72(6):561–569.

Reger MA, Smolenski DJ, Skopp NA, Metzger-Abamukang MJ, Kang HK, Bullman TA, Gahm GA. 2018. Suicides, homicides, accidents, and undetermined deaths in the U.S. military: comparisons to the U.S. population and by military separation status. Ann Epidemiol. 28(3):139–146.

Reger MA, Smolenski DJ, Skopp NA, Metzger-Abamukong MJ, Kang HK, Bullman TA, Gahm GA. 2017. Suicide risk among wounded U.S. service members. Suicide Life Threat Behav. 47(2):242–247.

Richardson DB, Cardis E, Daniels RD, Gillies M, Haylock R, Leuraud K, Laurier D, Moissonnier M, Schubauer-Berigan MK, Thierry-Chef I, et al. 2018. Site-specific solid cancer mortality after exposure to ionizing radiation: a cohort study of workers (INWORKS). Epidemiology. 29(1):31–40.

Robinette CD, Jablon S, Preston TL. 1985. Mortality of nuclear weapons test participants. Washington, DC: National Academy Press.

Ron E, Ikeda T, Preston DL, Tokuoka S. 2005. Male breast cancer incidence among atomic bomb survivors. J Natl Cancer Inst. 97(8): 603–605.

Schaeffer DM, Case DR, Ingraham JT, Blake PK. 2007. Plutonium bioassay testing of United States atmospheric nuclear test participants and occupation forces of Hiroshima and Nagasaki, Japan. Health Phys. 113:246–251.

Schubauer-Berigan MK, Daniels RD, Bertke SJ, Tseng CY, Richardson DB. 2015. Cancer mortality through 2005 among a pooled cohort of U.S. nuclear workers exposed to external ionizing radiation. Radiat Res. 183(6):620–631.

Schubauer-Berigan MK, Macievic GV, Utterback DF, Tseng CY, Flora JT. 2005. An epidemiologic study of mortality and radiation-related risk of cancer among workers at the Idaho National Engineering and Environmental Laboratory, a U.S. Department of Energy Facility. Health-Related Energy Research Branch. Cincinnati, OH: National Institute for Occupational Safety and Health(NIOSH). HHS/NIOSH Publication No. 2005-131. [accessed 2019 Oct 10]. https://www.cdc.gov/niosh/docs/2005-131/pdfs/2005-131.pdf.

Seltzer CC, Jablon S. 1974. Effects of selection on mortality. Am J Epidemiol. 100(5):367–372.

Seltzer CC, Jablon S. 1977. Army rank and subsequent mortality by cause: 23-year follow-up. Am J Epidemiol. 105(6):559–566.

Sher L, Braquehais MD, Casas M. 2012. Posttraumatic stress disorder, depression, and suicide in veterans. Cleve Clin J Med. 79(2):92–97.

Simon SL, Bailey SM, Beck HL, Boice JD, Bouville A, Brill AB, Cornforth MN, Inskip PD, McKenna MJ, Mumma MT, et al. 2019. Estimation of radiation doses to U.S. military test participants from nuclear testing: a comparison of historical film-badge measurements, dose Reconstruction and Retrospective Biodosimetry. Radiat Res. 191(4):297–310.

Smith TC, Jimenez DK, Smith B, Gray GC, Hooper TI, Gackstetter GD, Heller JM, Dalager NA, Kang HK, Hyams KC, et al. 2004. The postwar hospitalization experience of Gulf War veterans participating in US health registries. J Occup Environ Med. 46:386–397.

Stevens W, Thomas DC, Lyon JL, Till JE, Kerber RA, Simon SL, Lloyd RD, Elghany NA, Preston-Martin S. 1990. Leukemia in Utah and radioactive fallout from the Nevada test site. A case-control study. JAMA. 264(5):585–591.

Stram DO, Preston DL, Sokolnikov M, Napier B, Kopecky KJ, Boice J, Beck H, Till J, Bouville A. 2015. Shared dosimetry error in epidemiological dose-response analyses. PLoS One. 10(3):e0119418.

Thomas DB, Rosenblatt K, Jimenez LM, McTiernan A, Stalsberg H, Stemhagen A, Thompson WD, Curnen MG, Satariano W, Austin DF. 1994. Ionizing radiation and breast cancer in men (United States). Cancer Causes Control. 5(1):9–14.

Thomas TL, Kang HK, Dalager NA. 1991. Mortality among women Vietnam veterans, 1973-1987. Am J Epidemiol. 134(9):973–980.

Till JE, Beck HL, Aanenson JW, Grogan HA, Mohler HJ, Mohler SS, Voillequé PG. 2014. Military participants at U.S. atmospheric nuclear weapons testing-methodology for estimating dose and uncertainty. Radiat Res. 181(5):471–484.

Till JE, Beck HL, Aanenson JW, Grogan HA, Mohler HJ, Mohler SS, Voillequé PG. 2018a. Dosimetry associated with veterans who participated in nuclear weapons testing. Int J Radiat Biol. 1–9. DOI:10.1080/09553002.2018.1551639

Till JE, Beck HL, Boice JD, Jr, Mohler HJ, Mumma MT, Aanenson JW, Grogan HA. 2018b. Asbestos exposure and mesothelioma mortality among atomic veterans. Int J Rad Biol. 1–5. DOI:10.1080/09553002.2018.1551641

[DVA] U.S. Department of Veterans Affairs. 2011a. New study of atomic veterans who participated in above-ground nuclear testing. Ionizing Radiation Review Newsletter. 5 (1).

[DVA] U.S. Department of Veterans Affairs. 2011b. Veterans' Advisory Board on Dose Reconstruction (VBDR). A Special Supplement to the Ionizing Radiation Review Newsletter, February 2011. [accessed April 7, 2020]. https://www.publichealth.va.gov/docs/radiation/radiation-vbdr-insert-feb2011.pdf

[DVA] U.S. Department of Veterans Affairs. 2019. VA Mental Health Programs and Suicide Prevention Services. 2019 National Veteran Suicide Prevention Annual Report. Washington, DC: Department of Veterans Affairs. [accessed 2019 Oct 11]. https://www.mentalhealth.va.gov/docs/data-sheets/2019/2019_National_Veteran_Suicide_Prevention_Annual_Report_508.pdf

[UNSCEAR] United Nations Scientific Committee on the Effects of Atomic Radiation 2008. Effects of ionizing radiation. UNSCEAR 2006 report. Report to the general assembly. Volume 1. Scientific annex A. Epidemiological studies of radiation and cancer. Publication E.08.IX.6. New York, NY: United Nations.

[UNSCEAR] United Nations Scientific Committee on the Effects of Atomic Radiation. 2013. Sources, Effects and Risks of Ionizing Radiation. UNSCEAR 2013 report. Report to the general assembly. Volume II. Scientific annex B. Effects of radiation exposure of children. Publication E.14.IX.2. New York, NY: United Nations.

[VBDR] Veterans' Advisory Board on Dose Reconstruction. 2016. About the Veterans' Advisory Board on Dose Reconstruction. [accessed April 7, 2020] http://www.vbdr.org

Watanabe KK, Kang HK, Dalager NA. 1995. Cancer mortality risk among military participants of a 1958 atmospheric nuclear weapons test. Am J Public Health. 85(4):523–527.

Watanabe KK, Kang HK, Thomas TL. 1991. Mortality among Vietnam veterans: with methodological considerations. J Occup Med. 33(7): 780–785.

Watanabe KK, Kang HK. 1995. Military service in Vietnam and the risk of death from trauma and selected cancers. Ann Epidemiol. 5(5):407–412.

Weitz RL, Case DR, Chehata M, Egbert SD, Mason CL, Singer HA, Martinez DG, McKenzie-Carter MA, Shaw RS, Stiver JS. 2009. A Probabilistic approach to Uncertainty Analysis in NTPR Radiation Dose Assessments. McLean, VA: Science Applications International Corporation.

Yoder RC, Dauer L, Balter S, Boice JD, Jr, Grogan H, Mumma M, Passmore CN, Rothenberg LN, Vetter RJ. 2018. Dosimetry for the study of medical radiation workers with a focus on the mean absorbed dose to the lung, brain and other organs. Int J Radiat Biol. 1–12. DOI:10.1080/09553002.2018.1549756

Updated mortality analysis of the Mallinckrodt uranium processing workers, 1942–2012

Ashley P. Golden, Elizabeth D. Ellis, Sarah S. Cohen, Michael T. Mumma, Richard W. Leggett, Phillip W. Wallace, David J. Girardi, Janice P. Watkins, Roy E. Shore, and John D. Boice

ABSTRACT

Purpose: Mallinckrodt Chemical Works (MCW) was the earliest uranium processing facility in the United States, and in 1942 produced the uranium oxide used for the first sustained and controlled nuclear fission chain-reaction at the University of Chicago. A second follow-up through 2012 was conducted of 2514 White male workers employed 1942–1966 at the MCW for dose-response analyses for selected causes of death.

Materials and methods: Organ/tissue-specific dose reconstruction included both external (12,686 MCW film badge records, 210 other facility film badge records, and 31,297 occupational chest x-rays) and internal sources of uranium and radium (39,451 urine bioassays, 2341 breath radon measurements, and 6846 ambient radon measurements). Dust measurements from pitchblende facilitated quantitative risk estimates for non-radiogenic effects on the lung and kidney. Vital status was determined from multiple sources including the National Death Index and the Social Security Administration. Cox regression models were used for dose response analyses.

Results: Vital status was determined for 99% of the workers, of whom 75% had died. The mean lung dose from all sources of external and internal radiation combined was 69.9 mGy (maximum 885 mGy; percent workers >100 mGy, 10%) and there was no evidence for a dose response for lung cancer (Hazard Ratio (HR) of 0.95 (95% CI = 0.81–1.12) at 100 mGy). A significant association with radiation was found for kidney cancer (HR of 1.73 (95% CI = 1.04–2.79) at 100 mGy) and suggested for nonmalignant kidney diseases (HR of 1.30 (95% CI = 0.96–1.76) at 100 mGy). A non-radiation etiology could not be discounted, however, because of the possible renal toxicities of uranium, a heavy metal, and silica, a component of pitchblende dust. Non-significant HRs at 100 mGy for other sites of *a priori* interest were 0.36 (0.06–2.03) for leukemia other than CLL, 0.68 (0.17–2.77) for liver cancer, and 1.23 (0.79–1.90) for non-Hodgkin lymphoma. The HR at 100 mGy was 1.09 (0.99–1.20) for ischemic heart disease. An association was seen between dust and combined malignant and non-malignant lung disease, HR at 10 $mgm^{-3}year^{-1}$ of 1.01 (1.00–1.02).

Conclusions: A positive radiation dose response was observed for malignant and non-malignant kidney disease, and a negative dose response for malignant and non-malignant lung disease. Cumulative measures of dust were significantly associated with malignant and non-malignant lung disease and suggested for malignant and non-malignant kidney disease. Small numbers preclude definitive interpretations which will await the combination with similar studies of early uranium processing workers.

Introduction

The contribution of the Mallinckrodt Chemical Works (MCW) in the development of the atomic weapons that ended World War II was critical and often unappreciated. In 1942, Arthur Holly Compton, the Nobel-prize winning physicist, knew that ether, a dangerous and explosive solvent, would be required to refine the large amount of uranium ore needed to produce uranium metal and oxides (USACE 1996; Westbrook and Bloom 2010). He approached Edward

Mallinckrodt Jr, the president of MCW, to do this processing because MCW had a history of safely working with ether. Mallinckrodt agreed to support the war effort, and the uranium processing began in July 1942. On 2 December 1942, the entire 40 tons of uranium oxide produced at the MCW facility in St. Louis, MO, were used at the University of Chicago by Enrico Fermi to produce the first man-made sustained and controlled nuclear fission chain-reaction – and the atomic age was born.

Uranium was processed at MCW's St. Louis facilities from 1942 through 1957 and then moved to Weldon Spring, MO, where operations continued until shut down at the end of 1966. For MCW to produce the 40 tons of uranium oxide needed for the Manhattan Project, hundreds of tons of pitchblende from the then Belgian Congo (currently Democratic Republic of the Congo) were processed. In 1940, Belgium sent 1250 tons of pitchblende to Staten Island, New York, to prevent it from falling into the hands of Germany. During the early years of operation at MCW, pitchblende ore that was upwards of 60 percent pure uranium was processed.

The pitchblende ore contained up to 100 mg of radium per ton of ore, which increased to 300 mg per ton during processing. Both soluble and insoluble uranium compounds were produced during processing. Daily average dust concentrations of 100–200 times the maximum allowable concentration of 50 μgm^{-3} were measured in poorly ventilated processing areas. Throughout the processing steps, potential exposures were to internal and external radiation, as well as to silica and sulfuric acid (Eisenbud 1975). The chemical toxicity of uranium, a heavy metal, particularly with respect to damage to the kidneys, was also an industrial hygiene concern.

A previous study of 2514 White male workers at the MCW uranium processing plant employed 1942–1966 followed the workers through 1993 and considered only external radiation exposure (Dupree-Ellis et al. 2000). Significantly low standardized mortality ratios (SMR) were observed for all causes of death and for all circulatory diseases consistent with the healthy worker effect (Monson 1986). No cause of death was found to be significantly elevated. A dose response was indicated for kidney cancer with an excess relative risk (ERR) per Sv of 10.5 (90% confidence interval: 0.6, 57.4). Chronic nephritis was elevated but not significantly. Silica present in the pitchblende ore was suggested as a possible cause of kidney damage. Internal intakes of radium and uranium were not considered, nor were potential chemical exposures.

We extended and enhanced the original study of MCW uranium processing workers by adding 19 additional years of follow-up and by reconstructing organ/tissue-specific doses from intakes of radium and uranium using the latest biokinetic models. These organ/tissue doses were determined using 7 different sources of contemporary personal and area monitoring data: personal film badge readings during employment at MCW, annual chest X-rays required for employment, breath radon measurements, uranium urinalysis measurements, time-weighted job-specific analysis of exposure to dust, job-specific exposure to radon, and personal film badge data from pre- and post-employment at MCW (Watkins et al. 1997; Ellis et al. 2018). Using these sources, this study addresses directly the possible health effects from intakes of uranium and radium experienced about 60 years ago.

Outcomes of interest included the kidney because of the prior association with radiation (Dupree-Ellis et al. 2000) and because of potential damage from the chemical toxicity of uranium and silica present in the dust from pitchblende; the lung because of the inhalation of radium, uranium and silica; non-Hodgkin lymphoma (NHL) because the thoracic lymph nodes receive exposure from intakes of uranium and radium; liver and bone cancer because of exposure from internal emitters; leukemia other than CLL because the bone marrow is radiosensitive; ischemic heart disease because of current interest (Ozasa et al. 2017); brain cancer and combined dementia, Alzheimer's, Parkinson's, and motor neuron disease because of interest in alpha particle dose to brain (UNSCEAR 2017; Boice, Ellis, et al. 2018; Boice, Leggett, et al. 2018); and multiple myeloma, colon cancer, and mesothelioma because of reports in other uranium worker studies.

The MCW study is part of the Million US Worker and Veteran Study of low-dose health effects (MWS), sometimes called the Million Person Study. The MWS includes approximately 360,000 U.S. Department of Energy workers; 147,000 nuclear utility workers; 115,000 atomic veterans who participated in above-ground atmospheric tests; 170,000 radiologists and medical workers; 90,000 other workers including navy personnel; and 130,000 industrial radiographers (Bouville et al. 2015; Boice, Ellis, et al. 2018; NCRP 2018).

Materials and methods

Human subjects research approval was received from the Oak Ridge Site-wide Institutional Review Board

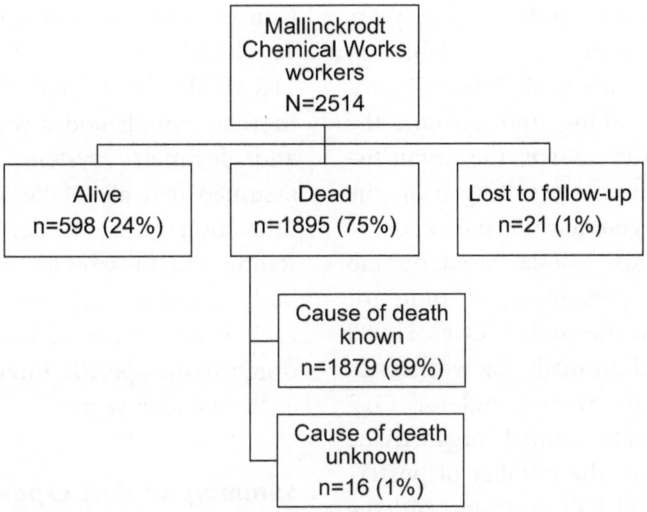

Figure 1. Vital status tracing of 2514 workers at the Mallinckrodt Chemical Works with follow-up through 31 December 2012.

and the Vanderbilt University Institutional Review Board.

Cohort definition

The study population was previously described by Dupree-Ellis et al. (2000). In brief, 2514 White males were employed on or after 1 January 1942 in the MCW Uranium Division located in St. Louis, MO and Weldon Spring, MO. All workers included in the study cohort were employed at least 30 days sometime between 1942 and 1966, which is the entire period of uranium processing operations at MCW.

Assignment of worker pay type (hourly or salary) was made using the first job title to establish a measure of socio-economic status (SES) as a surrogate for tobacco use (Sorensen et al. 2004). Each job title was assigned as hourly or salary first by identifying words in the job title that indicated either salary or hourly positions (e.g. manager, foreman, director were assigned salary; rigger, pipefitter, technician were assigned hourly), and then by individual inspection of job histories when identifying keywords were not present.

Vital status and outcome determination

The cohort had previously been traced for vital status through 1993 using the SSA Epidemiological Vital Status Service, Pension Benefit Information, and the National Death Index (NDI) (Dupree-Ellis et al. 2000). For this update, vital status as of 31 December 2012 was determined from linkages with the SSA Epidemiological Vital Status Service which confirmed that 598 workers (24%) were alive at the end of 2012 and NDI which confirmed that 1895 workers (75%) had died by the end of 2012 (Figure 1). In addition, the Centers for Disease Control and Prevention LinkPlus program, a probabilistic scoring system which does not require exact matches on all variables, was used to match the study roster against the SSA Death Master File and state mortality files (Campbell et al. 2008). Cause of death was determined for all but 16 (1%) of the 1895 workers who had died. Workers without a SSA, state mortality file or NDI match ($n = 21$, 1%) were assumed alive until the date of last contact which was usually the date of last employment at MCW.

Workers with serious renal disease were identified by linkage with the U.S. Renal Data System (1977–2008), which includes persons who received kidney dialysis or transplant (USRDS 2017).

Summary of radiation dose reconstruction

The methods for radiation dose reconstruction have been described (Ellis et al. 2018). Briefly, the approach to estimating occupational doses received by MCW employees followed the procedures outlined by Boice et al. (Boice, Cohen, et al. 2006; Boice, Leggett, et al. 2006; Boice et al. 2014) and the NCRP Report 178 (NCRP 2018). There were three sources for external dose estimates: personal film badge records and occupational chest X-rays from MCW, and individual film badge records for time periods of work at other facilities before or after MCW employment. Personal film badge records were obtained for 2024 workers included in the study (80% of the cohort) from the Manhattan Engineer District files and MCW occupational records from mid-1945 through 1966. Doses were imputed for those workers and working years were film badge records were not available as described previously (Dupree-Ellis et al. 2000). Organ/tissue doses were calculated taking into account the photon energy of the radiation source, the organ/tissue-specific conversion coefficients resulting from the energy (ICRP 2010), and the geometry of the exposure of the worker to the radiation source based on job groupings of the estimated percentage of time for anterior-posterior exposure to the source. Over 31,000 medical chest x-rays, required annually for workers in uranium processing operations, were found for 1995 workers and used to calculate annual organ/tissue doses, taking into account both the number of posterior-anterior (PA) and lateral (LAT) positions and the associated dose equivalents specific for the organ/tissue of interest given by Thomas (Thomas 2011). Two hundred and ten dosimetry records documenting radiation exposure received before or after employment at MCW were obtained for 43 workers from five additional sources: DOE Radiation Exposure Monitoring System (REMS) (DOE 2017); other historic DOE radiation exposure data not included in DOE REMS; the Nuclear Regulatory Commission Radiation Exposure Information and Reporting System (REIRS) (NRC 2018); Landauer, Inc.; and the Nuclear Test Personnel Review Program for military veterans who participated in the U.S. atmospheric nuclear tests from 1944 to 1962 or the occupation forces of Hiroshima and Nagasaki, Japan. All organ/tissue-specific doses from each source were added together to obtain the total organ/tissue-specific external dose received by each worker for each calendar year, as described by Ellis et al. (2018).

The important internal emitters at MCW were all naturally occurring radionuclides, primarily members of the ^{238}U and ^{235}U chains and in particular the elevated concentrations of airborne ^{238}U, ^{235}U, ^{234}U, ^{226}Ra, ^{222}Rn and its short-lived progeny. Detailed explanations for internal dose reconstructions are given by Ellis et al. (2018). Briefly, three sources of internal dosimetry were used to calculate organ/tissue-specific cumulative internal doses for MCW workers: 39,451 uranium urine samples available for 1925 workers from 1948 to 1966; 2341 radon breath samples available for 500 workers from 1948 to 1966, and 6846 ambient radon exposure measurements for 1392 workers available from 1944–1955. Workers without internal source records were considered unexposed. The biokinetic and dosimetric models used to estimate annual organ/tissue doses from internal sources in this study are those recommended in International Commission on Radiological Protection Publications 68 (ICRP 1994), 71 (ICRP 1995), 100 (ICRP 2006), 116 (ICRP 2010), and 137 (ICRP 2017). Dose to the thoracic lymph nodes was also based on ICRP models and dosimetry systems (ICRP 1994, 2006). It was assumed that all intakes of uranium or ^{226}Ra were via inhalation of moderately soluble (Type M) material. Radon breath samples were used to estimate radium body burden. All organ/tissue-specific doses from each source were added together to obtain the total organ/tissue-specific internal dose for each worker for each calendar year.

Summary of dust exposure

Because of the high levels of airborne dust in the uranium processing operations at MCW, there was concern about both the internal radiation exposure and the chemical toxicity of uranium and silica, a known important component of pitchblende, to organs such as the lung and kidney. General air and breathing zone sampling for dust were done frequently from 1942–1952, with worker-specific estimates of daily exposure to dust recorded in worksheets. These data were used to calculate job-specific daily time-weighted average dust exposures for 1091 of the 1457 workers employed during the time period that dust exposures were recorded as described in Ellis et al. (2018). Dust exposures were imputed based on job title and work history for the remaining 366 workers during this period. Because the internal intakes of uranium are best described by the available uranium bioassay data, the dust exposures were used to explore potential associations between lung and kidney outcomes and heavy metal toxicity.

Analytic methods

Standardized mortality ratio (SMR) analyses compared the numbers of deaths observed among MCW workers with the numbers expected based on general population rates in the United States for persons of the same age in five year groups, race (White) and sex (male) over the same 5-year time periods. The SMR analyses were based on the underlying cause of death. For all workers, person-time began 30 days after date of first hire at MCW or 1 January 1942, whichever was later, and ended at the date of death, age 95, date lost to follow-up, or 31 December 2012, whichever came first.

Within-cohort analyses (hereafter called internal analyses) were conducted using Cox proportional hazards models to compute risks across categories of estimated radiation dose to specific organ/tissues (Cox 1972). All Cox models were conducted using SAS/STAT software (SAS/STAT software, Version 9.4 of the SAS System for Windows, SAS Institute Inc., Cary, NC, USA). Excess Relative Risks at 100 mGy were estimated using the Peanuts program in Epicure (Preston et al. 2007). The internal analyses are considered more valid than the SMR analyses because potential biases associated with general population comparisons are eliminated and time-dependent internal radiation doses from intakes of radionuclides are incorporated. To increase statistical power, the cause-specific internal analyses included both the underlying and contributing causes of death obtained from the National Death Index and available death certificates for all sites except for lung cancer and ischemic heart disease. If a worker had both kidney cancer and non-malignant kidney disease, only kidney cancer was counted; similarly, for lung cancer and non-malignant respiratory disease excluding flu and pneumonia, only lung cancer was counted in the combined analyses. Year of birth and pay-type (hourly versus salary) were included in all models. Sensitivity analyses were conducted including adjustments for year of hire, duration of employment, socioeconomic status, and cumulative dust exposure.

For the internal cohort analyses, radiation dose was treated as a categorical measure, with categories defined based on the distribution of dose for each organ/tissue of interest. The dose category was then treated as a time-dependent measure, allowing workers to be assigned to increasingly higher dose categories over time as their individual radiation doses accrued. Doses accrued over time include both external doses and internal doses from the decay of radionuclides that had been retained in the body after inhalation or ingestion. Age was used as the underlying timescale for the hazard function. To allow for a possible latent period between radiation exposure and any effect consequent to it, doses were lagged, i.e. excluded if they occurred during some assumed interval prior to the event of interest, 10 years for solid cancers and 2 years for leukemia. Parameter estimates and standard errors for the dose categories in the Cox time-dependent regression models were used to obtain hazard ratios (HR) and 95% confidence intervals (CI) for death due to the cause under investigation compared with those in the referent group taken as the workers with low cumulative radiation dose. Trend tests treated the radiation dose as a single continuous measure, and two-sided p-values are presented. HRs at 100 mGy were computed for selected causes of death. In the primary analysis, a dose weighting factor of 1 was applied to alpha particle absorbed doses from uranium or radium-226, i.e. assuming that the absorbed dose was the relevant measure of biological effect. Additional analyses were performed assuming dose weighting factors of 10 or 20 as possible indicators of biological effectiveness for these alpha particle absorbed doses (data not shown). Excess relative risks (ERRs) also were estimated using the same model specifications as the Cox time-dependent regression model. ERRs at100 mGy are presented in the text and tables alongside the HR at 100 mGy.

Analyses of cumulative dust exposure in relation to kidney cancer, non-malignant kidney disease, lung cancer, and non-malignant lung disease were also conducted using Cox regression. Cumulative dust exposure was treated as a categorical measure using tertiles derived from the 1358 workers with dust measurements greater than zero; workers who had no dust exposure ($N=99$) were grouped with the lowest tertile of dust which was treated as the referent. Adjustment was made for pay type and year of birth in all models. Dust doses were not lagged but when adjustment was made for radiation dose to the kidney or lung, the radiation dose was lagged by 10 years. HRs and 95% CIs were calculated in the model using categories of dust, and at $10\,\text{mgm}^{-3}\text{year}^{-1}$ of dust exposure from a model treating dust as a continuous measure. ERR (95% CI) at $10\,\text{mgm}^{-3}\text{year}^{-1}$ were also estimated.

Results

All MCW workers included for study were White males and 83% were hourly wage earners. Demographic and occupational characteristics of the cohort are presented in Table 1. Many workers were born before 1920 (41%) and followed for more than 50 years (41%). Sixteen percent were hired between 1942 and 1945 (16%) and the mean duration of employment in the uranium division of MCW was 5.2 years. The mean age at start of follow-up was 30.3 years (range 16.1–65.3). Vital status was obtained for 99% of the population and 75% had died by the end of 2012 (Figure 1). The average follow-up was 43.3 years. Of the 2514 workers, 2459 (98%) had a film badge measurement greater than 0, and all who had chest X-rays were assumed to have exposures greater than 0. Of the 2160 workers for whom internal monitoring data were available, 1868 workers were monitored for uranium (1855 with dose >0), 487 for

Table 1. Demographic and occupational characteristics of 2514 workers employed during 1942–1966 at the Mallinckrodt Chemical Works.

	Total Population $N = 2{,}514$	
Characteristic	*n*	%
Sex-Male	2514	100%
Race-White	2514	100%
Pay type		
Hourly	2076	83%
Salary	438	17%
Year of Birth		
<1900	124	5%
1900-1909	264	11%
1910-1919	619	25%
1920-1929	858	34%
1930-1939	561	22%
1940-1947	88	4%
Year of Hire		
1942-1945	399	16%
1945-1954	1109	44%
1955-1966	1006	40%
Years of follow-up		
<20	232	9%
20–29	288	11%
30–39	395	16%
40–49	572	23%
50–59	809	32%
≥60	218	9%
Vital Status as of 12/31/2012		
Alive	598	24%
Dead	1895	75%
Lost to follow-up	21	1%
Radiation monitoring		
External - occupational (any)	2514	100%
External - medical (any)	2514	100%
Internal (any monitoring)	2160	86%
Internal (positive assay, dose >0)	2155	86%

radium (423 with dose >0), and 1356 for radon (all with dose >0) (Table 1). There were 43 (1.7%) workers monitored for radiation at other facilities either before or after employment at Mallinckrodt adding 575 mGy to the population dose.

Table 2 presents descriptive statistics for organ/tissue-specific radiation doses by different types of exposure. The mean external doses from film badges ranged from 27.9 mGy to 36.0 mGy for kidney and lung, respectively, while the highest external doses from occupational chest X-rays were for the heart, lung, and kidney. Thus, the highest combined external dose was observed in the lung (mean 46.7 mGy; maximum 778 mGy). Assuming a dose weighting factor of 1 for absorbed doses from internal exposures, the mean lung dose was 27.1 mGy (maximum 416 mGy), by far the highest internal exposure, primarily due to ambient radon. Except for the lung, the internal dose was a minor component of the total dose. The total combined external and internal doses are presented for dose weighting factors of 1, 10, and 20, where the highest organ doses are to the lung and thoracic lymph nodes.

Table 2. External and internal radiation doses to specific organs and tissues received by 2514 workers at the Mallinckrodt Chemical Works.

	workers				
Type of dose (mGy) Median	*N* Std	Max		Mean	
External - occupational[a]					
Lung	2514	36.0	11.2	68.9	724
Brain	2514	33.0	10.7	62.5	738
Heart	2514	35.6	11.0	68.4	711
Kidney	2514	27.9	8.9	53.1	604
Colon	2514	35.6	11.0	68.4	712
Red Bone Marrow	2514	32.9	10.3	62.9	676
Thoracic Lymph Nodes	2514	32.9	10.3	62.9	676
External - Medical x-ray (chest)					
Lung	2514	10.7	7.2	8.7	54
Brain	2514	4.2	3.0	3.6	20
Heart	2514	11.8	8.0	9.7	59
Kidney	2514	10.7	7.2	8.7	54
Colon	2514	2.3	1.8	1.9	12
Red Bone Marrow	2514	2.2	1.8	1.9	11
Thoracic Lymph Nodes	2514	10.0	6.7	8.1	50
Internal doses > 0[b]					
Lung	2155	27.1	11.2	39.2	416
Brain	2155	0.08	0.04	0.11	1
Heart	2155	0.09	0.04	0.12	1
Kidney	2155	0.73	0.41	0.91	12
Colon	2155	0.13	0.058	0.21	3
Red Bone Marrow	2155	0.57	0.18	1.05	10
Thoracic Lymph Nodes	2155	2.57	1.08	4.31	44
Total dose (external + internal)		Dose weighting factor = 1			
Lung	2514	69.9	33.1	103.0	885
Brain	2514	37.2	15.2	63.9	750
Heart	2514	47.5	23.3	73.0	738
Kidney	2514	39.2	20.2	57.9	637
Colon	2514	38.0	13.4	69.4	714
Red Bone Marrow	2514	35.6	12.9	64.3	683
Thoracic Lymph Nodes	2514	45.1	21.8	69.5	708
Total dose (external + internal)		Dose weighting factor = 10			
Lung	2514	277.4	102.4	426.6	4175
Brain	2514	37.7	15.6	64.4	750
Heart	2514	48.0	23.7	73.5	738
Kidney	2514	44.7	23.8	62.6	641
Colon	2514	38.7	13.9	70.6	717
Thoracic Lymph Nodes	2514	63.0	30.6	94.6	842
Total dose (external + internal)		Dose weighting factor = 20			
Lung	2514	508.0	178.0	796.7	8327
Brain	2514	38.3	16.1	65.0	750
Heart	2514	48.6	24.1	74.1	738
Kidney	2514	50.7	26.8	68.6	645
Colon	2514	49.0	24.3	74.6	736
Thoracic Lymph Nodes	2514	83.0	38.9	126.0	1138

[a]External occupational doses include film badge readings from Mallinckrodt and doses received at other facilities.
[b]2160 workers monitored for internal radiation. Only 5 workers who were monitored for internal radiation had 0 doses. Internal doses include uranium and radium-226 doses as well as radon and radon progeny doses.

Table 3 presents SMRs for all 2514 Mallinckrodt workers. Overall, the number of observed deaths was below that expected: 1894 deaths were observed and 2024.5 were expected (SMR 0.94; 95% CI 0.89–0.98). There were 488 deaths from cancer compared with 502.5 expected (SMR 0.97; 95% CI 0.89–1.06). The SMR for brain and central nervous system cancers was significantly elevated (SMR 1.85, 95% CI = 1.16–2.80) and the SMR for stomach cancer was

Table 3. Standardized mortality ratios (SMR) and 95% confidence intervals (CI) for cancer and other causes of death among 2514 Mallinckrodt Chemical Works workers with follow-up 1942–2012[a].

Cause of Death (ICD9)	Observed	Expected	SMR	95% CI
All Causes of Death (001–999)	1894	2,024.5	0.94*	0.89–0.98
All Malignant Neoplasms (140–208)	488	502.5	0.97	0.89–1.06
Buccal Cavity & Pharynx (140–149)	8	10.6	0.76	0.33–1.49
Esophagus (150)	15	13.7	1.09	0.61–1.80
Stomach (151)	7	14.9	0.47*	0.19–0.97
Colon (153)	46	42.2	1.09	0.80–1.45
Rectum (154)	8	9.0	0.89	0.38–1.76
Biliary Passages & Liver (155,156)	6	12.1	0.50	0.18–1.08
Pancreas (157)	29	25.7	1.13	0.76–1.62
Larynx (161)	5	5.7	0.87	0.28–2.04
Bronchus, Trachea, & Lung (162)	157	171.1	0.92	0.78–1.07
Breast (175)	–	0.6	0.00	0.00–5.90
Prostate (185)	48	48.3	0.99	0.73–1.32
Testes & Other Male Genital Organs (186, 187)	1	1.3	0.76	0.01–4.23
Kidney (189.0–189.2)	13	12.5	1.04	0.55–1.77
Bladder & Other Urinary (188, 189.3–189.9)	12	16.0	0.75	0.39–1.31
Melanoma of Skin (172)	10	7.8	1.28	0.61–2.36
Brain & CNS (191–192)	22	11.9	1.85*	1.16–2.80
Thyroid & Other Endocrine Glands (193–194)	1	1.4	0.70	0.01–3.89
Bone (170)	2	1.1	1.75	0.20–6.32
Connective & Other Soft Tissue (171)	3	2.4	1.23	0.25–3.59
All Lymphatic, Haematopoietic Tissue (200–208)	57	50.2	1.14	0.86–1.47
Non-Hodgkin Lymphoma (200; 202)	25	18.6	1.34	0.87–1.98
Hodgkin Lymphoma (201)	2	2.6	0.78	0.09–2.81
Leukemia & Aleukemia (204–208)	23	20.0	1.15	0.73–1.73
Chronic Lymphocytic Leukemia (CLL) (204.1)	5	4.7	1.07	0.35–2.50
Leukemia other than CLL	18	15.3	1.17	0.69–1.85
Multiple Myeloma (203)	6	8.5	0.71	0.26–1.54
Pleura & peritoneum (158.8, 158.9, 163) and mesothelioma (ICD10 C45)	2	2.1	0.95	0.11–3.41
AIDS (042–044, 795.8)	–	1.6	0.00	0.00–2.36
Smoking Related Cancers (140–150, 157, 161–162, 188–189)	239	255.4	0.94	0.82–1.06
Non-Smoking Related Cancers	249	247.1	1.01	0.89–1.14
Diabetes (250)	27	41.8	0.65*	0.43–0.94
Mental and Behavioral Disorders (290–319)	33	31.6	1.04	0.72–1.47
Diseases of the Nervous System (320–389)	72	63.3	1.14	0.89–1.43
Cerebrovascular Disease (430–438)	114	110.3	1.03	0.85–1.24
All Heart Disease (390–398, 404, 410–429)	648	728.4	0.89*	0.82–0.96
Ischemic Heart Disease (410–414)	521	567.0	0.92	0.84–1.00
Non-malignant Respiratory Disease (460–519)	176	187.3	0.94	0.81–1.09
Bronchitis, Emphysema, Asthma (490–493)	57	66.9	0.85	0.65–1.10
Cirrhosis of Liver (571)	32	35.3	0.91	0.62–1.28
Nephritis & Nephrosis (580–589)	31	24.4	1.27	0.86–1.80
All External Causes of Death (800–999)	98	119.2	0.82	0.67–1.00
Accidents (850–949)	72	80.1	0.90	0.70–1.13
Suicides (950–959)	21	29.6	0.71	0.44–1.09
Myelodysplastic Syndrome (238.7)	2	2.7	0.74	0.08–2.69
Polycythemia Vera (238.4)	–	0.4	0.00	0.00–8.73
Dementia & Alzheimer's Disease (290, 331)	50	42.4	1.18	0.88–1.55
Dementia, Alzheimer's, Parkinson's & Motor Neuron Diseases (290, 331, 332; 335.2)	71	60.7	1.17	0.91–1.48
Unknown Causes of Death	14	–	–	–.

[a]Total person-years of follow-up were 107,927; expected numbers for the referent population were based on the general population rates in the United States for persons of the same age, race (White) and sex (male) over the same time periods of follow-up.
*Indicates statistical significance at $p < .05$.

significantly below expectation (SMR 0.47, 95% CI = 0.19–0.97). The SMRs for cancer sites of *a priori* interest were not remarkable. Deaths due to all heart disease (SMR 0.89, 95% CI = 0.82–0.96) and ischemic heart disease (SMR 0.92, 95% CI = 0.84–1.00) were significantly below expectation. Rates of non-malignant respiratory disease and non-malignant renal disease were not significantly different from the general population, nor were dementia, Alzheimer's, Parkinson's, and motor neuron diseases.

Table 4 shows internal cohort dose-response analyses based on Cox proportional hazards models and excess relative risk models combining organ/tissue-specific doses from external radiation with internal radiation exposures from uranium and radium, as well as doses received before or after employment at MCW. All analyses presented in Table 4 assume a dose weighting factor of 1. Dose weighting factors of 10 and 20 for organ/tissue-specific alpha particle absorbed doses were also examined (results not

Table 4. Internal cohort dose-response analyses for select causes of death over categories of organ-specific radiation doses among 2514 workers at Mallinckrodt Chemical Works.

Dose (mGy)	Number of workers	Number of cases	HR	95% CI
Lung Cancer				
<10	703	46	1.00	REF
10–49	877	36	0.60	0.39, 0.94
50–99	422	37	1.05	0.68, 1.62
100–249	367	31	1.07	0.68, 1.70
≥250	145	7	0.59	0.26, 1.31
Total	2514	157	*p* for trend	*.56 (−)*
HR (95% CI) at 100 mGy				0.95 (0.81, 1.12)
ERR (95% CI) at 100 mGy				−0.06 (−0.18, 0.06)
Non-malignant respiratory disease excluding flu and pneumonia				
<10	703	52	1.00	REF
10–49	877	82	0.98	0.69, 1.39
50–99	422	48	0.91	0.61, 1.35
100–249	367	36	0.85	0.55, 1.31
≥250	145	20	1.26	0.75, 2.13
Total	2514	238	*p* for trend	*.30 (+)*
HR (95% CI) at 100 mGy				1.06 (0.95, 1.19)
ERR (95% CI) at 100 mGy				0.06 (−0.09, 0.21)
Combined Lung Disease (Cancer + Non-malignant respiratory disease)				
<10	703	98	1.00	REF
10–49	877	118	0.82	0.62, 1.07
50–99	422	85	0.95	0.71, 1.28
100–249	367	67	0.94	0.68, 1.28
≥250	145	27	0.96	0.62, 1.48
Total	2514	395	*p* for trend	*.72 (+)*
HR (95% CI) at 100 mGy				1.02 (0.93, 1.12)
ERR (95% CI) at 100 mGy				0.02 (−0,08, 0.12)
Brain Cancer				
<10	1066	12	1.00	REF
10–49	951	6	0.62	0.23, 1.67
≥50	497	4	0.86	0.26, 2.84
Total	2514	22	*p* for trend	.25 (-)
HR (95% CI) at 100 mGy				0.78 (0.29, 2.10)
ERR (95% CI) at 100 mGy				−0.13 (-0.55, 0.29)
Kidney cancer				
<10	885	4	*1.00*	REF
10–49	1079	8	*1.47*	0.44, 4.92
≥50	550	5	*1.93*	0.49, 7.58
Total	2514	17	*p* for trend	*.025 (+)*
HR (95% CI) at 100 mGy				1.73 (1.07, 2.79)
ERR (95% CI) at 100 mGy				Not presented due to lack of fit
Nephritis and Nephrosis				
<10	885	16	1.00	REF
10–49	1079	35	1.37	0.75, 2.48
50-99	319	12	1.13	0.53, 2.42
≥100	231	8	1.32	0.55, 3.15
Total	2514	71	*p* for trend	*.10 (+)*
HR (95% CI) at 100 mGy				1.30 (0.96, 1.76)
ERR (95% CI) at 100 mGy				0.34 (−0.31, 0.99)
Combined Kidney Disease (Cancer + Non-malignant kidney disease)				
<10	885	20	1.00	REF
10–49	1079	43	1.39	0.82, 2.37
50-99	319	13	1.08	0.53, 2.21
≥100	231	12	1.74	0.83, 3.65
Total	2514	88	*p* for trend	*.013 (+)*
HR (95% CI) at 100 mGy				1.39 (1.07, 1.80)
ERR (95% CI) at 100 mGy				0.52 (−0.19, 1.21)
Liver cancer				
<10	804	3	1.00	REF
10–49	1054	2	0.40	0.07-2.41
≥50	656	3	0.80	0.15-4.12
Total	2514	8	*p* for trend	*.59 (−)*
HR (95% CI) at 100 mGy				0.68 (0.17-2.77)
ERR (95% CI) at 100 mGy				−0.14 (−0.60, 0.32)
Colon cancer				
<10	1150	25	1.00	REF
10–49	859	23	1.01	0.57, 1.79
≥50	505	15	0.93	0.49, 1.78
Total	2514	63	*p* for trend	*.71 (+)*
HR (95% CI) at 100 mGy				1.06 (0.78, 1.43)

(Continued)

Table 4. Continued.

Dose (mGy)	Number of workers	Number of cases	HR	95% CI
ERR (95% CI) at 100 mGy				0.06 (−0.31, 0.43)
Ischemic Heart Disease				
<10	804	123	1.00	REF
10–49	1054	220	1.40	1.12, 1.76
50–99	361	86	1.15	0.87, 1.52
100–249	231	76	1.80	1.35, 2.40
≥250	64	16	1.24	0.74, 2.10
Total	2514	521	*p* for trend	*.07 (+)*
HR (95% CI) at 100 mGy				1.09 (0.99, 1.20)
ERR (95% CI) at 100 mGy				0.13 (−0.01, 0.28)
Alzheimer's/Dementia/Parkinson's Disease/Motor Neuron Disease				
<10	1066	30	1.00	REF
10–49	951	36	1.02	0.63, 1.66
50-99	263	17	1.23	0.67, 2.25
≥100	234	10	1.08	0.52, 2.25
Total	2514	93	*p* for trend	*.58 (−)*
HR (95% CI) at 100 mGy				0.91 (0.64, 1.29)
ERR (95% CI) at 100 mGy				−0.13 (−0.28, 0.02)
Leukemia other than CLL				
<10	1131	9	1.00	REF
10–49	901	6	0.76	0.26, 2.16
≥50	482	3	0.72	0.18, 2.81
Total	2514	18	*p* for trend	*.24 (−)*
HR (95% CI) at 100 mGy				0.36 (0.06, 2.03)
ERR (95% CI) at 100 mGy				−0.14 (−0.60, 0.33)
Non-Hodgkin Lymphoma				
<10	841	8	1.00	REF
10–49	1055	15	1.20	0.51, 2.84
≥50	618	7	1.00	0.34, 2.89
Total	2514	30	*p* for trend	*.36 (+)*
HR (95% CI) at 100 mGy				1.23 (0.79–1.90)
ERR (95% CI) at 100 mGy				0.20 (−0.23, 0.64)

All radiation doses have a dose weighting factor of 1 applied. Models include radiation doses lagged by 10 years for solid tumors and by 2 years for leukemia. Doses were analyzed using time-dependent covariates. Outcomes include deaths from underlying and contributing causes of death with the exception of ischemic heart disease and lung cancer. All models adjusted for pay type (hourly/salary) and year of birth. P-value for test for linear trend in the relative risk (i.e. hazard ratio) computed for continuous organ dose. *P*-values are two-sided. REF denotes the referent category. HR denotes Hazards Ratio. ERR denotes Excess Relative Risk.

shown) and minimally affected the estimates of risk; with the exception of lung cancer where the negative trends approached statistical significance. At 100 mGy, no statistically significant dose-response associations were seen for cancers of the lung, brain, or leukemia other than CLL. The risk estimates for ischemic heart disease were positive and of borderline significance. A positive and significant risk was observed for kidney cancer (HR 1.73, 95% CI = 1.07, 2.79), but risk estimates for nephritis and nephrosis were not significant. The ERR for kidney cancer is not presented in Table 4 due to significant lack of fit, such that the model would not converge. A combined analysis of both malignant and non-malignant kidney disease revealed an increased risk of borderline significance (HR 1.39, 95% CI = 1.07, 1.80). A similar combined analysis of both malignant and non-malignant lung disease was not remarkable. Because of the possibility that dust inhalation was a potential confounder, the models for kidney and lung cancer were adjusted for cumulative dust exposure. The risk for kidney cancer increased slightly (HR 1.85, 95% CI = 1.09, 3.14), while the risk for lung cancer decreased slightly (HR 0.93, 95% CI = 0.78, 1.11), though not significantly (data not shown).

The average annual cumulative dust exposure estimate was 15.08 $mgm^{-3}year^{-1}$, with the maximum average of 30.88 $mgm^{-3}year^{-1}$ in 1944 declining to 1.23 $mgm^{-3}year^{-1}$ in 1952. Internal cohort dose-response analyses with dust inhalation as the exposure of interest for kidney and lung outcomes are presented in Table 5. Dose-response relationships over cumulative dust categories for kidney cancer, non-malignant kidney disease, and a combined measure of both outcomes were not significant. The highest cumulative dust categories showed statistically significant elevations for non-malignant kidney disease and the combined malignant and non-malignant kidney disease category (Table 5). The dose responses for lung cancer, non-malignant disease, and the combined diseases were all positive, with the combined cancer and non-malignant disease category approaching statistical significance.

Linkages with the U.S. Renal Data System (USRDS) identified 23 former Mallinckrodt workers with end-stage renal disease (ESRD). The total radiation dose to

Table 5. Internal cohort dose-response analyses for kidney-related causes of death over categories of cumulative dust exposure among 1457 workers with potential for dust exposure at Mallinckrodt Chemical Works.

Cumulative dust exposure (mg/m^3-years)	Number of workers	Number of cases	HR	95% CI
Kidney Cancer				
0–3.77	551	1	1.00	REF
>3.77–23.61	452	2	2.75	0.24–31.38
>23.61	454	6	8.03	0.79–81.5
Total	1457	9	*p* for trend	*.72 (−)*
HR (95% CI) at 10 mg/m^3–year			*0.99 (0.93–1.06)*	
ERR (95% CI) at 10 mg/m^3–year			*0.22 (−0.70, 1.14)*	
Non-malignant kidney disease				
0–3.77	551	14	1.00	REF
>3.77–23.61	452	11	1.04	0.46–2.35
>23.61	454	12	2.52	1.14–5.60
Total	1457	37	*p* for trend	*.47 (−)*
HR (95% CI) at 10 mg/m^3–year			*0.99 (0.95–1.02)*	
ERR (95% CI) at 10 mg/m^3–year			*−0.04 (−0.25, 0.16)*	
Combined (Kidney cancer and Non-Malignant Kidney Disease)				
0–3.77	551	15	1.00	REF
>3.77–23.61	452	13	1.15	0.54–2.45
>23.61	454	18	2.86	1.36–5.98
Total	1457	46	*p* for trend	*.38 (−)*
HR (95% CI) at 10 mg/m^3–year			*0.99 (0.95-1.02)*	
ERR (95% CI) at 10 mg/m^3–year			*−0.001 (−0.24, 0.24)*	
Lung Cancer				
0–3.77	551	32	1.00	REF
>3.77–23.61	452	43	1.74	1.08–2.78
>23.61	454	27	1.29	0.72–2.30
Total	1457	102	*p* for trend	*.12 (+)*
HR (95% CI) at 10 mg/m^3–year			*1.01 (0.998–1.02)*	
ERR (95% CI) at 10 mg/m^3–year			*0.04 (−0.12, 0.21)*	
Non-malignant Lung disease				
0–3.77	551	47	1.00	REF
>3.77–23.61	452	50	1.35	0.89–2.04
>23.61	454	54	1.69	1.06–2.69
Total	1457	151	*p* for trend	*.24 (+)*
HR (95% CI) at 10 mg/m^3–year			*1.01 (0.995–1.02)*	
ERR (95% CI) at 10 mg/m^3–year			*0.15 (−0.06, 0.37)*	
Combined (Lung cancer and NMRD)				
0–3.77	551	79	1.00	REF
>3.77–23.61	452	93	1.50	1.10–2.05
>23.61	454	81	1.52	1.06–2.19
Total	1457	253	*p* for trend	*.06 (+)*
HR (95% CI) at 10 mg/m^3–year			*1.01 (1.00–1.02)*	
ERR (95% CI) at 10 mg/m^3–year			*0.09 (−0.05, 0.22)*	

Dust data not lagged. Doses were analyzed using time-dependent covariates. Outcomes include deaths from underlying and contributing causes. All models adjusted for SES, year of birth, and organ-specific (kidney or lung) radiation dose lagged 10 years. *p*-value for test for linear trend in the relative risk (i.e. hazard ratio) computed for continuous dust dose. *p*-values are two-sided. REF denotes reference category.

the kidney for these 23 workers ranged from 2.66 mGy to 183 mGy (dose weighting factor of 1). Nine of the 23 incident cases were attributed to hypertension, 6 to diabetes, 3 to glomeruleronephritis, 1 to cystic kidney disease, with 4 to a cause unknown. Among the 23 workers with ESRD, all but one had died by the end of study follow-up. Among these 22 deaths, no deaths from kidney cancer and nine deaths from non-malignant kidney disease were observed. Incidence rates from USRDS were used to calculate a standardized incidence ratio (SIR) of ESRD among the cases who were diagnosed from 1978 (the start of nearly full coverage of the USRDS registry) through end of follow-up ($N=22$ cases). The number of observed cases was 22 with 30.1 cases expected (SIR 0.73, 95% CI 0.46–1.10). A SIR analysis was also conducted by dust levels, as 11 of the 22 cases had measured dust level. However, no discernable pattern over dust concentration categories was observed. A SIR analysis was also conducted for ESRD attributed to glomerulonephritis; however, there were only 3 observed cases with 3.2 cases expected. The three cases of glomerulonephritis had estimated cumulative dust exposures of 0, 6.65, and 1055 mgm^{-3}year^{-1}.

Discussion

A second follow-up of 2514 White males employed at the MCW uranium processing facility between 1942–1966 was conducted with an extended period of observation of up to 70 years (mean 43.3 years) where over 75% of the workforce had died. Comparisons with the general population revealed a healthy workforce with overall death rates (SMR 0.94) significantly low, related in large part to the low death rate from heart disease (SMR 0.89). The overall cancer rates (SMR 0.97) were close to general population expectation and similar to the SMR computed at last follow-up (SMR 1.05) (Dupree-Ellis et al. 2000).

Few studies of uranium processing workers, including uranium millers, have incorporated organ/tissue-specific estimates of inhaled uranium and radium into analyses (UNSCEAR 2017). Most MCW workers had positive bioassay measures for uranium or radium (2,160 or 86%) which facilitated the computation of organ/tissue-specific doses from these internal alpha-particle emitters. The comprehensive dose-reconstruction incorporated external and internal exposures from processing pitchblende, medical x-rays for employment, and doses received before and after working at MCW. Statistically significant dose-response relationships were found for kidney cancer alone, and malignant and non-malignant kidney disease combined. Significant dose-response relationships were not found for malignant or non-malignant lung disease, non-CLL leukemia, brain and CNS cancer, non-Hodgkin lymphoma, ischemic heart disease (IHD) or dementia, Alzheimer's disease and other motor neuron diseases.

Comparisons with similar uranium processing facilities

Studies of workers exposed to uranium in underground mines, milling activities, gaseous diffusion plants, enrichment and other processing facilities have been recently reviewed (UNSCEAR 2017). Uranium has not been considered a human carcinogen (National Academy of Sciences 1988; IARC 2012; ATSDR 2013) and occupational limits are often based on its chemical toxicity, notably to the kidney (Leggett 1989; ICRP 1997). Human studies are limited, however, because of small sizes, relatively low estimates of radiation dose, and incomplete or absent measures of intakes of uranium. The discussion of and comparisons with previous uranium worker studies focuses on similar uranium processing plants or uranium mills and not on underground mines gaseous diffusion plants, enrichment or other facilities.

Four uranium processing plants operated in North America during the years of the Manhattan Project: MCW in Missouri (Dupree-Ellis et al. 2000), Middlesex Sampling Plant in New Jersey (Eisenbud 1975), Linde Air Products Company Ceramics Plant in New York (Dupree et al. 1987), and Port Hope Radium and Uranium Refining and Processing Plant in Canada. All processed high-grade uranium pitchblende ore obtained from the Belgian Congo which contained high concentrations of uranium, radium and silica. During the years 1958 through 1983, enriched uranium fuels were fabricated for research, space and power reactors at Rocketdyne (Atomics International) in California (Boice, Cohen, et al. 2006; Boice et al. 2011). Internationally, studies of the uranium processing workers and millers have been conducted in Germany (Kreuzer et al. 2015), France (Guseva Canu et al. 2010; Guseva Canu et al. 2011; Guseva Canu et al. 2012; Zhivin et al. 2016) and the United Kingdom (McGeoghegan and Binks 2006) but without individual dose reconstructions following intakes of uranium. A combined mortality analysis of Port Hope and German uranium processing workers was conducted of 7431 workers (Zablotska et al. 2018) and found no statistically significant elevations in lung cancer or cardiovascular disease as related to occupational ionizing radiation exposures. Studies of uranium millers include those of the Colorado Plateau (Pinkerton et al. 2004), Grants, New Mexico (Boice et al. 2008) and Uravan, Colorado (Boice et al. 2007). There were no consistent findings in these studies and any interpretation is hindered because of the inability to estimate organ/tissue doses related to the ingestion or inhalation of uranium processing materials.

Overall, these occupational studies suggested only a weak association between lung cancer with uranium work (UNSCEAR 2017), but the inconsistent results and imperfect dose reconstructions temper any conclusion that might be drawn. The three studies with comprehensive dose reconstructions for individual workers with intakes of uranium (the current MCW study (Ellis et al. 2018), Rocketdyne (Leggett et al. 2005; Boice, Cohen, et al. 2006; Boice et al. 2011) and Fernald Feed Materials Production Center (Anderson et al. 2012; Silver et al. 2013) studies, provide no evidence for a radiation-related lung cancer risk. The occupational studies also revealed no clear and consistent relationship between uranium work and any other cancer or non-cancer, including leukemia, lymphoma, digestive system cancers, kidney and other urological cancers, brain/CNS cancers, non-malignant respiratory and kidney disease and cardiovascular disease (UNSCEAR 2017).

The Fernald Feed Materials Production Center study included 6409 uranium processing workers employed 1951–1985 (Silver et al. 2013). Females and non-White employees were included. The dose reconstruction was comprehensive and organ/tissue-specific doses were estimated for external radiation, internal intakes of uranium based on urine samples, radon and its progeny (Hornung et al. 2008), and exposure received both before and after employment at Fernald (Anderson et al. 2012). A relationship between intestinal cancer (small intestine and colon) and organ dose was suggested. Leukemia (other than CLL) showed a significant negative dose response. No relationship was found for malignant or nonmalignant respiratory disease and uranium work. Salaried and hourly workers were identified, and an informative analysis showed that hourly workers, compared with the general population, were at significantly high risk for lung cancer and ischemic heart disease, whereas salaried workers were at significantly low risk for these causes of death. Since salaried workers have both a low exposure to radiation and a low risk of death to smoking-related conditions, not adjusting for pay type (or some other measure of socio-economic status) might produce spurious results. A major limitation is the low estimates of cumulative organ/tissue dose, i.e. the mean cumulative organ dose for the lung from internal uranium exposure was 1.1 mGy and 13.4 mGy from external exposures. The mean cumulative radon progeny exposure was 26 working level months (WLMs). Another limitation was that the urine analyses were based on single-void samples and not on 24-h urinary excretion collections, which introduces a large uncertainty in estimates of uranium intakes.

The Port Hope Radium and Uranium Refining and Processing Plant study in Canada included 3000 male and female uranium processing workers first employed during 1932–1980 (Zablotska et al. 2013). The workers processed pitchblende, were exposed to external gamma-rays and dust and had inhaled or ingested uranium and radium. Both mortality and cancer incidence were evaluated, and most analyses focused on the 2645 male workers. There were no significant findings, and the only elevation reported was a nonsignificant lung cancer risk following exposures to radon daughter products (mean WLM 15.9). The mean cumulative whole-body γ-ray dose was 134.4 mSv. Limitations include the small study size, incomplete dose reconstructions, lack of information on dust exposure and apparent lack of any measure of socio-economic status. To improve understanding of possible risks associated with intakes of uranium, the authors recommend further follow-up, improved dose reconstruction and pooling with other similar studies in North America (as listed above) and in Europe (Laurent et al. 2016; Gueguen et al. 2017).

Kidney disease

A significant association was seen in the present study between radiation kidney dose and malignant and nonmalignant kidney diseases. Other studies of uranium processing workers, however, fail to find a consistent link between kidney disease and radiation (UNSCEAR 2017). A recent study of workers at gaseous diffusion plants in the US, however, has reported the suggestion of a dose response for kidney cancer and non-malignant kidney disease, though not of statistical significance and the exposure scenarios were completely different from uranium processing facilities (Yiin et al. 2018). Further, of the 17 deaths due to cancer of the kidney and urinary organs among MCW workers, 16 were coded as cancer of the kidney parenchyma (ICD-9 code 189.0), i.e. a type of kidney cancer that has not been linked to radiation, except at high therapeutic doses (Preston et al. 2007; UNSCEAR 2008; Ozasa et al. 2012). Only cancers of the renal pelvis and ureter (i.e. primarily transitional cell carcinomas) have been increased following radiation. No cancer of the renal pelvis was observed among MCW workers, and only one was of the ureter (ICD-9 code 189.3). The previous studies of no association between radiation and dose to the kidney parenchyma, e.g. among atomic bomb survivors and other populations, raise questions as to whether radiation dose contributed to the increase in kidney cancer among MCW workers and may point to the possible influence of metal toxicity and silica exposure. The observed distribution by cell type among MCW workers is as expected from general population statistics which indicate that 90% of kidney cancer are adenocarcinomas of the renal parenchyma and only 10% are renal pelvis carcinomas (Chow et al. 2010).

Since radiation dose to the kidney appears an unlikely explanation for the observed "radiation dose-response", conceivably the increased risk might be related to uranium nephrotoxicity or damage from silica exposure. Radiation dose to the kidney is correlated with the inhalation of pitchblende dust which contains uranium and silica. Renal toxicity associated with high intakes of uranium has been demonstrated in animal experiments and ascribed to chemical damage to the kidneys (Leggett 1989; ATSDR 2013; Dinocourt et al. 2015; UNSCEAR 2017). The possible chemical toxicity of uranium is considered more important for human health than the risk of cancer from its radioactive properties.

Large intakes of uranium can cause kidney damage because the kidney both removes uranium from the body and is a site of uranium accumulation. However, renal damage observed in humans generally appears to be transient, and there have been no deaths attributable to uranium poisoning even among diabetic patients treated in the last century with large amounts of uranium (Kathren 2008). Further, there is no convincing evidence in human epidemiologic studies linking uranium intakes to deaths from kidney disease (Pinkerton et al. 2004; Boice et al. 2007; ATSDR 2013; UNSCEAR 2017). It is unclear, then, whether the direct uranium exposure to the kidney would have been sufficient to cause nonmalignant disease.

Dust inhalation has been linked to nonmalignant kidney disease, but not kidney cancer, in epidemiologic studies (Thun et al. 1985; Osorio et al. 1987; Calvert et al. 1997; Rapiti et al. 1999; Steenland et al. 2001; Steenland 2005; IARC 2012). The level of cumulative dust exposure among MCW workers was generally higher than those reported in studies of crystalline silica-exposed gold miners (Steenland and Brown 1995; Calvert et al. 1997), ceramics workers (Rapiti et al. 1999), workers in the industrial sand industry (Steenland et al. 2001) and most other nonmine occupations reporting significant increases of nonmalignant kidney disease (Steenland 2005). Although there was little evidence for an exposure-response over categories of cumulate dose, a significant risk was seen in the highest tertile (cumulative dust exposure $> 23.6\,\text{mgm}^{-3}\text{year}^{-1}$), suggesting the possibility of a dust effect. The ESRD registry has been used to reveal increases in renal toxicity, especially glomerulonephritis, from studies of workers exposed to silica dust (Calvert et al. 1997; Steenland et al. 2001; Steenland 2005). Linkage with the end stage renal disease registry identified 23 MCW workers. Three were diagnosed with glomerulonephritis with estimated cumulative dust exposures of 0, 6.65, and $1055.27\,\text{mgm}^{-3}\text{year}^{-1}$. It is of note that of the 1358 workers with dust measurements, one of the three workers in the ESRD registry with glomerulonephritis had the third highest cumulative dust exposure and fell in the 99 percentile of the exposure distribution. Silicosis was recorded as a contributing cause of death in two workers, providing limited support for a silica effect on other organs.

In summary, the significant dose response seen between radiation dose to the kidney and malignant and nonmalignant kidney diseases remains unexplained. Cancer of the kidney is not elevated in other studies of uranium workers nor is it considered a radiogenic cancer, except for cancer of the renal pelvis which occurs infrequently and not among MCW workers. Silica is considered a cause of nonmalignant kidney disease but not kidney cancer (ASTDR 2017). The observations, then, may have resulted from the combined effects of uranium and silica chemical nephrotoxicity from the dust and radiation dose or the play of chance.

Lung disease

There is mixed evidence as to whether lung cancer is associated with work at uranium processing facilities (UNSCEAR 2017). Studies in the United States (Dupree et al. 1987; Dupree-Ellis et al. 2000; Boice et al. 2011; Silver et al. 2013), Canada (Zablotska et al. 2013), and the United Kingdom (once cancer of the pleura was excluded) (McGeoghegan and Binks 2006) report no consistent elevations in risk. Studies in France and Germany are equivocal (UNSCEAR 2017). Studies of uranium millers in the United States find no increases in lung cancer related to uranium milling (Pinkerton et al. 2004; Boice et al. 2007, 2008). Limiting factors of these studies include incomplete or absent dose reconstructions (NRC 1983), and several studies make comparisons only with general population rates. The MCW workers provide no evidence to support an association between uranium work and lung cancer risk. We found no evidence of a dose response for lung cancer, nonmalignant lung disease, or malignant and nonmalignant lung disease combined. The absence of a lung cancer effect in these many studies of uranium processers and millers contrasts with the substantial increase in lung cancer mortality seen among underground uranium miners exposed to radon and radon progeny (Lubin et al. 1995).

There also is little to no evidence that non-malignant lung disease is associated with uranium or uranium work (UNSCEAR 2017), consistent with the MCW findings. The negative dose response for lung cancer over a broad range of absorbed doses and after applying dose weighting factors of 10 and 20 (data not shown) is peculiar and remains unexplained. There was suggestive evidence, however, of an association of malignant and nonmalignant lung disease with cumulative estimates of dust. This is consistent with studies of workers exposed to silica where significant increases have been reported for both lung cancer (Calvert et al. 2003) and nonmalignant lung disease (Osorio et al. 1987; Calvert et al. 1997; ASTDR 2017).

Ischemic heart disease

Studies of uranium processing workers in Canada (Zablotska et al. 2013), France and Germany (Guseva Canu et al. 2012; Zhivin et al. 2016; Zablotska et al. 2018) have suggested increases in ischemic heart disease, whereas studies in the United States have not (Dupree-Ellis et al. 2000; Silver et al. 2013). Studies of uranium millers also do not find elevations (Pinkerton et al. 2004; Boice et al. 2008). Most studies are limited by incomplete or no dose reconstruction, and risks in several studies were based only on comparisons with general population rates (SMRs). Overall, there is no convincing evidence that uranium exposure is associated with IHD (UNSCEAR 2017).

Similar to a comprehensive study of uranium processors in France (Guseva Canu et al. 2012), a dose response between IHD and heart dose of borderline significance was seen among MCW workers. However, the heart doses were very low, and related to external gamma ray and medical X-ray exposures (mean, 47.5 mGy) and not intakes of uranium or radium (mean, 0.09 mGy). Further, there is not sufficient evidence in human studies that absorbed doses in the heart below 500 mGy cause cardiovascular disease (CVD) (NCRP 2018; Shore et al. 2019).

If heart radiation is an unlikely cause of the IHD risk at these low heart doses, other explanations might include radiation effects on the kidney and inadequate adjustment for socioeconomic status. Damage to the kidney, which is a primary target organ for uranium intakes, might affect the risk of heart disease through the renin-angiotensin pathway (UNSCEAR 2017). For example, chronic kidney disease associated with radiation may have played a role in the development of cardiovascular disease (CVD) among atomic bomb survivors (Sera et al. 2013), i.e. radiation-induced renal dysfunction could be a mechanism that causes an increase in CVD (Adams et al. 2012). Malignant and nonmalignant kidney disease was increased among MCW workers, and thus damage to the kidney, whatever the cause/s, might be related to the observed increase in IHD.

In the absence of knowledge on smoking history, socioeconomic status (SES) such as pay type (salaried/hourly) is frequently used in occupational studies as a surrogate adjustment factor for tobacco use. This is because the smoking prevalence among hourly (blue-collar) workers is nearly double that among salaried (white-collar) workers (Sorensen et al. 2004). Workers at the Fernald Feed Materials Production Center were analyzed separately by salaried or hourly status. In comparison with general population rates, salaried worker had significantly low SMRs for IHD and lung cancer, whereas hourly workers had significantly high SMRs for both diseases (Silver et al. 2013), indirectly confirming the likelihood that pay type can predict smoking history. We conducted a similar SMR analysis with somewhat similar results: the SMRs for salaried workers were significantly low for both IHD (0.60) and lung cancer (0.48) but the SMRs for hourly workers were near population expectation: 0.99 for IHD and 1.03 for lung cancer (data not shown). We conducted analyses not adjusting for SES: the risk estimates for IHD increased slightly, and the dose-response trend changed from borderline significant (p trend=.07) to statistically significant (p trend=.029). There was little change in the lung cancer results. These analyses indicate the importance of adjusting for SES in the analyses, and the possibility of substantial confounding by smoking if not done.

Nonetheless, pay type is only a crude indicator for smoking status. Workers change from hourly to salaried positions, and we were only able to assign pay type at the time of first hire. Some skilled hourly workers have lifestyles similar to salaried workers, but this could not be determined. Only 17% of the MCW workers where salaried, but adjustment mattered for IHD. Conceivably inaccuracies in the pay type classification for a strong risk factor (smoking surrogate) could contribute to spurious results.

Brain

A significant SMR was seen for brain and CNS cancers. When conducting multiple statistical tests for numerous disease outcomes, some significantly elevated cancer and non-cancer risk estimates are expected to occur by chance alone and should be evaluated in context of findings from other studies as well as internal consistency. There was no association with radiation dose to the brain among Mallinckrodt workers. Further, the evidence that brain cancer is caused by radiation comes mainly from studies of children with inconsistent or nonexistent evidence following adult exposures (UNSCEAR 2008), especially in large worker studies where negative dose responses are reported (Cardis et al. 2007; Zablotska et al. 2014; Richardson et al. 2018). Animal experiments (Dinocourt et al. 2015; UNSCEAR 2017) and human autopsy evaluations provide evidence that intakes of uranium (Avtandilashvili et al. 2015) and radium (Schlenker et al. 1982) can deposit in the brain. The brain dose from alpha particles emitted from the intakes of uranium and radium was too small (median, 0.040 mGy; max 1.4 mGy) (Ellis et al. 2018), however, to expect an increase in brain and CNS cancers.

Non-Hodgkin lymphoma (NHL)

Because inhaled uranium deposited in the lung tends to migrate to the thoracic lymph nodes, lymphoma is a plausible adverse outcome (UNSCEAR 2017). Yet most studies of uranium processors (Silver et al. 2013; Zablotska et al. 2013) and uranium millers (Boice et al. 2008) do not find significant elevations of NHL, and current studies of all uranium workers do not provide clear evidence of uranium-related risks (UNSCEAR 2017). A positive, but not significant, dose-response for NHL was seen among MCW workers based on a 10-year radiation dose lagging period and dose to the active bone marrow which is often done as a surrogate for dose to the lymphatic system (data not shown). The dose response for NHL did not change and remained non-significant when the more relevant dose to the thoracic lymph nodes was used. External radiation has not generally been considered a cause of NHL except perhaps an unexplained excess risk among male but not female atomic bomb survivors (Hsu et al. 2013). Larger combined studies with improved dose reconstruction to the thoracic lymph nodes might confirm the absent or very low risk of NHL from uranium and radium depositions.

Dementia, Alzheimer's, Parkinson's and other motor neuron diseases

There is limited evidence to suggest that uranium exposures to brain tissue at sufficient levels might cause cognitive disturbances and behavior changes in experimental animals, most likely related to the chemical properties of uranium affecting the neural network (Dinocourt et al. 2015; UNSCEAR 2017). Such behavioral effects could not be evaluated among MCW workers. However, the possible effect of alpha particle exposure on brain disease could be evaluated with regard to death due to dementia and other motor neuron diseases.

There is no evidence that low-LET radiation increases the risk of dementia, Alzheimer's disease, Parkinson's disease, or motor neuron disease (Yamada et al. 2009). Experimental studies of animals briefly exposed to energetic heavy ions report behavioral changes as well as an acceleration in the occurrence of Alzheimer's disease raising the possibly that radiation exposure in space might increase the risk of Alzheimer's disease later in life (Cherry et al. 2012; NCRP 2016). The concern is whether galactic cosmic rays (GCR), the high-velocity heavy ions traveling through space after a super nova explodes (or as emanating from super massive black holes), might adversely affect the central nervous system (NCRP 2016; Boice, Ellis, et al. 2018).

There are no human exposures or analogs similar to GCR in space. Studies of workers with internal intakes of radioactive elements might serve as a possible, though imperfect, analog to such GCR exposures. Radiochemical analyses of tissues have demonstrated that uranium and radium can pass the blood-brain barrier, be deposited in brain tissue and expose brain tissue to high-LET alpha particles (Schlenker et al. 1982; Avtandilashvili et al. 2015; Kathren and Tolmachev 2015). Radiochemical analyses have also confirmed brain depositions from intakes of plutonium (James et al. 2003; James et al. 2007), americium (McInroy et al. 1985), as well as the more nearly uniformly distributed alpha particle emitter polonium (Nathwani et al. 2016). Preliminary analysis of workers at the Mound facility with intakes of polonium suggested an increased risk for the combined motor neuron diseases of dementia, Alzheimer's disease, Parkinson's disease and other motor neuron diseases (Boice 2017). A dose response for MCW workers with intakes of uranium and radium, however, was not apparent for the combined motor neuron diseases. Limitations of the MCW study includes small numbers and low brain doses from alpha particles (mean 0.08 mGy) and external radiation (mean ∼37 mGy). Combination of these data with similar cohorts in the MWS should enhance understanding as to whether such brain exposures might be related to dementia and associated conditions.

Other sites

Bone cancer, liver cancer and multiple myeloma were sites of interest because of the potential for exposure to intakes of uranium and radium. There was no evidence for an increased risk related to radiation, but the number of deaths from bone cancer ($n = 2$), liver cancer ($n = 6$) and multiple myeloma ($n = 6$) were too small for meaningful analyses. There was no evidence for an increase in mesothelioma in the current study based on only 2 cases, suggesting the absence of asbestos exposure at this early uranium processing facility. The Fernald study of uranium processing workers suggested an increased risk for intestinal cancer (colon and small intestine) (Silver et al. 2013); whereas we found no evidence for a radiation-related risk for colon cancer. Most studies of uranium workers do not find an association between multiple myeloma and radiation dose, except for one study of workers at gaseous diffusion plants in the United States (UNSCEAR 2008; Yiin et al. 2009; UNSCEAR 2017; Yiin et al. 2017; Yiin et al. 2018). The association reported between uranium bone marrow dose and multiple myeloma risk in the gaseous diffusion plant studies requires confirmation by other studies (UNSCEAR 2017); dose to the thoracic lymph nodes might be a better surrogate for dose to the lymphatic system than dose to bone marrow, although we found it made little difference.

Sensitivity analyses

A discussion of sensitivity analyses and uncertainty propagation techniques is found in Ellis et al. (2018). The relatively small size of the MCW study, small number of specific outcomes, relatively low organ/tissue-specific radiation doses, and complex exposure sources (intakes of radionuclides and dust, external exposures from chest x-rays and pitchblende, radon inhalation, and sources from other facilities) limited our ability to conduct meaningful uncertainty analyses. More substantive approaches for handling uncertainties are planned in future combined analyses of similar studies within the MWS, reducing the role of sampling variability and increasing statistical power and precision. The extensive collection of radiation exposure data available for the MCW uranium processing workers and the other similar studies provides a broad application of the methodology endorsed by the NCRP Publication 178 (NCRP 2018) for construction of organ/tissue doses for use in epidemiologic studies. Nonetheless, the limited approach to uncertainty is a study weakness.

Simple sensitivity analyses included making minimum and maximum assumptions for dose and evaluating the effect on dose-response analyses, which were minimal because of small number sampling variability. Assumptions were made on how to apply the dosimetric and biokinetic models needed to produce organ/tissue-specific doses. Doses for workers with missing or incomplete records were imputed. The radiation exposure data available for the 43 workers before and after employment at MCW did not contain the level of detail available for the same data collected while employed at MCW. The assumptions made, the methods used to impute missing data, and the decisions made to use the personal monitoring data collected outside of MCW employment may have affected individual worker dose. Fortunately, these numbers were small, and modifying the assumptions, imputation approaches, handling and adjusting for dust inhalation had little impact on the analyses and interpretations. Adjusting and not adjusting for potential confounding factors revealed the importance of socioeconomic status as a surrogate for tobacco use but mainly for IHD. The statistical power for all but a few organ/tissue sites was so low that the non-significant findings and patterns remained non-significant. More formal techniques for propagation of uncertainties in parameter values are planned in future combined analyses.

Strengths and limitations

Strengths of the study include the cohort design, the long and complete follow-up of up to 70 years, and the comprehensive dose reconstruction that included external and internal doses from six sources of exposure. Occupational doses received both before and after employment at MCW were captured, organ/tissue-specific doses from intakes of uranium and radium were computed, organ/tissue-specific doses from occupational medical X-rays required for employment were included, ambient radon levels in work areas were available, and dust measurements from pitchblende, including uranium and silica, were addressed in the analyses. The 39,451 urine bioassay samples (over 20 samples per worker) add to the quality of the information on dose reconstruction, as did the radon breathe measurements. Another strength was the estimation of radiation dose to the thoracic lymph nodes for NHL dose-response analyses, eliminating reliance on red bone marrow dose as a surrogate for dose to the lymphatic system. Linkages to identify serious renal disease within the U.S. Renal Data System added an extra dimension of understanding to the mortality findings. Other methodological strengths include the low percentage of workers who were lost to follow-up (0.8%), and the low percentage of deaths for which a specific cause was not available (0.8%).

Limitations of the study include the relatively small number of workers and the incomplete knowledge of confounding factors such as smoking history. Adjusting for pay type (hourly vs salaried) was an attempt to control indirectly for smoking and other lifestyle behaviors (Boice, Leggett, et al. 2006). Other uncertainties include the possibility of missing or incomplete measurements of radiation doses or dust exposure, and the assumptions of aerosol size and residence time in the lungs. Further, concomitant intakes of radionuclides such as actinium and protactinium could not be accounted for. The relatively low cumulative dose and dose range limited the ability of the study to detect radiation effects had there been any. Our inability to control for solvents and other laboratory chemical exposures in the workplace is a limitation.

Differences in smoking habits between MCW workers and the general population were not reflected in the SMRs for lung cancer (SMR 0.93) and smoking-related cancers (SMR 0.94), nor for non-malignant respiratory diseases (SMR 0.94). The overall SMR for lung cancer reflected an SMR of 1.03 for hourly workers and a significantly low SMR of 0.48 for salaried workers. Adjustment was made for job title in the intra-cohort analyses as a surrogate measure of socioeconomic class among radiation workers, and there was no evidence for a dose-response effect for lung cancer or non-malignant respiratory disease. The study examined mortality and not incidence of disease for which the number of events and quality of diagnoses would be expected to be higher. However, most of the diseases of interest, e.g. cancers of the lung, liver, and leukemia, however, have a high fatality rate over the years of study so that mortality would be expected to reflect incidence closely. A limitation is the small number of deaths for most individual outcomes of interest. Biomarkers for uranium in epidemiological studies might provide insights into low-dose effects (Hall et al. 2017).

Summary

The MCW workers have a singular place in history in producing the uranium oxide metal necessary for the first manmade sustained and controlled nuclear fission chain-reaction in 1942. Some 70 years later, 2514 workers were followed for health effects possibly related to the external and internal radiation doses received and the dust inhaled. The dose reconstruction took into account six sources of exposure and is perhaps one of the most comprehensive of any radiation epidemiologic study yet conducted. The MCW study is used as a model for other similar studies of uranium processing facilities within the MWS, such as Middlesex, NJ. The high quality of the available radiation and dust data at MCW was a testimony to the oversight provided by two of the renowned health physicists of the last century: Robley Evans and Merril Eisenbud. Although the cohort is relatively small, a significant radiation dose response was observed for malignant and non-malignant kidney disease. A non-radiation etiology could not be completely discounted, however, because of inconsistencies in the radiation literature and the likely influence of the toxic properties of uranium, a heavy metal, and silica, a component in the pitchblende dust inhaled. The negative findings for malignant and non-malignant lung disease are of note, as are the non-significant positive radiation association for ischemic heart disease. However, the small numbers preclude definitive interpretations and wait for combination with other studies following similar methodologies to refine and clarify radiation association following prolonged exposures to ionizing radiation and to evaluate more precisely risks from intakes of radioactive elements.

Disclosure statement

No potential conflict of interest was reported by the author(s).

Funding

This work was supported in part by a grant from the U.S. Nuclear Regulatory Commission (NRC-HQ-60-14-G-0011), a grant from the Centers for Disease Control and Prevention (5UE1EH000989), a grant from the National Aeronautics and Space Administration (Grant No. NNX15AU88G), and grants from the U.S. Department of Energy (Grant No. DE-SC0008944 and Grant No. DE-AU0000042 awarded to the National Council on Radiation Protection and Measurements, which included interagency support from the U.S. Nuclear Regulatory Commission, the U.S. Environmental Protection Agency and the National Aeronautics and Space Administration). Additional contract support was received by Oak Ridge National Laboratory from the Office of Radiation and Indoor Air, U.S. Environmental Protection Agency, under Interagency Agreement DOE No. 1824 S581-A1, under contract No. DE-AC05-00OR22725 with UT-Battelle; and contract support was received by Oak Ridge Associated Universities from the U.S. Department of Energy under contract No. DE-SC0014664.

ORCID

Ashley P. Golden http://orcid.org/0000-0002-9430-7702
Elizabeth D. Ellis http://orcid.org/0000-0003-2871-1587
Sarah S. Cohen http://orcid.org/0000-0003-0421-1983
Michael T. Mumma http://orcid.org/0000-0001-7506-8710
John D. Boice http://orcid.org/0000-0002-8755-1299

References

Adams MJ, Grant EJ, Kodama K, Shimizu Y, Kasagi F, Suyama A, Sakata R, Akahoshi M. 2012. Radiation dose associated with renal failure mortality: a potential pathway to partially explain increased cardiovascular disease mortality observed after whole-body irradiation. Radiat Res. 177:220–228.

Anderson JL, Daniels RD, Fleming DA, Tseng CY. 2012. Exposure assessment for a cohort of workers at a former uranium processing facility. J Expo Sci Environ Epidemiol. 22:324–330.

ASTDR 2017. Toxicological profile for silica. Draft for Public Comment. IUS Department of Health and Human Services Public Health Service Atlanta, GA: Agency for Toxic Substances and Disease Registry (ATSDR).

ATSDR 2013. Toxicological profile for uranium. US Department of Health and Human Services Public Health Service. Atlanta, GA: Agency for Toxic Substances and Disease Registry (ATSDR).

Avtandilashvili M, Puncher M, McComish SL, Tolmachev SY. 2015. US Transuranium and Uranium Registries case study on accidental exposure to uranium hexafluoride. J Radiol Prot. 35:129–151.

Boice JD. 2017. Space: the final frontier-research relevant to mars. Health Phys. 112:392–397.

Boice JD, Cohen SS, Mumma MT, Chadda B, Blot WJ. 2007. Mortality among residents of Uravan, Colorado who lived near a uranium mill, 1936–1984. J Radiol Prot. 27:299–319.

Boice JD, Cohen SS, Mumma MT, Chadda B, Blot WJ. 2008. A cohort study of uranium millers and miners of Grants, New Mexico, 1979–2005. J Radiol Prot. 28: 303–325.

Boice JD, Cohen SS, Mumma MT, Dupree-Ellis E, Eckerman KF, Leggett RW, Boecker B, Brill A, Henderson B. 2006. Mortality among radiation workers at Rocketdyne (Atomics International), 1948–1999. Radiat Res. 166:98–115.

Boice JD, Cohen SS, Mumma MT, Ellis ED, Cragle DL, Eckerman KF, Wallace PW, Chadda B, Sonderman JS, Wiggs LD. 2014. Mortality among mound workers exposed to polonium-210 and other sources of radiation, 1944–1979. Radiat Res. 181:208–228.

Boice JD, Cohen SS, Mumma MT, Ellis ED, Eckerman KF, Leggett RW, Boecker BB, Brill AB, Henderson BE. 2011. Updated mortality analysis of radiation workers at Rocketdyne (Atomics International), 1948–2008. Radiat Res. 176:244–258.

Boice JD, Ellis ED, Golden AP, Girardi DJ, Cohen SS, Chen H, Mumma MT, Shore RE, Leggett RW. 2018. The past informs the future: an overview of the Million Worker Study and the Mallinckrodt Chemical Works cohort. Health Phys. 114:381–385.

Boice JD, Leggett RW, Eckerman KF, Tolmachev SY, Woloschak GE, Golden AP, Ellis ED. 2018. Response to Mortazavi et al. on Detecting bone-seeking radionuclides in brain tissue. Health Phys. 115:389–390.

Boice JD, Leggett RW, Ellis ED, Wallace PW, Mumma MT, Cohen SS, Brill AB, Bandana C, Boecker BB, Yoder RC. 2006. A comprehensive dose reconstruction methodology for former Rocketdyne/Atomics International radiation workers. Health Phys. 90:409–430.

Bouville A, Toohey RE, Boice JD, Beck HL, Dauer LT, Eckerman KF, Hagemeyer D, Leggett RW, Mumma MT, Napier B, et al. 2015. Dose reconstruction for the Million Worker Study: status and guidelines. Health Phys. 108: 206–220.

Calvert GM, Rice FL, Boiano JM, Sheehy JW, Sanderson WT. 2003. Occupational silica exposure and risk of various diseases: an analysis using death certificates from 27 states of the United States. Occup Environ Med. 60: 122–129.

Calvert GM, Steenland K, Palu S. 1997. End-stage renal disease among silica-exposed gold miners. a new method for assessing incidence among epidemiologic cohorts. JAMA. 277:1219–1223.

Campbell K, Deck D, Krupski A. 2008. Record linkage software in the public domain: a comparison of Linke Plus, The Link King, and a 'basic' deterministic algorithm. Health Informatics J. 14:5–15.

Cardis E, Vrijheid M, Blettner M, Gilbert E, Hakama M, Hill C, Howe G, Kaldor J, Muirhead CR, Schubauer-Berigan M, et al. 2007. The 15-country collaborative study of cancer risk among radiation workers in the nuclear industry: estimates of radiation-related cancer risks. Radiat Res. 167:396–416.

Cherry JD, Liu B, Frost JL, Lemere CA, Williams JP, Olschowka JA, O'Banion MK. 2012. Galactic cosmic radiation leads to cognitive impairment and increased a beta plaque accumulation in a mouse model of Alzheimer's disease. PLoS One. 7:10.

Chow WH, Dong LM, Devesa SS. 2010. Epidemiology and risk factors for kidney cancer. Nat Rev Urol. 7:245–257.

Cox D. 1972. Regression models and life-tables. J R Stat Soc Ser B. 34:187–220.

Dinocourt C, Legrand M, Dublineau I, Lestaevel P. 2015. The neurotoxicology of uranium. Toxicology. 337:58–71.

DOE. 2017. DOE 2016 Occupational radiation exposure report. Oak Ridge, Tennessee: US Department of Energy, The Radiation Exposure Monitoring System (REMS). https://www.energy.gov/sites/prod/files/2017/12/f46/2016_Occupational_Radiation_Exposure_Report.pdf

Dupree-Ellis E, Watkins J, Ingle J, Phillips J. 2000. External radiation exposure and mortality in a cohort of uranium processing workers. Am J Epidemiol. 152:91–95.

Dupree E, Cragle D, McLain R, Crawford-Brown D, Teta M. 1987. Mortality among workers at a uranium processing facility, the Linde Air Products Company Ceramics Plant, 1943-1949. Scand J Work Environ Health. 13: 100–107.

Eisenbud M. 1975. Early occupational exposure experience with uranium processing. Conference on Occupational Health Experience with Uranium; Apr 28; Arlington, Virginia, Washington DC: ERDA-93 Energy Research and Development. [accessed 2019 Jan 28]. http://www.osti.gov/scitech/servlets/purl/4082757.

Ellis ED, Boice JD, Golden AP, Girardi DJ, Cohen SS, Mumma MT, Shore RE, Leggett RW, Kerr GD. 2018. Dosimetry is key to good epidemiology: workers at Mallinckrodt Chemical Works had seven different source exposures. Health Phys. 114:386–397.

Gueguen Y, Roy L, Hornhardt S, Badie C, Hall J, Baatout S, Laurent O, Ebrahimian T, Grison S, Ibanez C, et al. 2017. Biomarkers for uranium risk assessment for the development of a molecular epidemiological protocol. Toxicol Lett. 280:S300–S300.

Guseva Canu I, Cardis E, Metz-Flamant C, Caër-Lorho S, Auriol B, Wild P, Laurier D, Tirmarche M. 2010. French cohort of the uranium processing workers: mortality pattern after 30-year follow-up. Int Arch Occup Environ Health. 83:301–308.

Guseva Canu I, Garsi J, Cae¨r, Lorho S, Jacob S, Collomb P, Acker A, Laurier D. 2012. Does uranium induce circulatory diseases? First results from a French cohort of uranium workers. Occup Environ Med. 69:404–409.

Guseva Canu I, Jacob S, Cardis E, Wild P, Caer S, Auriol B, Garsi JP, Tirmarche M, Laurier D. 2011. Uranium carcinogenicity in humans might depend on the physical and chemical nature of uranium and its isotopic composition: results from pilot epidemiological study of French nuclear workers. Cancer Causes Control. 22:1563–1573.

Hall J, Jeggo PA, West C, Gomolka M, Quintens R, Badie C, Laurent O, Aerts A, Anastasov N, Azimzadeh O, et al. 2017. Ionizing radiation biomarkers in epidemiological studies - an update. Mutat Res. 771:59–84.

Hornung RW, Pinney SM, Lodwick J, Killough GG, Brewer DE, Nasuta J. 2008. Estimation of radon exposures to workers at the Fernald Feed Materials Production Center 1952–1988. J Expo Sci Environ Epidemiol. 18:512–523.

Hsu WL, Preston DL, Soda M, Sugiyama H, Funamoto S, Kodama K, Kimura A, Kamada N, Dohy H, Tomonaga M, et al. 2013. The incidence of leukemia, lymphoma and multiple myeloma among atomic bomb survivors: 1950–2001. Radiat Res. 179:361–382.

IARC 2012. A review of human carcinogens, Part D: radiation. Lyon, France: International Agency for Research on Cancer (IARC).

ICRP 1994. Dose coefficients for intakes of radionuclides by workers. ICRP Publication 68. Ann ICRP. 24(4):1–84.

ICRP 1995. Age dependent doses to members of the public from intake of radionuclides: part 4-inhalation dose coefficients. ICRP Publication 71. Ann ICRP. 25:3–4.

ICRP 1997. Individual monitoring for internal exposure of workers. ICRP Publication 78. Ann ICRP. 27:124–130.

ICRP 2006. Human alimentary tract model for radiological protection. ICRP Publication 100. Ann ICRP. 36(1–2): 69–111.

ICRP 2010. Conversion of coefficients for radiological protection quantities for external radiation exposures. Vol. 40. New York: Elsevier. (ICRP Publication 116, Ann ICRP; 2-5).

ICRP 2017. Occupational intakes of radionuclides: part 3. ICRP Publication 137. Ann ICRP. 46(3–4):297–486.

James AC, Filipy RE, Russell JJ, McInroy JF. 2003. USTUR Case 0259 whole body donation: a comprehensive test of the current ICRP models for the behavior of inhaled 238PuO$_2$ ceramic particles. US Transuranium and Uranium Registries. Health Phys. 84:2–33.

James AC, Sasser LB, Stuit DB, Wood TG, Glover SE, Lynch TP, Dagle GE. 2007. USTUR whole body case 0262: 33-y follow-up of PuO$_2$ in a skin wound and associated axillary node. Radiat Prot Dosimetry. 127:114–119.

Kathren RL, Burklin RK. 2008. Acute chemical toxicity of uranium. Health Phys. 94(2):170–179.

Kathren RL, Tolmachev SY. 2015. Natural uranium tissue content of three Caucasian males. Health Phys. 109: 187–197.

Kreuzer M, Dufey F, Laurier D, Nowak D, Marsh JW, Schnelzer M, Sogl M, Walsh L. 2015. Mortality from internal and external radiation exposure in a cohort of male German uranium millers, 1946–2008. Int Arch Occup Environ Health. 88:431–441.

Laurent O, Gomolka M, Haylock R, Blanchardon E, Giussani A, Atkinson W, Baatout S, Bingham D, Cardis E, Hall J, et al. 2016. Concerted Uranium Research in Europe (CURE): toward a collaborative project integrating dosimetry, epidemiology and radiobiology to study the effects of occupational uranium exposure. J Radiol Prot. 36:319–345.

Leggett RW. 1989. The behavior and chemical toxicity of U in the kidney: a reassessment. Health Phys. 57:365–383.

Leggett RW. 1994. Basis for the ICRP's age-specific biokinetic model for uranium. Health Phys. 67:589–610.

Leggett RW, Eckerman KF, Boice JD. 2005. A respiratory model for uranium aluminide based on occupational data. J Radiol Prot. 25:405–416.

Lubin J, Boice JJ, Edling C, Hornung RW, Howe G, Kunz E, Kusiak R, Morrison H, Radford E, Samet J. 1995. Lung cancer in radon-exposed miners and estimation of risk from indoor exposure. J Natl Cancer Inst. 87:817–827.

McGeoghegan D, Binks K. 2006. The mortality and cancer morbidity experience of workers at the Springfields uranium production facility, 1946–1995 (vol 20, pg 111, 2006) [Correction]. J Radiol Prot. 26:455–455.

McInroy J, Boyd H, Eutsler B, Romero D. 1985. The U.S. Transuranium Registry report of the 241Am content of a whole body. Part IV: preparation and analysis of the tissues and bones. Health Phys. 49:587–621.

Monson R. 1986. Observations on the healthy worker effect. J Occup Med. 28:425–433.

Nathwani AC, Down JF, Goldstone J, Yassin J, Dargan PI, Virchis A, Gent N, Lloyd D, Harrison JD. 2016. Polonium-210 poisoning: a first-hand account. Lancet. 388:1075–1080.

National Academy of Sciences. 1988. Health risks of radon and other internally deposited alpha-emitters (BEIR IV). Committee on the Biological Effects of Ionizing Radiations. Washington DC: National Academies Press.

NCRP. 2016. Commentary No. 25 - potential for central nervous system effects from radiation exposure during space activities phase 1: overview. Bethesda, MD: National Council on Radiation Protection and Measurements.

NCRP. 2018. Commentary No. 27 - implications of recent epidemiologic studies for the linear-nonthreshold model and radiation protection. Bethesda, MD: National Council on Radiation Protection and Measurements.

NRC. 1983. Health physics surveys in uranium mills. US Nuclear Regulatory Commission. Washington, DC: Regulatory guide No.: 8.30.

NRC 2018. Radiation exposure information and reporting system (REIRS). Washington, DC: US Nuclear Regulatory Commission.

Osorio A, Thun M, Novak R, Van Cura E, Avner E. 1987. Silica and glomerulonephritis: case report and review of the literature. Am J Kidney Dis. 9:224–230.

Ozasa K, Shimizu Y, Suyama A, Kasagi F, Soda M, Grant EJ, Sakata R, Sugiyama H, Kodama K. 2012. Studies of the mortality of atomic bomb survivors, report 14, 1950–2003: an overview of cancer and noncancer diseases. Radiat Res. 177:229–243.

Ozasa K, Takahashi I, Grant EJ, Kodama K. 2017. Cardiovascular disease among atomic bomb survivors. Int J Radiat Biol. 93:1145–1150.

Pinkerton LE, Bloom TF, Hein MJ, Ward EM. 2004. Mortality among a cohort of uranium mill workers: an update. Occup Environ Med. 61:57–64.

Preston DL, Ron E, Tokuoka S, Funamoto S, Nishi N, Soda M, Mabuchi K, Kodama K. 2007. Solid cancer incidence in atomic bomb survivors: 1958–1998. Radiat Res. 168: 1–64.

Rapiti E, Sperati A, Miceli M, Forastiere F, Di Lallo D, Cavariani F, Goldsmith DF, Perucci CA. 1999. End stage renal disease among ceramic workers exposed to silica. Occup Environ Med. 56:559–561.

Richardson DB, Cardis E, Daniels RD, Gillies M, Haylock R, Leuraud K, Laurier D, Moissonnier M, Schubauer-Berigan MK, Thierry-Chef I, et al. 2018. Site-specific solid cancer mortality after exposure to ionizing radiation: a cohort study of workers (INWORKS). Epidemiology. 29: 31–40.

Schlenker RA, Keane AT, Holtzman RB. 1982. The retention of 226Ra in human soft tissue and bone; implications for the ICRP 20 alkaline earth model. Health Phys. 42:671–693.

Sera N, Hida A, Imaizumi M, Nakashima E, Akahoshi M. 2013. The association between chronic kidney disease and cardiovascular disease risk factors in atomic bomb survivors. Radiat Res. 179:46–52.

Shore RE, Beck HL, Caffrey EA, Davis S, Grogan HA, Mettler FA, Preston RJ, Till JE, Wakeford R, Walsh L. 2019. Recent epidemiologic studies and the linear non-threshold model for radiation protection - considerations regarding NCRP Commentary No. 27. Health Phys. 116(2):235–246.

Silver SR, Bertke SJ, Hein MJ, Daniels RD, Fleming DA, Anderson JL, Pinney SM, Hornung RW, Tseng CY. 2013. Mortality and ionising radiation exposures among workers employed at the Fernald Feed Materials Production Center (1951–1985). Occup Environ Med. 70:453–463.

Sorensen G, Barbeau E, Hunt MK, Emmons K. 2004. Reducing social disparities in tobacco use: a social-contextual model for reducing tobacco use among blue-collar workers. Am J Public Health. 94:230–239.

Steenland K. 2005. One agent, many diseases: exposure-response data and comparative risks of different outcomes following silica exposure. Am J Ind Med. 48: 16–23.

Steenland K, Brown D. 1995. Mortality study of gold miners exposed to silica and nonasbestiform amphibole minerals: an update with 14 more years of follow-up. Am J Ind Med. 27:217–229.

Steenland K, Sanderson W, Calvert GM. 2001. Kidney disease and arthritis in a cohort study of workers exposed to silica. Epidemiology. 12:405–412.

Thomas E. 2011. Dose reconstruction from occupational medical X-Ray procedures. Oak Ridge, Tennessee: ORAU Team. OTIB-0006.

Thun MJ, Baker DB, Steenland K, Smith AB, Halperin W, Berl T. 1985. Renal toxicity in uranium mill workers. Scand J Work Environ Health. 11:83–90.

UNSCEAR. 2008. Effects of ionizing radiation. 2006 Report. Scientific annex A: epidemiological studies of radiation and cancer. New York: United Nations Scientific Committee on the Effects of Atomic Radiation.

UNSCEAR. 2017. Sources, effects and risks of ionizing radiation. 2016 Report to the General Assembly. Volume 1. Scientific annex D: biological effects of selected internal emitters – uraniaum. New York: United Nations Scientific Committee on the Effects of Atomic Radiation.

USACE. 1996. The history of the St. Louis uranium processing plant radioactive waste sites. http://www.mvs.usace.army.mil/Portals/54/docs/fusrap/docs/1996%20St%20Louis%20Sites%20Remediation%20Task-%20Force%20Report%202.pdf.

USRDS. 2017. USRDS 2017 annual data report: atlas of end-stage renal disease in the United States. Bethesda, MD: National Institutes of Health, National Institute of Diabetes and Digestive and Kidney Diseases; [accessed]. (http://www.usrds.org).

Watkins J, Cragle D, Frome E, Reagan J, West C, Crawford-Brown D, Tankersley W. 1997. Collection, validation, and treatment of data for a mortality study of nuclear industry workers. Appl Occup Environ Hygiene. 12:195–205.

Westbrook J, Bloom C. 2010. Basis for development of an exposure matrix for the Mallinckrodt Chemical Company St. Louis Downtown Site and the St. Louis Airport Site, St. Louis, Missouri. Oak Ridge, Tennessee: ORAU Team.

Yamada M, Kasagi F, Mimori Y, Miyachi T, Ohshita T, Sasaki H. 2009. Incidence of dementia among atomic-bomb survivors - radiation effects research foundation adult health study. J Neurol Sci. 281:11–14.

Yiin JH, Anderson JL, Bertke SJ, Tollerud DJ. 2018. Dose-response relationships between internally-deposited uranium and select health outcomes in gaseous diffusion plant workers, 1948–2011. Am J Ind Med. 61:605–614.

Yiin JH, Anderson JL, Daniels RD, Bertke SJ, Fleming DA, Tollerud DJ, Tseng CY, Chen PH, Waters KM. 2017. Mortality in a combined cohort of uranium enrichment workers. Am J Ind Med. 60:96–108.

Yiin JH, Anderson JL, Daniels RD, Seel EA, Fleming DA, Waters KM, Chen PH. 2009. A nested case-control study of multiple myeloma risk and uranium exposure among workers at the Oak Ridge Gaseous Diffusion Plant. Radiat Res. 171:637–645.

Zablotska LB, Fenske N, Schnelzer M, Zhivin S, Laurier D, Kreuzer M. 2018. Analysis of mortality in a pooled cohort of Canadian and German uranium processing workers with no mining experience. Int Arch Occup Environ Health. 91:91–103.

Zablotska LB, Lane RS, Frost SE. 2013. Mortality (1950–1999) and cancer incidence (1969–1999) of workers in the Port Hope cohort study exposed to a unique combination of radium, uranium and γ-ray doses. BMJ Open. 3:e002159.

Zablotska LB, Lane RS, Thompson PA. 2014. A reanalysis of cancer mortality in Canadian nuclear workers (1956–1994) based on revised exposure and cohort data. Br J Cancer. 110:214–223.

Zhivin S, Canu I, Samson E, Laurent O, Grellier J, Collomb P, Zablotska L, Laurier D. 2016. Mortality (1968-2008) in a french cohort of uranium enrichment workers potentially exposed to rapidly soluble uranium compounds. Occup Environ Med. 73:167–174.

Mortality among workers at the Los Alamos National Laboratory, 1943–2017

John D. Boice, Sarah S. Cohen, Michael T. Mumma, Ashley P. Golden, Sara C. Howard, David J. Girardi, Elizabeth D. Ellis, Michael B. Bellamy, Lawrence T. Dauer, Caleigh Samuels, Keith F. Eckerman, and Richard W. Leggett

ABSTRACT

Background: During World War II (WWII), the Manhattan Engineering District established a secret laboratory in the mountains of northern New Mexico. The mission was to design, construct and test the first atomic weapon, nicknamed 'The Gadget' that was detonated at the TRINITY site in Alamogordo, NM. After WWII, nuclear weapons research continued, and the laboratory became the Los Alamos National Laboratory (LANL).

Materials and methods: The mortality experience of 26,328 workers first employed between 1943 and 1980 at LANL was determined through 2017. Included were 6157 contract workers employed by the ZIA Company. Organ dose estimates for each worker considered all sources of exposure, notably photons, neutrons, tritium, ^{238}Pu and ^{239}Pu. Vital status determination included searches within the National Death Index, Social Security Administration and New Mexico State Mortality Files. Standardized Mortality Ratios (SMR) and Cox regression models were used in the analyses.

Results: Most workers (55%) were hired before 1960, 38% had a college degree, 25% were female, 81% white, 13% Hispanic and 60% had died. Vital status was complete, with only 0.1% lost to follow-up. The mean dose to the lung for the 17,053 workers monitored for radiation was 28.6 weighted-mGy (maximum 16.8 weighted-Gy) assuming a Dose Weighting Factor of 20 for alpha particle dose to lung. The Excess Relative Risk (ERR) at 100 weighted-mGy was 0.01 (95%CI −0.02, 0.03; $n = 839$) for lung cancer. The ERR at 100 mGy was −0.43 (95%CI −1.11, 0.24; $n = 160$) for leukemia other than chronic lymphocytic leukemia (CLL), −0.06 (95%CI −0.16, 0.04; $n = 3043$) for ischemic heart disease (IHD), and 0.29 (95%CI 0.02, 0.55; $n = 106$) for esophageal cancer. Among the 6499 workers with measurable intakes of plutonium, an increase in bone cancer (SMR 2.44; 95%CI 0.98, 5.03; $n = 7$) was related to dose. The SMR for berylliosis was significantly high, based on 4 deaths. SMRs for Hispanic workers were significantly high for cancers of the stomach and liver, cirrhosis of the liver, nonmalignant kidney disease and diabetes, but the excesses were not related to radiation dose.

Conclusions: There was little evidence that radiation increased the risk of lung cancer or leukemia. Esophageal cancer was associated with radiation, and plutonium intakes were linked to an increase of bone cancer. IHD was not associated with radiation dose. More precise evaluations will await the pooled analysis of workers with similar exposures such as at Rocky Flats, Savannah River and Hanford.

Introduction

The Million Person Study (MPS) of low-level radiation exposure was designed to evaluate the level of health effects when exposure is gradual over time and not delivered briefly. Over a million healthy American workers and veterans are being studied to evaluate cancer and non-cancer mortality following low-level low-LET as well as high-LET exposure (Boice, Cohen, et al. 2019; Boice, Quinn, et al. 2021). The large study size also enables the evaluation of radiation associations for rare cancers, intakes of radioactive elements, and differences between men and women. A major component of the MPS is the U.S. Department of Energy (DOE) workers employed during and after the Manhattan Project (Boice, Cohen, et al. 2006; Boice et al. 2011, 2014; Ellis, Boice, et al. 2018; Ellis, Girardi, et al. 2018; Boice, Cohen, et al. 2019; Golden et al. 2019), including workers at the Los Alamos National Laboratory (LANL).

During World War II (WWII), the Manhattan Engineering District established a secret laboratory in the mountains of northern New Mexico. Named Project Y, the mission was to design, construct, and test the first atomic weapon, nicknamed 'The Gadget', with the detonation code name of TRINITY. The plutonium device was successfully tested on 16 July 1945 in Alamogordo, New Mexico. After WWII, the laboratory continued with nuclear weapons research and design and became known as the Los Alamos Scientific Laboratory and eventually as LANL. Over the years, mortality studies were conducted of the early LANL workers exposed to external gamma-rays and intakes of plutonium (Hempelmann et al. 1973; Wiggs 1987; Wilkinson et al. 1987; Galke et al. 1992; Wiggs et al. 1994; Voelz et al. 1997; Daniels et al. 2006). The current investigation expands upon the previous work, extends the follow-up an additional 27 years, includes women for the first time, includes workers at Zia, a maintenance contractor (Galke et al. 1992; Ellis, Girardi, et al. 2018), and uses comprehensive dose reconstruction techniques to estimate organ-specific doses for each worker from photons, neutrons, tritium, ^{238}Pu, and ^{239}Pu.

Materials and methods

Human subjects research approval was received from the Oak Ridge Site-wide Institutional Review Board and the Vanderbilt University Institutional Review Board.

Study population

The LANL worker population was assembled from three previously studied cohorts: male LANL workers employed 1943 through 1977 (Wiggs et al.1994); female LANL workers employed 1943 through 1977 (Wiggs 1987); and male and female workers employed 1946 through 1977 by the Zia Company, a support services contractor to LANL (Galke et al. 1992; History Associates Incorporated 1997; Ellis, Girardi, et al. 2018). The Zia Company employed 14,326 workers but only the 6157 employees who were monitored for external and/or internal radiation while working at LANL are included. Throughout the paper, 'LANL workers' is used to refer to all LANL and ZIA employees who worked at LANL. The combined LANL population cohort initially consisted of 27,497 workers. After excluding 780 workers with insufficient or incorrect identifying information, 58 workers with missing or insufficient occupational data, and 331 workers employed for fewer than 30 days, 26,328 workers remained eligible for the study (Table 1, Figure 1). Worker information including demographics and employment dates were obtained from rosters utilized in the previous studies (Voelz and Hempelmann 1975; History Associates Incorporated 1997; Ellis, Girardi, et al. 2018). Educational attainment was available for 24,826 workers (94%) and was classified as less than high school, some high school, high school graduate, associate degree, college degree, or graduate degree. For 1116 workers (4.2%), education was imputed using residential address and Census data as described in Cohen et al. (2018). For $n = 387$ workers (1.5%), where an address history could not be obtained, the education level was assigned to be high school.

Table 1. Descriptive characteristics of the 26,328 LANL workers first employed 1943–1980 and followed through 2017.

Characteristic	*n*	%
Employer		
LANL	20,171	76.6
ZIA	4,237	16.1
Both LANL and ZIA	1,920	7.3
Sex		
Male	19,804	75.2
Female	6,524	24.8
Education		
<High school, some high school	3,823	14.5
High school graduate	2,097	8.0
Associates degree	10,364	39.4
College degree	5,750	21.8
Graduate degree	4,294	16.3
Race/ethnicity		
Asian	29	0.1
Black	102	0.4
Hispanic[a]	3,357	12.8
Native American	197	0.1
White	21,252	80.7
Unknown	1,391	5.3
Year of Birth		
Prior to 1900	934	3.6
1900–1919	6,143	23.3
1920–1939	11,818	44.9
1940–1962	7,433	28.2
Year of Hire		
1943–1949	8,368	31.8
1950–1959	6,180	23.5
1960–1969	5,188	19.7
1970–1980	6,592	25.0
Age at First Hire (years)		
16–19	3,079	11.7
20–24	6,738	25.6
25–29	6,146	23.3
30–39	6,641	25.2
40+	3,724	14.1
Age at End of Follow-up (Dec. 31, 2017) (years)		
<40	406	1.5
40–49	717	2.7
50–59	1,655	6.3
60–69	5,786	22.0
70–79	7,985	30.3
80–89	7,154	27.2
90–95+	2,625	10.0
Vital status as of Dec. 31, 2017		
Confirmed dead	15,737	59.8
Confirmed alive	10,517	40.0
Died outside U.S.	42	0.2
Lost to follow-up	32	0.1

[a]While Hispanic is an ethnicity rather than a race category, we chose to tabulate them separately.

Vital status and outcome determination

Vital status and cause of death for LANL workers as of 31 December 2017, was determined from linkages with the National Death Index (NDI); state mortality files; the Social Security Administration (SSA) Death Master File; the SSA Service to Epidemiological Researchers (which confirms alive status); and credit reporting agencies (Mumma et al. 2018). The Centers for Disease Control and Prevention LinkPlus program, which incorporates a probabilistic scoring system that does not require exact matches on all variables, was used for in-house matches (Campbell 2008). The ability to

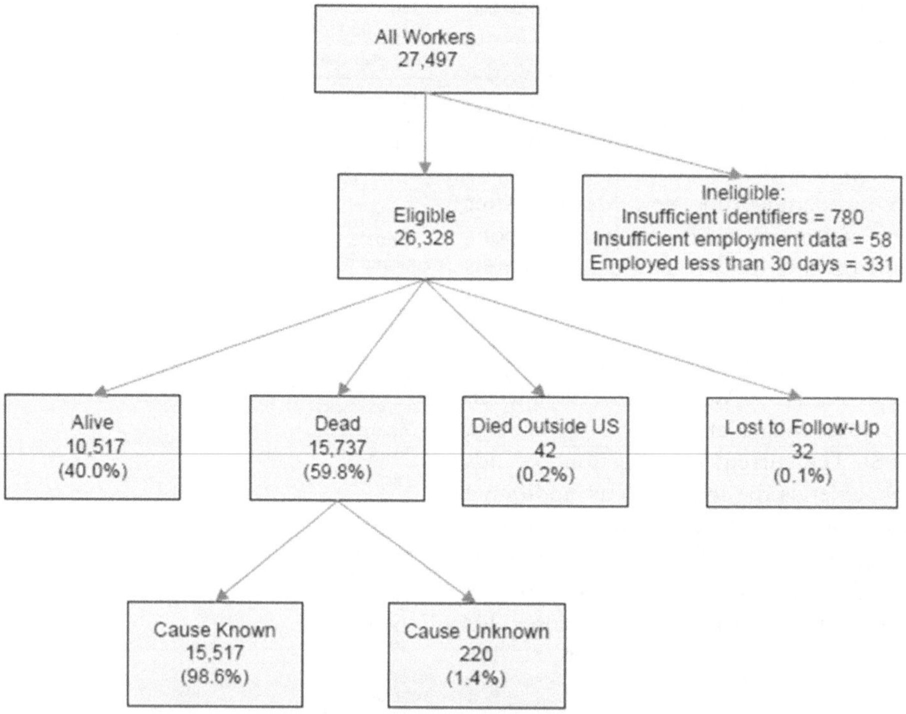

Figure 1. Schematic of the selection and vital status tracing results for the 26,328 workers at the Los Alamos National Laboratory (LANL) first employed between 1943 and 1980 and followed through 2017.

match worker rosters to state mortality data, including data from the New Mexico Department of Vital Status, was especially valuable in identifying deaths that occurred before 1979 when the NDI began.

SSA vital status files and other sources confirmed that 10,517 workers (40.0%) were alive in 2017 (Figure 1). The cause of death was determined for all but 220 (1.4%) of the 15,737 workers who had died. Workers without a match in SSA, a state mortality file, NDI, or if a credit header record could not be located ($n = 32$, 0.1%) were assumed alive until their date of last employment at LANL. Workers with serious renal disease were identified by linkage with the U.S. Renal Data System (1975–2018), which includes persons who received kidney dialysis or transplant (U.S. Renal Data System 2020), and their causes of death determined.

Radiation dose reconstructions for external sources of radiation and tritium

Measurements of gamma, neutron, and tritium exposures were obtained from data files submitted to the Department of Energy's Comprehensive Epidemiologic Data Resource (CEDR) (Department of Energy 2021) from previous studies (Wiggs 1987; Galke et al. 1992; Wiggs et al. 1994). Pocket chambers or film dosimeters were used for personnel monitoring from 1944 through 1980 when they were replaced with thermoluminescent dosimeters.

The weighted absorbed dose to the tissues of the body was estimated assuming an irradiation geometry and energy spectrum of incident photons and neutrons on the body. The irradiation geometry assumed half the radiation was incident of the body in an anterior-posterior (AP) manner and half in a rotational (ROT) manner as recommended in NCRP Report 178 (National Council on Radiation Protection and Measurements 2018b). The dose coefficients for such geometries are tabulated in ICRP Publication 116 (International Commission on Radiological Protection 2010). The recorded photon annual personal dose equivalent, Hp(10), was assumed to be associated with down-scattered photon energy spectra associated with ^{137}Cs. The distribution of the tissue doses is shown in Figure 2.

Annual personal dose equivalent values also were recorded for thermal, fast, and other neutron exposures. The fast neutron values were interpreted assuming the Watt spontaneous-fission spectra, while the thermal neutron values assume energy of 0.025 eV. The other neutron exposures assume a degraded fast neutron energy spectrum. The tissue equivalent annual doses for the neutron exposures assume a radiation weighting factor of 2.5 for the thermal neutrons (International Commission on Radiological Protection 2007, Figure 1). A radiation weighting factor of 16 for non-thermal neutrons, both fast neutrons and fission neutrons, was assumed based on the fast neutron spectrum-weighted average over the ICRP Publication 103 (International Commission on Radiological Protection 2007, Figure 1) continuous neutron radiation weighting factor function. The distribution of tissue doses for the fast neutrons is shown in Figure 2.

Tritium intakes were monitored by measurements of activity within the urine. The measurements were recorded without information on the exposure period. The pre-1968 analytical methods involved Geiger–Müller detectors which were replaced by liquid scintillation procedures in later years. The pre-1968 medium detectable tritium concentration was 1 μCi/L, and it was 0.1 μCi/L in later years. Considerable uncertainties exist in the estimated annual doses; however, these dose contributions are rather minor –

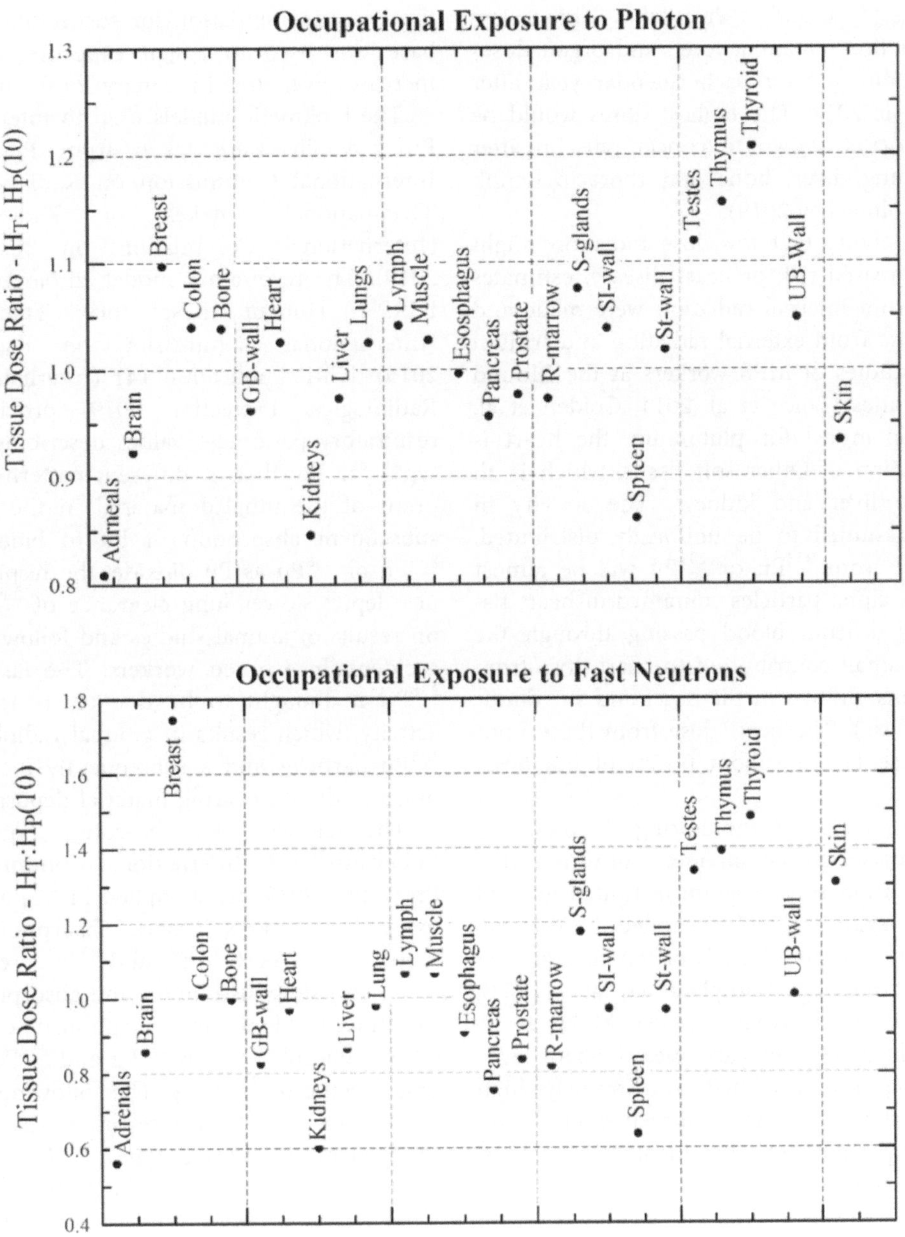

Figure 2. Ratio of photon tissue dose (top panel) to personnel equivalent dose and ratio of fast neutron tissue dose (bottom panel) to personnel equivalent dose for 25 tissues.

average lifetime tritium absorbed dose less than 3 mGy. The analyses assume a Dose Weighting Factor (DWF) of 1 for tritium, although higher values up to 3 are plausible (Cox et al. 2008; National Council on Radiation Protection and Measurements 2018c). Again, the tritium contribution to organ doses is so low that the DWF assumptions have a negligible effect. For example, the mean total dose to RBM for Pu (DWF = 1) and tritium (DWF = 1) is 12.4 mGy, and for a DWF of 2 for tritium, the mean dose to RBM increases only to 12.56 mGy while the maximum of 835 mGy remains the same.

Radiation dose reconstruction for internally deposited Pu

Many radionuclides were handled at LANL since the start of operations in 1943, but mostly in small quantities. By far the greatest potential for elevated doses from internal emitters arose from work with ^{238}Pu and ^{239}Pu, which were handled in relatively large quantities. Numerous short-term intakes of ^{239}Pu occurred over the years, particularly during early work involving reduction of Pu to metallic form and development of technology for use of Pu in a nuclear bomb. Later peacetime programs including the development of ^{238}Pu heat sources resulted in elevated intakes of ^{238}Pu. Since 1943, 6499 LANL workers were monitored for ^{239}Pu or ^{238}Pu and 5142 (79.1%) had positive bioassays. The number of Pu bioassay records available for evaluation was 158,222, that is, 24.3 records per worker on average. To reduce the number of time-consuming case-by-case dose reconstructions for relatively noninformative low intakes of plutonium, screening criteria were developed. Effective doses were available and 10 mSv was taken as the minimum criterion for conducting comprehensive individual dose reconstructions for the ^{238}Pu and ^{239}Pu intakes. There were 457 workers who met this screening level, and organ doses for 25 tissues were estimated for each calendar year after intake through the year 2017. The highest doses would be expected for those organs known to concentrate Pu after intake, that is, the lung, liver, bone and thoracic lymph nodes (Kathren and Tolmachev 2019).

Because of the possibility that low-dose radiation might be associated with increased risk of heart disease, estimates of dose to the heart from internal radiation were made and combined with the dose from external radiation and tritium as done in previous studies of MPS workers at the Mound and Mallinckrodt facilities (Boice et al. 2014; Golden et al. 2019). In the systemic model for plutonium, the heart is treated as a mass fraction of Other soft tissue, which is all soft tissue except the liver and kidneys. The activity in Other soft tissue is assumed to be uniformly distributed. The dose to the heart from ^{238}Pu or ^{239}Pu will be almost entirely self-dose from alpha particles contained in heart tissue plus alphas emitted from blood passing through the heart. There is a very small contribution to heart dose from other types of radiations emitted in the heart and surrounding tissues (crossfire dose). The heart dose from these non-alpha emissions represents only about 0.15% of the heart dose from alphas.

MPS dose reconstructions for monitoring Pu intake at LANL were based primarily on measurements of urinary Pu, the key component of a bioassay program instituted in 1944 for monitoring ^{239}Pu exposure and expanded in the late 1960s to monitor ^{238}Pu exposure. Details of the program are provided in publications by Campbell et al. (1972), Hempelmann et al. (1973), Lawrence (1978), Miller et al. (2008), and Efurd et al. (2008). The early measurement techniques involved sizable errors related to extremely high background counts and widely varying measurements over relatively short time periods for individual Pu workers. Continual improvements in urinary Pu measurements were made over time, and reasonably reliable techniques with relatively small errors appear to have been developed by 1957.

The bioassay data for Pu were supplemented with post-mortem measurements of ^{239}Pu in tissues of 28 LANL workers and ^{238}Pu measurements in tissues of 22 of these 28 workers (see for example McInroy et al. 1991; McInroy 1995). The autopsy data provided information on the relative contents of these Pu isotopes in different tissues including bone, liver, lungs, and tracheobronchial lymph nodes (TBLN). The relative content of Pu in lungs, TBLN, and bone plus liver is an indicator of the solubility of the inhaled material. The collective autopsy and bioassay data for these 28 workers provided checks on the reliability of the biokinetic model for systemic Pu used in the dose reconstructions for ^{238}Pu and ^{239}Pu (Leggett et al. 2005, 2018; International Commission on Radiological Protection 2019).

Additional information was provided by a LANL database of incidents involving potential intake of ^{238}Pu or ^{239}Pu. For each incident, the database records the date, worker identifier number, and abbreviations representing the nature of the exposure or reason for suspecting an intake. This database was used to assign exposure scenarios to observed increases over time in urinary Pu for the listed workers.

The biokinetic models used to interpret bioassay data for Pu generally were taken from Publication 141 of the International Commission on Radiation Protection ICRP), 'Occupational Intakes of Radionuclides: Part 4' (International Commission on Radiological Protection 2019). The respiratory model employed is an update of the ICRP's Human Respiratory Tract Model (HRTM) (International Commission on Radiological Protection 2015). ICRP Publication 141 (International Commission on Radiological Protection 2019) provides several sets of respiratory parameter values describing different absorption types for Pu, that is, different patterns over time of dissolution of the inhaled material in the respiratory tract and subsequent absorption of Pu to blood. For inhalation of ^{239}Pu or ^{238}Pu as Pu dioxide, the respiratory parameter values depict slower lung clearance of ^{239}Pu than ^{238}Pu based on results of animal studies and follow-up measurements on accidentally exposed workers. The faster lung clearance of ^{238}Pu is thought to be due to its relatively high specific activity, which results in gradual radiolytic fragmentation of ^{238}Pu particles and a subsequently increased rate of dissolution of the Pu-bearing material deposited in the lungs

The following Pu absorption types defined in ICRP Publication 141 (International Commission on Radiological Protection 2019) were applied in ^{239}Pu dose reconstructions for LANL workers: generic absorption types F, M, and S (applicable to both ^{238}Pu and ^{239}Pu) representing fast, moderate, and slow dissolution and absorption from the respiratory tract to blood; an absorption type-specific to Pu nitrate (also applicable to both ^{238}Pu and ^{239}Pu); and an absorption type-specific to $^{239}PuO_2$. The following Pu absorption types were applied in ^{238}Pu dose reconstructions for LANL workers: generic absorption types F, M, and S; an absorption type-specific to ceramic $^{238}PuO_2$ (used in the production of ^{238}Pu heat sources); an absorption type-specific to non-ceramic $^{238}PuO_2$ (representing all forms of $^{238}PuO_2$ other than ceramic $^{238}PuO_2$); and two variations of the ICRP's respiratory parameter values for ceramic $^{238}PuO_2$ depicting shorter peak urinary ^{238}Pu times (about 550 and 1100 days) than predicted by the parameter values for ceramic $^{238}PuO_2$ given in Publication 141 (about 1500 days). The indicated variations of the ICRP's model for ceramic $^{238}PuO_2$ allowed improved model fits urinary ^{238}Pu data for many LANL workers whose ^{238}Pu excretion patterns were qualitatively similar to that indicated by the ICRP's model for ceramic $^{238}PuO_2$ but showed earlier peaks than predicted by the ICRP's parameter values.

Incident or autopsy data were not available for the preponderance of LANL Pu workers with positive ^{238}Pu or ^{239}Pu bioassay. In the absence of such data, assignments of intake times and absorption types were based as feasible on comparisons of worker-specific patterns of time-dependent urinary Pu with patterns predicted by ICRP models for different forms of Pu. Differences in model-predicted patterns of 24-h or cumulative urinary ^{238}Pu or ^{239}Pu are illustrated

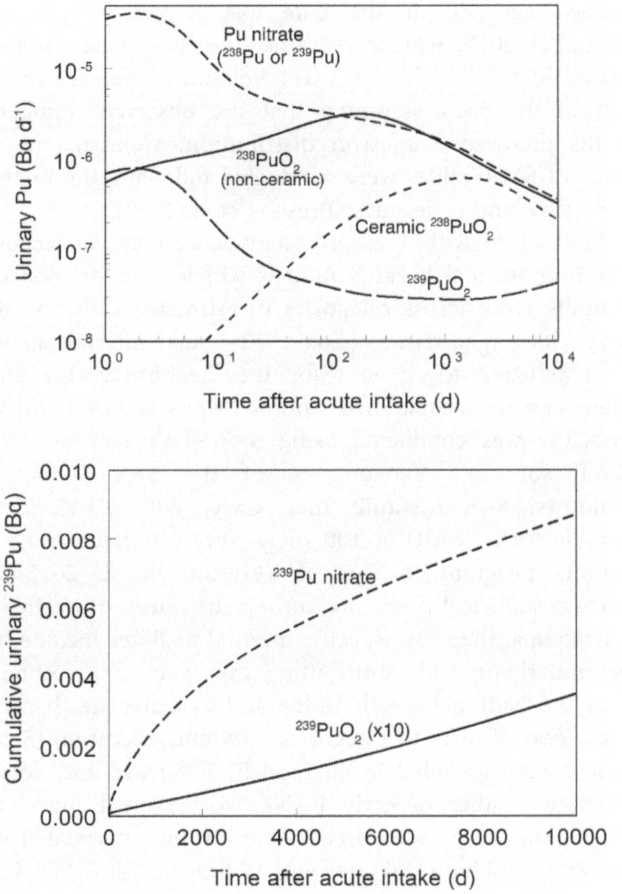

Figure 3. Illustrations of different patterns of daily and cumulative urinary Pu activity predicted for different absorption types considered in dose reconstructions for LANL workers. An intake of 1 Bq ^{238}Pu or ^{239}Pu of particle size 5 μm AMAD is assumed. In the lower panel, the cumulative activity curve for inhaled $^{239}PuO_2$ is raised 10-fold for ease of viewing.

in the upper panel of Figure 3 for selected forms of Pu. An intake of 1 Bq ^{238}Pu or ^{239}Pu of particle size 5 μm AMAD (Activity Median Aerodynamic Diameter) is assumed: 50% of the activity in the aerosol is associated with particles of aerodynamic diameter greater than the AMAD. For example, the predicted time post intake at which the peak urinary excretion rate occurs is a few days for $^{239}PuO_2$, ^{239}Pu nitrate, or ^{238}Pu nitrate; a few months for non-ceramic $^{238}PuO_2$; and a few years for ceramic $^{238}PuO_2$ (as depicted by the ICRP's model for ceramic $^{238}PuO_2$).

An indication of the solubility of the inhaled material often can be gained from inspection of the cumulative excretion curve inferred from the collective 24-h urinary Pu data. This is indicated in the lower panel of Figure 3 by model-predicted cumulative urinary ^{239}Pu curves for inhalation of the moderately soluble form ^{239}Pu nitrate and the relatively insoluble form $^{239}PuO_2$. In general, the curve for cumulative urinary ^{239}Pu for inhalation of a relatively soluble or moderately soluble form of ^{239}Pu will be concave downward, while the cumulative excretion curve for a relatively insoluble form will be a nearly straight line or may even be slightly concave upward.

Uncertainties in dose estimates for internally deposited Pu may be divided into the following categories:

Measurement of urinary Pu

Measurement errors for urinary Pu appear to have represented an important source of error in reconstructed doses for ^{239}Pu intake in the first year or two of operations, a smaller but still important source of error from the mid-1940s to 1957, and a generally less important source of error after the introduction of a relatively sensitive measurement technique in 1957.

Mode of Pu intake

The mode of intake of Pu by an individual worker was based on LANL's database of incidents when the worker and mode of intake were listed in the database and the listed date of exposure was consistent with increases in the worker's urinary Pu. In most cases of elevated urinary Pu, however, the time of increase in urinary Pu could not be matched to records in the incident database and intake via inhalation was assumed. This assumption could have resulted in order-of-magnitude overestimates of lung dose in many cases in which the preponderance of intake was via wounds. Estimated doses to tissues outside the respiratory and alimentary tract were only moderately sensitive to erroneous assignment of the mode of intake. This is because the latter tissue doses were back-calculated from the time course of urinary excretion of Pu, which largely reflects the time-dependent systemic behavior of Pu regardless of the pathway of entry of Pu into the systemic circulation.

Assignment of an absorption-type to inhaled Pu

Assignment of an absorption-type often was straightforward for ^{238}Pu because the urinary ^{238}Pu patterns frequently resembled the characteristic excretion curves for inhalation of either ceramic or non-ceramic $^{238}PuO_2$. For many workers with positive ^{239}Pu bioassay data and a smaller subset of workers with positive ^{238}Pu data, the data could be produced about equally well with different plausible exposure scenarios, that is, different combinations of intake time(s), intake mode(s), and/or respiratory absorption-type(s). In such cases, the exposure was assumed to be via inhalation, and absorption types representing intermediate residence times in the lungs were selected in an effort to avoid large overestimates or underestimates of lung dose. It is expected that the errors in lung dose estimates usually were less than a factor of three for inhaled ^{239}Pu and a factor of two for inhaled ^{238}Pu except in cases where Pu inhalation was erroneously assigned as the mode of intake.

Biokinetic model for systemic Pu

Autopsy data together with bioassay data for 28 workers allowed checks on the reliability of the biokinetic model for systemic Pu used in the dose reconstructions for ^{238}Pu and ^{239}Pu (Leggett et al. 2005; International Commission on Radiological Protection 2019). Model predictions based on urinary Pu measurements were found to be reasonably consistent with the postmortem systemic burdens, with the rate of urinary excretion of systemic Pu, and with observed mean or median ratios of bone Pu to liver Pu, although the relative Pu contents of liver and skeleton varied from worker to worker. Acute intake of Pu was assumed in all cases. Chronic exposure to Pu is an unlikely scenario for LANL and most other sites. The times of acute exposure were based on time-dependent urinary Pu patterns. These findings suggested that the systemic model for Pu typically was not a major source of error in dose estimates from internally deposited Pu but could represent a substantial (greater than a factor of 2) error in estimated doses to the liver, bone surface, or bone marrow for some individuals.

Career doses

In addition to the dosimetry records available for workers at LANL 'on site', dosimetry records documenting radiation exposure received before or after working at LANL were sought from eight additional sources: DOE Radiation Exposure Monitoring System (REMS) (Department of Energy 2018); other historic DOE radiation exposure data not included in DOE REMS; the Nuclear Regulatory Commission Radiation Exposure Information and Reporting System (REIRS) (Hagemeyer et al. 2018; Nuclear Regulatory Commission 2019); the US Navy Dosimetry System (Naval Dosimetry Center 2019) and US Army and US Air Force dosimetry files (Department of Army 2012); the DOD Nuclear Test Personnel Review program for military veterans who participated in the U.S. atmospheric nuclear tests from 1944 to 1962 (Till et al. 2014); and Landauer, Inc. dosimetry records (National Council on Radiation Protection and Measurements 2018b; Yoder et al. 2020). All nonoverlapping doses from each source were added together for each calendar year to obtain the total external dose received by each worker, following the procedures outlined in Boice, Leggett, et al. (2006) and Ellis, Boice, et al. (2018) which also describe how the calendar year organ-specific organ doses following intakes of radionuclides are incorporated into the individual dosimetry files.

Statistical analysis

Standardized Mortality Ratio (SMR) analyses were conducted by comparing the observed number of deaths with those expected based on mortality rates in the comparable general population of the United States. Observed numbers of deaths from cancers and other diseases were counted by age, sex, and calendar year for LANL workers overall and for subgroups defined by employer, education, year of birth, age, calendar year, ethnicity (e.g. Hispanic), type of radiation monitoring, and years of follow up. Expected values were based on general population rates for white males and females. Technically, Hispanic is an ethnicity and not a race. Hispanic ethnicity, when known, is often classified as white with white general population rates used for comparison. The SMR analyses were based on the underlying cause of death. The follow-up time for each worker began 30 days after the date of first employment at LANL, and ended at his or her date of death, or the date when the worker reached age 95, or the date lost to follow-up, or 31 December 2017, whichever came first. Statistical variability was evaluated by the 95% exact Poisson confidence interval (CI) of the SMR assuming that the observed number of deaths followed a Poisson distribution. Comparisons and ratios of SMR values were conducted following the methods of Breslow and colleagues (Breslow et al. 1983).

Internal, or within-cohort, analyses were conducted using Cox proportional hazards models which were developed to compute risks across categories of estimated radiation dose to specific organ/tissues (Cox 1972). Such internal analyses are conducted to account for the 'healthy worker effect' often seen in occupational studies (Monson 1986). All Cox modeling was conducted using SAS/STAT software (SAS/STAT software, Version 9.4 of the SAS System for Windows, SAS Institute Inc., Cary, NC, USA). Excess Relative Risks (ERR) at 100 mGy were estimated using the Peanuts program in Epicure (Preston et al. 2015). To increase statistical power for infrequent causes of death such as leukemia, the cause-specific internal analyses include both the underlying and contributing causes of death obtained from the National Death Index and available death certificates. Year of birth (4 categories), sex and education (5 categories) were included in all models. Extensive and detailed interview studies of early LANL workers had found that educational status was directly and strongly correlated with cigarette smoking (Mahoney and Wilkinson 1987; Wilkinson et al. 1987). Education for all workers was then used as a measure of socioeconomic status (SES) for all workers to account for smoking and other lifestyle factors (Cohen et al. 2018).

For the internal cohort analyses, radiation dose was treated as a categorical measure, with categories defined based on the distribution of dose for each organ/tissue of interest. The dose category was then treated as a time-dependent measure, allowing workers to be assigned to increasingly higher dose categories over time as their individual radiation doses accrued. Doses accrued over time include both external doses and internal doses from the decay of radionuclides that had been retained in the body after inhalation or ingestion. Age was used as the underlying timescale for the hazard function. To allow for a possible latent period between radiation exposure and any effect consequent to it, doses were lagged, that is, excluded if they occurred during some assumed interval prior to the event of interest. A 10-year lag was applied for solid cancers and other chronic conditions and a 2-year lag for leukemia, myelodysplastic syndrome (MDS) and bone cancer. Sensitivity analyses were conducted assuming a 5-year lag for selected cancers. Parameter estimates and standard errors for the dose categories in the Cox time-dependent regression models were used to obtain hazard ratios (HR) and 95% confidence intervals (CI) for death due to the cause under investigation compared with those in the referent group taken as the workers with low cumulative radiation dose (<5 mGy). Trend tests treated the radiation dose as a single continuous measure, and two-sided p-values are presented.

HRs at 100 mGy were computed for selected causes of death (Golden et al. 2018).

In the primary analysis for lung, liver and bone cancers, non-Hodgkin lymphoma (NHL) and liver cirrhosis, a dose weighting factor (DWF) of 20 was applied to alpha-particle absorbed doses from intakes of plutonium-239 and plutonium-238, that is, the biological effectiveness of the absorbed dose from Pu intakes was assumed to be 20 times that of the absorbed dose alone. The unit weighted-dose, for example, weighted-mGy, is used for such analyses. Additional analyses were performed assuming dose weighting factors of 1 or 10 as possible indicators of biological effectiveness for these alpha-particle absorbed doses. In the absence of a complete understanding of the biological effectiveness of alpha particles to cause-specific outcomes, the use of various DWFs provide a range of possibilities and boundary conditions. Other than for the lung, liver, thoracic lymph nodes (NHL) and bone, a DWF of 1 was used for other organs since Pu concentration was negligible, for example, the mean dose to the esophagus changed from 12.8 mGy (DWF = 1) to 12.9 weighted-mGy (DWF = 20).

Results

Descriptive characteristics of workers at LANL are shown in Table 1: 76.6% were employed only by LANL, 16.1% only by ZIA, and 7.3% by both sometime during their career. Overall, 75.2% were male, 80.7% white and 12.8% Hispanic. Fewer than a quarter of the workers had a high school education or less, and 38.1% had a college degree or higher. Approximately one-quarter of the study population was born before 1920. Workers were first hired in the 1940s, 1950s, 1960s and 1970–1984 for 31.8%, 23.5%, 19.7%, and 25.0% of the cohort, respectively. At the end of vital status tracing, 40% of the workers were alive, 59.9% had died, and only 32 (0.1%) were lost to follow-up (Figure 1). The cause of death was obtained for all but 1.2% of those who died. The mean duration of follow-up was 44.9 years.

Overall, 66.8% of workers at LANL were monitored for any type of radiation, 64.5% for external radiation, 43.6% for neutrons, 7.8% for tritium, 14.5% for ^{238}Pu and 24.7% for ^{239}Pu. (Table 2). There were 6499 workers who were monitored for either ^{238}Pu or ^{239}Pu of whom 1357 (21%) did not have a positive assay. For the 5142 workers with a positive assay, approximately 8.3% had an effective dose >10 mSv which triggered individual case-by-case dose reconstructions for 25 organs or tissues. For the remainder, 88% had effective doses <1 mSv; 1.7% between 1–5 mSv; and 2% between 5–10 mSv. Effective dose is a radiation protection unit and is not appropriate for epidemiologic analyses (Cox and Kellerer 2003). A crude characterization of the low organ doses received by workers not meeting the 10 mSv screening level, however, can be made by assuming various absorption types for the plutonium intakes contributing to the effective dose calculations, for example, 50% nitrate and 50% oxide. If done, crude characterizations of organ doses for lung, endosteal bone surface, and liver might be of the order of 3–4 times the effective dose and substantially less for other organs.

Table 2. Radiation monitoring history by source of exposure for the 26,328 LANL workers first employed 1943–1980 and followed through 2017.

Radiation monitoring	n	%
Any radiation monitoring (any type)		
Yes	17,588	66.8
None	8,740	33.2
Any external		
Yes	16,965	64.4
LANL only	10,792	41.0
LANL + Other facilities	5,288	20.1
Other facilities only	885	3.4
None	9,363	35.6
Any neutrons		
Yes	12,192	46.3
None	14,136	53.7
Any tritium		
Yes	2,075	7.9
None	24,253	82.1
Any ^{238}Pu		
Not monitored	22,514	85.5
Monitored, no positive assay	624	2.4
Monitored, positive Assay, <10 mSv[a]	2,965	2.4
Monitored, positive Assay, >10 mSv	225	0.8
Any ^{239}Pu		
Not Monitored	19,829	75.3
Monitored, no positive assay	1,301	4.9
Monitored, positive Assay, <10 mSv[a]	4,767	18.1
Monitored, positive Assay, >10 mSv	431	1.6
Any ^{238}Pu or ^{239}Pu		
Not monitored for either ^{238}Pu or ^{239}Pu	19,829	75.3
Monitored, no positive assay for either ^{238}Pu or ^{239}Pu	1,357	5.2
Monitored, positive Assay, <10 mSv[a] for ^{238}Pu and ^{239}Pu	4,685	17.8
Monitored, positive Assay, >10 mSv for either ^{238}Pu or ^{239}Pu	457	1.7

[a]10 mSv effective dose was taken as the minimum criterion for conducting comprehensive individual dose reconstructions for Pu intakes.

There were 6173 (23.4%) workers who were monitored for radiation other than at LANL (mean dose, 4.2 mGy, median dose 0.1 mGy, max dose 506 mGy), including 55 who were at the TRINITY detonation test and 335 who participated at other atmospheric nuclear weapons testing. Additional dosimetry information was obtained for 301 workers who served In the U.S. Navy, 138 in the U.S. Air Force and 19 in the U.S. Army. There were 554 who worked at a nuclear power plant, and 4500 who received doses at DOE facilities other than LANL: notably the Nevada National Security Site (~30%), Hanford (21%), Rocky Flats (7%), and Lawrence Livermore National Lab (6%). For the 8740 workers not monitored for radiation at LANL, 11 (0.1%) were monitored elsewhere (mean dose, 0.16 mGy, median 0.1, max dose 0.6 mGy). The nonmonitored LANL workers were more likely to be female, employed prior to 1960 and less educated than the monitored workers.

Table 3 provides summary statistics for 15 tissue-specific radiation absorbed doses (mGy) among LANL workers monitored for radiation. The organ-specific doses combine doses from photons, neutrons, tritium, ^{238}Pu and ^{239}Pu. Three different Dose Weighting Factors (DWF) were applied to the ^{238}Pu and ^{239}Pu alpha particle doses (1, 5, 10, and 20), and a DWF of 1 was applied to tritium doses. The weighting factors used for neutron doses were 2.5 (thermal) and 16 for nonthermal neutrons. Similar to the convention

Table 3. Tissue-specific radiation absorbed dose (weighted-mGy)[a], combining external and internal sources, for three Dose Weighting Factors (DWF) for Pu among the 17,053[b] LANL workers monitored for radiation.

Tissue	DWF	Mean	Median	Std.	Max.
Lung	1	13.9	0.90	48.7	1,248
	10	20.9	0.90	134.9	8,620
	20	28.6	0.90	251.6	16,811
Bone	1	13.8	0.89	47.4	924
	10	19.7	0.90	112.1	3,881
	20	26.4	0.90	204.6	7,701
Liver	1	13.5	0.87	46.8	864
	10	20.9	0.87	134.8	4,731
	20	29.2	0.87	253.9	9,413
Colon	1	13.6	0.87	47.3	906
	10	13.6	0.90	47.4	908
	20	13.7	0.91	47.5	909
Prostate	1	12.5	0.80	43.1	844
	10	12.6	0.85	43.2	846
	20	12.6	0.85	43.4	849
Esophagus	1	12.8	0.83	44.2	858
	10	12.8	0.86	44.3	860
	20	12.9	0.87	44.4	862
Brain	1	11.6	0.76	39.4	760
	10	11.7	0.79	39.5	762
	20	11.7	0.79	39.6	764
Stomach	1	13.3	0.86	46.5	893
	10	13.4	0.88	46.6	895
	20	13.4	0.90	46.7	897
Pancreas	1	12.1	0.80	41.7	815
	10	12.2	0.82	41.8	815
	20	12.2	0.83	41.9	815
Bladder	1	13.7	0.89	47.9	907
	10	13.8	0.91	48.0	909
	20	13.8	0.93	48.1	911
Kidneys	1	10.8	0.72	36.8	772
	10	11.0	0.76	37.1	772
	20	11.1	0.76	37.6	772
Breast	1	15.1	0.95	53.4	1,140
	10	15.1	0.99	53.5	1,140
	20	15.2	0.99	53.7	1,140
Heart	1	13.4	0.87	46.6	893
	10	13.4	0.90	46.7	895
	20	13.5	0.90	46.8	897
Red bone marrow	1	12.4	0.84	42.4	835
	5	12.7	0.85	43.3	845
Thoracic lymph nodes	1	36.5	0.89	419	25,600
	20	479.9	0.90	8,256	503,015

[a]Cumulative organ doses include doses from photons, neutrons, tritium, ^{238}Pu and ^{239}Pu. The weighting factors used for neutron doses were 2.5 (thermal), 16 (fast) and 16 (fission). Three different Dose Weighting Factors (DWF) were applied to the ^{238}Pu and ^{239}Pu doses (1, 10, and 20). A DWF of 1 was assumed for tritium doses.

[b]17,053 is the number of workers with dose information and differs from the 17,588 workers in Table 2 who were monitored for radiation. 535 workers were excluded who were only monitored for Pu and whose intake levels were very low and below the 10 mSv effective dose threshold for organ dose estimation.

used in the Life Span Study (LSS) of Japanese atomic bomb survivors, we refer to the neutron-weighted absorbed doses as weighted absorbed doses in Gy.

The mean and median doses were similar for most organs for DWFs of 1. For the lung (DWF of 1), the mean organ dose was 13.9 mGy and the range was broad and up to 1.25 Gy. For a DWF of 20, the mean lung dose was 28.6 weighted-mGy with a maximum of 16.8 weighted-Gy. The organs with a meaningful uptake of plutonium showed a doubling of the mean dose when a weighting factor of 20 is applied, specifically, the lung, bone (skeleton), liver and thoracic lymph nodes. The mean dose to the heart using a DWF of 1 was 13.4 mGy and the maximum was 893 mGy. For the heart and other organs/tissues with minimal uptake of plutonium isotopes, the mean and maximum doses were similar regardless of the DWF. DWFs of 1 and 5 are presented for red bone marrow since there does not appear to be much of increased biological effectiveness associated with alpha particle dose to the red bone marrow in causing leukemia in humans (Boice 1993; U.S. Environmental Protection Agency 1999; Daniels et al. 2006).

Table 4 shows the relative contribution of each radiation source to each mean tissue-specific dose among the 17,053 LANL workers monitored for radiation. External photons contribute about 84% of the dose for those organs that do not concentrate plutonium. But this percentage drops to below 50% for the organs that concentrate plutonium, that is, the lung, bone, liver and thoracic lymph nodes, and the predominant source comes from plutonium depositions.

Table 5 shows the SMRs for over 40 causes of death among the 26,328 LANL workers by sex. There were 1,181,472 person-years of follow-up, and the SMR for all causes of death was 0.76 (95%CI 0.75, 0.78). There were 839 deaths from lung cancer (SMR 0.54, 95%CI 0.51, 0.58). The SMRs for lung cancer did not differed by whether the workers were monitored for plutonium (SMR 0.55, 95%CI 0.51, 0.59; $n=612$) or not (SMR 0.53, 95%CI 0.46, 0.60; $n=227$). The SMR for leukemia other than CLL was 0.98 (95%CI 0.82, 1.16; $n=136$). Most causes of death were significantly below or close to expectation, for example, the liver (SMR 0.82; $n=110$), kidney (SMR 0.51; $n=62$) and NHL (SMR 0.85; $n=162$). The SMRs for liver cancer differed by whether the workers were monitored for plutonium (SMR 0.78, 95%CI 0.61, 0.98; $n=76$) or not (SMR 0.94, 95%CI 0.65, 1.32; $n=34$), but the difference was not statistically significant. Ischemic heart disease (IHD) also occurred below population expectation (SMR 0.60; $n=3043$) as did cerebrovascular disease (CeVD) (SMR 0.72; $n=871$). Cancer of the bone occurred close to expectation (12 deaths observed compared with 10.8 expected) as did liver cirrhosis (SMR 1.02; $n=319$). However, the SMRs for bone cancer differed by whether the workers were monitored for plutonium (SMR 2.44, 95%CI 0.98,5.03; $n=7$) or not (SMR 0.63, 95%CI 0.20, 1.48; $n=5$), and the ratio (3.86) was statistically significant (95%CI 1.22, 12.2). Cancers of the pleura and mesothelioma were not in excess (SMR 0.94; $n=20$). Causes of death that occurred significantly above expectation were suicides (SMR 1.16; $n=309$), notably among women, and beryllliosis (SMR 23.0; $n=4$), notably among men. Deaths due to dementia, Alzheimer's, Parkinson's, and other motor neuron disease were not increased (SMR 0.96, $n=973$), nor were deaths due to mental and behavioral disorders (SMR 0.99; $n=520$).

SMR analyses were computed by subgroups that included sex, race/ethnicity, education, employer (LANL, Zia) and others (data not shown). The all-cause SMR of death for women was significantly low (SMR 0.85, 95%CI 0.82, 0.88; $n=3545$) as it was for men (SMR 0.74, 95%CI 0.73, 0.76; $n=12,231$). There were no significantly high SMRs among women except for all external causes of death (SMR 1.17, 95%CI 1.00,1.35; $n=175$) which was attributed to a significantly high risk of suicide (SMR 1.92; $n=49$). For men,

Table 4. The relative contribution of radiation source to each mean tissue-specific dose (weighted-mGy)[a] among the 17,053 LANL workers monitored for radiation.

Tissue	Mean dose[a]	Percentage of photons (%)	Percentage of neutrons (%)	Percentage of tritium (%)	Percentage of plutonium (%)
Lung	28.6	81.8	14.1	1.3	2.8
Bone	26.4	82.1	13.9	1.3	2.8
Liver	29.2	81.6	14.2	1.3	2.9
Colon	13.7	82.8	15.0	1.3	0.9
Prostate	12.6	83.3	14.4	1.4	0.9
Esophagus	12.9	83.2	14.4	1.4	0.9
Brain	11.7	84.0	13.6	1.4	1.0
Stomach	13.4	82.8	15.0	1.3	0.9
Pancreas	12.2	83.7	14.1	1.4	0.9
Bladder	13.8	82.7	15.1	1.3	0.9
Kidneys	11.1	83.8	13.5	1.4	1.2
Breast	15.2	81.6	16.2	1.3	0.9
Heart	13.5	82.8	14.9	1.3	0.9
Red bone marrow	12.4	83.7	14.1	1.4	0.8
Thoracic lymph nodes	36.5	81.9	14.2	1.3	2.6

[a]Cumulative organ doses include the doses from photons, neutrons, tritium, ^{238}Pu and ^{239}Pu. The weighting factors used for neutron doses were 2.5 (thermal), 16 (fast) and 16 (fission). A Dose Weighting Factor of 20 was applied to the ^{238}Pu and ^{239}Pu doses in this table for all tissues except red bone marrow and thoracic lymph nodes which had a Dose Weighting Factor of 1 applied. A Dose Weighting Factor of 1 was used for tritium doses in this table.

only death due to berylliosis was significantly high (SMR 26.1, 95%CI 7.08, 66.8; $n = 4$). Among the Hispanic population, both sexes combined, the all-cause of death SMR also was significantly low (SMR 0.83, 95%CI 0.79, 0.86; $n = 1917$). However, there were a number of causes of death that were significantly high among the Hispanic population, that is, cancers of the stomach (SMR 2.40; $n = 39$) and liver (SMR 1.59; $n = 23$), mental and behavioral disorders (SMR 1.52; $n = 67$), cirrhosis of the liver (SMR 2.24; $n = 82$), nonmalignant kidney disease (SMR 1.68; $n = 49$), diabetes (SMR 1.44; $n = 71$) and accidents (SMR 1.73; $n = 157$).

Significantly high SMRs among whites were seen only for berylliosis (SMR 20.0; $n = 3$) and suicides (SMR 1.22; $n = 284$). The SMRs for lung cancer by educational status were highest for workers with a high school education (0.80; $n = 101$) or less (0.76; $n = 179$) and the lowest SMR were for those holding a college (0.44; $n = 164$) or graduate (0.22; $n = 68$) degree (data not shown). Both LANL (SMR 0.74; $n = 11{,}322$) and ZIA (SMR 0.90; $n = 4454$) workers had significantly low all-cause-of-death SMRs. Significantly high SMRs among LANL workers were seen for berylliosis (SMR 24.5; $n = 3$) and suicide (SMR 1.19; $n = 240$). Among Zia employees, significantly high SMRs were seen for mental and behavioral disorders (SMR 1.47; $n = 147$), cirrhosis of the liver (SMR 1.73; $n = 132$) and accidents (SMR 1.45; $n = 264$). Somewhat higher SMRs were observed for workers hired 1942–1949 where 3 of the 4 deaths due to berylliosis occurred.

Table 6 shows the hazard ratios (HR) and excess relative risks (ERRs) for lung cancer for DWF of 1, 10, and 20. There were 839 deaths due to lung cancer. The dose distribution was broad and up to 1.25 Gy for a DWF of 1 for plutonium intakes, and for a DWF of 20 it was up to 16.8 weighted-Gy. The ERR at 100 weighted-mGy was 0.01 (95%CI −0.02, 0.03), and there was no evidence for a dose-response (Figure 4). There was no statistically significant differences in lung cancer radiation risk estimates between men and women related in large part to the smaller number of women, their overall lower dose to lung, and the absence of any suggestion of a radiation association overall or for either sex. Assuming radiation weighting factors of 10 and 20 did not change the pattern of risk by dose or the risk coefficients.

Table 7 presents the HRs and ERRs for selected causes of death. A DWF of 20 was applied for those tissues that concentrate plutonium, that is, lung, liver, bone and thoracic lymph nodes (for NHL). A DWF of 1 was applied and presented for those tissues with minimal plutonium deposition since there were no meaningful variations in the patterns of risk for higher DWFs. The only significant dose-response was seen for esophageal cancer with the ERR at 100 mGy of 0.29 (95%CI 0.02, 0.55; $n = 106$). For sites with increased internal exposures to plutonium such as the liver and thoracic lymph nodes, there was no evidence for an association with radiation dose. The ERRs at 100 weighted-mGy were −0.01 (95%CI −0.14, 0.12; $n = 110$), −0.74 (95%CI −1.52, 0.03; $n = 319$), and −0.01 (95%CI −0.04, 0.02; $n = 162$) for liver cancer, liver cirrhosis and NHL, respectively. Neither was there a significant association seen for leukemia other than chronic lymphocytic leukemia (non-CLL) [ERR per 100 mGy −0.43 (95%CI −1.11, 0.24); $n = 160$)] nor for the combination of non-CLL leukemia and myelodysplastic syndrome (MDS) [(-0.35 (95%CI −0.92, 0.22)]. Cancers of the colorectum, stomach and bladder were not significantly increased. Multiple myeloma also failed to show a statistically meaningful radiation association. CLL was not increasing with the ERR per 100 mGy of 0.04 (95%CI −0.59, 0.66; $n = 58$). Other sites not convincingly or consistently linked to ionizing radiation also showed little evidence of a dose-response such as cancers of the prostate, brain (for adults) and pancreas, or ischemic heart disease. For ischemic heart disease, the ERR per 100 mGy was −0.06 (95%CI −0.16, 0.04; $n = 3043$) (Figure 5). For cerebrovascular disease (CeVD) the ERR per 100 mGy was −0.11 (95%CI −0.35, 0.12; $n = 871$). For prostate cancer, the ERR per 100 mGy was −0.01 (95%CI −0.24, 0.23; $n = 383$), and for brain cancer, it was 0.20 (95%CI −0.27, 0.67; $n = 94$). There was no evidence for a dose-response for the combined causes of

Table 5. Standardized Mortality Ratios (SMR) for 26,328 workers at LANL first employed 1943–1980 and followed through 2017 by sex.

	Males		Females		Total		
Persons-at-risk	19,808		6,520		26,328		
Person-years	867,912		313,561		1,181,472		
Cause of death (ICD9)[a]	Obs	SMR	Obs	SMR	Obs	SMR	95%CI
All causes of death (001–999)	12,231	0.74*	3,545	0.85*	15,776	0.76*	0.75–0.78
All malignant neoplasms (140–208)	2,886	0.72*	881	0.87*	3,767	0.75*	0.73–0.77
Buccal cavity & pharynx (140–149)	45	0.55*	9	0.80	54	0.58*	0.44–0.76
Esophagus (150)	95	0.81*	11	1.08	106	0.83	0.68–1.01
Stomach (151)	113	1.01	14	0.75	127	0.98	0.81–1.16
Colon (153)	228	0.70*	67	0.74*	295	0.71*	0.63–0.79
Rectum (154)	55	0.81	11	0.75	66	0.80	0.62–1.02
Biliary passages & liver (155,156)	87	0.79*	23	0.98	110	0.82*	0.68–0.99
Pancreas (157)	204	0.96	46	0.80	250	0.92	0.81–1.05
Pleura & peritoneum (158.8, 158.9, 163) and mesothelioma (ICD10 C45)[b]	17	0.88	3	1.63	20	0.94	0.57–1.45
Larynx (161)	23	0.53*	0	0.00	23	0.50*	0.31–0.74
Bronchus, trachea, & lung (162)	653	0.50*	186	0.78*	839	0.54*	0.51–0.58
Bone (170)	11	1.25	1	0.51	12	1.11	0.58–1.95
Connective & other soft tissue (171)	23	1.09	8	1.36	31	1.15	0.78–1.64
Female breast (174)			192	1.11	192	1.11	0.96–1.27
Male breast (175)	3	0.59			3	0.59	0.12–1.173
All uterine (Females only) (179–182)			34	0.74	34	0.74	0.51–1.03
Cervix Uteri (180)			10	0.52*	10	0.52*	0.25–0.96
Ovary and other female genital organs (183–184)			66	1.00	66	1.00	0.77–1.27
Prostate (185)	383	0.96			383	0.96	0.87–1.06
Testes & other male genital organs (186, 187)	11	1.12			11	1.12	0.56–2.00
Kidney (189.0–189.2)	49	0.48*	13	0.74	62	0.51*	0.39–0.66
Bladder & other urinary (188, 189.3–189.9)	93	0.67*	10	0.65	103	0.67*	0.55–0.81
Melanoma of skin (172)	70	1.03	15	1.29	85	1.07	0.86–1.32
Other malignant neoplasm of skin (173)	21	0.82	4	1.32	25	0.87	0.56–1.28
Eye (190)	3	1.33	0	0.00	3	1.06	0.21–3.10
Brain & central nervous system (191–192)	71	0.74*	23	0.98	94	0.79*	0.64–0.97
Thyroid & other endocrine glands (193–194)	10	0.86	1	0.24	11	0.69	0.35–1.24
All lymphatic, hematopoietic tissue (200–208)	376	0.92	77	0.80	453	0.90*	0.82–0.98
Non-Hodgkin lymphoma (200, 202)	126	0.83*	36	0.92	162	0.85*	0.72–0.99
Hodgkin lymphoma (201)	19	1.08	2	0.55	21	0.99	0.61–1.51
Leukemia (202.4, 204–208)	154	1.01	26	0.79	180	0.97	0.83–1.12
Chronic lymphocytic leukemia (CLL) (202.4, 204.1, 204.9)	34	0.89	9	1.29	43	0.95	0.69–1.28
Leukemia other than CLL	119	1.05	17	0.66	136	0.98	0.82–1.16
Acute lymphocytic leukemia (204.0)	8	1.41	0	0.00	8	1.13	0.49–2.23
Acute myeloid leukemia (205.0, 205.3, 206.0, 207.0, 207.2)	66	1.16	9	0.71	75	1.07	0.85–1.35
Chronic myeloid leukemia (205.1, 206.1)	16	1.23	3	1.02	19	1.19	0.71–1.85
Multiple myeloma (203)	68	0.96	11	0.62	79	0.89	0.71–1.11
Smoking related cancers (140–150, 157, 161–162, 188–189)	1,163	0.58*	275	0.78*	1,438	0.61*	0.58–0.64
Non-smoking related cancers	1,723	0.86*	606	0.91*	2,329	0.87*	0.84–0.91
Polycythemia Vera (238.4)	2	0.67	0	0.00	2	0.53	0.06–1.92
Myelodysplastic syndrome (238.7)	21	0.73	4	0.65	25	0.71	0.46–1.05
Diabetes (250)	263	0.72*	67	0.62*	330	0.70*	0.62–0.78
Mental and behavioral disorders (290–319)	361	1.04	159	0.90	520	0.99	0.75–0.78
Parkinson's disease (332)	156	1.12	37	1.38	193	1.16	1.00–1.34
Dementia and Alzheimer's disease(290.0–290.4, 331.0)	462	0.93	273	0.89	735	0.92*	0.85–0.98
Dementia, Alzheimer's, Parkinson's disease & other motor neuron (290.0–290.4; 331.0; 332; 335.2)	654	0.97	319	0.93	973	0.96	0.90–1.02
Diseases of the Nervous System (320–389)	580	0.95	235	1.02	815	0.97	0.90–1.03
All Heart Disease (390–398, 404, 410–429)	3,544	0.63*	822	0.69*	4,366	0.64*	0.62–0.66
Ischemic heart disease (410–414)	2,517	0.59*	526	0.68*	3,043	0.60*	0.58–0.63
Cerebrovascular disease (430–438)	625	0.70*	246	0.80*	871	0.72*	0.68–0.77
Nonmalignant respiratory disease (460–519)	1,209	0.77*	382	0.94	1,591	0.80*	0.76–0.84
Bronchitis, emphysema, asthma (490–493)	458	0.73*	170	0.95	628	0.78*	0.72–0.84
Berylliosis (503)	4	26.04*	0	0	4	23.04*	6.20–59.0
Cirrhosis of liver (571)	267	1.00	52	1.14	319	1.02	0.91–1.14
Nephritis & nephrosis (580–589)	164	0.73*	31	0.50*	195	0.68*	0.59–0.78
AIDS (042–044)	11	0.34*	0	0	11	0.32*	0.16–0.58
All external causes of death (800–999)	973	1.00	175	1.17*	1,148	1.02	0.96–1.08
Accidents (850–949)	647	1.00	117	1.05	764	1.00	0.93–1.08
Suicides (950–959)	260	1.08	49	1.92*	309	1.16*	1.03–1.30

[a]ICD denotes International Classification of Disease. ICD Revision 9 codes are shown; [b]mesothelioma did not have an explicit code in ICD9 but did in ICD10 as denoted.

*Indicates SMR is significantly different from 1.0; $p < 0.05$.

Table 6. Internal cohort dose-response analyses for lung cancer over categories of lung-specific radiation doses among 26,328 workers at LANL first employed 1943-1980 and followed through 2017 by sex and for different Dose Weighting Factors (DWF)

	DWF = 1 for Pu				DWF = 10 for Pu				DWF = 20 for Pu			
Dose (mGy)[a]	Number of Workers	Number of cases	HR	95% CI	Number of Workers	Number of cases	HR	95% CI	Number of Workers	Number of cases	HR	95% CI
Lung Cancer												
<5	21,728	694	1.0	REF	21,705	694	1.0	REF	21,704	694	1.0	REF
5 – <50	3,539	112	0.89	0.73; 1.10	3,403	104	0.87	0.70; 1.08	3,369	104	0.88	0.71; 1.09
50 - <100	503	14	0.77	0.45; 1.30	506	13	0.72	0.41; 1.24	473	12	0.71	0.40; 1.26
≥100	558	19	1.01	0.64; 1.59	714	28	1.08	0.74; 1.58	782	29	1.02	0.70; 1.49
P for trend		839		*0.87 (+)*		839		*0.65 (-)*		839		*0.64 (-)*
HR (95% CI) at 100 mGy			1.01	0.88; 1.19			1.01	0.96; 1.07			1.01	0.98; 1.04
ERR (95% CI) at 100 mGy			0.01	−0.15; 0.17			0.01	−0.04; 0.06			0.01	−0.02; 0.03

	MALES: DWF = 20 for Pu				FEMALES: DWF = 20 for Pu			
Dose (mGy)[a]	Number of Male Workers	Number of cases	HR	95% CI	Number of Female Workers	Number of cases	HR	95% CI
Lung Cancer								
<5	15,425	515	1.0	REF	6,279	179	1.0	REF
5 – <50	3,212	97	0.86	0.69; 1.07	157	7	*Categorical model doesn't converge*[b]	
50 - <100	445	12	0.75	0.43; 1.34	28	0		
≥100	722	29	1.11	0.51; 1.07	60	0		
P for trend		653		*0.52 (+)*		186		
P for heterogeneity for sex[b]								
HR (95% CI) at 100 mGy			1.01	0.98; 1.03				
ERR (95% CI) at 100 mGy			0.01	−0.02; 0.03				

Note: Models include radiation doses lagged by 10 years. Doses were analyzed as a time-dependent measure. Outcomes include deaths from underlying causes of death. All models adjusted for sex, education (5 categories, as in Table 1) and year of birth (4 categories, as in Table 1). P-value for test for linear trend in the relative risk (i.e., hazard ratio) computed for continuous lung dose. P-values are two-sided. REF denotes the referent category. HR denotes Hazards Ratio; CI denotes Confidence Interval; ERR denotes Excess Relative Risk.

[a]Cumulative weighted absorbed doses to the lung (weighted-mGy) include the contributions of photons, neutrons, tritium, ^{238}Pu and ^{239}Pu. The DWFs used for neutron doses are 2.5 (thermal), 16 (fast) and 16 (fission). Three different Dose Weighting Factors (DWF) are applied to the ^{238}Pu and ^{239}Pu doses (1, 10, and 20). A DWF of 1 was assumed for tritium doses.

[b]Small numbers of cases among females in the upper dose categories created an unstable model. A crude contrast of females with > 5 weighted-mGy to those with < 5 weighted-mGy yields a Risk Ratio of 1.00 (95% CI 0.47, 2.13). A similar crude contrast for males yields a Risk Ratio of 0.95 (95% CI 0.78, 1.14) (Rothman and Boice 1979). These estimates are consistent (p for heterogeneity = 0.88), but the number of female cases with high lung doses is small.

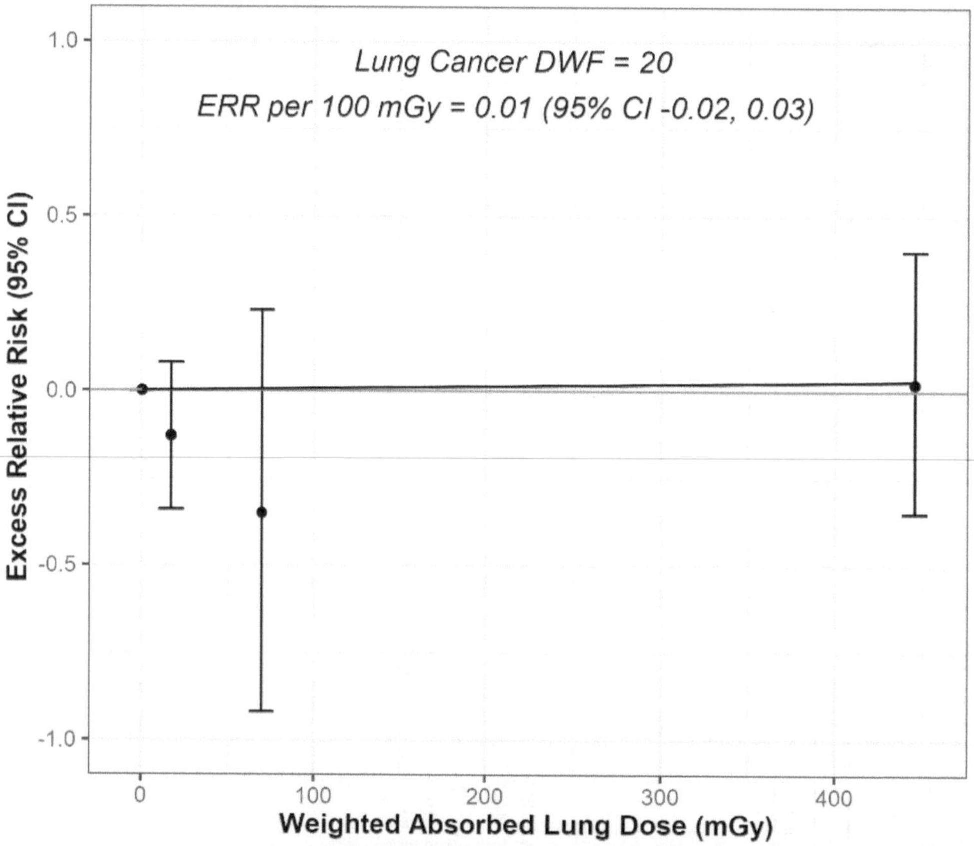

Figure 4. Excess Relative Risk (ERR) of lung cancer by weighted absorbed dose to lung among 26,328 workers at the Los Alamos National Laboratory (LANL) first employed 1943–1980 and followed through 2017. Cox proportional hazards models include radiation dose lagged by 10 years, adjusting for sex, education and year of birth. Dose was analyzed in a time-dependent manner. The ERR and 95% confidence intervals within selected dose categories are plotted at the mean dose within the category (Table 6). The linear trend line for a continuous measure of radiation dose to the lung is presented.

death from dementia, Alzheimer's disease, Parkinson's disease, and motor neuron disease with the ERR per 100 mGy of 0.01 (95%CI −0.19, 0.16; $n = 973$). There a suggested dose-response for Parkinson's disease, although not at the level of statistical significance: ERR per 100 mGy of 0.16 (95%CI −0.07, 0.40; $n = 273$). There was no evidence for a dose-response for deaths from suicide.

Table 8 shows the relative risk (hazards ratio) for lung cancer over categories of external dose and internal plutonium dose for 26,324 workers first employed at LANL between 1943–1980 and followed through 2017. External dose to the lung was taken as the contributions of dose from photons, neutrons, and tritium. The referent group was taken as the 21,745 workers with cumulative lung doses <5 mGy for both external lung dose and internal plutonium alpha-particle lung dose. Modeling the joint effects of external and internal exposures is imprecise since only 19 lung cancers occurred among the 447 workers with >5 weighted-mGy lung doses from plutonium intakes. Nonetheless, adjusting for sex, education, year of birth, and external dose to the lung, the HR at 100 weighted-mGy plutonium dose were 1.15 (95%CI 0.66,1.99; $n = 839$). The trend was not statistically significant ($p = .62$).

Linkage with the U.S. Renal Data System (USRDS) identified 334 former LANL workers with end-stage renal disease (ESRD). The earliest diagnosis was in 1975 and the last in 2018. Overall, 123 of the incident cases were attributable to diabetes, 83 to hypertension, 48 to glomerulonephritis, 8 to cystic kidney disease, 15 to other urologic disorders, and 57 to other or unknown conditions. Among the 6459 workers who were monitored for plutonium, there were 92 incident cases of ESRD (1.4%) of which 17 (18.5%) were attributed to glomerulonephritis. Among the larger number of 19,829 workers not monitored for plutonium, there were 242 incident cases of ESRD (1.3%) of which 31 (12.8%) were attributed to glomerulonephritis. Of the 334 workers who developed ESRD, 285 died by end of follow-up, including 2 from kidney cancer, 45 from nonmalignant kidney disease (nephritis or nephrosis) and 44 from diabetes. Both kidney cancers and 18 of the 45 deaths from nephritis or nephrosis were among the workers monitored for Pu.

Discussion

The study of 26,328 workers at the Los Alamos National Laboratory first employed between 1943 and 1980 is notable because of its size; comprehensive approach for individual dose reconstruction that incorporates doses from photons, neutrons, tritium and plutonium; inclusion of over 6000 female workers; complete and long-term follow-up of up to 75 years; and the inclusion of a Hispanic population of over 3000 workers. There were no statistically significant associations between radiation and lung cancer, liver cancer, NHL, leukemia other than CLL, or ischemic heart disease. The

Table 7. Internal cohort dose-response analyses for selected causes of death by organ-specific dose among 26,328 workers at LANL first employed 1943–1980 and followed through 2017[a,b].

Dose (weighted-mGy)	Workers	Cases	HR	95%CI
Esophageal cancer (ICD9 150)				
<5	21,901	80	1.0	REF
5 to <50	3,433	20	1.22	0.74; 2.01
≥50	994	6	1.24	0.54; 2.88
P for trend		106		0.03 (+)
HR (95%CI) at 100 mGy			1.33	1.02; 1.73
ERR (95%CI) at 100 mGy			0.29	0.02; 0.55
Stomach cancer (ICD9 (151)				
<5	21,832	105	1.0	REF
5 to <50	3,474	16	0.90	0.52; 1.54
50 to <100	484	4	1.60	0.58; 4.39
≥100	538	2	0.76	0.19; 3.10
P for trend		127		0.76 (−)
HR (95%CI) at 100 mGy			0.92	0.55; 1.56
ERR (95%CI) at 100 mGy			−0.08	−0.61; 0.45
Colorectal cancer (ICD9 153, 154)				
<5	21,789	303	1.0	REF
5 to <50	3,503	54	0.97	0.72; 1.31
50 to <100	488	7	0.86	0.41; 1.83
≥100	548	7	0.82	0.39; 1.75
P for trend		371		0.81 (−)
HR (95%CI) at 100 mGy			0.97	0.74; 1.27
ERR (95%CI) at 100 mGy			−0.03	−0.31; 0.24
Liver cancer (ICD9 155, 156)				
<5	21,797	91	1.0	REF
5 to <50	3,313	15	0.96	0.55; 1.69
≥50	1,218	2	0.69	0.25; 1.87
P for trend		110		0.88 (−)
HR (95%CI) at 100 mGy			0.99	0.87; 1.13
ERR (95%CI) at 100 mGy			−0.01	−0.14; 0.12
Pancreas cancer (ICD9 157)				
<5	22,000	207	1.0	REF
5 to <50	3,375	36	0.98	0.68; 1.40
≥50	953	7	0.69	0.32; 1.47
P for trend		250		0.54 (−)
HR (95%CI) at 100 mGy			0.87	0.57; 1.35
ERR (95%CI) at 100 mGy			−0.13	−0.57; 0.30
Female breast (ICD9 174)				
<5	6,285	183	1.0	REF
≥50	239	8	1.31	0.64; 2.68
P for trend		191		0.58 (+)
HR (95%CI) at 100 mGy			1.22	0.61; 2.42
ERR (95%CI) at 100 mGy			0.20	−0.49; 0.88
Prostate Cancer (ICD9 185)				
<5	15,647	294	1.0	REF
5 to <50	3,238	72	0.93	0.71; 1.20
≥50	923	17	0.71	0.43; 1.15
P for trend		383		0.95 (−)
HR (95%CI) at 100 mGy			0.99	0.79; 1.25
ERR (95%CI) at 100 mGy			−0.01	−0.24; 0.23
Bladder & other urinary (ICD9 188, 189.3–189.9)				
<5	21,769	73	1.0	REF
5 to <50	3,518	23	1.40	0.86; 2.26
50 to <100	487	4	1.60	0.58; 4.42
≥100	554	3	1.11	0.35; 3.56
P for trend		103		0.40 (+)
HR (95%CI) at 100 mGy			1.15	0.83; 1.58
ERR (95%CI) at 100 mGy			0.14	−0.18; 0.46
Brain cancer (ICD9 191–192)[c]				
<5	22,071	77	1.0	REF
5 to <50	3,330	11	0.86	0.45; 1.65
≥50	927	6	1.67	0.71; 3.89
P for trend		94		0.37 (+)
HR (95%CI) at 100 mGy			1.24	0.78; 1.98
ERR (95%CI) at 100 mGy			0.20	−0.27; 0.67
Non-Hodgkin lymphoma (ICD9 200, 202.0–202.1, 202.8–202.9)				
<5	21,709	130	1.0	REF
5 to <50	3,360	23	0.93	0.59; 1.49
50 to <100	443	3	0.88	0.28; 2.78
≥100	816	6	0.95	0.4; 2.18
P for trend		162		0.47 (−)
HR (95%CI) at 100 mGy			0.99	0.96; 1.02

(*continued*)

Table 7. Continued.

Dose (weighted-mGy)	Workers	Cases	HR	95%CI
ERR (95%CI) at 100 mGy			−0.01	−0.04; 0.02
Multiple myeloma (ICD9 203)				
<5	21,931	61	1.0	REF
≥5	4,397	18	1.08	0.63; 1.85
P for trend		79		0.19 (−)
HR (95%CI) at 100 mGy			0.36	0.08; 1.64
ERR (95%CI) at 100 mGy			−1.02	−2.54; 0.50
Non-CLL leukemia (ICD9 204.0, 204.2, 204.3–204.9, 205–208)				
<5	21,820	122	1.0	REF
5+	4,508	37	1.09	0.75; 1.59
P for trend		160		0.21 (−)
HR (95%CI) at 100 mGy			0.65	0.33; 1.28
ERR (95%CI) at 100 mGy			−0.43	−1.11; 0.24
CLL (ICD9 204.1, 202.4)				
<5	21,931	41	1.0	REF
5 to <50	3,425	12	1.32	0.68; 2.58
50 to <100	476	2	1.43	0.34; 5.99
≥100	496	3	2.01	0.61; 6.63
P for trend		58		0.91 (+)
HR (95%CI) at 100 mGy			1.04	0.56; 1.93
ERR (95%CI) at 100 mGy			0.04	−0.59; 0.66
Non-CLL leukemia + MDS (ICD9 204.0, 204.2, 204.3–204.9, 205–208, 238.7)				
<5	21,820	139	1.0	REF
≥5	4,508	46	1.13	0.80; 1.60
P for trend	185		0.23 (−)	
HR (95%CI) at 100 mGy		0.71	0.40; 1.25	
ERR (95%CI) at 100 mGy		−0.35−0.92; 0.22		
Dementia, Alzheimer's disease, Parkinson's disease, motor neuron disease (ICD9 290.0–290.4, 331.0, 332, 335.2)				
<5	22,071	786	1.0	REF
5 to <50	3,330	136	0.89	0.74; 1.08
50 to <100	458	25	1.07	0.72; 1.60
≥100	469	26	0.97	0.65; 1.44
P for trend		973		0.90 (−)
HR (95%CI) at 100 mGy			0.99	0.83; 1.18
ERR (95%CI) at 100 mGy			−0.01	−0.19; 0.16
Parkinson's Disease (ICD9 332)				
<5	22,071	213	1.0	REF
5 to <50	3,330	37	0.74	0.51; 1.05
50 to <100	458	14	1.83	1.06; 3.17
≥100	469	9	1.05	0.53; 2.06
P for trend		273		0.18 (+)
HR (95%CI) at 100 mGy			1.18	0.93; 1.49
ERR (95%CI) at 100 mGy			0.16	−0.07; 0.40
Ischemic heart disease (ICD9 410–414)				
<5	21,816	2,484	1.0	REF
5 to <50	3,489	429	0.96	0.87; 1.07
50 to <100	484	70	1.11	0.87; 1.41
≥100	539	60	0.91	0.70; 1.17
P for trend		3,043		0.23 (−)
HR (95%CI) at 100 mGy			0.94	0.85; 1.04
ERR (95%CI) at 100 mGy			−0.06	−0.16; 0.04
Cerebrovascular disease (ICD9 430–438)				
<5	22,071	742	1.0	REF
5 to <50	3,330	102	0.87	0.69; 1.06
50 to <100	458	15	0.90	0.54; 1.50
≥100	469	12	0.68	0.38; 1.21
P for trend		871		0.35 (−)
HR (95%CI) at 100 mGy			0.89	0.71; 1.13
ERR (95%CI) at 100 mGy			−0.11	−0.35; 0.12
Cirrhosis of liver (571)				
<5	21,821	279	1.0	REF
5 to <50	3,470	35	0.85	0.59; 1.22
50 to <100	498	4	0.72	0.27; 1.95
≥100	539	1	0.18	0.026; 1.31
P for trend		319		0.06 (−)
HR (95%CI) at 100 mGy			0.48	0.22; 1.03
ERR (95%CI) at 100 mGy			−0.74	−1.52; 0.03
Suicides (ICD9 950–959)				
<5	22,071	262	1.0	REF
5 to <50	3,330	40	1.15	0.82; 1.62
50 to <100	458	2	0.43	0.11; 1.72
≥100	469	5	1.15	0.47; 2.81
P for trend		309		0.79 (+)

(*continued*)

Table 7. Continued.

Dose (weighted-mGy)	Workers	Cases	HR	95%CI
HR (95%CI) at 100 mGy			1.05	0.72; 1.53
ERR (95%CI) at 100 mGy			0.05	−0.32; 0.43

MDS denotes Myelodysplastic Syndrome; CLL denotes Chronic lymphocytic leukemia; HR denotes Hazards Ratio; CI denotes Confidence Interval; REF denotes the referent category.

[a]Models are for radiation doses lagged by 10 years for all outcomes except Non-CLL and Non-CLL + MDS which uses radiation doses lagged by 2 years. Doses were analyzed as a time-dependent measure. Outcomes include deaths from underlying causes of death only, except for Non-CLL, CLL, and Parkinson's disease. Models adjusted for sex, education (5 categories, as in Table 1) and year of birth (4 categories, as in Table 1). *p*-Value for test for linear trend in the relative risk (i.e. hazard ratio) computed for continuous organ dose. *p*-Values are two-sided.

[b]Cumulative organ doses include dose contributions from photons, neutrons, tritium, ^{238}Pu and ^{239}Pu. The weighting factors used for neutron doses are 2.5 (thermal), 16 (fast) and 16 (fission). A DWF of 1 was assumed for tritium doses. For most organs with small plutonium depositions, a DWF = 1 was used since analyses with DWFs of 10 and 20 for Pu were essentially the same. For organs that concentrate plutonium after intake, a DWF of 20 for plutonium was used, that is, a DWF of 20 was assumed for cancers of the lung, liver and bone, for liver cirrhosis and for the thoracic lymph nodes. The cumulative organ doses are weighted absorbed doses.

[c]First two categories of Year of birth collapsed for model convergence

only significant radiation dose-response was for cancer of the esophagus. Measurable intakes of plutonium likely contributed to a small but significant excess of bone cancer. Plutonium may have contributed to a small but nonsignificant increase in lung cancer, but results were inconclusive. Workers in the 1940s showed a significant excess of deaths due to berylliosis related to beryllium exposure, based on 4 overall deaths. A dose-response of borderline significance was seen for Parkinson's disease which is consistent with a recent report among Mayak workers in Russia (Azizova et al. 2020). There were no meaningful differences between radiation risk for females compared with male workers, most notably for lung cancer, but the doses for women were low. The Hispanic population did not reveal any radiation dose-response relationships. Consistent with studies emerging from the Million Person Study, there was little evidence that radiation exposures received gradually over time significantly increased the risk of cancers and ischemic heart disease; however, the narrow dose ranges and associated sampling variability is such that radiation risks consistent with current understanding cannot be ruled out (Hauptmann et al. 2020).

Plutonium worker studies

Studies of plutonium workers in Russia have reported increased cancers of the lung, liver and bone for relatively high levels of plutonium intakes, (Koshurnikova et al. 2000; Gilbert et al. 2000, 2013; United Nations Scientific Committee on the Effects of Atomic Radiation 2008; ATSDR 2010; International Agency for Research on Cancer 2012; Sokolnikov et al 2008, 2015; Gillies, Kuznetsova, et al. 2017; Grellier et al. 2017). In some instances, intakes of plutonium were sufficiently large as to cause pulmonary sclerosis of the lung, a deterministic effect (Claycamp et al. 2000). There was little statistical evidence for radiation risks for other organs with lower levels of plutonium deposition (Shilnikova et al. 2003; Hunter et al. 2013; Sokolnikov et al. 2015; Kuznetsova et al. 2016). Studies of plutonium workers in the US and the UK involved much lower levels of plutonium, and radiation-related increases in cancer are not consistent (Wilkinson et al. 1987; Gilbert, Cragle, et al. 1993; Gilbert, Omohundro, et al. 1993; Wiggs et al. 1994; Voelz et al 1997; Omar et al. 1999; Brown et al. 2004; Wing et al. 2004; Wing and Richardson 2005; Agency for Toxic Substances and Disease Registry 2010; Boice et al. 2014).

Lung cancer

The mean cumulative lung dose for the 17,053 workers monitored for radiation was 13.9 mGy (maximum 1.2 Gy), with 1061 workers receiving >50 mGy. Applying a DWF of 20 for plutonium dose, the mean cumulative lung dose was 28.6 weighted-mGy (maximum 16.8 Gy). There were 836 deaths due to lung cancer, and no evidence for increasing risk with increasing radiation dose, regardless of applying radiation weighting factors for plutonium of 1, 10 or 20. ERRs per 100 weighted-mGy higher than 0.03 could be excluded with 95% confidence. There was no difference in the radiation-related lung cancer risks between men (19,808) and women (6520), but women had low radiation doses and number of cases were small. These findings for lung cancer and radiation are generally consistent with seven other cohort studies within the Million Person Study consisting of 385,778 workers (Boice et al. 2011, 2014; Boice, Ellis, et al. 2019; Golden et al. 2019; Boice et al. 2020; Boice, Cohen, Mumma, Hagemeyer, et al. 2021). A recently completed study of over 100,000 male and female medical radiation workers within the MPS found a statistically significant increase in lung cancer (Yoder et al. 2018, 2020; Boice, Cohen, et al. 2019; Boice, Cohen, Mumma, Yoder, et al. 2021), whereas a study of over 13,000 female workers at the Tennessee Eastman Corporation site did not find lung cancer to be associated with radiation lung doses from uranium dust inhalation (Boice, Cohen, Mumma, Golden, et al. 2021). Studies of Hanford workers reported a decrease in lung cancer risk with increasing radiation lung doses, though not statistically significant (Peterson et al. 1990). Studies of tuberculosis patients who received multiple chest fluoroscopies during the monitoring of lung collapse therapy found no evidence for radiation-related lung cancer risk, and there was no evidence that women had a higher risk than men (Boice, Ellis, et al. 2019). A recent study of medical radiological technicians also found little evidence for increasing risk of lung cancer with estimates of radiation dose (Velazquez-Kronen et al. 2020).

In contrast to the studies above, studies of the Japanese atomic bomb survivors exposed acutely to radiation in 1945 indicate a significant association between lung cancer and radiation dose and that women were at 2–3 times the excess relative risk as men (Cahoon et al. 2017). Studies of 22,374 Mayak workers also reported a significant difference between the female and male radiation risks for lung cancer, with females being 2–4 times higher than males based on incidence or mortality (Gillies, Kuznetsova, et al. 2017). The mean cumulative lung dose was 455 mGy (maximum 7.6 Gy) from gamma radiation and approximately 129–176 mGy (maximum 17–20 Gy) from ^{239}Pu. The ERR per Gy for lung cancer mortality for external lung dose was estimated to be about 0.38–0.40 (95%CI 0.22, 0.61) based on 158 lung cancer deaths. Similar to studies of Japanese atomic bomb survivors, smoking rates were very different in males (74%) and females (4%) for those with known smoking histories. When the analysis was restricted to workers who received less than about 200 mGy plutonium dose, there apparently was no difference in the male and female estimates of radiation risk for lung cancer. The overall lost-to-follow-up rate was relatively high at 22% (Preston et al. 2017). The estimates of plutonium dose were somewhat uncertain because few workers had urine bioassays during the time of exposure, and the average number of bioassays was less than 2 (Khokhryakov et al. 2013). Radiation estimates were based on incomplete coverage of only 42% of the workers employed at the plutonium facilities (Gillies, Kuznetsova, et al. 2017). A recent comprehensive dose uncertainty analysis found that accounting for uncertainty had a minor effect on the external gamma-ray dose-response, whereas it strikingly increased the width of the confidence intervals for the plutonium dose-response (Stram et al. 2021). Another large study involved the U.K. Sellafield workers which included 23,443 workers (Gillies, Kuznetsova, et al. 2017). The mean dose to the lung from external exposure was 73 mGy (maximum 1.8 Gy), and the dose from intakes of plutonium was of the order of 1.9–5.5 mGy (maximum 0.5–0.7 Gy) depending upon assumptions regarding solubility. These low doses were such that statistically significant associations between lung dose and lung cancer were not observed; however, pooled analyses with the higher-dose Mayak workers were informative, with the pooled ERR point estimates for lung cancer in relationship to plutonium lung dose within the range of 50–80 per 100 mGy. The risk coefficients for lung cancer for external exposure for LANL and Mayak workers are not statistically compatible, whereas they are compatible for plutonium dose although the LANL estimate is not statistically significant.

Whether there is a statistically meaningful difference in lung cancer risk between men and women is of importance to NASA (National Council on Radiation Protection and Measurements 2021). NASA bases its radiation protection standard for deep space exploration on individual lifetime cancer mortality predictions (National Aeronautics and Space Administration 2015, 2021). The lifetime cancer mortality models currently used are based on the atomic bomb survivor data, which limits the amount of time female astronauts can spend in space (National Academies/National Research Council 2012). This is because Japanese women exposed acutely to ionizing radiation were at a 2–3 times higher excess relative risk of lung cancer than men (National Council on Radiation Protection and Measurements 2016; Boice, Ellis, et al. 2019). The studies of plutonium workers exposed to mixed radiation fields (external low-LET photons and internal high-LET alpha particles) are one of the few terrestrial analogs capable of providing estimates of risks directly relevant for space radiation which consists of both low-LET radiation and high-LET high-energy heavy particle radiation (galactic cosmic rays (GCR)) (National Council on Radiation Protection and Measurements 2019; Boice, Ellis, et al. 2019; Boice 2019; Boice, Cohen, Mumma, Hagemeyer, et al. 2021). It might be noted that the exposures inside a space vehicle differ from those outside because of the interactions with shielding. Low energy alpha particles are a significant portion of the GCR exposure experienced inside a space vehicle, and much of the high-LET portion of radiation space exposure is actually from low energy alpha particles and protons (National Council on Radiation Protection and Measurements 2014; Slaba et al. 2016; Boice, Quinn et al. 2021).

Liver

The liver is similar to the lung, bone, and thoracic lymph nodes in being a primary site where plutonium would concentrate and be retained for extended periods (Voelz 1991; Agency for Toxic Substances and Disease Registry 2010; Kathren and Tolmachev 2019). Liver cancer has been found to be significantly increased among Japanese atomic bomb survivors for both males and females (Ozasa et al. 2012), whereas a negative dose-response was reported among INWORKS occupationally-exposed workers (Richardson et al. 2018). The radiation estimates among Japanese survivors are based on combined doses from external gamma and neutron radiation, whereas the INWORKS considered only external gamma radiation and excluded any concurrent worker doses from intakes of radionuclides or from external neutrons (National Council on Radiation Protection and Measurements 2018a). Significant excesses of liver cancer have been reported among the Mayak workers and appeared concentrated among workers with estimated plutonium absorbed doses >1 Gy (Gilbert et al. 2000; Sokolnikov et al. 2008). Among workers in the other MPS cohorts, no radiation-related increases of liver cancer were found among workers at the Rocketdyne (Boice, Cohen et al. 2006; Boice et al. 2011), Mound (Boice et al. 2014), or Mallinckrodt (Golden et al. 2019) facilities, nor among atomic veterans (Boice et al. 2020). Among LANL workers there was no evidence for an increased risk of liver cancer related to external or internal plutonium doses. The overall dose-response was negative, but only relatively high ERRs per 100 mGy >0.46 could be excluded with 95% confidence.

Bone

High bone doses from alpha-particle emitting radionuclides have been found to cause bone cancer, for example, among radium dial painters (Rowland et al. 1983; United Nations Scientific Committee on the Effects of Atomic Radiation 2008) and among Mayak plutonium workers (Koshurnikova et al. 2000, Sokolnikov et al. 2008). There is limited evidence that low-LET radiation causes bone cancer at relatively low doses below several Gy (United Nations Scientific

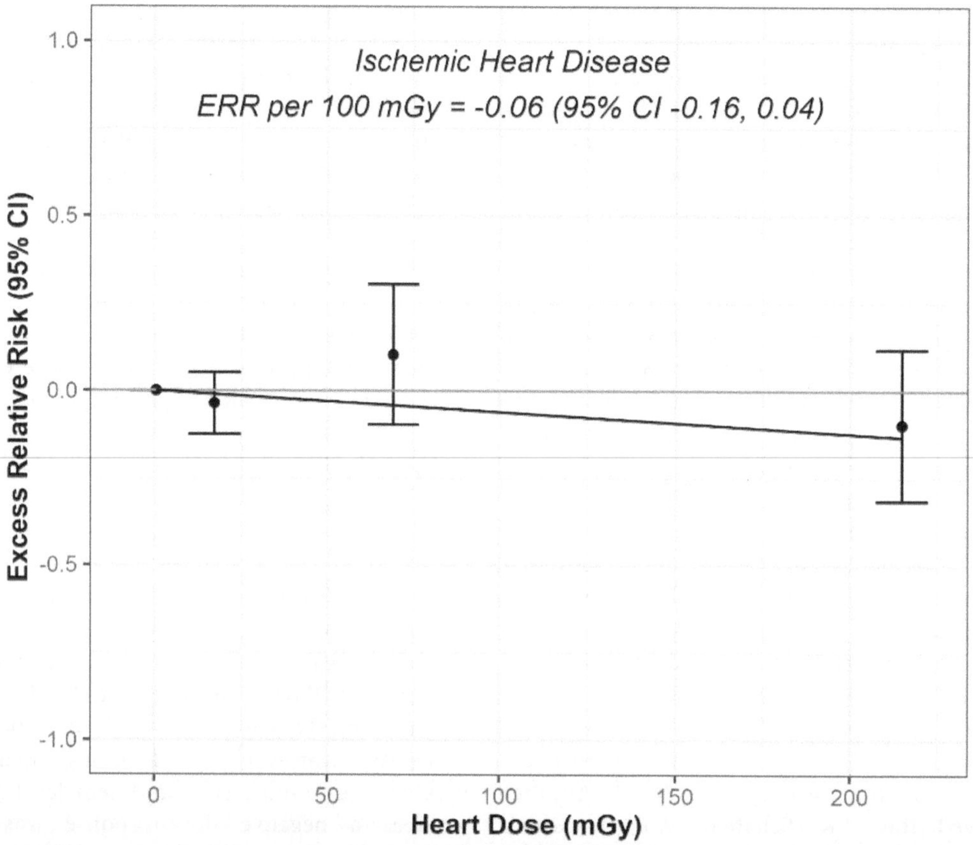

Figure 5. Excess Relative Risk (ERR) of ischemic heart disease by radiation dose among 26,328 workers at the Los Alamos National Laboratory (LANL) first employed 1943–1980 and followed through 2017. Cox proportional hazards models include radiation dose lagged by 10 years, adjusting for sex, education and year of birth. Dose was analyzed in a time-dependent manner. The ERR and 95% confidence intervals within selected dose categories are plotted at the mean dose within the category (Table 7). The linear trend line for a continuous measure of radiation dose to the heart is presented.

Table 8. Relative risks for lung cancer over categories of external[a] lung dose by internal plutonium lung dose among 26,328 workers at LANL first employed 1943–1980 and followed through 2017.

	External lung dose (mGy)[a]		
Internal plutonium lung dose (weighted-mGy)	<5	5–49	50+
<5			
No. of Workers	21,708	3337	836
No. of lung cancers	694	103	23
RR (95%CI)	1.0 (REF)	0.88 (0.71, 1.09)	0.83 (0.55, 1.26)
5–49			
No. of Workers	19	23	20
No. of lung cancers	0	2	1
RR (95%CI)	0.00 (0.00, 6.30)	1.86 (0.46, 7.46)	0.93 (0.13, 6.62)
50+			
No. of Workers	114	114	157
No. of lung cancers	3	6	7
RR (95%CI)	0.62 (0.20, 1.92)	1.26 (0.56, 2.82)	1.11 (0.53, 2.34)

[a]Absorbed doses to the lung are lagged 10 years. External dose to the lung is taken as the sum of photons, neutrons and tritium doses. The weighting factors used for neutron doses are 2.5 (thermal), 16 (fast) and 16 (fission). Internal dose to lung includes any intakes of ^{238}Pu or ^{239}Pu. Dose weighting factor (DWF) of 20 was applied to the ^{238}Pu and ^{239}Pu doses. RR denotes Relative Risk; CI denotes Confidence Interval; REF denotes Referent group

Committee on the Effects of Atomic Radiation 2008). The number of bone cancers occurring among the LANL workers was low and only 12 in total. There was a small but statistically significant increase in bone cancer among the workers monitored for plutonium, possibly attributable to prior plutonium intakes. It had been previously reported that one of the 26 earliest plutonium workers at LANL who were intensely studied clinically for over 50 years, had died of bone cancer (Hempelmann et al. 1973; Voelz 1991; Voelz and Lawrence 1991; Voelz et al. 1997). For this individual, our estimate of dose to the bone surface of about 55 mGy is 8 times lower than previous estimates of 440 mGy (Voelz 1991; Voelz and Lawrence 1991). This difference is related to differences in the ICRP biokinetic models used. In earlier models, the endosteum was assumed to be a 10-μm-thick layer that covered the surfaces of the bone, whereas new models assume the thickness to be 50-μm. The models predict the same activity in bone but the increase in mass results in a much lower dose. Assuming a DWF of 20, the dose to the bone for this worker is 1.10 weighted-Gy.

NHL

Because plutonium radionuclides inhaled in relatively soluble form tend to migrate to the thoracic lymph nodes, lymphoma is a plausible adverse health outcome (Kathren and Tolmachev 2019). Nearly 7000 LANL workers had positive assays reflecting intakes of ^{238}Pu and/or ^{239}Pu. This resulted in a contribution of dose to the thoracic lymph nodes that, on average, was about 3 times higher than the contribution from external sources of exposure. There was no evidence for a radiation-related excess of NHL among the LANL employees. The upper 95%CI about the ERR per 100 mGy of approximately 0.17 indicates that mortality risks higher than this level can be excluded with 95% confidence. A positive, but not significant, dose-response for NHL was reported among Mallinckrodt uranium processing workers for whom dose to the thoracic lymph nodes from intakes of uranium and radium also was evaluated (Golden et al. 2018). Studies of uranium processors (Silver et al. 2013; Zablotska et al. 2013) and uranium millers (Boice et al. 2008) do not find significant associations between radiation and NHL (United Nations Scientific Committee on the Effects of Atomic Radiation 2017). A relationship between radiation with either external gamma or internal plutonium for NHL was not seen among Mayak workers (Kuznetsova et al. 2016). External radiation has not generally been considered a cause of NHL (United Nations Scientific Committee on the Effects of Atomic Radiation 2008), although an excess was reported among male but not female atomic bomb survivors (Hsu et al. 2013). NHL was not associated with radiation in recent studies of nuclear weapons test participants (Boice et al. 2020) or US radiological technologists (Linet et al. 2020).

Leukemia, CLL, and MDS

Leukemia other than CLL is frequently found to be in excess among populations exposed to ionizing radiation. Excesses have been reported after 2 years from initial exposure, and the radiation risk coefficient is several times higher than observed for solid cancers. The mean dose to red bone marrow for LANL workers was 12.4 mGy and the maximum was 835 mGy. Leukemia other than CLL was not significantly elevated: ERR per 100 mGy −0.43 (95%CI −1.11, 0.24); $n = 130$. The upper 95%CI about the ERR per 100 mGy of 0.24 indicates that the data are consistent in a statistical sense with radiation risk estimates up to this level with 95% confidence; that is, the data are not inconsistent with those reported in studies of male atomic bomb survivors (Hsu et al. 2013) and in large occupational studies (Leuraud et al. 2015; National Council on Radiation Protection and Measurements 2018a; Gillies et al. 2019). Significant associations between non-CLL leukemia and radiation also have been seen in larger MPS populations of nuclear power plant workers and industrial radiographers (National Council on Radiation Protection and Measurements 2018a; Boice, Cohen, Mumma, Hagemeyer, et al. 2021). For Mayak workers, there was a significant risk of leukemia other than CLL associated with relatively high levels of gamma-ray exposures where nearly 20% of the exposed population receive greater than 3 Gy cumulative dose: ERR per Gy = 0.99 (90%CI 0.45; 2.12), n = 77, but there was no association with plutonium exposures (Shilnikova et al. 2003; Kuznetsova et al. 2016). Radiation associations among LANL workers were not found for CLL, consistent with studies of Mayak workers (Kuznetsova et al. 2016), and recent studies of nuclear weapons test participants (Boice et al. 2020) and US radiological technologists (Linet et al. 2020). MDS and MDS combined with AML also were not found to be associated with radiation, consistent with recent studies of nuclear weapons test participants (Boice et al. 2020) and US radiological technologists (Linet et al. 2020).

It is worth mentioning that MDS represents a grouping of heterogeneous diseases, and its association with radiation exposure is very uncertain (United Nations Scientific Committee on the Effects of Atomic Radiation 2020). The classification of MDS has evolved over the years, and in the past MDS cases were often misdiagnosed as AML. The United Nations Scientific Committee on the Effects of Atomic Radiation (2020) recently questioned whether MDS should be included in the assessment of leukemia risk. Among Japanese atomic bomb survivors included in the Life Span Study (LSS), MDS has been discussed but not included in the analyses of leukemia or lymphoma (Hsu et al. 2013). Radiation has been associated with MDS among long-term atomic bomb survivors at Nagasaki, a study which included survivors who were not in the LSS (Iwanaga et al. 2011). The results are difficult to interpret because of the extremely long latency following childhood exposures, up to 50 years which is inconsistent with the much shorter latencies seen for AML where a wave-like pattern of risk overtime is seen with peak excesses occurring 10 to 15 years after exposure and then risk diminishes over time.

Esophageal cancer

A significant association was seen between estimated dose to the esophagus and radiation among the LANL workers. The mean dose to esophagus was 12.8 mGy and the maximum was 858 mGy. Esophageal cancer also was significantly associated with radiation among Mayak workers (ERR per 100 mGy 0.13; 95%CI 0.04, 0.33) (Sokolnikov et al. 2015) and comparable to our ERR per 100 mGy of 0.29 (95%CI 0.02,0.55). Among workers at the Mound facility in Ohio, a significant positive dose-response trend for esophageal cancer was reported where the estimated excess relative risk at 100 mSv was 0.54 (05% CI 015–1.07) (Boice et al. 2014). A non-significant increase in esophageal cancer among nuclear weapons test participants was seen (ERR per 100 mGy 0.19; 95%CI −0.23, 0.60), although smoking habits among enlisted personnel may have been a contributing factor

(Boice et al. 2020). A significant excess risk of esophageal cancer related to radiation was seen in Japanese atomic bomb survivors for both cancer incidence (Preston et al. 2007) and mortality (Ozasa et al. 2012), although significance was apparent only among the female survivors. Increased ERRs per Gy were seen for esophageal cancer among the INWORKS consortium, but not at the level of statistical significance (Richardson et al. 2018).

Nephrotoxicity

Plutonium is an actinide heavy metal that has the potential to damage the kidneys (New Brunswick Laboratory 2015). There is little evidence, however, from human occupational studies or canine experimental studies for associations between plutonium intakes and malignant or nonmalignant kidney disorders (Agency for Toxic Substances and Disease Registry 2010). Similarly, there was no increase in nonmalignant kidney disease among all LANL workers, nor among the 6459 workers monitored for plutonium (SMR 0.86; 95%CI 0.67, 1.09). The ESRD registry has revealed increases in renal toxicity, especially glomerulonephritis in studies of workers exposed to silica dust and other non-radioactive materials (Steenland 2005). Linkage with the USRDS registry did not reveal increased numbers of glomerulonephritis cases among workers monitored for plutonium compared with those not monitored for plutonium. Additional evidence is thus provided that plutonium is an unlikely cause of serious kidney failure leading to either death or dialysis.

Ischemic heart disease

There was no evidence for a radiation-related association for ischemic heart disease based on 3043 confirmed deaths. The ERR per 100 mGy was −0.06 (95%CI 0.16, 0.04) indicating that risks higher than an ERR per 100 mGy of 0.04 could be rejected with 95% confidence. The absence of an association with ischemic heart disease is consistent with other MPS cohort studies that do not find increases in IHD over categories of radiation dose to the heart: Mallinckrodt (Golden et al. 2018), Rocketdyne (Boice, Cohen, et al. 2006), nuclear power plant workers and industrial radiographers (National Council on Radiation Protection and Measurements 2018a; Boice, Cohen, Mumma, Hagemeyer, et al. 2021), and atomic veterans (Boice et al. 2020). IHD was raised but not significantly so in a pooled analysis of Mayak and Sellafield nuclear workers: ERR per Sv of 0.22 (95%CI −0.06, 0.57) (Azizova, Batistatou, et al. 2018). A significant radiation risk for IHD was reported in the INWORKS study: ERR per 100 mSv = 0.017 (95%CI 0.0002, 0.036; $n = 17{,}279$). The INWORKS results need to be interpreted with caution, however, because of significant heterogeneity in the estimates of radiation risk, that is, significant radiation associations were seen for white-collar workers and not blue-collar workers and for female workers and not male workers (Gillies, Richardson, et al. 2017). The absence of information on internal radiation exposures and on potential confounders associated with lifestyle factors was also noted.

A recent report among the Japanese atomic bomb survivors found no association between radiation and IHD (Takahashi et al. 2017). Inconsistencies among studies may be related to the inability to control for potential confounding factors such as cigarette smoking and other lifestyle behaviors; medical conditions, such as diabetes and blood pressure; misclassification of death certificate causes of death; or whole-body radiation effects on renal function (Ozasa et al. 2017; de Vocht et al. 2020). Whether radiation doses to the heart <0.5 Gy can result in increased deaths due to IHD is an area of intense interest and study (Tran et al. 2017; National Council on Radiation Protection and Measurements 2018a; Wakeford 2019; Boice, Held, et al. 2019; EPRI 2020; Richardson et al. 2020; Tapio et al. 2021).

Cerebrovascular disease

There was no evidence for a radiation association for cerebrovascular (CeVD) disease based on 871 confirmed deaths among LANL workers. The ERR per 100 mGy was −0.11 (95%CI −0.35, 0.12), indicating that ERR per 100 mGy greater than 0.12 could be excluded with 95% confidence. Cerebrovascular disease was significantly raised among Mayak nuclear workers but only for incidence and not mortality (Azizova et al. 2011). The authors commented that differences in radiation risk between cerebrovascular disease incidence and mortality could not be ruled out, and discussed the possibility of surveillance bias which they concluded was unlikely. The total cumulative absorbed doses among Mayak were exceptionally large and reached up to 9 Gy. In a more recent mortality study that also included the Sellafield worker cohort, there were no radiation associations for CeVD (Azizova, Batistatou, et al. 2018). The dose-response was in relation to personal dose equivalent (Hp(10)). For Mayak, the ERR per 100 mSv for CeVD mortality was 0.00 (95%CI −0.006, 0.008) and for Sellafield it was 0.005 (95%CI −0.046, 0.079). The absence of a radiation association for CeVD mortality in both the Mayak and Sellafield workers is consistent with the LANL findings, although the statistical uncertainties are large.

Comparisons with the study of the Japanese atomic bomb survivors are not straightforward based on the published data because CeVD is not separated out from circulatory disease (Ozasa et al. 2012). However, Gillies, Richardson, et al. (2017, Table 7) obtained the LSS data on adult survivors and, using colon dose in the dose response, reported the CeVD mortality risk to be: ERR per 100 mSv = 0.008 (90% CI −0.0003, 0.018; $n = 7708$). Significant radiation associations for CeVD mortality were reported in the INWORKS study: ERR per 100 mSv = 0.049 (90% CI 0.011, 0.092; $n = 4399$) (Gillies, Richardson, et al. 2017). The authors raised caution in interpreting these findings because of the heterogeneity in the estimated radiation risks and the absence of information on internal radiation exposures as well as potential confounders associated with lifestyle factors. It was noted that a radiation risk was apparent only among blue-collar workers and not among white-collar workers. For the 15-Country Study using colon dose, the ERR per

100 mSv was 0.088 (95%CI −0.067, 0.316; $n = 1224$) (Gillies, Richardson, et al. 2017).

Berylliosis

Beryllium was used at LANL in the 1940s to develop and build implosion-assembly devices and more recently for operations related to nuclear reactors and weapons production, machining, fabrication, and testing of components (Stefaniak et al. 2003). At sufficiently high cumulative levels of intake, beryllium can cause chronic beryllium disease, lung cancer, and other adverse health effects (International Agency for Research on Cancer 2018). Excessive exposure to beryllium at LANL is evident by the significant increase in berylliosis, a disease that is caused only by beryllium exposure, although based on only 4 deaths. The risk of berylliosis and beryllium-associated diseases has not been comprehensively evaluated at the level of mortality within DOE facilities and future studies within the MPS will consider a broader evaluation by pooling results from other weapons facilities, including Mound, Rocky Flats, Fernald, Savannah River, Oak Ridge (K-25, Y-12, ORNL) and Hanford where beryllium exposure has occurred (Kreiss et al. 1993; Stange et al. 1996; Viet et al. 2000; Welch et al. 2013; Department of Energy 2019; Tolmachev et al. 2019; Boice, Quinn, et al. 2021).

Parkinson's disease

Parkinson's disease and other neurological conditions are generally not considered outcomes associated with ionizing radiation (Yamada et al. 2009; Pasqual et al. 2021). A recent study of Mayak workers, however, reported a significant correlation between low-LET external radiation and Parkinson's disease (Azizova et al. 2020). LANL workers also showed a significant increase in Parkinson's disease in comparison with the general population, SMR 1.16 (95%CI 1.00, 1.34), $n = 193$. A positive association with radiation was seen, ERR per 100 mGy of 0.16 (95%CI −0.07, 0.40), consistent with that among Mayak workers ERR per 100 mGy of 0.10 (95%CI 0.06, 0.17). There was no radiation-related increase seen for other neurological disorders including dementia, Alzheimer's disease, and motor neuron disease: ERR per 100 mGy of −0.01 (95%CI 0.19, 0.16; $n = 973$) for the combined neurological causes of death. Possible associations between neurological disorders and high-LET radiation as experienced in lengthy exposures during space exploration are of importance to NASA since animal experiments simulating galactic cosmic ray exposures find elevations in Alzheimer's disease as well as diminution of cognitive function (Cherry et al. 2012; National Council on Radiation Protection and Measurements 2016; Boice 2017, 2019; Simonsen et al. 2020; National Aeronautics and Space Administration 2021). There is also an interest in persistent depression-like behavior, including anxiety disorders, seen in animal experiments following exposures to analogs of spaceflight radiation (Clément et al. 2020).

Other sites

There was no evidence for significant radiation association for colorectal cancer (ERR per 100 mGy of −0.03, $n = 371$); brain cancer (0.20, $n = 94$); pancreatic cancer (−0.13, $n = 250$); prostate cancer (-0.01, $n = 383$); stomach cancer (−0.08, $n = 127$); bladder cancer (0.14, $n = 103$); liver cirrhosis (−0.74, $n = 319$) or suicide (0.05, $n = 309$). Similarly, there was no evidence for a radiation association for multiple myeloma, consistent with studies of Mayak workers (Kuznetsova et al. 2016) and recent studies of nuclear weapons test participants (Boice et al. 2020) and US radiological technologists (Linet et al. 2020). Japanese atomic bomb survivors (Hsu et al. 2013) and large occupational studies (Leuraud et al. 2015) provide little evidence for radiation associations with multiple myeloma. Melanoma among LANL workers was not associated with radiation consistent with previous LANL studies (Acquavella et al. 1983) and with Mayak studies (Azizova, Bannikova, et al. 2018). Brain cancer was not significantly associated with radiation consistent with recent studies of Mayak workers (Hunter et al. 2013; Sokolnikov et al. 2015), radiological technologists (Kitahara et al. 2017), INWORKS participants (Richardson et al. 2018), and atomic veterans (Boice et al. 2020). A recent study of CNS tumors in the Japanese LSS reported a significant radiation association for gliomas based on 67 incident cases: ERR per 100 mGy = 0.17 (95%CI 0.012, 0.53) (Brenner et al. 2020). The association was apparent only among males as females showed a negative dose response, though not statistically significant. Most of the survivors who developed glioma were under age 20 at exposure.

Plutonium dosimetry

There have been other approaches to plutonium dosimetry for LANL workers (Poudel et al. 2018) as well as for plutonium workers at other sites in the US, UK, and Russia (Leggett et al. 2005, 2018; Daniels et al. 2006; Birchall et al. 2010; Puncher and Riddell 2016; Bingham et al. 2017; Gregoratto and Bailey 2018). Differences in dose estimates based on the different approaches arise mainly from the application of different biokinetic models, assumptions concerning the form(s) of inhaled Pu, and data fitting methods. Other large uncertainties, known and unknown, might arise from missing or absent monitoring, different levels of detectability and contaminated urine samples. Reuse and contamination of glass collection bottles, for example, have been reported in the early operational years at a UK plutonium facility (Riddell et al. 2000).

As part of a case-control study of the association between external radiation exposures and leukemia mortality in workers from five US facilities including 4 DOE nuclear installations, Daniels et al. (2006) assessed the intake of Pu as a potential confounder. Of the 1269 workers studied in the five cohorts, 173 had worked at LANL, and 62 of the 173 had been monitored for Pu intake. The equivalent dose to active bone marrow was used as the exposure variable. Estimated equivalent doses to active marrow were based on the latest ICRP models at the time (since updated by the

ICRP), including the original Human Respiratory Tract Model of ICRP Publication 66 (International Commission on Radiological Protection 1994a), the gastrointestinal tract model of ICRP Publication 30 (International Commission on Radiological Protection 1979), and the systemic model for Pu applied in ICRP Publication 68 (International Commission on Radiological Protection 1994b). The intake of Pu was assumed to be via inhalation. For each monitored worker with positive urinary Pu measurements, a single inhalation intake was assumed to have occurred three days prior to the first positive measurement. The solubility of the inhaled material was characterized as 50% Type M (moderately soluble) and 50% Type S (relatively insoluble). The cumulative active marrow doses for the monitored LANL workers averaged about 15 mSv with a DWF of 20 applied for alpha particle dose. Daniels et al. (2006) addressed only a tiny portion of all LANL workers, old models were used and the only dose to active bone marrow was calculated.

Sensitivity analyses

An overview of the approach to sensitivity analyses and uncertainty propagation techniques is found in Ellis, Girardi, et al. (2018) for handling external and internal exposures at the Mallinckrodt Chemical Works that processed uranium and then in the comprehensive NCRP publications (Bouville et al. 2015; Dauer et al. 2018; National Council on Radiation Protection and Measurements 2018b) that were specific for the MPS cohorts. These publications provided a standard approach for developing organ doses useful in epidemiologic studies that incorporate the application of dose calculation methods in a wide range of investigations and exposures. They build upon other NCRP publications related to various aspects of uncertainty (National Council on Radiation Protection and Measurements 2009). Uncertainties surrounding the estimation of plutonium dose to tissue in LANL workers are described in the dosimetry section above and in papers by Leggett (2003) and Leggett et al. (2005, 2018).

Uncertainty analysis for external doses followed the approach outlined by Gilbert (1998), Gilbert and Fix (1996) and Gilbert et al. (1996) for evaluating errors in estimated external radiation doses for Hanford workers. There was a particular concern for the estimates of exposure for the earlier workers with the highest exposures and for which a greater uncertainty is possible (Wakeford 2018). Bias factors for recorded external photon doses as estimates of lung dose were assumed to be similar for LANL workers as for Hanford workers (see Table 1 in Gilbert 1998). Our analyses are generally consistent with the Hanford analyses, that is, even if very conservative assumptions about the magnitude of errors were made, they did not seriously distort the dose-response analyses. As discussed by Gilbert and Fix (1996) 'The reason for the lack of distortion appears to be that larger doses, which are the most influential in dose-response analyses, are the sum of independent measurements from many dosimeters; thus, the relative error in such doses is small'. The Hanford analyses were not able to account for doses from internally-deposited radionuclides, which we were able to do. A remaining uncertainty, however, is the possible underestimation of neutron doses in the early years. Nonetheless, the extensive industrial hygiene procedures in place for the Manhattan Project workers help minimize uncertainties, that is, the careful monitoring of external exposures, the extensive bioassay programs, the air sampling measurements, comprehensive job histories, and available medical monitoring data.

As discussed in the Lung Cancer section, Stram et al. (2021) recently conducted an innovative uncertainty analysis of plutonium and external gamma radiation doses to lung among Mayak workers in Russia. While Mayak workers received doses to the lung that were substantially greater than those received by Los Alamos workers, the Mayak findings are informative. Accounting for uncertainty in the gamma-ray doses had a minor effect on the dose-response, whereas it markedly increased the width of the confidence intervals for the plutonium dose-response. This is consistent with our uncertainty evaluation of external gamma-ray doses not affecting the dose-response. While the importance of the uncertainty in plutonium doses is noted and is an area of active examination, the mean cumulative contribution of total plutonium dose to LANL workers was only 2.8% (Table 4) so that any uncertainty likely would have minimal effect on the overall dose-response. It is also recognized that the relatively low plutonium doses and size of the Pu-exposed worker population limit the statistical ability to quantify any plutonium effect, which must await the future pooling of similar MPS cohort studies.

A number of analyses were conducted to evaluate the effects of using different model paraments, different DWFs and different approaches to the outcome selections. A 5-year lag instead of a 10-year lag was used for lung cancer without any change in the estimate of ERR per 100 mGy. DWFs of 1, 10 and 20 for plutonium organ doses were used for lung, liver, NHL (thoracic lymph nodes) and bone analyses with no change in the dose-response estimates or confidence limits. For the relatively infrequent outcomes, such as leukemia other than CLL, analyses based only on the underlying cause of death did not deviate from the finding based on combining the underlying and contributing causes of death, except that the confidence intervals were a bit wider. Assuming a DWF of 2 or 3 for tritium made no difference in the dose-response analyses since tritium was such a small contributor to the overall organ dose. For the 2085 workers with positive tritium readings, their mean absorbed dose was 1.46 mGy (max 157 mGy) for a DWF of 1, which increased to 4.39 mGy (max 473 mGy) for a DWF of 3.

Strengths and limitations

The strengths of the study of LANL workers include the cohort design, the long and complete follow-up of up to 75 years, and the relatively large number of workers eligible for the study, over 26,000 individuals, and the inclusion of women. Further, the comprehensive dose reconstruction included specific organ dose determinations from external photons and neutrons, and internal intakes of tritium and two plutonium radionuclides. Occupational doses received both before and after employment at LANL were obtained. The available 158,222 urine bioassay samples (24.3 samples per worker monitored) provided unusually detailed and high-quality information for dose reconstruction for plutonium intakes. Another strength is the estimation of radiation dose to thoracic lymph nodes for NHL dose-response analyses, removing reliance on red bone marrow dose as a surrogate for dose to the lymphatic system. Linkage to identify serious renal disease within the USRDS, while qualitative, added additional understanding to the mortality findings which failed to reveal an increase in malignant or nonmalignant kidney disease. Other methodological strengths include the low percentage of workers who were lost-to-follow-up (0.1%) and the low percentage of deaths for which a specific cause was not available (1.4%). Another strength was the ability to cost-effectively build upon and enhance three previous epidemiologic investigations of this workforce (Wiggs 1987; Galke et al. 1992; Wiggs et al. 1994; DOE 2021), while providing special attention to a small group of 26 early plutonium workers who have been clinically followed for over 50 years (Hempelmann et al. 1973; Voelz et al. 1997; Inkret and Miller 1995). The dose reconstruction incorporated autopsy data for 28 LANL workers who had donated their bodies for scientific research (McInroy 1995; McInroy et al. 1991); this provided valuable checks on the solubility of the inhaled material and the reliability of the biokinetic models applied in the dose reconstruction.

Another strength was that cigarette use and smoking patterns were directly assessed and correlated in an interview study of 5507 early LANL employees (Voelz and Hempelmann 1975; Mahoney and Wilkinson 1987; Wilkinson et al. 1987).

The importance of adjusting for SES as a surrogate for tobacco use and other lifestyle factors was evident in that the SMRs for IHD for workers who had a high school education or less was 0.71, whereas workers with a college or graduate degree had SMRs for IHD that were significantly lower, 0.52 and 0.30, respectively. Comparable findings were seen for lung cancer where the SMRs were about 0.80 for those with a high school education or less, whereas those with a college or graduate degree had significantly lower SMRs, 0.44 and 0.22, respectively. Education was known for practically all employees, and adjustment for education had been found to be a robust way to control indirectly smoking and other lifestyle behaviors (Cohen et al. 2018).

The ability to identify over 3000 workers who were Hispanic provided an opportunity to evaluate cancer on noncancer mortality for a population with different baseline rates of disease. There was no evidence for radiation associations, although the number of causes of death was significantly increased in comparison with the general White population, that is, cancers of the stomach and liver, cirrhosis of the liver, nonmalignant kidney disease, and diabetes. Such elevations are consistent with the role that lifestyle, genetics and infections play in disease occurrence among the Hispanic population (Dominguez et al. 2015; Miller et al. 2018).

Limitations of the study include the incomplete knowledge of concomitant exposures to toxic occupational hazards such as asbestos or beryllium. The absence of an increase in cancers of the pleura, mesothelioma and asbestosis, however, indicates a limited exposure to asbestos. The excess of berylliosis indicates a meaningful exposure to beryllium but the small number of deaths, only four, suggests that any associations with other causes of death are likely to be minimal. Nonetheless, our inability to control solvents and other laboratory chemical exposures in the workplace remains a limitation. Despite the comprehensive estimation of radiation dose and the completeness of the vital status follow-up, the statistical precision of risk estimates was limited. The 95% confidence limits were often consistent with both an absence of a radiation effect and an effect even 2 times higher than those reported in other investigations. Similarly, despite LANL workers experiencing some of the highest intakes of plutonium in the US, they were much lower than those experienced by Mayak workers in Russian, and not sufficient to convincingly discern radiation effects directly from plutonium dose to specific organs. The increase of bone and possibly lung cancer among plutonium workers at LANL indicates that combining similar MPS cohort studies, such as Rocky Flats and Hanford workers, should be informative and directly address these limitations of small numbers. Another limitation is the difficulty in assigning meaningful, defensible, quantitative uncertainties in dose estimates. While the approaches taken for external radiation (Gilbert et al. 1996) and for internal radiation (related to the systemic model used for Pu) did not suggest major sources of error in dose estimates, nonetheless, additional work is needed and planned (National Council on Radiation Protection and Measurements 2018b; Stram et al. 2021).

Conclusions

An extended mortality follow-up of over 26,000 workers at the Los Alamos National Laboratory failed to reveal significant associations between radiation dose and cancers of the lung, liver and NHL, sites that would have received the highest doses from intakes of plutonium, nor excesses due to leukemia or ischemic heart disease. A plutonium-related excess of bone cancer was plausible among workers with the highest intakes, but the numbers were small. An association between esophageal cancer and radiation is noted, and the association with Parkinson's disease is provocative given a recent investigation of workers at the Mayak facility in Russia. Beryllium exposure among early workers in the 1940s could be tied to a small number of deaths from berylliosis. The population of Hispanic workers could be evaluated separately but numbers were too small to evaluate any unique patterns of death following radiation exposure. Analyses focusing on plutonium dose were generally inconclusive, although the association with bone cancer was plausible. The LANL cohort study is notable in having nearly complete follow-up for up to 75 years, comprehensive external and internal organ dose determination, the inclusion of women, and relatively large numbers.

Acknowledgments

We are indebted to the many individuals who were instrumental in providing support and assistance throughout the conduct of the study of LANL workers employed 1943–1980. Greg Wilkinson (University of Texas, retired) provided advice and the unpublished manuscript by Warren Galke regarding a follow-up of the Zia workers who were included in our investigation. Drs. Luis Bertelli, John Klumpp and other colleagues at LANL were instrumental in providing updated plutonium bioassay urine results that we could incorporate into the dose reconstruction for individual workers. The Department of Energy Office of Environmental Health and Safety (Drs. Pat Worthington, Bonnie Richter and Joey Zhou) provided essential support for accessing DOE occupational records, including autopsy material from the USTUR. Drs. Sergei Tolmachev, Maia Avtandilashvili and Martin Šefl (USTUR) provided helpful insights into the organ doses received from plutonium among LANL workers as well as biological data on autopsy materials which helped validate the biokinetic models used. Dr. Mary Shepherd at the New Mexico State Health Department facilitated our independent linkages for vital status death determination. Linda Walsh (University of Zurich) provided helpful guidance on Poisson regression risk modeling and estimation. Laurie Wiggs (LANL, retired) provided important early guidance with regard to our follow-up of her previous studies. Ray Guilmette (formerly at Lovelace and LANL) provided critical early guidance with regard to the plutonium bioassay program at LANL. We are further indebted to Craig Yoder and Chris Passmore (Landauer, Inc.) for facilitating record linkage of career doses, Derek Hagemeyer (ORAU) for facilitating the career dose linkages for the REIRS and REMS data registries, and the U.S. Army Dosimetry Center (William S. Harris, Jr., CHP), the U.S. Air Force Radiation Dosimetry Laboratory (Ms. Linda Wilson), and the Naval Dosimetry Center (CAPT Thad Sharp and LCDR Nakima McCormack) for facilitating linkages with their respective military dosimetry files. The findings and conclusions in this paper are those of the authors. Its publication does not necessarily represent the official positions of or implied endorsement by the National Council on Radiation Protection and Measurements, Oak Ridge Associated Universities, Vanderbilt University Medical Center, or any of the acknowledged agencies.

Disclosure statement

The authors report no conflicts of interest. The authors alone are responsible for the content and writing of the paper.

Funding

The study of workers at the Los Alamos National Laboratory, a component of the Million Person Study, was financially supported in part by grants from the U.S. Department of Energy [DE-SC0008944] awarded to the National Council on Radiation Protection and Measurements, which included interagency support from the U.S. Nuclear Regulatory Commission, the U.S. Environmental Protection Agency, and the National Aeronautics and Space Administration; and more recent grants [DE-AU0000042, DE-AU0000046]; a grant from the U.S. Nuclear Regulatory Commission [NRC-HQ-60-14-G-0011]; grants from the Centers for Disease Control and Prevention [5UE1EH000989, 5NUE1EH001315]; and grants from the National Aeronautics and Space Administration [NNX15AU88G, 80NSSC17M0016, 80NSSC19M0161]. Contract support was received from the Naval Sea Systems Command [N00024-17-C-4322]. Further, contract support was received by Oak Ridge National Laboratory from the Office of Radiation and Indoor Air, U.S. Environmental Protection Agency, under Interagency Agreement DOE No. 1824 S581-A1, under contract No. DE-AC05-00OR22725 with UT-Battelle; and contract support was received by Oak Ridge Associated Universities from the U.S. Department of Energy under contract No. DE-SC0014664.

ORCID

John D. Boice http://orcid.org/0000-0002-8755-1299
Sarah S. Cohen http://orcid.org/0000-0003-0421-1983
Michael T. Mumma http://orcid.org/0000-0001-7506-8710
Ashley P. Golden http://orcid.org/0000-0002-9430-7702
David J. Girardi http://orcid.org/0000-0003-1253-4045
Elizabeth Dupree Ellis http://orcid.org/0000-0003-2871-1587
Michael B. Bellamy http://orcid.org/0000-0003-0892-7559
Lawrence T. Dauer http://orcid.org/0000-0002-5629-8462

References

Acquavella JF, Wilkinson GS, Tietjen GL, Key CR, Stebbings JH, Voelz GL. 1983. A melanoma case-control study at the Los Alamos National Laboratory. Health Phys. 45(3):587–592.

Agency for Toxic Substances and Disease Registry (ATSDR). 2010. Toxicological profile for plutonium. Atlanta (GA): ATSDR, Public Health Service, US Department of Health and Human Services; [accessed 2020 December 5]. https://www.atsdr.cdc.gov/toxprofiles/tp143.pdf

Azizova TV, Muirhead CR, Moseeva MB, Grigoryeva ES, Sumina MV, O'Hagan J, Zhang W, Haylock RJ, Hunter N. 2011. Cerebrovascular diseases in nuclear workers first employed at the Mayak PA in 1948-1972. Radiat Environ Biophys. 50(4):539–552.

Azizova TV, Bannikova MV, Grigoryeva ES, Rybkina VL. 2018. Risk of malignant skin neoplasms in a cohort of workers occupationally exposed to ionizing radiation at low dose rates. PLoS One. 13(10): e0205060.

Azizova TV, Batistatou E, Grigorieva ES, McNamee R, Wakeford R, Liu H, de Vocht F, Agius RM. 2018. An assessment of radiation-associated risks of mortality from circulatory disease in the cohorts of Mayak and Sellafield nuclear workers. Radiat Res. 189(4): 371–388.

Azizova TV, Bannikova MV, Grigoryeva ES, Rybkina VL, Hamada N. 2020. Occupational exposure to chronic ionizing radiation increases risk of Parkinson's disease incidence in Russian Mayak workers. Int J Epidemiol. 49(2):435–447.

Bingham D, Bérard P, Birchall A, Bull R, Cardis E, Challeton-de Vathaire C, Grellier J, Hurtgen C, Puncher M, Riddell A, et al. 2017. Reconstruction of internal doses for the alpha-risk case-control study of lung cancer and leukaemia among European nuclear workers. Radiat Prot Dosimetry. 174(4):485–494.

Birchall A, Puncher M, Harrison J, Riddell A, Bailey MR, Khokryakov V, Romanov S. 2010. Plutonium worker dosimetry. Radiat Environ Biophys. 49(2):203–212.

Boice JD. Jr. 1993. Leukemia risk in thorotrast patients. Radiat Res. 36(2):301–302.

Boice JD. 2017. Space: the final frontier-research relevant to mars. Health Phys. 112(4):392–397.

Boice JD. Jr. 2019. Relevance of the million person study to research needs for NASA and space exploration. Int J Radiat Biol. 1–9.

Boice JD, Cohen SS, Mumma MT, Dupree-Ellis E, Eckerman KF, Leggett RW, Boecker B, Brill A, Henderson B. 2006. Mortality among radiation workers at Rocketdyne (Atomics International), 1948-1999. Radiat Res. 166(1):98–115.

Boice JD, Leggett RW, Ellis ED, Wallace PW, Mumma MT, Cohen SS, Brill AB, Bandana C, Boecker BB, Yoder RC. 2006. A comprehensive dose reconstruction methodology for former Rocketdyne/Atomics International radiation workers. Health Phys. 90(5): 409–430.

Boice JD, Cohen SS, Mumma MT, Chadda B, Blot WJ. 2008. A cohort study of uranium millers and miners of Grants, New Mexico, 1979-2005. J Radiol Prot. 28(3):303–325.

Boice JD, Cohen SS, Mumma MT, Ellis ED, Eckerman KF, Leggett RW, Boecker BB, Brill AB, Henderson BE. 2011. Updated mortality analysis of radiation workers at Rocketdyne (Atomics International), 1948-2008. Radiat Res. 176(2):244–258.

Boice JD, Cohen SS, Mumma MT, Ellis ED, Cragle DL, Eckerman KF, Wallace PW, Chadda B, Sonderman JS, Wiggs LD, et al. 2014. Mortality among mound workers exposed to polonium-210 and other sources of radiation, 1944-1979. Radiat Res. 181(2):208–228.

Boice JD, Jr, Cohen SS, Mumma MT, Ellis ED. 2019. The Million Person Study, whence it came and why. Int J Radiat Biol. 1–14. DOI: 10.1080/09553002.2019.1589015

Boice JD, Jr, Ellis ED, Golden AP, Zablotska LB, Mumma MT, Cohen SS. 2019. Sex-specific lung cancer risk among radiation workers in the Million Person Study and among TB-fluoroscopy patients. Int J Radiat Biol. 1–12. DOI:10.1080/09553002.2018.1547441

Boice JD, Held KD, Shore RE. 2019. Radiation epidemiology and health effects following low-level radiation exposure. J Radiol Prot. 39(4): S14–S27.

Boice JD, Jr, Cohen SS, Mumma MT, Chen H, Golden AP, Beck HL, Till JE. 2020. Mortality among US military participants at eight aboveground nuclear weapons test series. Int J Radiat Biol. 1–22. DOI:10.1080/09553002.2020.1787543

Boice JD, Jr, Cohen SS, Mumma MT, Hagemeyer DA, Chen H, Golden AP, Yoder RC, Dauer LT. 2021. Mortality from leukemia, lung cancer and heart disease among U.S. nuclear power plant workers. Int J Radiat Biol.

Boice JD, Jr, Cohen SS, Mumma MT, Yoder RC, Dauer LT. 2021. Mortality among medical radiation workers in the United States, 1965–2016. Int J Radiat Biol.

Boice JD, Jr, Cohen SS, Mumma MT, Golden AP, Howard S, Girardi DJ, Dupree Ellis ED, Bellamy M, Dauer LT, Samuels C, et al. 2021. Mortality study of women and men working at a uranium processing facility during World War II., 1943-1947. The Tennessee Eastman Corporation (Y-12) Worker Study. Int J Radiat Biol.

Boice JD, Jr, Quinn B, Ansari A, Blake PK, Blattnig SR, Caffrey EA, Cohen SS, Golden AP, Held KD, Jokisch DW, et al. 2021. A million persons, a million dreams: A vision for a National Center of Radiation Epidemiology and Biology. Int J Radiat Biol.

Bouville A, Toohey RE, Boice JD, Beck HL, Dauer LT, Eckerman KF, Hagemeyer D, Leggett RW, Mumma MT, Napier B, et al. 2015. Dose reconstruction for the Million Worker Study: status and guidelines. Health Phys. 108(2):206–220.

Brenner AV, Sugiyama H, Preston DL, Sakata R, French B, Sadakane A, Cahoon EK, Utada M, Mabuchi K, Ozasa K. 2020. Radiation risk of central nervous system tumors in the Life Span Study of atomic bomb survivors, 1958-2009. Eur J Epidemiol. 35(6):591–600.

Breslow NE, Lubin JH, Marek P, Langholz D. 1983. Multiplicative models and cohort analysis. J Am Stat Assoc. 78(381):1–12.

Brown SC, Schonbeck MF, McClure D, Barón AE, Navidi WC, Byers T, Ruttenber AJ. 2004. Lung cancer and internal lung doses among plutonium workers at the rocky flats plant: a case-control study. Am J Epidemiol. 160(2):163–172.

Cahoon EK, Preston DL, Pierce DA, Grant E, Brenner AV, Mabuchi K, Utada M, Ozasa K. 2017. Lung, laryngeal and other respiratory cancer incidence among Japanese atomic bomb survivors: an updated analysis from 1958 through 2009. Radiat Res. 187(5): 538–548.

Campbell EE, Milligan MF, Moss WD, Schulte HF. 1972. History of the plutonium bioassay program at the Los Alamos Scientific Laboratory, 1944-1972. LA-5008, UC-41. October 1972. Los Alamos (NM): Los Alamos Scientific Laboratory; [accessed 2018 September 20]. https://www.osti.gov/servlets/purl/4587709

Campbell KM, Deck D, Krupski A. 2008. Record linkage software in the public domain: a comparison of Link Plus, The Link King, and a 'basic' deterministic algorithm. Health Informatics J. 14(1):5–15.

Cherry JD, Liu B, Frost JL, Lemere CA, Williams JP, Olschowka JA, O'Banion MK. 2012. Galactic cosmic radiation leads to cognitive impairment and increased a beta plaque accumulation in a mouse model of Alzheimer's disease. PLoS One. 7(12):1e53275.

Claycamp HG, Okladnikova ND, Azizova TV, Belyaeva ZD, Boecker BB, Pesternikova VS, Scott BR, Shekhter-Levin S, Sumina MV, Sussman NB, et al. 2000. Deterministic effects from occupational radiation exposures in a cohort of Mayak PA workers: data base description. Health Phys. 79(1):48–54.

Clément GR, Boyle RD, George KA, Nelson GA, Reschke MF, Williams TJ, Paloski WH. 2020. Challenges to the central nervous system during human spaceflight missions to Mars. J Neurophysiol. 123(5):2037–2063.

Cohen SS, Mumma MT, Ellis ED, Boice JD. Jr. 2018. Validating the use of census data on education as a measure of socioeconomic status in an occupational cohort. Int J Rad Biol. 1–6. DOI:10.1080/09553002.2018.1549758

Cox D. 1972. Regression models and life-tables. J R Stat Soc Ser B. 34(2):187–220.

Cox R, Kellerer AM. 2003. A current view in radiation weighting factors and effective dose. ICRP Publication 92. Ann ICRP. 33:1–4.

Cox R, Menzel HG, Preston J. 2008. Internal dosimetry and tritium–the ICRP position. J Radiol Prot. 28(2):131–135.

Daniels RD, Lodwick CJ, Schubauer-Berigan MK, Spitz HB. 2006. Assessment of plutonium exposures for an epidemiological study of US nuclear workers. Radiat Prot Dosimetry. 118(1):43–55.

Dauer LT, Bouville A, Toohey RE, Boice JD, Jr, Beck HL, Eckerman KF, Hagemeyer D, Leggett RW, Mumma MT, Napier B, et al. 2018. Dosimetry and uncertainty approaches for the million-worker study of radiation workers and veterans: overview of the recommendations in NCRP Report No. 178. A review. Int J Rad Biol. 1–10. DOI:10.1080/09553002.2018.1536299

Department of Army. 2012. Occupational dosimetry and dose recording for exposure to ionizing radiation. Pamphlet 385-25. Washington (DC): Department of Army; [Accessed 2021 June 4]. https://armypubs.army.mil/epubs/DR_pubs/DR_a/pdf/web/p385_25.pdf.

Department of Energy. 2018. DOE 2017 Occupational radiation exposure. Washington (DC): US Department of Energy, Office of Environment, Health, Safety and Security, The Radiation Exposure Monitoring System (REMS).

Department of Energy. 2019. Former worker medical screening program report. 2019 Annual Report. Washington (DC): Office of Environment, Health, Safety and Security; [accessed 2020 December 7]. https://www.energy.gov/sites/prod/files/2020/06/f75/2019-FW-Medical-Screening-Program-Annual-Report.pdf

Department of Energy. 2021. Comprehensive epidemiological data resource (CEDR). Washington (DC): Office of Environment, Health, Safety and Security; [accessed 2021 March 3]. https://oriseapps.orau.gov/cedr/default.aspx

de Vocht F, Hidajat M, Martin RM, Agius R, Wakeford R. 2020. Ischemic heart disease mortality and occupational radiation exposure in a nested matched case-control study of British nuclear fuel cycle workers: investigation of confounding by lifestyle, physiological traits and occupational exposures. Radiat Res. 194(4):431–444.

Dominguez K, Penman-Aguilar A, Chang MH, Moonesinghe R, Castellanos T, Rodriguez-Lainz A, Schieber R. 2015. Vital signs: leading causes of death, prevalence of diseases and risk factors, and use of health services among Hispanics in the United States - 2009-2013. MMWR Morb Mortal Wkly Rep. 64(17):469–478.

Efurd DW, Steiner RE, Lamont SP, Lewis D. 2008. History of the plutonium bioassay program at the Los Alamos National Laboratory, 1944-2006. J Radioanal Nucl Chem. 276(2):499–504.

Ellis ED, Boice JD, Golden AP, Girardi DJ, Cohen SS, Mumma MT, Shore RE, Leggett RW, Kerr GD. 2018. Dosimetry is key to good epidemiology: workers at Mallinckrodt Chemical Works had seven different source exposures. Health Phys. 114(4):386–397.

Ellis ED, Girardi D, Golden AP, Wallace PW, Phillips J, Cragle DL. 2018. Historical perspective on the Department of Energy mortality studies: focus on the collection and storage of individual worker data. Int J Radiat Biol. 1–8. DOI:10.1080/09553002.2018.1547851

EPRI. 2020. Cardiovascular risks from low dose radiation exposure: review and scientific appraisal of the literature. Palo Alto (CA): EPRI. [accessed 2020 December 19]. https://www.epri.com/research/products/000000003002018408

Galke WA, Johnson ER, Tietjen GL. 1992. Mortality in an ethnically diverse radiation exposed occupational cohort. Los Alamos (NM): Los Alamos National Laboratory. Occupational Medicine Group (HS-2).

Gilbert ES. 1998. Accounting for errors in dose estimates used in studies of workers exposed to external radiation. Health Phys. 74(1): 22–29.

Gilbert ES, Fix JJ. 1996. Laboratory measurement error in external dose estimates and its effects on dose-response analyses of Hanford worker mortality data. PNL-11289. Richland (WA): Pacific Northwest Laboratory. [accessed 2020 December 10]. https://www.osti.gov/servlets/purl/379945

Gilbert ES, Cragle DL, Wiggs LD. 1993. Updated analyses of combined mortality data for workers at the Hanford site, Oak Ridge National Laboratory, and Rocky Flats Weapons Plant. Radiat Res. 136(3): 408–421.

Gilbert ES, Omohundro E, Buchanan JA, Holter NA. 1993. Mortality of workers at the Hanford site: 1945-1986. Health Phys. 64(6): 577–590.

Gilbert ES, Fix JJ, Baumgartner WV. 1996. An approach to evaluating bias and uncertainty in estimates of external dose obtained from personal dosimeters. Health Phys. 70(3):336–345.

Gilbert ES, Koshurnikova NA, Sokolnikov M, Khokhryakov VF, Miller S, Preston DL, Romanov SA, Shilnikova NS, Suslova KG, Vostrotin VV. 2000. Liver cancers in Mayak workers. Radiat Res. 154(3): 246–252.

Gilbert ES, Sokolnikov ME, Preston DL, Schonfeld SJ, Schadilov AE, Vasilenko EK, Koshurnikova NA. 2013. Lung cancer risks from plutonium: an updated analysis of data from the Mayak worker cohort. Radiat Res. 179(3):332–342.

Gillies M, Kuznetsova I, Sokolnikov M, Haylock R, O'Hagan J, Tsareva Y, Labutina E. 2017. Lung cancer risk from plutonium: a pooled analysis of the Mayak and Sellafield worker cohorts. Radiat Res. 188(6):645–660.

Gillies M, Richardson DB, Cardis E, Daniels RD, O'Hagan JA, Haylock R, Laurier D, Leuraud K, Moissonnier M, Schubauer-Berigan MK, et al. 2017. Mortality from circulatory diseases and other non-cancer outcomes among nuclear workers in France, the United Kingdom and the United States (INWORKS). Radiat Res. 188(3):276–290.

Gillies M, Haylock R, Hunter N, Zhang W. 2019. Risk of leukemia associated with protracted low-dose radiation exposure: updated results from the National Registry for Radiation Workers Study. Radiat Res. 192(5):527–537.

Golden AP, Cohen SS, Chen H, Ellis ED, Boice JD. Jr. 2018. Evaluation of statistical modeling approaches for epidemiologic studies of low-dose radiation health effects. Int J Radiat Biol. 1–8. DOI:10.1080/09553002.2018.1554924

Golden AP, Ellis ED, Cohen SS, Mumma MT, Leggett RW, Wallace PW, Girardi DJ, Watkins JP, Shore RE, Boice JD. Jr. 2019. Updated mortality analysis of the Mallinckrodt uranium processing workers. Int J Radiat Biol. 1–17. DOI:10.1080/09553002.2019.1569773

Gregoratto D, Bailey MR. 2018. Estimation of lung absorption parameters for oxides of ^{238}Pu. J Radiol Prot. 38(2):831–853.

Grellier J, Atkinson W, Bérard P, Bingham D, Birchall A, Blanchardon E, Bull R, Guseva Canu I, Challeton-de Vathaire C, Cockerill R, et al. 2017. Risk of lung cancer mortality in nuclear workers from internal exposure to alpha particle-emitting radionuclides. Epidemiology. 28(5):675–684.

Hagemeyer D, Nichols G, Mumma MT, Boice JD, Jr, Brock TA. 2018. 50 years of the radiation exposure information and reporting system and importance to the Million Person Study. Int J Radiat Biol. 1–4. DOI:10.1080/09553002.2018.1540896

Hauptmann M, Daniels RD, Cardis E, Cullings HM, Kendall G, Laurier D, Linet MS, Little MP, Lubin JH, Preston DL, et al. 2020. Epidemiological studies of low-dose ionizing radiation and cancer: summary bias assessment and meta-analysis. J Natl Cancer Inst Monogr. 2020(56):188–200.

Hempelmann LH, Langham WH, Richmond CR, Voelz GL. 1973. Manhattan Project plutonium workers: a twenty-seven year follow-up study of selected cases. Health Phys. 25(5):461–479.

History Associates Incorporated. 1997. Los Alamos National Laboratory: a guide to records series supporting epidemiologic studies conducted for the Department of Energy. U.S. Department of Energy, Office of Epidemiology and Health Surveillance, prepared under Contract No. DE-AC01-93EH89246. Rockville (MD): HAI; [accessed 2020 December 4]. https://www.osti.gov/servlets/purl/477726

Hunter N, Kuznetsova IS, Labutina EV, Harrison JD. 2013. Solid cancer incidence other than lung, liver and bone in Mayak workers: 1948-2004. Br J Cancer. 109(7):1989–1996.

Hsu WL, Preston DL, Soda M, Sugiyama H, Funamoto S, Kodama K, Kimura A, Kamada N, Dohy H, Tomonaga M, et al. 2013. The incidence of leukemia, lymphoma and multiple myeloma among atomic bomb survivors: 1950-2001. Radiat Res. 179(3):361–382.

Inkret WC, Miller G. 1995. On the front lines: plutonium workers past and present share their experiences. Los Alamos Science. 23: 125–173.

International Agency for Research on Cancer. 2012. A review of human carcinogens. Part D: radiation. Vol 100D. Lyon (France): IARC Press.

International Agency for Research on Cancer. 2018. Beryllium and beryllium compounds. IARC Monogr Eval Carcinog Risks Hum. 100C:1–477.

International Commission on Radiological Protection. 1979. ICRP Publication 30, Part 1. Limits on intakes of radionuclides for workers. Ann ICRP. 2(3–4):1–116.

International Commission on Radiological Protection. 1994a. ICRP Publication 66. Human respiratory tract model for radiological protection. Ann ICRP. 24(1–3):1–482.

International Commission on Radiological Protection. 1994b. ICRP Publication 68. Dose coefficients for intake of radionuclides by workers. Ann ICRP. 24(4):1–83.

International Commission on Radiological Protection. 2007. ICRP publication 103. The 2007 recommendations of the international commission on radiological protection. Ann ICRP. 37(2–4):1–332.

International Commission on Radiological Protection. 2010. ICRP Publication 116. Conversion coefficients for radiological protection quantities for external radiation exposures. Ann ICRP. 40(2–5): 1–257.

International Commission on Radiological Protection. 2015. ICRP Publication 130. Occupational intake of radionuclides, Part 1. Ann ICRP. 44 (2):1–188.

International Commission on Radiological Protection. 2019. Occupational intakes of radionuclides: part 4. ICRP Publication 141. Ann ICRP. 48(1):1–514.

Iwanaga M, Hsu WL, Soda M, Takasaki Y, Tawara M, Joh T, Amenomori T, Yamamura M, Yoshida Y, Koba T, et al. 2011. Risk of myelodysplastic syndromes in people exposed to ionizing radiation: a retrospective cohort study of Nagasaki atomic bomb survivors. JCO. 29(4):428–434.

Kathren RL, Tolmachev SY. 2019. The US Transuranium and Uranium Registries (USTUR): a five-decade follow-up of plutonium and uranium workers. Health Phys. 117(2):118–132.

Khokhryakov VV, Khokhryakov VF, Suslova KG, Vostrotin VV, Vvedensky VE, Sokolova AB, Krahenbuhl MP, Birchall A, Miller SC, Schadilov AE, et al. 2013. Mayak Worker Dosimetry System 2008 (MWDS-2008): assessment of internal dose from measurement results of plutonium activity in urine. Health Phys. 104(4):366–378.

Kitahara CM, Linet MS, Balter S, Miller DL, Rajaraman P, Cahoon EK, Velazquez-Kronen R, Simon SL, Little MP, Doody MM, et al. 2017. Occupational radiation exposure and deaths from malignant intracranial neoplasms of the brain and CNS in U.S. Radiologic Technologists, 1983-2012. AJR Am J Roentgenol. 208(6):1278–1284.

Koshurnikova NA, Gilbert ES, Sokolnikov M, Khokhryakov VF, Miller S, Preston DL, Romanov SA, Shilnikova NS, Suslova KG, Vostrotin VV. 2000. Bone cancers in Mayak workers. Radiat Res. 154(3): 237–245.

Kreiss K, Mroz MM, Zhen B, Martyny JW, Newman LS. 1993. Epidemiology of beryllium sensitization and disease in nuclear workers. Am Rev Respir Dis. 148(4):985–991.

Kuznetsova IS, Labutina EV, Hunter N. 2016. Radiation risks of leukemia, lymphoma and multiple myeloma incidence in the Mayak cohort: 1948-2004. PLoS One. 11(9):e0162710.

Lawrence JNP. 1978. A history of PUQFUA – plutonium body burden (Q) from urine assays. LA-7403-H, History, UC-41. Los Alamos, NM: Los Alamos Scientific Laboratory. DOI:10.2172/6538507

Leggett RW. 2003. Reliability of the ICRP's dose coefficients for members of the public. III. Plutonium as a case study of uncertainties in the systemic biokinetics of radionuclides. Radiat Prot Dosimetry. 106(2):103–120.

Leggett RW, Eckerman KF, Khokhryakov VF, Suslova KG, Krahenbuhl MP, Miller SC. 2005. Mayak worker study: an improved biokinetic model for reconstructing doses from internally deposited plutonium. Radiat Res. 164(2):111–122.

Leggett RW, Eckerman KF, Bellamy M. 2018. MPS dose reconstruction for internal emitters: some site-specific issues and approaches. Int J Radiat Biol. 1–13. DOI:10.1080/09553002.2018.1558302

Leuraud K, Richardson DB, Cardis E, Daniels RD, Gillies M, O'Hagan JA, Hamra GB, Haylock R, Laurier D, Moissonnier M, et al. 2015. Ionising radiation and risk of death from leukaemia and lymphoma in radiation-monitored workers (INWORKS): an international cohort study. Lancet Haematol. 2(7):e276–e281.

Linet MS, Little MP, Kitahara CM, Cahoon EK, Doody MM, Simon SL, Alexander BH, Preston DL. 2020. Occupational radiation and haematopoietic malignancy mortality in the retrospective cohort study of US radiologic technologists, 1983-2012. Occup Environ Med. 77(12):822–831.

Mahoney MC, Wilkinson GS. 1987. Smoking patterns among Los Alamos National Laboratory employees. Report LA-10650. Los Alamos (NM): Los Alamos National Laboratory; [accessed 2020 December 4]. https://www.osti.gov/biblio/6403953-smoking-patterns-among-los-alamos-national-laboratory-employees

McInroy JF. 1995. A true measure of exposure - the human tissue analysis program at Los Alamos. Los Alamos Sc. 23:235–249.

McInroy JF, Kathren RL, Voelz GL, Swint MJ. 1991. U.S. Transuranium Registry report on the ^{239}Pu distribution in a human body. Health Phys. 60(3):307–333.

Miller G, Bertelli L, Guilmette R, McNaughton MW, Eisele WF. 2008. A study of early Los Alamos internal exposures to plutonium. Radiat Prot Dosim. 130(4):503–509.

Miller KD, Goding Sauer A, Ortiz AP, Fedewa SA, Pinheiro PS, Tortolero-Luna G, Martinez-Tyson D, Jemal A, Siegel RL. 2018. Cancer statistics for Hispanics/Latinos, 2018. CA Cancer J Clin. 68(6):425–445.

Monson R. 1986. Observations on the healthy worker effect. J Occup Med. 28(6):425–433.

Mumma MT, Cohen SS, Sirko JL, Ellis ED, Boice JD. Jr. 2018. Obtaining vital status and cause of death on a million persons. Int J Radiat Biol. 1–7. DOI:10.1080/09553002.2018.1539884

Naval Dosimetry Center. 2019. [accessed 2020 May 13]. Naval Dosimetry Center. https://www.cnic.navy.mil/regions/ndw/installations/nsa_bethesda/about/tenant_orgs/ndc.html

National Academies/National Research Council. 2012. Technical evaluation of the NASA model for cancer risk to astronauts due to space radiation. Washington (DC): National Academies Press. [accessed 2021 March 3]. https://www.nap.edu/catalog/13343/technical-evaluation-of-the-nasa-model-for-cancer-risk-to-astronauts-due-to-space-radiation

National Aeronautics and Space Administration. 2015. NASA space flight human-system standard, volume 1, revision A: crew health. NASA STD-3001. Houston (TX): NASA Johnson Space Center [accessed 2021 April 10]. https://standards.nasa.gov/standard/nasa/nasa-std-3001-vol-1

National Aeronautics and Space Administration. 2021. NASA human research program roadmap. Washington (DC): NASA; [accessed April 10]. https://humanresearchroadmap.nasa.gov/

National Council on Radiation Protection and Measurements. 2009. Report No. 164 - uncertainties in internal radiation dose assessment. Bethesda (MD): NCRP.

National Council on Radiation Protection and Measurements. 2014. Commentary No. 23 - radiation protection for space activities: supplement to previous recommendations. Bethesda (MD): NCRP.

National Council on Radiation Protection and Measurements. 2016. Commentary No. 25 - potential for central nervous system effects from radiation exposure during space activities phase 1: overview. Bethesda (MD): NCRP.

National Council on Radiation Protection and Measurements. 2018a. Commentary No. 27 - implications of recent epidemiologic studies for the linear-nonthreshold model and radiation protection. Bethesda (MD): NCRP.

National Council on Radiation Protection and Measurements. 2018b. Report No. 178 - deriving organ doses and their uncertainty for epidemiologic studies (with a focus on the One Million U.S. Workers and Veterans Study of Low-Dose Radiation Health Effects). Bethesda (MD): NCRP.

National Council on Radiation Protection and Measurements. 2018c. Report No. 181 - evaluation of the relative effectiveness of low-energy photons and electrons in inducing cancer in humans. Bethesda (MD): NCRP.

National Council on Radiation Protection and Measurements. 2019. Report No. 183 - radiation exposures in space and the potential for central nervous system effects: phase II. Bethesda (MD): NCRP.

National Council on Radiation Protection and Measurements. 2021. Scientific Committee 1-27: evaluation of sex-specific differences in lung cancer radiation risks and recommendations for use in transfer and projection models. Bethesda (MD): NCRP; [accessed 2021 April 10]. https://ncrponline.org/program-areas/sc-1-27-evaluation-of-sex-specific-differences-in-lung-cancer-radiation-risks-and-recommendations-for-use-in-transfer-and-projection-models/

New Brunswick Laboratory. 2015. Safety data sheet plutonium metal. Argonne (IL): US Department of Energy; [accessed 2020 December 2]. https://science.osti.gov/-/media/nbl/pdf/price-lists/SDS/SDS-Plutonium_Metal.pdf

Nuclear Regulatory Commission. 2019. Radiation exposure information and reporting system (REIRS). Rockville (MD): US Nuclear Regulatory Commission; [accessed 2020 February 7]. https://www.reirs.com/

Omar RZ, Barber JA, Smith PG. 1999. Cancer mortality and morbidity among plutonium workers at the Sellafield plant of British Nuclear Fuels. Br J Cancer. 79(7–8):1288–1301.

Ozasa K, Shimizu Y, Suyama A, Kasagi F, Soda M, Grant EJ, Sakata R, Sugiyama H, Kodama K. 2012. Studies of the mortality of atomic bomb survivors, report 14, 1950–2003: an overview of cancer and noncancer diseases. Radiat Res. 177(3):229–243.

Ozasa K, Takahashi I, Grant EJ, Kodama K. 2017. Cardiovascular disease among atomic bomb survivors. Int J Radiat Biol. 93(10): 1145–1150.

Pasqual E, Boussin F, Bazyka D, Nordenskjold A, Yamada M, Ozasa K, Pazzaglia S, Roy L, Thierry-Chef I, de Vathaire F, et al. 2021. Cognitive effects of low dose of ionizing radiation - Lessons learned and research gaps from epidemiological and biological studies. Environ Int. 147:106295.

Peterson GR, Gilbert ES, Buchanan JA, Stevens RG. 1990. A casecohort study of lung cancer, ionizing radiation, and tobacco smoking among males at the Hanford Site. Health Phys. 58(1):3–11.

Poudel D, Miller G, Klumpp JA, Bertelli L, Waters TL. 2018. Bayesian analysis of plutonium bioassay data at Los Alamos National Laboratory. Health Phys. 115(6):712–726. [accessed December 7. 2020]. https://www.epa.gov/sites/production/files/2015-05/documents/ornl-tm-2006-583.pdf.

Preston DL, Ron E, Tokuoka S, Funamoto S, Nishi N, Soda M, Mabuchi K, Kodama K. 2007. Solid cancer incidence in atomic bomb survivors: 1958-1998. Radiat Res. 168(1):1–64.

Preston DL, Lubin J, Pierce D, McConney M, Shilnikova N. 2015. Epicure risk regression and person-year computation software: command summary and user guide. Ottawa (CA): Risk Sciences International.

Preston DL, Sokolnikov ME, Krestinina LY, Stram DO. 2017. Estimates of radiation effects on cancer risks in the Mayak worker, Techa River and atomic bomb survivor studies. Radiat Prot Dosimetry. 173(1–3):26–31.

Puncher M, Riddell AE. 2016. A Bayesian analysis of plutonium exposures in Sellafield workers. J Radiol Prot. 36(1):1–19.

Richardson DB, Cardis E, Daniels RD, Gillies M, Haylock R, Leuraud K, Laurier D, Moissonnier M, Schubauer-Berigan MK, Thierry-Chef I. 2018. Site-specific solid cancer mortality after exposure to ionizing radiation: a cohort study of workers (INWORKS). Epidemiology. 29(1):31–40.

Richardson DB, Abalo K, Bernier MO, Rage E, Leuraud K, Laurier D, Keil AP, Little MP. 2020. Meta-analysis of published excess relative risk estimates. Radiat Environ Biophys. 59(4):631–641.

Riddell AE, Battersby WP, Peace MS, Strong R. 2000. The assessment of organ doses from plutonium for an epidemiological study of the Sellafield workforce. J Radiol Prot. 20(3):275–286.

Rowland RE, Stehney AF, Lucas HF. 1983. Dose-response relationships for radium-induced bone sarcomas. Health Phys. 44(Suppl (1): 15–31.

Shilnikova NS, Preston DL, Ron E, Gilbert ES, Vassilenko EK, Romanov SA, Kuznetsova IS, Sokolnikov ME, Okatenko PV, Kreslov VV, et al. 2003. Cancer mortality risk among workers at the Mayak nuclear complex. Radiat Res. 159(6):787–798.

Silver SR, Bertke SJ, Hein MJ, Daniels RD, Fleming DA, Anderson JL, Pinney SM, Hornung RW, Tseng CY. 2013. Mortality and ionising radiation exposures among workers employed at the Fernald Feed

Materials Production Center (1951-1985). Occup Environ Med. 70(7):453–463.

Simonsen LC, Slaba TC, Guida P, Rusek A. 2020. NASA's first ground-based galactic cosmic ray simulator: enabling a new era in space radiobiology research. PLoS Biol. 18(5):e3000669.

Slaba TC, Blattnig SR, Norbury JW, Rusek A, La Tessa C. 2016. Reference field specification and preliminary beam selection strategy for accelerator-based GCR simulation. Life Sci Space Res. 8:52–67.

Sokolnikov ME, Gilbert ES, Preston DL, Ron E, Shilnikova NS, Khokhryakov VV, Vasilenko EK, Koshurnikova NA. 2008. Lung, liver and bone cancer mortality in Mayak workers. Int J Cancer. 123(4):905–911.

Sokolnikov M, Preston D, Gilbert E, Schonfeld S, Koshurnikova N. 2015. Radiation effects on mortality from solid cancers other than lung, liver, and bone cancer in the Mayak worker cohort: 1948-2008. PLoS One. 10(2):e0117784.

Stange AW, Furman FJ, Hilmas DE. 1996. Rocky flats beryllium health surveillance. Environ Health Perspect. 104 Suppl 5(Suppl (5): 981–986.

Steenland K. 2005. One agent, many diseases: exposure response data and comparative risks of different outcomes following silica exposure. Am J Ind Med. 48(1):16–23.

Stefaniak AB, Weaver VM, Cadorette M, Puckett LG, Schwartz BS, Wiggs LD, Jankowski MD, Breysse PN. 2003. Summary of historical beryllium uses and airborne concentration levels at Los Alamos National Laboratory. Appl Occup Environ Hyg. 18(9):708–715.

Stram DO, Sokolnikov M, Napier BA, Vostrotin VV, Efimov A, Preston DL. 2021. Lung cancer in the Mayak workers cohort: risk estimation and uncertainty analysis. Radiat Res. 195(4):334–346.

Takahashi I, Shimizu Y, Grant EJ, Cologne J, Ozasa K, Kodama K. 2017. Heart disease mortality in the Life Span Study, 1950-2008. Radiat Res. 187(3):319–332.

Tapio S, Little MP, Kaiser JC, Impens N, Hamada N, Georgakilas AG, Simar D, Salomaa S. 2021. Ionizing radiation-induced circulatory and metabolic diseases. Environ Int. 146:106235.

Till JE, Beck HB, Aanenson JW, Grogan HA, Mohler HJ, Mohler SS, Voillequé PG. 2014. Military participants at the U.S. atmospheric nuclear weapons testing – methodology for estimating dose and uncertainty. Radiat Res. 181(5):471–484.

Tolmachev SY, Swint MJ, Bistline RW, McClellan RO, McInroy JF, Kathren RL, Filipy RE, Toohey RE. 2019. USTUR Special Session Roundtable-US Transuranium and Uranium Registries (USTUR): a five-decade follow-up of plutonium and uranium workers. Health Phys. 117(2):211–222.

Tran V, Zablotska LB, Brenner AV, Little MP. 2017. Radiation-associated circulatory disease mortality in a pooled analysis of 77,275 patients from the Massachusetts and Canadian tuberculosis fluoroscopy cohorts. Sci Rep. 7(1):1–10.

United Nations Scientific Committee on the Effects of Atomic Radiation. 2008. Effects of ionizing radiation. 2006 Report. Scientific Annex A: Epidemiological studies of radiation and cancer. New York (NY): UNSCEAR.

United Nations Scientific Committee on the Effects of Atomic Radiation. 2017. Sources, effects and risks of ionizing radiation. 2016 Report to the General Assembly. Volume 1. Scientific Annex D: Biological effects of selected internal emitters – uranium. New York (NY): UNSCEAR.

United Nations Scientific Committee on the Effects of Atomic Radiation. 2020. Sources, effects and risks of ionizing radiation. UNSCEAR 2019 Report to the General Assembly. Scientific Annex A: evaluation of selected health effects and inference of risk due to radiation exposure. New York (NY): UNSCEAR.

U.S. Environmental Protection Agency. 1999. Cancer risk coefficients for environmental exposure to radionuclides. Washington (DC): EPA. Federal Guidance Report No. 13, EPA-402-R99-001.

U.S. Renal Data System. 2020. USRDS 2019 annual data report: atlas of end-stage renal disease in the United States. Bethesda (MD): National Institutes of Health, National Institute of Diabetes and Digestive and Kidney Diseases.

Velazquez-Kronen R, Gilbert ES, Linet MS, Moysich KB, Freudenheim JL, Wactawski-Wende J, Simon SL, Cahoon EK, Alexander BH, Doody MM, et al. 2020. Lung cancer mortality associated with protracted low-dose occupational radiation exposures and smoking behaviors in U.S. radiologic technologists, 1983-2012. Int J Cancer. 147(11):3130–3138.

Viet SM, Torma-Krajewski J, Rogers J. 2000. Chronic beryllium disease and beryllium sensitization at Rocky Flats: a case-control study. AIHAJ. 61(2):244–254.

Voelz GL. 1991. Health considerations for workers exposed to plutonium. Occup Med. 6(4):681–694.

Voelz GL, Hempelmann LH. 1975. Health study of plutonium workers - protocol for a morbidity and mortality study. Los Alamos (NM): Los Alamos Scientific Laboratory.

Voelz GL, Lawrence JN. 1991. A 42-y medical follow-up of Manhattan Project plutonium workers. Health Phys. 61(2):181–190.

Voelz GL, Lawrence JN, Johnson ER. 1997. Fifty years of plutonium exposure to the Manhattan Project plutonium workers: an update. Health Phys. 73(4):611–619.

Wakeford R. 2018. The growing importance of radiation worker studies. Br J Cancer. 119(5):527–529.

Wakeford R. 2019. Does low-level exposure to ionizing radiation increase the risk of cardiovascular disease? Hypertension. 73(6): 1170–1171.

Welch LS, Ringen K, Dement J, Bingham E, Quinn P, Shorter J, Fisher M. 2013. Beryllium disease among construction trade workers at Department of Energy nuclear sites. Am J Ind Med. 56(10): 1125–1136.

Wiggs LD. 1987. Mortality among females employed by the Los Alamos National Laboratory: an epidemiologic investigation [dissertation]. Norman (OK): University of Oklahoma.

Wiggs LD, Johnson ER, Cox-DeVore CA, Voelz GL. 1994. Mortality through 1990 among white male workers at the Los Alamos National Laboratory: considering exposures to plutonium and external ionizing radiation. Health Phys. 67(6):577–588.

Wilkinson GS, Tietjen GL, Wiggs LD, Galke WA, Acquavella JF, Reyes M, Voelz GL, Waxweiler RJ. 1987. Mortality among plutonium and other radiation workers at a plutonium weapons facility. Am J Epidemiol. 125(2):231–250.

Wing S, Richardson D, Wolf S, Mihlan G. 2004. Plutonium-related work and cause-specific mortality at the United States department of energy Hanford site. Am J Ind Med. 45(2):153–164.

Wing S, Richardson DB. 2005. Age at exposure to ionising radiation and cancer mortality among Hanford workers: follow up through 1994. Occup Environ Med. 62(7):465–472.

Yamada M, Kasagi F, Mimori Y, Miyachi T, Ohshita T, Sasaki H. 2009. Incidence of dementia among atomic bomb survivors - radiation effects research foundation adult health study. J Neurol Sci. 281(1–2):11–14.

Yoder RC, Dauer L, Balter S, Boice JD, Jr, Grogan H, Mumma M, Passmore CN, Rothenberg LN, Vetter RJ. 2018. Dosimetry for the study of medical radiation workers with a focus on the mean absorbed dose to the lung, brain and other organs. Int J Radiat Biol. 1–12. DOI:10.1080/09553002.2018.1549756

Yoder RC, Balter S, Boice JD, Jr, Grogan H, Mumma M, Rothenberg L, Passmore C, Vetter R, Dauer L. 2020. Using personal monitoring data to derive organ doses for medical radiation workers in the Million Person Study – considerations regarding NCRP Commentary No. 30. J Radiol Prot 1–15. DOI:10.1088/1361-6498/abcfcb.

Zablotska LB, Lane RS, Frost SE. 2013. Mortality (1950–1999) and cancer incidence (1969–1999) of workers in the Port Hope cohort study exposed to a unique combination of radium, uranium and c-ray doses. BMJ Open. 3(2):e002159.

Mortality among Tennessee Eastman Corporation (TEC) uranium processing workers

John D. Boice, Sarah S. Cohen, Michael T. Mumma, Ashley P. Golden, Sara C. Howard, David J. Girardi, Elizabeth D. Ellis, Michael B. Bellamy, Lawrence T. Dauer, Keith F. Eckerman, and Richard W. Leggett

ABSTRACT

Background: There are few occupational studies of women exposed to ionizing radiation. During World War II, the Tennessee Eastman Corporation (TEC) operated an electromagnetic field separation facility of 1152 calutrons to obtain enriched uranium (^{235}U) used for the Hiroshima atomic bomb. Thousands of women were involved in these operations.

Materials and methods: A new study was conducted of 13,951 women and 12,699 men employed at TEC between 1943 and 1947 for at least 90 days. Comprehensive dose reconstruction techniques were used to estimate lung doses from the inhalation of uranium dust based on airborne measurements. Vital status through 2018/2019 was obtained from the National Death Index, Social Security Death Index, Tennessee death records and online public record databases. Analyses included standardized mortality ratios (SMRs) and Cox proportional hazards models.

Results: Most workers were hourly (77.7%), white (95.6%), born before 1920 (58.3%), worked in dusty environments (57.0%), and had died (94.9%). Vital status was confirmed for 97.4% of the workers. Women were younger than men when first employed: mean ages 25.0 years and 33.0 years, respectively. The estimated mean absorbed dose to the lung was 32.7 mGy (max 1048 mGy) for women and 18.9 mGy (max 501 mGy) for men. The mean dose to thoracic lymph nodes (TLNs) was 127 mGy. Statistically significant SMRs were observed for lung cancer (SMR 1.25; 95% CI 1.19, 1.31; $n = 1654$), nonmalignant respiratory diseases (NMRDs) (1.23; 95% CI 1.19, 1.28; $n = 2585$), and cerebrovascular disease (CeVD) (1.13; 95% CI 1.08, 1.18; $n = 1945$). For lung cancer, the excess relative rate (ERR) at 100 mGy (95% CI) was 0.01 (−0.10, 0.12; $n = 652$) among women, and −0.15 (−0.38, 0.07; $n = 1002$) among men based on a preferred model for men with lung doses <300 mGy. NMRD and non-Hodgkin lymphoma were not associated with estimated absorbed dose to the lung or TLN.

Conclusions: There was little evidence that radiation increased the risk of lung cancer, suggesting that inhalation of uranium dust and the associated high-LET alpha particle exposure to lung tissue experienced over a few years is less effective in causing lung cancer than other types of exposures. There was no statistically significant difference in the lung cancer risk estimates between men and women. The elevation of certain causes of death such as CeVD is unexplained and will require additional scrutiny of workplace or lifestyle factors given that radiation is an unlikely contributor since only the lung and lymph nodes received appreciable dose.

Introduction

From June 1943 to May 1947, the Tennessee Eastman Corporation (TEC) operated a uranium processing facility that employed nearly 40,000 workers in Oak Ridge, TN. Y-12 was the wartime code name given to the facility. Enriched uranium (^{235}U) was obtained by electromagnetic processing and was then sent to Los Alamos National Laboratory (LANL) and used to develop atomic weapons, including the Hiroshima atomic bomb (Polednak and Frome 1981). The mass spectrometer separation process resulted in many TEC workers being exposed to varying amounts of ^{238}U, ^{235}U and their progeny. These radionuclides are primarily alpha-particle emitters, and the primary health concern was from the inhalation of uranium dust. The exposure to external radiation was minimal (Polednak et al. 1982; Beck et al. 1983; Dupree et al. 1995). There are few occupational studies of women exposed to ionizing radiation and even fewer investigations that have evaluated intakes of inhaled radionuclides or uranium dust (ATSDR 2013; UNSCEAR 2017). Thus, the TEC operation offers a unique opportunity to study a sizable number of women who have never been studied before. The women employed at this facility were popularized in the book '*The Girls of Atomic City*' (Kiernan 2013).

The electromagnetic field separation of the uranium isotopes was done by 1152 calutrons, named for the *Cal*ifornia *U*niversity cyclo*tron* and designed by Ernest O. Lawrence. The calutrons were housed in nine buildings. Five buildings were designated as 'alpha' buildings and four as 'beta' buildings. The designations were not related to radioactive emissions and were just building names.

TEC operated the electromagnetic separation process from 1943 to 1947 and employed the largest number of workers of all the early uranium processing plants involved in the Manhattan Project. The plant converted uranium oxide (UO_3) received from the Mallinckrodt Chemical Works in Missouri to uranium chloride (UCl_4). The UCl_4 was then enriched, i.e. the percentage of ^{235}U was increased by the calutron (mass spectrometer) electromagnetic separation process. The separation process involved two stages ('alpha' and 'beta'), each consisting of multiple calutron separators. The alpha stage was discontinued in late 1945, when enriched uranium fluoride (UF_6) was received from the Oak Ridge gaseous diffusion plant (K-25), converted to UF_4, enriched further by the beta calutrons, and shipped to LANL for development of nuclear weapons.

White male workers at TEC were previously studied through 1973 (Polednak and Frome 1981; Cookfair 1982; Cookfair et al. 1983), and as part of a four-site case-control study of lung cancer, including female workers, based on follow-up through 1982 (Dupree et al. 1995). The initial studies suggested an association between lung cancer and lung dose (Polednak et al. 1982; Cookfair et al. 1983), but this was not confirmed in a later study that also accounted for tobacco use (Dupree et al. 1995). The current investigation builds upon these previous studies, extends the follow-up of male workers through 2018 (for an additional 45 years), improves upon the estimation of lung dose from uranium dust inhalation, enhances the evaluation of tobacco use, includes a female TEC cohort for the first time, and includes a separate analysis of black women. As occupational exposure is from inhalation of airborne uranium dust with essentially no external radiation, the focus is on lung cancer, NMRD and NHL as possible late-occurring health outcomes. Herein, we report lifetime risks for all causes of death and sex-specific radiation risk estimates for these outcomes among workers employed between 1943 and 1947 for more than 90 days at the TEC uranium conversion and enrichment plant. The TEC cohort is a component of the Million Person Study (MPS) of U.S. Radiation Workers and Veterans evaluating health effects following low-level chronic (low-dose rate) exposures (Boice, Cohen, Mumma, & Ellis 2022; Boice, Quinn, Al-Nabulsi, Ansari, et al. 2022).

Materials and methods

Human subjects research approval was received from the Oak Ridge Site-Wide Institutional Review Board and the Vanderbilt University Institutional Review Board.

Study population

The previous TEC studies restricted the cohort to white males who were employed only at the TEC facility and not subsequently at any other Oak Ridge facility (Polednak 1980; Polednak and Frome 1981; Polednak et al. 1982). The current study extended the follow-up of this white male population; and includes all women, white and black, who were employed at the TEC facility. The study does not exclude women who transferred to the newer Y-12 plant operated by the Union Carbide Corporation or who transferred to any other Oak Ridge facility such as Oak Ridge National Laboratory (X-10) or the gaseous diffusion plant (K-25). Overall, there were 20,434 women first employed at TEC between 1943 and 1947. Women employed for fewer than 90 days ($n = 6483$) were excluded, leaving an eligible population of 13,951 female workers. The initial male population consisted of 18,869 workers. After excluding those who worked fewer than 90 days ($n = 6170$), 12,699 male workers remained eligible for study. In total, the study population comprised 26,650 workers.

Vital status and outcome determination

The procedures described in Mumma et al. (2022) were followed to determine vital status as of 31 December 2019 for the TEC women and 31 December 2018 for the TEC men. Probabilistic matching programs, such as LinkPlus (Campbell et al. 2008), were used to obtain date of death and cause of death from state and national sources of vital status information, including the National Death Index (NDI), the California Death Statistical Master File, and the Social Security Administration (SSA) Death Master File. Alive status was confirmed from the SSA Service for Epidemiological Researchers. To confirm or correct identifying information used in the matching procedures (e.g. name, social security number and date of birth), TLO, a subscription service of TransUnion (www.tlo.com) was used. Individual one-on-one searches also were employed to locate study subjects using resources such as Ancestry.com and Google obituary search. Unique databases in Tennessee were also searched on a one-on-one basis because of the difficulty in tracing young unmarried women who worked in the 1940s.

Radiation dose reconstruction

The radiation dose reconstruction approach followed the guidance set forth in NCRP Report 178 (2018) for deriving organ doses and their uncertainty specifically for the MPS. The manual nature of the operations at the TEC plant resulted in many workers being exposed to varying amounts of uranium compounds of ^{234}U, ^{235}U, ^{238}U and their progeny. Since these uranium isotopes are chiefly alpha-radiation emitters and inhalation was the primary mode of exposure, internal doses to the lung and thoracic lymph nodes (TLNs) were most important when considering health risks. External radiation was not expected to be consequential due to relatively low energies of photons emitted by natural uranium isotopes, and no film badges or other measurements were used. The internal radiation dosimetry for TEC workers built upon the foundational work of previous dose reconstructions (Polednak and Frome 1981; Cookfair 1982; Polednak et al. 1982; Beck et al. 1983; Cookfair et al. 1983; Dupree et al. 1995). The present dose estimations were based on air monitoring data for locations near primary personnel work areas at the two processing facilities (i.e. the alpha and beta buildings), and accounted for the likely chemical and physical forms of uranium. The highest average concentrations of airborne uranium dust (300 and 250 $\mu g/m^3$) were measured in the chemistry department in the alpha building and were associated with the sublimation (i.e. the transition from a solid to a gaseous state) and bottle filling areas, respectively. Concentrations of airborne uranium in several beta chemistry areas averaged 50 μ g/m^3 (Polednak and Frome 1981).

An estimate of annual radiation absorbed doses to the lung from inhaled uranium was obtained for each worker. First, to obtain estimates of the uranium intakes, and recognizing that many TEC employees worked in more than one department (and some changed jobs within departments), an employee's total work experience was divided into groups according to specific job factors: job title, job code, department, location, and calendar date range. Second, these work experiences were linked to appropriate average airborne uranium dust concentration measurements (Dupree et al. 1995) accounting for the changes in enrichment percentages that occurred over time (e.g. for beta areas the uranium enrichment changed from 13% before August 1945 to 17.8% by October 1945, and ultimately to 25% thereafter). Third, the job-specific exposure concentrations were multiplied by days worked in that specific job to obtain daily integrated inhaled activity of the uranium isotopic mixture. Finally, these activities were converted to a lung absorbed dose in accordance with current ICRP internal dosimetry methodology for ^{234}U, ^{235}U, and ^{238}U (ICRP 2007, 2017) and were based on daily intake estimates and job start and end times. The analyses assumed an aerosol size with an activity median aerodynamic diameter (AMAD) of 5 μm, a breathing rate of 27.3 m^3/d for females and 28.8 m^3/d for males based on ICRP Publication 66 (ICRP 1994), absorption type M/S, and 6 h/d of exposed time. Thus, the female and male daily air intakes while at work would be 10.4 and 12.6 m^3, respectively. Similar procedures were used to estimate individual doses to TLNs.

AMAD is the particle size in an aerosol where 50% of the activity is associated with particles of aerodynamic diameter greater than the AMAD. Clearance of the deposited chemical form of the inhaled material is characterized by the absorption type parameter. Compounds that are readily absorbed are referred to as type F compounds. No uranium compounds were designated as F at any of the TEC facilities. Type M compounds have intermediate rates of absorption into blood, and type S compounds are relatively insoluble and slowly absorbed. This analyses utilized an absorption type M/S in accord with ICRP Publication 137 that notes that ^{238}U and U dioxide compounds have absorption rates between those of default type M and S. Sensitivity analyses were conducted to consider the impact of these assumptions (i.e. considering alternate assumptions of AMAD of 1 or 10 μm; breathing rate of 14.5 or 51.1 m^3/d; absorption type S or M, and exposure times of 4 h or 8 h per day).

Career doses

Dosimetry records documenting radiation exposure received before or after employment at TEC (1943–1947) were obtained following the procedures outlined in Boice, Leggett, et al. (2006) and Ellis et al. (2018). There were only 315 women and 333 men who received occupational exposures elsewhere, and the additional contribution to lung dose, primarily from gamma radiation, was small.

Smoking history

Tobacco use for individual workers is often unavailable or too expensive to obtain in retrospective epidemiologic studies. Smoking data, however, were available in the TEC occupational medical records and were abstracted and considered in two previous overlapping studies that included TEC workers (Cookfair et al. 1983; Dupree et al. 1995).

Smoking histories were sought for 641 women who died of lung cancer and for a comparison group of 707 women who did not die of lung cancer (selected to be 10% larger than the case group to ensure adequate data collection for missing medical records). The level of detail of smoking information was 'tobacco use Y/N.' During the abstraction process, information on chest X-ray examinations also was collected.

A similar evaluation of tobacco use was conducted among 1002 men who died of lung cancer and an approximate 10% sample of the 12,699 men who did not die of lung cancer as comparison subjects ($n = 1107$). The number of male lung cancer deaths was larger than in previous studies because our follow-up was longer by up to 35 years through 2018 compared with 1983 (Dupree et al. 1995), and was nearly to the end of life as 98.6% of the men had died. Information on chest X-ray examinations was not collected because the results from the women indicated only one chest X-ray on average.

Statistical analysis

The start of follow-up began 90 days after date of first hire at TEC (1943–1947), and ended at the date of death, age 95, date lost to follow-up, or 31 December 2019 for women and 31 December 2018 for men, whichever came first. Standardized mortality ratios (SMRs) for over 50 causes of death were computed as follows. The observed numbers of cause-specific deaths were compared with the expected numbers based on U.S. mortality rates accounting for age and calendar year in five year categories, and sex. SMRs were further computed within pay category (hourly/salary) work area (alpha, beta, other), age at hire, race and years of follow-up. Exact 95% Poisson confidence intervals (CIs) for the SMRs were computed (Rothman and Boice 1979).

Cox proportional hazards models were used in the dose–response analyses to compute organ-specific radiation risks (Cox 1972). Such internal cohort analyses are conducted to address the 'healthy worker effect' often seen in occupational studies (Monson 1986). Cox models were conducted using SAS/STAT software (SAS/STAT software, Version 9.4 of the SAS System for Windows, SAS Institute Inc., Cary, NC). Excess relative rates (ERRs) at 100 mGy were estimated using the Peanuts program in Epicure (Preston et al. 2015). Year of birth and a measure of socioeconomic status (SES), i.e. pay category (hourly/salary), were included in all models as was race which was available for women. Pay category was selected as the measure of SES to indirectly account for smoking and other life choices. Previous studies of TEC workers found that pay category was correlated with tobacco use (Dupree et al. 1995). Questionnaire interviews at the time of employment were available in medical records to validate the use of pay category as a reasonable surrogate for tobacco use. Internal dose–response analyses were conducted for lung cancer among the 1168 black women who worked at TEC. For female workers, both race and pay category were included as adjustment factors in all models. Conceivably, collinearity might exist which could distort results. To address this possibility, a sensitivity analysis was conducted only adjusting for pay category and then only adjusting for race. Results were not meaningfully different so both factors were kept as adjustments in the dose–response analyses.

For the internal cohort analyses, radiation dose was treated as a categorical measure, with categories based on the distribution of dose to lung. The dose category was treated as a time-dependent measure, allowing workers to be assigned to increasingly higher dose categories over time as their individual organ-specific radiation doses accrued. To allow for a possible latent period between radiation exposure and any consequent effect, doses were lagged, i.e. excluded if they occurred during some assumed interval prior to the event of interest, 10 years for lung cancer and also for non-malignant respiratory disease (NMRD) excluding flu and pneumonia (NMRD), and for NHL. Parameter estimates and standard errors for the dose categories in the Cox time-dependent regression models were used to obtain hazard ratios (HRs) and 95% CIs for death due to lung cancer, NMRD and NHL compared with those in the referent group taken as the workers with low cumulative radiation dose. Trend tests treated the radiation dose as a single continuous time-dependent measure, and two-sided p values are presented. Dose weighting factors (DWFs) of 1, 10 and 20 were applied to the alpha-particle absorbed doses as possible indicators of biological effectiveness. ERRs were estimated using the same model specifications as the Cox time-dependent regression model.

Sensitivity analyses addressed changing the minimum duration of employment required for inclusion; excluding workers with lung dose = 0 and using >0 to <5 mGy as the referent category; restricting the analysis of male workers to those with lung doses <300 mGy; examining different dose-lag intervals; and varying the assumptions with regard to the uranium dust parameters used in the absorbed lung dose calculations. Malignant and nonmalignant lung disease and NHL were the primary outcomes of interest because of the substantial contribution of airborne uranium dust to lung dose and TLN dose. Due to the nature of the uranium conversion and enrichment processes relying on mass spectrometry, there was little external radiation exposure, and no other organs received more than a tiny dose from intakes of uranium (Polednak and Frome 1981).

Results

Descriptive characteristics of the TEC workers are shown in Table 1. There were 26,650 workers employed 1943–1947 at TEC for at least 90 days. Follow-up was through 2019 for the 13,951 female workers (52.3%) and through 2018 for the 12,699 male workers (47.7%). The total person-years of observation were 1,199,423 for an average of 45.0 years. The majority of workers were hourly workers (77.7%), white (95.6%), born between 1910 and 1929 (73.5%), under age 30 when first hired (60.4%), worked for more than one year (55.5%), and had died (94.9%). Fifty-seven percent of workers were employed in buildings with dusty environments (alpha or beta processing, or alpha chemistry). Through the end of follow-up, 672 workers (2.5%) were confirmed alive, and cause of death was determined for all but 473 (1.9%) of the 25,288 workers who had died. Only 690 (2.6%) of the workers were lost to follow-up (Figure 1).

Table 2 presents the cumulative lung doses and the TLN doses among the female and male workers. These dose estimates were used in the main analysis of dose response for lung disease and non-Hodgkin lymphoma (NHL). The preferred analyses assumed a particle size of 5 μm AMAD, breathing rate of 27.3 m^3/d for females and 28.8 m^3/d for males (ICRP 1994), solubility class of M/S, and 6 h/d of exposed time. For a DWF of 1, the mean dose to lung was 32.7 mGy and the maximum was 1.05 Gy for women. Assuming a DWF of 20, the mean is 591 weighted-mGy lung dose with a maximum of 18.5 weighted-Gy lung dose. For men, the mean absorbed dose to lung was 18.9 mGy (maximum 0.50 Gy). For a DWF of 20, the mean is 352 weighted-mGy lung dose with a maximum of 8.91 weighted-Gy dose to lung. The mean dose for DWF = 20 is a bit lower than what would be computed by multiplying the mean dose for DWF = 1 by 20 because the total dose computations also incorporate photon doses which do not change when the uranium dose is multiplied by any DWF. Thus, the mean dose for DWF = 20 will always be lower than the product of 20 times the mean dose with DWF = 1 given concomitant photon contributions to dose. The mean dose to TLN was 127 mGy, and the patterns between women and men were similar to those for lung dose, except

Table 1. Descriptive characteristics of the TEC cohort of 26,650 workers employed 1943–1947 for at least 90 days and followed through 2019 for the 13,951 female and through 2018 for the 12,699 male workers.

Sex	Female $N = 13,951$		Male $N = 12,699$		Total $N = 26,650$	
Characteristic	N	%	N	%	N	%
Pay category						
Hourly	10,775	77.2%	9921	78.1%	20,696	77.7%
Salary	3176	22.8%	2778	21.9%	5954	22.3%
Race						
Black	1168	8.4%	0	0%	1168	4.4%
White	12,783	91.6%	12,699	100%	25,482	95.6%
Year of birth						
1885–1899	312	2.2%	1968	15.5%	2280	8.6%
1900–1909	1486	10.7%	3295	26.0%	4781	17.9%
1910–1919	3908	28.0%	4563	35.9%	8471	31.8%
1920–1930	8245	59.1%	2873	22.6%	11,118	41.7%
Mean YOB	1919		1911		1915	
Year of hire						
1943	1908	13.7%	2105	16.6%	4013	15.1%
1944	7811	56.0%	7132	56.2%	14,943	56.1%
1945	3970	28.5%	3129	24.6%	7099	26.6%
1946	262	1.9%	333	2.6%	595	2.2%
Mean YOH	1944		1944		1944	
Age at hire (years)						
15 to <20	3651	26.2%	739	5.8%	4390	16.5%
20 to <25	4807	34.5%	2251	17.7%	7058	26.5%
25 to <30	2314	16.6%	2323	18.3%	4637	17.4%
30 to <45	2893	20.7%	5533	43.6%	8426	31.6%
45+	286	2.1%	1853	14.6%	2139	8.0%
Mean age at hire	25.0		33.0		28.8	
Processing category						
Minimal u dust exposure	5211	37.4%	6241	49.2%	11,452	43.0%
Alpha process or beta	7109	51.0%	5960	46.9%	13,069	49.0%
Alpha chemistry	1631	11.7%	498	3.9%	2129	8.0%
Duration of employment						
>90 days to 1 year	6560	47.0%	5312	41.8%	11,872	44.6%
1–2 year	4611	33.1%	4306	33.9%	8917	33.5%
2+ years	2780	19.9%	3081	24.3%	5861	22.0%
Mean (d)	457.2		493.4		474.4	
Age at end of follow-up (years)						
39–44	832	6.0%	556	4.4%	1388	5.2%
45–49	195	1.4%	410	3.2%	605	2.3%
50–54	308	2.2%	637	5.0%	945	3.6%
55–59	435	3.1%	886	7.0%	1321	5.0%
60–64	663	4.8%	1229	9.7%	1892	7.1%
65–69	933	6.7%	1482	11.7%	2415	9.1%
70–74	1361	9.8%	1703	13.4%	3064	11.5%
75–79	1727	12.4%	1778	14.0%	3505	13.2%
≥80	7497	53.7%	4018	31.6%	11,515	43.2%
Mean age (years)	77.0		71.8		74.5	
Vital status[a]						
Alive	592	4.2%	80	0.6%	672	2.5%
Dead	12,773	91.6%	12,515	98.6%	25,288	94.9%
Lost to follow-up	586	4.2%	104	0.8%	690	2.6%

[a]Vital status determined as of 31 December 2019 for women and as of 31 December 2018 for men.

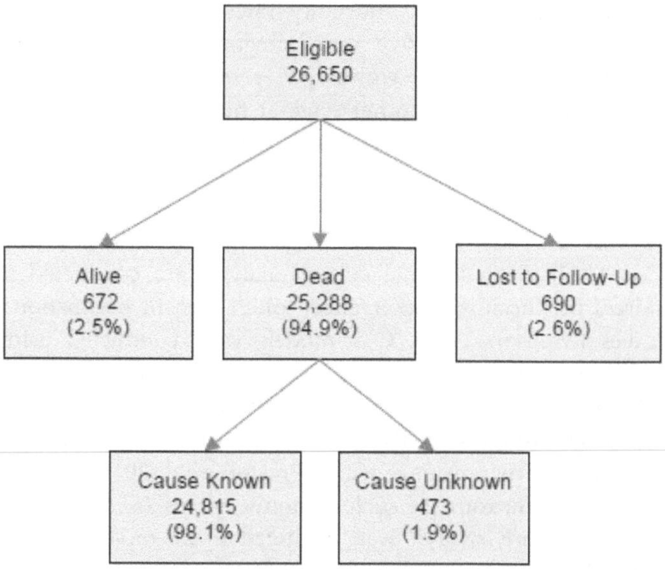

Figure 1. Vital status tracing results for the 26,650 Tennessee Eastman Corporation (Y-12) workers employed 1943–1947 for at least 90 days and followed through 2019 for the 13,951 females and through 2018 for the 12,699 males.

that the TLN measures of dose were about five times that of the lung doses. Women had higher estimates of cumulative dose in large part because they worked longer in dusty environments than their male counterparts.

Sensitivity analyses were conducted using different assumptions for the four dose parameters used in the dose estimation procedure. Realistic low-sided and high-sided estimates for the four parameters were applied for particle size, breathing rate, duration of daily exposure, and the solubility of the inhaled uranium (Supplemental Table 1). For example, compared with the AMAD particle size of 5 μm used in the main analysis, the dose to lung would be two times higher for a 1 μm assumption and 50% lower for a 10 μm assumption.

The SMRs for 59 causes of death among the 13,951 TEC women and the 12,699 men are shown in Table 3. The SMRs were similar for women and men. All causes of death among the 26,650 TEC workers were significantly increased based on 25,288 deaths (SMR 1.09; 95% CI 1.07, 1.10), related to significant elevations of lung cancer (SMR 1.25; 95% CI 1.19, 1.31; $n = 1654$), all NMRD (SMR 1.23; 95% CI 1.19, 1.28; $n = 2585$), NMRD excluding flu and pneumonia (SMR 1.28; 95% CI 1.22, 1.34; $n = 1798$), mental and behavioral disorders (SMR 1.36; 95% CI 1.27, 1.45; $n = 855$), cerebrovascular disease (CeVD) (SMR 1.13; 95% CI 1.08, 1.18; $n = 1945$), and accidents (SMR 1.20; 95% CI 1.12, 1.28; $n = 843$). There were significant deficits of cancers of the colon (SMR 0.83; 95% CI 0.75, 0.91; $n = 405$), stomach (SMR 0.83; 95% CI 0.70, 0.99; $n = 132$), and nonsmoking related cancers (SMR 0.94; 95% CI 0.90, 0.97; $n = 2866$). There was an elevated mortality from cancer of the uterine cervix based on 80 cases (SMR 1.23; 95% CI 0.98, 1.53) and of NHL based on 207 cases (SMR 1.11; 95% CI 0.96, 1.27). Other causes of death of interest included ischemic heart disease (IHD) (SMR 0.99; 95% CI 0.96, 1.01; $n = 5994$), kidney cancer (SMR 0.89; 95% CI 0.72, 1.09; $n = 91$), nonmalignant kidney disease (SMR 1.01; 95% CI 0.90, 1.14; $n = 297$), cirrhosis of the liver (SMR 0.84; 95% CI 0.74, 0.95; $n = 256$), dementia and Alzheimer's disease (SMR 1.42; 95% CI 1.35, 1.49; $n = 1464$), Parkinson's disease (SMR 1.13; 95% CI 0.96, 1.32; $n = 153$), diabetes (SMR 0.92; 95% CI 0.84, 1.00; $n = 467$), leukemia other than chronic lymphocytic leukemia (CLL) (SMR 0.93; 95% CI 0.76, 1.12; $n = 109$), CLL (SMR 1.38; 95% CI 1.04, 1.79; $n = 57$), cancer of the prostate (SMR 0.96; 95% CI 0.85,1.08; $n = 265$), and cancer of the eye (SMR 2.77; 95% CI 1.26, 5.25; $n = 9$).

SMR analyses by pay category (hourly/salary) and sex were also conducted (Supplemental Table 2). The 20,696

Table 2. Dose to the lung and dose to the thoracic lymph nodes (TLNs) by sex for different dose weighting factors (DWF) of 1, 10 and 20 for TEC workers employed 1943–1947 for at least 90 d and followed through 2019 (women) or 2018 (men).

		TEC women (N = 13,948)				TEC men (N = 12,695)				All TEC workers (N = 26,643)			
Organ	DWF	Mean (mGy)	Median (mGy)	Std	Max (Gy)	Mean (mGy)	Median (mGy)	Std	Max (Gy)	Mean (mGy)	Median (mGy)	Std	Max (Gy)
Lung	1	32.7	7.6	73	1.0	18.9	6.3	35	0.5	26.3	6.9	59.0	1.0
	10	297	69	654	9.3	177	58	321	4.5	240	63.4	526.1	9.3
	20	591	137	1301	18.5	352	115	638	8.9	478	125.9	1046	18.5
TLN	1	162	37	359	5.1	88	29	160	2.3	127	33	284	5.1
	10	1556	360	3429	49.0	845	276	1533	21.5	1217	319	2721	49.0
	20	3105	720	6841	97.6	1687	552	3060	42.9	2429	637	5428	97.6

Cumulative organ doses include the sum of the internal doses received at TEC and any external doses received at other facilities. For DWF = 1, the lung dose is in units of absorbed dose, mGy or Gy. For DWF = 10 and 20, the lung dose is in units of weighted-lung dose, weighted-mGy or weighted-Gy. Doses are unlagged. There were 7 TEC workers with no available dose records who are not included in these tabulations.

hourly workers had higher SMRs than the 5954 salaried workers for most specific causes of death. For all causes of death, the SMRs were 1.12 ($n = 9871$) vs. 0.96 ($n = 2902$) for female hourly workers compared with salaried workers and were 1.18 ($n = 9785$) vs. 0.89 ($n = 2730$) for male workers. For lung cancer, the SMRs were 1.18 ($n = 477$) vs. 1.25 ($n = 175$) for female hourly workers compared with salaried workers and were 1.44 ($n = 832$) vs. 0.84 ($n = 170$) for male workers. IHD was significantly higher among hourly workers compared with salaried workers as were deaths from accidents. For smoking-related cancers, the salaried male workers had a significantly low SMR (0.80; $n = 251$) in contrast to significantly high SMRs among hourly male workers (1.27; $n = 1123$), hourly female workers (1.14; $n = 722$), and salaried female workers (1.17; $n = 253$). This pattern was particularly noticeable for lung cancer, smoking-related cancers, and NMRD. Hourly female workers had a significantly low risk of breast cancer (0.89; $n = 303$) and a significantly high risk of cervical cancer (1.32; $n = 68$). These comparisons indicate the importance of adjusting for pay category in the internal dose–response analyses.

Table 4 presents the internal dose–response analysis for lung cancer over categories of radiation dose to the lung by sex for three different DWFs of 1, 10, and 20. The estimates of risk, ERR per 100 weighted-mGy dose to lung, did not change when the DWF was increased from 1 to 20. However, for DWF = 20, the number of workers in the high-dose category (>250 weighted-mGy lung dose) did increase appreciably by 9137 and the number of lung cancer deaths by 595. There was no evidence for an increased risk of lung cancer among either women or men. For a DWF of 1, the ERR per 100 mGy lung dose (95% CI) was 0.01 (−0.10; 0.12) for females, −0.22 (−0.43; −0.01) for males, and −0.09 (−0.19, 0.02) for all workers. Graphical representation of the dose response for all workers is presented in Figure 2, for women in Figure 3, and for men in Figure 4 for DWFs 1 and 20. Figure 4(c) presents an analysis of male workers with lung absorbed doses <300 mGy, i.e. it excludes 25 workers (0.2% of the male population) with very high lung absorbed doses, >300 mGy. This was done in part because of concern that bias in the dust to dose conversion process for a few workers in the highest dose categories might have overestimated the lung doses and disproportionately influence the dose response which was significantly negative. Excluding these few workers (none who died of lung cancer) increased the estimated ERR per 100 mGy from −0.22 (95% CI −0.43; −0.01; $n = 1002$; p for trend =.042) to −0.15 (95% CI −0.38, 0.07; $n = 1002$; p for trend =.18) and the trend was no longer significantly negative.

Three previous studies of TEC workers reported a radiation-related increase of lung cancer among workers first hired over the age of 45 years, although not statistically significant (Polednak et al. 1982; Cookfair et al. 1983; Dupree et al. 1995). The current study included of 2139 workers over age 45 years when first hired; 286 were women and 1853 were men (Table 1). The SMR for lung cancer among all workers hired at age 45 years or older was 1.16 (95% CI 0.94, 1.43) based on 90 lung cancer deaths. For workers hired at ages 15–24 years ($n = 11,448$) and 25–44 years ($n = 10,565$), lung cancer SMRs were 1.28 (95% CI 1.14, 1.32; $n = 737$) and 1.28 (95% CI 1.20, 1.37; $n = 827$), respectively (data not shown). An internal dose–response analysis for male workers first hired after age 45 years found the ERR (95% CI) at 100 mGy to be 0.30 (95% CI −0.43, 1.02). The ERRs (95% CI) at 100 mGy for male workers hired at ages 15–24 years and 25–44 years were −0.41 (95% CI −0.85, 0.04) and −0.19 (95% CI −0.45, 0.06), respectively (Supplemental Table 3). The number of female workers hired after age 45 years was only 286 and too small for any meaningful analyses.

Table 5 presents the internal dose–response analysis for NMRD, excluding flu and pneumonia, over categories of radiation dose to the lung by sex for three different DWFs of 1, 10, and 20. Similar to the estimates of radiation risk for lung cancer, the estimates for NMRD did not vary by DWF. There was no evidence for a radiation-related increased risk of NMRD among either women or men. For all workers, the sex-adjusted NMRD ERR per 100 mGy lung dose (95% CI) for a DWF of 1 was −0.07 (−0.16, 0.02; $n = 1798$); it was −0.05 (−0.15, 0.05; $n = 881$) for females and −0.02 (−0.20, 0.17; $n = 917$) for males.

In the past, there were few occupational studies that included women and even fewer that included nonwhite women. This was true in large part because there were relatively few women in radiation jobs, the likelihood of receiving meaningful radiation exposures was limited, and women were more challenging to trace than men. The number of nonwhite workers was a small component of the TEC female radiation workforce. Given the limited numbers of studies including black women, a dose–response evaluation

Table 3. Standardized mortality ratios (SMRs) for the 26,650 TEC workers employed 1943–1947 for at least 90 days and followed through 2018 for males and 2019 for females, by sex.

	Women		Men		Total		
Persons-at-risk	13,951		12,699		26,650		
Person-years	716,143		483,280		1,199,423		
Cause of death (ICD9)[a]	Obs	SMR	Obs	SMR	Obs	SMR	95% CI
All causes of death (001–999)	12,773	1.08*	12,515	1.10*	25,288	1.09*	1.07; 1.10
All malignant neoplasms (140–208)	2656	1.01	2560	1.03	5216	1.02	0.99; 1.05
Buccal cavity and pharynx (140–149)	33	1.06	54	0.95	87	0.99	0.79; 1.22
Esophagus (150)	28	1.00	45	0.77	73	0.84	0.66; 1.06
Stomach (151)	51	0.86	81	0.82	132	0.83*	0.70; 0.99
Colon (153)	220	0.84*	185	0.81*	405	0.83*	0.75; 0.91
Rectum (154)	37	0.86	41	0.71*	78	0.77*	0.61; 0.97
Biliary passages and liver (155,156)	56	0.92	43	0.82	99	0.88	0.71; 1.07
Pancreas (157)	161	1.08	132	1.05	293	1.07	0.95; 1.20
Pleura and peritoneum (158.8, 158.9, 163) and mesothelioma (ICD10 C45)[b]	2	0.46	3	0.43	5	0.45	0.14; 1.04
Larynx (161)	11	1.36	17	0.56*	28	0.73	0.49; 1.06
Bronchus, trachea, and lung (162)	652	1.20*	1002	1.28*	1654	1.25*	1.19; 1.31
Bone (170)	6	1.03	6	0.79	12	0.90	0.46; 1.56
Connective and other soft tissue (171)	11	0.79	5	0.47	16	0.65	0.37; 1.06
Breast (174, 175)	420	0.93	1	0.31	421	0.92	0.84; 1.02
All uterine (females only) (179–182)	135	0.96	–	–	135	0.96	0.80; 1.14
Cervix uteri (180)	80	1.23	–	–	80	1.23	0.98; 1.53
Ovary and other female genital organs (183–184)	158	0.93	–	–	158	0.93	0.79; 1.09
Prostate (185)	–	–	265	0.96	265	0.96	0.85; 1.08
Testes and other male genital organs (186, 187)	–	–	8	1.14	8	1.14	0.49; 2.25
Kidney (189.0–189.2)	46	1.04	45	0.78	91	0.89	0.72; 1.09
Bladder and other urinary (188, 189.3–189.9)	43	1.01	79	0.90	122	0.94	0.78; 1.12
Melanoma of skin (172)	31	1.19	39	1.32	70	1.26	0.98; 1.59
Other malignant skin neoplasm (173)	12	1.45	8	0.57	20	0.90	0.55; 1.38
Eye (190)	3	1.93	6	3.54*	9	2.77*	1.26; 5.25
Brain and central nervous system (191–192)	55	1.00	44	0.82	99	0.91	0.74; 1.11
Thyroid and other endocrine glands (193–194)	8	0.73	5	0.70	13	0.72	0.38; 1.22
All lymphatic, hematopoietic tissue (200–208)	270	1.06	250	1.04	520	1.05	0.96; 1.15
Non-Hodgkin lymphoma (200, 202)	117	1.14	90	1.07	207	1.11	0.96; 1.27
Hodgkin lymphoma (201)	4	0.38*	13	0.87	17	0.67	0.39; 1.07
Multiple myeloma (203)	62	1.29	38	0.98	100	1.15	0.94; 1.40
All leukemia (202.4, 204–208)	81	0.98	86	1.10	167	1.04	0.89; 1.21
Acute lymphocytic leukemia (204.0)	5	1.66	0	0.00	5	0.87	0.28; 2.03
Acute myeloid leukemia (205.0, 205.3, 206.0, 207.0, 207.2)	25	0.88	27	1.09	52	0.98	0.73; 1.28
Chronic myeloid leukemia (205.1, 206.1)	7	0.89	7	0.91	14	0.90	0.49; 1.51
Chronic lymphocytic leukemia (CLL) (202.4, 204.1, 204.9)	25	1.27	32	1.47*	57	1.38*	1.04; 1.79
Leukemia other than CLL	55	0.89	54	0.97	109	0.93	0.76; 1.12
Smoking related cancers (140–150, 157, 161–162, 188–189)	975	1.15*	1374	1.15*	2349	1.15*	1.10; 1.20
Nonsmoking related cancers	1681	0.94*	1185	0.93*	2866	0.94*	0.90; 0.97
Polycythemia vera (238.4)	2	0.85	2	0.69	4	0.76	0.20; 1.94
Myelodysplastic syndrome (238.7)	14	0.93	10	1.07	24	0.99	0.63; 1.47
Diabetes (250)	299	0.96	168	0.86*	467	0.92	0.84; 1.00
Mental and behavioral disorders (290–319)	656	1.36*	199	1.36*	855	1.36*	1.27; 1.45
Dementia, Alzheimer's, Parkinson's disease and other motor neuron (290.0–290.4, 331.0, 332, 335.2)	1277	1.37*	375	1.38*	1652	1.37*	1.31; 1.44
Dementia and Alzheimer's disease (290.0–290.4, 331.0)	1180	1.40*	284	1.50*	1464	1.42*	1.35; 1.49
Parkinson's disease (332)	80	1.13	73	1.12	153	1.13	0.96; 1.32
Diseases of the nervous system (320–389)	814	1.25*	447	1.15*	1261	1.22*	1.15; 1.28
All heart disease (390–398, 404, 410–429)	3508	0.96*	4635	1.02	8143	0.99	0.97; 1.01
Rheumatic heart disease (393–398)	53	0.71*	66	1.03	119	0.86	0.71; 1.03
Ischemic heart disease (410–414)	2294	0.95*	3700	1.01	5994	0.99	0.96; 1.01
Cerebrovascular disease (430–438)	1081	1.11*	864	1.15*	1945	1.13*	1.08; 1.18
Nonmalignant respiratory disease (460–519)	1265	1.18*	1320	1.29*	2585	1.23*	1.19; 1.28
Bronchitis, emphysema, asthma (490–493)	481	1.21*	408	1.36*	889	1.27*	1.19; 1.36
Cirrhosis of liver (571)	109	0.88	147	0.82*	256	0.84*	0.74; 0.95
Nephritis and nephrosis (580–589)	182	1.05	115	0.96	297	1.01	0.90; 1.14
AIDS (042–044)	2	2.64	0	0.00	2	1.01	0.11; 3.65
All external causes of death (800–999)	410	1.07	755	1.27*	1165	1.19*	1.12; 1.26
Accidents (850–949)	328	1.12*	515	1.25*	843	1.20*	1.12; 1.28
Suicides (950–959)	47	0.8	172	1.17	219	1.06	0.93; 1.21
Unknown causes of death	283	–	190	–	473	–	–

*SMR is significantly different from 1.0 ($p<.05$).
[a]ICD: International Classification of Disease. Revision 9 codes are shown.
[b]Mesothelioma did not have an explicit code in ICD9 but did in ICD10 as denoted.

Table 4. Internal cohort dose–response analyses for lung cancer over categories of lung dose among 13,951 women and 12,699 men employed for more than 90 days and followed through 2019 (women) or 2018 (men).

	DWF = 1				DWF = 10				DWF = 20			
Lung dose (mGy)[a]	No. of persons	No. of cases	HR	95% CI	No. of persons	No. of cases	HR	95% CI	No. of persons	No. of cases	HR	95% CI
All workers												
0	3758	160	1.0	Ref.	3758	160	1.0	Ref.	3758	160	1.0	Ref.
>0 to <10	11,919	796	0.96	0.80; 1.14	3339	244	1.11	0.90; 1.36	1744	123	1.13	0.89; 1.44
10 to <50	7840	522	0.86	0.71; 1.03	5829	381	0.92	0.68; 1.12	4368	292	0.95	0.78; 1.17
50 to <100	1816	106	0.77	0.60; 0.99	3140	196	0.83	0.67; 1.04	3088	210	0.96	0.77; 1.18
100 to <250	863	52	0.90	0.65; 1.24	4677	316	0.86	0.71; 1.05	4101	256	0.82	0.67; 1.01
250+	454	18	0.64	0.39; 1.04	5907	357	0.81	0.66; 0.98	9591	613	0.84	0.70; 1.01
Total	26,650	1654			26,650	1654			26,650	1654		
p for trend				.10 (–)				.080 (–)				.078 (–)
HR (95% CI) at 100 mGy			0.92	0.83; 1.02			0.99	0.98; 1.001			0.995	0.99; 1.001
ERR (95% CI) at 100 mGy			–0.09	–0.19; 0.02			–0.01	–0.02; 0.001			–0.005	–0.01; 0.001
Females												
0	2502	83	1.0	Ref.	2502	83	1.0	Ref.	2502	83	1.0	Ref.
>0 to <10	5420	272	0.99	0.77; 1.28	1546	94	1.21	0.89; 1.63	926	59	1.23	0.88; 1.72
10 to <50	4068	203	0.94	0.71; 1.24	2466	111	0.88	0.65; 1.18	1655	80	0.99	0.72; 1.35
50 to <100	953	45	0.93	0.63; 1.36	1609	77	0.89	0.68; 1.24	1447	66	0.87	0.62; 1.22
100 to <250	603	32	1.10	0.72; 1.69	2377	119	0.92	0.68; 1.24	2088	99	0.89	0.65; 1.21
250+	405	17	0.88	0.52; 1.59	3450	168	0.92	0.68; 1.22	5333	265	0.93	0.71; 1.22
Total	13,951	652			13,951	652			13,951	652		
p for trend				.87 (+)				.91 (+)				.91 (+)
HR (95% CI) at 100 mGy			1.01	0.90; 1.13			1.001	0.99; 1.01			1.00	0.99; 1.01
ERR (95% CI) at 100 mGy			0.01	–0.10; 0.12			0.001	–0.01; 0.01			0.00	–0.01; 0.01
Males												
0	1256	77	1.0	Ref.	1256	77	1.0	Ref.	1256	77	1.0	Ref.
>0 to <10	6499	524	0.96	0.74; 1.23	1793	150	1.03	0.77; 1.36	818	64	1.00	0.71; 1.40
10 to <50	3772	319	0.89	0.68; 1.15	3363	270	0.96	0.74; 1.25	2713	212	0.94	0.72; 1.24
50 to <100	863	61	0.74	0.52; 1.05	1531	119	0.84	0.63; 1.14	1641	144	1.04	0.78; 1.39
100 to <250	260	20	0.83	0.50; 1.37	2299	197	0.89	0.68; 1.18	2013	157	0.84	0.63; 1.11
250+	49	1	0.19	0.03; 1.36	2457	189	0.81	0.61; 1.07	4258	348	0.86	0.66; 1.11
Total	12,699	1002			12,699	1002			12,699	1002		
p for trend				.042 (–)				.033 (–)				.032 (–)
HR (95% CI) at 100 mGy			0.80	0.65; 0.99			0.98	0.95; 0.998			0.99	0.98; 0.999
ERR (95% CI) at 100 mGy			–0.22	–0.43; –0.01			–0.03	–0.05; –0.002			–0.01	–0.02; –0.001

HR: hazard ratio; CI: confidence interval; ERR: excess relative rate.

Risk estimates are presented by sex and for different dose weighting factors (DWF). Death is from underlying cause only. Model adjusted for year of birth (categories), pay (hourly/salary) and, for women, race (white/black). Underlying timescale is age.

[a]Cumulative lung doses include the sum of the internal doses received at TEC and any external doses received at other facilities. Three different dose weighting factors (DWF) are applied to the high-LET component of the internal doses (1, 10, and 20). For DWF = 10 and 20, the lung dose is in units of weighted-lung dose.

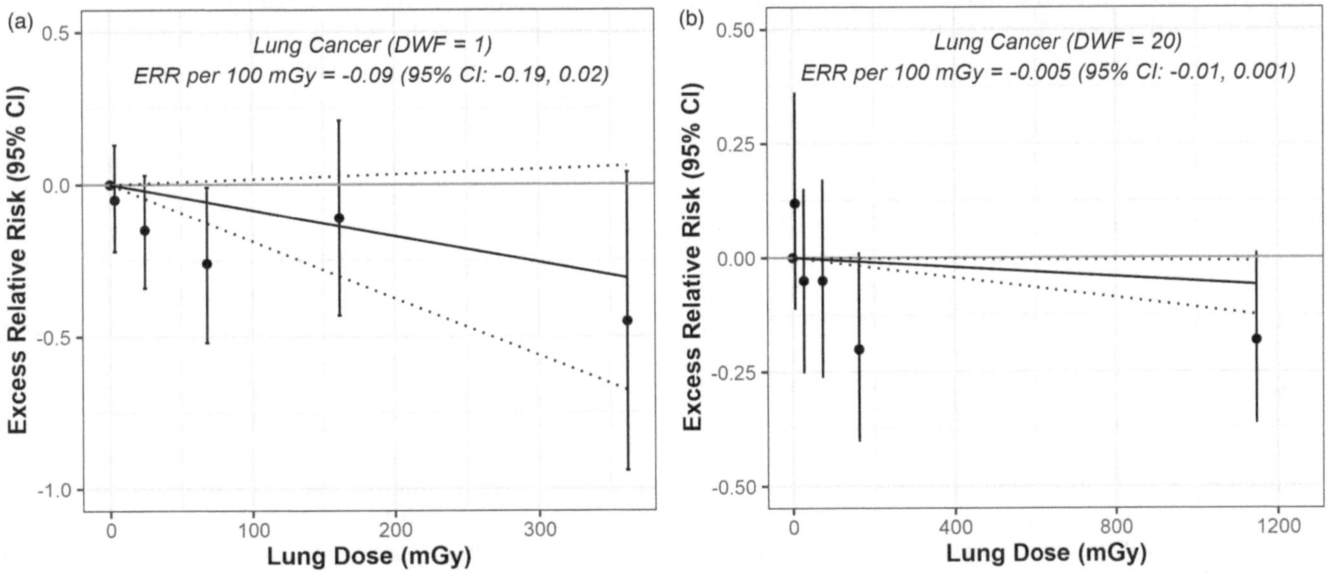

Figure 2. Excess relative rate (ERR) of lung cancer by radiation dose to lung among all 26,650 workers employed 1943–1947 for at least 90 d at Tennessee Eastman Corporation (TEC), Oak Ridge, and followed through 2019 for the 13,951 females and through 2018 for the 12,699 males. A linear model is presented (solid line) based on a continuous measure of dose, and 95% confidence bands (dotted lines) presented about the regression line. Doses were lagged by 10 years, a dose weighting factor of 1 was applied (a), and adjustments were made for sex, race (white, black), year of birth, and socioeconomic status (pay category). ERRs and 95% confidence limits are calculated for selected categories of dose to the lung and displayed at the mean lung dose in each category. A dose weighting factor of 20 was applied for (b). (a) All TEC Workers DWF = 1. (b) All TEC Workers DWF = 20.

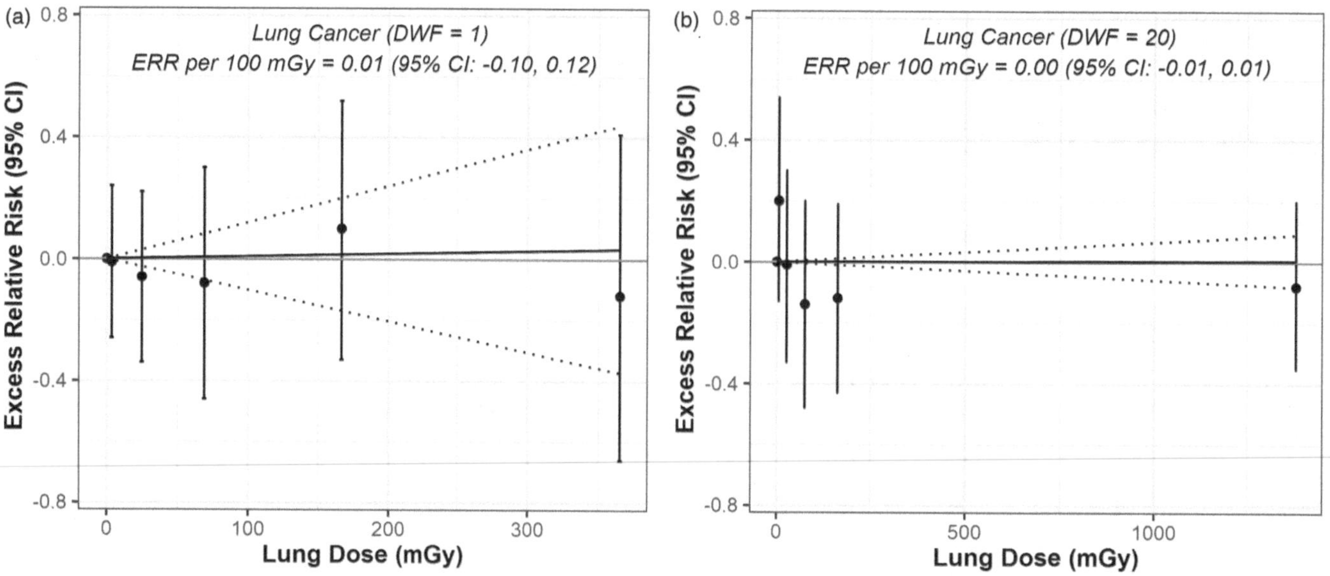

Figure 3. Excess relative rate (ERR) of lung cancer by radiation dose to lung among 13,951 female workers employed 1943–1947 for at least 90 d at Tennessee Eastman Corporation (TEC), Oak Ridge, and followed through 2019. A linear model is presented (solid line) based on a continuous measure of dose, and 95% confidence bands (dotted lines) presented about the regression line. Doses were lagged by 10 years, a dose weighting factor of 1 was applied (a), and adjustments were made for race (white, black), year of birth, and socioeconomic status (pay category). ERRs and 95% confidence limits are calculated for selected categories of dose to the lung and displayed at the mean lung dose in each category. A dose weighting factor of 20 was applied for (b). (a) Female TEC Workers DWF = 1. (b) Female TEC Workers DWF = 20.

for the 1168 black women who worked at TEC was conducted, recognizing the small numbers and statistical uncertainties (Supplemental Table 4). There were 34 women who died of lung cancer and their ERR (95% CI) at 100 weighted-mGy lung dose was estimated as 0.13 (95% CI −0.03, 0.28). A DWF of 10 was applied because a DWF = 1 resulted in a narrow dose range that precluded model convergence. Future analyses will combine black workers from other nuclear facilities (Boice, Quinn, Al-Nabulsi, Ansari, et al. 2022), such as the Savannah River site (Wartenberg et al. 2001).

The highest dose from the inhalation of uranium dust was to the TLNs: the mean dose was 127 mGy and the maximum dose was 5.1 Gy (Table 2). An internal dose–response analysis was conducted of the 207 workers who died of NHL. For a DWF = 1, the ERR (95% CI) at 100 mGy for NHL was estimated as 0.03 (95% CI −0.02, 0.07) (Table 6).

Table 7 presents a sensitivity analysis on the influence of the assumptions made to estimate lung dose on the lung cancer dose response estimates of risk. The four key parameters to estimate lung dose from inhaled uranium dust included the particle size of the dust inhaled, the breathing rate of the worker, the daily exposure during time on the job, and the solubility factor of the uranium whether it was absorbed into the blood at a slow rate (*S*) or a moderate rate (*M*). The low and high level parameter assumptions were judged to be realistic, though somewhat extreme, possibilities. Dose response evaluations were made for best estimate (Table 4), realistic low estimate, and realistic high estimate (Table 7). The Hazard ratio minus one (HR-1) is presented as an estimate of ERR. For female workers, the HR-1 estimates at 100 mGy lung dose ranged between 0.00 and 0.03 and all were consistent with the results from the main analysis (ERR per 100 mGy 0.01; 95% CI −0.10, 0.12). For male workers, however, there was a broader range in the estimates of HR-1 at 100 mGy of −0.59 to −0.09, which were not as consistent with the main analysis (ERR per 100 mGy estimate of −0.22; 95% CI −0.43, −0.01). The solubility assumptions (totally *S* or totally *M*) for male workers result in the most inconsistent risk estimates which might be expected because the inhaled uranium dust was known to be of various solubilities and not totally *S* or totally *M*.

Table 8 presents additional sensitivity analyses of lung cancer risk considering employment duration, dose lagging intervals, a different referent category, and a modified high dose category. For the 14,778 workers employed for more than 1 year, the ERR estimate was essentially the same as the analysis based on all workers: ERR per 100 mGy of −0.09 (95% CI −0.19; 0.03; $n = 871$) compared with −0.09 (95% CI −0.19; 0.02; $n = 1654$). Similarly, the lagging intervals made little difference: for the 5 years lag and 20 years lag intervals the ERRs per 100 mGy were −0.09 (95% CI −0.19; 0.01) and −0.06 (95% CI −0.17; 0.04), respectively. Related in part to the unexpected statistically significant negative dose response for lung cancer among the 12,699 male workers (ERR per 100 mGy of −0.22 (95% CI −0.43, −0.01; $n = 1002$); p for trend =.042) (Table 4), and the possible disproportional influence of a few very high dose workers on the dose response, two additional analyses were conducted: one excluding the 1256 male workers with 0 lung dose, and the other excluding the 25 male workers with estimated lung dose >300 mGy. For both analyses, the ERR estimate increased, and the negative dose response was no longer statistically significant. Due to the apparent influence of a handful of high dose workers perhaps related to bias in the dust to dose conversion for the dustiest environments, the preferred ERR per 100 mGy was taken as -0.14

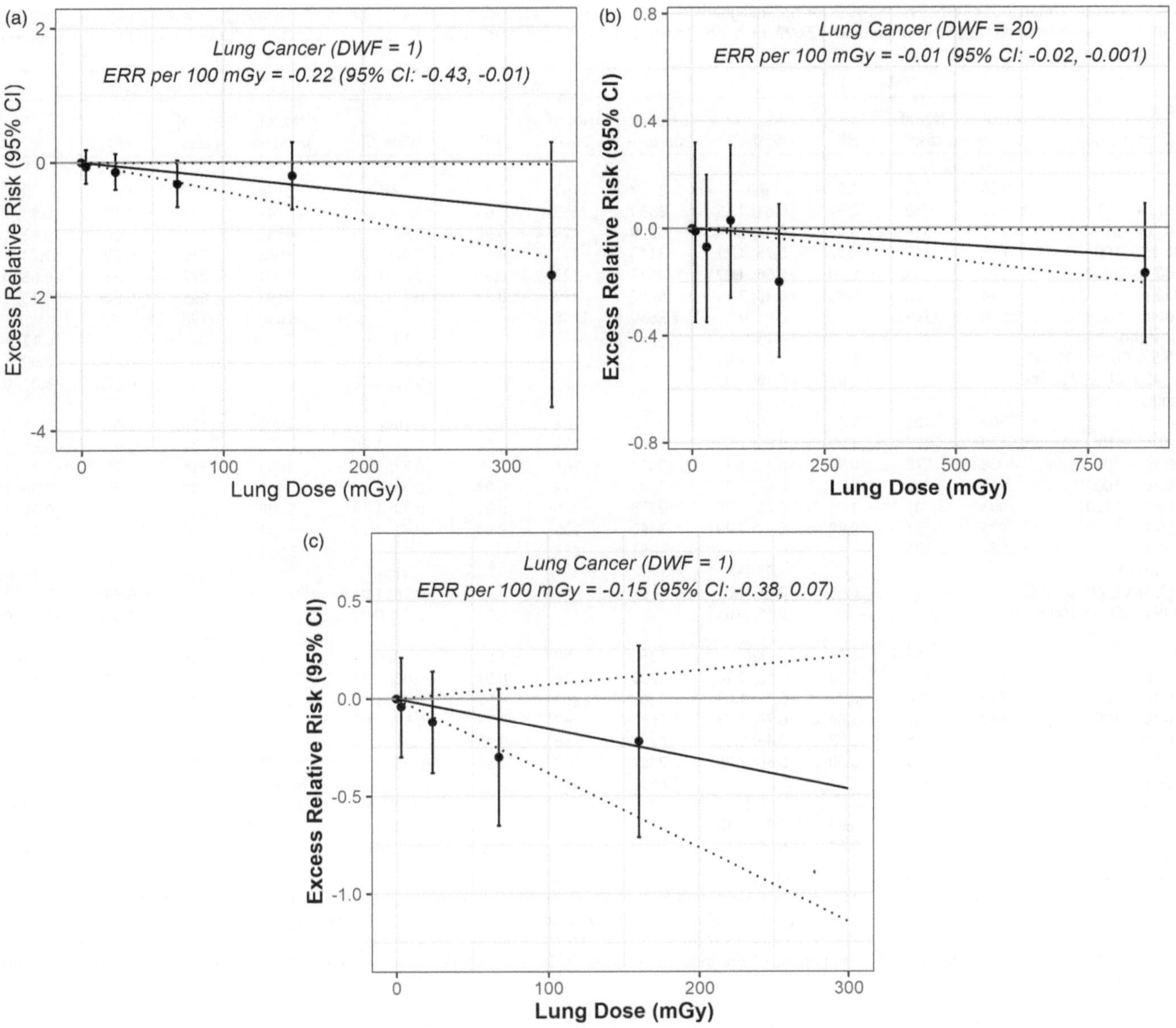

Figure 4. Excess relative rate (ERR) of lung cancer by radiation dose to lung among 12,699 male workers employed 1943–1947 at Tennessee Eastman Corporation (TEC), Oak Ridge, for at least 90 d and followed through 2018. A linear model is presented (solid line) based on a continuous measure of dose, and 95% confidence bands (dotted lines) presented about the regression line. Doses were lagged by 10 years, a dose weighting factor of 1 was applied (a), and adjustments were made for year of birth and socioeconomic status (pay category). ERRs and 95% confidence limits are calculated for selected categories of dose to the lung and displayed at the mean lung dose in each category. A dose weighting factor of 20 was applied for (b). (c) The dose–response model as described above for a dose weighting factor of 1 and for all male workers with lung dose <300 mGy. (a) Male TEC Workers DWF=1. (b) Male TEC Workers DWF=20. (c) Male TEC Workers (DWF=1) with lung dose <300 mGy.

(95% CI −0.32, 0.08; $n = 1002$; p for trend =.18) for male workers (Figure 4(c)).

Tobacco use

Smoking information was sought from TEC medical records on all female lung cancer cases ($n = 641$) and a sample of comparison women who did not die from lung cancer ($n = 707$). Medical records were found for 1235 (91.6%) of these 1348 women: 224 (18.1%) women had used tobacco, 374 (30.3%) had not used tobacco, and 637 (51.6%) had no information on tobacco use. For the 598 women for whom tobacco use was recorded, either yes or no, a higher proportion of tobacco use was seen among the 294 lung cancer cases compared with the 304 comparison women who did not die of lung cancer. Of the 294 women who died of lung cancer, 146 (49.7%) used tobacco, and 78 (25.7%) of the 304 comparison women who did not die of lung cancer used tobacco. These percentages are similar to those reported in an earlier study (Dupree et al. 1995). A dose–response analysis directly adjusting for tobacco use (Supplemental Table 5) did not modify the ERR estimate of lung cancer risk compared with the estimate adjusting for pay category as a surrogate for smoking (Table 4). A small number of black women ($n = 16$) had information on tobacco use, and their percentage who used tobacco (31.3%) was similar to the percentage for the 582 white women (37.6%).

Smoking information also was sought for all male lung cancer cases ($n = 1002$) and a comparison sample of men

Table 5. Internal cohort dose–response analyses for nonmalignant respiratory disease (ICD9 codes 460–478, 490–519) excluding flu and pneumonia over categories of lung doses among 13,951 women and 12,699 men employed 1943–1947 for more than 90 days at TEC, and followed through 2019 (women) or 2018 (men).

	DWF = 1				DWF = 10				DWF = 20			
Lung dose (mGy)[a]	No. of persons	No. of cases	HR	95% CI	No. of persons	No. of cases	HR	95% CI	No. of persons	No. of cases	HR	95% CI
All workers												
0	3758	162	1.0	Ref.	3758	162	1.0	Ref.	3758	162	1.0	Ref.
>0 to <10	11,919	850	0.97	0.81; 1.15	3339	237	1.04	0.85; 1.27	1744	135	1.17	0.93; 1.48
10 to <50	7840	554	0.82	0.68; 0.99	5829	429	0.98	0.82; 1.19	4368	316	1.00	0.82; 1.21
50 to <100	1816	147	0.96	0.76; 1.21	3140	211	0.83	0.67; 1.03	3088	218	0.92	0.74; 1.13
100 to <250	863	59	0.90	0.66; 1.22	4677	322	0.80	0.66; 0.98	4101	272	0.81	0.66; 0.99
250+	454	26	0.73	0.48; 1.11	5907	437	0.87	0.721; 1.05	9591	695	0.84	0.70; 1.01
Total	26,650	1798			26,650	1798			26,650	1798		
p for trend				.13 (–)				.13 (–)				.12 (–)
HR (95% CI) at 100 mGy			0.93	0.85; 1.02			0.99	0.98; 1.002			0.996	0.99; 1.001
ERR (95% CI) at 100 mGy			−0.07	−0.16; 0.02			−0.01	−0.02; 0.002			−0.004	−0.01; 0.001
Females												
0	2502	104	1.0	Ref.	2502	104	1.0	Ref.	2502	104	1.0	Ref.
>0 to <10	5420	377	1.03	0.82; 1.29	1546	113	1.12	0.85; 1.46	926	72	1.17	0.86; 1.58
10 to <50	4068	276	0.94	0.73; 1.20	2466	166	0.99	0.77; 1.28	1655	118	1.10	0.84; 1.44
50 to <100	953	62	0.91	0.65; 1.26	1609	112	0.97	0.73; 1.29	1447	90	0.89	0.66; 1.19
100 to <250	603	41	1.06	0.73; 1.55	2378	156	0.91	0.70; 1.19	2088	137	0.91	0.69; 1.19
250+	405	21	0.80	0.49; 1.29	3450	230	0.94	0.72; 1.19	5333	360	0.93	0.73; 1.18
Total	13,951	881			13,951	881			13,951	881		
p for trend				.33 (–)				.31 (–)				.31 (–)
HR (95% CI) at 100 mGy			0.95	0.86; 1.05			0.99	0.98; 1.01			0.997	0.99; 1.003
ERR (95% CI) at 100 mGy			−0.05	−0.15; 0.05			−0.01	−0.02; 0.01			0.003	−0.01; 0.003
Males												
0	1256	58	1.0	Ref.	1256	58	1.0	Ref.	1256	58	1.0	Ref.
>0 to <10	6499	473	0.90	0.68; 1.20	1793	124	0.91	0.66; 1.26	818	63	1.09	0.76; 1.56
10 to <50	3772	278	0.77	0.57; 1.03	3363	263	0.97	0.72; 1.30	2713	198	0.91	0.67; 1.23
50 to <100	863	85	1.04	0.74; 1.47	1531	99	0.73	0.52; 1.02	1641	128	0.95	0.69; 1.31
100 to <250	260	18	0.78	0.45; 1.33	2299	166	0.75	0.55; 1.02	2013	135	0.74	0.54; 1.02
250+	49	5	0.90	0.36; 2.25	2457	207	0.88	0.65; 1.19	4258	335	0.81	0.61; 1.09
Total	12,699	917			12,699	917			12,699	917		
p for trend				.87 (–)				.90 (–)				.90 (–)
HR (95% CI) at 100 mGy			0.99	0.82; 1.18			0.999	0.98; 1.02			0.999	0.99; 1.01
ERR (95% CI) at 100 mGy			−0.02	−0.20; 0.17			−0.001	−0.02; 0.02			−0.001	−0.01; 0.01

HR: hazard ratio; CI: confidence interval; ERR: excess relative rate.

Risk estimates are presented by sex and for different dose weighting factors (DWF). Death is from underlying cause only. Model adjusted for year of birth (categories) and pay (hourly/salary). For all workers, adjustment is also made for sex, and for female workers adjustment included race. Underlying timescale is age.

[a]Cumulative organ doses include the sum of the internal doses received at TEC and any external doses received at other facilities. Three different dose weighting factors (DWF) are applied to the high-LET component of the internal U doses (1, 10, and 20). For DWF = 10 and 20, the lung dose is in units of weighted-lung dose.

who did not die from lung cancer ($n = 1107$). Medical records were found for 1905 (94.3%) of these 2109 men: 782 (41.0%) men had used tobacco, 181 (9.5%) had not used tobacco, and 942 (49.4%) had no information on tobacco use. For the 963 men for whom tobacco use was recorded, a higher proportion of tobacco use was seen among the 489 lung cancer cases compared with the 474 comparison men who did not die of lung cancer. Of the 489 men who died of lung cancer, 437 (89.4%) used tobacco, and 345 (72.8%) of the 474 comparison men who did not die of lung cancer used tobacco. A dose–response analysis directly adjusting for tobacco use did not appreciably modify the ERR estimate of lung cancer risk compared with the estimate adjusting for pay category as a surrogate for smoking (Table 4), although the negative trend was no longer statistically significant (ERR per 100 mGy −0.13 (95% CI −0.45, 0.19; $n = 489$; p trend =.41)) (Supplemental Table 5).

Medical X-rays

Chest X-ray information was sought for the 1234 women included in the tobacco use medical record abstraction study. Chest X-ray information was found for 78% of the women, and there was no difference between the number of X-rays among lung cancer cases and the comparison women. Among the 585 women who died from lung cancer, 270 (46.1%) received only one X-ray and only 13 women received more than three X-rays. The numbers and proportions of X-rays received by the comparison women were similar. The dose to the lung from a chest X-ray was very low and of the order of a mGy, similar to other cohorts (Ellis et al. 2018). An evaluation of low-dose chest X-rays for men was not deemed necessary so was not included in the medical record abstraction.

Discussion

The study of 26,650 workers employed 1943–1947 at the TEC-Y12 electromagnetic field uranium processing facility is unique because of the large number of women who received higher exposures than the male workers, the follow-up for mortality was up to 75 years and 45 years on average, and practically all workers had died (94.9%). There was little exposure to external radiation in contrast to relatively large

Table 6. Internal cohort dose–response analyses for non-Hodgkin lymphoma (ICD9 categories 200 and 202) over categories of dose to thoracic lymph nodes (TLNs) among 13,951 women and 12,699 men employed for more than 90 days and followed through 2019 (women) or 2018 (men).

	DWF = 1				DWF = 10				DWF = 20			
Thoracic lymph node dose (mGy)[a]	No. of persons	No. of cases	HR	95% CI	No. of persons	No. of cases	HR	95% CI	No. of persons	No. of cases	HR	95% CI
All workers												
0	3753	17	1.0	Ref.	3753	17	1.0	Ref.	3753	17	1.0	Ref.
>0 to <10	5781	38	1.17	0.65; 2.09	965	8	1.41	0.61; 3.26	612	2	0.58	0.13; 2.50
10 to <50	6790	61	1.54	0.88; 2.70	2980	22	1.29	0.68; 2.46	1695	16	1.58	0.79; 3.15
50 to <100	3242	20	0.94	0.48; 1.84	1993	10	0.91	0.41; 2.01	1651	12	1.32	0.62; 2.79
100 to <250	4259	41	1.49	0.82; 2.71	3701	33	1.51	0.82; 2.77	2769	21	1.37	0.71; 2.63
250+	2825	30	1.80	0.96; 3.38	13,258	117	1.38	0.80; 2.37	16,170	139	1.37	0.80; 2.35
Total	26,650	207			26,650	207			26,650	207		
p for trend				.23 (+)				.24 (+)				.24 (+)
HR (95% CI) at 100 mGy			1.03	0.98; 1.07			1.003	0.998; 1.01			1.001	0.999; 1.004
ERR (95% CI) at 100 mGy			0.03	−0.02; 0.07			0.003	−0.002; 0.01			0.001	−0.001; 0.004

HR: hazard ratio; CI: confidence interval; DWF: dose weighting factor; ERR: excess relative rate.

Death is from underlying cause only. Model adjusted for sex, year of birth (categories) and pay (hourly/salary). Underlying timescale is age. Estimates of risk are presented for different dose weighting factors (DWF).

[a]Cumulative organ doses include the sum of the internal doses received at TEC and external doses received at other facilities. Three different dose weighting factors (DWF) are applied to the high-LET component of the internal doses (1, 10, and 20). For DWF = 10 and 20, the lung dose is in units of weighted-lung dose (weighted-mGy).

Table 7. Sensitivity analyses based on varying assumptions regarding uranium dust characteristics used in the calculation of internal dose to lung for DWF = 1.

Females

Lung dose (mGy)[a]	Number of women	Number lung cases	HR	95% CI	Number of women	Number lung cases	HR	95% CI
	AMAD high				AMAD low			
0	2502	83	1.0	Ref.	2502	83	1.0	Ref.
>0 to <10	7190	356	0.97	0.76; 1.25	3903	199	1.01	0.78; 1.32
10 to <50	3293	167	0.97	0.73; 1.30	3985	195	0.93	0.70; 1.22
50 to <100	440	20	0.94	0.57; 1.56	1695	87	0.97	0.70; 1.33
100 to <250	498	24	1.03	0.64; 1.65	1033	49	0.94	0.64; 1.36
250+	28	2	1.59	0.39; 6.51	833	39	0.97	0.65; 1.45
Total	13,951	652			13,951	652		
p for trend				.88 (+)				.87 (+)
HR-1 (95% CI) at 100 mGy			0.02	−0.20; 0.29			0.01	−0.05; 0.07
	Breathing rate high				Breathing rate low			
0	2502	83	1.0	Ref.	2502	83	1.0	Ref.
>0 to <10	3940	201	1.01	0.78; 1.32	6880	341	0.97	0.76; 1.25
10 to <50	3989	195	0.93	0.70; 1.22	3508	177	0.97	0.73; 1.28
50 to <100	1670	87	0.98	0.71; 1.35	441	20	0.92	0.56; 1.53
100 to <250	1024	47	0.91	0.62; 1.32	553	29	1.12	0.72; 1.73
250+	826	39	0.98	0.66; 1.47	67	2	0.71	0.17; 2.89
Total	13,951	652			13,951	652		
p for trend				.87 (+)				.87 (+)
HR-1 (95% CI) at 100 mGy			0.01	−0.05; 0.07			0.02	−0.17; 0.25
	4 hour workday				8 hour workday			
0	2502	83	1.0	Ref.	2502	83	1.0	Ref.
>0 to <10	6353	313	0.97	0.75; 1.25	4749	251	1.04	0.80; 1.34
10 to <50	3777	194	0.98	0.74; 1.29	4120	190	0.86	0.65; 1.14
50 to <100	557	27	1.01	0.64; 1.59	1272	67	0.98	0.70; 1.38
100 to <250	620	26	0.90	0.57; 1.42	690	30	0.89	0.58; 1.38
250+	142	9	1.40	0.69; 2.81	618	31	1.05	0.68; 1.62
Total	13,951	652			13,951	652		
p for trend				.87 (+)				.87 (+)
HR-1 (95% CI) at 100 mGy			0.01	−0.14; 0.20			0.01	−0.07; 0.09
	Type M solubility				Type S solubility			
0	2502	83	1.0	Ref.	2502	83	1.0	Ref.
>0 to <10	8889	445	0.98	0.76; 1.25	3336	172	1.01	0.77; 1.32
10 to <50	1975	95	0.95	0.69; 1.31	3878	189	0.92	0.70; 1.22
50 to <100	462	22	1.00	0.62; 1.63	1963	98	0.94	0.69; 1.29
100 to <250	123	7	1.34	1.61; 2.92	1300	65	1.02	0.72; 1.44
250+	–	–	–	–	972	45	0.98	0.67; 1.45
Total	13,951	652			13,951	652		
p for trend				.90 (+)				.85 (–)
HR-1 (95% CI) at 100 mGy			0.03	−0.34; 0.60			0.00	−0.04; 0.05

(*continued*)

Table 7. Continued.

Males

Lung dose (mGy)[a]	Number of men	Number lung cases	HR	95% CI	Number of women	Number lung cases	HR	95% CI
		AMAD high				AMAD low		
0	1256	77	1.0	Ref.	1256	77	1.0	Ref.
>0 to <10	8277	677	0.94	0.73; 1.21	5090	415	0.99	0.77; 1.27
10 to <50	2878	228	0.84	0.64; 1.11	3880	319	0.87	0.67; 1.13
50 to <100	215	16	0.78	0.45; 1.35	1411	115	0.86	0.64; 1.16
100 to <250	73	4	0.55	0.20; 1.51	868	60	0.71	0.50; 1.01
250+	–	–	–	–	194	16	0.88	0.51; 1.51
Total	12,699	1002			12,699	1002		
p for trend				.04 (–)				.04 (–)
HR-1 (95% CI) at 100 mGy			–0.37	–0.60; –0.02			–0.11	–0.20; –0.00[b]
		Breathing rate high				Breathing rate low		
0	1256	77	1.0	Ref.	1256	77	1.0	Ref.
>0 to <10	5101	417	0.99	0.77; 1.27	7990	650	0.94	0.73; 1.21
10 to <50	3886	319	0.87	0.67; 1.14	3105	250	0.86	0.65; 1.12
50 to <100	1402	114	0.86	0.63; 1.16	254	18	0.75	0.45; 1.26
100 to <250	864	59	0.71	0.50; 1.01	94	7	0.81	0.37; 1.76
250+	190	16	0.89	0.52; 1.54	–	–	–	–
Total	12,699	1002			12,699	1002		
p for trend				.04 (–)				.04 (–)
HR-1 (95% CI) at 100 mGy			–0.11	–0.2; –0.00[b]			–0.34	–0.56; –0.01
		4 hour workday				8 hour workday		
0	1256	77	1.0	Ref.	1256	77	1.0	Ref.
>0 to <10	7435	597	0.94	0.73;1.20	5838	472	0.97	0.75; 1.25
10 to <50	3474	289	0.87	0.67; 1.14	3911	325	0.88	0.67; 1.14
50 to <100	371	28	0.79	0.51; 1.23	1160	87	0.78	0.57; 1.07
100 to <250	163	13	0.96	0.53; 1.74	438	34	0.81	0.54; 1.23
250+	–	–	–	–	96	7	0.79	0.36; 1.72
Total	12,699	1002			12,699	1002		
p for trend				.04 (–)				.04 (–)
HR-1 (95% CI) at 100 mGy			–0.28	–0.47; –0.01			–0.15	–0.28; –0.01
		Type M solubility				Type S solubility		
0	1256	77	1.0	Ref.	1256	77	1.0	Ref.
>0 to <10	9712	794	0.93	0.73; 1.19	4997	402	0.97	0.75; 1.24
10 to <50	1636	125	0.80	0.60; 1.08	3721	318	0.90	0.69; 1.16
50 to <100	95	6	0.80	0.35; 1.85	1551	128	0.88	0.65; 1.18
100 to <250	–	–	–	–	936	59	0.68	0.48; 0.96
250+	–	–	–	–	238	18	0.83	0.49; 1.40
Total	12,699	1002			12,699	1002		
p for trend				.03 (–)				.06 (–)
HR-1 (95% CI) at 100 mGy			–0.59	–0.81; –0.09			–0.09	–0.18; 0.01

HR: hazard ratio; CI: confidence interval; DWF: dose weighting factor; ERR: excess relative rate.

Particle size, breathing rate, daily exposure, and uranium solubility factors were modified to evaluate their effect on the lung cancer dose response and risk estimates. Analyses are for the 13,951 women and 12,699 men employed at TEC 1943–1947 for at least 90 days and followed through 2019 (women) or 2018 (men). Death is from underlying cause only. Model adjusted for year of birth (categories), race (white/black) for females, and pay category (hourly/salary). Underlying timescale is age.

[a]Cumulative lung doses include the sum of the internal doses received at TEC and any external doses received at other facilities.

[b]Rounded to two significant digits. The negative sign indicates that the HR-1 was between 0.000 and –0.005.

doses to lung from inhalation of airborne uranium dust, i.e. the mean absorbed dose to lung for all workers was 26.3 mGy and the maximum was 1.05 Gy. The high doses and large numbers of deaths from lung cancer ($n = 1654$) resulted in high statistical power to reveal any radiation-related lung cancer risks and to distinguish any differences in risk estimates between men and women had they occurred. The data, however, did not support an association between radiation dose to the lung from inhaled uranium dust and lung cancer or NMRD. There was no difference in the lung cancer risk by sex. There was no significant radiation association for NHL for doses to the TLNs that were higher than the lung doses.

The unexpected significant negative dose response for male lung cancer was addressed by conducting several sensitivity analyses (Table 8). First, male workers who had no lung dose were removed as the referent group, after recognizing that they were not comparable in job categories, potential for dust exposure, or tobacco use as other workers. This was reflected in the SMR analyses where salaried male workers had significantly low risks of lung cancer whereas the hourly males, hourly females and salaried females all had significantly high lung cancer SMRs. The dose response was no longer significantly negative after these zero-dose workers were excluded.

Second, concern over possible measurement assessments that may have been exceptionally high among a few high dose workers, was confirmed in an analysis that excluded all male workers with cumulative absorbed lung doses >300 mGy ($n = 25$ and no lung cancer deaths). The ERR

Table 8. Sensitivity analyses for lung cancer considering (1) only the 14,778 TEC workers employed for more than 1 year; (2) applying different lag periods of 5 years and 20 years for the 26,650 workers employed 1943–1947 for more than 90 days; (3) analyses of male workers excluding those with lung dose = 0 mGy (primarily salaried workers) and using >0 to <5 mGy as the referent category; (4) and analyses of male workers with lung doses <300 mGy (excluding the 25 workers with lung dose >300 mGy).

(1) All workers employed >1 year

Lung dose (mGy)[a]	No. of persons	No. of cases	HR	95% CI
All workers				
0	1621	63	1.0	Ref.
>0 to <10	4844	299	1.04	0.78; 1.38
10 to <50	5953	385	0.97	0.73; 1.29
50 to <100	1496	87	0.86	0.61; 1.21
100 to <250	455	23	0.86	0.53; 1.40
250+	409	14	0.64	0.35; 1.16
Total	14,778	871		
p for trend				.15 (–)
HR (95% CI) at 100 mGy			0.91	0.81; 1.03
ERR (95% CI) at 100 mGy			–0.09	–0.21; 0.03

(2) All workers – lagging scenarios of 5 years and 20 years

	Lag 5 years				Lag 20 years			
Lung dose (mGy)[a]	No. of persons	No. of cases	HR	95% CI	No. of persons	No. of cases	HR	95% CI
All workers								
0	3372	142	1.0	Ref.	5018	220	1.0	Ref.
>0 to <10	12,109	802	0.94	0.78; 1.13	11,259	772	1.17	0.99; 1.38
10 to <50	7981	533	0.86	0.71; 1.04	7460	498	1.03	0.86; 1.24
50 to <100	1847	105	0.74	0.57; 0.96	1653	96	0.92	0.71; 1.19
100 to <250	881	54	0.91	0.66; 1.26	821	51	1.12	0.81; 1.54
250+	460	18	0.62	0.38; 1.02	439	17	0.77	0.47; 1.28
Total	26,650	1,654			26,650	1654		
p for trend				.09 (–)				.23 (–)
HR (95% CI) at 100 mGy			0.92	0.83; 1.02			0.94	0.85; 1.04
ERR (95% CI) at 100 mGy			–0.09	–0.19; 0.01			–0.06	–0.17; 0.04

(3) Male workers excluding those with lung dose = 0 mGy

Lung dose (mGy)[a]	No. of persons	No. of cases	HR	95% CI
Males				
>0 to <5	4956	404	1.0	Ref.
5 to <10	1543	120	0.96	0.75; 1.13
10 to <50	3772	319	0.95	0.82; 1.11
50 to <100	863	61	0.79	0.60; 1.03
100 to <250	260	20	0.88	0.56; 1.39
250+	49	1	0.20	0.03; 1.42
Total	11,443	925		
p for trend				.11 (–)
HR (95% CI) at 100 mGy			0.84	0.69; 1.04
ERR (95% CI) at 100 mGy			–0.17	–0.38; 0.04

(4) Male workers with lung dose <300 mGy (Figure 4(c))

Lung dose (mGy)[a]	No. of persons	No. of cases	HR	95% CI
Males				
0	1256	77	1.0	Ref.
>0 to <10	6499	524	0.96	0.74; 1.23
10 to <50	3772	319	0.89	0.68; 1.15
50 to <100	863	61	0.74	0.52; 1.05
100 to <300	284	20	0.80	0.49; 1.31
Total	12,674	1002		
p for trend				.18 (–)
HR-1 (95% CI) at 100 mGy			–0.14	–0.32; 0.08
ERR (95% CI) at 100 mGy			–0.15	–0.38; 0.07

HR: hazard ratio; CI: confidence interval; ERR: excess relative rate; DWF: dose weighting factor.

Death is from underlying cause only. Model adjusted for year of birth (categories), sex (for all worker analyses), and pay category (hourly/salary). Underlying timescale is age. DWF = 1 for all analyses.

[a]Cumulative lung doses include the sum of the internal doses received at TEC and any external doses received at other facilities.

(95% CI) per 100 mGy became −0.14 (95% CI −0.32, 0.08; $n = 1002$; *p* for trend =.18). Removal of this small number of apparent outliers resulted in a dose response that was no longer statistically significant and the ERR estimate was compatible with the TEC female estimate and with other MPS studies with intakes of radionuclides, i.e. workers at LANL (Boice, Cohen, Mumma, Golden, et al. 2022) and at the Mallinckrodt Chemical Workers (Golden et al. 2022). Additional analyses were conducted removing 16 male workers with lung doses >350 mGy (ERR= −0.17) and 11 workers >400 mGy (ERR= −0.18) and neither dose response trend was statistically significant. Although speculative as to why the dose assessment may have been biased, conceivably it might be related to unknown use of respirators in the dustier environments. Earlier assessments of the TEC male cohort noted a possible limitation if respiratory protection was used during certain operations involving high uranium air concentrations (Polednak et al. 1982). Unfortunately, we and Polednak et al. (1982) did not identify any specific documents on the use of respirators. If, however, respirators were worn and were fully operational in the dustier workplaces, i.e. worn and properly cleaned, actual exposures would probably be only 25% of the measured average concentration (Lippmann 1959). Accordingly, the analysis of workers with lung doses <300 mGy (i.e. excluding 25 workers with doses >300 mGy and no lung cancer deaths) was taken as the preferred ERR estimate for male workers (Table 8, Figure 4(c)).

Comparison with previous TEC studies

In a subgroup analysis of men who were 45 years of age or older when first hired, Cookfair et al. (1983) reported a significant smoking-adjusted lung cancer risk among those with a cumulative lung dose of greater than 200 mGy. There were 330 lung cancer deaths evaluated by Cookfair et al. (1983) and follow-up was to 1 July 1973. The earlier study of white males at TEC also noted an elevation of lung cancer ($n = 324$ total) for those over age 45 years when hired (lung cancers, $n = 97$) and among a subset of 'chemical' workers (lung cancers, $n = 7$) who had the highest potential for uranium dust exposure (Polednak and Frome 1981). The larger four-facility study of 787 lung cancer cases (567 among workers at TEC), included contributing causes of death and follow-up was through 1982 (Dupree et al. 1995). The findings resembled those of Cookfair et al. (1983) but did not reach a level of statistical significance after adjusting for tobacco use (Dupree et al. 1995). Among 1853 male workers over age 45 years when first hired in our study, the ERR (95% CI) at 100 mGy was increased, 0.30 (95% CI −0.43, 1.02), but not at the level of statistical significance. In contrast, for workers 15–24 years and 25–44 years at first hire, the ERRs were all negative, but also not statistically significant. This pattern of decreasing ERRs with increasing age at hire might partially reflect the increase background rates for lung cancer with age at hire. Future combined cohorts of MPS cohorts will evaluate further the possible influence of age at exposure and subsequent lung cancer risk.

Comparison with other uranium processing cohorts

A comprehensive study of uranium processing workers who inhaled uranium dust was conducted of 2514 white males who worked at the Mallinckrodt Chemical Works (Boice et al. 2018; Ellis et al. 2018; Golden et al. 2022). The exposure assessment was extensive and included 39,451 uranium urine bioassays, 2341 breath radon measurements, and 15,518 occupational medical X-rays in addition to film badge measurements of external gamma radiation exposures. The estimated mean lung absorbed dose from all sources of exposure was 69.9 mGy (maximum 0.89 Gy). The mean lung dose from uranium inhalation alone was 27.1 mGy (max 0.42 Gy) which is similar to our mean lung dose of 26.3 mGy (max 1.0 Gy) for all workers. Similar to our findings that are based only on airborne uranium measurements, there was no evidence of a dose response for either lung cancer ($n = 157$) or NMRD excluding flu and pneumonia ($n = 238$). It remains equivocal whether exposure to uranium is a cause of lung cancer (UNSCEAR 2017), and occupational guidelines are based in large part on the chemical toxicity of uranium, notably to the kidney (ATSDR 2013). An extensive review of the literature in 2013 concluded that 'no definitive evidence has been found in epidemiologic studies that links human deaths to uranium exposure' (ATSDR 2013).

Smoking considerations

For the 1561 workers (598 women and 963 men) with available information on tobacco use obtained from medical questionnaires completed during employment at TEC, internal cohort dose–response analyses were conducted for lung cancer. The estimates of radiation risk (ERRs) with direct adjustment for tobacco use were comparable to those computed for the full cohort adjusting for pay category (salary/hourly) (Supplemental Table 5). For all workers, the lung cancer ERR per 100 mGy with direct adjustment for tobacco use was −0.03 (95% CI −0.20, 0.13; $n = 784$; p trend .70). The lung cancer ERR per 100 mGy with adjustment for pay category was −0.09 (95% CI −0.19, 0.02; $n = 1654$; p trend .10) and statistically indistinguishable. For males, the ERR per 100 mGy approached the null, increasing from −0.22 to −0.13, and the dose response was no longer statistically significant. These analyses suggest that the absence of a lung cancer radiation effect among TEC workers is unlikely to be related to confounding by cigarette smoking. Further, that adjusting for pay category can be assumed to be closely equivalent to directly adjusting for tobacco use in this population. To support this conclusion that pay category is an adequate surrogate for cigarette smoking, we evaluated the risk of lung cancer among hourly workers (SMR = 1.32; 95% CI 1.25–1.40; $n = 1299$) and salaried workers (SMR = 1.01; 95% CI 0.90–1.12; $n = 344$) (Supplemental Table 2). The ratio of SMRs, 1.31, was statistically significant (95% CI 1.16, 1.47), as was the SMR for lung cancer among the hourly workers (but not salaried workers).

Non-Hodgkin lymphoma

Death from NHL was increased overall based on comparison with the general population (SMR 1.11; 95% CI 0.96, 1.27; $n = 207$). The TLNs received a higher mean dose (127 mGy) than the lung (26.3 mGy) (Table 2). An internal analysis revealed little evidence for a radiation association with NHL (ERR at 100 mGy 0.03; 95% CI −0.02, 0.07) (Table 6). This finding is consistent with previous MPS studies of workers with intakes of radionuclides at LANL in New Mexico (Boice, Cohen, Mumma, Golden, et al. 2022) and at the Mallinckrodt Chemical Works in Missouri (Golden et al. 2022). No radiation association for NHL was observed following external radiation among nuclear power plant workers in the MPS (Boice, Cohen, Mumma, Hagemeyer, et al. 2022). In contrast, a recent report from the UK National Registry of Radiation Workers found a significant dose response for NHL which appeared concentrated among workers who received cumulative external doses >0.5 Sv (Hunter and Haylock 2021).

Skin cancer

Polednak and Frome (1981) commented that the dose to the skin from beta-emitting progeny of ^{238}U might be a potential hazard but provided no dosimetric details. Melanoma of the skin (SMR 1.26; 95% CI 0.98, 1.59; $n = 90$) and cancer of the eye (SMR 2.77; 95% CI 1.26, 5.25; $n = 9$) were elevated which prompted us to look carefully at the possible range of doses to the skin from the uranium dust. We found that the skin dose was miniscule (mean dose <0.05 mGy; max ~1.0 mGy) and was lower than the lung dose by a factor of 1000.

Parkinson's disease

A study of Mayak plutonium workers in Russia recently reported a dose–response relationship between low-LET radiation and Parkinson's disease (Azizova et al. 2020). Similar positive associations have been reported among MPS cohorts (Boice and Dauer 2021; Boice, Cohen, Mumma, Golden, et al. 2022; Boice, Cohen, Mumma, Hagemeyer et al. 2022; Boice, Cohen, Mumma, Howard, et al. 2022; Boice, Quinn, Al-Nabulsi, Ansari, et al. 2022). The SMR for Parkinson's disease among TEC workers was slightly elevated at 1.13 (95% CI 0.96, 1.32; $n = 153$). As workers had little radiation dose to brain, dose–response analyses were precluded. However, we were able to evaluate whether smoking cigarettes was related to a protective effect of Parkinson's disease as seen in epidemiologic studies of smokers, and most recently in the 65-year follow-up of British doctors (Mappin-Kasirer et al. 2020). Assuming that pay category is an adequate surrogate for tobacco use, we evaluated the SMRs of Parkinson's disease among hourly workers (SMR = 1.03; 95% CI 0.85, 1.26; $n = 97$) and among salaried workers (SMR = 1.30; 95% CI 0.97, 1.71; $n = 52$) (Supplemental Table 2). These results are as predicted and consistent with a protective effect of cigarette smoking for Parkinson's disease, i.e. hourly workers had a lower SMR than salaried workers. The ratio of SMRs, 1.26, was close to statistical significance (95% CI 0.90, 1.76), as was the increased SMR among the small number of salaried workers.

Interpretation of SMR analysis and future evaluations

Interpretation of SMR analysis must be done cautiously for a number of reasons, including non-comparability of the general population with a working population, and the ability to make hundreds of comparisons when evaluating many causes of death by many categories. There were nearly 400 SMRs computed for the male and female TEC workers which resulted in many significantly high and significantly low estimates (Table 3, Supplemental Table 2). By chance alone about 20 statistically significant estimates might be expected. Thus, specific hypotheses are important beforehand to discourage, in part, attempts to explain every significant observation or subgroup analysis that may have been serendipitous and not of *a priori* interest. In our study of uranium processing the highest radiation doses were anticipated for the lung and TLNs, so that focusing on malignant and nonmalignant lung disease and perhaps NHL and kidney disease seemed appropriate. A number of causes of death had statistically significant SMRs that are not easily interpreted with respect to radiation exposure because the organ-specific doses were very small. For example, there was a significant excess of deaths from CeVD in contrast to a nearly significant deficit of deaths from IHD. Similarly, a significantly high excess of CLL was seen whereas there was no increase in deaths from leukemia excluding CLL; the latter is frequently found in excess following external (although not internal) radiation exposures whereas the former is not. SMR elevations were seen for cancer of the eye, melanoma, accidents, and other sites suggesting that the comparison population, the general population of the United States, is not ideal for radiation risk indications. Perhaps equally important, the TEC WWII worker population may have had unique lifestyle and demographic features that are not readily characterized. The population smoked heavily, for example, as seen by the significant excesses from smoking-related cancers and the significant deficits for nonsmoking related cancers.

The WWII population of men and women working in Oak Ridge on a secret Manhattan Project program could very well differ in many ways from the general population of the United States. Accordingly, we computed SMRs based on population rates of the state of Tennessee. The overall SMR for all causes of death decreased to 1.01 (95% CI 1.00, 1.03), and there were other changes observed such as the deficit of IHD became significantly low (SMR 0.92; 95% CI 0.89, 0.94). These examples point to the challenges in interpreting SMRs even when best attempts are made to obtain an adequate general population. Thus, internal dose–response relationships considering only the workers themselves and making comparisons over categories of radiation dose with adjustment for potential confounders is optimum. Such an approach addresses the so-called healthy worker effect where occupational groups are healthier and less likely to die during the years after first hire than segments of the general population that are unable to work because of illnesses or other infirmities. However, in circumstances such as TEC there are few causes of death for which *radiation* dose–response evaluations can be conducted because only the lung and lymph nodes received appreciable radiation dose, and other occupational exposures are either unknown or difficult to estimate. Future evaluations should consider, if possible, any chemical exposures (e.g. phosgene, Cl_2, HNO, NO_X); and any medical conditions such as tuberculosis, other lung diseases, or other conditions that may be available in medical records of the TEC population. For TEC workers alive in 1999 and after, we will obtain detailed information on disease occurrence, such as cancer incidence, and risk factors, such as smoking, with linkages to Medicare claims data and other extensive datasets available from the Centers for Medicare and Medicaid Services (CMS) (Boice, Quinn, Al-Nabulsi, Ansari, et al. 2022). There were 9995 TEC workers alive in 1999, 7624 women and 2371 men, who are eligible for the CMS linkages.

Quality of death information for those dying after age 85 years

Analyses of MPS cohorts commonly end the follow-up of workers not known to have died when they reach the age of 95 years. This is done because of perceived concern that the causes of death recorded on death certificates for the very old would be nonspecific and noninformative, for example 'old age', and that the rates of mortality for cause-specific deaths for ages greater than 95 years accordingly would be uncertain.

We conducted a sensitivity analysis (Supplemental Table 6) where we ended the follow-up when a worker reached the age of 85 and compared the resulting SMRs with those who died at older ages (Table 3). There were essentially no differences in the SMRs in relation to the ages when follow-up ended. For all causes of death, the SMRs using an 85 years cutoff was 1.06 (95% CI 1.05, 1.08; $n = 18{,}327$) compared with 1.07 (95% CI 1.06, 1.08; $n = 25{,}282$) for using the later cutoff. There were 6,955 additional deaths that occurred after age 85, and the force of mortality was not seen to change, i.e. the women did not die at a higher or lower rate as compared with the general population, nor did the SMRs change. A key issue, however, is whether the distribution of the causes of death changed over these advanced ages of follow-up. In brief, they did not. Based on large numbers of deaths from all cancers, the SMRs at cutoff age 85 and at ages >85 years were the same at 1.01 with an additional 750 cancer deaths observed after age 85. For lung cancer, the SMRs at cutoff age 85 and at ages >85 years were also the same at 1.24 with an additional 164 lung cancer deaths observed after age 85. Similar patterns are seen for the other major causes of death including breast cancer, prostate cancer, leukemia, mental and behavioral disorders, all heart disease including IHD, CeVD, NMRD, nephritis and nephrosis, and all external causes of death. There were some differences seen for non-Hodgkin lymphoma and diabetes, but based on small numbers and, conceivably, might indicate better diagnoses of these causes of death than in the general population of elderly persons.

Medical diagnoses and treatment for the elderly and disabled in the United States improved markedly since 1965 with the expansion of Medicare programs that ensured access to and payment for health care (CMS 2015). These programs were accompanied with an increased accuracy of disease diagnoses and cause of death determination compared with years past, e.g. the top five causes of death among centenarians in 2014 were heart disease, Alzheimer's disease, stroke, cancer, and influenza and pneumonia and similar to death distributions for those younger but considered elderly (Xu 2016). We conclude that using an age at follow-up cutoff of 95 years is reasonable, and does not bias or dilute the analyses. Future analyses will consider the ramifications of extending the follow-up without using any age cutoff and recording deaths until the end of life, e.g. up to age 105 years.

High-LET dose to lung

Table 9 presents lung cancer risk estimates for MPS cohorts with intakes of radionuclides (Boice, Cohen, Mumma, Ellis 2022). The workers in these cohorts had intakes of uranium, polonium, plutonium, or other radionuclides, and exhibited little evidence for any increase in lung cancer risk. The over 35,000 workers with intakes of radionuclides were employed at Rocketdyne (Leggett et al. 2005; Boice, Cohen, et al. 2006; Boice, Leggett, et al. 2006; Boice et al. 2011), Mound (Boice et al. 2014), Mallinckrodt Chemical Works (Golden et al. 2022), LANL (Boice, Cohen, Mumma, Golden, et al. 2022), and Rocky Flats (Boice, Quinn, Al-Nabulsi, Ansari, et al.

Table 9. Lung cancer radiation risk estimates in Million Person Study cohorts with intakes of radionuclides used in the computation of lung dose.

Cohort[a]	Number of workers	No. of monitored for radionuclides and dose to lung computed	Primary radionuclide	Number lung cancers	Study metric	Estimated ERR per 100 mGy (95% CI)
TEC men[b]	12,699	11,443	Uranium	1002	ERR at 100 mGy	−0.14 (−0.32; 0.08)
TEC women	13,951	11,449	Uranium	652	ERR at 100 mGy	0.01 (−0.10; 0.12)
Rocketdyne	5801	2232	Uranium	151	RR at 100 mSv[c]	−0.01 (−0.14; 0.13)
Mound	4672	3082	Polonium	203	HR at 100 mSv[c]	0.00 (−0.03; 0.04)
Mallinckrodt men	2514	2155	Uranium	157	ERR at 100 mGy	−0.06 (−0.18; 0.06)
Los Alamos	26,328	5142	Plutonium	839	ERR at 100 mGy	0.01 (−0.15; 0.17)
Rocky Flats	9397	4812	Plutonium	324	ERR at 100 mGy	0.01 (−0.10; 0.11)

When a different metric was used to estimate risk in a particular study, an approximation to an estimated ERR is presented. The mGy doses are unweighted absorbed lung doses (Dose Weighting Factor of 1) and include external doses.

[a]Rocketdyne (Boice et al. 2011); Mound (Boice et al. 2014); Mallinckrodt (Golden et al. 2022); Los Alamos (Boice, Cohen, Mumma, Golden, et al. 2022). Summaries also found in Boice, Quinn, Al-Nabulsi, Ansari, et al. (2022).

[b]Male workers with lung doses <300 mGy.

[c]These values were converted to an approximate ERR per 100 mGy.

2022). Other than the male TEC estimate, the ERRs per 100 mGy lung dose ranged from −0.06 to 0.01, which can be compared with the estimate of 0.02 (95% CI −0.03, 0.07) for low-LET exposure in the combined cohorts of nuclear power plant workers, industrial radiographers and medical radiation workers ($n = 367{,}722$) (Boice, Cohen, Mumma, Howard, et al. 2022; Boice, Cohen, Mumma, Hagemeyer, et al. 2022; Boice, Ellis, Golden, Zablotska, et al. 2022).

Our study of high-LET alpha particle exposure to lung from inhaled uranium dust is relevant to concerns for astronauts embarking on long missions outside of Earth orbit. The possible health effects of galactic cosmic rays (GCRs), i.e. the very high-LET heavy ions traveling near the speed of light, are an area of intense investigation (Boice 2017, 2022; NASA 2021; NASEM 2021; NCRP 2021a, 2021b; Boice, Ellis, Golden, Zablotska, et al. 2022). There are no circumstances on Earth that can represent GCR so that animal studies and GCR simulations are used to estimate the risk of cancer and nonmalignant diseases, including cognitive function. The studies of the Japanese atomic bomb survivors exposed acutely to radiation in 1945 indicate that women are at 2–3 times higher risk than men on a relative scale for lung cancer (NCRP 2014). Until recently (NASA 2022) based its radiation protection guidance on individual lifetime risk estimates and relies heavily upon the Japanese data. Accordingly, women were limited in their flight hours compared with men because of the differences in radiation-related lung cancer mortality risk. Although an imperfect analog to GCR, the alpha particle exposure to lung from intakes of radionuclides is from high-LET particles and of high dose from protracted exposure over years (Boice, Quinn, Al-Nabulsi, Ansari, et al. 2022). The dose distributions, however, are not similar within the lung, the track structure differs, and there is the possible effect of inhomogeneous distributions of the heavy metal uranium within the lung as well as any effect of metal toxicity. Nonetheless, it provides another set of human data that could be considered for radiation protection guidance, including the combination with other informative data sets (Simonsen and Slaba 2021).

Strengths and limitations

The study of the TEC workers employed 1943–1947, has a number of epidemiologic strengths as mentioned: a large number of workers, including female workers, near complete lifetime follow-up for mortality, information on individual use of tobacco from medical questionnaires, and a comprehensive dose-reconstruction process. Further, a wide range of sensitivity analyses were conducted to help interpret observations such as varying the input parameters required to estimate lung dose from uranium air monitoring data, evaluating different dose lagging procedures, examining the effect of different inclusion and exclusion criteria for conducting dose response (such as excluding 25 highest dose male workers and limiting the analyses to worker with <300 mGy cumulative lung dose). The number of women was larger than the number of men and their lung doses were larger; this is unusual in occupational studies and a strength when conducting sex-specific comparisons of lung cancer risks. The reason for the higher doses among the women is that they remained in the 'dusty environments' for longer employment periods than the men.

A major limitation in most occupational studies of lung cancer is the limited information available on smoking histories for individual workers. During WWII, smoking was not uncommon among men or women. Pay category (hourly/salary) was used as the measure of SES and as a surrogate for tobacco use. Medical record abstraction on the use of tobacco among TEC men and women confirmed that pay category was directly correlated with tobacco use. Nearly, 50% of the workers who died of lung cancer had information on tobacco use available in medical questionnaires. Analyses adjusting directly for tobacco use among those workers with information available in their medical records (both lung cancer cases and comparison workers who did not die of lung cancer) were similar to analyses of the full cohort adjusting for pay category as a surrogate for tobacco use, although the negative dose response for male workers was no longer statistically significant. The current study is generally consistent with previous MPS studies evaluating lung cancer risk following intakes of radionuclides (Table 9).

Another limitation is the possibility of inaccuracies within the dose reconstruction process since the lung doses are based almost entirely on air monitoring measurements in the various calutron (alpha and beta processing) areas and other TEC buildings, and no urine samples or other bioassays measurements were taken. Also, it was not known whether or when respiratory protections may have been used by individual workers which could bias the estimates of inhaled airborne dust and overestimate the lung doses. Conceivably, this bias might have been possible for workers engaged in very dusty environments where respirators might have been available and recommended. Separately, inaccuracies or misclassifications could be in the assignment of exposure for job-specific categories, the duration of exposure, uranium solubility assumptions, and particle size assumptions. While a comprehensive effort was made to convert air measurements to worker dose based on likely airborne dust exposure for specific job locations (alpha building, beta building, chemistry lab, other), this process was nonetheless based on a series of assumptions. The sensitivity analysis associated with varying the parameters needed for dose reconstruction did not indicate any appreciable difference in risk estimates for female workers, but the male workers were more affected by changes in the parameter values (Table 7).

Our previous studies found that air monitoring measurements alone are subject to uncertainties that are difficult to address (Boice, Leggett, et al. 2006; Golden et al. 2022). Though not as optimal as lung-dose estimates based on periodic individual bioassay measurements over time, air sampling measurements for particulate radioactivity can help to determine the intake of airborne radioactivity for individual workers or groups (Marshall and Stevens 1980; Polednak and Frome 1981; Keane and Polednak 1983). In addition, sampling at TEC included collection of dust settled on surfaces (Polednak and Frome 1981), and later studies of Oak Ridge workers exposed to enriched airborne compounds demonstrated positive correlations between surface contamination levels and employee uranium excretion rates (Schultz and Becher 1963). Although based on only a few postmortem examinations of TEC male workers employed for several years in environments with high airborne uranium dust levels, the fact that high uranium retention in the lungs and TLNs occurred is not questioned (Keane and Polednak 1983). It is the uncertainty surrounding accurate estimates of lung dose that is a limitation, particularly for individual workers. Static (area) air monitoring may provide reasonable estimates of average airborne activity in a work area but less accurate estimates of intake by workers at individual workstations within that area. In general, estimates of intake based on air monitoring alone are prone to sizable errors for individual workers even if personal air samplers are worn (ICRP 2015).

State of the art approaches were used in the conversion processes to estimate absorbed dose to the lung, and account was taken of the number of environmental samples and the estimates of uranium dust over the years of TEC operation, including the changing enrichment factors and solubility factors. The absence of a dose response in previous studies at four uranium processing plants, including TEC, suggests that with the possible exception of male workers errors in measurements are likely to be small (Dupree et al. 1995) as does the consistency with previous MPS studies of lung cancer following intakes of uranium (Boice et al. 2011; Golden et al. 2022) and other radionuclides (Boice et al. 2014; Boice, Cohen, Mumma, Golden, et al. 2022) (Table 9). Finally, a standard approach was followed for estimating organ doses after intakes of radionuclides for epidemiologic studies, specific to the MPS investigations, and described in detail in NCRP Report 178 (NCRP 2018) and in other publications (Bouville et al. 2015; Ellis et al. 2018; Dauer et al. 2022).

Conclusions

The study of 26,650 men and women who inhaled uranium dust during the years 1943–1947 found no evidence that high-LET alpha particle exposures increased the risk of lung cancer, NHL, or NMRD in either men or women. The absorbed doses to lung and to TLN were broad and reached over 1 Gy. The large population of female workers at this uranium conversion and enrichment plant had not been studied before, their exposures were greater than those for their male counterparts, and sex-specific comparisons indicated no significant difference in the lung cancer risk estimates between male and female TEC workers.

Acknowledgements

We gratefully acknowledge the advice and discussions from Dr. Diane Cookfair for her insights into the conduct of her comprehensive case-control study of lung cancers among TEC male workers; Mr. Jack Beck for his insights into the dose reconstructions used in the Cookfair et al. (1983) studies; Dr. Lynn Sowder for discussions related to the complex algorithms used in the past to convert measurement of uranium dose into estimates of absorbed dose to lung; and Ms. Emily Sirko for extensive tracing support. The findings and conclusions in this paper are those of the authors. Its publication does not necessarily represent the official position of or imply endorsement by the National Council on Radiation Protection and Measurements, Oak Ridge National Laboratory, Oak Ridge Associated Universities, Memorial Sloan Kettering Cancer Center, Vanderbilt University, or any of the acknowledged agencies.

Disclosure statement

The authors report no conflicts of interest. The authors alone are responsible for the content and writing of the paper.

Funding

The study of workers at the Tennessee Eastman Corporation (Y-12, 1943–1947), a component of the Million Person Study, was supported in part by grants from the U.S. Department of Energy (Grant No. DE-SC0008944 awarded to the National Council on Radiation Protection and Measurements, which included interagency support from the U.S. Nuclear Regulatory Commission, the U.S. Environmental Protection Agency, and the National Aeronautics and Space Administration; and more recent Grants No. DE-AU0000042 and DE-AU0000046); a grant from the U.S. Nuclear Regulatory Commission (NRC-HQ-60-14-G-0011); grants from the Centers for Disease Control and Prevention (5UE1EH000989, 5NUE1EH001315); and grants from the National Aeronautics and Space Administration (NNX15AU88G, 80NSSC17M0016). Contract support was received from the Naval Sea Systems Command (N00024-17-C-4322). Further, contract support was received by Oak Ridge National Laboratory from the Office of Radiation and Indoor Air, U.S. Environmental Protection Agency, under Interagency Agreement DOE No. 1824 S581-A1, under contract No. DE-AC05-00OR22725 with UT-Battelle; and contract support was received by Oak Ridge Associated Universities from the U.S. Department of Energy under contract No. DE-SC0014664.

ORCID

John D. Boice http://orcid.org/0000-0002-8755-1299
Sarah S. Cohen http://orcid.org/0000-0003-0421-1983
Michael T. Mumma http://orcid.org/0000-0001-7506-8710
Ashley P. Golden http://orcid.org/0000-0002-9430-7702
Sara C. Howard http://orcid.org/0000-0002-9290-4319
David J. Girardi http://orcid.org/0000-0003-1253-4045
Elizabeth D. Ellis http://orcid.org/0000-0003-2871-1587
Michael B. Bellamy http://orcid.org/0000-0003-0892-7559
Lawrence T. Dauer http://orcid.org/0000-0002-5629-8462

References

[ATSDR] Agency for Toxic Substances and Disease Registry. 2013. Toxicological profile for uranium. Atlanta (GA): U.S. Department of Health and Human Services, Public Health Service.

Azizova TV, Bannikova MV, Grigoryeva ES, Rybkina VL, Hamada N. 2020. Occupational exposure to chronic ionizing radiation increases risk of Parkinson's disease incidence in Russian Mayak workers. Int J Epidemiol. 49(2):435–447.

Beck WL, Cookfair DL, Sowder CL, Gissel K. 1983. Radiation dosimetry for epidemiologic lung cancer study. In: Epidemiology Applied to Health Physics: Sixteenth Midyear Topical Meeting of the Health Physics Society; Jan 9–13; Albuquerque, NM. CONF-830101. National Technical Information Service. Springfield (VA): U.S. Department of Commerce. p. 436–443.

Boice JD Jr. 2017. Space: the final frontier—research relevant to Mars. Health Phys. 112(4):392–397.

Boice JD Jr. 2022. The Million Person Study relevance to space exploration and Mars. Int J Radiat Biol. 98(4):551–559.

Boice JD Jr, Cohen SS, Mumma MT, Ellis ED. 2022. The Million Person Study, whence it came and why. Int J Radiat Biol. 98(4):537–550.

Boice JD, Cohen SS, Mumma MT, Ellis ED, Cragle DL, Eckerman KF, Wallace PW, Chadda B, Sonderman JS, Wiggs LD, et al. 2014. Mortality among Mound workers exposed to polonium-210 and other sources of radiation, 1944–1979. Radiat Res. 181(2):208–228.

Boice JD Jr, Cohen SS, Mumma MT, Ellis ED, Eckerman KF, Leggett RW, Boecker BB, Brill AB, Henderson BE. 2011. Updated mortality analysis of radiation workers at Rocketdyne (Atomics International), 1948–2008. Radiat Res. 176(2):244–258.

Boice JD, Cohen SS, Mumma MT, Ellis ED, Eckerman KF, Leggett RW, Boecker B, Brill A, Henderson B. 2006. Mortality among radiation workers at Rocketdyne (Atomics International), 1948–1999. Radiat Res. 166(1 Pt 1):98–115.

Boice JD Jr, Cohen SS, Mumma MT, Golden AP, Howard S, Girardi DJ, Ellis ED, Bellamy M, Dauer LT, Samuels C, et al. 2022. Mortality among workers at the Los Alamos National Laboratory, 1943–2017. Int J Radiat Biol. 98(4):722–749.

Boice JD Jr, Cohen SS, Mumma MT, Hagemeyer DA, Chen H, Golden AP, Yoder RC, Dauer LT. 2022. Mortality from leukemia, cancer and heart disease among U.S. nuclear power plant workers, 1957–2011. Int J Radiat Biol. 98(4):657–678.

Boice JD Jr, Cohen SS, Mumma MT, Howard SC, Yoder RC, Dauer LT. 2022. Mortality among medical radiation workers in the United States, 1965–2016. Int J Radiat Biol. 1–25.

Boice JD Jr, Dauer LT. 2021. Million Person Study of low-dose radiation health effects. Trans Am Nucl Soc. 125(1):1068–1071.

Boice JD Jr, Ellis ED, Golden AP, Girardi DJ, Cohen SS, Chen H, Mumma MT, Shore RE, Leggett RW. 2018. The past informs the future: an overview of the million worker study and the Mallinckrodt Chemical Works cohort. Health Phys. 114(4):381–385.

Boice JD Jr, Ellis ED, Golden AP, Zablotska LB, Mumma MT, Cohen SS. 2022. Sex-specific lung cancer risk among radiation workers in the million person study and patients TB-fluoroscopy. Int J Radiat Biol. 98(4):769–780.

Boice JD Jr, Leggett RW, Ellis ED, Wallace PW, Mumma M, Cohen SS, Brill AB, Chadda B, Boecker BB, Yoder RC, et al. 2006. A comprehensive dose reconstruction methodology for former Rocketdyne/Atomics International radiation workers. Health Phys. 90(5):409–430.

Boice JD Jr, Quinn B, Al-Nabulsi I, Ansari A, Blake PK, Blattnig SR, Caffrey EA, Cohen SS, Golden AP, Held KD, et al. 2022. A million persons, a million dreams: a vision for a National Center of Radiation Epidemiology and Biology. Int J Radiat Biol. 98(4):795–821.

Bouville A, Toohey RE, Boice JD, Beck HL, Dauer LT, Eckerman KF, Hagemeyer D, Leggett RW, Mumma MT, Napier B, et al. 2015. Dose reconstruction for the million worker study: status and guidelines. Health Phys. 108(2):206–220.

Campbell KM, Deck D, Krupski A. 2008. Record linkage software in the public domain: a comparison of Link Plus, The Link King, and a 'basic' deterministic algorithm. Health Informatics J. 14(1):5–15.

CMS. 2015. Centers for Medicare and Medicaid Services. Milestones, 1937–2015, July 2015; [accessed 2022 Apr 25]. https://www.cms.gov/About-CMS/Agency-Information/History.

Cookfair DL. 1982. A case control study of lung cancer among workers at a uranium processing plant [dissertation]. Chapel Hill (NC): University of North Carolina, Chapel Hill.

Cookfair DL, Beck WL, Shy CM, Lushbaugh CC, Sowder CL. 1983. Lung cancer among workers at a uranium processing plant. In: Epidemiology Applied to Health Physics: Sixteenth Midyear Topical Meeting of the Health Physics Society; Jan 9–13; Albuquerque (NM). CONF-830101. Springfield (VA): National Technical Information Service, U.S. Department of Commerce. p. 398–406.

Cox D. 1972. Regression models and life-tables. J R Stat Soc Ser B. 34(2):187–220.

Dauer LT, Bouville A, Toohey RE, Boice JD Jr, Beck HL, Eckerman KF, Hagemeyer D, Leggett RW, Mumma MT, Napier B, et al. 2022.

Dosimetry and uncertainty approaches for the million person study of low-dose radiation health effects: overview of the recommendations in NCRP report no. 178 . Int J Radiat Biol. 98(4):600–609.

Dupree EA, Watkins JP, Ingle JN, Wallace PW, West CM, Tankersley WG. 1995. Uranium dust exposure and lung cancer risk in four uranium processing operations. Epidemiology. 6(4):370–375.

Ellis ED, Boice JD, Golden AP, Girardi DJ, Cohen SS, Mumma MT, Shore RE, Leggett RW, Kerr GD. 2018. Dosimetry is key to good epidemiology: workers at Mallinckrodt Chemical Works had seven different source exposures. Health Phys. 114(4):386–397.

Golden AP, Ellis ED, Cohen SS, Mumma MT, Leggett RW, Wallace PW, Girardi DJ, Watkins JP, Shore RE, Boice JD Jr. 2022. Updated mortality analysis of the Mallinckrodt uranium processing workers, 1942–2012. Int J Radiat Biol. 98(4):572–579.

Hunter N, Haylock R. 2022. Radiation risks of lymphoma and multiple myeloma incidence in the updated NRRW-3 cohort in the UK: 1955–2011. J Radiol Prot. 42(1).

[ICRP] International Commission on Radiological Protection. 1994. ICRP publication 66. Human respiratory tract model for radiological protection. Ann ICRP. 24(1–3):1–482.

[ICRP] International Commission on Radiological Protection. 2007. ICRP publication 103. The 2007 recommendations of the international commission on radiological protection. Ann ICRP. 37(2–4):1–332.

[ICRP] International Commission on Radiological Protection. 2015. ICRP publication 130. Occupational intakes of radionuclides: part 1. Ann ICRP. 44(2):98–99.

[ICRP] International Commission on Radiological Protection. 2017. ICRP publication 137. Occupational intakes of radionuclides: part 3. Ann ICRP. 46(3/4):1–486.

Keane AT, Polednak AP. 1983. Retention of uranium in the chest: implications of findings in vivo and postmortem. Health Phys. 44(Suppl. 1):391–402.

Kiernan D. 2013. The girls of atomic city: the untold story of the women who helped win World War II. New York (NY): Atria Books.

Leggett RW, Eckerman KF, Boice JD Jr. 2005. A respiratory model for uranium aluminide based on occupational data. J Radiol Prot. 25(4):405–416.

Lippmann M. 1959. Environmental exposure to uranium compounds; air urine, and medical findings. AMA Arch Ind Health. 20:211–226.

Mappin-Kasirer B, Pan H, Lewington S, Kizza J, Gray R, Clarke R, Peto R. 2020. Tobacco smoking and the risk of Parkinson disease: a 65-year follow-up of 30,000 male British doctors. Neurology. 94(20):e2132–e2138.

Marshall M, Stevens DC. 1980. The purposes, methods and accuracy of sampling for airborne particulate radioactive materials. Health Phys. 39(3): 409–423.

Monson RR. 1986. Observations on the healthy worker effect. J Occup Med. 28(6):425–433.

Mumma MT, Cohen SS, Sirko JL, Ellis ED, Boice JD Jr. 2022. Obtaining vital status and cause of death on a million persons. Int J Radiat Biol. 98(4):580–586.

[NASA] National Aeronauts and Space Administration. 2021. Human research roadmap. A risk reduction strategy for human space exploration; [accessed 2021 Aug 1]. https://humanresearchroadmap.nasa.gov/.

[NASA] NASA space flight human-system standard, volume 1, revision B: crew health. NASA STD-3001; [accessed 2022 Jun 7] https://standards.nasa.gov/sites/default/files/standards/NASA/B/2022-01-05-NASA-STD-3001-Vol1-Rev-B-Final-Draft-Signature-010522.pdf .

[NASEM] National Academies of Sciences, Engineering, and Medicine. 2021. Space radiation and astronaut health: managing and communicating cancer risks. Washington (DC): The National Academies Press.

[NCRP] National Council on Radiation Protection and Measurements. 2014. Radiation protection for space activities: supplement to previous recommendations. NCRP commentary 23. Bethesda (MD): NCRP.

[NCRP] National Council on Radiation Protection and Measurements. 2018. Deriving organ doses and their uncertainty for epidemiologic studies (with a focus on the one million US workers and veteran study of low-dose radiation health effects). NCRP report 178. Bethesda (MD): NCRP.

[NCRP] National Council on Radiation Protection and Measurements. 2021a. Development of kinetic and anatomical models for brain dosimetry for internally deposited radionuclides. NCRP commentary 31. Bethesda (MD): NCRP.

[NCRP] National Council on Radiation Protection and Measurements. 2021b. Scientific Committee 1-27: evaluation of sex-specific differences in lung cancer radiation risks and recommendations for use in transfer and projection models; [accessed 2021 Dec 22]. https://ncrponline.org/program-areas/sc-1-27-evaluation-of-sex-specific-differences-in-lung-cancer-radiation-risks-and-recommendations-for-use-in-transfer-and-projection-models/.

Polednak AP, Frome EL. 1981. Mortality among men employed between 1943 and 1947 at a uranium processing plant. J Occup Med. 23(3):169–178.

Polednak AP, Keane AT, Beck WL. 1982. Estimation of radiation doses to the lungs of early uranium processing plant workers. Environ Res. 28(2):313–328.

Polednak AP. 1980. Long-range studies of uranium workers and the Oak Ridge radiation-worker population. In: Hubner KF, Fry SA, editors. The medical basis for radiation accident preparedness. New York (NY): Elsevier/North-Holland.

Preston DL, Lubin J, Pierce D, McConney M, Shilnikova N. 2015. Epicure risk regression and person-year computation software: command summary and user guide. Ottawa (CA): Risk Sciences International.

Rothman KJ, Boice JD Jr. 1979. Epidemiologic analysis with a programmable calculator. NIH Publication 79-1649. Washington (DC): US Government Printing Office.

Schultz NB, Becher AF. 1963. Correlation of uranium alpha surface contamination, air-borne concentrations, and urinary excretion rates. Health Phys. 9(9):901–909.

Simonsen LC, Slaba TC. 2021. Improving astronaut cancer risk assessment from space radiation with an ensemble model framework. Life Sci Space Res. 31:14–28.

[UNSCEAR] United Nations Scientific Committee on the Effects of Atomic Radiation. 2017. Sources, effects and risks of ionizing radiation. UNSCEAR 2016 Report to the general assembly. Volume 1. Scientific annex D. Biological effects of selected internal emitters – uranium. Publication E.17.IX.1. New York (NY): United Nations.

Wartenberg D, Brown S, Mohr S, Cragle D, Friedlander B. 2001. Are African-American nuclear workers at lower mortality risk than Caucasians? J Occup Environ Med. 43(10):861–871.

Xu J. 2016. Mortality among centenarians in the United States, 2000–2014. NCHS Data Brief. 233:1–8.

Mortality among medical radiation workers in the United States, 1965–2016

John D. Boice, Sarah S. Cohen, Michael T. Mumma, Sara C. Howard, R. Craig Yoder, and Lawrence T. Dauer

ABSTRACT

Background: Estimates of radiation risks following prolonged exposures at low doses and low-dose rates are uncertain. Medical radiation workers are a major component of the Million Person Study (MPS) of low-dose health effects. Annual personal dose equivalents, $H_P(10)$, for individual workers are available to facilitate dose-response analyses for lung cancer, leukemia, ischemic heart disease (IHD) and other causes of death.

Materials and methods: The Landauer, Inc. dosimetry database identified 109,019 medical and associated radiation workers first monitored 1965–1994. Vital status and cause of death were determined through 2016. Mean absorbed doses to red bone marrow (RBM), lung, heart, and other organs were estimated by adjusting the recorded $H_P(10)$ for each worker by scaling factors, accounting for exposure geometry, the energy of the incident photon radiation, sex of the worker and whether an apron was worn. There were 4 exposure scenarios: general radiology characterized by low-energy x-ray exposure with no lead apron use, interventional radiologists/cardiologists who wore aprons, nuclear medicine personnel and radiation oncologists exposed to high-energy photon radiation, and other workers. Standardized mortality ratio (SMR) analyses were performed. Cox proportional hazards models were used to estimate organ-specific radiation risks.

Results: Overall, 11,433 deaths occurred (SMR 0.60; 95%CI 0.59,0.61), 126 from leukemia other than chronic lymphocytic leukemia (CLL), 850 from lung cancer, and 1654 from IHD. The mean duration of monitoring was 23.7 y. The excess relative rate (ERR) per 100 mGy was estimated as 0.10 (95% CI −0.34, 0.54) for leukemia other than CLL, 0.15 (0.02, 0.27) for lung cancer, and −0.10 (−0.27, 0.06) for IHD. The ERR for lung cancer was 0.16 (0.01, 0.32) among the 55,218 male workers and 0.09 (−0.19, 0.36) among the 53,801 female workers; a difference that was not statistically significant (p-value = 0.23).

Conclusions: Medical radiation workers were at increased risk for lung cancer that was higher among men than women, although this difference was not statistically significant. In contrast, the study of Japanese atomic bomb survivors exposed briefly to radiation in 1945 found females to be nearly 3 times the radiation risk of lung cancer compared with males on a relative scale. For medical workers, there were no statistically significant radiation associations with leukemia excluding CLL, IHD or other specific causes of death. Combining these data with other cohorts within the MPS, such as nuclear power plant workers and nuclear submariners, will enable more precise estimates of radiation risks at relatively low cumulative doses.

Introduction

The first cancer attributed to ionizing radiation was reported on the hand of a radiologist in 1902 (Yoshinaga et al. 2004; Boice 2006). Radiation-related leukemia was first linked to chronic exposures in studies of U.S. radiologists (March 1944) and subsequently confirmed in studies of pioneering radiologists in the United States (Matanoski et al. 1975, 1984), United Kingdom (Court-Brown and Doll 1958; Smith and Doll 1981; Berrington et al. 2001) and China (Wang et al. 1988, 1990, 2002; Sun et al. 2016). These radiation workers practiced during the first half of the 20th century and before radiation protection guidelines were commonplace (Boice, Dauer et al. 2020). The studies provide convincing evidence that leukemia and skin cancer can result from repeated, presumably small, radiation exposures received over a period of many years if the cumulative dose is sufficiently high. The cumulative doses were very high and estimated to range between 1 and 8 Gy (Smith and Doll 1981; Sherwood 2001; Boice 2006; Linet et al. 2010). Many pioneering radiologists experienced deterministic effects, e.g. burn-like dermatitis, ulcerations, reduction in white blood cell counts, and testicular shrinkage, leading to amputations and early death (DiSantis 1986). The absence of dose measurements for individual workers, however, limits the usefulness of these data for radiation risk quantification. Subsequent studies of radiologists, radiologic technologists and medical radiation workers who began employment after the early era have been conducted. But again, the near absence of reliable data on individual exposures remains a shortcoming that hinders adequate quantification of radiation risks (Jablon and Miller 1978; Yoshinaga et al. 1999, 2004; Boice 2006; Linet et al. 2010, 2020; Liu et al. 2014; Preston et al. 2016; NCRP 2018b).

The current study of U.S. medical workers was designed to address recognized limitations of previous studies, and in particular to provide individual radiation dose estimates based entirely on badge dose measurements, i.e. personal dose equivalents. Another goal was to include a large number of female workers, an understudied group in occupational studies.

The study of U.S. medical workers is a major component of the Million Persons Study (MPS) and was specifically designed to examine sex-specific lung cancer risks following low-dose fractionated exposures that accumulated to a relatively high lung dose (Bouville et al. 2015; Dauer et al. 2018; Yoder et al. 2018, 2021; NCRP 2018a, 2020). Because NASA guidance for radiation protection is based upon individual lifetime cancer mortality projections, any differences in lung cancer risks between men and women are of particular importance (NCRP 2014, 2021; Boice, Cohen et al. 2019, Boice, Ellis et al. 2019; National Academies of Sciences et al. 2021). Projections based on the acute exposure received by atomic bomb survivors currently limit the amount of time female flight crews can spend in space because their lung cancer risks, on a relative scale, are estimated to be 2–3 times higher than for men (NCRP 2014). The design of the current study was based on the 'film badge dose', $H_P(10)$, which showed a broad dose distribution with substantial statistical power to identify lung cancer risks had they occurred (Table 1). The comprehensive dose reconstruction that accounted for the use of lead aprons among interventional fluoroscopists narrowed the dose distribution and subsequently reduced statistical power (NCRP 2020; Yoder et al. 2021).

Herein we report cause-specific mortality risks for over 40 causes of death and radiation risk estimates for leukemia, lung cancer, all solid cancer, ischemic heart disease (IHD), Parkinson's disease, and selected causes of death among medical radiation workers first monitored between 1965 and 1994 in the United States. This is the first use of the Landauer, Inc. dosimetry database to form the basis of an entire epidemiologic population for radiation research.

Materials and methods

Human subjects' approval for the study was received from Vanderbilt University.

Cohort definition

Landauer, Inc., a large supplier of personal dosimetry services, has provided services to hospitals, clinics, and private medical practices since its founding in 1954 (NCRP 2020; Yoder et al. 2018, 2021). In the early 1980s, Landauer created an electronic database to support an epidemiologic study of radiologic technologists (Boice et al. 1992; Simon et al. 2006; Yoder et al. 2018). The digital database was periodically updated and currently includes annual badge doses for over 5 million workers monitored since 1977, and lifetime cumulative doses extending back into the mid-1960s. A potentially eligible population for the study was selected from a stratified sample of 1.7 million medical workers first monitored before 1995 with sufficient identifying information (SSN) to allow vital status tracing. The selection and sampling criteria changed somewhat over the years which accounts for the difference in numbers previously reported (Bouville et al. 2015; NCRP 2018a; Boice, Cohen et al. 2019). Short-term workers and those with very low cumulative badge doses were excluded because the added costs would not be commensurate with any epidemiologic information gained. Workers were excluded from the study if they were monitored for less than 2 y, had a cumulative badge dose of less than 10 mSv, were already in another MPS study, or had invalid identifiers (Figure 1). The final study population consisted of 109,019 medical workers, 55,218 males (50.6%) and 53,801 females (49.4%). The medical worker cohort includes physicians and technologists (general radiology, interventional radiology/cardiology, nuclear medicine and radiation oncology), nuclear pharmacists, medical and radiation physicists, nurses, veterinarians, chiropractors, dentists and allied healthcare support workers monitored for radiation in similar environments. Overall, 7.5% of the cohort members had doctoral titles (Dr or MD). Approximately 5.6% of the cohort did not fall into a distinct medical specialty and were classified as 'other'. Although nearly 4% of the other workers had doctoral titles, they were typically in nonmedical professions and had not been included in any previous MPS study.

Vital status and outcome determination

Vital status as of 31 December 2016 was determined following the approach described in Mumma et al. (2018). Briefly, probabilistic linkages of the study roster to the National Death Index (NDI); the California Death Statistical Master File and other state mortality files; and the Social Security Administration (SSA) Death Master File were conducted using SAS and LinkPlus (Campbell et al. 2008) to obtain a date and/or a cause of death. The SSA Service for Epidemiological Researchers was used to confirms alive status among those not found to have died. TLO, a subscription service of TransUnion (www.tlo.com) was used to confirm and correct key identifiers (name, SSN, DOB). Workers without a confirmed vital status, but who were located via TLO with no indication of death, were assumed alive through the end of follow-up. In addition to this extensive armamentarium of tracing assets, individual one-on-one

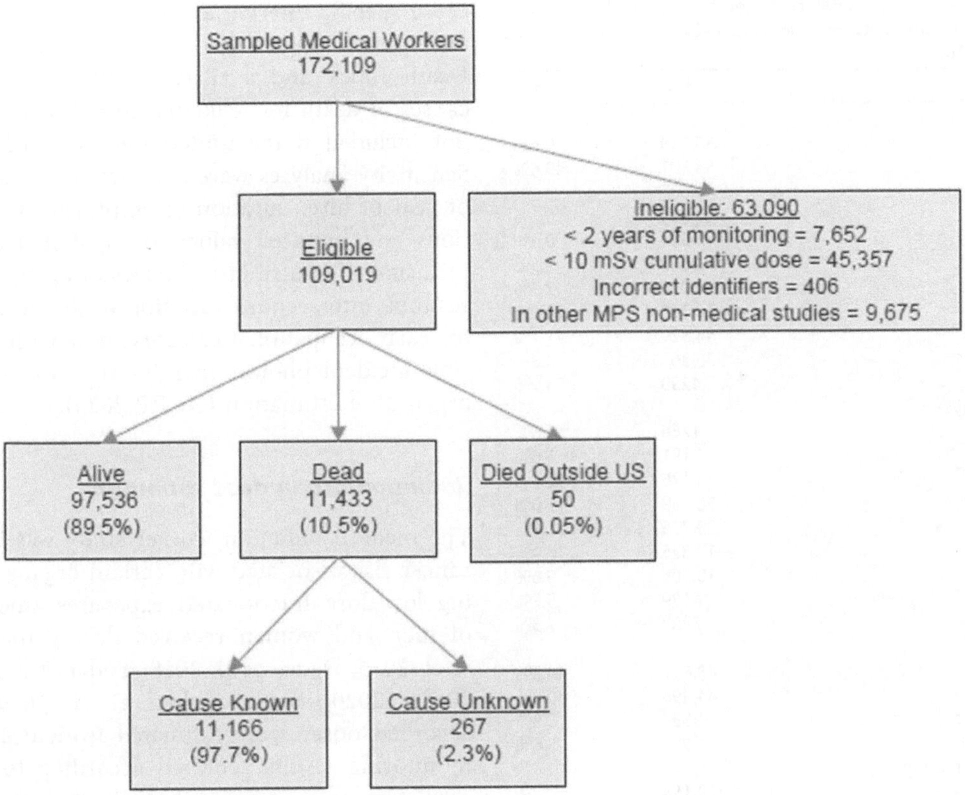

Figure 1. Schematic of the selection and vital status of 109,019 medical radiation workers first monitored between 1965 and 1994 in the United States and followed through 2016.

approaches also were employed to locate study subjects using resources such as Ancestry.com and Google obituary search. These obituary sources are simply adjuncts to our comprehensive approach (Mumma et al. 2018). Through the end of follow-up in 2016, 86,689 workers (79.6%) were confirmed to be alive, and 10,847 (9.9%) workers were assumed to be alive (Table 1, Figure 1). Cause of death was determined for all but 267 (2.3%) of the 11,433 (10.5%) workers who had died.

Analytic methods

External standardized mortality ratio (SMR) analyses were conducted by comparing the observed number of deaths with those expected based on mortality rates in the general population of the United States. Rates for white men and women were calculated using mortality data from the National Center for Health Statistics. Evaluation of NDI returns indicated that over 90% of the study population was white. Observed numbers of deaths from cancers and other diseases were tabulated by age in 5-year age groups, sex, and 5-year calendar year groups for all workers and subgroups defined by occupational category (general radiology, interventional radiology/cardiology, nuclear medicine and radiation oncology, and other). The SMR analyses were based on the underlying cause of death. The start of follow-up began on the later of January 1 of the second year of Landauer annual records, the date cumulative badge dose was at least 10 mSv, or 1 January 1977. Follow-up ended at the date of death, age 95 or 31 December 2016, which ever came first. Statistical variability was evaluated by calculating the 95% exact confidence interval (CI) of the SMR assuming that the observed number of deaths followed a Poisson distribution (Rothman and Boice 1979). The ratio of two SMRs and the associated 95% exact confidence limits were computed following the methods of Breslow and Day (1987) and Armstrong (1995).

Within-cohort analyses were conducted to account in part for the healthy worker effect often seen in occupational studies and as evident when comparisons are made with the general population (Monson 1986; Buckley et al. 2015). Internal within-cohort analyses used Cox proportional hazards models to compute hazards ratios (HR) and corresponding 95% confidence intervals (CI) for a continuous cumulative measure of radiation dose to the lung and other organs in a time-dependent manner (Cox 1972). Age was used as the underlying timescale for the hazard function. A possible latent period between radiation exposure and any effect consequent to it was examined by lagging doses 10 y for solid cancers and other chronic conditions or 2 y for leukemia other than CLL and myelodysplastic syndrome (MDS). Two-sided p-values are presented. HRs at 100 mGy were computed for selected causes of death. Dose-response functions for continuous measurement of radiation organ dose were modeled as a linear function (Golden et al. 2018). Tests for linear trend in the dose response treated the cumulative organ dose as a single continuous measure. All Cox models were conducted using SAS/STAT software (SAS/STAT software, Version 9.4 of the SAS System for Windows, SAS Institute Inc., Cary, NC, USA). Excess

Table 1. Demographic and occupational characteristics of 109,019 medical radiation workers first monitored between 1965 and 1994 in the United States and followed through 2016.

Characteristic	N	%
Sex		
Male	55,218	50.6%
Female	53,801	49.4%
Year of birth		
1896–1909	36	<0.1%
1910–1919	564	0.5%
1920–1929	2682	2.5%
1930–1939	7911	7.3%
1940–1949	23,386	21.5%
1950–1959	44,829	41.1%
1960–1969	25,361	23.3%
1970–1977	4250	3.9%
Age at end of follow-up (years)		
39–44	1766	1.6%
45–49	7191	6.6%
50–54	13,176	12.1%
55–59	20,739	19.0%
60–64	23,725	21.8%
65–69	17,825	16.4%
70–74	10,506	9.6%
75–79	6189	5.7%
≥80	7902	7.3%
Occupational Category		
General Radiology	48,837	44.8%
Interventional Radiology	44,498	40.8%
Nuclear Medicine and Radiation Oncology	9597	8.8%
Other	6087	5.6%
Age at first Landauer monitoring		
<20	12,158	11.2%
20–24	25,660	23.5%
25–29	24,971	22.9%
30–34	20,979	19.2%
35–39	12,052	11.1%
40–45	7469	6.9%
≥45	5730	5.3%
Year of first Landauer monitoring		
1965–1969	1422	1.3%
1970–1974	8085	7.4%
1975–1979	29,291	26.9%
1980–1984	28,404	26.1%
1985–1989	21,535	19.8%
1990–1994	20,282	18.6%
Cumulative badge dose (mSv)		
10–<25	38,862	35.7%
25–<50	26,313	24.1%
50–<100	25,939	23.8%
100–<250	14,729	13.5%
250–<500	2540	2.3%
≥500	636	0.6%
Vital status as of 31 December 2016		
Confirmed alive	86,689	79.6%
Assumed alive	10,847	9.9%
Dead inside the US	11,433	10.5%
Died outside the US	50	0.05%

Relative Rates (ERR) per 100 mGy along with corresponding CIs were estimated using the Peanuts program in Epicure (Preston et al. 2015). All models were adjusted for sex, year of birth and occupational category (categories as shown in Table 1). The occupational category was taken as a measure of socioeconomic status (SES) and adjustment was made in all analyses.

Nested models with and without an interaction term between sex and radiation dose were compared using the likelihood ratio test (LRT) to test whether sex was an effect modifier, i.e. whether risk differed by sex (see also Boice, Ellis et al. 2019). To increase statistical power for a few infrequent causes of death such as leukemia, the cause-specific internal analyses included both the underlying and contributing causes of death obtained from the National Death Index and available death certificates. Contributing causes of death for leukemia ($n = 9$) and MDS ($n = 17$) were not included if the underlying cause of death was cancer. Sensitivity analyses were conducted to examine the influence of year of hire, duration of employment, dosimetry assumptions and imputed values, occupational categories, and the inclusion of contributing causes of death. Uncertainty analyses took into account variation in the dosimetry assumptions for each occupational category, notably for the range of possible incident photon energies and concomitant variation in organ dose estimation (NCRP 2020).

Radiation organ dose estimation

The medical radiation worker study was initiated to address cancer risk associated with certain organs and tissues following low-dose fractionated exposures among large numbers of men and women received during their career (Bouville et al. 2015; Dauer et al. 2018; Yoder et al. 2018, 2021; NCRP 2018a, 2020; Boice, Cohen et al. 2019). Average organ absorbed doses were estimated from occupational personal monitoring results acquired according to Federal and State regulatory requirements that directed workers to wear personal dosimeters at points on the body likely to receive the highest doses. Relating such personal monitoring results to organ doses received by workers during their career (lifetime dose) requires a series of dose conversions outlined in NCRP Report 178 (NCRP 2018a). The dose conversions first normalized the different monitoring quantities measured from the 1950s through today to a common quantity, the personal dose equivalent, $H_P(10)$. Tabulated dose conversion coefficients published in NCRP Report 178 (NCRP 2018a) enabled average organ absorbed doses to be estimated from $H_P(10)$ after accounting for the conditions of radiation exposure (e.g. radiation quality, exposure geometry, sex/morphology of the worker and presence of lead aprons).

Approach to classify radiation workers into exposure categories

For the medical worker cohort, radiation exposures are typical to x- and gamma-rays (collectively called photons) with energies from about 0.02 MeV to slightly greater than 1 MeV. The relationship between a personal monitoring result expressed as $H_P(10)$ in mSv and the mean organ dose such as the lung in mGy is relatively constant for photon energies above 0.1 MeV but rapidly changes for energies less than 0.1 MeV. Therefore, the dose reconstruction process identified and separated those subjects exposed to the higher energies from those exposed to the lower energies. This was accomplished through the use of institutional worker or department identifiers and dose levels known to be indicative of the photon energy regime; namely, radiology settings where photons with lower energies prevail and nuclear medicine and radiation therapy settings where higher photon energies predominate.

Table 2. Radiation monitoring histories for 109,019 medical workers identified within the Landauer, Inc. dosimetry database.

Gaps in monitoring histories (y)	Males (N = 55,218)		Females (N = 53,801)		Total (N = 109,019)	
	N	%	N	%	N	%
None	25,409	46%	24,727	46%	50,136	46%
1–4	22,863	41%	22,137	41%	45,000	41%
5–9	7574	14%	7863	14%	15,437	14%
≥10	3850	7%	3853	7%	7703	7%

Duration and gaps in monitoring coverage by sex are presented. Duration of monitoring is taken as the date of last monitoring minus the date of first monitoring as recorded in the digital dosimetry database. The mean durations of monitoring were 23.7 y overall, 23.9 y for men and 23.5 y for women.

The NCRP has extensively reviewed radiation dose management for fluoroscopy and fluoroscopically guided interventional (FGI) medical staff protection considerations and associated medical radiation work practices and exposure scenarios noting that the use of protective aprons was mandatory for such workers (NCRP 2010). Therefore, individuals identified as fluoroscopy and FGI workers were assumed to use radio-protective aprons (NCRP 2020). The use of radio-protective aprons markedly reduced the dose to the organs located in the torso by an average factor of 20 or more (NCRP 2020). Due to regulatory requirements, dosimeters were typically worn over the apron so as to measure the dose to the part of the body receiving the highest exposure (i.e. the unshielded head and upper extremities are considered to be part of the whole body) (USNRC 1991; NCRP 2020). Therefore, the dose reconstruction analysis was required to estimate the dose existing under the apron from measurements made over the apron. Radio-protective aprons are not used to shield workers from higher energy photons due to the inability to significantly reduce the torso doses.

The result of these dose reconstruction steps was to place a medical worker into one of three exposure groups: exposure to low energy photons with no use of a protective apron; exposure to low energy photons with the use of a protective apron; and exposure to higher energy photons. NCRP Scientific Committee 6–11 has tabulated conversion factors that relate the personal monitoring results, after correction for errant conditions and the effects of protective aprons, to the mean organ dose for various organs and tissues (NCRP 2018a, 2020). For a few organs (namely the breast and sexual reproductive organs) the dose conversion factors vary according to sex, but for most organs sex differences are minimal.

The decision to assign a given worker with the use of a protective apron relied on several pieces of information available in the dosimetry database. Primary was the use of department codes indicating radiology activities certain to require protective aprons such as interventional radiology, fluoroscopy and 'cardiac cath lab'. When the department code was ambiguous, the annual dose value was used to indicate the probability that an apron would have been used. For example, many annual doses would have exceeded regulatory limits had an apron not been used; however, actual over-exposures are rare, and the 'apparent' over-exposures were indicative of the use of a dosimeter worn over a protective apron. Finally, name fields that indicated the user's title (e.g. Dr or MD) that were linked to high doses were taken to indicate high-dose situations when a physician would have been present, such as interventional radiology or cardiac catheterization. For those situations when a dosimeter was worn under an apron, the dose level would be extremely low and no correction for the use of an apron would have been appropriate.

Evaluation of the completeness of the Landauer database in terms of long-term monitoring of occupational exposures

An extensive 'gap analysis' was performed to evaluate whether there were intervals in the Landauer occupational radiation monitoring histories not covered by Landauer dose records (Table 2). The duration of a worker's occupational history was taken as the difference between the date of last monitoring and first monitoring within the Landauer database. The mean duration of monitoring was 23.7 y, and some workers were monitored for almost 60 y. The results were reassuring in that 87% of all cohort employment was covered by an annual dose record. In addition, nearly 50% of all medical workers had no gaps in their dose histories, and 90% had only small periods without monitoring, i.e. less than 5 y. All patterns were similar between men and women in the length of monitoring, the number of annual dose records and gaps in coverage. As described in the methods section, imputed dose values for gaps of less than 5 y were taken as the mean annual dose. Dose-response analyses with or without annual dose imputation had little effect on the estimates of radiation associations.

Additional analyses were conducted by length of Landauer monitoring (<5, 5–9, 10–19, 20–29, and ≥30 y). Even among participants with occupational histories of 30 or more years, annual dose records were available for 87% of this time, and only 7% had gaps of 10 y or more. It is noted that a 'gap' does not necessarily mean that occupational doses of consequence were missed, e.g. a worker may not have been working during that time or may have been working in an occupation where radiation monitoring was not required. The additional evaluation of hospital records described below for a large number of radiation workers, provides additional assurance as to the general validity of the assumptions made regarding dose reconstruction.

In addition to the overall gap analysis of the entire cohort, a quality control validation of the completeness of the Landauer database was conducted by comparing both the Landauer doses and study population to the independent Memorial Sloan Kettering Cancer Center (MSKCC) dosimetry database (Dauer et al. 2019). Among the 555 workers in the Landauer defined cohort study who were employed at MSKCC, 4,968 annual worker doses (covering years of employment from 1967 to 2015) matched nearly perfectly, with more than 99% of the Landauer dosimetry file being confirmed in the MSKCC records. For the <1% with slight differences, most were related to the recording of very low annual badge doses (e.g. 0.001 or 0.0001 mSv). There was no difference in the annualized doses nor were there any

Table 3. Standardized mortality ratios (SMR) and 95% confidence intervals (CI) for cancers and other causes of death among 109,019 medical radiation workers first monitored between 1965 and 1994 in the United States and followed through 2016, by sex.

	Males		Females		Total		
Persons at risk	55,218		53,801		109,019		
Person years at risk	1,406,960		1,372,878		2,779,838		
Cause of death (ICD9[a])	Obs	SMR	Obs	SMR	Obs	SMR	95% CI
All causes of death (001–999)	7,123	0.55[c]	4,310	0.70[c]	11,433	0.60[c]	0.59–0.61
All malignant neoplasms (140–208)	2,047	0.57[c]	1,579	0.75[c]	3,626	0.64[c]	0.62–0.66
Buccal cavity & pharynx (140–149)	32	0.39[c]	16	0.81	48	0.47[c]	0.35–0.63
Esophagus (150)	75	0.51[c]	16	0.82	91	0.55[c]	0.44–0.67
Stomach (151)	61	0.78	25	0.86	86	0.80[c]	0.64–0.99
Colon (153)	155	0.61[c]	106	0.81[c]	261	0.68[c]	0.60–0.77
Rectum (154)	26	0.48[c]	11	0.42[c]	37	0.46[c]	0.32–0.63
Biliary passages & liver (155,156)	79	0.49[c]	32	0.65[c]	111	0.52[c]	0.43–0.63
Pancreas (157)	167	0.76[c]	104	0.92	271	0.81[c]	0.72–0.91
Pleura & peritoneum (158.8,158.9,163) and mesothelioma (ICD10 C45)[b]	6	0.30[c]	9	2.30[c]	15	0.62	0.35–1.03
Larynx (161)	8	0.20[c]	2	0.30	10	0.22[c]	0.10–0.40
Bronchus, trachea, & lung (162)	480	0.42[c]	370	0.68[c]	850	0.50[c]	0.47–0.54
Bone (170)	5	0.64	1	0.25	6	0.51	0.19–1.11
Connective & other soft tissue (171)	21	0.81	16	0.90	37	0.84	0.59–1.16
Breast (174, 175)	5	1.04	311	0.79[c]	316	0.79[c]	0.70–0.88
All Uterine (Females only) (179–182)	–	–	57	0.52[c]	57	0.52[c]	0.39–0.67
Cervix uteri (180)	–	–	26	0.52[c]	26	0.52[c]	0.34–0.76
Ovary and other female genital organs (183–184)	–	–	119	0.84	119	0.84	0.70–1.01
Prostate (185)	172	0.81[c]	–	–	172	0.81[c]	0.69–0.94
Testes & other male genital organs (186, 187)	5	0.56	–	–	5	0.56	0.18–1.31
Kidney (189.0–189.2)	72	0.65[c]	26	0.75	98	0.68[c]	0.55–0.82
Bladder & other urinary (188, 189.3–189.9)	58	0.57[c]	8	0.38[c]	66	0.54[c]	0.41–0.68
Melanoma of skin (172)	61	0.71[c]	15	0.45[c]	76	0.64[c]	0.50–0.80
Other malignant neoplasm of skin (173)	8	0.32[c]	1	0.22	9	0.30[c]	0.14–0.58
Eye (190)	1	0.49	1	0.84	2	0.62	0.07–2.23
Brain & central nervous system (191–192)	113	0.93	54	0.82	167	0.89	0.76–1.04
Thyroid & other endocrine glands (193–194)	13	1.03	8	0.91	21	0.98	0.61–1.50
All lymphatic, hematopoietic tissue (200–208)	270	0.76[c]	165	1.05	435	0.85[c]	0.77–0.93
Non-Hodgkin lymphoma (200, 202)	95	0.68[c]	59	0.98	154	0.77[c]	0.65–0.90
Hodgkin lymphoma (201)	9	0.81	6	1.02	15	0.88	0.49–1.46
Multiple myeloma (203)	61	0.95	42	1.44[c]	103	1.10	0.90–1.34
Leukemia (202.4, 204–208)	103	0.74[c]	58	0.94	161	0.80[c]	0.68–0.93
Acute lymphocytic leukemia (204.0)	2	0.24[c]	4	0.77	6	0.45[c]	0.16–0.98
Acute myeloid leukemia (205.0, 205.3, 206.0, 207.0, 207.2)	49	0.79	32	1.03	81	0.87	0.69–1.08
Chronic myeloid leukemia (205.1, 206.1)	10	0.97	4	0.83	14	0.92	0.50–1.55
Chronic lymphocytic leukemia (CLL) (204.1)	23	0.83	10	1.38	33	0.94	0.65–1.32
Leukemia other than CLL	78	0.71[c]	48	0.89	126	0.77[c]	0.64–0.91
Smoking related cancers (140–150, 157, 161–162, 188–189)	892	0.48[c]	542	0.72[c]	1,434	0.55[c]	0.52–0.58
Non-smoking related cancers	1,155	0.67[c]	1,037	0.77[c]	2,192	0.71[c]	0.68–0.74
Polycythemia vera (238.4)	0	0.00	0	0.00	0	0.00	0.00–1.94
Myelodysplastic syndrome (238.7)	36	1.49[c]	9	1.11	45	1.40[c]	1.02–1.87
Diabetes (250)	194	0.51[c]	106	0.56[c]	300	0.53[c]	0.47–0.59
Mental and behavioral disorders (290–319)	148	0.53[c]	98	0.68[c]	246	0.58[c]	0.51–0.66
Dementia, Alzheimer's, Parkinson's disease & motor neuron (290.0–290.4, 331.0, 332, 335.2)	308	0.77[c]	168	0.69[c]	476	0.74[c]	0.68–0.81
Dementia and Alzheimer's disease (290.0–290.4, 331.0)	192	0.72[c]	134	0.69[c]	326	0.70[c]	0.63–0.79
Parkinson's disease (332)	75	0.90	12	0.53[c]	84	0.82[c]	0.66–1.02
Diseases of the nervous system (320–389)	307	0.75[c]	164	0.66[c]	471	0.72[c]	0.65–0.78
All heart disease (390–398, 404, 410–429)	1778	0.52[c]	741	0.66[c]	2,519	0.56[c]	0.54–0.58
Rheumatic heart disease (393–398)	8	0.57	11	0.77	19	0.67	0.40–1.04
Ischemic heart disease (410–414)	1230	0.51[c]	424	0.62[c]	1654	0.53[c]	0.51–0.56
Cerebrovascular disease (430–438)	261	0.56[c]	201	0.73[c]	462	0.62[c]	0.57–0.68
Non-malignant respiratory disease (460–519)	439	0.42[c]	352	0.63[c]	791	0.49[c]	0.46–0.53
Bronchitis, emphysema, Asthma (490–493)	204	0.36[c]	210	0.62[c]	414	0.46[c]	0.42–0.51
Cirrhosis of liver (571)	117	0.32[c]	73	0.55[c]	190	0.39[c]	0.33–0.44
Nephritis & nephrosis (580–589)	97	0.56[c]	71	0.81	168	0.64[c]	0.55–0.75
AIDS (042–044, 795.8)	131	0.80[c]	4	0.18[c]	135	0.73[c]	0.61–0.86
All external causes of death (800–999)	718	0.54[c]	324	0.65[c]	1042	0.57[c]	0.53–0.60
Accidents (850–949)	395	0.49[c]	193	0.60[c]	588	0.52[c]	0.48–0.57
Suicides (950–959)	256	0.66[c]	92	0.82	348	0.70[c]	0.63–0.77
Unknown causes of death	168		99		267		

[a]ICD = International Classification of Disease. Revision 9 codes are shown.
[b]Mesothelioma did not have an explicit code in ICD9 but did in ICD10 as denoted.
[c]Indicates SMR is significantly different from 1.0.

differences with regard to gaps in monitoring, i.e. they were comparable and few. Further, the highest doses were confirmed to be interventional radiologists, then nuclear medicine technologists/nurses, followed by a mix of others. Finally, for all of the MSKCC interventional radiologists, the study value recorded for annual dose was noted to be based on badges worn outside of the lead apron/collar, confirming assumptions with regard to apron correction application.

Identification and treatment of errant doses

Personal monitoring rests on the assumption that the dose received by the monitor (dosimeter) is highly correlated with that received by the subject – a condition that exists when the dosimeter is worn by the worker during periods of exposure. The human condition, however, may create situations when this assumption is not valid; for example when a dosimeter is left in a treatment room attached to a lab coat or protective apron (NCRP 2020). Errant doses were discovered during a manual review of the 57,768 annual doses received by 1,757 medical workers who had initial estimates of lifetime doses greater than 375 mSv. These workers exhibited a nearly uninterrupted sequence of annual doses from which dose accumulation patterns could be discerned and abnormal patterns identified. For example, a single annual dose that accounted for the majority of the lifetime dose estimate would create suspicions of an errant dose having been measured for one or more of the dosimeters issued to the subject during the year. In addition, annual doses that greatly exceeded regulatory limits were also suspect, particularly those annual doses that would generate acute tissue reactions. The manual review of the high-dose workers led to the rule that any annual dose greater than 200 mSv would be considered an errant dose. Errant doses totaled 456 occurrences with a median value of 682 mSv and a mean value of 1,302 mSv denoting a skewed distribution. This skewed distribution reflected a combination of measurement technology traits, maximum dose reporting rules employed by the dosimetry service and client attention to correcting errant doses in the subject's dose record. When film dosimeters are not properly inserted into the dosimeter holder and exposed to the x-ray energies encountered in the interventional fluoroscopy room, the measured dose can be as much as 15 times greater than it should be. Errant doses identified for each worker were removed from the annual and lifetime dose records and replaced by the average annual dose received in the years prior to and following the year of the errant dose.

Career dose estimation

Dosimetry records documenting radiation exposure received before or after employment as a medical radiation worker were obtained for 4,482 workers from four additional sources: DOE Radiation Exposure Monitoring System (REMS) (U.S. Department of Energy (DOE) 2021) including other historic DOE radiation exposure data not included in DOE REMS ($n = 734$); the Nuclear Regulatory Commission Radiation Exposure Information and Reporting System (REIRS) ($n = 1{,}999$) (USNRC 2019); the U.S. Navy Dosimetry System ($n = 2{,}110$) (NDC 2021); and the Nuclear Test Personnel Review Program (DTRA 2020; Boice, Cohen et al. 2020). All organ/tissue-specific doses from each source were added together to obtain the total organ/tissue-specific external dose received by each worker for each calendar year, following the procedures outlined in Boice et al. (2006) and Ellis et al. (2018).

Results

Descriptive characteristics of the 109,019 medical radiation workers are presented in Table 1. The numbers of males ($n = 55{,}218$) and females ($n = 53{,}801$) were nearly equal, most workers were born after 1950 (68.8%), and just over 10% were born before 1940. Radiology was the most common specialty comprising 85.6% of the cohort: general radiology (44.8%) and interventional radiology/cardiology (40.8%). The years of first monitoring within the Landauer database were from 1965 to 1994 (mean 1983). The age at first monitoring was 29 y on average, and the age at the end of follow-up was 63 y on average. The mean and median cumulative badge doses, i.e. $H_P(10)$ personal dose equivalents, were 63 mSv and 37 mSv, respectively. and ranged to over 500 mSv. There were 2,779,838 person-years of follow-up; 25.5 y on average. At the end of the follow-up, 89.5% of the cohort were determined to be alive (Figure 1). Cause of death was determined for all but 267 (2.3%) of the 11,433 (10.5%) workers who had died.

Table 3 presents the SMRs for all 109,019 medical workers by sex. There was a statistically significant deficit of all causes of death (SMR 0.60; 95% CI 0.59, 0.61: $n = 11{,}433$). Similarly, significant deficits were seen for all malignant neoplasms (SMR 0.64; $n = 3{,}626$), all solid cancer (SMR 0.62; 95% CI .060, .064; $n = 3191$), and all heart disease (SMR 0.56; 95% CI 0.64, 0.58; $n = 2{,}519$). Cancer of the lung was significantly low (SMR 0.50; 95% CI 0.47, 0.54; $n = 850$), as were deaths due to leukemia other than CLL (SMR 0.77; 95% CI 0.64,0.91; $n = 126$). Cancers of the brain and central nervous system deaths were also below population expectation, but not significantly (SMR 0.89; 95% CI 0.76–1.04; $n = 167$). Deaths due to dementia, Alzheimer's, Parkinson's, and motor neuron disease were all significantly low in both sexes. The only significantly elevated cause of death in comparison with the general population was myelodysplastic syndrome (MDS) (SMR 1.40; 95% CI 1.02–1.87; $n = 45$), where the increase appeared concentrated among the male medical workers (SMR 1.49; 95% CI 1.04, 2.06; $n = 36$). The general pattern of significantly low SMRs for practically all specific causes of death was similar among both males and females, although the overall male deficit (SMR 0.55; 95% CI 0.54, 0.56) was significantly lower than that for the females (SMR 0.70; 95% CI 0.68, 0.72), i.e. the ratio of SMRs was 0.79 (95%CI 0.76, 0.82). The SMR for cancer of the prostate was significantly low (0.81; 95% CI 0.69–0.94; $n = 172$) as was the SMR for female breast cancer (0.79; 95% CI 0.69–0.94; $n = 311$).

Table 4 presents the SMRs for each of the four occupational categories. The three medical specialties of general radiography, interventional fluoroscopy, and nuclear medicine and radiation oncology showed similar significant deficits for practically all causes of death. A typical pattern was that the *other* worker category had the highest SMRs (although most were below 1.0), followed by general

Table 4. Standardized mortality ratios (SMR) for cancers and other causes of death among 109,019 medical radiation workers first monitored between 1965 and 1994 in the United States and followed through 2016, by occupational category.

	General radiography		Interventional fluoroscopy		Nuclear medicine and radiation oncology		Other	
Persons at risk	48,837		44,498		9,597		6,087	
Person years at risk	1,153,327		1,212,652		235,474		178,384	
Cause of death (ICD9[a])	Obs	SMR	Obs	SMR	Obs	SMR	Obs	SMR
All causes of death (001–999)	4,955	0.64[c]	4,191	0.51[c]	561	0.45[c]	1,726	0.91[c]
All Malignant Neoplasms (140–208)	1,585	0.67[c]	1,366	0.57[c]	179	0.48[c]	496	0.93
Buccal Cavity & Pharynx (140–149)	18	0.45[c]	21	0.46[c]	2	0.29	7	0.74
Esophagus (150)	45	0.70[c]	32	0.42[c]	3	0.26[c]	11	0.72
Stomach (151)	33	0.77	37	0.79	7	1.01	9	0.84
Colon (153)	123	0.78[c]	92	0.56[c]	11	0.45[c]	35	0.90
Rectum (154)	13	0.40[c]	12	0.35[c]	4	0.73	8	1.08
Biliary passages & liver (155,156)	45	0.53[c]	44	0.47[c]	8	0.53	14	0.78
Pancreas (157)	127	0.92	98	0.69[c]	13	0.59	33	1.08
Pleura & peritoneum (158.8,158.9,163) and mesothelioma (ICD10 C45)[b]	6	0.62	8	0.76	0	0.00	1	0.42
Larynx (161)	2	0.11[c]	5	0.24[c]	0	0.00	3	0.67
Bronchus, trachea, & lung (162)	357	0.51[c]	307	0.43[c]	38	0.36[c]	148	0.90
Bone (170)	1	0.21	2	0.39	1	1.17	2	1.99
Connective & other soft tissue (171)	14	0.77	15	0.80	3	0.94	5	1.41
Breast (174, 175)	159	0.82[c]	94	0.62[c]	20	0.69	43	1.57[c]
All uterine (females only) (179–182)	21	0.39[c]	25	0.61[c]	6	0.72	5	0.74
Cervix uteri (180)	8	0.33[c]	12	0.63	4	0.99	2	0.75
Ovary and other female genital organs (183–184)	63	0.91	39	0.75	9	0.90	8	0.79
Prostate (185)	63	0.77[c]	66	0.71[c]	8	0.77	35	1.24
Testes & other male genital organs (186, 187)	1	0.32	3	0.70	1	1.43	0	0.00
Kidney (189.0–189.2)	29	0.50[c]	51	0.80	6	0.62	12	0.89
Bladder & other urinary (188, 189.3–189.9)	27	0.55[c]	26	0.49[c]	4	0.56	9	0.67
Melanoma of skin (172)	35	0.74	32	0.61[c]	2	0.24[c]	7	0.69
Other malignant neoplasm of skin (173)	4	0.34[c]	4	0.30[c]	0	0.00	1	0.34
Eye (190)	0	0.00	1	0.72	1	4.45	0	0.00
Brain & central nervous system (191–192)	81	1.06	67	0.82	5	0.37[c]	14	0.93
Thyroid & other endocrine glands (193–194)	5	0.56	12	1.32	1	0.67	3	1.64
All lymphatic, hematopoietic tissue (200–208)	200	0.96	175	0.79[c]	13	0.40[c]	47	0.92
Non-Hodgkin lymphoma (200, 202)	74	0.92	59	0.68[c]	5	0.40[c]	16	0.79
Hodgkin lymphoma (201)	6	0.90	7	0.92	0	0.00	2	1.36
Multiple myeloma (203)	43	1.12	45	1.13	3	0.52	12	1.28
Leukemia (202.4, 204–208)	76	0.93	63	0.73[c]	5	0.39[c]	17	0.86
Acute lymphocytic leukemia (204.0)	4	0.73	2	0.34	0	0.00	0	0.00
Acute myeloid leukemia (205.0, 205.3, 206.0, 207.0, 207.2)	39	1.01	28	0.70	5	0.80	9	1.07
Chronic myeloid leukemia (205.1, 206.1)	4	0.67	6	0.90	0	0.00	4	2.52
Chronic lymphocytic leukemia (CLL) (204.1)	15	1.07	16	1.06	0	0.00	2	0.49
Leukemia other than CLL	60	0.89	47	0.66[c]	5	0.47	14	0.90
Smoking related cancers (140-150, 157, 161-162, 188-189)	605	0.57[c]	540	0.48[c]	66	0.40[c]	223	0.89
Non-smoking related cancers	980	0.76[c]	826	0.64[c]	113	0.55[c]	273	0.98
Polycythemia vera (238.4)	0	0.00	0	0.00	0	0.00	0	0.00
Myelodysplastic syndrome (238.7)	17	1.28	18	1.33	2	1.14	8	2.21
Diabetes (250)	139	0.59[c]	104	0.43[c]	13	0.34[c]	44	0.83
Mental and behavioral disorders (290–319)	105	0.61[c]	93	0.52[c]	6	0.24[c]	42	0.83
Dementia, Alzheimer's, Parkinson's Disease & motor neuron disease (290.0–290.4, 331.0, 332, 335.2)	200	0.75[c]	182	0.71[c]	22	0.74	72	0.81
Dementia and Alzheimer's disease (290.0–290.4, 331.0)	139	0.72[c]	122	0.68[c]	12	0.61	53	0.77
Parkinson's disease (332)	38	0.88	29	0.67	7	1.46	13	0.93
Diseases of the nervous system (320–389)	218	0.79[c]	171	0.63[c]	24	0.63[c]	58	0.79
All heart disease (390–398, 404, 410–429)	1,041	0.58[c]	911	0.47[c]	114	0.42[c]	453	0.88[c]
Rheumatic heart disease (393–398)	8	0.68	7	0.59	1	0.61	3	0.90
Ischemic heart disease (410–414)	691	0.57[c]	594	0.44[c]	71	0.38[c]	298	0.83[c]
Cerebrovascular disease (430–438)	191	0.63[c]	169	0.55[c]	21	0.49[c]	81	0.93
Non-malignant respiratory disease (460–519)	347	0.52[c]	286	0.43[c]	29	0.31[c]	129	0.73[c]
Bronchitis, emphysema, asthma (490–493)	190	0.50[c]	138	0.37[c]	14	0.26[c]	72	0.81
Cirrhosis of liver (571)	77	0.40[c]	68	0.31[c]	16	0.42[c]	29	0.77
Nephritis & nephrosis (580–589)	65	0.60[c]	67	0.61[c]	6	0.38[c]	30	1.09
AIDS (042–044, 795.8)	81	1.25	34	0.38[c]	7	0.43[c]	13	0.94
All external causes of death (800–999)	432	0.61[c]	401	0.48[c]	71	0.46[c]	138	1.00
Accidents (850–949)	235	0.54[c]	227	0.45[c]	40	0.43[c]	86	1.00
Suicides (950–959)	149	0.78[c]	139	0.60[c]	22	0.52[c]	38	1.03
Unknown causes of death	126		90		11		40	

[a]ICD: International Classification of Disease. Revision 9 codes are shown.
[b]Mesothelioma did not have an explicit code in ICD9 but did in ICD10 as denoted.
[c]Indicates SMR is significantly different from 1.0.

Table 5. Internal cohort dose-response analyses for lung cancer over categories of radiation absorbed dose to the lung among 109,019 medical radiation workers first monitored between 1965 and 1994 in the United States and followed through 2016, by sex[a].

Cause of death	Workers	Cases	HR	95% CI
Lung cancer (both sexes)				
Lung dose (mGy)				
<5	48,391	386	1.0	REF
5–<10	28,464	196	0.98	0.80;1.22
10–<25	22,151	160	0.98	0.77;1.24
25–<50	6183	56	1.21	0.86;1.71
50–<100	2902	35	1.38	0.92;2.08
≥100	928	17	1.42	0.83;2.43
Totals and *p* for trend	109,019	850		0.02 (+)
HR (95% CI) at 100 mGy			1.16 (1.02;1.31)	
ERR (95% CI) at 100 mGy			0.15 (0.02;0.27)	
Male lung cancer				
Lung dose (mGy)				
<5	25,488	205	1.0	REF
5–<10	12,978	98	0.91	0.68;1.22
10–<25	10,902	105	1.09	0.81;1.48
25–<50	3353	31	1.04	0.66;1.62
50 –<100	1765	26	1.41	0.87;2.29
≥100	732	15	1.52	0.84;2.76
Totals and *p* for trend	55,218	480		0.035 (+)
HR (95% CI) at 100 mGy			1.18 (1.01;1.37)	
ERR (95% CI) at 100 mGy			0.16 (0.01;0.32)	
Female lung cancer				
Lung dose (mGy)				
<5	22,903	181	1.0	REF
5–<10	15,486	98	1.14	0.82;1.58
10–<25	11,249	55	0.87	0.59;1.28
25–<50	2830	25	1.60	0.94;2.75
50–<100	1137	9	1.27	0.59;2.75
≥100	196	2	0.79	0.19;3.35
Totals and *p* for trend	53,801	370		0.53 (+)
HR (95% CI) at 100 mGy			1.09 (0.83;1.44)	
ERR (95% CI) at 100 mGy			0.09 (−0.19;0.36)	
p for interaction between sex and dose[b]			0.23	

[a]Models include radiation doses lagged by 10 y. Doses were analyzed as a time-dependent measure. Outcomes include deaths from underlying causes of death only. All models adjusted for sex, year of birth (categories as shown in Table 1), and occupational category (categories as shown in Table 1) taken as a measure of socioeconomic status. *p*-value for test for linear trend in the hazards ratio computed for continuous organ dose. *p*-values are two-sided. Abbreviations: REF: referent category; HR: hazards ratio; CI: confidence interval; ERR: excess relative rate.

[b]*p*-value from likelihood ratio test (LRT) for interaction between sex and dose in a Cox proportional hazards model including males and females.

radiography, including technologists, then the interventional fluoroscopists followed by the nuclear medicine personnel and radiation oncologists. The pattern of lung cancer SMRs, for example, was 0.90, 0.51, 0.43, and 0.36, respectively, over these occupational categories. The 44,498 interventional radiologists/cardiologists were not at significantly different mortality risk from any cause of death compared with the *other* workers. The SMR for brain cancer, for example, was 0.82 (95% CI 0.64, 1.04; $n=67$) and lower but statistically comparable with the SMR seen among general radiologists/technologists of 1.06 (95 CL 0.84, 1.32; $n=81$). The 6,087 workers classified as *other* had a higher proportion of workers who had died, over 28%, compared with about 9.4% for the medical specialties combined, reflecting an older age distribution and earlier years of first monitoring on average. Of the over 100 SMRs presented in Table 4, the only SMR that was significantly elevated was breast cancer among the 1,963 *other* workers (SMR 1.57; 95% CI 1.10, 2.09; $n = 43$). AIDS among the general radiography workers was elevated (SMR 1.25; 95% CI 0.99,1.55; $n=81$). The mean cumulative absorbed dose to the breast for the *other* female workers was 40.2 mGy (maximum 1.59 Gy; percent >100 mGy 6.6%) and over twice that of the full cohort (16.6 mGy).

Table 5 presents the internal cohort dose-response analysis for lung cancer. The mean cumulative absorbed dose to the lung was 13.0 mGy (maximum 1.27 Gy; percent >100 mGy 1.21%). The mean absorbed dose was somewhat lower for female workers 11.5 mGy (maximum 1.27 Gy; percent >100 mGy 0.68%) compared with male workers 14.5 mGy (maximum 1.11 Gy; percent >100 mGy 1.73%). Based on 840 lung cancer deaths, the sex-adjusted dose-response trend was statistically significant ($p=0.02$) and the ERR at 100 mGy was 0.15 (95% CI 0.02, 0.27) (Figure 2). A significant trend was seen among the 55,218 male medical workers with an ERR at 100 mGy equal to 0.16 (95% CI 0.01, 0.32) $n=480$, but the trend for the 53,801 female medical workers was not statistically significant [ERR at 100 mGy equal to 0.09 (95% CI −0.19, 0.36 $n=370$] (Figure 2). The ratio of the male to female ERRs at 100 mGy was 1.78 ($p_{\text{interaction}} =$ 0.23) which was similar to the ratio of the male to female SMRs for lung cancer of 1.62 (95% CI 1.41, 1.85).

There was heterogeneity among the four occupational categories with regard to the mean lung absorbed doses. The mean lung dose (max) in mGy for general radiology, interventional radiology, nuclear medicine and radiation oncology, and *other* were 11.6 (312.5), 3.0 (65.2), 50.8 (1106), and 38.6 (1272), respectively (Boice, Quinn et al. 2021). The lower doses among the interventionalists were because of their use of lead aprons.

Table 6 presents the internal cohort dose-response analysis for leukemia other than CLL, CLL, MDS, multiple myeloma, breast cancer, brain cancer, stomach cancer, esophageal cancer, prostate cancer, dementia and other neuronal diseases, Parkinson's disease, IHD and cerebrovascular disease. The mean cumulative absorbed dose to RBM was 12.9 mGy (maximum 1.19 Gy; percent >100 mGy 1.01%). Based on 139 cases of leukemia other than CLL, there was limited statistical evidence for an increasing trend over estimated doses to the RBM with the ERR at 100 mGy equal to 0.10 (95% CI −0.34, 0.54) (Figure 3). Nor was there evidence for a significant dose-response for CLL: ERR at 100 mGy 0.19 (95% CI −0.53, 0.90), $n=38$. The dose-response for MDS was negative based on 66 cases but not statistically significant. The dose-response for MDS in combination with acute myelogenous leukemia (AML) was positive based on 126 cases but also not statistically significant.

Breast cancer mortality among the 53,801 female medical workers was not related to radiation dose to the breast with the ERR per 100 mGy equal to −0.11 (95% CI −0.62, 0.40) $n=311$ (Table 6). The mean cumulative absorbed dose to the breast was 16.6 mGy (maximum 1.59 Gy; percent >100 mGy 1.43%). The internal dose-response analysis for brain cancer, based on 165 deaths, showed a positive relationship with radiation dose but the trend was not significant: ERR at 100 mGy was 0.20 (95% CI −0.30, 0.71). The mean

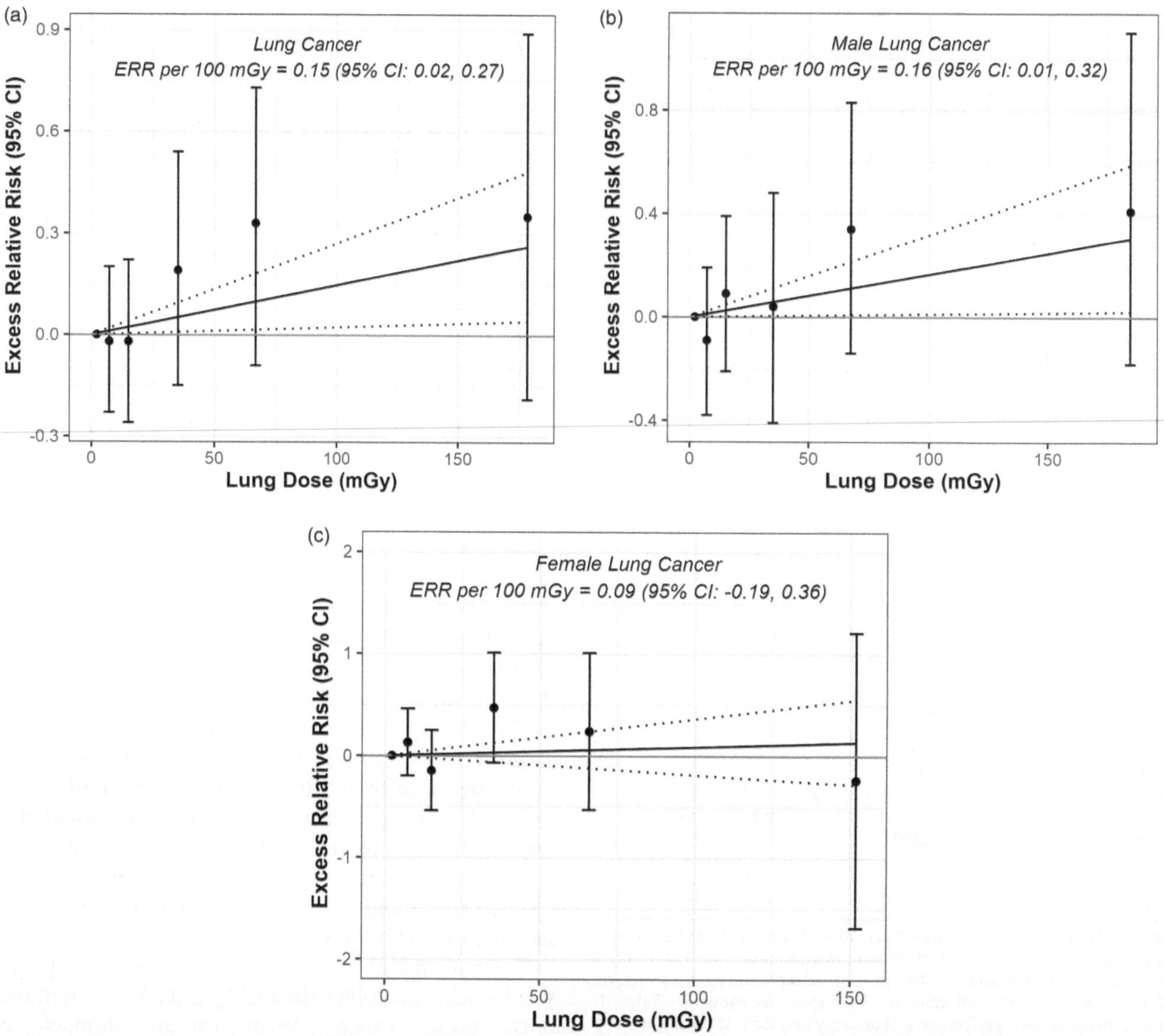

Figure 2. Excess Relative Risk (ERR) of lung cancer by radiation dose to lung among 109,019 medical radiation workers first monitored between 1965–1994 in the United States and followed through 2016. A linear model is presented (solid line) based on a continuous measure of dose, and 95% confidence bands (dotted lines) presented about the regression line. Doses were lagged by 10 y and adjustment was made for year of birth and socioeconomic status (taken as occupational category). Figure 2a presents the sex-adjusted dose response. Figures 2b and 2c present the male and female dose responses, respectively. ERRs and 95% confidence limits are calculated for selected categories of dose to the lung and displayed at the mean lung dose in each category. Categories <5 (referent), 5-, 10-, 25-, 50-, and >100 with respective mean doses of 2.2, 7.2, 14.9, 35.1, 66.8, and 178.1 mGy for all workers; 2.6, 7.2, 14.8, 35.0, 65.9, and 151.7 mGy for females; and 2.2, 7.2, 15.1, 35.1, 67.4, and 185.1 mGy for males.

cumulative absorbed dose to the brain was 18.9 mGy (maximum 1.08 Gy; percent >100 mGy 1.81%), The dose-response analysis for dementia, Alzheimer's, Parkinson's, and motor neuron disease was based on 476 cases and showed no significant evidence of radiation-related increased risk with the ERR at 100 mGy equal to 0.05 (95% CI −0.13, 0.23). Parkinson's disease was evaluated separately, and findings indicated a positive dose-response based on 87 underlying causes of death: ERR at 100 mGy equal to 0.17 (95% CI −0.20, 0.54) but was not statistically significant. The greatest number of deaths for a specific cause was for IHD ($n = 1{,}655$). The mean cumulative absorbed dose to the heart was 14.6 mGy (maximum 1.27 Gy; percent >100 mGy 1.48%). The association with radiation was negative but not statistically significant. The ERR at 100 mGy was −0.10 (95% CI −0.27, 0.06) (Figure 4). Cerebrovascular disease was not significantly related to radiation dose: the ERR at 100 mGy was 0.04 (95% CI −0.16, 0.23), $n = 462$.

Positive dose-response trends were seen for multiple myeloma and cancers of the stomach and esophagus, but none were statistically significant. The dose–response for prostate cancer was negative but also not statistically significant.

Internal cohort dose-response analyses for all solid cancers (ICD9 140–199) with and without the inclusion of lung cancer (ICD9 162) over categories of radiation dose to the colon are presented in Table 7. The absorbed dose to the colon was taken as an estimate of whole-body dose. For all solid cancers, the ERR (95% CI) at 100 mGy was 0.07 (-0.01, 0.15), $n = 3{,}191$. For all solid cancers excluding lung cancer, the ERR (95% CI) at 100 mGy was reduced and estimated as 0.03 (-0.08, 0.14), $n = 2{,}341$.

Table 6. Internal cohort dose-response analyses for selected causes of death over categories of radiation absorbed dose to specific organs or tissue among 109,019 medical workers first monitored between 1965 and 1994 in the United States and followed through 2016[a].

Cause of death	Workers	Cases	HR	95% CI
Leukemia other than CLL** (ICD9 204.0,204.2, 204.3–204.9,205–208)				
RBM dose (mGy)				
<5	40,186	47	1.0	REF
5–<10	32,231	40	1.11	0.62;1.99
10–<25	25,355	33	1.23	0.65;2.35
25–<50	6426	11	1.83	0.77;4.35
50–<100	3683	5	1.73	0.56;5.38
≥100	1138	3	2.42	0.62;9.49
Totals and *p* for trend	109,019	139		0.64 (+)
HR (95% CI) at 100 mGy				1.11 (0.72; 1.72)
ERR (95% CI) at 100 mGy				0.10 (−0.34; 0.54)
CLL** (ICD9 204.1, 202.4)				
RBM dose (mGy)				
<5	47,154	18	1.0	REF
5–<10	29,723	7	0.96	0.34;2.73
10–<25	22,409	10	2.23	0.79;6.33
25–<50	6204	1	1.38	0.15;12.30
≥50	3529	2	4.78	0.75;30.58
Totals and *p* for trend	109,019	38		0.62 (+)
HR (95% CI) at 100 mGy				1.20 (0.58; 2.48)
ERR (95% CI) at 100 mGy				0.19 (−0.53;0.90)
MDS (ICD9 238.7)				
RBM dose (mGy)				
<5	40,186	18	1.0	REF
5–<10	32,231	17	1.17	0.46; 2.97
10–<25	25,355	5	0.42	0.12; 1.49
≥50	11,247	9	1.50	0.41; 5.56
Totals and *p* for trend	109,019	49		*0.74 (–)*
HR (95% CI) at 100 mGy				0.83 (0.27; 2.57)
ERR (95% CI) at 100 mGy				−0.19 (−1.32; 0.94)
AML + MDS (205.0, 238.7)				
RBM dose (mGy)				
<5	40,186	40	1.0	REF
5–<10	32,231	41	1.29	0.70;2.38
10–<25	25,355	24	0.99	0.49;2.00
25–<50	6426	10	1.74	0.70;4.34
50–<100	3683	8	2.72	0.97;7.60
≥100	1138	3	2.27	0.57;9.11
Totals and *p* for trend	109,019	126		0.61 (+)
HR (95% CI) at 100 mGy				1.11 (0.75; 1.65)
ERR (95% CI) at 100 mGy				0.10 (−0.29; 0.50)
Multiple myeloma (ICD9 203)				
RBM dose (mGy)				
<5	47,154	51	1.0	REF
5–<10	29,723	30	0.93	0.52; 1.64
10–<25	22,409	16	0.62	0.31; 1.26
25–<50	6204	2	0.28	0.06; 1.30
≥50	3529	4	0.74	0.22; 2.54
Totals and *p* for trend	109,019	103		0.50 (+)
HR (95% CI) at 100 mGy				1.16 (0.76; 1.76)
ERR (95% CI) at 100 mGy				0.14 (−0.28; 0.56)
Stomach cancer (ICD9 151)				
Stomach dose (mGy)				
<5	40,114	38	1.0	REF
5–<10	27,893	20	1.17	0.61; 2.24
10–<25	28,489	20	1.12	0.55; 2.29
25–<50	7696	2	0.39	0.08; 1.79
≥50	4827	6	1.46	0.48; 4.43
Totals and *p for trend*	109,019	86		0.66 (+)
HR (95% CI) at 100 mGy				0.80 (0.30; 2.15)
ERR (95% CI) at 100 mGy				−0.22 (−1.21; 0.77)
Esophageal cancer (ICD9 150)				
Esophagus dose (mGy)				
<5	48,457	42	1.0	REF
5–<10	28,416	30	1.39	0.78; 2.46
10–<25	21,892	14	0.76	0.37; 1.58
25–<50	6060	3	0.59	0.16; 2.17

(continued)

Table 6. Continued.

Cause of death	Workers	Cases	HR	95% CI
≥50	4194	2	0.43	0.09; 2.07
Totals and *p* for trend	109,019	91		0.79 (+)
HR (95% CI) at 100 mGy				1.08 (0.61; 1.94)
ERR (95% CI) at 100 mGy				0.08 (−0.50; 0.66)
Female breast cancer[b] (ICD9 174)				
Breast dose (mGy)				
<5	18,559	123	1.0	REF
5–<10	9,552	48	0.90	0.59–1.35
10–<25	18,167	98	0.98	0.67–1.44
25–<50	5214	22	0.79	0.46–1.37
50–<100	1899	17	1.48	0.79–2.78
≥100	410	3	0.85	0.25–2.83
Totals and *p* for trend	53,801	311		0.67 (-)
HR (95% CI) at 100 mGy				0.90 (0.54; 1.50)
ERR (95% CI) at 100 mGy				−0.11 (−0.62; 0.40)
Prostate cancer[c] (ICD9 185)				
Prostate dose (mGy)				
<5	30,904	87	1.0	REF
5–<10	11,825	39	0.99	0.65; 1.50
10–<25	7283	25	0.79	0.46; 1.38
25–<50	2767	10	0.77	0.35; 1.72
50–<100	1688	7	0.79	0.32; 1.98
≥100	751	4	0.67	0.22; 2.06
Totals and *p* for trend	55,218	172		0.32 (-)
HR (95% CI) at 100 mGy				0.72 (0.38; 1.36)
ERR (95% CI) at 100 mGy				−0.32 (−0.96; 0.31)
Brain cancer (cranial) (ICD9 191, 192.0–192.1)				
Brain dose (mGy)				
<5	41,483	81	1.0	REF
5–<10	18,259	19	0.79	0.14–1.34
10–<25	29,474	45	1.47	0.88–2.45
25–<50	14,084	10	0.72	0.33–1.55
50–<100	4464	7	1.38	0.58–3.31
≥100	1255	3	1.96	0.57–6.75
Totals and *p* for trend	109,019	165		0.43 (+)
HR (95% CI) at 100 mGy				1.23 (0.74; 2.03)
ERR (95% CI) at 100 mGy				0.20 (−0.30; 0.71)
Dementia/Alzheimer's/Parkinson's/motor neuron disease (ICD9 290.0–290.4, 331.0, 332, 335.2)				
Brain dose (mGy)				
<5	41,483	157	1.0	REF
5–<10	18,259	75	1.08	0.80–1.46
10–<25	29,474	125	1.11	0.77–1.60
25–<50	14,084	73	1.20	0.80–1.80
50–<100	4464	34	1.24	0.77–1.99
≥100	1255	12	1.12	0.58–2.18
Totals and *p* for trend	109,019	476		0.58 (+)
HR (95% CI) at 100 mGy				1.05 (0.88; 1.25)
ERR (95% CI) at 100 mGy				0.05 (−0.13; 0.23)
Parkinson's disease (ICD9 332)				
Brain dose (mGy)				
<5	41,483	27	1.0	REF
5–<10	18,259	14	1.30	0.65; 2.59
10–<25	29,474	17	1.32	0.55; 3.18
25–<50	14,084	16	2.16	0.85; 5.44
≥50	5719	13	2.59	0.98; 6.88
Totals and *p* for trend	109,019	87		0.37 (+)
HR (95% CI) at 100 mGy				1.18 (0.82; 1.71)
ERR (95% CI) at 100 mGy				0.17 (−0.20; 0.54)
Ischemic heart disease (ICD9 410–414)				
Heart dose (mGy)				
<5	42,678	617	1.0	REF
5–<10	28,251	425	1.22	1.04–1.43
10–<25	26,526	400	1.14	0.96–1.35
25–<50	7,199	116	1.13	0.89–1.44
50–<100	3,311	70	1.29	0.96–1.72
≥100	1,054	27	0.96	0.63–1.46
Totals and *p* for trend	109,019	1,655		0.21 (-)
HR (95% CI) at 100 mGy				0.90 (0.77; 1.06)
ERR (95% CI) at 100 mGy				−0.10 (−0.27; 0.06)

(continued)

Table 6. Continued.

Cause of death	Workers	Cases	HR	95% CI
Cerebrovascular disease (ICD9 430–438)				
Brain dose (mGy)				
<5	41,483	167	1.0	REF
5–<10	18,259	63	1.04	0.76; 1.43
10–<25	29,474	113	1.36	0.96; 1.92
25–<50	14,084	82	1.97	1.35; 2.88
50–<100	4464	26	1.44	0.88; 2.38
≥100	1255	11	1.59	0.81; 3.13
Totals and *p* for trend	109,019	462		0.70 (+)
HR (95% CI) at 100 mGy				1.04 (0.86; 1.26)
ERR (95% CI) at 100 mGy				0.04 (−0.16; 0.23)

[a]Models include radiation doses lagged by 2 y for leukemia other than CLL, AML, and MDS; otherwise, 10 y lags were applied. Doses were analyzed as a time-dependent measure. Outcomes include deaths from underlying causes of death unless noted as ** in which case contributing causes of death were also included. All models adjusted for sex, year of birth (categories as shown in Table 1), and occupational category (categories as shown in Table 1) taken as a measure of socioeconomic status. *p*-value for test for linear trend in the hazards ratio computed for continuous organ dose. *p*-values are two-sided. REF: referent category; HR: hazards ratio; CI: confidence interval; ERR: excess relative rate.
[b]Limited to females only.
[c]Limited to males only.

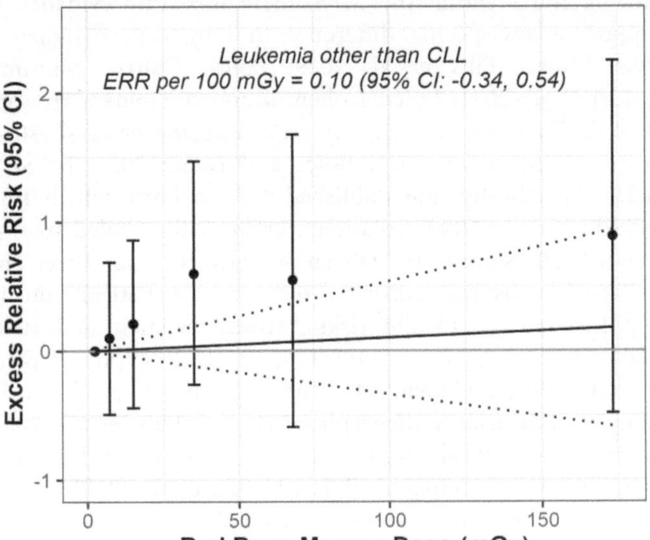

Figure 3. Excess Relative Risk (ERR) of leukemia other than chronic lymphocytic leukemia (CLL) by radiation dose to red bone marrow (RBM) among 109,019 medical radiation workers first monitored between 1965-1994 in the United States and followed through 2016. A linear model is presented (solid line) based on a continuous measure of dose, and 95% confidence bands (dotted lines) presented about the regression line. Doses were lagged by 2 y and adjustments were made for sex, year of birth, and socioeconomic status (taken as occupational category). ERRs and 95% confidence limits are calculated for selected categories of dose to the RBM and displayed at the mean RBM dose in each category. Categories <5 (referent), 5-, 10-, 25-, 50-, and ≥100 with respective mean doses of 2.5, 7.2, 15.0, 35.1, 68.1, and 173.3 mGy.

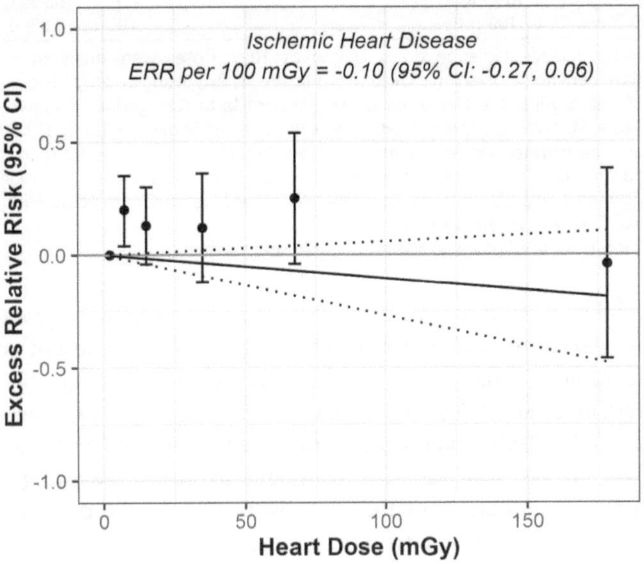

Figure 4. Excess Relative Risk (ERR) of ischemic heart disease (IHD) by radiation dose to the heart among 109,019 medical radiation workers first monitored between 1965–1994 in the United States and followed through 2016. A linear model is presented (solid line) based on a continuous measure of dose, and 95% confidence bands (dotted lines) presented about the regression line. Doses were lagged by 10 y and adjustment was made for sex, year of birth and socioeconomic status (taken as occupational category). ERRs and 95% confidence limits are calculated for selected categories of dose to the heart and displayed at the mean heart dose in each category. Categories <5 (referent), 5-, 10-, 25-, 50-, and ≥100 with respective mean doses of 2.2, 7.3, 15.2, 34.8, 67.4, and 178.1 mGy.

Discussion

Previous studies of radiologists and other medical radiation workers have reported significant increases in leukemia, multiple myeloma, and cancers of the breast, thyroid, bone, pancreas, lung and skin (Yoshinaga et al. 2004; Boice 2006; Linet et al. 2010). Except for leukemia and skin cancer among those first employed before 1950, there is little consistency in reported findings. Inconsistencies may be related to methodological deficiencies in design and conduct, small cohort sizes, short and incomplete follow-up, the influence of unknown confounding factors, differences in the cumulative radiation exposures received, and/or inadequate measures of radiation exposure. It is the absence of individual measures of radiation exposure over employment years that severely limits the interpretation of these early studies, i.e. using measures such as years worked as a surrogate for cumulative radiation exposure, the uncertainties and possible systematic errors in estimating doses in the absence of badge readings for individual workers, especially before 1960, and

Table 7. Internal cohort dose-response analyses for all solid cancers (ICD9 140–199) with and without inclusion of lung cancer over categories of radiation absorbed dose to the colon among 109,019 medical workers first monitored between 1965 and 1994 in the United States and followed through 2016[a].

Cause of Death	Workers	Cases	HR	95% CI
All solid cancers				
Colon dose (mGy)[b]				
<5	41,210	1304	1.0	REF
5–<10	25,594	679	1.02	0.91; 1.15
10–<25	29,256	801	1.03	0.91; 1.16
25–<50	8117	214	1.03	0.86; 1.22
50–<100	3623	131	1.28	1.03; 1.58
≥100	1219	62	1.25	0.94; 1.66
Totals and *p* for trend	109,019	3191		0.07 (+)
HR (95% CI) at 100 mGy			1.08	0.99; 1.16
ERR (95% CI) at 100 mGy			0.07	−0.01; 0.15
All solid cancers excluding lung cancer				
Colon dose (mGy)[b]				
<5	41,210	973	1.0	REF
5–<10	25,594	496	1.01	0.88; 1.15
10–<25	29,256	605	1.06	0.92; 1.22
25–<50	8117	139	0.91	0.73; 1.12
50–<100	3623	87	1.17	0.90; 1.51
≥100	1219	41	1.16	0.82; 1.63
Totals and *p* for trend	109,019	2341		0.58 (+)
HR (95% CI) at 100 mGy			1.03	0.92; 1.15
ERR (95% CI) at 100 mGy			0.03	−0.08; 0.14

[a]Models include radiation doses lagged by 10 y. Doses were analyzed as a time-dependent measure. Outcomes include deaths from underlying causes of death. All models adjusted for sex, year of birth (categories as shown in Table 1), and occupational category (categories as shown in Table 1) taken as a measure of socioeconomic status. *p*-value for test for linear trend in the hazards ratio computed for continuous organ dose. *p*-values are two-sided. Abbreviations: REF: referent category; HR: hazards ratio; CI: confidence interval; ERR: excess relative rate.
[b]Colon dose used as a surrogate for whole-body dose.

difficulties in accounting for the changing radiologic practices overtime when dosimetric measures are absent or incomplete (Boice 2006; Linet et al. 2010; Liu et al. 2014; Preston et al. 2016; NCRP 2018b; Karellas 2020; Villoing et al. 2021). Individual dosimetric badge readings are more likely to be available and complete for those working in more recent years, but these exposures are low and provide little statistical power to uncover radiation associations. The current study of U.S. medical workers was designed to address the recognized limitations of previous studies, and in particular to provide individual radiation dose estimates based on badge dose measurements, i.e. personal dose equivalents, over all years of employment. Notably, the duration of monitoring for medical workers within the Landauer database ranged from 2 to 60 y and was 23.7 y on average.

The current study of 109,019 medical radiation workers in the United States with equal numbers of men and women found a significant correlation between radiation dose and lung cancer mortality with the risk among males nearly 2 times higher than among females on a relative scale. Although this difference was not statistically significant, there was no evidence to indicate that females were at higher risk than males. There were no statistically significant associations found for deaths from leukemia other than CLL, CLL, MDS, MDS in combination with acute myelogenous leukemia, cancers of the breast and brain, dementia including Parkinson's disease, cerebrovascular disease, IHD and several other cancers. For leukemia other than CLL, and several other cancers, the sampling variability associated with narrow dose ranges and relatively small number of outcomes was such that radiation risks consistent with current understanding could not be ruled out (Hauptmann et al. 2020; Gilbert et al. 2020). The relatively young age of the cohort with nearly 90% alive at last follow-up points to the value of continuing the follow-up to improve the estimates of radiation risk and to clarify any sex-specific differences in radiation associations.

Lung cancer

Lung cancer mortality was significantly associated with occupational exposure as a medical radiation worker. The radiation risk was highest among the male workers (ERR per 100 mGy of 0.16) compared with the female workers (ERR per 100 mGy of 0.09) but the difference was not statistically significant. There was no evidence that the dose-response relationships for lung cancer differed among the four occupational categories ($p_{\text{interaction}} = .62$). Similar to other cohorts within the MPS, there was little evidence to support a sex-specific difference in lung cancer radiation risks (Boice, Ellis et al. 2019; Boice, Cohen, Mumma, Golden et al. 2021; Boice, Cohen, Mumma, Golden, Howard et al. 2021; Boice, Cohen, Mumma, Hagemeyer et al. 2021; Boice, Quinn et al. 2021; Boice and Dauer 2021; NASEM 2021). Specifically, the published MPS cohorts which now include over 500,000 radiation workers either failed to find a statistically significant association between lung cancer and radiation exposures and/or failed to find a statistically meaningful difference in the risks between women and men. These MPS cohorts include 135,193 nuclear power plant workers (Boice, Cohen, Mumma, Hagemeyer et al. 2021), 123,556 industrial radiographers (Boice, Ellis et al. 2019; Boice, Quinn et al. 2021), 114,270 above-ground nuclear weapons test participants (Boice, Cohen et al. 2020), 26,328 workers at the Los Alamos National Laboratory (Boice, Cohen, Mumma, Golden et al. 2021), 4,954 workers at the Mound facility in Ohio (Boice et al. 2014), 2,514 workers at the Mallinckrodt Chemical Works in Ohio (Boice, Ellis et al. 2019), 5,801 workers at the Rocketdyne (Atomics International) complex in California (Boice et al. 2006, 2011), and 7,000 workers at two early uranium processing facilities in the U.S.: the Fernald Feed Materials Production Center in Ohio and the Middlesex Sampling Plant in New Jersey (Golden et al. 2021). Additional large cohorts will be evaluated within the next two years including nuclear submariners (Charpentier et al. 1993; Boice, Cohen et al. 2019; Friedman-Jimenez et al. 2021).

The sex-adjusted ERR per 100 mGy of 0.15 (95% CI 0.02, 0.27) in the current study is consistent with the INWORKS occupational study [ERR per 100 mGy of 0.05 (90% CI 0.00, 1.09)] (Richardson et al. 2018). INWORKS did not provide sex-specific estimates of cancer risks.

Our study results differ from the U.S. radiologic technologist study of 106,068 questionnaire respondents where the ERR per 100 mGy unadjusted for smoking was −0.02 (95% CI <0, 0.13), n=1090 deaths; and the ERR per 100 mGy adjusted for smoking was 0.16 (95% CI <0, 0,46), (Velazquez-Kronen et al. 2020). There were more female (n=80,180) than male (n=25, 888) workers in the U.S. radiologic technologist study. Our study of 109,019 U.S. medical radiation workers had near equal numbers by sex, fewer technologists, and predominantly interventional and general radiologists, nuclear medicine and radiation oncology workers. The mean cumulative absorbed dose to the lung in the U.S. radiologic technologist study was 25 mGy (maximum 810 mGy), with little difference between the men and women. The radiation risk was highest among the female workers (ERR per 100 mGy of 0.06; 95% CI <0,0.23; n=711) compared with the male workers (ERR per 100 mGy of −0.14; 95% CI <0, 0.09; n=379) but neither estimate was statistically significant nor was the difference in the ERR estimates (p–value for heterogeneity = 0.18).

Cumulative individual doses in the U.S. radiologic technologist study were estimated in a comprehensive manner (Simon 2011; Simon et al. 2014), but the incomplete individual dosimetry data, especially for the early workers when exposures were highest, is a shortcoming that tempers the interpretation of the radiation risk estimates (NCRP 2018b). Recognized limitations by the authors include the uncertainties and possible systematic errors in estimating doses in the absence of badge readings for individual workers, especially before 1960, relying on self-reported work history data and historical data, and the absence of information on nearly 40,000 (27%) of the technologist selected for study who died before the mailed questionnaire was sent in the early 1980 or who failed to respond (Liu et al. 2014; Preston et al. 2016; NCRP 2018b; Linet et al. 2020). Nonetheless, the study has many strengths including large numbers and long follow-up, large numbers of women (n = 80,180), a comprehensive approach to dosimetry including chromosomal analyses, and the ability to assess, based on self-reported questionnaire responses, the use of lead aprons, cigarette smoking histories, and other occupational and behavioral characteristics. While there was not a radiation association with lung cancer overall, significant associations with radiation dose and lung cancer were seen for nonsmokers and light smokers but not for heavy smokers. Interestingly, heavy smokers also were at much lower radiation risk than nonsmokers or light smokers among Japanese atomic bomb survivors who were acutely exposed to higher levels of radiation (Furukawa et al. 2010, 2017; Cahoon et al. 2017).

Our study of 109,019 U.S. medical radiation workers first monitored 1965–1985 is not consistent with the study of Japanese atomic bomb survivors exposed briefly in 1945. While we find males to be at 1.8 times higher risk than females on a relative scale, the Japanese Life Span Study reported females to be nearly 3 times the lifetime risk of dying from lung cancer than males (Ozasa et al. 2012; NCRP 2014). Overall, the sex-averaged ERR per 100 mGy was 0.075 (95% CI 0.051, 0.103) and the ERR per 100 mGy was greater for women (0.11; 95% CI 0.068, 0.16) than for men (0.040; 95% CI 0.017, 0.067). NCRP is currently assembling and evaluating many relevant data sets to provide guidance to NASA on sex-specific lung cancer risks used in lifetime mortality risk projections for astronaut protection (Boice 2019b; National Academies of Sciences et al. 2021; NCRP 2021).

Leukemia and MDS

A statistically significant radiation association was not apparent for death from leukemia other than CLL which is consistent with a recent report from the U.S. radiologic technologist study (Linet et al. 2020) but inconsistent with previous studies of early radiologists and technologists (Jablon and Miller 1978; Matanoski et al. 1984; Berrington et al. 2001; Wang et al. 2002; Yoshinaga et al. 2004) and with large-scale occupational studies of radiation workers within the MPS and around the world (Boice 2017; NCRP 2018b; Shore et al. 2019; Boice, Cohen, Mumma, Hagemeyer et al. 2021; Boice, Quinn et al. 2021). The ERR per 100 mGy of 0.10 (95% CI −0.34, 0.54) is based on a rather narrow dose distribution with only 1% of the workers receiving >100 mGy to the active bone marrow and relatively small numbers of leukemia deaths, n = 139. The wide confidence limits, i.e. the upper bound of 0.54, indicates that the data are statistically compatible with studies with broader dose ranges including that of the Japanese atomic bomb survivors (Hsu et al. 2013). The consistency with the non-overlapping U.S. radiologic technologist study, recognizing the differences in study design and dose reconstruction mentioned above, suggests that occupational exposures in the medical workplace in more recent years have been greatly reduced compared with the turn of the last century, reflecting radiation protection awareness and implementation of safety guidance (Boice, Dauer et al. 2020). An association with CLL was also not observed, as expected since CLL is not considered inducible by radiation or that the risk is so low that it is not detectable by epidemiologic methods. Also based on rather narrow dose ranges to RBM, there was no statistical evidence for an association with MDS, or MDS combined with acute myelogenous leukemia, consistent with larger studies within the MPS (Boice, Cohen, Mumma, Hagemeyer et al. 2021). Recently, the possibility of a causal association between radiation and MDS has been questioned (UNSCEAR 2020).

Breast cancer

Female breast cancer mortality was not linked to radiation exposure [ERR per 100 mGy −0.11 (95% CI −0.62, 0.40)], but the data are consistent with relatively high risks, i.e. ERRs per 100 mGy up to 0.40 are compatible with the data at 95% confidence. Despite the relatively large number of women studied nearly 54,000, and the 311 deaths from breast cancer, the narrow dose distribution coupled with the ages of the women when exposed likely contributed to the inconclusive findings. Among the subset of *other* workers, however, the dose distribution was broader and follow-up longer, and 13 of the 20 breast cancer among women who received >50 mGy occurred among this smaller group. The SMR of 1.54 was also statistically significant but the small numbers overall precluded convergence of the dose-response models. The female breast is highly susceptible to radiation-induced carcinogenesis and radiogenic risk is seen to decrease with increasing age at exposure (Preston et al. 2002; UNSCEAR 2008). Nearly 40% of the women were monitored after age 35 when radiation risk is lower than at younger ages. Nearly 90% of the women were alive at last follow-up in 2016 and the majority have not reached the ages later in life when breast cancer rates are high and radiation-related excesses appear.

The early studies of the pioneering radiologists and medical workers included few female workers. It was only in the study of Chinese medical workers where breast cancer was significantly elevated, but the absence of dose estimates hinders interpretation (Wang et al. 2002). A more recent report included dose reconstructions for the colon for use in all solid cancer evaluations, but not specifically for the breast or other organs (Sun et al. 2016). There were several reports evaluating breast cancer within the U.S. radiologic technologist study that indicated elevated risks for those who began employment before 1950, but radiation absorbed dose to the breast was not estimated (Boice et al. 1995; Mohan et al. 2002; Doody et al. 2006). A recent report incorporating individual estimates of the radiation-absorbed dose was based on 83,523 female radiologic technologists who were alive in 1983 and completed a mail questionnaire (Preston et al. 2016). The estimated mean cumulative breast dose was 37 mGy and ranged from 6 mGy for those first employed after 1980 to 1,168 mGy for women employment before 1930. The ERR at 100 mGy was estimated to be 0.07 (95% CI −0.005, 0.19) for incidence ($n = 1{,}922$) and 0.31 (95% CI 0.11, 0.67) for mortality ($n = 586$). Conclusions are tempered because of the limitations previously mentioned, i.e. relying on estimates of radiation exposure without measurements for practically all of the early workers, the criteria that workers had to be alive in 1983 and to have responded to a mailed questionnaire, the reliance on self-reported incident cases of breast cancer (which does not affect the mortality analysis), and concerns as described in NCRP Commentary 27 (NCRP 2018b). The authors also recommended caution in interpreting the results because of the uncertainties and possible systematic errors in estimating doses and the absence of badge readings for individual workers, especially before 1960 (Preston et al. 2016). One strong point is being able to account for important risk factors for breast cancer, such as nulliparity and age at menarche. The risk estimates also are compatible with the study of Japanese atomic bomb survivors, and with a pooled analysis of two Massachusetts tuberculosis-fluoroscopy cohorts and a U.S. thymic irradiation cohort as reported in Preston et al. (2016). These similarities of the ERRs per Gy with the tuberculosis and thymus cohorts are somewhat unexpected since the dose rates were much higher than those received occupationally as a technologist, the background rates for breast cancer are so different and the sources of exposure so varied, e.g. infants given radiotherapy for thymic enlargement. There is little doubt that women under the age of 35 when first exposed to radiation are at notable increased risk for developing breast cancer later in life. However, more work is needed to quantify with precision radiation risks following frequent yet low-level exposures cumulating to 100 mGy.

Brain cancer

The evidence and the degree to which radiation at low levels might increase the occurrence of the brain and central nervous system (CNS) tumors among adults is uncertain (UNSCEAR 2008) and for some lethal cancers such as glioma, the risk appears to largely disappear at ages of exposure after about 20 y (UNSCEAR 2013). There have been concerns, however, of elevated occurrences of brain tumors among interventional radiologists who perform fluoroscopically-guided procedures (Picano et al. 2012; Cousins et al. 2013; Roguin et al. 2013). Overall, the study of medical radiation workers in the United States did not find that radiation was significantly associated with brain cancers [ERR at 100 mGy was 0.20 (95%CI −0.30, 0.71) $n = 165$] following a mean cumulative absorbed dose to the brain of 18.9 mGy, maximum 1.08 Gy. Neither was their an increase among the interventional fluoroscopists, although the numbers were smaller. These findings are generally consistent with the U.S. radiologic technologist study [ERR per 100 mGy of 0.1 (95% CI <−0.3, 1.5); $n = 193$] following an estimated cumulative mean absorbed brain dose of 12 mGy, maximum 290 mGy (Kitahara et al. 2017), although the estimate for radiation dose is uncertain as described above (NCRP 2018b). No increases in brain cancer have been reported in other MPS studies of following occupational radiation exposure.

Parkinson's disease and neurological disorders

In light of a recent report among Russian workers at the Mayak nuclear facility indicating a correlation between low-LET radiation and Parkinson's disease (Azizova et al. 2020) and the recent interest in cognitive impairment following low-level radiation exposure (Pasqual et al. 2021), an evaluation was similarly conducted. There was no evidence for a radiation association for the combined categories of dementia, Alzheimer's, Parkinson's, and motor neuron disease. However, the dose-response for Parkinson's disease alone was positive (ERR per 100 mGy 0.17; 95% CI −0.20, 0.54) (Figure 5) and consistent with the estimate of risk among Mayak workers: ERR per 100 mGy 0.102; 95% CI 0.059, 0.163; $n = 300$ diagnoses. Associations with Parkinson's disease have been reported in other MPS studies including Los Alamos National Laboratory workers (Boice, Cohen, Mumma, Golden et al. 2021), workers at the Mound Facility in Ohio (Boice et al. 2014), nuclear power plant workers (Boice, Cohen, Mumma, Hagemeyer et al. 2021), and industrial radiographers (Boice and Dauer 2021; Boice, Quinn et al. 2021) but not among Mallinckrodt Chemical workers (Golden et al. 2019). The future pooling of MPS cohorts will confirm and improve estimates of radiation mortality risks following prolonged exposures for Parkinson's disease and other neurological disorders.

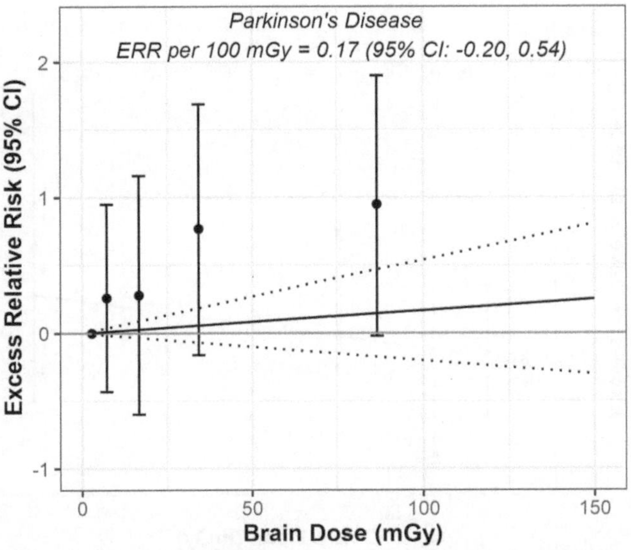

Figure 5. Excess Relative Risk (ERR) of Parkinson's disease by radiation dose to brain among 109,019 medical radiation workers first monitored between 1965–1994 in the United States and followed through 2016. A linear model is presented (solid line) based on a continuous measure of dose, and 95% confidence bands (dotted lines) presented about the regression line. Doses were lagged by 10 y and adjustment was made for sex, year of birth and socioeconomic status (taken as occupational category). ERRs and 95% confidence limits are calculated for selected categories of dose to the brain and displayed at the mean brain dose in each category. Categories <5 (referent), 5-, 10-, 25-, and ≥50 with respective mean doses of 2.8, 7.1, 16.6, 34.3, and 86.7 mGy.

Ischemic heart disease

Whether IHD is a consequence of low-level radiation exposures accumulating under 500 mGy remains a question of great scientific and public health importance (NCRP 2014, 2018b; Tran et al. 2017; Cutter et al. 2019; Wakeford 2019; Boice, Held et al. 2019, Boice, Cohen, Mumma, Hagemeyer et al. 2021; EPRI 2020; Richardson et al. 2020; Anderson et al. 2021; Tapio et al. 2021). Even a small risk might equate to a large absolute number of excess cases because heart disease is such a prevalent cause of death in the United States and around the world. Consistent with the MPS studies published to date, the medical radiation worker analyses failed to reveal a significant increase in IHD mortality over categories of dose to the heart (Boice et al. 2011; 2014; Boice, Cohen, Mumma, Golden et al. 2021; Boice, Cohen, Mumma, Hagemeyer et al. 2021; NCRP 2018b; Golden et al. 2019; Boice, Quinn et al. 2021). Based on 1,655 deaths attributed to IHD, the ERR at 100 mGy was −0.10 (95% CI −0.27, 0.06). The sampling variability was such, however, that an ERR per 100 mGy of 0.06 is statistically compatible with these data at 95% confidence. Thus, the study is consistent with the overall estimate of IHD reported in INWORKS (ERR per 100 mSv 0.017; 90% CI 0.0002, 0.036), although there were notable differences in that INWORKS risks were significantly higher for females than males and for white collar than blue collar workers (Gillies et al. 2017). Significant radiation associations for heart disease mortality have been reported among the Japanese atomic bomb survivors (Ozasa et al. 2012). Recent reports on IHD, however, indicate that the risks were no longer statistically significant and that explanations other than radiation for the increase were plausible (Ozasa et al. 2017; Takahashi et al. 2017).

The previous studies of early radiologists and medical radiation workers in the United States (Matanoski et al. 1984), the United Kingdom (Berrington et al. 2001), and China (Wang et al. 2002) did not provide estimates of radiation dose which were likely to be substantial given the significant excesses of leukemia and skin cancer reported. Only for the U.S. pioneering radiologists, however, were diseases of the circulatory system reported to be significantly elevated. In contrast, UK radiologists showed significant deficits of circulatory disease deaths. The data within the U.S. radiologic technologist study also did not find associations for IHD with estimated radiation doses to the heart, there were but suggested increased risks among those who first worked before 1950 when compared with technologists who began employment in 1960 or later (Hauptmann et al. 2003). In a more recent evaluation among 22,209 radiologic technologists who reported that they had performed fluoroscopically-guided interventional procedures, there was little indication of an increased risk for IHD in comparison with technologists who did not report similar participation with fluoroscopically-guided procedures [HR 1.04 (95% CI 0.87, 1.26)' m=166))] (Rajaraman et al. 2016). In another subgroup analysis involving radiologic technologists who reported performing medical procedures involving radionuclides, there was an increased incidence for self-reported myocardial infarction when comparisons were made with radiologic technologists who never performed these procedures but there was no increase seen for IHD [HR 1.05 (95% CI 0.83, 1.33); $n = 93$] (Kitahara et al. 2015). These recent publications are difficult to interpret because of the absence of individual radiation dose estimates to the heart and the lack of dose-response relationships, the reliance on self-reported work histories and outcomes, and the comparisons with other technologists who reported not performing specific procedures but likely performed others. The reliance on self-reported diagnoses of circulatory disease that was not confirmed adds another caution in interpreting these findings (Kitahara et al. 2015) as well as the concerns previously mentioned regarding the dose reconstruction gaps (NCRP 2018b). Nonetheless, the medical radiation worker studies in the U.S. provide little consistent evidence for a risk of IHD at the cumulative dose levels evaluated. The combined cohort data within the MPS will provide a more powerful evaluation of the possible level of IHD risk following low-level radiation exposures received in the occupational setting over many years.

Other cancers

Esophageal cancer was increased in several MPS cohorts, including studies of Mound workers (Boice et al. 2014), nuclear weapons test participants (Boice, Cohen et al. 2020), workers at Los Alamos National Laboratory (Boice, Cohen, Mumma, Golden et al. 2021), and nuclear power plant workers (Boice, Cohen, Mumma, Hagemeyer et al. 2021). There was a positive dose-response among medical workers but not statistically significant and based on a small number

Table 8. Sensitivity analysis adjusting for missing dose for leukemia other than CLL and for lung cancer.

Cause of death	Workers	Cases	HR	95% CI
Leukemia other than CLL				
RBM dose (mGy)				
<5	37,373	42	1.0	REF
5–<10	24,929	34	1.24	0.65; 2.38
10–<25	34,028	44	1.52	0.75; 3.08
25–<50	7479	11	1.85	0.73; 4.67
50–<100	4016	5	1.88	0.58; 6.16
≥100	1194	3	2.68	0.66; 10.9
Totals and *p* for trend	109,019	139		0.69 (+)
HR (95% CI) at 100 mGy			1.10	0.70; 1.73
ERR (95% CI) at 100 mGy			0.09	−0.36; 0.55
Lung cancer				
Lung dose (mGy)				
<5	43,090	359	1.0	REF
5–<10	26,259	190	0.94	0.75; 1.19
10–<25	28,629	182	0.87	0.68; 1.12
25–<50	6,977	67	1.27	0.91; 1.79
50–<100	3,108	35	1.29	0.85; 1.96
≥100	956	17	1.34	0.78; 2.32
Totals and *p* for trend	109,019	850		0.026 (+)
HR (95% CI) at 100 mGy			1.15	1.02; 1.31
ERR (95% CI) at 100 mGy			0.14	0.02; 0.27

Adjustments to annual doses to account for the potential for missed doses were performed by increasing each annual dose to a minimum value of 0.4 mSv and adding 0.2 mSv for annual doses that are between 0.4 and 1 mSv (NCRP 2020). Internal cohort dose-response analyses were then conducted for leukemia other than CLL and for lung cancer. Models include radiation doses lagged by 2 y for leukemia other than CLL and 10 y for lung cancer. Doses were analyzed as a time-dependent measure. Outcomes include deaths from underlying and, for leukemia, the contributing causes of death. All models adjusted for sex, year of birth (categories as shown in Table 1), and occupational category (categories as shown in Table 1) taken as a measure of socioeconomic status. *p*-value for test for linear trend in the hazards ratio computed for continuous organ dose. *p*-values are two-sided. Abbreviations: REF: referent category; HR: hazards ratio; CI: confidence interval; ERR: excess relative rate.

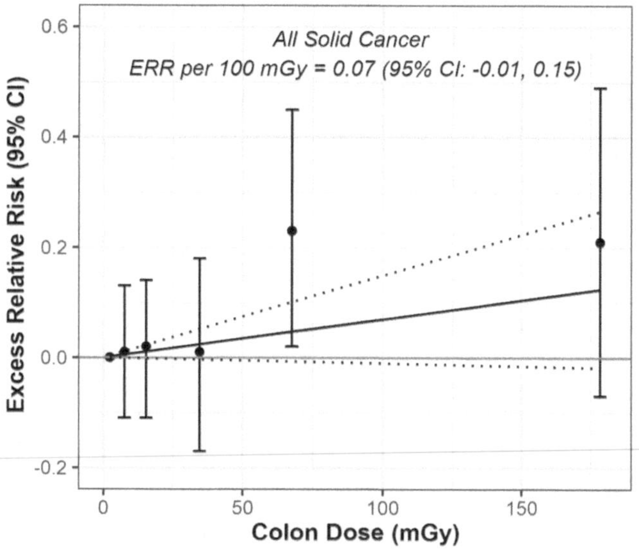

Figure 6. Excess Relative Risk (ERR) of All Solid Cancers by radiation dose to colon among 109,019 medical radiation workers first monitored between 1965–1994 in the United States and followed through 2016. A linear model is presented (solid line) based on a continuous measure of dose, and 95% confidence bands (dotted lines) presented about the regression line. Doses were lagged by 10 y and adjustment was made for sex, year of birth and socioeconomic status (taken as occupational category). ERRs and 95% confidence limits are calculated for selected categories of dose to the colon and displayed at the mean brain dose in each category. Categories <5 (referent), 5-, 10-, 25-, 50-, and ≥100 with respective mean doses of 2.2, 7.5, 15.4, 34.6, 67.7, and 178.5 mGy

of deaths, $n = 91$: the ERR per 100 mGy was 0.08 (95% CI −0.50, 0.66). Consistent with recent MPS studies (Boice, Cohen et al. 2020; Boice, Cohen, Mumma, Golden et al. 2021; Boice, Cohen, Mumma, Hagemeyer et al. 2021) and with the INWORKS investigation (Richardson et al. 2018), there was little evidence of a dose-response for cancer of the prostate. In contrast, a positive dose-response was reported among incident prostate cancers among Japanese atomic bomb survivors (Mabuchi et al. 2021). There were no significant dose responses seen for multiple myeloma or stomach cancers, but the numbers of deaths were relatively small, and the confidence limits were wide so that only relatively high levels of risk could be rejected with 95% confidence.

All solid cancers

The ERR per 100 mGy colon dose was 0.07 (95%CI −0.01, 0.15) for all solid cancers (Figure 6), and was 0.03 (95%CI −0.08, 0.14) with the exclusion of lung cancer. These estimates are consistent with the estimate of 0.042 (95% CI 0.032, 0.053) reported among the Japanese atomic bomb survivors (Ozasa et al. 2012). Interpreting the combined outcome of all solid cancers is problematic in part because of the heterogeneity of background rates for individual cancers and the differences seen in the dose-response when selected cancers are removed from combined cancer analyses (Cologne et al. 2019). Biological plausibility is another reason to be cautious because of the many differences in etiologies for individual cancer such as smoking for lung cancer, hormonal factors for breast cancer, viruses for cervical cancers; the differences between cancers in response to age at exposure and attained age; and the combining of cancers that are not consistently increased following radiation or are seen only after high, therapeutic, doses (NCRP 2012). With these cautions in mind, solid cancer risk estimates are often used in summarizing overall risks in exposed populations (Hauptmann et al. 2000).

Occupational categories

The four occupational categories covered a broad range of professions, including physicians and technologists, nuclear pharmacists, medical radiation physicists, nurses and allied healthcare support workers. A small number of workers, 6,087 (5.6%), did not fall into a distinct medical specialty and were classified as 'other'. Although some held medical-affiliated jobs such as radiopharmaceutical delivery and nearly 4% had doctoral titles (Dr or MD), many were in nonmedical professions. They had to satisfy the same inclusion criteria as the medical professionals, i.e. monitored for more than 2 y within the Landauer database with a cumulative badge dose of more than 10 mSv and not in another MPS study. The other workers, however, were older on average than the medical workers, had longer coverage within the Landauer database and thus a slightly broader dose distribution, and had higher SMRs in part because of the longer follow-up and associated diminution of the healthy worker effect over time since first employment. These differences were particularly notable for breast cancer. The rationale for inclusion was to increase statistical precision in the analyses, and adjusting for the occupational category was made to address any dissimilarities. The dose-response relationships and estimates of radiation risk for lung cancer among the four occupational categories were compatible.

Analyses conducted excluding the other workers were consistent with analyses for the full cohort, although the risk estimates were less precise (wider confidence intervals). For lung cancer, the ERR per 100 mGy for the full cohort was 0.15 (95% CI 0.02,0.27) and 0.38 (0.04,0.72) when the other workers were excluded. For leukemia other than CLL, the ERR per 100 mGy for the full cohort was 0.10 (95% CL −0.34,0.54) and −0.09 (95% CI −1.39,1.21) when the other workers were excluded.

Strengths and limitations

The study of U.S. medical radiation workers monitored between 1965 and 1994 and followed through 2016 has a number of strengths. The study size of 109,019 healthy workers is large, and the over 50,000 female workers comprise one of the largest occupational studies conducted to evaluate and quantify radiation effects among women. The duration of radiation monitoring with personal dosimeters within the Landauer database was long, 23.7 y on average and ranged up to 60 y, providing exceptional documentation of annual exposure measurements. The comprehensive program to develop organ-specific estimates of radiation dose from badge doses available for all individual workers is another study strength.

The mortality follow-up was complete and was up to 50 y. The determination of causes of death from the NDI and state mortality indices assured accurate, unbiased and consistent measures. The cause of death was known for 97.7% of all workers who had died. A comprehensive dosimetry program coordinated by the NCRP provided high-quality and precise organ dose estimates for medical radiation workers (Dauer et al. 2018; Yoder et al. 2018, 2021; NCRP 2018a, 2020). The importance of the organ-specific dosimetry programs, and in particular accounting for the use of lead aprons and the energy of the incident radiation, is seen by the substantial reduction in the organ-specific dose distributions compared with the uncorrected distribution of badge doses, i.e. $H_P(10)$. There was essentially no confounding influence related to possible neutron exposures or intakes of radionuclides. The cross-evaluation of the dosimetry data from a large medical center validated the accuracy of the information obtained from the Landauer digital dosimetry files that formed the basis of the epidemiologic study (cf. Dauer et al. 2019). There was a near-complete determination of occupational doses from all places of employment other than as recorded in the Landauer dosimetry database.

One advantage over previous studies of medical radiation workers is the ability to estimate organ-specific doses based on nearly complete annual records of badge doses for all workers. The selection of the cohort was dependent upon a worker having been monitored as a Landauer client for many years which assured that reasonably complete exposure measurements were available. This was confirmed in gap analyses of dosimetry coverage where the mean number of years with dosimetry measurements was 23.7 y. In a sensitivity analysis, doses that might have been missed because of incomplete coverage were imputed but these additional doses made little difference in the dose-response analyses and estimates of risk. In another sensitivity analysis, we evaluated whether adjusting for duration of monitoring made a difference in the radiation risk estimates or the precision of the estimates; there were no meaningful differences.

Limitations include the absence of individual risk information on important live style factors such as cigarette smoking; a relatively narrow distribution of radiation dose that reduced the statistical ability to estimate radiation organ-specific risks with precision (Gilbert et al. 2020); the possible impact of measurement errors and missing doses on risk estimation; the relatively young age of the population at last follow-up; and the need for additional and more rigorous uncertainty analyses.

SES adjustment. The importance of adjusting for SES as a surrogate for tobacco use and other lifestyle behaviors was reflected in the SMRs for lung cancer. Education is a strong predictor of smoking habits, and adjustment is considered a robust way to control indirectly for smoking (Cohen et al. 2018). Physicians and medical workers, in general, have more college and graduate degrees compared with the general population, and thus would be expected to smoke less and have lower rates of smoking-related deaths such as lung cancer. A ranking of educational achievements from highest to lowest by occupational category would be medical doctor and trained specialists (interventional fluoroscopy and nuclear medicine & radiation oncology), general radiography (the preponderance being technologists), and others. The corresponding SMRs for lung cancer are 0.36 and 0.43, 0.51 and 0.90 (Table 4) and suggest a credible correlation between occupational category and SES status, and the adequacy of the adjustment for tobacco use. A significant dose-response was found for lung cancer and the estimates of radiation risk did not differ among the occupational categories. Similarly, the SMRs for IHD revealed the same pattern: 0.38 and 0.44 (medical doctor and specialists), 0.57 (general radiograph), and 0.83 (other).

Reduced statistical power. Accounting for the wearing of lead aprons among interventional radiologists/cardiologists and the unanticipated large number of errant doses that had to be excluded were the reasons for the reduction in individual doses and accordingly in statistical power. Nonetheless, a statistically significant association between cumulative absorbed dose to the lung and lung cancer risk was observed; reflecting the large number of workers studied, the accuracy of the organ-specific doses, and the relatively large number of deaths attributed to lung cancer. The lower doses received by women compared with men reduced the ability to demonstrate convincingly a difference in sex-specific risk estimates for lung cancer, and the interaction did not reach the level of statistical significance ($p = .062$). The pooling of sex-specific lung cancer risks in other MPS studies will help clarify any differences between men and women as to their radiation risk for lung cancer (Boice, Ellis et al. 2019).

Young cohort. Another limitation is the relatively young age of the cohort, i.e. 89.9% of the medical workers were alive at the last follow-up and nearly 40% were under age 60 y at the last follow-up. Additional follow-up will provide a more comprehensive evaluation of lifetime risks.

Possible unrecorded doses and gap analysis. While the approach to dosimetry was sophisticated and carefully crafted, relying on a dosimetry database whose primary use was commercial raises the possibility of missed doses, incorrect doses and inaccurate doses. To evaluate the possibility of unrecorded doses, we conducted gap analyses indicating that within the Landauer, Inc. database, there were few workers with gaps in occupational monitoring data and the differences did not vary between men and women. The accuracy of coverage was confirmed in part with the consistency between the Landauer digital database with the dosimetry records of a large medical facility with records going back to 1970. Compared with studies of workers at nuclear installations where the adequacy of recorded doses received in the early years of work in the 1940s and 1950s may be a concern (Wakeford 2021), the potential for missed photon doses appears less likely among medical workers first monitored in 1965 with a mean number of years with dosimetry measurements of 23.7 y.

Adjustment for missed dose. Any adjustments for 'missed dose', i.e. the dose a worker may have received but was less than the minimum detection or recording level of the dosimetry system, would primarily impact the least exposed members of the cohort because they have the largest number of measurements that yielded results less than the minimum detection or recording limit (NCRP 2020). Following the guidance of NCRP Commentary No. 30 (NCRP 2020), sensitivity analyses that included additions to annual doses to account for the potential for such missed dose were performed by increasing each annual dose to a minimum value of 0.4 mSv and adding 0.2 mSv for annual doses that are between 0.4 and 1 mSv. Such adjustments did not influence the dose responses reported in this study (Table 8). For lung cancer, the ERR per 100 mGy was 0.15 (95% CI 0.02;0.27) which became 0.14 (95% CI 0.02;0.27) after adjusting for the missed dose. For leukemia other than CLL, the ERR per 100 mGy was 0.10 (95% CI −0.34; 0.54) which became 0.09 (95% CI (−0.36; 0.55)) after adjusting for the missed dose.

Intentionally not wearing dosimeters. Another possible source of missed dose readings is intentional acts of not wearing dosimeters in order to remain below radiation protection limits. Conceivably, this might occur among physicians who needed to perform as many fluoroscopic procedures as possible to reach specialist qualification status. While this is a possibility noted by some institutions, the effect would not be expected to significantly affect the study findings. Such a habit would reduce the observed lifetime dose values, placing the physician incorrectly into a lower dose group. This placement would minimize the chance that the physician would be included in our cohort, either because his or her lifetime dose was below that necessary for inclusion based on the eligibility criteria or that the lifetime dose level was such that inclusion would have arisen strictly on the probability of being randomly selected. As the number of physicians (about 8%) was less than the number of non-physicians (about 92%), the probability of inclusion would be small.

While worker scenarios were developed that accounted for the energy of the incident photons, geometric orientation, the use of a lead apron, missed and unrecorded doses and other dosimetric assessment issues, additional and more rigorous uncertainty analyses are warranted. Most of the dosimetric uncertainty arises from the conversion of measurements made from a dosimeter with relatively known amounts of uncertainty to the organ dose that depends more on knowledge regarding the attributes of the person and conditions of exposure than on the measurement per se. For example, the assumption that worker morphology conforms to that assumed by the ICRP mathematical models used to derive conversion coefficients would be another shared error.

Nonoccupational sources of radiation exposure. The associations presented are with occupational radiation exposure and do not consider all sources of exposure such as from ubiquitous natural background radiation or personal exposures to medical radiation. It is assumed that such unknown exposures do not vary by occupational dose category, but this is nearly impossible to verify. The NCRP (Report 160, NCRP 2009) has estimated that natural background radiation and medical exposures contribute about 6 mSv effective dose annually to the general population further challenging the conduct and interpretation of future epidemiologic studies of low-dose exposures that might be initiated.

Value of cancer incidence data. High-quality cancer incidence data, when available, would be preferable to mortality data because the number of cases and the quality of the diagnoses would be higher. Unfortunately, the U.S. does not have a national cancer incidence registry that would cover workers employed in the radiation professions over the years 1965–2016. Many of the cancers of interest, e.g. of the lung and leukemia, have a high fatality rate over the years of study so that mortality might reflect incidence closely, as seen generally in studies of the Japanese atomic bomb survivors (Ron et al. 1994) and workers employed at the Mound nuclear facility in Ohio (Boice et al. 2014). Cancers such as the thyroid that have a low fatality rate can be evaluated in mortality studies, but the statistical power to identify significant increases would be low. Among atomic bomb survivors, the estimates of risk were generally higher for incidence than mortality, and it was recommended that both types of analyses be conducted since the end points provided complementary information for risk assessment (Ron et al. 1994). In the future, comprehensive cancer incidence data is planned to be obtained for all radiation workers within the MPS through linkage to the Centers for Medicine and Medicaid Services (CMS) claims data (Cooper et al. 1999; Mues et al. 2017). While these data are essentially population-based with minimal selection bias with respect to socioeconomic status or other demographic features, the coverage begins only in 1999 and is only for those who have reached the age of 65 y and qualify in most part for Medicare coverage.

Summary

The study of 109,019 medical radiation workers in the United States was designed to evaluate any differences in the risk of radiation-related lung cancer between women and men. Overall, radiation was correlated with an increased lung cancer risk with the association statistically strongest for male workers. Although male workers had nearly 2 times the radiation risk of lung cancer compared with female workers on a relative scale, the difference was not statistically significant. These data indicating no significant sex difference in radiation-related lung cancer risk, in combination with other large-scale studies conducted within the MPS, should provide direction to NASA with regard to their lifetime radiation risk projection models currently used in providing protection guidance for flight crews on long-duration space missions. The absence of a significant radiation association for leukemia other than CLL is consistent with recent studies of medical radiation workers but not earlier studies of pioneering radiologists or other large-scale occupational studies within the MPS. The differences may reflect the narrow and lower dose distributions for medical radiation workers which limit statistical stability or, perhaps, point to a lower risk following cumulative exposures under 100 mGy than predicted based on a linear model. The absence of radiation risk for IHD is consistent with previous MPS cohorts with broader dose distributions. The larger number of deaths attributed to IHD in the current and previous studies within the MPS at low doses received gradually over time supports the notion that any radiation risk, if it exists, at these dose levels may be very small. The MPS will be able to provide substantial information on the level of radiation risk for specific causes of death as the individual cohorts are combined.

Acknowledgments

Special thanks to Chris Passmore, Mark Salasky and colleagues at Landauer Inc. who provided essential advice and guidance in understanding, using and accessing the dosimetry database used for defining the cohort selected for study, including the recording of personal dose equivalents (badge doses) over the more than 50 y of individual dosimetry data available for dose reconstruction. We are grateful for the detailed expert guidance on the dosimetric aspects of this work as provided by the members of the National Council on Radiation Protection and Measurements Scientific Committees SC 6-9 (NCRP, Report 178) and SC 6-11 (NCRP Commentary 30). We especially thank Daniel Miowdownik, Memorial Sloan Kettering Cancer Center for his assistance in the validation efforts. We also are indebted to the Department of Energy (Nimi Rao), the Nuclear Regulatory Commission (Doris Lewis), Oak Ridge Associated Universities (Derek A. Hagemeyer) and the Naval Dosimetry Center (CAPT Thad Sharp and LCDR Nakima McCormack) for facilitating linkages with their respective dosimetry files. The results presented herein represent the conclusions and opinions solely of the authors. Its publication does not imply endorsement by the National Council on Radiation Protection and Measurements, Vanderbilt University or any of the acknowledged agencies.

Disclosure statement

Most of the data used to reconstruct doses for the medical radiation worker cohort arose from measurements made by Landauer, Inc. and its predecessors over a period in excess of 50 y. Dr. R. Craig Yoder, a former long-term employee of Landauer, Inc and now retired was chair of the NCRP Scientific Committee that recently completed Commentary 30 on dosimetry guidance for medical workers. He contributed to the assembling, evaluation and interpretation of the recorded doses used in the analyses as well as assuring the historical accuracy of both technical and administrative data, His participation does not reflect any endorsement of the commercial offerings of Landauer by the NCRP, and he received no compensation from Landauer with regard to any aspect of the research reported herein. Further, he attests that his former associations had no influence on the scientific accuracy, or any other aspect of the work reported here. The other authors report no conflicts of interest. The authors alone are responsible for the content and writing of the paper.

Funding

The study of the medical workers, a component of the Million Person Study, was supported in part by a research grant from the U.S. Department of Energy (DOE) [Grant No. DE-SC0008944] awarded to the National Council on Radiation Protection and Measurements, which included interagency support from the U.S. Nuclear Regulatory Commission (NRC), the U.S. Environmental Protection Agency and the National Aeronautics and Space Administration (NASA); and two more recent DOE Grants [No. DE-AU0000042 and DE-AU0000046]. Additional support included grants from the NRC [NRC-HQ-60-14-G-0011]; the Centers for Disease Control and Prevention [5UE1EH000989, 5NUE1EH001315]; NASA [NNX15AU88G, 80NSSC17M0016]. Contract support also was received from the Naval Sea Systems Command [N00024-17-C-4322] for dosimetry linkages.

ORCID

John D. Boice http://orcid.org/0000-0002-8755-1299
Sarah S. Cohen http://orcid.org/0000-0003-0421-1983
Michael T. Mumma http://orcid.org/0000-0001-7506-8710
Sara C. Howard http://orcid.org/0000-0002-9290-4319
R. Craig Yoder http://orcid.org/0000-0002-5132-967X
Lawrence T. Dauer http://orcid.org/0000-0002-5629-8462

References

Anderson JL, Bertke SJ, Yiin J, Kelly-Reif K, Daniels RD. 2021. Ischaemic heart and cerebrovascular disease mortality in uranium enrichment workers. Occup Environ Med. 78(2):105–111.

Armstrong BG. 1995. Comparing standardized mortality ratios. Ann Epidemiol. 5(1):60–64.

Azizova TV, Bannikova MV, Grigoryeva ES, Rybkina VL, Hamada N. 2020. Occupational exposure to chronic ionizing radiation increases risk of Parkinson's disease incidence in Russian Mayak workers. Int J Epidemiol. 49(2):435–447.

Berrington A, Darby SC, Weiss HA, Doll R. 2001. 100 years of observation on British radiologists: mortality from cancer and other causes 1897–1997. Br J Radiol. 74(882):507–519.

Boice JD, Jr. 2006. Ionizing radiation. In Schottenfeld D, Fraumeni JF Jr, editors. Cancer epidemiology and prevention. 3rd ed. New York (NY): Oxford University Press; p. 259–293.

Boice JD, Jr. 2017. The linear nonthreshold (LNT) model as used in radiation protection: an NCRP update. Int J Radiat Biol. 93(10): 1079–1092.

Boice JD, Jr. 2019a. The Million Person Study relevance to space exploration and Mars. Int J Radiat Biol. 1–9. DOI:10.1080/09553002.2019.1589020.

Boice JD, Jr. 2019b. NCRP vision for the future and program area committee activities in 2018. Health Phys. 116(2):282–294.

Boice JD, Jr, Cohen SS, Mumma MT, Chen H, Golden AP, Beck HL, Till JE. 2020. Mortality among US military participants at eight aboveground nuclear weapons test series. Int J Radiat Biol. 1–22. DOI:10.1080/09553002.2020.1787543.

Boice JD, Jr, Cohen SS, Mumma MT, Ellis ED. 2019. The Million Person Study, whence it came and why. Int J Radiat Biol. 1–14. DOI:10.1080/09553002.2019.1589015.

Boice JD, Jr, Cohen SS, Mumma MT, Ellis ED, Cragle DL, Eckerman KF, Wallace PW, Chadda B, Sonderman JS, Wiggs LD, et al. 2014. Mortality among Mound workers exposed to polonium-210 and other sources of radiation, 1944–1979. Radiat Res. 181(2):208–228.

Boice JD, Jr, Cohen SS, Mumma MT, Ellis ED, Eckerman KF, Leggett RW, Boecker BB, Brill AB, Henderson BE. 2011. Updated mortality analysis of radiation workers at Rocketdyne (Atomics International), 1948–2008. Radiat Res. 176(2):244–258.

Boice JD, Jr, Cohen SS, Mumma MT, Golden AP, Howard S, Girardi DJ, Ellis ED, Bellamy M, Dauer LT, Samuels C, et al. 2021. Mortality among workers at the Los Alamos National Laboratory, 1943–2017. Int J Radiat Biol. 1–28. DOI:10.1080/09553002.2021.1917784.

Boice JD, Jr, Cohen SS, Mumma MT, Golden AP, Howard SC, Girardi DJ, Ellis ED, Bellamy MB, Dauer LT, Eckerman KF, Leggett RW. 2021. Mortality among Tennessee Eastman Corporation (TEC) uranium processing workers, 1943–2017. Int J Radiat Biol. (in press).

Boice JD, Jr, Cohen SS, Mumma MT, Hagemeyer DA, Chen H, Golden AP, Yoder RC, Dauer LT. 2021. Mortality from leukemia, cancer and heart disease among U.S. nuclear power plant workers. Int J Radiat Biol. 1957–2011. DOI:10.1080/09553002.2021.1967507.

Boice JD, Jr, Dauer LT. 2021. Million Person Study of low-dose radiation health effects. Trans Am Nucl Soc. (in press).

Boice JD, Jr, Dauer LT, Kase KR, Mettler FA, Jr, Vetter RJ. 2020. Evolution of radiation protection for medical workers. Br J Radiol. 93(1112):20200282.

Boice JD, Jr, Ellis ED, Golden AP, Zablotska LB, Mumma MT, Cohen SS. 2019. Sex-specific lung cancer risk among radiation workers in the million-person study and patients TB-Fluoroscopy. Int J Radiat Biol. 1–12. DOI:10.1080/09553002.2018.1547441.

Boice JD, Held KD, Shore RE. 2019. Radiation epidemiology and health effects following low-level radiation exposure. J Radiol Prot. 39(4): S14–S27.

Boice JD, Jr, Leggett RW, Ellis ED, Wallace PW, Mumma M, Cohen SS, Brill AB, Chadda B, Boecker BB, Yoder RC, et al. 2006. A comprehensive dose reconstruction methodology for former Rocketdyne/Atomics International radiation workers. Health Phys. 90(5):409–430.

Boice JD, Jr, Mandel JS, Doody MM. 1995. Breast cancer among radiologic technologists. JAMA. 274(5):394–401.

Boice JD, Jr, Mandel JS, Doody MM, Yoder RC, McGowan R. 1992. A health survey of radiologic technologists. Cancer. 69(2):586–598.

Boice JD, Jr, Quinn B, Ansari A, Blake PK, Blattnig SR, Caffrey EA, Cohen SS, Golden AP, Held KD, Jokisch DW, et al. 2021. A million persons, a million dreams: a vision for a National Center of Radiation Epidemiology and Biology. Int J Radiat Biol. DOI:10.1080/09553002.2021.1988183.

Bouville A, Toohey RE, Boice JD Jr, Beck HL, Dauer LT, Eckerman KF, Hagemeyer D, Leggett RW, Mumma MT, Napier B, et al. 2015. Dose reconstruction for the Million Worker Study: status and guidelines. Health Phys. 108(2):206–220.

Breslow NE, Day NE. 1987. Statistical methods in cancer research. Vol. 2. The design and analysis of cohort studies. Publication 82. Lyon (FR): International Agency for Research on Cancer.

Buckley JP, Keil AP, McGrath LJ, Edwards JK. 2015. Evolving methods for inference in the presence of healthy worker survivor bias. Epidemiology. 26(2):204–212.

Cahoon EK, Preston DL, Pierce DA, Grant E, Brenner AV, Mabuchi K, Utada M, Ozasa K. 2017. Lung, laryngeal and other respiratory cancer incidence among Japanese atomic bomb survivors: an updated analysis from 1958 through 2009. Radiat Res. 187(5): 538–548.

Campbell KM, Deck D, Krupski A. 2008. Record linkage software in the public domain: a comparison of Link Plus, The Link King, and a 'basic' deterministic algorithm. Health Informatics J. 14(1):5–15.

Charpentier P, Ostfeld AM, Hadjimichael OC, Hester R. 1993. The mortality of US nuclear submariners, 1969–1982. J Occup Med. 35: 501–509.

Cohen SS, Mumma MT, Ellis ED, Boice JD, Jr. 2018. Validating the use of census data on education as a measure of socioeconomic status in an occupational cohort. Int J Radiat Biol. 1–10 DOI:10.1080/09553002.2018.1549758.

Cologne J, Kim J, Sugiyama H, French B, Cullings HM, Preston DL, Mabuchi K, Ozasa K. 2019. Effect of heterogeneity in background incidence on inference about the solid-cancer radiation dose response in atomic bomb survivors. Radiat Res. 192(4):388–398.

Cooper GS, Yuan Z, Stange KC, Dennis LK, Amini SB, Rimm AA. 1999. The sensitivity of Medicare claims data for case ascertainment of six common cancers. Med Care. 37(5):436–444.

Court-Brown WM, Doll R. 1958. Expectation of life and mortality from cancer among British radiologists. Br Med J. 2(5090):181–187.

Cousins C, Miller DL, Bernardi G, Rehani MM, Schofield P, Vañó E, Einstein AJ, Geiger B, Heintz P, Padovani R, et al. 2013. ICRP publication 120: radiological protection in cardiology. Ann ICRP. 42(1): 1–125.

Cox D. 1972. Regression models and life-tables. J R Stat Soc Ser B. 34(2):187–220.

Cutter DJ, Taylor CW, Rahimi K, McGale P, Ferreira V, Burrage M, Vivekananda S, Hawkins M, Darby S. 2019. Effects of radiation therapy on the cardiovascular system. In: Ewer MS, editor. Cancer and the heart. 3rd ed. Cary (NC): PMPH USA; p. 119–138.

Dauer LT, Bouville A, Toohey RE, Boice JD, Jr, Beck HL, Eckerman KF, Hagemeyer D, Leggett RW, Mumma MT, Napier B, et al. 2018. Dosimetry and uncertainty approaches for the million person study of low-dose radiation health effects: overview of the recommendations in NCRP Report No. 178. Int J Radiat Biol. 1–10. DOI:10.1080/09553002.2018.1536299.

Dauer LT, Woods M, Miodownik D, Serencsits B, Quinn B, Bellamy M, Yoder C, Liang X, Boice JD, Jr, Bernstein J. 2019. Cohort profile - MSK radiation workers: a feasibility study to establish a deceased worker sub-cohort as part of a multicenter medical radiation worker component in the Million Person Study of Low-Dose Radiation Health Effects. Int J Radiat Biol. 1–7. DOI:10.1080/09553002.2019.1587194

Defense Threat Reduction Agency (DTRA). 2020. Design Aspects of DTRA's NuTRIS-Web Software Application. Fort Belvoir (VA): DTRA-TR-19-039. [accessed 2021 Jan 23]. https://apps.dtic.mil/sti/pdfs/AD1093055.pdf

DiSantis DJ. 1986. Early American Radiology: the pioneer years. AJR Am J Roentgenol. 147(4):850–853.

Doody MM, Freedman DM, Alexander BH, Hauptmann M, Miller JS, Rao RS, Mabuchi K, Ron E, Sigurdson AJ, Linet MS. 2006. Breast cancer incidence in U.S. radiologic technologists. Cancer. 106(12): 2707–2715.

Electric Power Research Institute (EPRI). 2020. Cardiovascular risks from low dose radiation exposure: review and scientific appraisal of the literature. Palo Alto(CA): EPRI. [accessed 2021 Jul 31]. https://www.epri.com/research/products/000000003002018408

Ellis ED, Boice JD, Golden AP, Girardi DJ, Cohen SS, Mumma MT, Shore RE, Leggett RW, Kerr GD. 2018. Dosimetry is key to good epidemiology: workers at Mallinckrodt Chemical Works had seven different source exposures. Health Phys. 114(4):386–397.

Friedman-Jimenez G, Kato I, Factor-Litvak P, Shore R. 2021. Mortality of enlisted men who served on nuclear-powered submarines in the United States Navy. J Occup Environ Med. DOI:10.1097/JOM.0000000000002364.

Furukawa K, Preston DL, Lonn S, Funamoto S, Yonehara S, Matsuo T, Egawa H, Tokuoka S, Ozasa K, Kasagi F, et al. 2010. Radiation and smoking effects on lung cancer incidence among atomic bomb survivors. Radiat Res. 174(1):72–82.

Furukawa K, Preston DL, Misumi M, Cullings HM. 2017. Handling incomplete smoking history data in survival analysis. Stat Methods Med Res. 26(2):707–723.

Gilbert ES, Little MP, Preston DL, Stram DO. 2020. Issues in interpreting epidemiologic studies of populations exposed to low-dose, high-energy photon radiation. J Natl Cancer Inst Monogr. 2020(56):176–187.

Gillies M, Kuznetsova I, Sokolnikov M, Haylock R, O'Hagan J, Tsareva Y, Labutina E. 2017. Lung cancer risk from plutonium: a pooled analysis of the Mayak and Sellafield worker cohorts. Radiat Res. 188(6):645–660.

Golden AP, Cohen SS, Chen H, Ellis ED, Boice JD, Jr. 2018. Evaluation of statistical modeling approaches for epidemiologic studies of low dose radiation health effects . Int J Radiat Biol. 1–26 DOI:10.1080/09553002.2018.1554924.

Golden AP, Ellis ED, Cohen SS, Mumma MT, Leggett RW, Wallace PW, Girardi DJ, Watkins JP, Shore RE, Boice JD, Jr. 2019. Updated mortality analysis of the Mallinckrodt uranium processing workers, 1942–2012. Int J Radiat Biol. 1–21 DOI:10.1080/09553002.2019.1569773.

Golden AP, Milder CM, Ellis ED, Anderson JL, Boice JD, Jr, Bertke SJ, Zablotska LB. 2021. Cohort profile: four early uranium processing facilities in the US and Canada. Int J Radiat Biol. 97(6):833–847.

Hauptmann M, Daniels RD, Cardis E, Cullings HM, Kendall G, Laurier D, Linet MS, Little MP, Lubin JH, Preston DL, et al. 2020. Epidemiological studies of low-dose ionizing radiation and cancer: summary bias assessment and meta-analysis. J Natl Cancer Inst Monogr. 2020(56):188–200.

Hauptmann M, Mohan AK, Doody MM, Linet MS, Mabuchi K. 2003. Mortality from diseases of the circulatory system in radiologic technologists in the United States. Am J Epidemiol. 157(3):239–248.

Hsu WL, Preston DL, Soda M, Sugiyama H, Funamoto S, Kodama K, Dohy H, Tomonaga M, Iwanaga M, Miyazaki Y, et al. 2013. The incidence of leukemia, lymphoma and multiple myeloma among atomic bomb survivors: 1950–2001. Radiat Res. 179(3):361–382.

Jablon S, Miller R. 1978. Army technologists: 29-year follow up for cause of death. Radiology. 126(3):677–679.

Karellas A. 2020. Challenges in dosimetry and radiation dose trends. Radiology. 294(2):360–361.

Kitahara CM, Linet MS, Balter S, Miller DL, Rajaraman P, Cahoon EK, Velazquez-Kronen R, Simon SL, Little MP, Doody MM, et al. 2017. Occupational radiation exposure and deaths from malignant intracranial neoplasms of the brain and CNS in U.S. radiologic technologists 1983–2012. AJR Am J Roentgenol. 208(6):1278–1284.

Kitahara CM, Linet MS, Drozdovitch V, Alexander BH, Preston DL, Simon SL, Freedman DM, Brill AB, Miller JS, Little MP, et al. 2015. Cancer and circulatory disease risks in US radiologic technologists associated with performing procedures involving radionuclides. Occup Environ Med. 72(11):770–776.

Linet MS, Kim KP, Miller DL, Kleinerman RA, Simon SL, Berrington de Gonzalez A. 2010. Historical review of occupational exposures and cancer risks in medical radiation workers. Radiat Res. 174(6): 793–808.

Linet MS, Little MP, Kitahara CM, Cahoon EK, Doody MM, Simon SL, Alexander BH, Preston DL. 2020. Occupational radiation and haematopoietic malignancy mortality in the retrospective cohort study of US radiologic technologists, 1983–2012. Occup Environ Med. 77(12):822–831.

Liu JJ, Freedman DM, Little MP, Doody MM, Alexander BH, Kitahara CM, Lee T, Rajaraman P, Miller JS, Kampa DM, et al. 2014. Work history and mortality risks in 90,268 US radiological technologists. Occup Environ Med. 71(12):819–835.

Mabuchi K, Preston DL, Brenner AV, Sugiyama H, Utada M, Sakata R, Sadakane A, Grant EJ, French B, Cahoon EK, et al. 2021. Risk of prostate cancer incidence among atomic bomb survivors: 1958–2009. Radiat Res. 195(1):66–76.

March H. 1944. Leukemia in radiologists. Radiology. 43(3):275–278.

Matanoski GM, Sartwell PE Elliott Ea Tonascia J, Sternberg A. 1984. Cancer risks in radiologists and radiation workers. In: Boice JD Jr, Fraumeni JF Jr, editors. Radiation carcinogenesis: epidemiology and biological significance. New York (NY): Raven Press; p. 83–96.

Matanoski GM, Seltser R, Sartwell PE, Diamond EL, Elliott EA. 1975. The current mortality rates of radiologists and other physician specialists: deaths from all causes and from cancer. Am J Epidemiol. 101(3):188–198.

Mohan AK, Hauptmann M, Linet MS, Ron E, Lubin JH, Freedman DM, Alexander BH, Boice JD, Jr, Doody MM, Matanoski GM. 2002. Breast cancer mortality among female radiologic technologists in the United States. J Natl Cancer Inst. 94(12):943–948.

Monson R. 1986. Observations on the healthy worker effect. J Occup Med. 28(6):425–433.

Mues KE, Liede A, Liu J, Wetmore JB, Zaha R, Bradbury BD, Collins AJ, Gilbertson DT. 2017. Use of the Medicare database in epidemiologic and health services research: a valuable source of real-world evidence on the older and disabled populations in the US. CLEP. Volume 9:267–277.

Mumma MT, Cohen SS, Sirko JL, Ellis ED, Boice JD Jr. 2018. Obtaining vital status and cause of death on a million persons. Int J Radiat Biol. 1–21. DOI:10.1080/09553002.2018.1539884.

National Academies of Sciences, Engineering, and Medicine (NASEM). 2021. Space radiation and astronaut health: managing and communicating cancer risks. Washington (DC): The National Academies Press. https://doi.org/10.17226/26155.

National Council on Radiation Protection and Measurements (NCRP). 2009. Ionizing radiation exposure of the population of the United States. Bethesda (MD): NCRP. NCRP Report 160.

National Council on Radiation Protection and Measurements (NCRP). 2010. Radiation dose management for fluoroscopically-guided interventional medical procedures. Bethesda (MD): NCRP. NCRP Report 168.

National Council on Radiation Protection and Measurements (NCRP). 2012. Uncertainties in the estimation of radiation risks and probability of disease causation. Bethesda (MD): NCRP. NCRP Report No. 171.

National Council on Radiation Protection and Measurements (NCRP). 2014. Radiation protection for space activities: supplement to previous recommendations. Bethesda (MD): NCRP. NCRP Commentary No. 23.

National Council on Radiation Protection and Measurements (NCRP). 2018a. Deriving organ doses and their uncertainty for epidemiologic studies (with a focus on the one million U.S. workers and veterans study of low-dose radiation health effects). Bethesda (MD): NCRP. NCRP Report 178.

National Council on Radiation Protection and Measurements (NCRP). 2018b. Implications of recent epidemiologic studies for the linear nonthreshold model and radiation protection. Bethesda (MD): NCRP. NCRP Commentary 27.

National Council on Radiation Protection and Measurements (NCRP). 2020. Dosimetry guidance for medical radiation workers with a focus on lung dose reconstruction. Bethesda (MD): NCRP. NCRP Commentary 30.

National Council on Radiation Protection and Measurements (NCRP). 2021. Scientific Committee 1-27: evaluation of sex-specific differences in lung cancer radiation risks and recommendations for use in transfer and projection models. [accessed 2021 Jul 31]. https://ncrponline.org/program-areas/sc-1-27-evaluation-of-sex-specific-differences-in-lung-cancer-radiation-risks-and-recommendations-for-use-in-transfer-and-projection-models/

Naval Dosimetry Center (NDC). 2021. [accessed 2021 Nov 15]. https://www.cnic.navy.mil/regions/ndw/installations/nsa_bethesda/about/tenant_orgs/ndc.html

Ozasa K, Shimizu Y, Suyama A, Kasagi F, Soda M, Grant EJ, Sakata R, Sugiyama H, Kodama K. 2012. Studies of the mortality of atomic bomb survivors, report 14, 1950–2003: an overview of cancer and noncancer diseases. Radiat Res. 177(3):229–243.

Ozasa K, Takahashi I, Grant EJ, Kodama K. 2017. Cardiovascular disease among atomic bomb survivors. Int J Radiat Biol. 93(10): 1145–1150.

Pasqual E, Boussin F, Bazyka D, Nordenskjold A, Yamada M, Ozasa K, Pazzaglia S, Roy L, Thierry-Chef I, de Vathaire F, et al. 2021. Cognitive effects of low dose of ionizing radiation – lessons learned and research gaps from epidemiological and biological studies. Environ Int. 147:106295.

Picano E, Vañó E, Domenici L, Bottai M, Thierry-Chef I. 2012. Cancer and non-cancer brain and eye effects of chronic low-dose ionizing radiation exposure. BMC Cancer. 12(1):157.

Preston DL, Kitahara CM, Freedman DM, Sigurdson AJ, Simon SL, Little MP, Cahoon EK, Rajaraman P, Miller JS, Alexander BH, et al. 2016. Breast cancer risk and protracted low-to-moderate dose occupational radiation exposure in the US radiologic technologists cohort, 1983–2008. Br J Cancer. 115(9):1105–1112.

Preston DL, Lubin J, Pierce D, McConney M, Shilnikova N. 2015. Epicure risk regression and person-year computation software: command summary and user guide. Ottawa (CA): Risk Sciences International.

Preston DL, Mattsson A, Holmberg E, Shore R, Hildreth NG, Boice JD Jr. 2002. Radiation effects on breast cancer risk: a pooled analysis of eight cohorts. Radiat Res. 158(2):220–235.

Rajaraman P, Doody MM, Yu CL, Preston DL, Miller JS, Sigurdson AJ, Freedman DM, Alexander BH, Little MP, Miller DL, et al. 2016. Incidence and mortality risks for circulatory diseases in US radiologic technologists who worked with fluoroscopically guided interventional procedures, 1994–2008. Occup Environ Med. 73(1):21–27.

Richardson DB, Abalo K, Bernier MO, Rage E, Leuraud K, Laurier D, Keil AP, Little MP. 2020. Meta-analysis of published excess relative risk estimates. Radiat Environ Biophys. 59(4):631–641.

Richardson DB, Cardis E, Daniels RD, Gillies M, Haylock R, Leuraud K, Laurier D, Moissonnier M, Schubauer-Berigan MK, Thierry-Chef I, et al. 2018. Site-specific solid cancer mortality after exposure to ionizing radiation: a cohort study of workers (INWORKS). Epidemiology. 29(1):31–40.

Roguin A, Goldstein J, Bar O, Goldstein JA. 2013. Brain and neck tumors among physicians performing interventional procedures. Am J Cardiol. 111(9):1368–1372.

Ron E, Preston DL, Mabuchi K, Thompson DE, Soda M. 1994. Cancer incidence in atomic bomb survivors. Part IV: comparison of cancer incidence and mortality. Radiat Res. 137(2):S98–S112.

Rothman KJ, Boice JD. 1979. Epidemiologic analysis with a programmable calculator. NIH Publication 79-1649. Washington (DC): US Government Printing Office.

Sherwood T. 2001. 100 years' observation of risks from radiation for British (male) radiologists. Lancet. 358(9282):604.

Shore RE, Beck HL, Boice JD Jr, Caffrey EA, Davis S, Grogan HA, Mettler FA Jr, Preston RJ, Till JE, Wakeford R, et al. 2019. Recent epidemiologic studies and the linear no-threshold model for radiation protection-considerations regarding NCRP commentary 27. Health Phys. 116(2):235–246.

Simon SL. 2011. Organ-specific external dose coefficients and protective apron transmission factors for historical dose reconstruction for medical personnel. Health Phys. 101(1):13–27.

Simon SL, Preston DL, Linet MS, Miller JS, Sigurdson AJ, Alexander BH, Kwon D, Yoder RC, Bhatti P, Little MP, et al. 2014. Radiation organ doses received in a nationwide cohort of U.S. radiologic technologists: methods and findings. Radiat Res. 182(5):507–528.

Simon SL, Weinstock RM, Doody MM, Neton J, Wenzl T, Stewart P, Mohan AK, Yoder RC, Hauptmann M, Freedman DM, et al. 2006.

Estimating historical radiation doses to a cohort of U.S. radiologic technologists. Radiat Res. 166(1 Pt 2):174–192.

Smith PG, Doll R. 1981. Mortality from cancer and all causes among British radiologists. Br J Radiol. 54(639):187–194.

Sun Z, Inskip PD, Wang J, Kwon D, Zhao Y, Zhang L, Wang Q, Fan S. 2016. Solid cancer incidence among Chinese medical diagnostic x-ray workers, 1950–1995: estimation of radiation-related risks. Int J Cancer. 138(12):2875–2883.

Takahashi I, Shimizu Y, Grant EJ, Cologne J, Ozasa K, Kodama K. 2017. Heart disease mortality in the Life Span Study, 1950–2008. Radiat Res. 187(3):319–332.

Tapio S, Little MP, Kaiser JC, Impens N, Hamada N, Georgakilas AG, Simar D, Salomaa S. 2021. Ionizing radiation-induced circulatory and metabolic diseases. Environ Int. 146:106235.

Tran V, Zablotska LB, Brenner AV, Little MP. 2017. Radiation-associated circulatory disease mortality in a pooled analysis of 77,275 patients from the Massachusetts and Canadian tuberculosis fluoroscopy cohorts. Sci Rep. 7(1):44147.

U.S. Department of Energy (DOE). 2021. DOE 2019 Occupational radiation exposure. US Department of Energy, Office of Environment, Health, Safety and Security, The Radiation Exposure Monitoring System (REMS). [accessed 2021 Jul 31]. http://energy.gov/ehss/occupational-radiation-exposure

U.S. Nuclear Regulatory Commission (NRC). 1991. US Nuclear Regulatory Commission Regulations. Title 10 Code of Federal Regulations Part 20, Subpart C – Occupational dose limits, 20120(c), Occupational dose limits for adults. 56 FR 23396. [accessed 2021 Nov 15]. https://www.nrc.gov/reading-rm/doc-collections/cfr/part020/part020-1201.html

U.S. Nuclear Regulatory Commission (NRC). 2019. Radiation exposure information and reporting system (REIRS). US Nuclear Regulatory Commission. [accessed 2021 Jul 31]. https://www.reirs.com/

United Nations Scientific Committee on the Effects of Atomic Radiation (UNSCEAR). 2008. Effects of ionizing radiation. 2006 Report. Scientific annex A: epidemiological studies of radiation and cancer. New York (NY): UNSCEAR.

United Nations Scientific Committee on Atomic Radiation (UNSCEAR). 2013. Sources, effects and risks of ionizing radiation. UNSCEAR Report 2013 to the General Assembly with Scientific annexes. Volume II, Scientific annex B: Effects of radiation exposure of children. E.14.IX.2. New York (NY): UNSCEAR. [accessed 2021 Oct 23]. http://www.unscear.org/docs/reports/2013/UNSCEAR2013Report_AnnexB_Children_13-87320_Ebook_web.pdf.

United Nations Scientific Committee on the Effects of Atomic Radiation (UNSCEAR). 2020. Sources, effects and risks of ionizing radiation. UNSCEAR 2019 Report to the General Assembly. Scientific Annex A: evaluation of selected health effects and inference of risk due to radiation exposure. New York (NY): UNSCEAR.

Velazquez-Kronen R, Gilbert ES, Linet MS, Moysich KB, Freudenheim JL, Wactawski-Wende J, Simon SL, Cahoon EK, Alexander BH, Doody MM, et al. 2020. Lung cancer mortality associated with protracted low-dose occupational radiation exposures and smoking behaviors in U.S. radiologic technologists, 1983–2012. Int J Cancer. 147(11):3130–3138.

Villoing D, Borrego D, Preston DL, Alexander BH, Rose A, Salasky M, Linet MS, Lee C, Kitahara CM. 2021. Trends in occupational radiation doses for U.S. radiologic technologists performing general radiologic and nuclear medicine procedures, 1980–2015. Radiology. 300(3):605–612.

Wakeford R. 2019. Does low-level exposure to ionizing radiation increase the risk of cardiovascular disease? Hypertension. 73(6): 1170–1171.

Wakeford R. 2021. Overview of epidemiological studies of nuclear workers: opportunities, expectations, and limitations. J Radiol Prot. DOI:10.1088/1361-6498/ac0df4.

Wang JX, Boice JD Jr, Li BX, Zhang JY, Fraumeni JF Jr. 1988. Cancer among medical diagnostic x-ray workers in China. J Natl Cancer Inst. 80(5):344–350.

Wang J-X, Inskip PD, Boice JD, Li B-X, Zhang J-Y, Fraumeni JF Jr. 1990. Cancer incidence among medical diagnostic X-ray workers in China, 1950 to 1985. Int J Cancer. 455(5):889–895.

Wang JX, Zhang LA, Li BX, Zhao YC, Wang ZQ, Zhang JY, Aoyama T. 2002. Cancer incidence and risk estimation among medical x-ray workers in China, 1950–1995. Health Phys. 82(4):455–466.

Yoder RC, Balter S, Boice JD Jr, Grogan H, Mumma M, Rothenberg L, Passmore C, Vetter R, Dauer L. 2021. Using personal monitoring data to derive organ doses for medical radiation workers in the Million Person Study – considerations regarding NCRP Commentary No. 30. J Radiol Prot. 41(1):118–128.

Yoder RC, Dauer L, Balter S, Boice JD Jr, Grogan H, Mumma M, Passmore CN, Rothenberg LN, Vetter RJ. 2018. Dosimetry for the study of medical radiation workers with a focus on the mean absorbed dose to the lung, brain and other organs. Int J Radiat Biol. 1–36. DOI:10.1080/09553002.2018.1549756.

Yoshinaga S, Aoyama T, Yoshimoto Y, Sugahara T. 1999. Cancer mortality among radiological technologists in Japan: updated analysis of follow-up data from 1969 to 1993. J Epidemiol. 9(2):61–72.

Yoshinaga S, Mabuchi K, Sigurdson AJ, Doody MM, Ron E. 2004. Cancer risks among radiologists and radiologic technologists: review of epidemiologic studies. Radiology. 233(2):313–321.

Radium dial workers: back to the future

Nicole E. Martinez, Derek W. Jokisch, Lawrence T. Dauer, Keith F. Eckerman, Ronald E. Goans, John D. Brockman, Sergey Y. Tolmachev, Maia Avtandilashvili, Michael T. Mumma, John D. Boice, and Richard W. Leggett

ABSTRACT

Purpose: This paper reviews the history of the radium dial workers in the United States, summarizes the scientific progress made since the last evaluation in the early 1990s, and discusses current progress in updating the epidemiologic cohort and applying new dosimetric models for radiation risk assessment.

Background: The discoveries of radiation and radioactivity led quickly to medical and commercial applications at the turn of the 20th century, including the development of radioluminescent paint, made by combining radium with phosphorescent material and adhesive. Workers involved with the painting of dials and instruments included painters, handlers, ancillary workers, and chemists who fabricated the paint. Dial painters were primarily women and, prior to the mid to late 1920s, would use their lips to give the brush a fine point, resulting in high intakes of radium. The tragic experience of the dial painters had a significant impact on industrial safety standards, including protection measures taken during the Manhattan Project. The dial workers study has formed the basis for radiation protection standards for intakes of radionuclides by workers and the public.

Epidemiologic approach: The mortality experience of 3,276 radium dial painters and handlers employed between 1913 and 1949 is being determined through 2019. The last epidemiologic follow-up was 30 years ago when most of these workers were still alive. Nearly 65% were born before 1920, 37.5% were teenagers when first hired, and nearly 50% were hired before 1930 when the habit of placing brushes in mouths essentially stopped. Comprehensive dose reconstruction techniques are being applied to estimate organ doses for each worker related to the intake of ^{226}Ra, ^{228}Ra, and associated photon exposures. Time dependent dose-response analyses will estimate lifetime risks for specific causes of death.

Discussion: The study of radium dial workers is part of the Million Person Study of low-dose health effects that is designed to evaluate radiation risks among healthy American workers and veterans. Despite being one of the most important and influential radiation effects studies ever conducted, shifting programmatic responsibilities and declining funding led to the termination of the radium program of studies in the early 1990s. Renewed interest and opportunity have arisen. With scientific progress made in dosimetric methodology and models, the ability to perform a study over the entire life span, and the potential applicability to other scenarios such as medicine, environmental contamination and space exploration, the radium dial workers have once again come to the forefront.

Introduction

The epidemiologic investigation of United States radium dial workers is one of the most important and influential radiation effects studies ever conducted (Rowland 1994; Fry 1998). The last epidemiological follow-up was conducted more than 30 years ago, at which time most of the radium dial workers were still alive (Stebbings et al. 1984; Rowland 1994; Department of Energy (DOE) 2021). Ongoing extended follow-up will provide new information on the lifetime risk of cancer and other adverse effects of ionizing radiation among women following intakes of radium. The study of radium dial workers is part of the Million Person Study (MPS) of low-dose health effects that is designed to evaluate radiation risks among healthy American workers

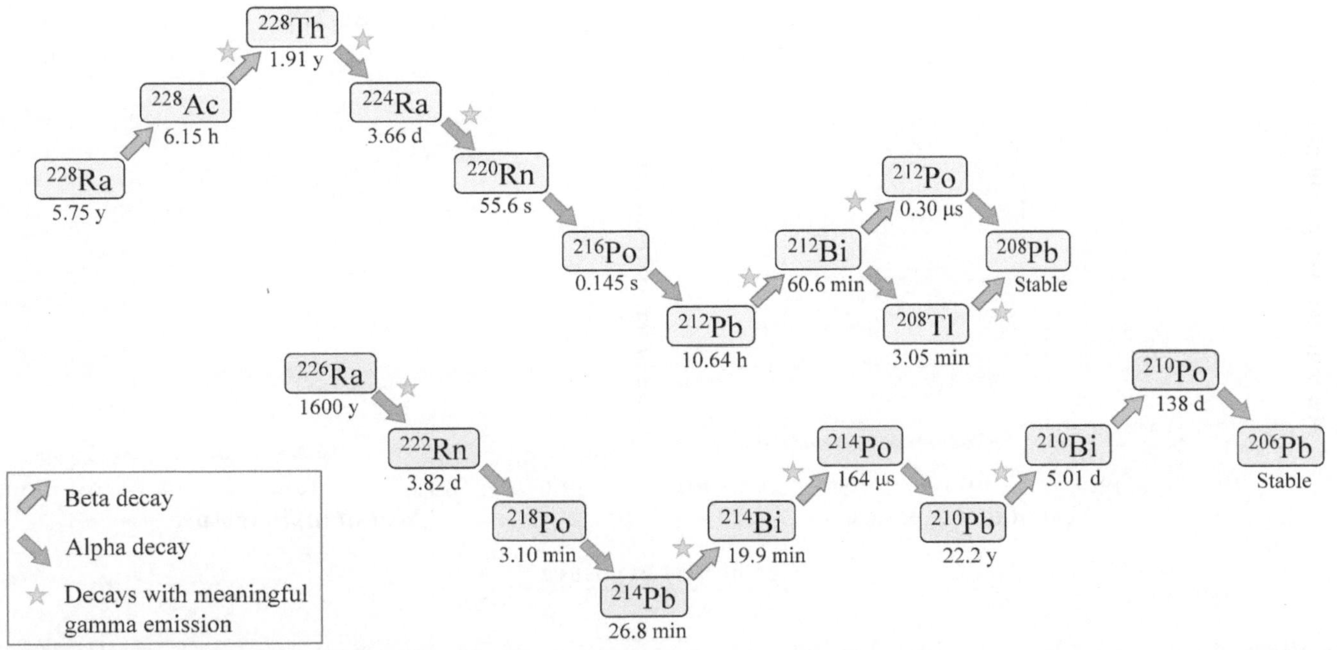

Figure 1. Decay chains for ^{228}Ra (top) and ^{226}Ra (bottom). Arrows pointing up and to the right indicate beta decay. Arrows pointing down and to the right indicate alpha decay. Stars indicate decays that are associated with meaningful gamma emission. Radiological half-lives are listed below each nuclide (International Commission on Radiological Protection (ICRP) 2008).

and veterans (Boice et al. 2019). This paper briefly reviews the history of the radium dial workers in the United States, summarizes the scientific progress made since the last epidemiologic analysis of this cohort (Rowland 1994), and discusses current progress in expanding and applying updated models to the original cohort.

Historical context

There are numerous excellent reviews and books that discuss the history of the dial painters (e.g. Martland 1929; Sharpe 1978; Rowland 1994; Fry 1998; Gunderman and Gonda 2015), including Kate Moore's novel *The Radium Girls* (2017) that focuses on the women's experiences. From this tragedy, information was learned about radiation-induced osteosarcomas as well as mastoid and paranasal sinus carcinomas. Here we provide a summary to provide context for the discussion that follows.

Marie and Pierre Curie discovered radium in 1898 and soon after radium was being marketed as a medicinal cure-all (Cothern and Smith 1987). The early 1900s also brought the recognition that radium could be combined with phosphorescent material (e.g. zinc sulfide) to make self-luminous paint; several proprietary formulas were eventually developed. Dr. Sabin von Sochocky, who studied both medicine and atomic physics (the latter under Dr. Ernest Rutherford), is credited with inventing a radioluminous paint used widely in the United States that was cheaper than that developed in Europe (NYT 1928; Sharpe 1978). In 1913, seven years after immigrating to the United States (US), he began selling radioluminescent watches commercially and two years later became an original founder of what would become the US Radium Corporation (USRC); he left the company in 1922. Dial-painting enterprises prospered early on due to the wartime demand for radioluminescent dials (Rowland 1994).

^{226}Ra ($t_{1/2} = 1600$ y), an alpha-emitter, was used in dial paint through the summer of 1919, at which point some facilities, notably USRC in New Jersey, also began to use ^{228}Ra ($t_{1/2} = 5.75$ y), a beta-emitting decay product of ^{232}Th commonly referred to as mesothorium (Keane et al. 1994). Figure 1 contains decay schemes for these two radium isotopes. Despite being chemically identical, ^{228}Ra was cheaper than ^{226}Ra. The processing of thorium ore results in the production of thorium nitrate, which was used at the time in the manufacture of incandescent gas mantles. ^{228}Ra was a byproduct of this process that could be obtained locally, used after a year or two following extraction to allow ingrowth of alpha-emitting progeny, necessary to achieve reasonable luminescence. Additionally, with a higher specific activity, ^{228}Ra also had a 'greater practical luminosity' than ^{226}Ra and was thus used in some locations to supplement ^{226}Ra in dial paint (MLR 1926; Stewart 1929; Sharpe 1978). Retrospective analysis of USRC dial paint determined an average ratio of ^{228}Ra to ^{226}Ra of about 8.4 for paints used between 1919 and 1925, with other years' paint likely ^{226}Ra only (Keane et al. 1994).

Workers in dial painting facilities included dial painters, dial handlers, chemists, and other ancillary workers. Making up the largest percentage (90.2%) of the current epidemiological cohort are the dial painters. Thousands of workers in the current study, mainly women (96.4%), painted dials and instruments with radium paint, using their lips to finely point their camel hair paint brushes to paint delicate pieces quickly and precisely. Being a dial painter was considered glamorous and patriotic, and early on little if any concern was expressed as to worker health and safety; many individuals believed radium exposure was 'good for you.'

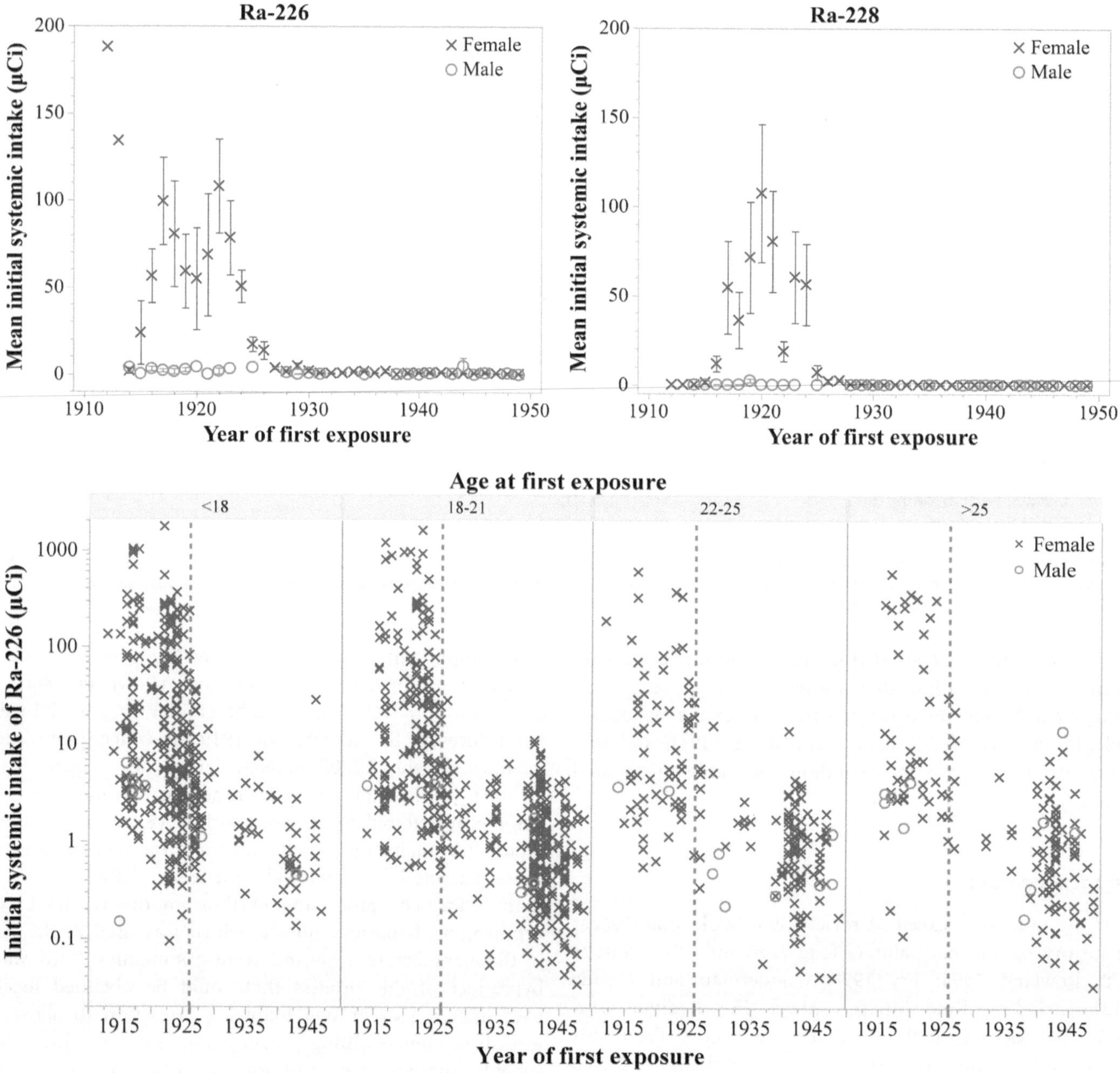

Figure 2. Available initial systemic intakes of radium (μCi) for individuals in the current cohort by year of first exposure, generally taken to be the year an individual was first hired. (×) represent females and (o) represent males. Data were drawn from the Comprehensive Epidemiologic Data Resource (DOE 2021). (Top panel) Mean initial systemic intakes (linear scale) of ^{226}Ra (left) and ^{228}Ra (right). Error bars represent the standard error of the mean in reported intakes and do not include consideration of measurement or modeling uncertainty. (Bottom panel) Individual initial systemic intakes (log scale) of ^{226}Ra grouped by age at first exposure. Vertical dashed lines represent the year 1926.

Interestingly, Moore reports an anecdote from 1918, where a New Jersey dial painter remembers von Sochocky, as he quickly passed through the room, telling her not to put the brush in her mouth because she would get sick (Moore 2017). In 1919, cloths were provided to help the painters shape the brushes but were removed due to their 'waste' of paint (Moore 2017). Despite this passing concern, it appears supervisors and managers remained either unconvinced or unconcerned that radium had negative health consequences, and most dial painters continued to moisten and/or point the brush with their mouth for several years following.

It wasn't until 1923 that Dr. Theodor Blum, a dentist in Orange, New Jersey, first publicly reported that osteomyelitis of the mandible and maxilla had occurred in a young painter of luminous dials, a condition termed 'radium jaw' (Blum 1924; Fry 1998), although Dr. Martin Szamatolski, consulting chemist for New Jersey Department of Labor, is often credited with the earliest written suspicion (January 1923) that radium was the source of this occupational disease (Stewart 1929; Sharpe 1978). Research on health effects of internal radium contamination began in earnest in the mid-1920s when unusual occurrences of bone and other conditions began appearing among New Jersey radium workers (Martland 1929; Martland and Humphries 1929; Aub et al. 1952; Rowland 1994; Stebbings 2001). A report by Harvard Medical School health and safety experts Drs. W.B.

Castle, Katherine Drinker, and Cecil Drinker was written for USRC in June 1924 and published in 1925 despite objections from the company (Castle et al. 1925; Rowland 1994). This report was one of the first studies to link exposure to radium with blood changes and jaw necrosis observed in dial painters (Castle et al. 1925; Rowland 1994). Dr. Frederick Hoffman, a statistician by education, published independent observations the same year, concluding that detrimental effects observed in dial painters were most likely due to direct contact with radium in the paint via lip pointing, although effects were attributed to ^{228}Ra (Hoffman 1925; Sharpe 1978). This assumption was seemingly based on the fact that, at the time, affected women were from USRC and believed to have painted with only ^{228}Ra-containing paint, contrasted with other facilities using ^{226}Ra paint. Ultimately this resulted in ^{228}Ra no longer being used in dial paint (Stewart 1929), which is supported by the retrospective paint analysis mentioned above (Keane et al. 1994). Interestingly, one difference between facilities that had an impact on radium intake was the type of adhesive used. Compared to an oil-varnish adhesive, paint applied with a water-based adhesive usually resulted in more frequent lip-pointing as water tends to separate brush hairs and was also less objectionable to put in the mouth. Also, paint was easier to apply with a stylus of some kind (e.g. glass rod or metal pen) with oil-varnish adhesive, which is likely why European dial painters did not exhibit the same effects as early American dial painters (Stewart 1929, Sharpe 1978).

The top panel of Figure 2 shows yearly averages in previously computed initial systemic intakes of ^{226}Ra (left) and ^{228}Ra (right) for individuals in the current cohort; non-zero intakes have been reported for dial workers through 1929 for ^{228}Ra and 1949 for ^{226}Ra (Department of Energy (DOE) 2021; Rowland 1994). Error bars represent the standard error of the mean and do not include consideration of measurement or modeling uncertainty. Estimates of initial systemic intake are available for 1,558 individuals in the current cohort of 3,276 radium dial workers (see the following section). Initial systemic intake, or the amount of radium that entered systemic circulation during an individual's exposure period, was calculated based on measurements of body burden, or residual radium content in the body (Rowland 1994). It was, however, sometimes necessary to estimate an individual's ^{228}Ra intake from colleagues' results or from measurement of exposure materials (Department of Energy (DOE) 2021). Initial systemic intake was found to be a useful metric for developing dose-response type of relationships as it was time-invariant and involved no assumptions as to the critical tissues at risk (Rowland and Lucas 1982).

Many dial-painting facilities reported prohibiting lip-pointing around 1925 (Martland 1929), although some seem to have delayed implementing the rule (Stewart 1929) and reports from painters themselves describe continuing to lip-point for a year or two more (Rowland and Lucas 1982; Moore 2017). Despite some uncertainty in when and to what extent lip-pointing was discontinued, calculated intakes of radium by dial workers were much lower in the years following 1926 (Figure 2).

The bottom panel of Figure 2 contains individual initial systemic intakes of ^{226}Ra for those in the current cohort for whom estimates are available (Department of Energy (DOE) 2021; Rowland 1994). Widespread publicity of the hazards of dial-painting and tragic consequences to early dial workers is attributed to the reduction in number of employees following 1925, with renewed interest in the 1940s associated with the wartime demand for luminous dials (Rowland and Lucas 1982). Figure 2 also highlights the overlap of age and year of first exposure. Prior to the 1926 benchmark, a large number of dial workers were under 18 years old. Looking forward to the 1940s, there are comparatively fewer teenage workers. This may be in part due to the Fair Labor Standards Act, first passed in 1938 (29 USC Chapter 8; Department of Labor (DOL) 2011).

Perhaps the most extensive of the early research into the effects observed in dial workers was the body of work led by Dr. Harrison Martland, a pathologist at Newark City Hospital who became the medical examiner for Essex County, New Jersey in 1925 (Martland et al. 1925; Sharpe 1978; ; Rowland 1994; Fry 1998). He reports in a series of papers detailed exposure history, symptoms, pathology, and prognosis of the radium-induced diseases seen in dial workers (e.g. jaw necrosis, aplastic anemia, and osteosarcomas) along with corresponding radiological measurements (Martland et al. 1925; Martland 1926, 1929, 1931; Martland and Humphries 1929; Aub et al. 1952). Martland and colleagues, including von Sochocky, were among the first to develop and use techniques for *in vivo* measurement of radioactivity (Martland et al. 1925; Martland 1929). Martland was an advocate for the dial workers, and his papers frequently included sociolegal aspects of the occupational circumstances (e.g. Martland 1929; Fry 1998).

The early health studies were revived by the Atomic Energy Commission following World War II, partly because of the importance of radium studies in predicting the health effects of plutonium, a new bone-seeking alpha-emitting radionuclide. In 1969, the three major human studies of radium were centralized at Argonne National Laboratory, following an initial proposal made by Dr. Robley D. Evans as to the need for a National Center of Human Radiobiology. Evans was a physicist at the Massachusetts Institute of Technology who made substantial contributions to the radium studies starting in the early 1930s, fresh out of graduate school, through his retirement in 1972. The Argonne program was terminated in the early 1990s and materials were transferred to Washington State University and stored at the National Human Radiobiology Tissue Repository (NHRTR) in Richland WA (Rowland 1994). The United States Transuranium and Uranium Registries (USTUR) research program is a federal-grant-funded human tissue research program providing long-term study of actinide biokinetics in former nuclear workers with accidental internal depositions of these elements. The USTUR conducts autopsies and performs radiochemical analyses of voluntarily donated tissue samples (Kathren and Tolmachev 2019;

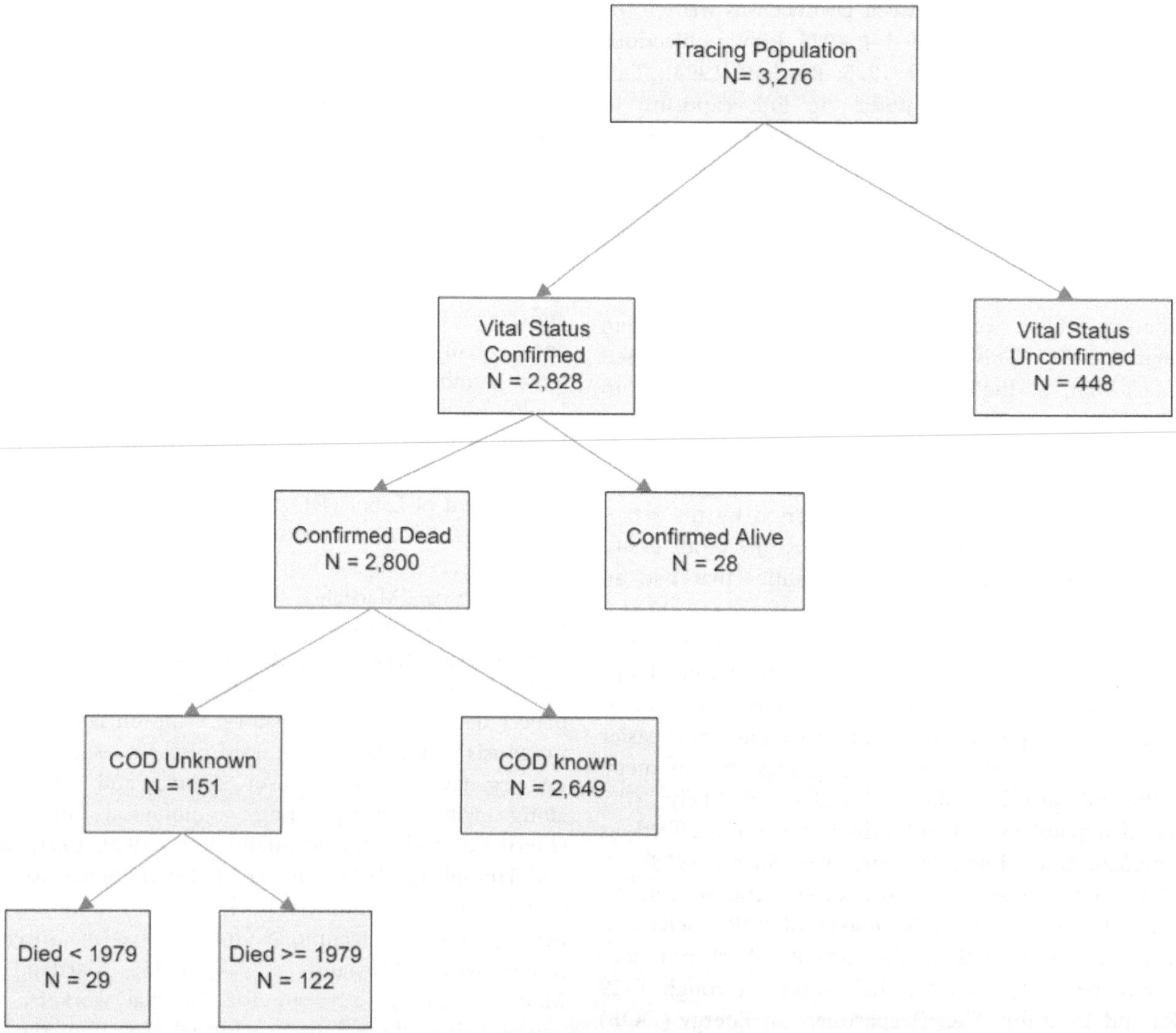

Figure 3. Schematic of the selection and vital status tracing results as of 31 December 2019 for the study population of 3,276 radium dial painters and radium dial handlers employed prior to 1950. COD indicates Cause of Death.

Tolmachev et al. 2019). NHRTR holds all tissues donated to the USTUR, along with specimens acquired from the US Radium Worker Studies (Rowland 1994). The USTUR/NHRTR is a unique resource for retrospective analyses and distribution studies of plutonium, uranium, americium, radium, and barium in the human whole body, as well as in specific tissues and organs. In fact, the USTUR repository contains 1000s of specimens from the radium dial workers and has been accessed to help inform dosimetric models, e.g. radiochemical determination of radium in brain tissue of a painter (Leggett et al. 2018; Kathren and Tolmachev 2019; Tolmachev et al. 2019)

Cohort definition

Over the years, there have been several epidemiologic studies and analyses of the radium dial workers (Polednak et al. 1978; Polednak 1978a, 1978b; Rowland, Stehney, Brues, et al. 1978; Rowland, Stehney, Lucas 1978; Rowland and Lucas 1982, 1984; Rowland et al. 1983, 1989; Stebbings et al. 1984; Rowland 1993, 1994). The current population is composed of the dial painters studied by Polednak et al. 1978 (900 dial painters first hired prior to 1930) and Stebbings et al. 1984 (approximately 2,600 dial painters hired prior to 1950) supplemented with workers available from the comprehensive dataset from Argonne National Laboratory described by Rowland (1994) and digitally archived at the US Department of Energy (DOE) Comprehensive Epidemiologic Data Resource (CEDR). This dataset contains information on about 6,000 individuals with radium exposure, including not only dial workers, but also radium chemists, patients treated with radium, individuals who were known to have ingested radium water (e.g. *Radithor*), and other miscellaneous exposures (Department of Energy (DOE) 2021; Rowland 1994). While the information available in CEDR and in Rowland (1994) helps define the study population, a key source of information is the 100,000 s of pages of individual microfilm and microfiche records that were converted to optical character read format. Detailed clinical data, dosimetry data, and follow-up data for

Table 1. Descriptive characteristics of the 3,276 radium dial painters and handlers first employed 1910–1949 that comprise the epidemiologic cohort being studied.

Characteristic	*N*	%
Primary employer		
Bendix	266	8.1
Elgin National Watch Company	52	1.6
Ingraham Clock Company	78	2.4
Luminous Engineering/Processes	978	29.9
Radium Dial Company	396	12.1
Standard Chemical	83	2.5
US Radium Company	594	18.1
Waltham Watch Company	88	2.7
Waterbury Clock Company	457	13.9
Other	284	8.7
Job category		
Dial painter	2,955	90.2
Dial handler	321	9.8
Sex		
Male	119	3.6
Female	3,157	96.4
Year of birth		
1856–1899	473	14.4
1900–1919	1,637	50.0
1920–1932	711	21.7
Unknown	455	13.9
Year of first hire		
1912–1919	529	16.1
1920–1924	546	16.7
1925–1929	445	13.6
1930–1939	172	5.3
1940–1949	1,481	45.2
Unknown	103	3.1
Age at first hire		
10–15 years	183	5.6
16–19 years	1,044	31.9
20–24 years	822	25.1
25–29 years	334	10.2
30–39 years	263	8.0
40 years or older	121	3.7
Unknown	509	15.5
Vital status as of 31 December 2019		
Confirmed dead	2,800	85.5
Confirmed alive	28	0.9
Currently being traced	448	13.7
Age at death or EOFU (Dec. 31, 2019)-VS confirmed		
<40 years	155	5.5
40–49 years	89	3.1
50–59 years	171	6.0
60–69 years	348	12.3
70–79 years	571	20.2
80–89 years	829	29.3
90 year and older	539	19.1
Unknown	126	4.5

EOFU: end of follow up; VS: vital status.

individuals provides a treasure trove of information to supplement and enhance the epidemiologic data. These data were made available through the DOE USTUR.

The population selection and the incremental cost for data abstraction and tracing were balanced against a small benefit from including a substantially expanded set of radium dial workers who had very low or minimal radium intakes (e.g. those first employed after about 1925, and certainly after 1950). The final defined population includes all dial painters (DP) and dial handlers (DH) employed prior to 1950. The final population size consists of 3,276 workers, including a small number ($n = 119$) of male painters and handlers.

Vital status and cause of death determination

Vital status, date and cause of death as of 31 December 2019 (aka vital status tracing) are being sought from linkages with the National Death Index (NDI); state mortality files; the Social Security Administration (SSA) Death Master File; the SSA Service to Epidemiological Researchers (which confirms alive status); and credit reporting agencies using the methods outlined in Mumma et al. (2018). The Centers for Disease Control and Prevention LinkPlus program, which incorporates a probabilistic scoring system that does not require exact matches on all variables, was used for in-house matches (Campbell 2008). Online ancestry providers (Ancestry.com) and credit record providers (Transunion) are important sources to help complete and correct key demographic and linking data, such as Social Security Number, last names (which often changed since employment), and dates of birth and death. Vital status (VS) tracing (Figure 3) continues, and preliminary results are presented in Table 1. End-of-follow-up (EOFU) in Table 1 refers to the date a person is no longer considered at-risk for analytic purposes. It is their date of death, 95th birthday or December 31 of the calendar year. We anticipate that most of the study participants still being traced will be deceased as of December 2019.

Career doses

Dosimetry records documenting radiation exposure received after employment at dial painting facilities will be sought from additional sources: the DOE Radiation Exposure Monitoring System (REMS), including historic DOE radiation exposure data not included in REMS; the Nuclear Regulatory Commission (NRC) Radiation Exposure Information and Reporting System (REIRS) and Landauer, Inc. dosimetry records (Department of Energy (DOE) 2018; National Council on Radiation Protection and Measurements (NCRP) 2018; Nuclear Regulatory Commission (NRC) 2019; Yoder et al. 2021). Based on the age structure of the population, we do not anticipate many additional external dose records. All organ/tissue-specific doses from each source will be added together to obtain the total organ/tissue-specific external and internal dose received by each worker for each calendar year, following the procedures outlined in Boice et al. (2006) and Ellis et al. (2018).

Integration of radiation biology, dosimetry, and epidemiology for the radium dial worker cohort

Epidemiology and health outcomes

As above, the mortality experience of a new cohort of 3,276 radium dial painters and handlers employed between 1913 and 1949 is being examined through 2019. Nearly 65% were born prior to 1920, 37.5% were teenagers when first hired, 96.4% were female, 90.2% were dial painters, and nearly 50% were hired before 1930 when the habit of placing brushes in mouths essentially stopped (Table 1). The large number of dial workers first employed 1940–1949 (46.7%),

Table 2. Distribution of bone sarcomas and head carcinomas* in radium-exposed persons as of 1990 by exposure circumstance (radium dial worker or other**), measurement status*** and sex, modified from Rowland (1994)

	Number of persons			Osteosarcomas (%)			Head carcinomas (%)		
Radium-Exposure Circumstance	Female	Male	All	Female	Male	All	Female	Male	All
Persons measured									
Dial worker	1747	161	1908	46 (2.6%)	0 (0.0%)	46 (2.4%)	19 (1.1%)	0 (0.0%)	19 (1.0%)
Other	156	319	475	15 (9.6%)	3 (1.4%)	18 (3.8%)	8 (5.1%)	5 (1.6%)	13 (2.7%)
Total	1903	480	2383	61 (3.2%)	3 (0.6%)	64 (2.7%)	27 (1.4%)	5 (1.0%)	32 (1.3%)
Persons not measured									
Dial worker	1910	315	2225	18 (0.9%)	0 (0.0%)	18 (0.8%)	5 (0.3%)	0 (0.0%)	5 (0.2%)
Other	871	1196	2067	1 (0.1%)	2 (0.2%)	3 (0.1%)	0 (0.0%)	0 (0.0%)	0 (0.0%)
Total	2781	1511	4292	19 (0.7%)	2 (0.1%)	21 (0.5%)	5 (0.2%)	0 (0.0%)	5 (0.1%)
All persons									
Dial worker	3657	476	4133	64 (1.8%)	0 (0.0%)	64 (1.5%)	24 (0.7%)	0 (0.0%)	24 (0.6%)
Other	1027	1515	2542	16 (1.6%)	5 (0.3%)	21 (0.8%)	8 (0.8%)	5 (0.3%)	13 (0.5%)
Total	4684	1991	6675	80 (1.7%)	5 (0.3%)	85 (1.3%)	32 (0.7%)	5 (0.3%)	37 (0.6%)

*Carcinomas originating in the paranasal sinuses or mastoids.
**Other includes chemists, physicists, laboratory technicians, nurses, offspring born to female subjects who had been exposed to radium, persons and patients injected with radium, and persons who drank *Radithor* (a form of radium available to the public in the 1920s which included bottled drinking water spiked with radium and sold over the counter or by mail).
***Whether or not measurements were made to determine radium body burden. The highest measurements were found among workers in the dial industry, patients treated with radium and from self-administered radium, e.g. drinking *Radithor*.

represents a relatively large low-dose group for comparison. The cohort was assembled over the years from over 10 different companies located primarily in New Jersey, Connecticut, and Illinois (Table 1) (Polednak et al. 1978; Rowland 1994). A comprehensive approach to ascertaining vital status (Mumma et al. 2018) has already confirmed to date that 86% have died; 99% of those with known vital status have died. Comprehensive dose reconstruction techniques are being applied to estimate organ doses for each worker related to the intake of ^{226}Ra, ^{228}Ra, and associated photon exposures. Time dependent dose-response analyses will estimate lifetime risks for specific causes of death, with a particular focus on osteosarcoma, mastoid and paranasal sinus carcinoma; leukemia; cancers of the lung, breast, and brain; ischemic heart disease; and dementia, Alzheimer's, Parkinson's, and motor neuron disease.

A comprehensive review of the published epidemiological studies of radium workers can be found in Rowland (1994); Dr. Robert E. Rowland, a biophysicist, was the first director of the Center for Radiation Biology at Argonne National Laboratory and heavily involved in the radium studies. Updates and informative reviews also are available in the Biological Effects of Ionizing Radiations (BEIR) IV Report (National Academies/National Research Council (NA/NRC) 1988) and articles by Fry (1998) and Stebbings (2001). Over the years, different epidemiologic cohorts were defined, e.g. women first employed before 1930 (Polednak et al. 1978), women first employed before 1950 (Stebbings et al. 1984) and women and men first employed before 1950 as radium workers (Rowland and Lucas 1984). Further, some publications focused primarily on dose-response relationships and included only women with measured intakes of radium (Rowland, Stehney, and Lucas 1978), and some included other types of radium workers in addition to dial painters (Rowland 1994). As mentioned above, the current cohort under study includes radium dial painters and a few associated workers, e.g. dial handlers, first employed before 1950 and previously studied in large part by Stebbings et al. (1984) and supplemented as described in Rowland (1994).

Briefly, osteosarcomas and head carcinomas (mastoid and paranasal sinus carcinomas) have been convincingly associated with internal radium exposure. The distribution of these cancers in radium-exposed persons as of 1990 is shown in Table 2; note that five individuals were diagnosed with both osteosarcomas and head carcinomas. About 1.5% (64 individuals) of dial workers were diagnosed with osteosarcomas, a cancer which was found to be more effectively induced by ^{228}Ra compared with ^{226}Ra (Rowland, Stehney, and Lucas 1978). About 0.6% (24 individuals) of dial workers were diagnosed with head carcinomas, attributed to the accumulation of ^{222}Rn in the sinus cavities (National Academies/National Research Council (NA/NRC) 1988; Rowland 1994). All these cancers were observed in female dial painters, attributed primarily to their higher levels of intake compared to dial handlers or male dial painters; few men painted dials (Rowland et al. 1983) and those who did generally had measured intake levels of ^{226}Ra that were much lower than those among women (Rowland 1994; see also Figure 2).

Other previously studied health outcomes included leukemia (Spiers et al. 1983), breast cancer (Adams and Brues 1980; Stebbings 2001), fertility (Polednak 1980; Schieve et al. 1997), multiple myeloma (Stebbings et al. 1984; Stebbings 2001), thyroid tumors (Polednak 1986) and cataracts (Adams et al. 1983). Although increased risks were reported for some of these conditions, the associations were either not statistically meaningful and/or were not convincingly related to estimated intakes of radium.

The current work will expand upon these previous observations and will provide a complete assessment of lifetime risks related to radium ingestion. The health outcomes considered will include the previously studied cancer and non-cancer outcomes and have been broadened to include cognitive deficits possibly related to radiation exposure to brain tissue (Marazziti et al. 2012; Parihar et al. 2016, 2020;

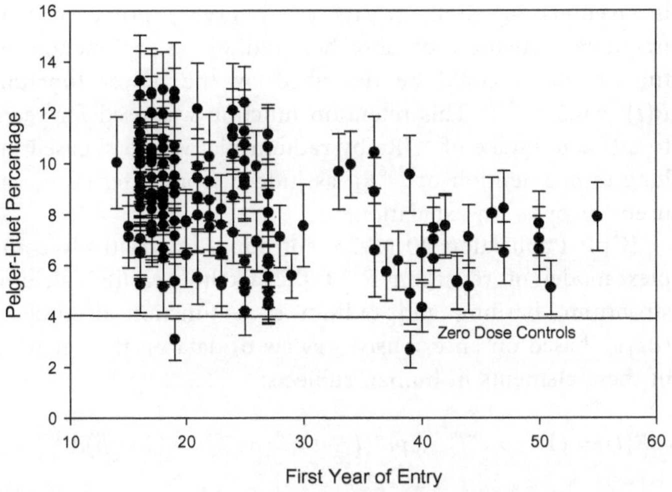

Figure 4. Pelger Huët percent (mean % ± SEM) observed in a cohort of 166 radium dial painters and ancillary workers versus date of entry into the workforce.

Azizova et al. 2020; Pasqual et al. 2021). Interestingly, NASA is concerned about possible behavioral and cognitive impairments from high energy heavy ions in space [galactic cosmic radiation (GCR)] that might jeopardize long missions, and, possibly, lead to Alzheimer's, Parkinson's or dementia later in life (Boice 2019; National Council on Radiation Protection and Measurements (NCRP) 2019). There are no human exposure circumstances similar to GCR in space that can provide direct information on cognition or neurological diseases following such high-LET exposure to brain tissue. The intake of radium can result in meaningful exposure to brain tissue from alpha particles, and the medical records of the dial painters are substantial, going back as early as the 1920s. Although an imperfect analog, the study will be able to address the likelihood that high-LET exposures to brain might cause cognitive impairment and provide some guidance as to the seriousness of this threat for space exploration.

Uncertainties associated with this work are those typically encountered in epidemiological studies evaluating radiation-induced health effects, although the current study has the unique benefit of documented radiological measurements for about half of the population, the continued monitoring and clinical visits of individuals in the population, and the current availability of thousands of tissue samples, and bones, of the dial painters available at USTUR. Dose reconstructions and analysis will follow current best practices with respect to evaluation of uncertainty (e.g. National Council on Radiation Protection and Measurements (NCRP) 2009, 2018), including consideration of the new approaches to address uncertainty in worker studies following intakes of plutonium, another bone-seeker (Stram et al. 2021).

Cellular dosimetry

Peripheral blood slides have been reviewed from a cohort of 166 radium dial painters and ancillary workers (Goans et al. 2019). The blood slides were prepared in 1960-1975 during medical follow-up and were made available in collaboration with the USTUR. The cohort contained 107 dial painters, 22 dial handlers, 19 radium chemists, and other personnel dealing with radium. Members of the cohort had ingestion of ^{226}Ra and ^{228}Ra at an early age (average age 20.6 ± 5.4 y; range 13–40 y) during the years 1914–1955. Exposure duration ranged from 1 to 1,820 weeks with red marrow dose 1.5–6,750 mGy.

The cell of interest in the peripheral smear is the pseudo-Pelger Huët cell (PH). PH is characterized in neutrophils by a bi-lobed nucleus whose lobes are joined by a thin chromatin bridge. PH in this case is caused by a radiation-induced decrease in the amount of the Lamin B receptor (LBR). The gene that encodes the LBR is known to be located on the long arm of chromosome 1, 1q42.12.

PH has been described as a novel, permanent, radiation-induced biomarker in circulating neutrophils (Goans et al. 2015, 2017). In studies involving a group of workers from the Y-12 criticality accident (1958) and from controlled primate studies at the Armed Forces Radiobiology Research Institute (AFRRI), PH has shown a linear dose response for mixed dose in red marrow from 1-10 Gy. In the radium dial painter cohort, PH expressed as a percentage of total neutrophils has been shown to rise in a sigmoidal fashion over five decades of red marrow dose, best fit with a sigmoid function and suggestive of a threshold effect (Goans et al. 2019). These results are consistent with health outcome findings (discussed above) in that no bone sarcomas were observed in the radium dial cohort at total systemic intakes below 100 μCi (Rowland 1994). Thus, PH percentages from peripheral blood tracks alpha dose to bone marrow and have the potential to be a useful metric for supporting dose estimates. Figure 4 shows a plot of PH percent (mean % ± standard error of the mean) from this cohort versus date of entry into the workforce. A decrease toward control values is seen for entry into the workplace after 1930.

Radium biokinetics and associated models

Overview of radium kinetics in the body

Radium belongs to the alkaline earth family (Group 2) of the periodic table and is a physiological analogue of the lighter alkaline earth metals calcium, strontium, and barium. The rates of uptake and removal by tissues differs from one alkaline earth to another due to discrimination by biological membranes and hydroxyapatite crystals of bone (Leggett, 1992; International Commission on Radiological Protection (ICRP) 1993). The biokinetics of radium in the human body resembles that of barium more closely than that of calcium or strontium.

Biokinetic data for radium in adult humans were reviewed in International Commission on Radiological Protection (ICRP) Publication 137 (International Commission on Radiological Protection (ICRP) 2017). Briefly, the biokinetics of the individual alkaline earth elements and the comparative behaviors of different pairs of these elements have been studied extensively in human subjects and laboratory animals. Based on controlled studies on adult human subjects it is estimated that about a third of radium atoms leaving blood deposit in excretion pathways, predominantly in the colon. Soft tissues initially accumulate a substantial portion of retained systemic radium but lose most of the deposited activity within a few days. Bone soon becomes the primary systemic repository of radium after its acute uptake to blood. Radium and other alkaline earths entering bone initially deposit on bone surfaces, from which they are removed over a period of hours or days back to blood and to a lesser extent to a bone volume pool referred to as exchangeable bone volume. The rate of loss of alkaline earth elements from exchangeable bone over the first few months after uptake to blood increases in the order radium > barium > strontium > calcium. A portion of radium, barium, strontium, or calcium entering exchangeable bone volume returns to blood over a period of months and a smaller portion becomes firmly fixed in bone crystals and is retained there until removed by bone restructuring processes. Calcium, strontium, barium, and radium are all about equally likely to transfer from bone surface to exchangeable bone volume, but the likelihood of becoming firmly fixed in bone crystal decreases in the order calcium > strontium > barium > radium. The rate at which the non-exchangeable (firmly fixed) alkaline earth elements are removed from bone volume to blood appears to depend completely on the rate of turnover of the bone type (trabecular or cortical bone) and thus is independent of the element. The portion of acutely injected radium in bone of a mature adult human typically is about 20–40% after 1 day, 6–12% after 1 month and 2–4% after 1 year.

Information is available on the systemic behavior of radium in immature humans (International Commission on Radiological Protection (ICRP) 1973; Parks et al., 1978; Muth and Glöbel, 1983; Parks and Keane, 1983; Keane and Schlenker, 1987). More detailed data on the age-specific behavior of systemic radium are available for laboratory animals, particularly dogs (Lloyd, Mays, and Atherton 1976; Lloyd, Mays, Atherton, et al. 1976; Lloyd et al. 1982; Bruenger et al. 1983; Lloyd, Bruenger, Jones et al. 1983; Lloyd, Bruenger, Mays et al. 1983; Lloyd, Jones, et al. 1983; Lloyd, Taylor, et al. 1983; Bruenger and Lloyd 1989). Differences with age in the systemic behavior of radium are consistent with findings for the other alkaline earth elements. That is, retention of radium is greater in growing bone than in mature bone; changes with age in uptake of radium by the skeleton are roughly proportional to the age-specific rate of calcium addition to bone from bone growth plus bone remodeling; at times remote from exposure, skeletal burdens acquired during periods of growth tend to remain higher than those acquired by mature skeletons except for skeletal burdens acquired during or soon after infancy; and both deposition and removal of radium appear to be greater in areas of bone undergoing rapid remodeling than in areas of relatively slow remodeling.

Biokinetic models for radium previously applied to dial workers

Mainly on the basis of follow-up data for Elgin State Hospital patients administered known amounts of ^{226}Ra via intravenous injection, Norris et al. (1955) proposed that fractional retention of absorbed radium as a function of time t (days) could be described by the power function $R(t) = 0.54t^{-0.52}$. This retention function was used for years to estimate intake of ^{226}Ra by radium dial workers, based on long-term retention of ^{226}Ra as judged from external measurement or ^{222}Rn exhalation.

ICRP Publication 20 (1973) introduced a relatively complex model of retention R of the alkaline earths calcium, strontium, barium, and radium as a function of time, t (days), based on an extensive review of data on the behavior of these elements in human subjects:

$$R(t) = (1-p)e^{-mt} + p\varepsilon^{b}(t+\varepsilon)^{-b}\left[\beta e^{-r\lambda t} + (1-\beta)e^{-\sigma r\lambda t}\right] \quad (1)$$

Some of the parameters in this retention function are element dependent and others represent physiological or unknown, element-independent processes:

λ is the rate of resorption of compact bone,
σ is the ratio of turnover rates of trabecular and compact bone,
β is the fraction of bone volume activity deposited in compact bone,
m is a rate constant representing an early exponential process,
p is the fraction of retention not in the early exponential process,
ε (element-specific) is related to fast turnover of an initial pool,
b (element-specific) is related to diffusion of exchangeable activity from bone,
r (element specific) is related to redeposition of activity at bone-forming sites.

In ICRP Publication 20, retention in soft tissues was calculated as the difference between total-body retention $R(t)$ and components of the model interpreted as representing activity in bone and blood.

Schlenker et al. (1982) concluded from a review of the distribution and retention of radium in soft tissues that the model of ICRP Publication 20 did not accurately depict the time-dependent distribution of radium between bone and soft tissue. They modified selected parameter values for radium to obtain a better fit to their collected data.

As described by Rowland (1993, 1994), measurements of retention of ^{226}Ra in the dial painters made 30-60 years after exposure indicated faster loss of radium from the body than predicted by the model of ICRP Publication 20. To address this issue, Rowland (1993) modified Schlenker's revision of the ICRP model to incorporate a higher bone turnover rate. It was later observed that case-specific modifications of the Schlenker model may sometimes be needed because the rate of bone resorption may be greatly reduced in cases of extremely high intake of ^{226}Ra due to radiation damage to bone (Rowland 1994).

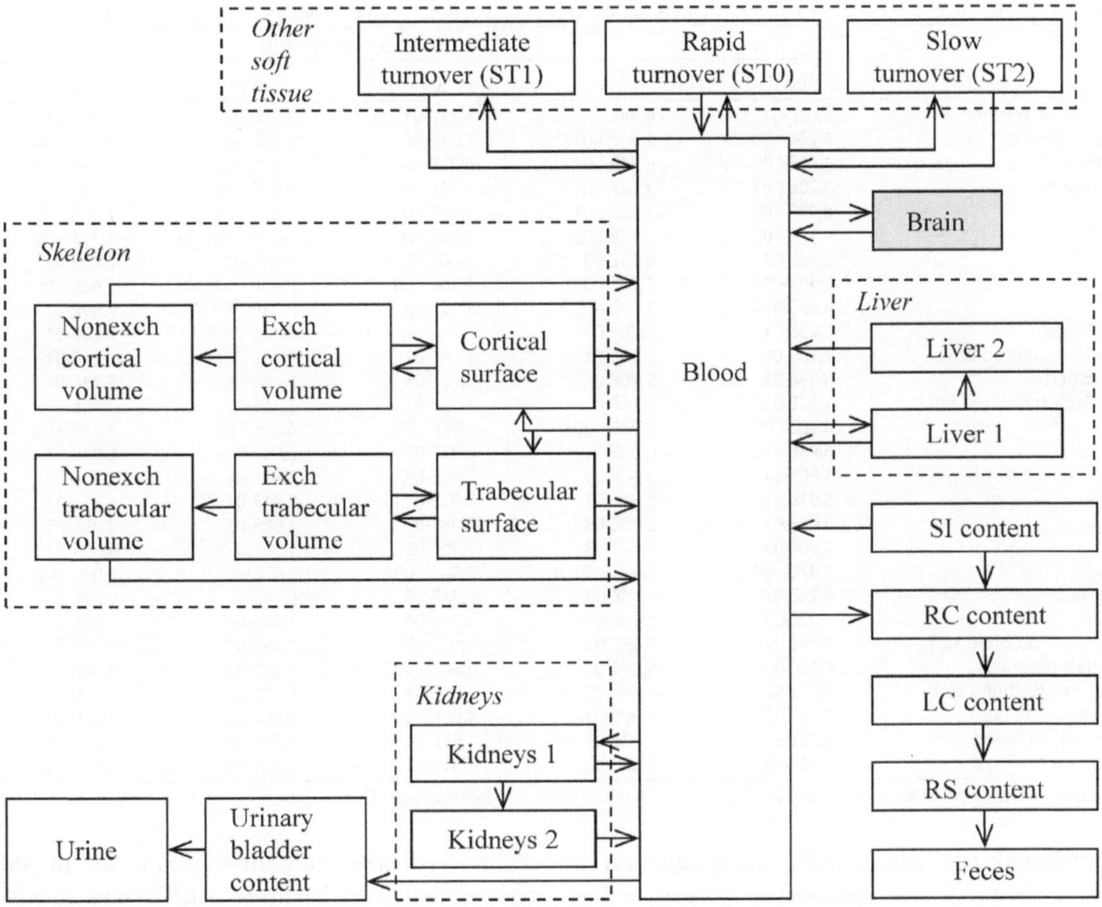

Figure 5. Structure of the ICRP's current biokinetic model for systemic radium (International Commission on Radiological Protection (ICRP) 2017) with the addition of the brain (shaded).

Biokinetic model for systemic radium applied in the present study

The latest version of the ICRP's biokinetic model for systemic radium will be used as the starting place for reconstructing intake of ^{226}Ra and ^{228}Ra by radium dial painters. As described below, it is expected that parameter values of the ICRP model will be modified where feasible to improve estimates of the intake and subsequent behavior of radium in individual dial painters or subgroups of dial painters.

The ICRP's latest biokinetic model for a reference adult is described in Publication 137 (2017), which is Part 3 of an ICRP series of reports on occupational intake of radionuclides (OIR). An age-specific version of that model has been developed and will be described and applied in an upcoming series of ICRP reports on environmental intake of radionuclides (EIR) by members of the public. The structure of the model is shown in Figure 5. Reference age-specific transfer coefficients are listed in Table 3. The following modifications of the radium model described in Figure 5 and Table 3 are planned for application to the radium dial painters:

- *Age- and sex-specific biokinetics of radium and progeny throughout life.* The ICRP's biokinetic model for radium addresses differences with age in radium kinetics only through age 25 y and does not address gender differences at any age. For application to the radium dial painters, the ICRP model will be modified to address age and gender differences in radium biokinetics throughout life. For example, the model will address age and gender differences in bone remodeling rates throughout life and in rates of net bone loss starting in the fifth decade of life. The ICRP model will also be modified to address differences with gender in the rates of bone maturation during adolescence.
- The relatively simple model applied by the ICRP to radon produced in vivo by decay of radium will be replaced with the more sophisticated ICRP model for radon as a parent.
- The structure of the ICRP's current biokinetic model for systemic radium (Figure 5) will be modified, insofar as allowed by available biokinetic data, to improve dose estimates for tissues of interest that are not explicitly depicted in the ICRP model. For example, a compartment representing the brain will be added. Parameter values describing uptake and removal of radium by the brain will be taken from an earlier paper (Leggett et al. 2018).

As in the ICRP's OIR and EIR series, the assumption of independent kinetics of radioactive progeny will be applied in dose reconstructions for radium dial painters. That is, radioactive progeny produced in the body after intake of ^{226}Ra or ^{228}Ra will be assumed to follow the characteristic biological behaviors of those elements (as opposed to

Table 3. ICRP's age-specific transfer coefficients for radium (adopted for use and to be published in the environmental intake of radionuclide [EIR] series).

	Transfer coefficient (d^{-1})					
Path[a]	100 d	1 y	5 y	10 y	15 y	Adult
Blood to urinary bladder content	2.02E-01	4.44E-01	4.85E-01	3.56E-01	2.10E-01	6.06E-01
Blood to right colon content	7.26E+00	1.60E+01	1.74E+01	1.28E+01	7.55E+00	2.18E+01
Blood to Trab bone surf	1.05E+01	6.30E+00	6.23E+00	9.87E+00	1.44E+01	9.72E+00
Blood to Cort bone surf	4.20E+01	2.52E+01	2.18E+01	2.93E+01	3.74E+01	7.78E+00
Blood to other 3	6.98E+00	1.53E+01	1.67E+01	1.23E+01	7.26E+00	2.09E+01
Blood to other 4	1.17E+00	2.57E+00	2.80E+00	2.05E+00	1.21E+00	3.50E+00
Blood to other 5	2.33E-02	5.13E-02	5.60E-02	4.11E-02	2.43E-02	7.00E-02
Blood to liver 1	1.40E+00	3.08E+00	3.36E+00	2.46E+00	1.46E+00	4.20E+00
Blood to kidneys 1	4.67E-01	1.03E+00	1.12E+00	8.21E-01	4.85E-01	1.40E+00
Trab bone surf to blood	5.78E-01	5.78E-01	5.78E-01	5.78E-01	5.78E-01	5.78E-01
Trab bone surf to Exch trab bone vol	1.16E-01	1.16E-01	1.16E-01	1.16E-01	1.16E-01	1.16E-01
Cort bone surf to Blood	5.78E-01	5.78E-01	5.78E-01	5.78E-01	5.78E-01	5.78E-01
Cort bone surf to Exch trab bone vol	1.16E-01	1.16E-01	1.16E-01	1.16E-01	1.16E-01	1.16E-01
Other 3 to blood	6.98E+00	6.98E+00	6.98E+00	6.98E+00	6.98E+00	6.98E+00
Other 4 to blood	6.93E-01	6.93E-01	6.93E-01	6.93E-01	6.93E-01	6.93E-01
Other 5 to blood	3.80E-04	3.80E-04	3.80E-04	3.80E-04	3.80E-04	3.80E-04
Liver 1 to blood	6.91E-01	6.91E-01	6.91E-01	6.91E-01	6.91E-01	6.91E-01
Liver 1 to liver 2	2.08E-03	2.08E-03	2.08E-03	2.08E-03	2.08E-03	2.08E-03
Liver 2 to blood	1.90E-03	1.90E-03	1.90E-03	1.90E-03	1.90E-03	1.90E-03
Kidneys 1 to blood	2.07E+00	2.07E+00	2.07E+00	2.07E+00	2.07E+00	2.07E+00
Kidneys 1 to kidneys 2	6.24E-03	6.24E-03	6.24E-03	6.24E-03	6.24E-03	6.24E-03
Kidneys 2 to blood	1.90E-03	1.90E-03	1.90E-03	1.90E-03	1.90E-03	1.90E-03
Exch trab bone vol to Trab bone surf	1.85E-02	1.85E-02	1.85E-02	1.85E-02	1.85E-02	1.85E-02
Exch to Nonexch trab bone vol	4.60E-03	4.60E-03	4.60E-03	4.60E-03	4.60E-03	4.60E-03
Exch cort bone vol to Cort bone surf	1.85E-02	1.85E-02	1.85E-02	1.85E-02	1.85E-02	1.85E-02
Exch to Nonexch cort bone vol	4.60E-03	4.60E-03	4.60E-03	4.60E-03	4.60E-03	4.60E-03
Nonexch cort bone vol to Blood	8.22E-03	2.88E-03	1.53E-03	9.04E-04	5.21E-04	8.21E-05
Nonexch trab bone vol to Blood	8.22E-03	2.88E-03	1.81E-03	1.32E-03	9.59E-04	4.93E-04

[a]Trab: trabecular; Cort: cortical; surf: surface; vol: volume; Exch: exchangeable; Nonexch: nonexchangeable.

following the biokinetics of radium) following production in soft tissues or on bone surface. Radioactive progeny other than radon produced in bone volume will be assumed to be removed from bone volume only through bone remodeling or net bone loss.

Radium distribution in brain

Analysis of brain tissue of a female individual occupationally exposed to radium was conducted at USTUR recently to study the distribution of ^{226}Ra. The concentrations of ^{226}Ra were measured with inductively coupled plasma mass spectrometry (ICP-MS) at the University of Missouri in the corpus callosum, the white and gray matter of the cerebrum lobes, the cerebellum, and brainstem segments of the brain. Preliminary results indicate that ^{226}Ra concentration in the white matter (18.3 ± 3.0 Bq kg^{-1}) was about 3.5 times higher than the average of all other brain segments (ranged 4.9–5.7 Bq kg^{-1}). With only one case studied, current preliminary results suggest non-uniformly distribution of radium in the human brain. In the future, this finding might have an impact on biokinetic modeling of internally deposited radionuclides in the brain as well as on the assessment of radiation doses to the brain. Current systemic biokinetic models, recommended by the ICRP assume a uniform distribution in the brain for any specific element because at present, it is not feasible to characterize with much confidence (1) the distribution of most individual elements among different regions of the brain or (2) element-specific biokinetics in individual regions of the brain. Thus, the assumption still would be that the element is uniformly distributed in the entire brain. We recognize, however, the importance of different distributions of radium within different brain regions and would address such issues moving forward as new information becomes available (Boice et al. 2021; NCRP 2021).

Overview of methodology for internal dose calculation

The starting point for internal dose calculation will be measurement-based radium body burden estimates performed in the past. The timing, number, and type of radiological measurement(s) made to estimate body burdens in dial workers varied by individual and were dependent on the available instrumentation and technology. For example, thallium-activated sodium iodide (NaI(Tl)) crystal was reported to be an effective scintillation material in 1948 (Hine 1977), but neither useful crystals, nor appropriate electronic circuitry to collect and interpret the light signal from a NaI(Tl) crystal, were readily available at the time. This limited the practical application of the NaI(Tl) detector for the next few years. Common types of measurements made, though, included radon breath measurements, whole-body gamma-ray measurements, and autopsy or other posthumous measurements (e.g. autoradiography of bones) (Rowland 1994).

Body burden estimates provide the activity of ^{226}Ra and/or ^{228}Ra in the body at the time of measurement, which can then be coupled with biokinetic models and the time of first intake from an individual's work history to develop a time series of relevant activity levels in the body. In other words, the latest biokinetic models described above will use previously determined body burdens to develop estimates of activity versus time in each pertinent source region in the body for radium and its progeny. The temporal activity data

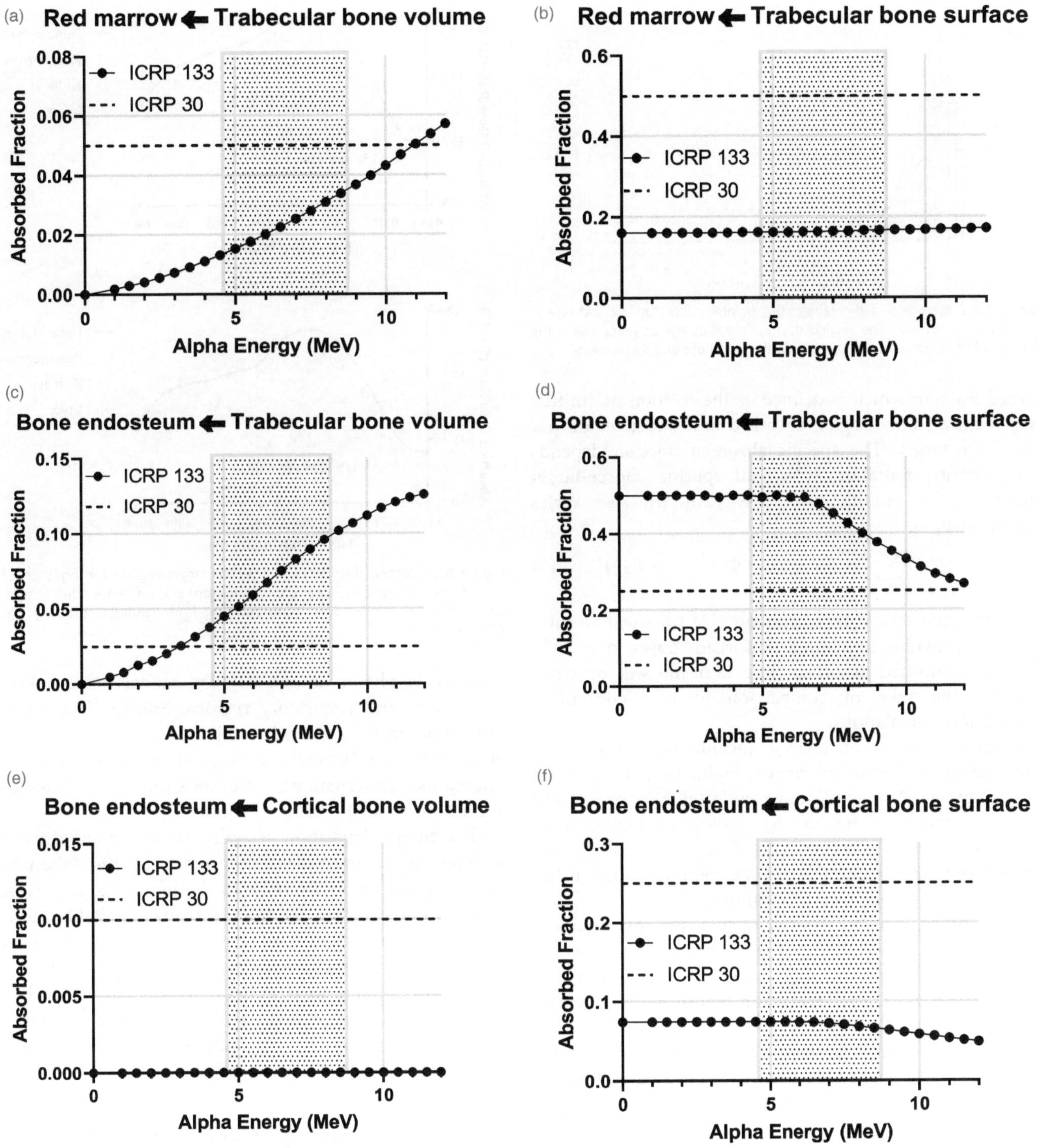

Figure 6. Plots of the fraction of the energy absorbed in a target tissue per alpha energy emitted from a skeletal source region. Data points (circles) are from ICRP Publication 133 (2016) and dashed lines are the values in ICRP Publication 30 (1979). The shaded region represents the range of alpha particle energies emitted in the ^{226}Ra and ^{228}Ra decay chains.

will then be customized to each dial worker based on their age and duration of employment. This work will also treat the intake of radium as chronic over their employment. Other factors, such as whether a dial painter's work bridged a transition from mouth tipping of brushes to cessation of this practice (somewhere around 1926) can be folded into the chronic intake model.

The absorbed dose rate, $\dot{D}_T$, to a particular target region, T, is computed using Equation (2) where A_S is the activity at a given time in a source region, S, and $S(r_T \leftarrow r_S)$ is the S-coefficient, defined as the absorbed dose to a target region, r_T, per nuclear transformation taking place in a source region, r_S.

$$\dot{D}_T = \sum_{r_S} A_S \cdot S(r_T \leftarrow r_S) \tag{2}$$

The S-coefficient is computed as shown in Equation (3) and depends on the energy, $E_{R,i}$, and yield, $Y_{R,i}$, of emission, i, of radiation type, R, from a given radionuclide. The last term in Equation (3), $\Phi(r_T \leftarrow r_S, E_{R,i})$, is the specific

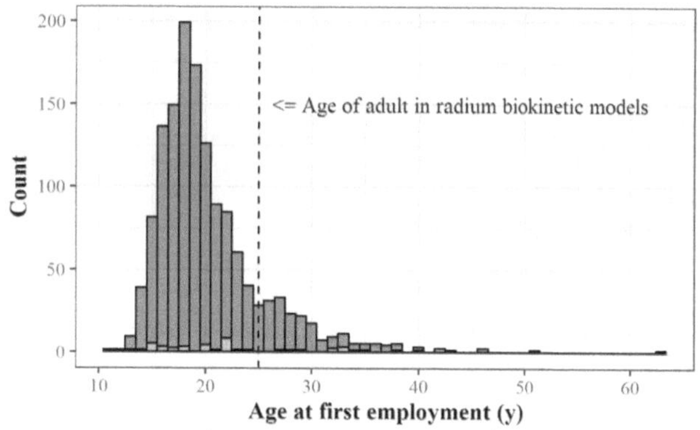

Figure 7. 1,558 members of the radium dial painter cohort by sex and age at time of first employment. The vertical dashed line is at age 25 years and is the age of the adult in systemic biokinetic models for radium and its progeny.

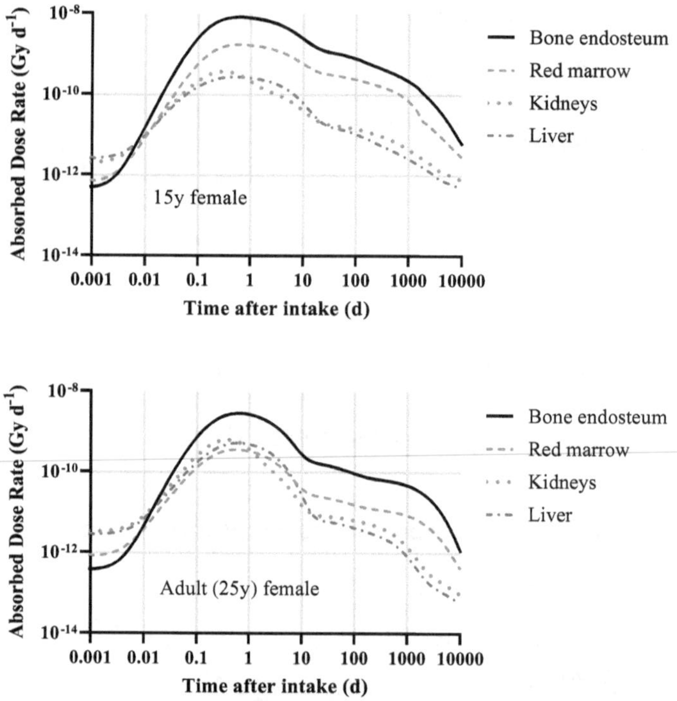

Figure 8. Absorbed dose rates to selected target regions for ingestion of 1 Bq of ^{226}Ra in a reference 15-year-old female (top) and reference adult female (bottom). The plots include contributions from radium progeny created post-ingestion.

absorbed fraction which is defined as the fraction of emitted energy from a source region absorbed in a target region per mass of the target. The specific absorbed fraction depends on the energy, radiation type, and specific source-target geometry. Before adulthood, the S-coefficient also varies with respect to age.

$$S(r_T \leftarrow r_S) = \sum_R \sum_i E_{R,i} \cdot Y_{R,i} \cdot \Phi(r_T \leftarrow r_S, E_{R,i}) \quad (3)$$

The absorbed dose rates as functions of time will be integrated to provide annualized absorbed doses over life. Similarly, committed absorbed dose over life will be computed for the sake of comparison to past committed absorbed dose calculations.

Dosimetric targets of interest include the bone endosteum, red (active) marrow, breast, brain, liver, lung, heart wall, and others. While the bone endosteum is considered the current target of interest for radiogenic bone cancer (Gossner et al. 2000, Gossner 2003, Bolch et al. 2007), dose to the entire bone volume will also be computed since it has been used in past studies on this cohort.

The absorbed dose to the red (active) marrow resulting from alpha particles emitted in the bone surface will not be uniform across the marrow cavity due to the short range of the alpha particles. The current definition of the bone endosteum target is the first 50 µm of marrow space adjacent to the bone surface. This region contains red marrow and, for the case of radiations emitted from the bone, the endosteum dose is equivalent to the shallow red marrow dose. The difference between the shallow marrow dose and the marrow dose averaged over the entire cavity may provide some insight into deterministic and stochastic hematopoietic response, or the lack thereof as seen in previous reports of radium dial painters (Spiers et al. 1983; Priest 1989).

Use of the latest energy absorption data

Energies and yields of the various radionuclide emissions will be taken from ICRP Publication 107 (International Commission on Radiological Protection (ICRP) 2008). In 2016, the ICRP published new specific absorbed fractions for reference adults (International Commission on Radiological Protection (ICRP) 2016). These specific absorbed fractions were computed using the latest whole-body voxel phantoms and detailed models of the skeleton, alimentary, and respiratory regions. Similar data for reference children were recently published by the ICRP (Schwarz et al. 2021a, 2021b) using an age-dependent set of reference phantoms (International Commission on Radiological Protection (ICRP) 2020a) and models.

The energy deposition of alpha particles emitted from the skeleton is of particular importance to the ingestion of radium given its high uptake to the bone. Figure 6 contains plots of the absorbed fractions provided in Publication 133 compared to values in Publication 30 (International Commission on Radiological Protection (ICRP) 1979). The differences are due to new definitions of source and target regions and an improved capability to perform radiation transport calculations in complex geometries such as the skeleton.

Age and sex dependency of dosimetric calculations

As referenced earlier, the updated dosimetry calculations will include consideration of the age and sex of the workers at the time of their ingestion of radium. This will impact both the biokinetic, or activity term in Equation (2) and the energy absorption term.

Figure 7 is a histogram of the ages of dial workers when they began their work. Most dial workers were teenagers at the time of first employment (see also Figure 2) and therefore at the time they were ingesting radium. Figure 8 gives the absorbed dose rates for selected target tissues due to ingestion of 1 Bq of ^{226}Ra as a function of time after intake. Due to significant differences in skeletal growth at age 15 compared to age 25 (adult), intake of radium as a teenager

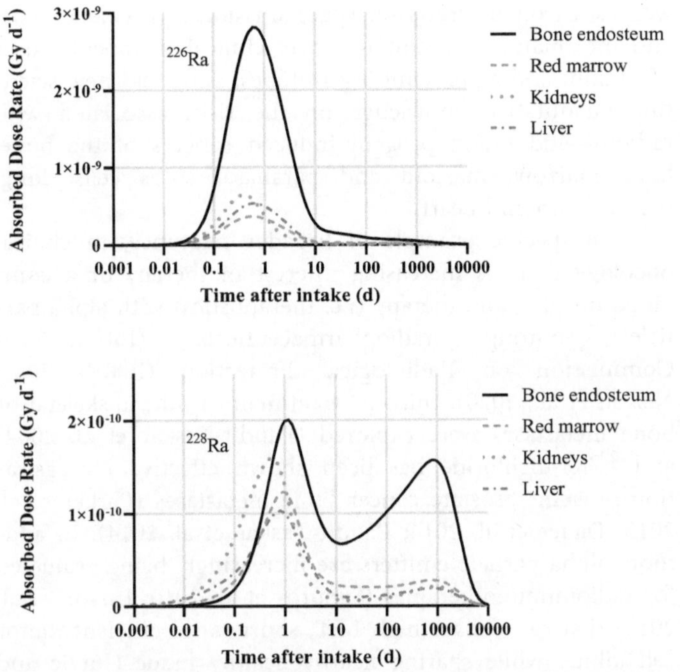

Figure 9. Absorbed dose rate to selected targets versus time following the ingestion of 1 Bq of ^{226}Ra (top) and ^{228}Ra (bottom) including contributions from radium progeny created post-ingestion.

will result in extended retention of radium and its progeny in the skeleton.

Also of note is that age and sex influence body weight. ICRP Publication 23 (International Commission on Radiological Protection (ICRP) 1975) lists weight of the reference total body as 70 kg for men and 58 kg for women. Current ICRP reference total body masses are 60 kg for adult females and 53 kg for 15-year-old females (International Commission on Radiological Protection (ICRP) 2009, 2020a). The United States did not start collecting detailed, comprehensive data on heights and weights of Americans until 1960 (Stoudt et al. 1960; Flegal 2009), although pockets of data were collected as early as 1858 for adults (Hathaway and Foard 1960) and 1877 for children and teenagers (Hathaway 1957). Interestingly, a 1923 dataset drawing from 12 schools in the North Eastern and North Central states (about 55,000 girls) determined a standard weight for 15-year-old females (depending on height) to be about 53 kg (Hathaway 1957), giving additional confidence in applying the current ICRP model to teenagers in our cohort.

Unique temporal aspects of dose resulting from ^{226}Ra and ^{228}Ra

The two radium decay chains (Figure 1) have several differences in their half-lives and elemental constituents. As mentioned previously, ^{226}Ra has a significantly longer half-life than ^{228}Ra and is itself an alpha emitter unlike the beta-emitting ^{228}Ra. However, the comparatively short half-life of ^{228}Ra allows for in-growth of the alpha-emitting progeny ^{228}Th and ^{224}Ra which are important dose contributors. Figure 9 contains plots of the absorbed dose rate to selected target tissues at times following the ingestion of 1 Bq of the two radium isotopes. Note that the dose rate due to ^{226}Ra and progeny peaks at under 1-day following ingestion. Conversely, the absorbed dose rate due to ^{228}Ra and progeny reaches has two peaks. The early peak at 1 day is due to dose from the beta-emitting ^{228}Ra and removal of the non-absorbed radium via the alimentary tract. The late peak occurring just before 1000 days after ingestion is due to the in-growth of alpha-emitting progeny ^{228}Th, ^{224}Ra and others from the fraction of radium which was absorbed into the skeleton and systemic tissue. It takes time for the in-growth of important ^{228}Ra progeny in the body.

By computing dose rate and annualized doses, the current study will allow for the important temporal contributions of the two radium isotopes and their progeny to be included in the dosimetry. This is particularly important since radiogenic disease occurred at varying times in members of the cohort.

Rowland, Stehney and Lucas (1978), among others, previously described various strategies for weighting ^{228}Ra compared to ^{226}Ra for the purpose of assessing dose-effect relationships, ultimately assigning different values for different biological effects. For example, Rowland, Stehney and Lucas (1978) describe three different ways to weight ^{228}Ra with respect to the induction of osteosarcomas: comparative effective alpha energy per decay, comparative energy deposited over the average time to sarcoma appearance, and comparison of dose-response curves between two groups with intakes of predominantly one isotope. The average value of these approaches, 2.5, was taken for comparative effectiveness of ^{228}Ra in the induction of osteosarcomas. Corresponding dose-response relationships (where 'dose' is taken to be systemic intake) were developed and analyzed for a total systemic intake of ^{226}Ra activity plus 2.5 times ^{228}Ra activity.

Comparatively, only ^{226}Ra intake was considered in developing dose-response relationships for the induction of mastoid and paranasal sinus carcinomas. None of these types of cancers were observed in those whose intake was primarily ^{228}Ra; this has been attributed to the importance of radon progeny accumulating in the sinus cavities, as touched on previously (Rowland 1994). ^{220}Rn ($t_{1/2} = 55.6$ s), fifth progeny of ^{228}Ra, does not have time to migrate to the sinus cavities, compared to ^{222}Rn ($t_{1/2} = 3.82$ d), first progeny of ^{226}Ra (Figure 1).

Missing body burden measurements

A job exposure matrix will be developed to provide body burden estimates, based on work history, for those members of the cohort without body burden measurements. This matrix will use the ingestion rate for other cohort members performing the same task in the same workplace during the same period of time. A model will be developed to provide a probability distribution of possible doses to these individuals.

Additional routes of exposure

Dial workers were exposed to gamma radiations emitted from the paint and inhalation of airborne activity. The magnitude of these exposures depends on the working conditions, e.g. number of coworkers, ventilation, and work practices. Bloomfield and Knowles (1933) surveyed several facilities and reported that painters were daily handling 50–500 μg (1.9–19 MBq) of radium. Measured external dose rates ranged from 0.7 to 46 cGy/y (0.8–5.3 μGy/h). Airborne activity of radium and radon were also observed. The average airborne activity concentration of radon was 0.051 μCi/m^3 (1.8 kBq/m^3) and that of radium was 260 pCi/m^3 (9.6 Bq/m^3). Estimates of the dose to the female breast and other tissues can be derived from these measurements and applying the dose rate coefficients of ICRP Publication 144 (International Commission on Radiological Protection (ICRP) 2020b). The dose contribution due to inhalation of airborne radium would largely follow the above methodology with the respiratory tract being the entrance into the body. The dose contribution of the airborne radon can be derived using the methods of ICRP Publication 137 (International Commission on Radiological Protection (ICRP) 2017) with due consideration to the workplace air ventilation and potential unattached radon short-lived progeny. Tissue dose coefficients can be derived from the measured external dose rates using the dose rate coefficients of ICRP Publication 144 (International Commission on Radiological Protection (ICRP) 2020b), which are based on the computational phantoms of ICRP Publication 110 (International Commission on Radiological Protection (ICRP) 2009), in the manner outlined in Appendix E of NCRP Report 178 (National Council on Radiation Protection and Measurements (NCRP) 2018).

These contributions are in addition to the potential contribution of internally deposited radium. Rowland et al. (1989) concluded that the elevated breast cancer risk among the dial painters cannot be attributed to the external dose and questioned the potential contribution of internally deposited radium. This conclusion is consistent with the studies in the United Kingdom (UK) where the habit of licking brushes was not seen, and any radiation exposure was primarily from external gamma radiation (mean absorbed breast dose 330 mGy) (Baverstock and Papworth 1989). The possibility that external radiation as well as radium intake among young women might be related to breast cancer (Stebbings 2001) will be reexamined, addressing personal characteristics, such as nulliparity, which are related to breast cancer risk (Adams and Brues 1980; Schieve et al. 1997).

Applicability to current exposure circumstances

The current work ultimately seeks to address female-specific health risks following intakes of radium; provide information on public and worker health risks relevant to environmental cleanups of former nuclear facilities and weapons testing; provide information on cognitive function following brain exposure relevant to flight crews at high altitude as well as astronaut crews on space missions; provide insights and information relevant to novel clinical therapeutic uses of radium and alpha emitting isotopes; and yield new scientific quantitative knowledge on the risks associated with radium- and radon progeny-induced cancers of the bone, bone marrow, mastoid and paranasal sinus cells, lung, breast, brain and heart.

As a specific example of broader relevance, in clinical oncology there is increasing interest in therapy or a combined imaging and therapy (i.e. theranostics) with alpha particle emitting radiopharmaceuticals (International Commission on Radiological Protection (ICRP) 2019; Nelson et al., 2020). Initially, treatment of diffuse skeletal or bone metastases were explored (Pandit-Taskar et al. 2004) and ^{223}Ra-dichloride has been shown effective for castration-resistant prostate cancer bone metastases (Parker et al. 2013; Dauer et al. 2014; Pandit-Taskar et al. 2014). In addition, alpha particle emitters are increasingly being evaluated for radioimmunotherapies (Sgouros et al. 2010; Larson et al. 2015) that can deliver high LET, short range efficient tumor cell killing while sparing nearby healthy tissue (Jurcic and Rosenblat 2014). As these clinical applications are moving forward, there is interest in addressing uncertainties associated with actual local absorbed doses (especially in the bone) and the associated selection of the most appropriate radiation weighting factors for alpha dosimetry (Sgouros et al. 2010; Lassmann and Nosske, 2013). This updated study of the radium dial worker cohort and the associated improvements in the understanding of the dosimetry of alpha emitters (along with their progeny) in the body will be informative for the ongoing clinical development of effective and optimal treatment protocols using radium or other alpha particles.

Conclusion

In the 1967 report proposing a National Center of Human Radiobiology (Rowland 1994), Evans made arguments still relevant today for continuing the study of the dial workers, e.g.: the nuclear era necessitates valid radiation protection criteria; human protection criteria are best derived from human evidence and, despite tragic outcomes for many of the exposed and their families, the dial workers represent a unique scientific opportunity not likely to be repeated; and gaining information from this group will benefit current and future generations, with progressing results likely applicable to other exposure scenarios not yet envisioned (such as in medicine, environmental contamination and space exploration).

Acknowledgements

We are tremendously indebted to innumerable individuals, including the scientists, physicians, epidemiologists, statisticians, technicians, librarians, etc. who have studied these workers and who have published and otherwise preserved associated information since the 1920s. We also acknowledge that in many cases, radium dial workers suffered tremendously and died, and that this suffering and loss was also acutely felt by their loved ones and their communities.

Disclosure statement

No potential conflict of interest was reported by the author(s).

Funding

The study of radium dial painters, a component of the Million Person Study, is supported in part by grants from the US Department of Energy [Grant No. DE-AU0000042 and DE-AU0000046] awarded to the National Council on Radiation Protection and Measurements, and a grant from the National Aeronautics and Space Administration [80NSSC17M0016]. Further, contract support was received by Oak Ridge National Laboratory from the Office of Radiation and Indoor Air, US Environmental Protection Agency, under Interagency Agreement DOE No. 1824 S581-A1, under contract No. DE-AC05-00OR22725 with UT-Battelle. The United States Transuranium and Uranium Registries is funded by US Department of Energy, Office of Domestic and International Health Studies (AU-13), under grant award DE-HS0000073 to the College of Pharmacy and Pharmaceutical Sciences at Washington State University.

ORCID

Nicole E. Martinez http://orcid.org/0000-0002-7184-3043
Derek W. Jokisch http://orcid.org/0000-0002-1567-5268
Lawrence T. Dauer http://orcid.org/0000-0002-5629-8462
John D. Brockman http://orcid.org/0000-0001-7419-5558
Sergey Y. Tolmachev http://orcid.org/0000-0003-0077-106X
Maia Avtandilashvili http://orcid.org/0000-0001-9400-8617
Michael T. Mumma http://orcid.org/0000-0001-7506-8710
John D. Boice http://orcid.org/0000-0002-8755-1299

References

Adams EE, Brues AM. 1980. Breast cancer in female radium dial workers first employed before 1930. J Occup Med. 22:583–587.

Adams EE, Brues AM, Anast GA. 1983. Survey of ocular cataracts in radium dial workers. Health Phys. 44(Suppl 1):73–79.

Aub JC, Evans RD, Hemplelmann LH, Martland HS. 1952. The late effects of internally-deposited radioactive materials in man. Medicine. 31(3):221–329.

Azizova TV, Bannikova MV, Grigoryeva ES, Rybkina VL, Hamada N. 2020. Occupational exposure to chronic ionizing radiation increases risk of Parkinson's disease incidence in Russian Mayak workers. Int J Epidemiol. 49(2):435–447.

Baverstock KF, Papworth DG. 1989. The UK radium luminizer survey. In: Taylor DM, Mays CW, Gerber GB, Thomas RG, editors. BIR Report 21: Risks from Radium and Thorotrast. London (UK): British Institute of Radiology; p. 72–76.

Bloomfield JJ, Knowles FL. 1933. Health aspects of radium dial painting. II. Occupational environment. J Indust Hygiene. 15:368–382.

Blum T. 1924. Osteomyelitis of the mandible and maxilla. J Am Dent Assoc. 11:802–805.

Boice JD, Cohen SS, Mumma MT, Dupree-Ellis E, Eckerman KF, Leggett RW, Boecker B, Brill A, Henderson B. 2006. Mortality among radiation workers at Rocketdyne (Atomics International), 1948–1999. Radiat Res. 166(1):98–115.

Boice JD Jr, Cohen SS, Mumma MT, Ellis ED. 2019. The Million Person Study, whence it came and why. Int J Radiat Biol. 1–14. doi:10.1080/09553002.2019.1589015.

Boice JD Jr, Quinn B, Ansari A, Blake PK, Blattnig SR, Caffrey EA, Cohen SS, Golden AP, Held KD, Jokisch DW, Mumma MT, Samuels C, Till JE, Tolmachev SY, Yoder RC, Zhou J, Dauer LT. 2021. A million persons, a million dreams: the study of one million radiation workers and veterans. Health Phys. (submitted)

Bolch WE, Shah AP, Watchman CJ, Jokisch DW, Patton PW, Rajon DA, Zankl M, Petoussi-Henss N, Eckerman KF. 2007. Skeletal absorbed fractions for electrons in the adult male: considerations of a revised 50-μm definition of the bone endosteum. Radiat Prot Dosim. 127(1–4):169–173.

Bruenger FW, Lloyd RD. 1989. The influence of age at time of exposure to Ra-226 or Pu-239 on distribution, retention, post-injection survival and bone tumor induction in beagle dogs. In: Thirty-fourth Annual Meeting of the Health Physics Society, Abstracts of Papers Presented at the Meeting Vol. 56 (Suppl. l), 27; June 25–29, 1989; Albuquerque, NM: Albuquerque Convention Center; New York (NY): Pergamon Press.

Bruenger FW, Smith JM, Atherton DR, Jee WSS, Lloyd RD, Stevens W. 1983. Skeletal retention and distribution of Ra-226 and Pu-239 in beagles injected at ages ranging from 2 days to 5 years. Health Phys. 44:513–527.

Campbell KM, Deck D, Krupski A. 2008. Record linkage software in the public domain: a comparison of Link Plus, The Link King, and a 'basic' deterministic algorithm. Health Informatics J. 14(1):5–15.

Castle WB, Drinker KR, Drinker CK. 1925. Necrosis of the jaw in workers employed in applying a luminous paint containing radium. J Ind Hyg. 8:371–382.

Cothern CR, Smith JE Jr. 1987. Environmental radon. New York (NY): Springer.

Dauer LT, Williamson MJ, Humm J, O'Donoghue J, Ghani R, Awadallah R, Carrasquillo J, Pandit-Taskar N, Aksnes A-K, Biggin C, et al. 2014. Radiation safety considerations for the use of $^{112}RaCl_2$ in men with castration-resistant prostate cancer. Health Phys. 106(4):494–504.

Department of Energy (DOE). 2018. DOE 2017 occupational radiation exposure. Washington (DC): US Department of Energy, Office of Environment, Health, Safety and Security, The Radiation Exposure Monitoring System (REMS).

Department of Energy (DOE). 2021. Comprehensive epidemiological data resource (CEDR). [accessed 2020 Dec 18]. https://oriseapps.orau.gov/cedr/default.aspx.

Department of Labor (DOL). 2011. The Fair Labor Standards Act of 1938, as amended. Washington (DC): US Deptartment of Labor, Wage and Hour Division. [accessed 2021 Jan 12]. http://purl.fdlp.gov/GPO/gpo24709.

Ellis ED, Boice JD, Golden AP, Girardi DJ, Cohen SS, Mumma MT, Shore RE, Leggett RW, Kerr GD. 2018. Dosimetry is key to good epidemiology: workers at Mallinckrodt Chemical Works had seven different source exposures. Health Phys. 114(4):386–397.

Flegal KM.2009. Trends in body weight and overweight in the U.S. population. Nutr Rev. 54(4):S97–S100.

Fry SA. 1998. Studies of U.S. radium dial workers: an epidemiological classic. Radiat Res. 150(5):S21–S29.

Goans RE, Iddins CJ, Christensen D, Wiley A, Dainiak N. 2015. Appearance of Pseudo Pelger Huet anomaly after exposure to ionizing radiation *in vivo*. Health Phys. 108(3):303–307.

Goans RE, Iddins CJ, Ossetrova NI, Ney PH, Dainiak N. 2017. The Pseudo-Pelger Huët cell – a new permanent radiation biomarker. Health Phys. 112(3):252–257.

Goans RE, Toohey RE, Iddins CJ, McComish SL, Tolmachev SY, Dainiak N. 2019. The pseudo-Pelger Huët cell as a retrospective dosimeter: analysis of a radium dial painter cohort. Health Phys. 117(2):143–148.

Gossner W. 2003. Target cells in internal dosimetry. Radiat Prot Dosim. 105: 39–42.

Gossner, W., Masse, R., Stather, J.W., 2000. Cells at risk for dosimetric modeling relevant to bone tumour induction. Radiat Prot Dosim, 1. 92:209–213.

Gunderman RB, Gonda AS. 2015. Radium girls. Radiology. 274(2): 314–318.

Hathaway ML. 1957. Heights and weights of children and youth in the United States. home economics research report No. 2. Washington (DC): US Department of Agriculture.

Hathaway ML, Foard ED. 1960. Heights and weights of adults in the United States. Home Economics Research Report No. 10. Washington (DC): US Department of Agriculture.

Hine GJ. 1977. The inception of photoelectric scintillation detection commemorated after three decades. J Nucl Med. 18(9):867–871.

Hoffman FL. 1925. Radium (mesothorium) necrosis.JAMA. 85(13): 961–965.

International Commission on Radiological Protection (ICRP). 1973. Alkaline earth metabolism in adult man (ICRP Publication 20). Health Phys. 24(2):125–221.

International Commission on Radiological Protection (ICRP). 1975. Report of the task group on reference man (ICRP Publication 23). Oxford (UK): Pergamon Press.

International Commission on Radiological Protection (ICRP). 1979. Limits for intakes of radionuclides by workers (ICRP Publication 30). Ann ICRP. 2(3/4):1–116.

International Commission on Radiological Protection (ICRP). 1993. Age-dependent doses to members of the public from intake of radionuclides: Part 2, ingestion dose coefficients (ICRP Publication 67). Ann ICRP. 23(3/4):1–167.

International Commission on Radiological Protection (ICRP). 2008. Nuclear decay data for dosimetric calculations (ICRP Publication 107). Ann ICRP. 38(3):1–96.

International Commission on Radiological Protection (ICRP). 2009. Adult reference computational phantoms (ICRP Publication 110). Ann ICRP. 39(2):1–110.

International Commission on Radiological Protection (ICRP). 2016. The ICRP computational framework for internal dose assessment for reference adults: specific absorbed fractions (ICRP Publication 133). Ann ICRP. 45(2):1–74.

International Commission on Radiological Protection (ICRP). 2017. Occupational intakes of radionuclides: Part 3 (ICRP Publication 137). Ann ICRP. 46(3/4):1–486.

International Commission on Radiological Protection (ICRP). 2019. Radiological protection in therapy with radiopharmaceuticals (Publication 140). Ann ICRP. 48(1):1–102.

International Commission on Radiological Protection (ICRP). 2020a. Paediatric reference computational phantoms (ICRP Publication 143). Ann ICRP. 49(1):1–299.

International Commission on Radiological Protection (ICRP). 2020b. Dose coefficients for external exposures to environmental sources (ICRP Publication 144). Ann ICRP. 49(2):1–147.

Jurcic JG, Rosenblat TL. 2014. Targeted alpha-particle immunotherapy for acute myeloid leukemia. Am Soc Clin Oncol Educ Book, 34. e126–131, e131.

Kathren RL, Tolmachev SY. 2019. The US Transuranium and Uranium Registries (USTUR): a five-decade follow-up of plutonium and uranium workers. Health Phys. 117(2):118–132.

Keane AT, Holtzmann RB, Rundo J. 1994. Prevalence of technical mesothorium in self-luminous compounds used by New Jersey radium dial workers (ANL/ER/CP-81143). Argonne (IL): Argonne National Laboratory.

Keane AT, Schlenker RA. 1987. Long-term loss of radium in 63 subjects exposed at ages 6–46. In: Gerger H, Metivier H, Smith H, editors. Age-related factors in radionuclide metabolism and dosimetry. Proceedings of a workshop in Angers, November 26-28, 1986. Dordrecht (NL): Martinus Nijhoff Publishers; p. 127–135.

Larson SM, Carrasquillo JA, Cheung N-KV, Press OW. 2015. Radioimmunotherapy of human tumours. Nat Rev Cancer. 15(6): 347–360.

Lassmann M, Nosske D. 2013. Dosimetry of 223Ra-chloride: dose to normal organs and tissues.Eur J Nucl Med Mol Imaging. 40(2): 207–212.

Leggett RW. 1992. A generic age-specific biokinetic model for calcium-like elements. Rad Prot Dosim. 41:183–198.

Leggett RW, Tolmachev SY, Boice JD, Jr. 2018. Potential improvements in brain dose estimates for internal emitters. Int J Rad Biol. 4:1–54. doi:10.1080/09553002.2018.1554923.

Lloyd RD, Bruenger FW, Jones CW, Taylor GN, Mays CW. 1983a. Retention in mature beagles injected at 5 years of age. Radiat Res. 94(1):210–216.

Lloyd RD, Bruenger FW, Mays CW, Jones CW. 1983b. Skeletal radon-to-radium ratios in neonatal, juvenile and mature beagles and in adult St. Bernards. Health Phys. 44:61–63.

Lloyd RD, Jones CW, Bruenger FW, Atherton DR, Mays CW. 1983c. Radium retention and dosimetry in juvenile beagles. Radiat Res. 94(2):295–304.

Lloyd RD, Taylor GN, Jones CW, Mays CW. 1983d. Radium retention and dosimetry in the St. Bernard. Radiat Res. 95(1):150–157.

Lloyd RD, Mays CW, Atherton DR. 1976a. Distribution of injected Ra-226 and Sr-90 in the beagle skeleton. Health Phys. 30:183–189.

Lloyd RD, Mays CW, Atherton DR, Taylor GN, Van Dilla MA. 1976b. Retention and skeletal dosimetry of injected Ra-226, Ra-228, and Sr-90 in beagles. Radiat Res. 66(2):274–287.

Lloyd RD, Mays CW, Taylor GN, Atherton DR, Bruenger FW, Jones CW. 1982. Radium-224 retention, distribution, and dosimetry in beagles. Radiat Res. 92(2):280–295.

Marazziti D, Baroni S, Catena-Dell'Osso M, Schiavi E, Ceresoli D, Conversano C, Dell'Osso L, Picano E. 2012. Cognitive, psychological and psychiatric effects of ionizing radiation exposure.CMC. 19(12): 1864–1869.

Martland HS. 1926. Histopathology of certain anemias due to radioactivity. Proc NY Pathological Soc NS. 26:1–5.

Martland HS. 1929. Occupational poisoning in manufacture of luminous watch dials. JAMA. 92(6):466.

Martland HS. 1931. The occurrence of malignancy in radioactive persons. Am J Cancer. 15:2435.

Martland HS, Conlon P, Knef JP. 1925. Some unrecognized dangers in the use and handling of radioactive substances: with especial reference to the storage of insoluble products of radium and mesothorium in the reticulo-endothelial system.JAMA. 85(23):1769–1776.

Martland HS, Humphries RE. 1929. Osteogenic sarcoma in dial painters using luminous paint. Arch Path Lab Med. 7:406–417.

Monthly Labor Review (MLR). 1926. Effects of use of radioactive substances on the health of workers. MLR. 22(5):18–31.

Moore K. 2017. The radium girls: the dark story of America's shining women. Naperville (IL): Sourcebooks.

Mumma MT, Cohen SS, Sirko JL, Ellis ED, Boice JD Jr. 2018. Obtaining vital status and cause of death on a million persons. Int J Radiat Biol. 9:1–21. doi:10.1080/09553002.2018.1539884.

Muth H, Glöbel B. 1983 Age dependent concentration of Ra-226 in human bone and some transfer factors from diet to human tissues. Health Phys. 44:113–121.

National Academies/National Research Council (NA/NRC). 1988. Health risks of radon and other internally deposited alpha-emitters: BEIR IV. Washington (DC): National Academies Press.

National Council on Radiation Protection and Measurements (NCRP). 2009. Report No. 164: uncertainties in internal radiation dose assessment. Bethesda (MD): NCRP.

National Council on Radiation Protection and Measurements (NCRP). 2018. Report No. 178: deriving organ doses and their uncertainty for epidemiologic studies (with a focus on the one million U.S. workers and veterans study of low-dose radiation health effects). Bethesda (MD): NCRP.

National Council on Radiation Protection and Measurements (NCRP). 2019. Report No. 183: Radiation exposures in space and the potential for central nervous system effects: phase II. Bethesda (MD): NCRP.

National Council on Radiation Protection and Measurements (NCRP). 2021. NCRP Draft SC 6-12 Commentary: development of kinetic and anatomical models for brain dosimetry for internally deposited radionuclides. Bethesda (MD): NCRP. [accessed 2021 February 20]. https://ncrponline.org/wp-content/themes/ncrp/Docs_in_Review/SC6-12_Council%20Review%20Draft_2021%2002%2016.pdf

Nelson BJ, Andersson JD, Wuest F.2020. Targeted alpha therapy: progress in radionuclide production, radiochemistry, and applications. Pharmaceuticals. 13(1):49.

New York Times. 1928. Radium paint takes its inventor's life. Obituary. *New York Times TimesMachine*, 15 November 1928, p. 29, column 5. [accessed 2021 Mar 11]. https://timesmachine.nytimes.com/

Norris WP, Speckman TW, Gustafson PF. 1955. Studies of the metabolism of radium in man. Am J Roentgen. 73:785–802.

Nuclear Regulatory Commission (NRC). 2019. Radiation exposure information and reporting system (REIRS). US Nuclear Regulatory Commission. [accessed 2020 Feb 7]. https://www.reirs.com/.

Pandit-Taskar N, Batraki M, Divgi CR. 2004. Radiopharmaceutical therapy for palliation of bone pain from osseous metastases. J Nucl Med. 45(8):1358–1365.

Pandit-Taskar N, Larson SM, Carrasquillo JA. 2014. Bone-seeking radiopharmaceuticals for treatment of osseous metastases, Part 1: alpha therapy with 223Ra-dichloride. J Nucl Med. 55(2):268–274.

Parihar VK, Allen BD, Caressi C, Kwok S, Chu E, Tran KK, Chmielewski NN, Giedzinski E, Acharya MM, Britten RA, Baulch JE, Limoli CL. 2016. Cosmic radiation exposure and persistent cognitive dysfunction. Sci Rep, 16:34774.

Parihar VK, Angulo MC, Allen BD, Syage A, Usmani MT, Passerat de la Chapelle E, Amin AN, Flores L, Lin X, Giedzinski E, Limoli CL. 2020. Sex-specific cognitive deficits following space radiation exposure. Front Behav Neurosci. 14:535885.

Parker, C., Nilsson, S., Heinrich, D., Helle, S.I., O'Sullivan, J.M., Fosså, S.D., Chodacki, A., Wiechno, P., Logue, J., Seke, M., Widmark, A., 2013. Alpha emitter radium-223 and survival in metastatic prostate cancer.N Engl J Med. 369(3):213–223.

Parks NJ, Keane AT. 1983. Consideration of age-dependent radium retention in people on the basis of the beagle model. Health Phys. 44(Suppl 1):103–112.

Parks NJ, Pool RR, Williams JR, Wolf HG. 1978. Age and dosage-level dependence of radium retention in beagles. Radiat Res. 75(3): 617–632.

Pasqual E, Boussin F, Bazyka D, Nordenskjold A, Yamada M, Ozasa K, Pazzaglia S, Roy L, Thierry-Chef I, de Vathaire F, Benotmane MA, Cardis E. 2021. Cognitive effects of low dose of ionizing radiation – lessons learned and research gaps from epidemiological and biological studies. Environ Int. 147:106295.

Polednak AP. 1978a. Bone cancer among female radium dial workers. Latency periods and incidence rates by time after exposure: brief communication. J Natl Cancer Inst. 60(1):77–82.

Polednak AP. 1978b. Long-term effects of radium exposure in female dial workers: differential white blood cell count. Environ Res. 15(2):252–261.

Polednak AP. 1980. Fertility of women after exposure to internal and external radiation. J Environ Pathol Toxicol. 4:457–470.

Polednak AP. 1986. Thyroid tumors and thyroid function in women exposed to internal and external radiation. J Environ Pathol Toxicol Oncol. 7:53–64.

Polednak AP, Stehney AF, Rowland RE. 1978. Mortality among women first employed before 1930 in the U.S. radium dial-painting painting industry. A group ascertained from employment lists. Am J Epidemiol. 107(3):179–195.

Priest ND. 1989. Alpha-emitters in the skeleton: an evaluation of the risk of leukaemia following intakes of plutonium 239. In: Taylor DM, Mays CW, Gerber GH, Thomas RG, editors. BIR report 21: risks from radium and thorotrast. London (UK): British Institute of Radiology; p. 159–165.

Rowland RE. 1993. Low-level radium retention by the human body: a modification of the ICRP publication 20 retention equation. Health Phys. 65(5):507–513.

Rowland RE. 1994. Radium in humans. A review of U.S. studies. Report ANUER-3. Argonne (IL): Argonne National Laboratory.

Rowland RE, Lucas HF Jr. 1982. Radium dial workers. Conference on radiation carcinogenesis: epidemiology and biological significance, Washington, DC, USA, 24 May 1982.

Rowland RE, Lucas HF Jr. 1984. Radium-dial workers. In: Boice JD, Jr. and Fraumeni JF Jr., editors. Radiation carcinogenesis: epidemiology and biological significance. New York (NY): Raven Press; p. 231–240.

Rowland RE, Lucas HF, Schlenker RA. 1989. External radiation doses received by female radium dial painters. In: Taylor DM, Mays CW, Gerber GB, Thomas RG, editors. BIR report 21: risks from radium and thorotrast. London (UK): British Institute of Radiology; p. 67–72.

Rowland RE, Stehney AF, Brues AM, Littman MS, Keane AT, Patten BC, Shanahan MM. 1978. Current status of the study of 226Ra and 228Ra in humans at the Center for Human Radiobiology. Health Phys. 35:159–166.

Rowland RE, Stehney AF, Lucas HF Jr. 1978. Dose-response relationships for female radium dial workers. Radiat Res. 76(2):368–383.

Rowland RE, Stehney AF, Lucas HF. 1983. Dose-response relationships for radium induced bone sarcomas. Health Phys. 44(Suppl I):15–31.

Schieve LA, Davis F, Roeske J, Handler A, Freels S, Stinchcomb T, Keane A. 1997. Evaluation of internal alpha-particle radiation exposure and subsequent fertility among a cohort of women formerly employed in the radium dial industry. Radiat Res, 2147:236–244.

Schlenker RA, Keane AT, Holtzman RB. 1982. The retention of ^{226}Ra in human soft tissue and bone: Implications for the ICRP 20 alkaline earth model. Health Phys. 42:671–693.

Schwarz BC, Godwin WJ, Wayson MB, Dewji SA, Jokisch DW, Lee C, Bolch WE. 2021a. Specific absorbed fractions for a revised series of the UF/NCI pediatric reference phantoms: internal electron sources. Phys Med Biol. 66:035005.

Schwarz BC, Godwin WJ, Wayson MB, Dewji SA, Jokisch DW, Lee C, Bolch WE. 2021b. Specific absorbed fractions for a revised series of the UF/NCI pediatric reference phantoms: internal photon sources. Phys Med Biol. 66:035006.

Sgouros G, Roeske JC, McDevitt MR, Palm S, Allen BJ, Fisher DR, Brill AB, Song H, Howell RW, Akabani G, In oration with the SNM MIRD Committee: Wesley E. Bolch, A. Bertrand Brill, Darrell R. Fisher, Roger W. Howell, Ruby F. Meredith, George Sgouros (Chair), Barry W. Wessels, and Pat B. Zanzonico. 2010. MIRD pamphlet No. 22 (abridged): radiobiology and dosimetry of alpha-particle emitters for targeted radionuclide therapy. J Nucl Med. 51(2):311–328.

Sharpe WD. 1978. The New Jersey radium dial painters: a classic in occupational carcinogenesis. Bull Hist Med. 52(4):560–570.

Spiers FW, Lucas HF, Rundo J, Anast GA. 1983. Leukemia incidence in the U.S. dial workers. Health Phys. 44(Suppl. 1):65–72.

Stebbings JH. 2001. Health risks from radium in workplaces: an unfinished story. Occup Med. 16:259–270.

Stebbings JH, Lucas HF, Stehney AF. 1984. Mortality from cancers of major sites in female radium dial workers.Am J Ind Med. 5(6): 435–459.

Stewart E. 1929. Radium poisoning. Industrial poisoning from radioactive substances. MLR. 28(6):20–95.

Stoudt HW, Damon A, McFarland RA. 1960. Heights and weights of white Americans. Hum Biol. 32(4):331–341.

Stram DO, Sokolnikov M, Napier BA, Vostrotin VV, Efimov A, Preston DL. 2021. Lung cancer in the Mayak workers cohort: risk estimation and uncertainty analysis. Radiat Res. 195:334–346.

Tolmachev SY, Swint MJ, Bistline RW, McClellan RO, McInroy JF, Kathren RL, Filipy RE, Toohey RE. 2019. USTUR special sessions roundtable: United States Transuranium and Uranium Registries (USTUR): a five-decade follow-up of plutonium and uranium workers. Health Phys. 117(2):211–222.

Yoder RC, Balter S, Boice JD Jr, Grogan H, Mumma M, Rothenberg L, Passmore C, Vetter R, Dauer L. 2021. Using personal monitoring data to derive organ doses for medical radiation workers in the Million Person Study – considerations regarding NCRP Commentary No. 30. J Radiol Prot. 41:118–128.

Sex-specific lung cancer risk among radiation workers in the Million Person Study and among TB-fluoroscopy patients

John D. Boice, Elizabeth D. Ellis, Ashley P. Golden, Lydia B. Zablotska, Michael T. Mumma, and Sarah S. Cohen

ABSTRACT

Background: The study of Japanese atomic bomb survivors, exposed briefly to radiation, finds the risk of radiation-induced lung cancer to be nearly three times greater for women than for men. Because protection standards for astronauts are based on individual lifetime risk projections, this sex-specific difference limits the time women can spend in space. Populations exposed to chronic or fractionated radiation were evaluated to learn whether similar differences exist when exposures occur gradually over years.

Methods and materials: Five occupational cohorts within the Million Person Study of Low-Dose Health Effects (MPS) and a Canadian Fluoroscopy Cohort Study (CFCS) of tuberculosis patients who underwent frequent chest fluoroscopic examinations are evaluated. Included are male and female workers at the Mound nuclear facility, nuclear power plants (NPP), and industrial radiographers (IR). Workers at the Mallinckrodt Chemical Works and military participants at aboveground nuclear weapons tests provide information on the risk among males. Cox proportional hazards and Poisson regression models were used to estimate sex-specific radiation risks for lung cancer and to compare any differences.

Results: Overall, 15,065 lung cancers occurred among the 443,684 subjects studied: 50,111 women and 395,573 men. The mean cumulative dose to the lung was 166.3 mGy (range 6 to 1,055 mGy) with the highest among the TB-fluoroscopy patients (mean 1,055 mGy). Mean lung dose for women in the worker cohorts was generally 4 times lower than for men. Of the 12 estimates of radiation-related risk, only one, for male IRs, showed a significant elevation (ERR 0.09; 95% CI 0.02–0.16, at 100 mGy). In contrast, the dose response for male NPP workers was negative (ERR −0.05; 95% CI −0.10, 0.01, at 100 mGy). Combined, these two cohorts provided little evidence for a radiation effect among males (ERR 0.01; 95% CI −0.04, 0.06, at 100 mGy). There was no significant dose-response among females within any cohort. There was no difference in the sex-specific estimates of lung cancer risk.

Conclusions: There was little evidence that chronic or fractionated exposures increased the risk of lung cancer. There were no differences in the risks of lung cancer between men and women. However, the sex-specific analyses are limited because of small numbers of women and relatively low doses. A more definitive study is ongoing of medical radiation workers which include 85,000 women and 85,000 men (overall mean dose 82 mGy, max 1,140 mGy). Additional understanding will come from the ongoing follow-up of the CFCS.

Introduction

The evidence that low-LET radiation causes lung cancer comes mainly from acute high-dose-rate exposures and at moderate radiation doses (UNSCEAR 2008). The Japanese atomic bomb survivor study provides evidence for an increased radiation risk of lung cancer but also indicates that the risk of radiation-induced lung cancer is nearly three times greater for women than for men (Ozasa et al. 2012; NCRP 2014; Cahoon et al. 2017). In contrast, studies of tuberculosis patients undergoing frequent chest fluoroscopic examination to monitor lung collapse therapy in Canada and in Massachusetts have failed to find an increased risk of lung cancer despite lung doses that exceed those of the Japanese atomic bomb survivors (Davis et al. 1989; Howe 1995). It has not been clear why such differences exist. One possibility is that fractionation of dose over time results in lower risk than when the exposure is received in a short period of time (Fry 1996). This difference in sex-specific radiation lung cancer risk is of special interest to NASA which bases its protection standards on lifetime risk estimates for individuals based on the atomic bomb survivor study (NA/NRC 2012). Since women have a much higher risk of lung cancer on a relative scale than men, their lifetime risk estimates per unit dose are much higher and thus their time in space for long-term missions is limited.

Several radiation cohorts within the MPS have sufficient numbers of female workers to be able to contrast the risks among males with those among females. The studies evaluated here include the Mound facility workers in Ohio (Boice et al. 2014), nuclear power plant (NPP) workers (Boice, Cohen, Mumma, Hagemeyer, Chen et al. 2019), and industrial radiographers (IR) (Boice, Cohen, Mumma, Hagemeyer, Golden et al. 2019). These studies were analyzed in a similar fashion and radiation-related estimates of lung cancer mortality can be made for both males and females. Male workers at the Mallinckrodt Chemical Works (MCW) and participants at above-ground nuclear weapons tests (Boice, Till et al. 2019) were also included to provide additional estimates of lung cancer risk among males, and because a component of the lung dose among MCW workers comes from intakes of high-LET emitters (uranium and radium) (Ellis et al. 2018; Golden et al. 2019).

There are few studies that have focused on sex-specific differences in lung cancer radiation risks. The six studies selected for analysis in this paper were conducted by the authors; thus, the datasets could be analyzed for sex-specific lung cancer risks using the same methods. The studies preferentially included available cohorts within the MPS, the focus of this special issue. The total numbers, nearly 444,000 persons and broad dose distributions were recognized. Seven important studies could not be included, e.g. the study of Mayak workers, and are described in the discussion section. Several studies not explicitly discussed can be found in an informative summary of sex-specific differences by the German Commission on Radiological Protection (Strahlenschutzkommission 2009). Finally, since lung cancer has such a high case-fatality rate, the patterns for incidence would be expected to be similar to mortality investigations.

Materials and methods

Statistical methods

For the MPS cohorts, Cox proportional hazards models were used to compute hazards ratios (HR) and corresponding 95% confidence intervals (CI) for a continuous measure of radiation dose to the lung (Cox 1972). Age was used as the underlying timescale for the hazard function. To allow for a possible latent period between radiation exposure and any effect consequent to it, doses were lagged 10 years. All Cox models were conducted using SAS/STAT software (SAS/STAT software, Version 9.4 of the SAS System for Windows, SAS Institute Inc., Cary, NC, USA). Excess Relative Risks per Gy (ERR Gy^{-1}) along with corresponding CIs were estimated using the Peanuts program in Epicure (Preston et al. 2018). All models were adjusted for year of birth and a measure of socioeconomic status which varied by cohort and was either education level or pay type, such as hourly versus salary.

Further details of the analytic methods for the Canadian Fluoroscopy Cohort Study (CFCS) are available in Howe (Howe 1995). Briefly, standard Poisson regression models (Breslow and Day 1987) were used to compute relative risks and corresponding CIs for a continuous dose to the lung with a 10-year lag applied to account for a latent period. The Epicure program (Preston et al. 2018) was used to compute linear excess relative risks per Gy. All models included adjustment for age at risk, calendar year at risk, and sex by stratification.

For all cohorts, analyses were conducted within cohort except for the NPP and IR cohorts which were examined both individually as well as combined. This could be done with minimal bias because both populations were identified from the same databases, had the same covariates, and used the same procedures for obtaining and adjusting for socioeconomic status. Models were first stratified by sex and then combined over sex with further adjustment for male versus female. To test whether sex was an effect modifier, i.e. whether risk differed by sex, nested models with and without an interaction term between sex and radiation dose were compared using the likelihood ratio test (LRT).

Results and study descriptions

The occupational studies within the MPS and the CFCS studies are described below and in Tables 1–3.

Mound nuclear facility

The Mound Nuclear Facility near Dayton, OH, employed 7,270 workers during the years 1944–1971, including 4,954 workers monitored for radiation exposure (Boice et al. 2014; Tables 1–3). Among females, mean lung dose was 24.9 mGy and the maximum was nearly 1.5 Gy. There was no evidence for an increase in lung cancer deaths related to radiation (Tables 2 and 3). There was no increased risk in lung cancer mortality among the 971 female workers (ERR at 100 mGy= −0.01; 95% CI −0.07, 0.07) nor among the 3983 male workers (ERR at 100 mGy = 0.01; 95% CI −0.02, 0.04) (Table 2), and accordingly no difference in the sex-specific lung cancer risk estimates (LRT $p = .72$).

Nuclear power plant workers (NPP)

Over 135,000 workers first monitored at nuclear power plants in the United States from 1957–1984 were evaluated through 2011 (Boice, Till et al. 2019; Tables 1–3). There was no evidence for an increase in radiation-related lung cancer death, and the sex-adjusted dose response was negative and of borderline significance (ERR at 100 mGy= −0.05; 95% CI −0.10, 0.001). The mean dose to the lung was 40.5 mGy and ranged up to 1085 mGy. Overall, there were 4420 female workers and 130,773 male workers (Table 1). A radiation-related dose response was not seen for the female workers (ERR at 100 mGy = 0.80; 95% CI −0.96, 2.56) nor among the male workers (ERR at 100 mGy= −0.05; 95% CI −0.10, 0.01). The estimates of radiation risk for lung cancer

Table 1. Subject counts, lung cancer deaths and descriptive measures of dose, by cohort and sex.

Cohort	Females					Males				
	N subjects	N cases lung cancer deaths	Person-Years	Mean [Median] Dose, mGy	95th percentile [Maximum] Dose, mGy	*N* subjects	*N* cases lung cancer deaths	Person-Years	Mean [Median] Dose, mGy	95th percentile [Maximum] Dose, mGy
Mallinckrodt (Golden et al. 2019)	na	na	na	na	na	2514	157	107,920	67.3 [30.7]	273 [829]
Mound (Boice et al. 2014)	971	21	41,750	24.9 [1.6]	132 [1,460]	3983	182	159,771	112.9 [14.7]	555 [17,478]
Atomic Veterans (Boice, Till et al. 2019)	na	na	na	na	na	113,806	8027	5,350,400	6.21 [2.28]	25 [972]
Nuclear Power Plant (NPP) (Boice, Cohen, Mumma, Hagemeyer, Chen et al. 2019)	4420	48	131,642	17.9 [7.7]	75.7 [466]	130,773	3337	3,947,599	41.3 [21.4]	152 [1085]
Industrial Radiographers (IR) (Cohen, Mumma, Hagemeyer, Golden et al. 2019)	12,933	55	210,913	2.0 [0.02]	8.3 [302]	110,577	2060	2,427,345	12.0 [0.4]	67 [1,435]
Combined NPP and IR	17,353	103	342,555	6.1 [0.09]	35 [466]	241,350	5397	6,374,944	27.8 [9.8]	123 [1435]
TB Fluoroscopy – Canadian[a] (Howe 1995)	31,787	266	883,172	1072 [622]	4115 [9640]	31,920	912	725,318	1038 [617]	4210 [9280]

[a]Excludes person-years for age at risk <20 years and deaths and person-years at risk within 10 years of exposure.

mortality were not significantly different between females and males (LRT $p = .37$).

Industrial radiographers (IR)

Over 123,000 industrial radiographers in the United States who were first monitored 1939–2011 were studied (Boice, Cohen, Mumma, Hagemeyer, Chen et al. 2019; Tables 1–3). The workers over the years were exposed only to external radiation and mainly iridium-192 and cobalt-60. The mean lung dose was 10.9 mGy and ranged up to 1435 mGy. The sex-averaged lung cancer risk estimate for mortality was significantly elevated (ERR at 100 mGy = 0.09; 95% CI 0.04, 0.19). There were 12,933 female industrial radiographers and 110,577 male radiographers (Table 1). The mean dose to the lung was 2.0 mGy for females and 12.0 mGy for males. The radiation risk estimates were non-significantly negative for females (ERR at 100 mGy= −0.33; 95% CI −0.45, 0.21) and significantly positive for males (ERR at 100 mGy = 0.09; 95% CI 0.02, 0.16). While a significant dose response was observed for radiation-induced lung cancer mortality among male radiographers, the difference in risk estimates between female and male workers was not significant (LRT $p = .43$).

The NPP and the IR worker studies were constructed using the same methodologies. The NRC REIRS database (NRC 2018) and the Landauer dosimetry records were used to select the study cohorts and the analytical and dosimetry approaches were exactly the same, including the adjustments for socioeconomic status based on areawide educational categories. A pooled analysis of these two cohorts was conducted to increase statistical power. The pooled analysis included 17,353 females and 241,350 males (Tables 1–3). There was no significant increase in the sex-adjusted radiation risk for lung cancer mortality (ERR at 100 mGy = 0.02; 95% CI −0.03, 0.07) overall, nor among females (ERR at 100 mGy = 0.16; 95% CI −0.49, 0.81) or males (ERR at 100 mGy = 0.01; 95% CI −0.04, 0.06). The pooled analysis did not support a radiation-related risk for lung cancer nor a statistically significant difference in sex-specific radiation lung cancer risk estimates (LRT $p = .86$).

Mallinckrodt chemical works

The Mallinckrodt Chemical Works (MCW) study evaluated 2514 white male workers employed 1942–1966 (Boice et al. 2018; Ellis et al. 2018; Golden et al. 2019; Tables 1–3). The workers over the years were exposed to external gamma radiation plus internal radiation from inhalation and ingestion of pitchblende dust (uranium, radium, and silica). Lung dose from occupational medical x-rays was included in the individual dose reconstructions as were ambient levels of radon and radon progeny. The mean dose to the lung was 67.3 mGy and ranged up to 829 mGy (Table 3). There was no evidence for a radiation dose response for lung cancer mortality (ERR at 100 mGy= −0.06; 95% CI −0.18, 0.06) and the negative trend was nearly significant. Because only males were included, the MCW study could not evaluate sex-specific radiation risks. It did suggest that chronic

Table 2. Sex-specific excess relative risks (ERR) and hazard ratios (HR) at 100 mGy for lung cancer deaths, by cohort and sex.

	Females		Males	
Cohort	ERR at 100 mGy (95% CI)	HR at 100 mGy (95% CI)	ERR at 100 mGy (95% CI)	HR at 100 mGy (95% CI)
Mallinckrodt	na	na	−0.06 (−0.18, 0.06)	0.95 (0.81, 1.12)
Mound	−0.01 (−0.07, 0.07)	0.86 (0.48, 1.55)	0.01 (−0.02, 0.04)	1.01 (0.99, 1.03)
Atomic Veterans	na	Na	0.08 (−0.06, 0.22)	1.04 (0.90, 1.21)
Nuclear Power Plant (NPP)	0.80 (−0.96, 2.56)	1.80 (0.86, 3.78)	−0.05 (−0.10, 0.01)	0.96 (0.90, 1.02)
Industrial Radiographers (IR)	−0.33 (−0.45, 0.21)	0.69 (0.14, 3.49)	0.09 (0.02, 0.16)	1.09 (1.02, 1.17)
Combined NPP and IR	0.16 (−0.49,0.81)	1.17 (0.61, 2.24)	0.01 (−0.04, 0.06)	1.01 (0.97, 1.06)
TB Fluoroscopy – Canadian	−0.007 (−0.015, 0.002)	0.97 (0.81, 1.15)	0.002 (−0.003, 0.008)	0.96 (0.87, 1.057)

Table 3. Sex-adjusted ERR and HR at 100 mGy for lung cancer deaths by cohort, combined over sex.

Cohort	Total Subjects both sexes	Total Lung cancers deaths both sexes	Mean [Median] dose, mGy	95th percentile [Maximum] dose, mGy	Sex-adjusted ERR at 100 mGy (95%CI)	Sex-adjusted HR at 100 mGy (95%CI)	p^a
Mallinckrodt	2514	157	67.3 [30.7]	273 [829]	−0.06 (−0.18, 0.06)	0.95 (0.81, 1.12)	na
Mound	4954	203	95.6 [9.1]	442 [17,478]	0.01 (−0.03, 0.05)	1.01 (0.99, 1.03)	.72
Atomic Veterans	113,806	8027	6.21 [2.28]	25 [972]	0.08 (−0.06, 0.22)	1.04 (0.90–1.21)	na
Nuclear Power Plant (NPP)	135,193	3385	40.5 [20.9]	150 [1,085]	−0.05 (−0.10, 0.001)	0.96 (0.90, 1.02)	.37
Industrial Radiographers (IR)	123,510	2115	10.9 [0.3]	61 [1435]	0.09 (0.04, 0.19)	1.09 (1.02, 1.17)	.43
Combined NPP and IR	258,703	5500	26.4 [8.6]	119 [1435]	0.02 (−0.03, 0.07)	1.01 (0.97–1.06)	.86
TB Fluoroscopy - Canadian	63,707	1178	1055 [620]	4146 [9640]	0.00 (−0.005, 0.005)	0.96 (0.88, 1.05)	.097

[a] *p*-value from likelihood ratio test (LRT) for interaction between sex and dose in a Cox proportional hazards model with males and females.

radiation exposures, including that from intakes of radionuclides that provide high-LET dose to lung tissue, is not associated with a detectable increase of lung cancer risk at the radiation doses received. A negative radiation dose response for non-malignant lung disease (Golden et al. 2019) provided some additional evidence that low-dose radiation exposures to the lung experienced over a period of years may be less carcinogenic than when the doses are received all at once and at a high level. The small numbers in this study, however, temper the strength of conclusions that might be drawn.

Nuclear weapons test participants (atomic veterans)

The United States conducted over 200 above-ground atmospheric weapons tests during the Cold War and many involved military personnel at the Nevada Test Site (NTS) and the Pacific Proving Grounds (e.g. Bikini Atolls) (Till et al. 2014; Caldwell et al. 2016; Beck et al. 2017; Tables 1–3).

The epidemiologic study of above-ground nuclear weapons test participants (atomic veterans) included 113,806 male military personnel who participated in weapons testing between 1945 and 1962. The atomic veterans are included as a component of the MPS that provides information on the radiation risk of death from lung cancer among males. Vital status was obtained for nearly 97% of the veterans and over 70% had died. There were 8027 deaths due to lung cancer. Overall, radiation doses to the lung were small, mean 6.21 mGy with a maximum of 972 mGy (Table 1). There was no statistical evidence for a radiation dose response over the very narrow range of lung doses (ERR 0.08; 95% CI −0.06, 0.22; at 100 mGy).

Canadian fluoroscopy cohort study (CFCS)

A lung cancer mortality study was conducted between 1950–1987 among 64,172 Canadian tuberculosis patients who received multiple chest fluoroscopic examinations during the monitoring of lung collapse therapy (Sherman et al. 1978; Howe and Yaffe 1992; Howe 1995; Tables 1–3). The current analyses are based on the previously published data (Howe 1995) using Poisson and Cox regression models. There were 31,917 female patients and 32,255 male patients (Table 1). The mean dose to the lung was 1055 mGy and the maximum was 9640 mGy (Table 3). Overall there was no evidence for a radiation-related increase in lung cancer among either females (ERR at 100 mGy= −0.007; 95% CI −0.015, 0.002) or males (ERR at 100 mGy = 0.002; 95% CI −0.003, 0.008) (Table 2). There was no difference between the sexes in their risk of radiation-induced lung cancer (LRT $p = .097$). These data are consistent with the Massachusetts TB-fluoroscopy study which also did not find a radiation-related risk for lung cancer (Davis et al. 1989; Brenner et al. 2007).

Socioeconomic status (SES)

SES was adjusted for in the analysis as a surrogate for demographic and lifestyle factors such as cigarette smoking (Wakeford 2005). SES was defined for each cohort as follows: for Mallinckrodt, it was pay code (Hourly vs. Salary); for Mound, it was self-reported educational achievement; for NPP and IR it was educational achievement based on census block group data; for Atomic Veterans, it was rank (Enlisted vs. Officer). The importance of this adjustment was clear in that significant SMRs for lung cancer were seen among workers categorized in the lowest SES levels and deficits were seen in the highest SES categories (e.g. Boice, Till et al. 2019; Golden et al. 2019).

Analyses were conducted (not shown) of the HRs at 100 mGy by SES status to learn whether there was a tendency for radiation risks to differ by SES status overall and by sex. There were no statistically meaningful differences by SES status or by sex, but the statistical variability was large. More robust analyses will await the completion of the medical radiation worker study, and the Los Alamos National Laboratory and Rocky Flats cohort studies that have substantial numbers of workers, broad dose distributions and much larger numbers of female workers.

Discussion

There are few studies, that have focused on sex-specific differences in lung cancer radiation risks. Seven studies with estimates of sex-specific lung cancer risks are briefly summarized below: Massachusetts TB- fluoroscopy patients (Boice et al. 1978; Davis et al. 1989; Brenner et al. 2007); scoliosis patients (Ronckers et al. 2010); Hodgkin lymphoma patients (Gilbert et al. 2003); and workers within the Mayak (Gilbert et al. 2004, 2013; Gillies et al. 2017), INWORKS (Richardson et al. 2018) and Rocketdyne (Boice et al. 2011) studies.

Summaries of other published studies

TB- Fluoroscopy Massachusetts study

The Massachusetts TB-fluoroscopy mortality study included 13,385 patients treated between 1925–1954 (Boice et al. 1978; Davis et al. 1989; Brenner et al. 2007; Little 2018). The patients received an average of 77 x-ray fluoroscopic examinations, and the mean dose to lung tissue was 840 mGy. Risks were not reported for women and men separately. Lung cancer (SMR 0.8, $n = 69$) was not elevated despite a range of doses to the lung that reached over 8000 mGy; there was no evidence of a dose response. Adjustments for smoking and the amount of lung tissue at risk did not appreciably modify these observations. The authors concluded that frequent exposures to low doses of radiation over a period of several years did not increase the risk of lung cancer and that fractionated exposure appeared less effective in causing lung cancer than a single exposure of the same total dose.

U.S. Scoliosis study

The U.S. Scoliosis Study includes 5573 female patients monitored with frequent spinal x-rays for curvature changes during the adolescent growth spurt and before 20 years of age, 1912–1965 (Hoffman et al. 1989; Ronckers et al. 2010). The average number of x-rays was 22.9 and ranged up to 553. The average cumulative lung dose was 41 mGy and the maximum was 676 mGy. Breast cancer mortality was significantly elevated. Lung cancer mortality (SMR 0.8, $n = 57$), however, was not elevated; and the dose response was negative (ERR Gy^{-1} −1.4; 95% CI −7.1, 3.1). These results are identical to the Canadian and Massachusetts TB-fluoroscopy studies where significant positive dose-response relationships for breast cancer were observed (Boice et al. 1979) but evidence for increased lung cancer risk was absent (Davis et al. 1989; Howe 1995). These three studies of patients monitored with diagnostic radiation examinations for many years are not consistent with the Japanese study of atomic bomb survivors exposed acutely to radiation where strong relationships where found for both breast and lung cancer (Preston et al. 2002; Cahoon et al. 2017). Possible explanations raised by the authors of the scoliosis studies for these discrepancies in lung cancer findings included a low-dose threshold or nonlinear radiation dose response when doses are fractionated, or failure to adequately control for the carcinogenic effects of smoking (Ronckers et al. 2010).

Rocketdyne workers

The Rocketdyne (Atomics International) study involved 46,970 workers employed 1948–1999 (Boice et al. 2011). Overall, 5801 workers were monitored for radiation activities (Boice et al. 2006). The mean dose to lung from external and internal radiation combined was 19 mSv (maximum 3600 mSv). Overall, there was no evidence for a radiation-associated risk for lung cancer mortality, i.e. the trend for lung cancer risk over categories of cumulative dose to the lung was negative but not significant. The RR at 100 mSv for lung cancer was estimated to be 1.01 (95% CI 0.89, 1.16). There was limited ability to evaluate sex-specific lung cancer radiation risk because only 8% of the population was female. The Rocketdyne study provides little evidence that fractionated exposures to the lung experienced over years is related to lung cancer risk. The study is limited by small numbers, but the dose range was broad and much of the high lung doses among workers came from intakes of radionuclides emitting high-LET alpha particles.

Hodgkin lymphoma patients

Cancer incidence studies of patients with Hodgkin lymphoma, treated with chest radiotherapy (Travis et al. 2002; Dores et al. 2002; Gilbert et al. 2003), provide evidence that high-dose radiotherapy causes lung cancer but not for risk differences by sex. The data for lung cancer were obtained within a population-based cohort of 19,046 Hodgkin lymphoma patients diagnosed from 1965–1994. Radiation dose was determined to the specific location in the lung where cancer developed, and tobacco use was adjusted for in the analyses. The mean dose to the lung was 36 Gy from the mantle radiotherapy received. Practically all radiotherapy doses to the lung were >5 Gy. Seventy-five percent of the patients with lung cancer within the nested case-control study were male and 25% were female. The radiation-related risk of lung cancer for the patients receiving >5 Gy was four times higher for males (RR 7.2; 95% CI 3.0, 18.6) than for females (RR 2.1; 95% CI 0.6, 11.5) but this difference was not statistically significant. This comprehensive study of lung cancer risk following radiotherapy at high doses provides no evidence for a difference in the sex-specific risk for radiation-induced lung cancer. The actuarial risk of lung cancer at 25 years after treatment was significantly higher in males (6.2%; 95% CI 5%, 7.4%) than females (3.2%; 95% CI 2.2%, 4.2%), although the radiation component was not separated from the entire cohort (Dores et al. 2002). The relevance of such very high doses to low-dose risk assessment is unclear.

UK national registry for radiation workers (NRRW)

The UK NRRW study includes 167,003 workers followed for 32 years and recently updated (Haylock et al. 2018). The overall mean lifetime external dose was 25.3 mSv, and higher among the 150,566 male workers (27.5 mSv) than among the 16,437 female workers (5.6 mSv). Nearly 25% of the workers were monitored for intakes of radionuclides such as plutonium, uranium, and tritium, but only the fact of monitoring was known, and internal doses were not included in the analyses. There were no statistically meaningful risks of lung cancer seen either for mortality (ERR $Sv^{-1}=0.028$; 95% CI −0.44, 0.63; $n=3058$) or for incidence (ERR $Sv^{-1}=0.13$; 95%CI −0.35, 0.72; $n=3263$). Cancers of the brain and mouth were significantly low whereas cancer of the rectum and bladder were significantly high. Cancer of the pleura (likely mesothelioma) was significantly based on incidence data. No sex-specific analyses were presented. There was no statistical evidence that cancer of the lung was associated with occupational external radiation, however the inability to take intakes of radionuclides into account tempers the strength of this observation.

Inworks

The INWORKS occupational mortality study consists of 308,297 nuclear workers in France, the United Kingdom, and the United States (Richardson et al. 2018). Overall there were 268,262 male workers and 40,035 female workers. The mean (and 95th percentile) cumulative lung doses were 22.8 mGy (106.3 mGy) for men and 4.8 mGy (18.8 mGy) for women. There were 5802 lung cancer deaths. The maximum likelihood estimate of excess relative risk per Gy for lung cancer mortality was 0.51 (90% CI 0.00, 1.09). A 10-year lag was applied; the risk was sex-adjusted but sex-specific estimates were not presented. A hierarchical Bayesian analytical method was also applied to estimate ERR Gy^{-1} with slightly higher risk estimates for lung cancer than provided by the maximum likelihood approach. The individual site-specific estimates in INWORKS are difficult to interpret because very high and statistically significant risks were seen for sites not found or rarely found to be increased following radiation exposures such as cancer of the rectum, peritoneum, pleura and mesothelioma, testis, larynx, and skin (e.g. melanoma), whereas sites that are frequently increased following radiation exposure were not, such as cancers of the colon, liver, bladder, and kidney (UNSCEAR 2008).

15-Country study

For completion, the 15-Country study reported an ERR $Sv^{-1}=1.86$ (90% CI 0.49, 3.63) overall, and an ERR $Sv^{-1}=1.87$ (90%CI 0.50, 3.66) for male workers and an ERR $Sv^{-1}=-1.04$ (90% CI <0, 11.1) for female workers (Cardis et al. 2007). Confounding by asbestos exposure is a concern as when cancer of the pleura was excluded from the analyses, the ERR Sv^{-1} was no longer significantly different from zero. Other uncertainties have tempered the conclusions that might be drawn (Wakeford 2005; NCRP 2012). Further, the large U.S. and U.K. components within the 15-Country study are also included in INWORKS, so results are not independent.

Mayak nuclear facility workers

A mortality study of lung cancer was conducted among 14,621 Mayak workers who were hired between 1948-1982 (Gilbert et al. 2004, 2013; Gillies et al. 2017). Significant differences between the female and male ERR Gy^{-1} for lung cancer for intakes from plutonium were reported with females being 4 times higher for mortality (Gillies et al. 2017). For internal dose, the excess absolute risk (EAR) for females was less than half that for males (Gilbert et al. 2004). The lung cancer risk estimates from gamma exposure were, on the other hand, 2–3 times lower than comparable risks in the Japanese atomic bomb survivor study (Preston et al. 2017). Interpretation, however, is challenging for several reasons including the differences in smoking prevalence: 74% of male workers smoked and only 5% of female worker smoked and ERR Gy^{-1} risks varied significantly by smoking status (Gillies et al. 2017); the overall lost-to-follow-up rate was high at 22% (Preston et al. 2017); it is challenging to separate out a gamma ray component to risk for workers who also had significant plutonium intakes; and the estimates of plutonium dose were based on incomplete coverage of only 42% of the workers employed at the plutonium facilities (Gillies et al. 2017). Further, 'A particularly important limitation of our lung cancer risk estimates relates to intrinsic difficulties in estimating doses from plutonium exposure. Despite extensive efforts by Russian and non-Russian dosimetrists, these doses are subject to many uncertainties including especially those resulting from imprecision in the urine measurements. Only 33% of the 6540 workers with estimates of positive plutonium doses had more than two measurements, although this percentage was higher (51%) for workers with estimated plutonium doses exceeding 0.2 Gy. For most workers, these measurements were taken many years after the exposure occurred, since routine plutonium monitoring did not begin until about 1970. Such measurement errors, often referred to as classical errors, are known to bias the dose-response coefficient toward zero and might also distort the shape of the dose-response' (Gilbert et al. 2013). In a comparable but much less powerful study of Sellafield workers, there was no difference in the male and female risk estimates (Gillies et al. 2017).

When the analysis of Mayak workers was restricted to workers who received <200 mGy, there was no difference in the male and female estimates of risk (Gillies et al.

2017). The mean dose to gamma-rays was 455 mGy (maximum 7595 mGy). The mean dose from plutonium-239 was 129-176 mGy (maximum 16,500-19,700 mGy). The differences related to the plutonium doses depends on whether the biokinetic models assumed either a slow clearance or a fast clearance from the lung, that is related to solubility. When the ERR Gy^{-1} analyses were allowed to depend on smoking status, the plutonium ERR Gy^{-1} for non-smokers was significantly greater than that for smokers which suggested a sub-multiplicative relationship. While the ERR Gy^{-1} for non-smokers was estimated to be four times that for smokers, the female/male ratio was estimated to be 1.0 (95% CI 0.3, 3.4). That is, when the plutonium ERR Gy^{-1} was allowed to depend on smoking, the ratio of female and male risk estimates was reduced from 3.3 to 1.0. The study does raise the issue as to the most appropriate scale for comparisons: the excess relative risk which depends on the background rates or the absolute excess risk which depends less so. This is a remarkably comprehensive and important study regarding the risks of high-dose plutonium and high-dose gamma exposures, but there remains some uncertainty in the estimated radiation risks and in the risks of sex-specific radiation-induced lung cancer.

Japanese atomic bomb survivor study

The study of Japanese atomic bomb survivors provides strong evidence that acute exposures to ionizing radiation can lead to an increased risk of lung cancer. Females are reported to be at 2–3 times higher risk of radiation-induced lung cancer than males (Preston et al. 2007; Furukawa et al. 2010; Ozasa et al. 2012; Cahoon et al. 2017). The lung cancer incidence analyses showed that lung cancer risk strongly depended on smoking patterns, with radiation risk highest for those smoking 10 cigarettes per day and approaching no risk at 20 cigarettes per day. Most men smoked whereas most women did not. The female to male incidence ratio of excess risk decreased with increasing radiation dose. Application of these risk estimates to American workers, and to the NASA astronaut corps (NCRP 2014), is somewhat problematic because of the different population characteristics and circumstances: the Japanese population was exposed briefly in 1945 to the atomic bombings in Hiroshima and Nagasaki and lived in a war-torn country for many years after (NCRP 2014; Boice 2017).

The most recent analyses evaluated cancer incidence from 1958–2009 (Cahoon et al. 2017). There were 105,444 survivors evaluated of whom a little over 86,000 had estimates of radiation dose. For non-smokers, the ERR Gy^{-1} estimate for lung cancer was 0.81 (95% CI 0.15, 1.18) and the female to male ratio was 2.83. The estimate of risk was significantly high for low-to-moderate smokers and no radiation risk was apparent for heavy smokers. Smoking information was available for 64,465 subjects (81%). For those with information on smoking, 86% of men and 18% of women indicated that they had ever smoked. The authors mentioned that the results should be interpreted in the context of several limitations, e.g. smoking history was unknown for ≈40% of the lung cancer cases and ≈60% of the follow-up time. It was also peculiar that no or very low radiation risks were seen for those who smoked heavily whereas moderate smokers had the highest radiation risk for lung cancer, as previously reported (Furukawa et al. 2010, 2017).

The mortality data (Ozasa et al. 2012) are more comparable to our mortality studies of worker populations and TB-fluoroscopy patients, although not of the diagnostic quality of that for incidence data. The Japanese mortality data are more complete than the incidence data in the sense that the entire country of Japan after 1950 was covered for death determinations, whereas the two cancer registries in Hiroshima and Nagasaki began in 1958 and could not include the entire country for incidence determinations. The mortality analyses focused on the 86,611 lifespan study cohort members with dose estimates. The sex-averaged ERR Gy^{-1} was 0.75 (95% CI 0.51,1.03). The ERR Gy^{-1} was greater among females (1.1; 95% CI 0.68, 1.6) compared with males (0.40; 95% CI 0.17, 0.67). The male to female ratio was 2.7 (95% CI 1.3, 6.8). The excess absolute risk (EAR) ratio of females to males was 0.98 (95% CI 0.40, 1.8), indicating the importance of background rates of disease when making inferences on 'risk differences'.

Animal Experiments

An informative review of experimental studies on sex-specific difference can be found in Strahlenschutzkommission (2009). Only a brief overview is presented here. There is clear evidence in experimental studies that low dose-rate exposures are less effective in causing lung cancer than are high dose-rate exposures (Ullrich and Storer 1979a, 1979b; Ullrich et al. 1987; Fry 1996; NRC 2006, Figure 10b-1). The evidence for high-LET exposures, such as neutrons, is not as apparent (Ullrich et al. 1976, 1977; Storer et al. 1979; Storer and Ullrich 1983; Storer and Fry 1995; Heidenreich et al. 2006). Interestingly, the relative risk estimates for man and mouse for lung tumors were not significantly different (Storer et al. 1988). Apparently, for low-LET radiation, low-dose rates consistently result in dose-response relationships and risks significantly lower than that observed for acute or higher-dose rate exposures (Ullrich et al. 1976, 1987; Ullrich and Storer 1979a, 1979b; Ullrich et al. 1987, 1976). However, the evidence that male and female risk estimates differ is not clear and depended upon mouse strain, dose rate, and radiation quality. To evaluate more fully the information available from animal experiments as well as in humans, NCRP has assembled scientific committee (SC 1-27), entitled 'Sex-Specific Differences in Lung Cancer Risks and Guidance on Modification of Lifetime Risk Projections.'

The current study computed sex-specific lung cancer risks for occupational cohorts within the MPS and for TB patients who received frequent chest fluoroscopic examinations. Insights and preliminary conclusions can be drawn as to sex-specific radiation-related lung cancer risks when exposure is protracted or fractionated over time.

Six cohort analyses of occupational and patient exposures showed little evidence that prolonged or fractionated exposures increase the risk of lung cancer. Further, there were no significant differences in the radiation risk estimates between men and women, although the small numbers of females temper the strength of this conclusion. These data for low-LET radiation are consistent with studies of Hodgkin lymphoma patients and Rocketdyne workers. They are also consistent with animal studies showing a marked reduction in risk when low-LET radiation dose is delivered at a low rate, i.e. spread over time. The studies providing evidence for a sex-specific difference in lung cancer risk are those of the Japanese atomic bomb survivors and the study of Mayak workers exposed to exceptionally high doses of plutonium and to relatively high doses from gamma-rays. Animal studies do not consistently find an ameliorating effect of protracted radiation on lung cancer risk when the exposures are to neutrons.

These data suggest that for low-LET radiation, chronic or fractionated exposures to radiation have little effect on increasing lung cancer risk among either men or women. Further, the new analyses of the occupational cohorts and the TB-fluoroscopy patients find no evidence for any differences in lung cancer risk between men and women. The reduction in lung cancer risk with protracted exposures is not seen in studies of radiation-induced breast cancer (Preston et al. 2002) and suggests that different tissues respond differently to the carcinogenic effects of chronic or fractionated exposures. Interestingly, the TB-fluoroscopy studies showing minimal radiation risk for lung cancer (Davis et al. 1989; Howe 1995) are the same studies showing, on an absolute scale, that breast cancer risk is not appreciably influenced by fractionation of dose (Miller et al. 1989; Boice et al. 1991). Similar observations of a significant breast cancer risk and the absence of a lung cancer risk are seen in the U.S. Scoliosis Study (Ronckers et al. 2010).

Of the six cohorts evaluated, the most informative on sex-specific lung cancer risk is the Canadian TB-fluoroscopy study because of the large numbers of females ($n = 31{,}917$) and the high mean cumulative dose to lung, 1072 mGy (Table 1). In contrast, the number of women in the other cohorts ($n = 18{,}324$) is 40% fewer and the mean lung doses much lower (from 2 to 24.9 mGy). The lower mean lung dose is related in part to the higher percentages of low dose subjects. The broad dose distributions are not reflected in the low mean dose as women received up to 300–1470 mGy; dose-response analyses could be conducted and risk estimates per 100 mGy could be made. Tables 2 and 3 indicate the consistency of the radiation risk estimates among the CFCS and the other five cohorts. The male-specific radiation risk estimates are spot on with few discernable differences. The female-specific differences are not as clear, in part because of the wide confidence intervals about the point estimates. Although the confidence intervals are narrower for the CFCS compared with the other studies, all female lung cancer radiation risks estimates are statistically compatible with each other. Nonetheless, while it seems justified to suggest that there may be a lowering of lung cancer radiation-related risk when the dose rate is chronic and not acute, the sex-specific differences are not so convincing, in part due to small numbers of women and the associated generally lower lung doses. Thus, strong conclusions cannot be made about whether there are or are not any sex-specific differences in lung cancer mortality based on these studies. These data, however, will be strengthened when several ongoing MPS cohorts are completed, specifically the medical radiation workers and the worker cohorts at the Los Alamos National Laboratory and the Rocky Flats plant for which large numbers of women ($n > 100{,}000$) and broad dose distributions are currently under study.

Strengths and weaknesses

The studies included in these analyses have been comprehensive, and the epidemiologic methods and approaches for dose reconstruction described (NCRP 2018a; NCRP 2018b). The follow-up rate was over 95% overall and the cause of death determination nearly complete. The causes of death were obtained in the same manner by the same investigators relying upon state and national death information in the United States for the occupational cohorts (Mumma et al. 2018). The numbers are large, with 443,684 individuals and 15,065 lung cancer deaths. The dose distribution was broad with mean lung doses ranging from 6 mGy to >1,000 mGy. The dose reconstruction procedures for each occupational cohort followed guidance found in NCRP Report 178 (NCRP 2018a) by essentially the same team of investigators. The TB-fluoroscopy studies have been ongoing for over 40 years with a focus on dose reconstruction, follow-up and death ascertainment. The findings of the TB-fluoroscopy studies with regard to breast cancer have been influential in understanding radiation risks and guiding protection guidance (Upton et al. 1977; NCRP 2018b).

Nonetheless, there are limitations that temper the strength of conclusions that can be drawn. First, despite the larger number of women (50,111) and their high mean cumulative dose to lung (677.1 mGy), the statistical power to discern sex-specific differences is low, in part because only one of the 12 sex-specific estimates of risks was significantly different from the null and tests to compare 'no effect' to 'no effect' are not very informative. Secondly, information on smoking histories was incomplete or not available. In the absence of individual data on tobacco use, it is important to have good measures of SES that can be used in the analyses as an adjustment for likely cigarette use. We used the best measures of SES available for each cohort. Nonetheless, such an adjustment cannot be expected to be completely successful (Wakeford 2005). For the occupational studies, education or job category were taken as measures of socioeconomic status and as an indirect way to account for lifestyle and behavior factors, but there remains some uncertainty as to the effectiveness of this adjustment to adequately control for smoking. A small interview study of Rocketdyne workers did confirm that hourly workers compared with salaried workers were significantly more likely to have smoked cigarettes, to start at a younger age, to quit at an older age, and to smoke for more years (Boice et al. 2006). Cigarette use was available for a fraction of subjects in the TB-fluoroscopy studies. In the CFCS, there was little variation in the prevalence of smoking by dose category, and thus smoking was not considered a confounder of the radiation risks of lung cancer (Howe 1995). Thirdly, because of the relatively small numbers exposed to intakes of radionuclides, the studies could not powerfully evaluate lung cancer risk following high-LET exposures except perhaps for the Mound study which involved intakes of polonium and relatively high doses to lung. Other studies initiated or planned within the MPS will be able to address high-LET exposures more powerfully, such as the plutonium workers at the Los Alamos National Laboratory (Wiggs et al. 1994), the Rocky Flats Nuclear Weapons Plant and the Hanford site (Gilbert et al. 1993), and the uranium workers at the Fernald Feed Materials Production Center (Silver et al. 2013) and the Linde Air Products Company Ceramics Plant (Dupree et al. 1987).

The study of Japanese atomic bomb survivors continues to show a risk of radiation-induced lung cancer and a substantial difference between male and female risks. These observations could be causally related to acute exposures or perhaps related to uncontrolled confounding or incomplete smoking ascertainment, perhaps because few Japanese women smoked whereas most men smoked. Consistent with animal experiments a very brief exposure, or high dose-rate exposure, appears to carry more risk than a similar dose to the lung that is received gradually over time, or at a low dose rate. The Mayak study is exceptionally powerful and comprehensive but interpretations are with caution given the uncertainties in the dose reconstructions and to the possible influence of the very high lung doses that caused deterministic effects (tissue reactions), e.g., pulmonary fibrosis (termed plutonium pneumosclerosis) (Okladnikova et al. 1994). Conceivably the influence of such high doses might be different than that of lower doses as suggested in the companion study of the Sellafield workers who received very much lower doses and had little evidence of a lung cancer risk (Gillies et al. 2017).

More work could be done with regard to evaluating the excess absolute risks (EAR Gy^{-1}) in addition to the excessive relative risks (ERR Gy^{-1}) per unit radiation dose. When differences in radiation risks between men and women are reported, they usually are based on a relative risk scale which is related more strongly to the background disease rate than an absolute risk scale.

Conclusion

These analyses and evaluations provide little evidence that chronic or fractionated exposures to radiation increase the risk of radiation-induced lung cancer. Further, there was no evidence that radiation-related lung cancer risks differed between women and men. However, these conclusions are preliminary and limited by relatively small numbers of women who received relatively high doses. The findings, however, are supported by other epidemiologic studies that also fail to find a statistically meaningful increase in lung cancer risk: workers at Rocketdyne, in the UK NRRW, INWORKS, 15-Country Study, and patients receiving medical radiation for scoliosis, Hodgkin lymphoma, and to monitor lung collapse among Massachusetts TB patients. Lung cancer risk following high-LET exposures was evaluated in three MPS cohorts, i.e. Mound, Mallinckrodt and Rocketdyne, but numbers and doses were not sufficient to make strong conclusions. Ongoing studies will be more informative. The ongoing study of medical radiation workers within the MPS includes 85,000 women and 85,000 men and will provide substantial information as to the effectiveness of gradual exposures over time to cause lung cancer and to determine whether such exposure results in different radiation lung cancer risks between men and women (Boice 2017; Yoder et al. 2018). The CFCS of 64,000 patients includes equal numbers of men and women exposed at all ages. It is currently being updated for mortality and incidence and should provide essentially lifespan follow-up of the cohort.

Disclosure statement

The authors report no conflicts of interest. The authors alone are responsible for the content and writing of the paper.

Funding

The Million Person Study of Low-Dose Health effects is supported in part by a grant from the U.S. Nuclear Regulatory Commission (NRC-HQ-60-14-G-0011), grants from the Centers for Disease Control and Prevention (5UE1EH000989, 5NUE1EH001315), grants from the National Aeronautics and Space Administration (NNX15AU88G, 80NSSC17M0016), grants from the U.S. Department of Energy (Grant No. DE-SC0008944 awarded to the National Council on Radiation Protection and Measurements, which included interagency support from the U.S. Nuclear Regulatory Commission, the U.S. Environmental Protection Agency, and the National Aeronautics and Space Administration; and more recent Grants No. DE-AU0000042 and DE-AU0000046, a grant from the National Cancer Institute (Grant No. U01 CA137026) and a Discovery Grant from the Vanderbilt-Ingram Cancer Center (Center no. 404-357-9682). Contract support was also received by Oak Ridge National Laboratory from the Office of Radiation and Indoor Air, U.S. Environmental Protection Agency, under Interagency Agreement DOE No. 1824 S581-A1, under contract No. DE-AC05-00OR22725 with UT-Battelle; and contract support was received by Oak Ridge Associated Universities from the U.S. Department of Energy under contract No. DE-SC0014664. Dr. Zablotska was supported by National Cancer Institute and National Institutes of Health (Grant No. R01CA197422).

ORCID

John D. Boice http://orcid.org/0000-0002-8755-1299
Elizabeth D. Ellis http://orcid.org/0000-0003-2871-1587
Ashley P. Golden http://orcid.org/0000-0002-9430-7702
Michael T. Mumma http://orcid.org/0000-0001-7506-8710
Sarah S. Cohen http://orcid.org/0000-0003-0421-1983

References

Beck HL, Till JE, Grogan HA, Aanenson JW, Mohler HJ, Mohler SS, Voillequé PG. 2017. Red bone marrow and male breast doses for a cohort of atomic veterans. Radiat Res. 187:221–228.

Boice JD. Jr. 2017. Space: the final frontier-research relevant to mars. Health Phys. 112:392–397.

Boice JD, Jr, Cohen SS, Mumma MT, Ellis ED, Cragle DL, Eckerman KF, Wallace PW, Chadda B, Sonderman JS, Wiggs LD, et al. 2014. Mortality among mound workers exposed to polonium-210 and other sources of radiation, 1944–1979. Radiat Res. 181:208–228.

Boice JD, Jr., Cohen SS, Mumma MT, Ellis ED, Eckerman KF, Leggett RW, Boecker BB, Brill AB, Henderson BE. 2011. Updated mortality analysis of radiation workers at Rocketdyne (Atomics International), 1948–2008. Radiat Res. 176:244–258.

Boice JD, Jr, Ellis ED, Golden AP, Girardi DJ, Cohen SS, Chen H, Mumma MT, Shore RE, Leggett RW. 2018. The past informs the future: an overview of the million worker study and the mallinckrodt chemical works cohort. Health Phys. 114:381–385.

Boice JD, Jr, Land CE, Shore RE, Norman JE, Tokunaga M. 1979. Risk of breast cancer following low-dose radiation exposure. Radiology. 131:589–597.

Boice JD, Jr, Leggett RW, Ellis ED, Wallace PW, Mumma M, Cohen SS, Brill AB, Chadda B, Boecker BB, Yoder RC, et al. 2006. A comprehensive dose reconstruction methodology for former Rocketdyne/Atomics International radiation workers. Health Phys. 90:409–430.

Boice JD, Jr, Preston D, Davis FG, Monson RR. 1991. Frequent chest X-ray fluoroscopy and breast cancer incidence among tuberculosis patients in Massachusetts. Radiat Res. 125:214–222.

Boice JD, Jr., Rosenstein M, Trout ED. 1978. Estimation of breast doses and breast cancer risk associated with repeated fluoroscopic chest examinations of women with tuberculosis. Radiat Res. 73:373–390.

Boice JD, Jr, Cohen SS, Mumma MT, Hagemeyer D, Chen H, Yoder RC, Dauer LT. 2019. Leukemia among early nuclear power plant workers employed between 1957 and 1984 in the United States. Int J Radiat Biol. (Submitted).

Boice JD, Jr, Cohen S, Mumma M, Hegemeyer D, Golden A, Yoder RC, Dauer L, et al. 2019. Leukemia among Industrial Radiographers. Int J Radiat Biol. (Submitted)

Boice JD, Jr, Till J, Mumma M, Chen H, et al. 2019. Mortality among US military participants at seven aboveground nuclear weapons test series. Int J Radiat Biol. (Submitted).

Brenner AV, Gilbert ES, Land CE. 2007. Lung cancer mortality after exposure to fractionated ionizing radiation in a cohort of Massachusetts tuberculosis patients. The 13th International Congress of Radiation Research (ICRR 2007); Jul 8–12; San Francisco, CA, USA, (poster PS1134: Abstract Book p. 82).

Breslow NE, Day NE. 1987. Statistical Methods in Cancer Research, Vol. 2. Lyon: IARC. The design and analysis of cohort studies; 131–135.

Cahoon EK, Preston DL, Pierce DA, Grant E, Brenner AV, Mabuchi K, Utada M, Ozasa K. 2017. Lung, laryngeal and other respiratory cancer incidence among Japanese atomic bomb survivors: an updated analysis from 1958 through 2009. Radiat Res. 187:538–548.

Caldwell GG, Zack MM, Mumma MT, Falk H, Heath CW, Till JE, Chen H, Boice JD. 2016. Mortality among military participants at the 1957 PLUMBBOB nuclear weapons test series and from leukemia among participants at the SMOKY test. J Radiol Prot. 36: 474–489.

Cardis E, Vrijheid M, Blettner M, Gilbert E, Hakama M, Hill C, Howe G, Kaldor J, Muirhead CR, Schubauer-Berigan M, et al. 2007. The 15-country collaborative study of cancer risk among radiation workers in the nuclear industry: estimates of radiation-related cancer risks. Radiat Res. 167:396–416.

Cox DR. 1972. Regression models and life-tables. J R Stat Soc B. 34: 187–220.

Davis FG, Boice JD, Jr., Hrubec Z, Monson RR. 1989. Cancer mortality in a radiation-exposed cohort of Massachusetts tuberculosis patients. Cancer Res. 49:6130–6136.

Dores GM, Metayer C, Curtis RE, Lynch CF, Clarke EA, Glimelius B, Storm H, Pukkala E, van Leeuwen FE, Holowaty EJ, et al. 2002. Second malignant neoplasms among long-term survivors of Hodgkin's disease: a population-based evaluation over 25 years. JCO. 20:3484–3494.

Dupree EA, Cragle DL, McLain RW, Crawford-Brown DJ, Teta MJ. 1987. Mortality among workers at a uranium processing facility, the Linde Air Products Company Ceramics Plant, 1943–1949. Scand J Work Environ Health. 13:100–107.

Ellis ED, Boice JD, Jr, Golden AP, Girardi DJ, Cohen SS, Mumma MT, Shore RE, Leggett RW, Kerr GD. 2018. Dosimetry is key to good epidemiology: workers at Mallinckrodt Chemical Works had seven different source exposures. Health Phys. 114:386–397.

Furukawa K, Preston DL, Lonn S, Funamoto S, Yonehara S, Matsuo T, Egawa H, Tokuoka S, Ozasa K, Kasagi F, et al. 2010. Radiation and smoking effects on lung cancer incidence among atomic bomb survivors. Radiat Res. 174:72–82.

Furukawa K, Preston DL, Misumi M, Cullings HM. 2017. Handling incomplete smoking history data in survival analysis. Stat Methods Med Res. 26:707–723.

Gilbert ES, Cragle DL, Wiggs LD. 1993. Updated analyses of combined mortality data for workers at the Hanford Site, Oak Ridge National Laboratory, and Rocky Flats Weapons Plant. Radiat Res. 136: 408–421.

Gilbert ES, Koshurnikova NA, Sokolnikov ME, Shilnikova NS, Preston DL, Ron E, Okatenko PV, Khokhryakov VF, Vasilenko EK, Miller S, et al. 2004. Lung cancer in Mayak workers. Radiat Res. 162:505–516.

Gilbert ES, Sokolnikov ME, Preston DL, Schonfeld SJ, Schadilov AE, Vasilenko EK, Koshurnikova NA. 2013. Lung cancer risks from plutonium: an updated analysis of data from the Mayak worker cohort. Radiat Res. 179:332–342.

Gilbert ES, Stovall M, Gospodarowicz M, Van Leeuwen FE, Andersson M, Glimelius B, Joensuu T, Lynch CF, Curtis RE, Holowaty E, et al. 2003. Lung cancer after treatment for Hodgkin's disease: focus on radiation effects. Radiat Res. 159:161–173.

Gillies M, Kuznetsova I, Sokolnikov M, Haylock R, O'Hagan J, Tsareva Y, Labutina E. 2017. Lung cancer risk from plutonium: a pooled analysis of the Mayak and Sellafield worker cohorts. Radiat Res. 188: 645–660.

Golden AP, Ellis ED, Cohen SS, Mumma MT, Leggett RW, Wallace PW, Girardi D, Watkins JP, Shore R, Boice JD. Jr. 2019. Updated mortality analysis of the Mallinckrodt uranium processing workers, 1942-2012. Int J Radiat Biol. https://doi.org/10.1080/09553002.2019.1569773

Haylock RGE, Gillies M, Hunter N, Zhang W, Phillipson M. 2018. Cancer mortality and incidence following external occupational radiation exposure: an update of the 3rd analysis of the UK national registry for radiation workers. Br J Cancer. 119:631–637.

Heidenreich WF, Carnes BA, Paretzke HG. 2006. Lung cancer risk in mice: analysis of fractionation effects and neutron RBE with a biologically motivated model. Radiat Res. 166:794–801.

Hoffman DA, Lonstein JE, Morin MM, Visscher W, Harris BS, 3rd, Boice JD. Jr. 1989. Breast cancer in women with scoliosis exposed to multiple diagnostic x rays. J Natl Cancer Inst. 81:1307–1312.

Howe GR. 1995. Lung cancer mortality between 1950 and 1987 after exposure to fractionated moderate-dose-rate ionizing radiation in the Canadian Fluoroscopy Cohort Study and a comparison with lung cancer mortality in the Atomic Bomb survivors study. Radiat Res. 142:295–304.

Howe GR, Yaffe M. 1992. Estimation of lung tissue doses following exposure to low-LET radiation in the Canadian study of cancer following multiple fluoroscopies, INFO-0405. Ottawa, Canada: Atomic Energy Control Board. [accessed 2018 Aug 27]. http://www.iaea.org/inis/collection/NCLCollectionStore/_Public/23/078/23078790.pdf?r=1

Fry RJ. 1996. Effects of low doses of radiation. Health Phys. 70: 823–827.

Little M. 2018. Tuberculosis fluoroscopy. National Cancer Institute, Division of Cancer Epidemiology and Genetics. [accessed 2018 Aug 27]. https://dceg.cancer.gov/research/what-we-study/tb-fluoroscopy

Miller AB, Howe GR, Sherman GJ, Lindsay JP, Yaffe MJ, Dinner PJ, Risch HA, Preston DL. 1989. Mortality from breast cancer after irradiation during fluoroscopic examinations in patients being treated for tuberculosis. N Engl J Med. 321:1285–1289.

Mumma MT, Cohen SS, Sirko JL, Ellis ED, Boice JD Jr. 2018. Obtaining vital status and cause of death on a million persons. Int J Radiat Biol. https://doi.org/10.1080/09553002.2018.1539884

National Council on Radiation Protection and Measurements (NCRP). 2012. Uncertainties in the Estimation of Radiation Risks and Probability of Disease Causation. NCRP Report 171. Bethesda, MD: NCRP.

National Council on Radiation Protection and Measurements (NCRP). 2014. Commentary No. 23, Radiation Protection for Space Activities: Supplement to Previous Recommendations. Bethesda, MD: National Council on Radiation Protection and Measurements.

National Council on Radiation Protection and Measurements (NCRP). 2018a. Deriving organ doses and their uncertainty for epidemiologic studies (With a focus on the one million U.S. workers and veterans study of low-dose radiation health effects. NCRP Report 178. Bethesda, MD: NCRP.

National Council on Radiation Protection and Measurements (NCRP). 2018b. Implications of recent epidemiologic studies for the linear nonthreshold model and radiation protection. NCRP Commentary 27. Bethesda, MD: NCRP.

National Academies/National Research Council (NA/NRC). 2012. Technical evaluation of the NASA model for cancer risk to astronauts due to space radiation. Washington, DC: National Academies Press.

National Research Council (NRC), Committee on the Biological Effects of Radiation. 2006. Health risks from exposure to low levels of ionizing radiation (BEIR VII Phase 2). Washington, DC: National Academies Press.

Okladnikova ND, Pesternikova VS, Sumina MV, Doshchenko VN. 1994. Occupational diseases from radiation exposure at the first nuclear plant in the USSR. Sci Total Environ. 142:9–17.

Ozasa K, Shimizu Y, Suyama A, Kasagi F, Soda M, Grant EJ, Sakata R, Sugiyama H, Kodama K. 2012. Studies of the mortality of atomic bomb survivors, Report 14, 1950-2003: an overview of cancer and noncancer diseases. Radiat Res. 177:229–243.

Preston DL, Mattsson A, Holmberg E, Shore R, Hildreth NG, Boice JD. Jr. 2002. Radiation effects on breast cancer risk: a pooled analysis of eight cohorts. Radiat Res 158:220–235.

Preston DL, Ron E, Tokuoka S, Funamoto S, Nishi N, Soda M, Mabuchi K, Kodama K. 2007. Solid cancer incidence in atomic bomb survivors: 1958–1998. Radiat Res. 168:1–64.

Preston DL, Sokolnikov ME, Krestinina LY, Stram DO. 2017. Estimates of radiation effects on cancer risks in the mayak worker, techa river and atomic bomb survivor studies. Radiat Prot Dosimetry. 173: 26–31.

Preston DL, Lubin JH, Pierce DA, McConney ME, Shilnikova NS. 2018. Epicure risk regression and person-year computation software: command summary and user guide (version 2.00.03) [Software]; Ottawa, Canada: Risk Sciences International. [accessed 2018 Aug 27]. http://epicurehelp.risksciences.com

Richardson DB, Cardis E, Daniels RD, Gillies M, Haylock R, Leuraud K, Laurier D, Moissonnier M, Schubauer-Berigan MK, Thierry-Chef I, et al. 2018. Site-specific solid cancer mortality after exposure to ionizing radiation: a cohort study of workers (INWORKS). Epidemiology. 29:31–40.

Ronckers CM, Land CE, Miller JS, Stovall M, Lonstein JE, Doody MM. 2010. Cancer mortality among women frequently exposed to radiographic examinations for spinal disorders. Radiat Res. 174:83–90.

Sherman GJ, Howe GR, Miller AB, Rosenstein M. 1978. Organ dose per unit exposure resulting from fluoroscopy for artificial pneumothorax. Health Phys. 35:259–269.

Silver SR, Bertke SJ, Hein MJ, Daniels RD, Fleming DA, Anderson JL, Pinney SM, Hornung RW, Tseng CY. 2013. Mortality and ionising radiation exposures among workers employed at the Fernald Feed Materials Production Center (1951-1985). Occup Environ Med. 70: 453–463.

Storer JB, Fry RJ. 1995. On the shape of neutron dose-effect curves for radiogenic cancers and life shortening in mice. Radiat Environ Biophys. 34:21–27.

Storer JB, Mitchell TJ, Fry RJ. 1988. Extrapolation of the relative risk of radiogenic neoplasms across mouse strains and to man. Radiat Res. 114. 114:331–353.

Storer JB, Serrano LJ, Darden EB, Jr, Jernigan MC, Ullrich RL. 1979. Life shortening in RFM and BALB/c mice as a function of radiation quality, dose, and dose rate. Radiat Res. 78:122–161.

Storer JB, Ullrich RL. 1983. Life shortening in BALB/c mice following brief, protracted, or fractionated exposures to neutrons. Radiat Res. 96:335–347.

Strahlenschutzkommission. 2009. Statement by the german commission on radiological protection (English translation). Sex-specific differences in radiation sensitivity – epidemiological, clinical and biological studies. Geschäftsstelle der Strahlenschutzkommission, Postfach 12 06 29, D-53048 Bonn. [accessed 2018 Aug 20]. https://www.ssk.de/SharedDocs/Beratungsergebnisse_PDF/2009/Sex_Specific_Differences_in_Radiation_Sensitivity.pdf?__blob=publicationFile

Till JE, Beck HL, Aanenson JW, Grogan HA, Mohler HJ, Mohler SS, Voillequé PG. 2014. Military participants at U.S. Atmospheric nuclear weapons testing-methodology for estimating dose and uncertainty. Radiat Res. 181:471–484.

Travis LB, Gospodarowicz M, Curtis RE, Clarke EA, Andersson M, Glimelius B, Joensuu T, Lynch CF, van Leeuwen FE, Holowaty E, et al. 2002. Lung cancer following chemotherapy and radiotherapy for Hodgkin's disease. J Natl Cancer Inst. 94:182–192.

Ullrich RL, Jernigan MC, Cosgrove GE, Satterfield LC, Bowles ND, Storer JB. 1976. The influence of dose and dose rate on the incidence of neoplastic disease in RFM mice after neutron irradiation. Radiat Res. 68:115–131.

Ullrich RL, Jernigan MC, Storer JB. 1977. Neutron carcinogenesis. Dose and dose-rate effects in BALB/c mice. Radiat Res. 72:487–498.

Ullrich RL, Jernigan MC, Satterfield LC, Bowles ND. 1987. Radiation carcinogenesis: time-dose relationships. Radiat Res. 111:179–184.

Ullrich RL, Storer JB. 1979a. Influence of gamma irradiation on the development of neoplastic disease in mice. II. Solid tumors. Radiat Res. 80:317–324.

Ullrich RL, Storer JB. 1979b. Influence of gamma irradiation on the development of neoplastic disease in mice. III. Dose-rate effects. Radiat Res. 80:325–342.

United Nations Scientific Committee on the Effects of Atomic Radiation (UNSCEAR). 2008. Effects of ionizing radiation. UNSCEAR 2006 report. Report to the general assembly. Volume 1. Scientific annex A. Epidemiological studies of radiation and cancer. Publication E.08.IX.6. New York, NY: United Nations.

Upton AC, Beebe GW, Brown JM, Quimby EH, Shellabarger C. 1977. Report of NCI Ad Hoc Working Group on the Risks Associated with Mammography in Mass Screening for the Detection of Breast Cancer. J Natl Cancer Inst. 59:480–493.

U.S. Nuclear Regulatory Commission (NRC). 2018. Radiation exposure information and reporting system (REIRS). [accessed 2018 Aug 31]. http://www.nrc.gov/reading-rm/doc-collections/gils/rad-exp.html

Wakeford R. 2005. Cancer risk among nuclear workers. J Radiol Prot. 25:225–228.

Wiggs LD, Johnson ER, Cox-DeVore CA, Voelz GL. 1994. Mortality through 1990 among white male workers at the Los Alamos National Laboratory: considering exposures to plutonium and external ionizing radiation. Health Phys. 67:577–588.

Yoder RC, Dauer L, Balter S, Boice JD, Jr, Grogan H, Mumma M, Passmore CN, Rothenberg LN, Vetter RJ. 2018. Dosimetry for the study of medical radiation workers with a focus on the mean absorbed dose to the lung, brain and other organs. Int J Radiat Biol. https://doi.org/10.1080/09553002.2018.1549756

Asbestos exposure and mesothelioma mortality among atomic veterans

John E. Till, Harold L. Beck, John D. Boice, H. Justin Mohler, Michael T. Mumma,
Jill W. Aanenson, and Helen A. Grogan

ABSTRACT

Background: The United States (U.S.) conducted 230 above-ground atmospheric nuclear weapons tests between 1945 and 1962 involving over 250,000 military personnel. This is the first quantitative assessment of asbestos-related mesothelioma, including cancers of the pleura and peritoneum, among military personnel who participated in above-ground nuclear weapons testing.

Methods: Approximately 114,000 atomic veterans were selected for an epidemiological study because they were in one of eight series of weapons tests that were associated with somewhat higher personnel exposures than the other tests and because they have been previously studied. We were able to categorize specific jobs into potential for asbestos exposure based on a detailed database of the military activities of the atomic veterans. Standardized mortality ratios (SMR) were calculated by service, rank (officer/enlisted) and ratings (occupation code and work location aboard ship) after 65 years of follow-up.

Results: Mesothelioma deaths were significantly increased overall (SMR 1.56; 95% CI 1.32–1.82; $n = 153$). This increase was seen only among those serving in the PPG (SMR 1.97; 95% CI 1.65–2.34; $n = 134$), enlisted men (SMR 1.81; 95% CI 1.53–2.13; $n = 145$), and the 70,309 navy personnel (SMR 2.15; 95% CI 1.80–2.56; $n = 130$). No increased mortality rates were seen among the other services: army (SMR 0.45), air force (SMR 0.85), or marines (SMR 0.75). Job categories with the highest potential for asbestos exposure (machinist's mates, boiler technicians, water tender, pipe fitters, and fireman) had an of SMR 6.47. Job categories with lower potential (SMR =1.35) or no potential (SMR =1.28) for asbestos exposure had non-significantly elevated mesothelioma mortality.

Conclusions: The large excess of mesothelioma deaths seen among atomic veterans was explained by asbestos exposure among enlisted naval personnel. The sources of exposure were determined to be on navy ships in areas (or with materials) with known asbestos content. No excess of mesothelioma was observed in other services or among naval personnel with minimal exposure to asbestos in this low-dose radiation exposed cohort.

Background

Military veterans who participated in atmospheric nuclear weapons testing conducted by the United States between 1945 and 1963 are being studied for late-occurring health effects. More than 250,000 veterans participated in these tests. The cohort being investigated consists of ~114,000 individuals who were present in one or more of eight test series: CASTLE, CROSSROADS, GREENHOUSE, REDWING, and HARDTACK I conducted at the Bikini and Enewetak Atolls and Johnston Island in the Pacific Proving Grounds (PPG); UPSHOT-KNOTHOLE and PLUMBBOB conducted at the Nevada Test Site (NTS); and TRINITY, the first nuclear test, conducted in New Mexico. We refer to our analysis of this collection of tests as the Eight Series Study which is a subset of the Million Person Study (NCRP 2018; Boice 2019).

Prior to 1980, all navy ships used asbestos extensively for boilers, steam pipe insulation, hull insulation, floors, gaskets, cables, fireproofing, and many other applications. Although most interior areas contained some asbestos material, areas especially susceptible to exposure were engineering spaces where when making repairs, the asbestos insulation had to be removed to allow access and replaced when a job was completed. Over time the asbestos insulation would deteriorate and become brittle making exposure even more likely. Since 1980 the navy has taken far-reaching steps to remove and reduce the use of asbestos and to find other types of material as a substitute.

The purpose of this study was to evaluate an observed excess of mesothelioma in the entire cohort of 114,270 nuclear weapons test participants compared with the US general population. The excess was seen only among Navy

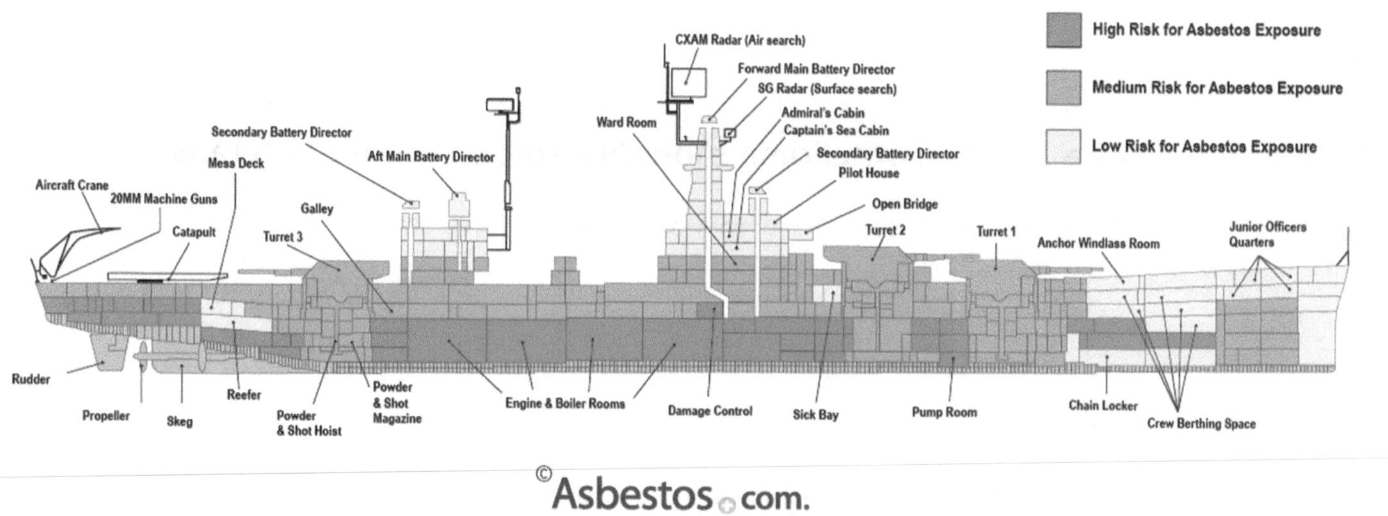

Figure 1. Diagram of naval vessel showing levels of risk for asbestos exposure (figure used with permission and acknowledgement of asbestos.com).

personnel. We applied basic methods previously described in Till et al. (2014) to identify specific jobs with high potential for asbestos exposure, and accordingly, the personnel who served in areas aboard ship where asbestos exposure would be likely. These efforts were concomitant with those used to estimate cumulative radiation exposure associated with the weapons testing.

Methods

Identification of exposed populations

Determination of which personnel aboard ship were most likely to be exposed to asbestos was achieved by identifying the individuals' job aboard ship and the locations on the ship where asbestos is located. The U.S. Navy employs a system of job descriptions among enlisted personnel that describes specifically what the person does while on duty. These specialties are known as ratings[1]. For example, boiler technicians spent most of their time below deck in the boiler room; boatswain's mates would generally be topside focusing on the material condition of the ship above decks; machinist's mates would generally be in the engineering spaces focusing on the propulsion system; and electrician's mates could be at almost any location aboard ship working on electrical components. Officers are more difficult to assign a specific location because they move from one leadership position to another over time, but as a general rule, they would not encounter the same level of asbestos exposure as enlisted personnel.

In addition to ratings for navy personnel that describe an individual's technical specialty, there are also pay grades known as rates. The most junior enlisted rates are seaman or fireman recruit (E-1), seaman or fireman apprentice (E-2), and seaman or fireman (E-3). The characterization of seaman or fireman describes the sailor's preference to 'strike' for a rating more related to either topside duties (e.g. boatswain's mate) or engineering duties (e.g. engineman). Among these junior rates, the 'fireman' designation would, therefore, be expected to be more highly exposed than the seaman and was assigned to a number of different duties aboard ship, including maintenance work where they may even provide the bulk of the manpower. Therefore, as we evaluated those most likely to be exposed, we considered the E-1 through E-3 fireman rates to be in the high-risk exposure category.

Figure 1 is a ship drawing illustrating where the highest levels of exposure to asbestos would be anticipated to occur. There were many types and sizes of navy ships involved in testing at the PPG and this figure is for illustrative purposes only, but the same fundamental compartments such as the engine room and boiler room, existed on essentially all of the ships stationed at the PPG. There are three grades of potential exposure risk—high, medium, and low. As indicated in the figure, the lower levels of the ship both fore and aft as well as the boiler and engine rooms present the highest exposure risk. Medium and low exposure would occur on decks above the lower levels and risk decreases within spaces as they ascend farther above the waterline.

If our conclusions about who would be most exposed were correct, we would expect to see the most junior rates (E-1 through E-3) and petty officers (E-4 through E-6) associated with the engine room, boiler room, and other engineering ratings as most likely to contract mesothelioma.

To support the Eight Series Study, a database was developed to manage the review of existing information for the atomic veterans, including details of their participation and duties (i.e. ratings) in the different test series at the NTS and the PPG (Till et al. 2014). Resulting exposures and doses were incorporated, along with information about the exposure location, activity, and source. The database facilitates review of existing information for the entire cohort, or for only the cases and controls in a specific unit, and it was used to develop the percentage of mesothelioma cases associated with different ratings.

The study population was selected from databases within the Department of Defense (DoD) Nuclear Test Review (NTPR) program. Military personnel who participated in one of eight test series (CASTLE, CROSSROADS, GREENHOUSE, REDWING, HARDTACK I, UPSHOT-KNOTHOLE, PLUMBBOB, and TRINITY) at any point were selected for inclusion in the study. These test series

Table 1. Standardized mortality analyses (SMR)[a] of mesothelioma[b] among nuclear weapons test participants by population characteristics.

Population characteristic	No. of atomic veterans	No. mesothelioma deaths[b]	SMR[a]	95% CI
All	114,270	153	1.56*	1.32–1.82
Rank				
Enlisted	96,071	145	1.81*	1.53–2.13
Officer	18,199	8	0.44*	0.19–0.87
Theatre				
Nevada Test Site	28,589	14	0.57*	0.31–0.95
Pacific Proving Ground	79,313	134	1.97*	1.65–2.34
Both	6368	5	0.89	0.29–2.07
Service				
Navy	70,309	130	2.15*	1.80–2.56
Officers	6494	4	0.61	0.16–1.56
Enlisted	63,815	126	2.34*	1.95–2.79
All other services[c]	43,961	23	0.61*	0.38–0.91
Army	26,091	10	0.45*	0.22–0.83
Air Force	12,848	10	0.85	0.41–1.56
Marines	5003	3	0.75	0.15–2.19

[a]The standardized mortality ratio is the observed number of deaths divided by the expected number of deaths computed on the basis of rates in the general population for males of similar age, race, and calendar years of life.
[b]Taken as cancers of the pleura and peritoneum (ICD5: 201; ICD6: 162.2; ICD7: 158,162.2; ICD8: 158.9,163.0; ICD9: 158.8,158.9,163) and mesothelioma (ICD10: C45).
[c]Nineteen atomic veterans were not in these four branches.
* $P < 0.05$.

were selected because of the availability of high-quality personnel radiation dosimetry records and epidemiologic data from previous investigations (for all series except TRINITY) (Johnson et al. 1996; Thaul et al. 2000; Watanabe et al. 1995). Available information included name, military service number, date of birth, sex, pay grade, rank or rating, unit membership, and dates during participation, and permanent unit.

Vital status determination

Ascertainment of vital status (alive or dead) and cause of death was determined by accessing various mortality databases, including the US Department of Veterans Affairs (VA) BIRLS (Beneficiary Identification Record Location Subsystem), the National Death Index, the Social Security Administration (Death Master File and Service for Epidemiologic Researchers), and death certificates requested and provided by the VA and state departments of health. See Mumma et al. (2018) for a detailed description of vital status and cause of death ascertainment. Deaths from mesothelioma, including cancers of the pleura and peritoneum, were coded according to the International Classification of Disease (ICD) revision being used at the time of death. Only the underlying cause of death was used in the analyses. Because an ICD code specific to malignant mesothelioma did not appear until ICD revision 10 in 1999 (C45), we identified probable deaths from mesothelioma, for example, cancers of the pleura and peritoneum, prior to ICD 10 using the following codes: ICD5: 201; ICD6: 162.2; ICD7: 158,162.2; ICD8: 158.9,163.0; ICD9: 158.8,158.9,163.

Statistical analyses

Standardized mortality ratios (SMRs) were computed comparing the numbers of deaths observed from mesothelioma and other causes among the workers with the numbers expected based on mortality rates in the general US population. US population mortality rates by age, calendar year, and sex were applied to the corresponding person-years of follow-up to obtain the expected number of deaths from mesothelioma had the study population experienced the same force of mortality as that of the U.S. population as a whole. The end of follow-up was taken as the date of death, the 95th birthday, or 31 December 2010, whichever came first. Observed and expected numbers of deaths were examined by test area (NTS, PPG, both areas), service (navy, army, air force, marines, and rank (enlisted, officer). Statistical variability was evaluated by the 95% exact Poisson confidence interval (CI) of the SMR assuming that the observed number of deaths followed a Poisson distribution.

Results

The SMR analyses revealed significant excess in mesothelioma deaths, overall (SMR 1.56; 95% CI 1.32–1.82; $n = 153$) and among certain subgroups (Table 1). Statistically significant elevations were seen among enlisted men (SMR 1.81; $n = 145$), navy personnel (SMR 2.15; $n = 130$), and those who served in the PPG (SMR 1.97; $n = 134$). Statistically significant deficits were seen among officers (SMR 0.44; $n = 8$), army personnel (SMR 0.45; $n = 10$) and those who served at the NTS (SMR 0.57; $n = 14$).

The SMR analyses showed a clear concentration of mesothelioma deaths among navy personnel and no increased mortality rates were seen among the other services: army (SMR 0.45), air force (SMR 0.85), or marines (SMR 0.75). In fact, the naval personnel had a significant excess of mesothelioma deaths (SMR 1.53), whereas the other three services combined had a significant deficit of mesothelioma deaths (SMR 0.61; 95% CI 0.38, 0.91; n-23). Thus the database developed to support the Eight Series Study was also used to evaluate the number and percentage of mesothelioma deaths for each naval rating and to determine whether or not the

Table 2. Standardized mortality ratio of mesothelioma by selected ratings and potential for asbestos exposure among 63,815 enlisted naval personnel (sailors).

Asbestos potential and rating	No. of sailors[a]	Observed	Expected	SMR	95% CI
High Potential					
Boiler technician	1002	8	0.72	11.1	4.78–21.9
Pipe fitter	283	3	0.23	13.0	2.62–38.1
Machinist's mate	3809	29	3.2	9.18	6.15–13.2
Water tender	845	2	0.65	3.06	0.34–11.1
Fireman	7302	30	6.2	4.85	3.27–6.92
Total	12,880	69	10.7	6.47	5.03–8.19
Low/Medium Potential					
Boatswain's mate	2938	2	2.40	0.83	0.09–3.01
Hull technician	2	0	<.01	n/a	n/a
Seaman	21,144	25	17.7	1.42	0.92–2.09
Total	23,936	27	19.9	1.35	0.89–1.97
Any Potential (Low to High)	36,639	96	30.5	3.15	2.55–3.85
No to minimal potential	27,176	30	23.4	1.28	0.87–1.83

[a]Six hundred and eighty-two naval personnel have more than one rating.

data support our conclusion that certain naval ratings would have a greater risk of asbestos exposure and of developing mesothelioma due to the areas they worked aboard ship.

If asbestos exposures aboard naval ships were sufficient to cause excess mesothelioma deaths, then it might be expected that deaths due to asbestosis also would be increased. A subsequent SMR analyses confirmed this to be the case. For all atomic veterans, 65 deaths from asbestosis occurred reflecting a doubling of risk (SMR 2.13; 95% CI 1.64-2.71). The increased mortality was concentrated entirely among navy personnel (SMR 3.22; 95% CI 2.45-4.14; $n=60$) and all other services had SMRs less than 1. Furthermore, Navy enlisted sailors with job categories with the highest potential for asbestos exposure had very high rates of asbestosis (SMR 10.0; 95% CI 6.90-14.2; $n=32$), whereas those with no/minimal potential had no significant excess (SMR 1.78, 95% CI 0.95-3.05, $n=13$).

Within naval personnel, we looked at the difference in the number of mesothelioma deaths between officers and enlisted. Officers had a nearly fourfold lower SMR from mesothelioma than enlisted personnel. To further investigate the deaths due to mesothelioma among different ratings within the U.S. Navy, we calculated SMRs within the ratings by the potential for asbestos exposure. Boiler technician, pipe fitter, machinist's mate, water tender, and fireman have the highest potential for exposure based on where they worked on the ship and job duties. Boatswain's mate, hull technician, and seaman have lower potential.

Table 2 shows the number and SMR for mesothelioma deaths by ratings based on the potential for exposure to asbestos among the 63,815 enlisted naval personnel. Boiler technician, pipe fitter, and machinist's mate all have SMRs exceeding 9. Likewise, fireman has an SMR of nearly 5. Ratings with low to medium potential have a slightly elevated non-significant SMR. Naval personnel with minimal potential for asbestos exposure have an SMR of 1.28 (95 CI 0.87–1.83).

Discussion

A key objective of this analysis was to determine if the methodology developed to determine radiation doses to navy military personnel who were exposed during atmospheric testing of nuclear weapons could be applied to investigate mortality from mesothelioma among the same group. The process of estimating radiation dose was based on knowing an individual's general location aboard ship using the navy's system of rates and ratings for enlisted personnel and rank for officers. The results of the statistical analysis revealed that mesothelioma deaths occurred at higher rates and at locations aboard ship where asbestos exposure would have been greater with respect to other parts of the ship. Therefore, the method of using the navy job classification system of enlisted rates and ratings and rank for officers provides a viable method for evaluating asbestos exposure among navy personnel.

A secondary benefit to the study is the confirmation of the methodology to estimate radiation dose described in Till et al. (2014). In that study, higher radiation exposures were found to be in locations where personnel were stationed near contaminated seawater used for condenser cooling and at other locations around the ship such as berthing quarters below decks that were near the hull.

Although sailors who died from mesothelioma were not included in the subset of veterans whose radiation exposure was reconstructed in great detail (Till et al. 2014), their whole-body radiation doses could be accurately estimated from military records (Beck et al. 2017). Absorbed doses to the lung were low (mean 6.21 mGy with a maximum of 972 mGy) and the doses for the high-risk asbestos-mesothelioma rating groups did not differ meaningfully from the doses to the members of the other naval cohort rating groups nor the doses to members of the entire Navy cohort.

Conclusions

Over half of the mesothelioma deaths among 63,815 enlisted navy personnel were for sailors with jobs as machinist's mate, pipe fitters, boiler technicians, water tender, and fireman, even though these jobs were held by only 20% of the naval enlisted population. These jobs all had high potential for exposure to asbestos products, which would explain the significant increase in risk of death from mesothelioma in the navy (SMR 2.15) compared with the significant deficit seen among other services (SMR 0.61) which had low potential for asbestos exposure. Sailors who did not work in these job categories with high potential for asbestos exposure, were not found to be at excess risk for mesothelioma mortality compared to the general US population.

Our evaluation and analyses of the comprehensive data sets available for military personnel support the conclusion that the high mortality from mesothelioma among sailors in the U.S. Navy during the atmospheric nuclear testing at the PPG was related to a high potential for exposure to asbestos. In contrast, the military personnel serving in other services had a low potential exposure to asbestos as evidenced by their significant low mortality rate from mesothelioma. Certain naval ratings were at a greater risk for exposure to asbestos than others. There is also strong evidence that officers have a lower risk of exposure than enlisted personnel, and accordingly a lower risk of death from mesothelioma. This analysis demonstrates the importance of considering an individual's job or duty on a ship and its relationship to risk of exposure to asbestos. Similar relationships were established between ratings, rates, and officers aboard ships in our estimation of radiation dose to the atomic veteran cohort (Till et al. 2014; Caldwell et al. 2016; Beck et al. 2017). In the absence of detailed knowledge on the potential exposure to a powerful carcinogen, such as asbestos, in a low-dose radiation study, misleading conclusions might be drawn. Thus, this investigation rules in asbestos as the cause of the mesothelioma excess, and accordingly rules out low-dose radiation.

Notes

1. The system of ratings and their abbreviations used by the U.S. Navy evolved over more than 200 years. The rating defines the job specialty of enlisted personnel and characterizes where aboard ship (or ashore) they work and what they do. Some ratings no longer exist, and new ratings have been introduced due to changes in technology with time.

Acknowledgements

Dr. John D. Boice, Jr. is the principal investigator for the Million Person Study. This support is acknowledged and appreciated. We also acknowledge Dr. Paul Blake, Director, Chief, Nuclear Test Personnel Review, Defense Threat Reduction Agency, Department of Defense and his staff for their technical support of the project.

Disclosure statement

The authors declare no conflict of interest.

Funding

This research was supported in part by grants from the National Cancer Institute (Grant No. U01 CA137026), the U.S. Department of Energy (Grant no. DE-SC0008944 awarded to the National Council on Radiation Protection and Measurement which included interagency support from the U.S. Nuclear Regulatory Commission, the U.S. Environmental Protection Agency, and the National Aeronautics and Space Administration, and Grant No. DE-AU0000042), and a Discovery Grant from the Vanderbilt-Ingram Cancer Center (Center no. 404-357-9682).

ORCID

John E. Till http://orcid.org/0000-0003-3666-8806
Michael T. Mumma http://orcid.org/0000-0001-7506-8710

References

Beck HB, Till JE, Grogan HA, Aanenson JW, Mohler HJ, Mohler SS, Voillequé PG. 2017. Red bone marrow and male breast doses for a cohort of atomic veterans. Radiat Res. 187:221–228.

Boice JD. 2019. The Million Person Study, whence it came and why. Int J Radiat Biol (Submitted for this special issue).

Caldwell GG, Zack M, Mumma M, Falk H, Heath CW, Till JE, Chen H, Boice JD. Jr. 2016. Mortality among military participants at the 1957 PLUMBBOB nuclear weapons test series and from leukaemia among participants at the SMOKY test. J Radiol Prot. 36:474–489.

Johnson JC, Thaul S, Page WF, Crawford H. 1996. Mortality of veteran participants in the CROSSROADS nuclear test. Washington (DC): National Academies Press.

Mumma MT, Cohen SS, Sirko JL, Ellis ED, Boice JD. 2018. Obtaining vital status and cause of death on a million persons. Int J Radiat Biol. 1–21. DOI: 10.1080/09553002.2018.1539884. [Epub ahead of print]

National Council on Radiation Protection and Measurements (NCRP). 2018. Deriving organ doses and their uncertainty for epidemiologic studies (With a focus on the one million U.S. workers and veterans study of low-dose radiation health effects. NCRP Report 178. Bethesda (MD): NCRP.

Thaul S, Page WF, Crawford H, O'Maonaigh H. 2000. The five series study: mortality of military participants in U.S. nuclear weapons tests. Washington (DC): National Academies Press.

Till JE, Beck HB, Aanenson JW, Grogan HA, Mohler HJ, Mohler SS, Voillequé PG. 2014. Military participants at the U.S. atmospheric nuclear weapons testing—methodology for estimating dose and uncertainty. Radiat Res. 181:471–484.

Watanabe KK, Kang HK, Dalager NA. 1995. Cancer mortality risk among military participants of a 1958 atmospheric nuclear weapons test. Am J Public Health. 85:523–527.

Mesothelioma mortality within two radiation monitored occupational cohorts

Michael T. Mumma, Jennifer L. Sirko, John D. Boice, and William J. Blot

ABSTRACT

Purpose: The risk of mesothelioma, including cancers of the pleura and peritoneum, was examined within two large cohorts of workers monitored for exposure to ionizing radiation.

Methods and materials: Mortality was assessed among 253,632 workers routinely monitored for external radiation, including 30,724 industrial radiographers (IR) at shipyards, 142,583 workers at nuclear power plants (NPP), and 83,441 IR who had not worked at an NPP or shipyard. Follow-up was from 1969 through 2011. Standardized mortality ratios (SMRs) and 95% confidence intervals (CIs) were computed; observed numbers of deaths from mesothelioma (including cancers of the pleura and peritoneum) and asbestosis were compared with numbers expected based on age-, sex-, and calendar year-specific national mortality rates. Job history and quantitative asbestos exposure data were unavailable, but work at a shipyard was taken as a surrogate for the likelihood of exposure. Cox proportional hazards models were used to estimate hazard ratios (HRs) for mesothelioma in relation to estimated cumulative radiation exposure to the lung.

Results: The mean duration of follow-up was 25.3 years (max 42 years). The mean cumulative lung dose was 28.6 mGy (7.3% > 250 mGy). Nearly 20% of the workers had died by 2011. A total of 421 mesothelioma deaths were found (75% occurring after 1999) with increased SMRs among workers monitored in shipyards (SMR 9.97; 95% CI 8.50–11.63) and for NPP workers (SMR 5.55; 95% CI 4.88–6.29), but not for IR who had not worked in shipyards (SMR 1.15; 95% CI 0.53–2.19). Likewise, deaths from asbestosis ($n = 189$) were also increased for shipyard and NPP workers (SMR = 18.1 and 9.2, respectively), but not among workers who never worked at a shipyard or NPP (SMR = 0.70; $n = 1$). Radiation dose to the lung was not associated with a statistically meaningful dose-response trend for mesothelioma in the combined cohorts (HR at 100 mGy = 1.10; 95% CI 0.96–1.27; $p = .18$), nor was mesothelioma risk associated with radiation exposure among IR who had not worked in a shipyard and assumed minimally exposed to asbestos.

Conclusions: An elevated rate of death from mesothelioma was observed in two radiation-exposed occupational groups with potential for asbestos exposure. The increased risk of death from asbestosis, combined with little evidence of a rising trend in mesothelioma mortality with increasing radiation exposure, suggests that the mesothelioma (and asbestosis) excess in these workers was due to asbestos exposure in shipyards and power plants and not to occupational low-dose radiation.

Introduction

Mesothelioma is a rare and highly lethal cancer originating in the mesothelial cells that line the pleura, peritoneal, heart, and other cavities. Most cases (81%) of mesothelioma arise in the pleural lining of the lung and chest cavity, while 9% arise in the peritoneal lining of the abdominal cavity and a smaller percentage in the pericardium (ACS 2017). Asbestos exposure in the workplace is the major cause of mesothelioma, accounting for up to 80% of cases (Crew et al. 2005; Boffetta and Stayner 2006; International Agency for Research on Cancer (IARC) 2012; Carbone et al. 2012; Carbone and Yang 2017). Asbestos was widely used in heat-resistant materials and insulation from the 1940s through the 1970s in the United States, primarily in the construction, textile, auto, aircraft and shipbuilding industries, and in electric power plants including nuclear power plants (NPP) (Selikoff and Seidman 1991). Incidence rates of mesothelioma in the United States increased until the mid-1990s, remained level for several years, and began to decline in 2005 in men while remaining stable in women (ACS 2017). These trends and sex-specific differences reflect in large part the reduction in workplace exposure to asbestos in jobs that were traditionally male-oriented professions.

Suspected non-asbestos risk factors for malignant mesothelioma include ionizing radiation (high-dose radiotherapy and Thorotrast (i.e. radioactive thorium dioxide that emits alpha particles)), viral infections (e.g. SV40), genetic predisposition (e.g. germline mutations of the BRCA1-associated protein 1), chronic pleural inflammation, and erionite (a naturally occurring fibrous material similar to asbestos). These factors may act alone or as co-factors together with asbestos or with 'naturally occurring asbestos-like material' such as erionite, to cause mesothelioma (Crew et al. 2005; Carbone et al. 2012, 2016; Attanoos et al. 2018). Animal experiments have shown the co-carcinogenic properties of high-dose radiation with asbestos (Warren et al. 1981). The polyomavirus simian virus 40 (SV40) has been reported to induce mesothelioma and to be a co-factor for asbestos carcinogenesis in experimental animals; however, the evidence for a causal relationship between SV40 and mesothelioma in humans is lacking (Carbone et al. 2012). Chronic inflammation appears to promote the occurrence of asbestos-related mesothelioma (Carbone and Yang 2017). Mesothelioma may thus be a multifactorial disease and not all non-asbestos causes may have been identified nor have all co-factor mechanisms.

Low-dose ionization radiation has been suggested as a possible risk factor for mesothelioma (Goodman et al. 2009), but the evidence is limited (Metz-Flammant et al. 2011). Practically all worker studies of low-dose radiation are unable to rule out asbestos used in thermal insulation at the worksite or used in shipyards as a contributing causal factor. Studies with information on potential asbestos exposure (e.g. based on job category and work environment, shipyard work, and clinical examinations) discount a radiation association (Schubauer-Berigan et al. 2005, 2015; Richardson et al. 2007; Till et al. 2018). It has been suggested that exposure to asbestos may have been positively correlated with radiation dose within a pooled analysis of five U.S. cohorts (Schubauer-Berigan et al. 2005), suggesting that any radiation association with mesothelioma may be confounded by unknown or uncontrolled exposure to asbestos. That is, jobs linked with asbestos work also have a high potential for radiation exposure. Furthermore, there is no evidence that cancer of the pleura or peritoneum is increased among Japanese atomic bomb survivors (Grant et al. 2017).

There have been few if any occupational studies of radiation-exposed groups with large enough size, duration of follow-up, and assessment of asbestos to evaluate convincingly any association between radiation exposure and mesothelioma (Metz-Flammant et al. 2011). Accordingly, we evaluated the risk of mesothelioma, including cancer of the pleura and peritoneum, by combining two large occupational cohorts monitored for external radiation, that is, NPP workers and industrial radiographers (IR). We sought to evaluate (1) the risk of mesothelioma overall; (2) the risk of mesothelioma among the subgroups of these workers who had been employed in shipyards and/or power plants, where there is high potential for asbestos exposure (Selikoff and Hammond 1978; Selikoff et al. 1979); (3) the risk of asbestosis, due entirely to asbestos exposure, to help assess confounding between asbestos and radiation exposures; and (4) any association between mesothelioma and cumulative low-dose radiation exposure overall and specifically among IRs not known to have worked at a shipyard.

Methods

Population identification. The study population was formed by combining two occupational cohorts: NPP workers ($n = 135,193$) and IRs ($n = 123,510$) within the Million Person Study of Low-Dose Health Effects (MPS) (NCRP 2018; Boice et al. 2019). Both cohorts were drawn from either the U.S. Nuclear Regulatory Commission (NRC) Radiation Exposure Information and Reporting System (REIRS) database or the electronic records of Landauer, Inc. (Hagemeyer et al. 2018). The NPP workers were identified by selecting workers monitored for radiation exposure between 1969 and 31 December 1984 at an NPP. Because of the very large numbers of NPP workers with a career badge dose of less than 10 mSv, a 10% random sample was selected for cost efficiency with minimal loss of statistical power. All NPP workers with career badge dose greater than 10 mSv were included in the cohort. Individuals were considered eligible for study if sufficient personal identifiers (either a Social Security Number or name and date of birth) were available to allow for vital status follow-up and had at least 30 days of radiation monitoring as an IR or at an NPP facility. There were 5071 workers in both the IR and NPP cohorts, resulting in a combined study population of 253,632.

Mortality follow-up. Mortality status as of 31 December 2011 as well as cause of death were determined from various mortality databases, including the National Death Index, the Social Security Administration (Death Master File and Service for Epidemiologic Researchers), and death certificates provided by state departments of health (Mumma et al. 2018). Underlying cause of death was coded according to the International Classification of Diseases (ICD) revision in effect at the time of death. An ICD code specific to malignant mesothelioma (C45) did not appear until 1999 when the 10th revision of the ICD was issued (ICD10). For earlier years we combined deaths from malignant cancers of the pleura (ICD8 163.0; ICD9 163) and peritoneum (ICD8 158.9; ICD9 158.8–158.9). Asbestosis mortality was indicated by the following codes: ICD8 515.2, ICD9 501, and ICD10 J61.

Radiation dose estimation. Estimation of occupational career radiation dose to the lung followed the procedures outlined by Boice et al. (2006), and described more fully in NCRP (2018) and by Dauer et al. (2018). Briefly, radiation doses from all places of employment were sought by linking the study population to the U.S. Department of Energy Radiation Exposure Monitoring Systems dosimetry database, historical DOE study databases as well as U.S. military service dosimetry databases. These doses, in addition to the doses from the REIRS and Landauer databases, were combined to create annual dose estimates of all occupational external radiation. There were minimal intakes of radionuclides in these occupations studied and thus no radiation from internal emitters was included in the annual dose estimates. Cumulative doses were obtained by summing the annual doses across all years monitored for radiation exposure. The procedures outlined in NCRP (2018) were then used to convert the measured personal dose equivalents

Table 1. Characteristics of the combined industrial radiographer (IR) and nuclear power plant (NPP) worker study populations ($n = 253{,}632$).

Characteristic	*n*	%
Total	253,632	100.0
Sex		
Female	17,312	6.8
Male	236,320	93.2
Occupational category		
Only monitored for radiation in shipyard	27,608	10.9
Only monitored for radiation at NPP	139,467	55.0
Monitored at both at shipyard and NPP	3116	1.2
Never monitored at shipyard or NPP (IR workers only)	83,441	32.9
Year of Birth		
1904–1929	26,280	10.4
1930–1939	25,629	10.1
1940–1949	59,038	23.3
1950–1959	78,507	31.0
1960–1993	64,178	25.3
Mean YOB	*1951*	
Start of follow-up		
1969–1979	103,760	40.9
1980–1989	88,251	34.8
1990–2011	61,621	24.3
Mean year at start of follow-up	*1983*	
Mean age at start of follow-up	*31.8*	
Years of follow-up		
0–9	28,736	11.3
10–19	42,058	16.6
20–29	73,798	29.1
30–39	93,168	36.7
40–42	15,872	6.3
Mean duration of follow-up (years)	*25.3*	
Career badge dose		
Less than 1 mSv	73,418	28.9
1–<10 mSv	41,635	16.4
10–<50 mSv	86,355	34.0
50–<100 mSv	27,542	10.9
100–<250 mSv	20,126	7.9
250 mSv or greater	4556	1.8
Mean career badge dose (mSv)	*35.4*	
Mean number of years monitored	*9.80*	
Vital status as of 31 December 2011		
Alive	204,206	80.5
Dead	49,426	19.5
Mean age at end of follow-up	*57.7*	

[Hp (10)] into individual lung doses using the appropriate dose conversion factors for the incident gamma ray energies. The primary source of exposure at NPPs was to radioactive cobalt (^{60}Co and ^{58}Co) and to some extent radioactive cesium (^{137}Cs). The primary source of exposure to IRs was to radioactive iridium (^{192}Ir) and to some extent radioactive cobalt (^{60}Co).

Asbestos exposure. Personal or work area measurements on asbestos exposures were not available. To investigate the potential of occupational asbestos exposure, each worker's exposure history within the REIRS and Landauer dosimetry databases was used to determine if they were ever monitored for radiation at either a shipyard or a U.S. naval facility (for IRs) or at an NPP, occupational settings with increased likelihood of asbestos exposure (Boffetta and Stayner 2006; IARC 2012). Keyword searches of the employer name (within REIRS) and account name (within Landauer) were used to identify shipyards, marinas, dry docks, and naval facilities. NRC license codes and Landauer account names were used to identify nuclear power facilities. Employment at a shipyard carried a strong potential for asbestos exposure (Selikoff and Hammond 1978; Selikoff et al. 1979; Tagnon et al. 1980; Matanoski 1991; Matanoski et al. 2008). Within nuclear and other power plants, asbestos was used for insulation of pipes, boilers, generators, and electric wirings (Boffetta et al. 1991). Individual workers within NPPs, however, could not be identified as to their potential for asbestos exposure. To address confounding by potential asbestos exposure, we created three non-overlapping groups of workers among whom risk of mesothelioma and asbestosis were assessed: (1) those ever monitored at shipyards (predominantly IR); (2) those who worked only at NPP but not at a shipyard; and (3) IR who had not been monitored at a shipyard or NPP.

Statistical analysis. Standardized mortality ratios (SMRs) were computed comparing the numbers of deaths observed from mesothelioma, asbestosis, and other causes among the workers with the numbers expected based on mortality rates in the general U.S. population. U.S. mortality rates by age, calendar year, and sex were applied to the corresponding person-years of follow-up to obtain the expected number of deaths (overall and for other conditions, including mesothelioma and asbestosis). The end of follow-up was taken as the date of death, the 95th birthday, or 31 December 2011, whichever came first. Observed and expected numbers of deaths were examined by sex, potential occupation exposure to asbestos (as judged by ever being monitored at a shipyard and/or a NPP), year of first radiation monitoring, time since first monitoring, and duration of follow-up. Confidence intervals (CI) for the SMRs were computed assuming that the observed number of deaths followed a Poisson distribution (Bailar and Ederer 1964). A 95% CI that excluded 1.0 was considered statistically significant at the two-sided significance level of .05.

Within-cohort (internal) analyses were conducted using Cox proportional hazards models to compute hazard ratios (HRs) and corresponding 95% CI. Age at follow-up was used as the underlying timescale for the analysis. Covariates in the models included sex, year of birth, potential occupational exposure to asbestos (ever monitored at a shipyard), year of first radiation monitoring, and socioeconomic status (SES), with the exposure variable of interest being cumulative radiation exposure to the lung. Direct measures of individual-level SES were not available; thus, to derive a surrogate measure for missing smoking and individual-level SES data, we obtained and geocoded residential address histories for all workers, linked them to area-level SES measures from the United States Census, and used the mean educational attainment within the Census Block Group as a measure of SES (Cohen et al. 2018). For the internal analyses, radiation exposure was lagged 10 years to be consistent with observations from reports of radiation therapy and secondary mesothelioma, where mesotheliomas occurred an average of 18 years (range 5–>40 years) after high-dose radiation treatment (Cavazza et al. 1996; Crew et al. 2005; Hodgson et al. 2007; Goodman et al. 2009; Carbone et al. 2012; Farioli et al. 2016) and with the estimated minimum latency period between initial asbestos exposure and mesothelioma onset (Boffetta and Stayner 2006). Furthermore, radiation studies that include workers potentially exposed to asbestos have reported latencies up to 40 years (Schubauer-

Table 2. Standardized mortality ratios (SMR) and 95% confidence intervals (CIs) for all causes of death and deaths due to mesothelioma and asbestosis among the combined industrial radiographer (IR) and nuclear power plant (NPP) worker cohorts ($n = 253{,}632$), 1969–2011.

	All causes of death			Mesothelioma[a]			Asbestosis		
	N	SMR	95% C.I.	*N*	SMR	95% C.I.	*N*	SMR	95% CI
Total	49,275	0.88*	0.88–0.89	421	6.10*	5.53–6.71	189	11.4*	9.85–13.2
Sex									
Male	48,338	0.88*	0.87–0.89	421	6.15*	5.58–6.77	87	9.15*	7.34–11.3
Female	937	0.97	0.91–1.04	0	0.00	0.00–6.53	103	14.0*	11.4–16.9
Occupational category									
Ever monitored at shipyard	13,474	0.86*	0.85–0.88	163	9.97*	8.50–11.6	100	18.1*	14.8–22.1
Only monitored at NPP	29,535	0.89*	0.88–0.90	249	5.55*	4.88–6.29	88	9.16*	7.35–11.3
Never at shipyard or NPP	6266	0.91*	0.89–0.93	9	1.15	0.53–2.19	1	0.70	0.01–3.89
Socioeconomic status									
Low	17,040	1.04*	1.03–1.06	108	5.48*	4.50–6.62	84	16.5*	13.2–20.4
Medium	25,178	0.91*	0.90–0.92	236	6.90*	6.05–7.84	90	11.0*	8.84–13.5
High	7057	0.60*	0.58–0.61	77	5.10*	4.02–6.37	15	4.60*	2.58–7.60
Year first monitored									
1940–1959	24,351	0.88*	0.87–0.89	276	8.37*	7.42–9.42	135	14.0*	11.7–16.5
1960–1990	23,185	0.89*	0.88–0.90	141	4.18*	3.52–4.93	54	8.19*	6.15–10.7
1991–2011	1739	0.81*	0.47–4.47	4	1.75	0.47–4.48	0	0	0.00–13.5
Time since first monitor									
Less than 10 years	5,175	0.67*	0.65–0.69	17	4.88*	2.84–7.81	4	6.79*	1.83–17.4
10–19 years	10,614	0.85*	0.84–0.87	34	3.66*	2.53–5.11	21	8.97*	5.55–13.7
20–29 years	16,779	0.95*	0.93–0.96	129	4.86*	4.06–5.78	65	11.7*	9.00–14.9
30 years or more	16,707	0.94*	0.92–0.95	241	8.12*	7.13–9.21	99	12.3*	10.0–15.0
Time period of follow-up									
1969–1979	1861	0.56*	0.53–0.59	4	2.73	0.73–6.99	4	20.9*	5.63–53.5
1980–1999	20,744	0.85*	0.84–0.86	102	7.47*	6.09–9.07	89	16.2*	13.0–20.0
2000–2011	26,670	0.95*	0.94–0.96	315	5.85*	5.22–6.53	96	8.84*	7.16–10.8

*$p < .05$.
[a]Pleura and peritoneum (ICD8: 158.9,163.0; ICD9: 158.8-158.9,) and mesothelioma (ICD10 C45).

Berigan et al. 2015; Yiin et al. 2017). Analyses showed an SMR of 4.9 ($n = 17$) within the first 10 years of first radiation monitoring (suggestive of asbestos exposure prior to radiation monitoring); therefore, follow-up within the first 10 years of radiation exposure were excluded in the internal Cox analyses. For consistency and comparisons with other cohorts, we also computed HRs adjusting for duration of monitoring at an IR or NPP facility (<1, 1–4, 5–9, 10–19, 20–29, ≥30 years). Special attention was paid to IR who had not worked in shipyards, a group with low likelihood of asbestos exposure, but small numbers of mesotheliomas ($n = 9$) among this group precluded informative Cox analyses.

Analyses were conducted with SAS/STAT software (version 9.4 of the SAS System for Windows, SAS Institute Inc., Cary, NC). Human subjects research approval was received from Vanderbilt University Institutional Review Board.

Results

Demographic and occupational characteristics of the combined study cohorts are shown in Table 1. Among 253,632 workers, 12% had ever worked in shipyards or naval facilities (90% as IR), 55% were NPP workers only, and 33% were IR who had not been monitored at shipyards or NPP. Most workers (93%) were male, the mean year of birth was 1951, and the mean duration of follow-up was 25 years. Over 49,000 workers, or 20%, had died as of 31 December 2011. The mean career badge dose was 35 mSv. The mean career lung dose was 28.6 mGy. During nearly 6,500,000 person-years of follow-up there were 421 deaths from mesothelioma and 189 deaths from asbestosis, all among men.

Total mortality was significantly below expectation in the combined study population (SMR 0.88), consistent with a healthy worker effect (Table 2). The numbers of total cancers (SMR 1.02; CI 1.00–1.03, $n = 15{,}105$) and smoking-related cancers (SMR 1.03; CI 1.01–1.05, $n = 7826$) were close to those expected (not shown). However, there were large and significant excesses of mesothelioma among shipyard workers (SMR 9.97; $n = 163$) and NPP workers (SMR 5.55; $n = 249$). The 83,441 IR workers who never worked at a shipyard did not show a significant increased rate of mesothelioma (SMR = 1.15; 95% CI 0.53–2.19; $n = 9$). Parallel findings were seen for deaths from asbestosis, with SMRs reaching 18.14 (95% CI 14.8–22.1) and 9.16 (95% CI 7.50–11.4) for shipyard and NPP workers, respectively, but 0.70 for IR workers not employed at shipyards (95% CI 0.01–3.89). Increasing SMRs from high to low SES were seen for total mortality and asbestosis but not for mesothelioma, possibly due to higher smoking prevalence among those in lower socioeconomic groups. SMRs were higher for mesothelioma prior to 2000, but the increased rates persisted beyond 2000 (SMR = 5.85) with 315 deaths (75% of all mesothelioma deaths) observed during the latest follow-up time period. SMRs also were highest when follow-up time exceeded 30 or more years from first radiation monitoring.

Table 3 shows the HRs and 95% CI derived from the Cox models for mesothelioma mortality in the combined cohorts by cumulative lung dose categories (including a lag of 10 years for all radiation doses as well as exclusion of the first 10 years of follow-up for all individuals) adjusted for sex, year of first monitoring, year of birth, ever monitored at a shipyard and SES. Because workers with <1 mGy cumulative dose included disproportionate numbers of office and other workers with characteristics different from the

Table 3. Hazard ratios (HR) and 95% confidence intervals (CI) derived from Cox proportional hazards models for mesothelioma mortality, follow-up starting 10 years after first radiation monitoring and dose lagged 10 years for the combined IR and NPP cohorts ($n = 157,855$ workers), 1969–2011.

			Not adjusted for duration of radiation monitoring		Adjusted for duration of radiation monitoring	
Lung dose (mGy)	Number of workers	Number of cases	HR[a]	95% CI	HR[a]	95% CI
1–<10	40,739	81	1.00	REF	1.00	REF
10–<50	78,819	172	1.22	0.92–1.60	1.21	0.92–1.59
50–<100	22,123	37	0.98	0.66–1.46	0.96	0.64–1.44
≥100	1	40	1.36	0.92–2.00	1.31	0.89–1.95
	p for trend[b]		*.18*		*.24*	
	Hazard ratio (95% CI) at 100 mGy		*1.10 (0.96–1.27)*		*1.09 (0.95–1.26)*	

[a]Adjusted for sex, year of first monitoring, year of birth, SES, and ever monitored at a shipyard.
[b]*p*-Value for test for linear trend in the HRs computed using a measure of continuous dose.

other monitored workers, the referent category was taken as 1 - <10 mGy for internal analyses. None of the HRs for dose categories above the 1–9 mGy reference category were significantly elevated. From a model treating dose as a continuous (linear) variable, the estimated HR at 100 mGy was 1.10 (95% CI 0.96–1.27) and there was no significant increase in the HR with increasing dose ($p = .18$). When adjusted for duration of monitoring, the HRs were essentially the same, again with no evidence of a linear trend.

Because of apparent confounding from asbestos exposure, that is, jobs with high potential for asbestos exposure also having high potential for radiation exposure, we carried out an internal analysis for radiation dose effects among the 83,441 IR workers who were not known to have been employed in shipyards. These workers also had lower radiation exposure (mean and 95th percentile lung doses of 10.2 and 53.7 mGy, respectively, compared with 23.3 and 116.5 mGy for IR workers who had been employed in shipyards). Due to small numbers of mesothelioma deaths (only nine) among non-shipyard working IR, hazard ratios by dose categories are not shown. A test for linear trend over increasing cumulative lung dose using a continuous measure of dose was not significant ($p = .56$).

Discussion

Several epidemiologic studies have reported on risks of mesothelioma among radiation-exposed workers. We briefly review the existing literature before discussing results and implications from this study.

Radiotherapy and Thorotrast patients

Several follow-up studies of cancer patients treated with high-dose ionizing radiation have reported increased rates of mesothelioma, suggesting that radiotherapy is a probable, though rare, cause of mesothelioma (Crew et al. 2005; Goodman et al. 2009; Carbone et al. 2012; Farioli et al. 2016). In addition, Thorotrast (colloidal $^{232}ThO_2$), a diagnostic radiographic contrast medium injected into the body for cerebral angiography, has been linked to small excesses of cancers of the peritoneum and pleura in epidemiologic studies in Denmark, Sweden, Portugal, Germany, and the US (Andersson et al. 1995; Travis et al. 2003; Becker et al. 2008). The colloidal Thorotrast would aggregate into 'hot spots', remain in the body for life irradiating mesothelial cells to high doses of alpha particles and cause chronic inflammation (NRC 1988). The non-radiation properties of Thorotrast, a heavy metal, could not be completely ruled out as contributing to the mesothelioma increase (Andersson et al. 1995).

Atomic bomb survivors

Among the nearly 100,000 atomic bomb survivor cohort, no significant increases in cancer of the pleura (one case in men) or peritoneum (three cases in women) were reported (Grant et al. 2017), nor have there been reports of increased mesothelioma. A case report of mesothelioma among a Nagasaki survivor concluded that work at a shipyard and inhalation of asbestos fibers was the likely cause (Mizuki et al. 1997).

Occupational groups with chronic low-dose radiation exposure

Metz-Flammant et al. (2011) reviewed 17 studies of nuclear workers and noted that, while 15 showed elevations of malignant pleural mesothelioma, of the 12 studies conducting radiation dose response analyses, only 1 found a significant association with mesothelioma. The reviewers concluded that asbestos was an important confounder in most studies, and that ionizing radiation was not associated with mesothelioma mortality. More recent large-scale occupational studies of workers exposed to chronic low dose radiation also tend not to find significant associations between radiation and mesothelioma mortality (Habib et al. 2006; Gillies and Haylock 2014; Schubauer-Berigan et al. 2015; Haylock et al. 2018; Richardson et al. 2018). Asbestos exposures were present in all cohorts, with mesothelioma elevations interpreted to be most likely due to asbestos rather than radiation exposure (Muirhead et al. 2009).

There were four notable studies of radiation-exposed workers with significant elevations in mesothelioma deaths that were able to evaluate the potential for asbestos exposures based on job categories or work in areas with known asbestos exposure. These were studies of workers at the Savannah River Site (SRS) (Makie et al. 2005; Richardson et al. 2007), the Idaho National Laboratory (INL) (Schubauer-Berigan et al. 2005), the Portsmouth Naval Shipyard (PNS) (Rinsky et al. 1988; Silver et al. 2004; Zaebst et al. 2009; Schubauer-Berigan et al. 2015), and military participants at above-ground nuclear weapons tests (Till et al. 2018).

Savannah River Site (SRS)

Among 15,264 male workers at the SRS, a significant SMR of 4.7 was found for cancer of the pleura which was notable among hourly-paid workers (Richardson et al. 2007). Industrial hygiene reports from the 1970s found increased asbestos exposure for some activities based on area-wide sampling, and medical evaluations, including chest radiographs among 1368 SRS workers as part of the SRS Former Production Worker Health Project (Makie et al. 2005). Pleural abnormalities were higher in former SRS workers than in the general population, and were associated with jobs with likely occupational exposure to asbestos, for example, 'Electrical construction work (including asbestos workers engaged in insulation)'.

Idaho National Laboratory (INL)

Among 63,561 workers at INL, significantly increased SMRs were found for asbestosis (SMR 3.67, $n = 10$) and respiratory cancers likely to be mesothelioma (SMR 1.74, $n = 22$) (Schubauer-Berigan et al. 2005). The association with asbestosis indicates that high levels of asbestos were present, since asbestos is the only cause of asbestosis (Silver et al. 2004). Significant excess asbestos-related deaths occurred among 36,534 workers monitored for radiation exposure: asbestosis (SMR 3.21, $n = 6$) and likely mesothelioma (SMR 2.59, $n = 15$), but there was no association with radiation dose. Industrial hygiene data were collected at INL and 2741 workers with high likelihood of asbestos exposure were identified based on job titles such as asbestos mechanic, asbestos foreman, engineer asbestos, asbestos worker, insulation worker, boiler maker, and pipefitter. Standardized rate ratios for asbestos workers were notable and significantly high for asbestosis (SRR 25.6; 95% CI 6.25, 105) and likely mesothelioma (SRR 4.28; 95% CI 1.19, 15.5). The cases were restricted to workers with jobs related to maintenance and construction activities, and were highest in the group of jobs with likely asbestos exposure.

Portsmouth Naval Shipyard (PNS)

Among 37,853 workers at PNS, significantly high SMRs for asbestosis (SMR 5.71, $n = 12$) again indicated that high levels of asbestos were present. An expert panel also was able to estimate the potential for asbestos exposure for all workers on the basis of shop, job title, and time-period available from each worker's employment history (Zaebst et al. 2009). The significant elevations in asbestosis mortality occurred among the 13,468 workers monitored for radiation, whereas asbestosis occurred below expectation among the 24,385 non-monitored workers (SMR 0.79). Mesothelioma and pleural cancers combined were significantly elevated in a pooled analysis of five U.S. nuclear facilities (SMR 2.80; 95% CI 2.27, 3.42, $n = 96$) with most of the excess coming from the PNS cohort (Schubauer-Berigan et al. 2015). A radiation dose-response analysis for mesothelioma and pleural cancer deaths combined, revealed a positive but not statistically significant trend. The significantly high SMR for asbestosis among monitored workers and the deficit of asbestosis deaths among non-monitored workers indicated a strong correlation between radiation work and asbestos work. That is, radiation workers were more heavily exposed to asbestos than were other workers, and that asbestos was the most likely cause of mesothelioma and pleural cancers.

For completion, increased rates of mesothelioma also were observed in combined cohorts of uranium enrichment works, including Oak Ridge Gaseous Diffusion Plant (also known as K-25) in Oak Ridge, Tennessee; the Portsmouth Gaseous Diffusion Plant in Piketon, Ohio; and the Paducah Gaseous Diffusion Plant Paducah, Kentucky (Yiin et al. 2017). A non-significant association between radiation and mesothelioma was reported. Unfortunately, asbestos had not been identified *a priori* as an important workplace exposure and was not assessed.

Atomic veterans

A study of 114,270 military participants at above-ground nuclear weapons tests was able to explain a notably high increase in deaths due to mesothelioma by exposure to asbestos and not to low-dose radiation (Till et al. 2018). Mesothelioma mortality was significantly increased in the entire study population (SMR 1.56; 95% CI 1.32–1.82; $n = 153$). The increase was seen, however, only among 70,309 navy personnel (SMR 2.15; 95% CI 1.80–2.56; $n = 130$), and not among the other services: army (SMR 0.45), air force (SMR 0.85), or marine corps (SMR 0.75). Asbestos was used in naval vessels for insulation in the boiler room, engine room, and other areas. The 12,880 sailors with job categories having the highest potential for asbestos exposure, that is, machinist's mates, boiler technicians, water tenders, pipe fitters, and firemen, had the highest mortality increase (SMR 6.47; 95% CI 5.03–8.19; $n = 69$). The 27,176 sailors with no potential for asbestos exposure did not have a significant increase in mesothelioma mortality (SMR 1.28; 95% CI 0.87–1.83; $n = 30$). Similarly, 65 deaths from asbestosis occurred (SMR 2.13; 95% CI 1.64–2.71), and the increased mortality was concentrated entirely among navy personnel (SMR 3.22; 95% CI 2.45–4.14; $n = 60$). Furthermore, Navy enlisted sailors with job categories with the highest potential for asbestos exposure had very high rates of asbestosis (SMR 10.0; 95% CI 6.90–14.2; $n = 32$). Asbestos exposure among naval personnel serving on ships during nuclear weapon testing explained the large excess of mesothelioma and asbestosis deaths seen among atomic veterans in this low-dose radiation cohort.

Current study

This study confirmed and quantified high mortality rates of mesothelioma and asbestosis among workers at NPP and in shipyards. The excess rate among IR workers appeared only for those who had worked in shipyards, a setting with known asbestos exposure in ship building and repair (Tagnon et al. 1980; IARC 2012). No increase in mesothelioma or asbestosis was seen among radiation-exposed IR who had not been employed in shipyards or NPP, thus providing evidence that mesothelioma is not caused by low-level radiation when confounding by asbestos can be ruled out or at least minimized. Similarly, the sixfold increase in mesothelioma among NPP workers can be attributable to an increased probability of on-site exposure to asbestos used as insulation materials in the plants. A notable observation is that mesothelioma mortality remained high after 30 years of follow-up and was increased sixfold during 2000–2011, indicating both the very long latency of mesothelioma and suggesting the need for continued monitoring for these occupational groups. The very high SMRs for asbestosis among the NPP workers and the shipyard workers and the absence of a risk among IR workers who did not work at a shipyard confirms these observations as well as the presence of asbestos in NPP and shipyards.

The ability to evaluate the risk of mesothelioma in relation to ionizing radiation in these occupational settings was aided by the availability of annual and individual radiation monitoring data for the IR and NPP cohorts. Radiation doses to the lung were broad and cumulative doses were over 100 mGy for nearly 3000 workers. Radiation dose to the lung was not associated with a statistically significant dose-response trend for mesothelioma (HR at 100 mGy = 1.10; 95% CI 0.96–1.27; $p = .18$).

The difficulty in estimating an independent effect of ionizing radiation on malignant mesothelioma in any occupational setting with known asbestos exposure cannot be overstated. While some have questioned the potential for confounding by asbestos (Matanoski et al. 2008; Goodman et al. 2009), it has been demonstrated in several cohorts that job titles with a higher likelihood of ionizing radiation exposure also had a higher likelihood of asbestos exposure (Matanoski et al. 2008; Schubauer-Berigan et al. 2005, 2015). The pooled study of U.S. nuclear workers (Schubauer-Berigan et al. 2015), which included the PNS study (Rinsky et al. 1988; Silver et al. 2004; Yiin et al. 2007), suggested that exposure to asbestos was positively correlated with radiation dose within the pooled cohort and that asbestos was a confounder at PNS for lung cancer. Furthermore, the strong correlation between radiation and deaths due to asbestosis (which is only caused by asbestos) seen at INL signifies a strong link between radiation work and asbestos work (Schubauer-Berigan et al. 2015). Our results are consistent with and strengthen these prior findings indicating the key role of asbestos in inducing mesothelioma in NPP and IR workers. Other large occupational cohorts within the MPS such as the 170,000 medical radiation workers (where exposure to asbestos in the workplace is unlikely) should help further elucidate the association between long-term radiation exposure and mesothelioma (Yoder et al. 2018).

Strengths and limitations

The strengths of this study include the large size of the radiation cohorts, the comprehensive follow-up covering 70 years, the comprehensive estimation of cumulative radiation dose to lung, and the large number of mesothelioma and asbestos deaths observed. The ability to identify over 83,000 workers who had only minimal likelihood of asbestos exposure provided an opportunity to evaluate a low-dose radiation association with mesothelioma with minimal confounding by asbestos. Radiation doses were recorded for each worker according to reporting regulations and reasonably accurate radiation doses to the lung were reconstructed. Internal cohort analyses compared workers to workers over radiation dose categories and thus minimized a 'healthy worker effect' which can occur when comparisons are made with the general population.

Limitations included the inability to estimate the potential for asbestos exposure for individual workers, the assumption that cancers of the pleura and peritoneum are mainly mesotheliomas, the limited number of female workers, the absence of data on smoking, the absence of specific job categories which involved asbestos exposure and asbestos exposure measurements. Prior to the use of ICD-10 codes in mortality data, an ICD code specific to malignant mesothelioma did not exist. ICD codes for malignant cancers of the pleura and peritoneum likely underestimate the true number of deaths from mesothelioma (Davis et al. 1992).

Confounding by asbestos for radiation associations was demonstrable by the correlation between radiation and asbestosis, which is only caused by asbestos. IR workers at shipyards who had a high potential for asbestos exposure could be identified and separated, allowing analyses of IR workers with likely and unlikely exposures to asbestos. However, we did not have access to job titles or any other data to quantify asbestos exposure among individual workers. Residual confounding by asbestos for radiation dose-response analyses could not be discounted.

Conclusion

Asbestos. Our study of 253,632 workers monitored for occupational radiation exposure as IRs or NPP workers confirms and quantifies the high rates of mesothelioma mortality (SMR = 6.1; $n = 421$) and asbestosis mortality (SMR = 11.4; $n = 189$) in these professions. The high mortality rates have persisted for over 30 years and increases remained near sixfold through 2011, reflecting the long latencies for these asbestos-related conditions. While the absolute number of asbestos-related deaths were small in comparison with the large number of workers, an awareness and need for monitoring is clearly indicated.

Radiation. Therapeutic doses of radiation and injection with Thorotrast, an alpha-particle emitting colloidal solution of thorium dioxide used as a diagnostic radiographic contrast medium, are likely causes of mesothelioma, including cancers of the pleura and peritoneum. However, cancer of the pleura is not increased among atomic bomb survivors, and large-scale studies of occupational groups exposed to low radiation levels are either negative or the increases are attributed to asbestos exposure. Associations between radiation and asbestosis (which is only caused by asbestos) indicate that jobs with relatively high radiation exposures are highly correlated with jobs with relatively high asbestos exposure. Over 83,000 IRs not employed at shipyards did not have an excess of mesothelioma or asbestosis, and there was no clear evidence of a radiation dose-response relationship in the overall cohort. Low-dose radiation received in these occupational setting was not found to be associated with mesothelioma.

Disclosure statement

M. T. Mumma and W. J. Blot have previously consulted with a former manufacturer of products containing asbestos on matters related to litigation. Other authors report no conflict of interest. The authors alone are responsible for the content and writing of the paper.

Funding

This work was supported in part by a grant from the U.S. Nuclear Regulatory Commission [NRC-HQ-60-14-G-0011], a grant from the Centers for Disease Control and Prevention [5UE1EH000989], a grant from the National Aeronautics and Space Administration [Grant No. NNX15AU88G], and grants from the U.S. Department of Energy [Grant No. DE-SC0008944 and Grant No. DE-AU0000042 awarded to the National Council on Radiation Protection and Measurements, which included interagency support from the U.S. Nuclear Regulatory Commission, the U.S. Environmental Protection Agency, and the National Aeronautics and Space Administration]. Furthermore, contract support was received by Oak Ridge National Laboratory from the Office of Radiation and Indoor Air, U.S. Environmental Protection Agency, under Interagency Agreement DOE No. 1824 S581-A1, under contract No. DE-AC05-00OR22725 with UT-Battelle; and contract support was received by Oak Ridge Associated Universities from the U.S. Department of Energy under contract No. DE-SC0014664.

ORCID

Michael T. Mumma http://orcid.org/0000-0001-7506-8710
John D. Boice http://orcid.org/0000-0002-8755-1299

References

American Cancer Society (ACS). 2017. Cancer facts & figures 2017. Special section: rare cancers in adults. Atlanta (GA): American Cancer Society; [accessed 2018 Oct 22]. https://www.cancer.org/content/dam/cancer-org/research/cancer-facts-and-statistics/annual-cancer-facts-and-figures/2017/cancer-facts-and-figures-2017.pdf

Andersson M, Wallin H, Jönsson M, Nielsen LL, Visfeldt J, Vyberg M, Bennett WP, De Benedetti VM, Travis LB, Storm HH. 1995. Lung carcinoma and malignant mesothelioma in patients exposed to Thorotrast: incidence, histology and p53 status. Int J Cancer. 63: 330–336.

Attanoos RL, Churg A, Galateau-Salle F, Gibbs AR, Roggli VL. 2018. Malignant mesothelioma and its non-asbestos causes. Arch Pathol Lab Med. 142:753–760.

Bailar III JC, Ederer F. 1964. Note: significance factors for the ratio of a Poisson variable to its expectation. Biometrics. 20:639–643.

Becker N, Liebermann D, Wesch H, Van Kaick G. 2008. Mortality among Thorotrast-exposed patients and an unexposed comparison group in the German Thorotrast study. Eur J Cancer. 44:1259–1268.

Boffetta P, Cardis H, Vainio H, Coleman P, Kogevinas M, Nordberg G, Parkin DM, Partensky C, Shuker D, Tomatis L. 1991. Cancer risks related to electricity production. Eur J Cancer. 27:1504–1519.

Boffetta P, Stayner R. 2006. Pleural and peritoneal cancers. In: Scottenfeld D, Fraumeni JF Jr, editors. Cancer epidemiology and prevention. 3rd ed. New York (NY): Oxford University Press; p. 659–673.

Boice JD, Jr, Cohen SS, Mumma MT, Ellis ED. 2019. The Million Person Study, from whence it came and why. Int J Radiat Biol. 4: 1589015. [Epub ahead of print] [PMID

Boice JD, Jr., Leggett RW, Ellis ED, Wallace PW, Mumma M, Cohen SS, Brill AB, Chadda B, Boecker BB, Yoder RC, et al. 2006. A comprehensive dose reconstruction methodology for former Rocketdyne/Atomics International radiation workers. Health Phys. 90:409–430.

Carbone M, Kanodia S, Chao A, Miller A, Wali A, Weissman D, Adjei A, Baumann F, Boffetta P, Buck B. 2016. Consensus Report of the 2015 Weinman International Conference on Mesothelioma. J Thorac Oncol. 11:1246–1262.

Carbone M, Ly BH, Dodson RF, Pagano I, Morris PT, Dogan UA, Gazdar AF, Pass HI, Yang H. 2012. Malignant mesothelioma: facts, myths, and hypotheses. J Cell Physiol. 227:44–58.

Carbone M, Yang H. 2017. Mesothelioma: recent highlights. Ann Transl Med. 5:238.

Cavazza A, Travis LB, Travis WD, Wolfe JT, Foo ML, Gillespie DJ, Weidner N, Colby TV. 1996. Post-irradiation malignant mesothelioma. Cancer. 77:1379–1385.

Cohen SS, Mumma MT, Ellis ED, Boice JD. Jr. 2018. Validating the use of census data on education as a measure of socioeconomic status in an occupational cohort. Int J Rad Biol. DOI:10.1080/09553002.2018.1549758

Crew KD, Neugut AI, Antman KH. 2005. Malignant mesothelioma following radiation: advances in pathogenesis, diagnosis and translational therapies. In: Pass HI, Vogelzang NJ, Carbone M, eds. Malignant mesothelioma. New York (NY): Springer; p. 350–363.

Dauer LT, Bouville A, Toohey RE, Boice JD, Jr, Beck HL, Eckerman KF, Hagemeyer D, Leggett RW, Mumma MT, Napier B. 2018. Dosimetry and uncertainty approaches for the million person study of radiation workers and veterans: overview of the recommendations in NCRP Report No. 178. Review. Int J Rad Biol. DOI:10.1080/09553002.2018.1536299

Davis LK, Martin TR, Klinger B. 1992. Use of death certificates for mesothelioma surveillance. Public Health Rep. 107:481–483.

Farioli A, Ottone M, Morganti AG, Compagnone G, Romani F, Cammelli S, Mattioli S, Violante FS. 2016. Radiation-induced mesothelioma among long-term solid cancer survivors: a longitudinal analysis of SEER database. Cancer Med. 5:950–959.

Gillies M, Haylock R. 2014. The cancer mortality and incidence experience of workers at British Nuclear Fuels plc, 1946-2005. J Radiol Prot. 34:595–623.

Goodman JE, Nascarella MA, Valberg PA. 2009. Ionizing radiation: a risk factor for mesothelioma. Cancer Causes Control. 20:1237–1254.

Grant EJ, Brenner A, Sugiyama H, Sakata R, Sadakane A, Utada M, Cahoon EK, Milder CM, Soda M, Cullings HM. 2017. Solid cancer incidence among the life span study of atomic bomb survivors: 1958-2009. Radiat Res. 187:513–537.

Habib RR, Abdallah SM, Law M, Kaldor J. 2006. Cancer incidence among Australian nuclear industry workers. J Occup Health. 48: 358–365.

Hagemeyer D, Nichols G, Mumma MT, Boice JD, Jr, Brock TA. 2018. 50 years of the radiation exposure information and reporting system and importance to the Million Person Study. Int J Radiat Biol. DOI: 10.1080/09553002.2018.1540896

Haylock RGE, Gillies M, Hunter N, Zhang W, Phillipson M. 2018. Cancer mortality and incidence following external occupational radiation exposure: an update of the 3rd analysis of the UK National Registry for Radiation Workers. Br J Cancer. 119:631–637.

Hodgson DC, Gilbert ES, Dores GM, Schonfeld SJ, Lynch CF, Storm H, Hall P, Langmark F, Pukkala E, Andersson M, et al. 2007. Long-term solid cancer risk among 5-year survivors of Hodgkin's lymphoma. J Clin Oncol. 25:1489–1497.

International Agency for Research on Cancer (IARC). 2012. Asbestos (chrysotile, amosite, crocidolite, tremolite, actinolite, and anthophlyte). IARC Monograph 100. Lyon (France): IARC.

Makie T, Adcock D, Lackland DT, Hoel DG. 2005. Pulmonary abnormalities associated with occupational exposures at the Savannah River Site. Am J Ind Med. 48:365–372.

Matanoski GM, Tonascia JA, Correa-Villaseñor A, Yates YC, Fink N, Elliott E, Sanders B, Lantry D. 2008. Cancer risks and low-level radiation in US shipyard workers. JRR. 49:83–91.

Matanoski GM. 1991. Health effects of low-level radiation in shipyard workers. Final report:[Draft]. DOE Contract Number DE-AC02-79EV10095; [accessed 2019 Jan 12] https://www.osti.gov/biblio/10103020

Metz-Flammant C, Guseva Canu I, Laurier D. 2011. Malignant pleural mesothelioma risk among nuclear workers: a review. J Radiol Prot. 31:9–23.

Mizuki M, Yukishige K, Abe Y, Tsuda T. 1997. A case of malignant pleural mesothelioma following exposure to atomic radiation in Nagasaki. Respirology. 2:201–205.

Muirhead CR, O'Hagan JA, Haylock RG, Phillipson MA, Willcock T, Berridge GL, Zhang W. 2009. Mortality and cancer incidence following occupational radiation exposure: third analysis of the National Registry for Radiation Workers. Br J Cancer. 100:206–212.

Mumma MT, Cohen SS, Sirko JL, Ellis ED, Boice JD. 2018. Obtaining vital status and cause of death on a million persons. Int J Radiat Biol. DOI:10.1080/09553002.2018.1539884

National Council on Radiation Protection and Measurements (NCRP). 2018. Deriving organ doses and their uncertainty for epidemiologic studies (with a focus on the one million U.S. workers and veterans study of low-dose radiation health effects). NCRP Report No. 178. Bethesda (MD): NCRP.

National Research Council (NRC). 1988. Committee on the biological effects of ionizing radiations. Health risks of radon and other internally deposited alpha-emitters: BEIR IV. Washington (DC): National Academy Press.

Richardson DB, Cardis E, Daniels RD, Gillies M, Haylock R, Leuraud K, Laurier D, Moissonnier M, Schubauer-Berigan MK, Thierry-Chef I, et al. 2018. Site-specific solid cancer mortality after exposure to ionizing radiation: a cohort study of workers (INWORKS). Epidemiology. 29:31–40.

Richardson DB, Wing S, Wolf S. 2007. Mortality among workers at the Savannah River Site. Am J Ind Med. 50:881–891.

Rinsky RA, Melius JM, Hornung RW, Zumwalde RD, Waxweiler RJ, Landrigan PJ, Bierbaum PJ, Murray WE. Jr. 1988. Case-control study of lung cancer in civilian employees at the Portsmouth Naval Shipyard, Kittery, Maine. Am J Epidemiol. 127:55–64.

Schubauer-Berigan MK, Daniels RD, Bertke SJ, Tseng CY, Richardson DB. 2015. Cancer mortality through 2005 among a pooled cohort of U.S. nuclear workers exposed to external ionizing radiation. Radiat Res. 183:620–631.

Schubauer-Berigan MK, Macievic GV, Utterback DF, Tseng CY, Flora JT. 2005. an epidemiologic study of mortality and radiation-related risk of cancer among workers at the Idaho National Engineering and Environmental Laboratory, a U.S. Department of Energy Facility. Health-Related Energy Research Branch. HHS/NIOSH Publication No. 2005-131, National Institute for Occupational Safety and Health, Cincinnati, OH; [accessed 2019 Jan 12]. https://www.cdc.gov/niosh/docs/2005-131/pdfs/2005-131.pdf

Selikoff IJ, Hammond EC. 1978. Asbestos-associated disease in United States shipyards. CA Cancer J Clin. 28:87–99.

Selikoff IJ, Lilis R, Nicholson WJ. 1979. Asbestos disease in United States shipyards. Ann N Y Acad Sci. 330:295–311.

Selikoff IJ, Seidman H. 1991. Asbestos-associated deaths among insulation workers in the United States and Canada, 1967-1987. Ann N Y Acad Sci. 643:1–14.

Silver SR, Daniels RD, Taulbee TD, Zaebst DD, Kinnes GM, Couch JR, Kubale TL, Yiin JH, Schubauer-Berigan MK, Chen PH. 2004. Differences in mortality by radiation monitoring status in an expanded cohort of Portsmouth Naval Shipyard workers. J Occup Environ Med. 46:677–690.

Tagnon I, Blot WJ, Stroube RB, Day NE, Morris LE, Peace BB, Fraumeni JF. 1980. Mesothelioma associated with the shipbuilding industry in coastal Virginia. Cancer Res. 40:3875–3879.

Till JE, Beck HL, Boice JD, Mohler HJ, Mumma MT, Aanenson JW, Grogan HA. 2018. Asbestos exposure and mesothelioma mortality among atomic veterans. Int J Radiat Biol. 4:1–15. [Epub ahead of print]

Travis LB, Hauptmann M, Gaul LK, Storm HH, Goldman MB, Nyberg U, Berger E, Janower ML, Hall P, Monson RR, et al. 2003. Site-specific cancer incidence and mortality after cerebral angiography with radioactive thorotrast. Radiat Res. 160:691–706.

Warren S, Brown CE, Chute RN, Federman M. 1981. Mesothelioma relative to asbestos, radiation, and methylcholanthrene. Arch Pathol Lab Med. 105:305–312.

Yiin JH, Anderson JL, Daniels RD, Bertke SJ, Fleming DA, Tollerud DJ, Tseng CY, Chen PH, Waters KM. 2017. Mortality in a combined cohort of uranium enrichment workers. Am J Ind Med. 60:96–108.

Yiin JH, Silver SR, Daniels RD, Zaebst DD, Seel EA, Kubale TL. 2007. A nested case-control study of lung cancer risk and ionizing radiation exposure at the Portsmouth Naval Shipyard. Radiat Res. 168: 341–348.

Yoder RC, Dauer L, Balter S, Boice JD, Jr, Grogan H, Mumma M, Passmore CN, Rothenberg LN, Vetter RJ. 2018. Dosimetry for the study of medical radiation workers with a focus on the mean absorbed dose to the lung, brain and other organs. Int J Radiat Biol. DOI:10.1080/09553002.2018.1549756

Zaebst DD, Seel EA, Yiin JH, Nowlin SJ, Chen P. 2009. Summary of retrospective asbestos and welding fume exposure estimates for a nuclear naval shipyard and their correlation with radiation exposure estimates. J Occup Environ Hyg. 6:404–414.

Meeting report

A million persons, a million dreams: a vision for a National Center for Radiation Epidemiology and Biology

John D. Boice, Brian Quinn, Isaf Al-Nabulsi, Armin Ansari, Paul K. Blake, Steve R. Blattnig, Emily A. Caffrey, Sarah S. Cohen, Ashley P. Golden, Kathryn D. Held, Derek W. Jokisch, Richard W. Leggett, Michael T. Mumma, Caleigh Samuels, John E. Till, Sergey Y. Tolmachev, R. Craig Yoder, Joey Y. Zhou, and Lawrence T. Dauer

ABSTRACT

Background: Epidemiologic studies of radiation-exposed populations form the basis for human safety standards. They also help shape public health policy and evidence-based health practices by identifying and quantifying health risks of exposure in defined populations. For more than a century, epidemiologists have studied the consequences of radiation exposures, yet the health effects of low levels delivered at a low-dose rate remain equivocal.

Materials and Methods: The Million Person Study (MPS) of U.S. Radiation Workers and Veterans was designed to examine health effects following chronic exposures in contrast with brief exposures as experienced by the Japanese atomic bomb survivors. Radiation associations for rare cancers, intakes of radionuclides, and differences between men and women are being evaluated, as well as noncancers such as cardiovascular disease and conditions such as dementia and cognitive function. The first international symposium, held November 6, 2020, provided a broad overview of the MPS. Representatives from four U.S. government agencies addressed the importance of this research for their respective missions: U.S. Department of Energy (DOE), the Centers for Disease Control and Prevention (CDC), the U.S. Department of Defense (DOD), and the National Aeronautics and Space Administration (NASA). The major components of the MPS were discussed and recent findings summarized. The importance of radiation dosimetry, an essential feature of each MPS investigation, was emphasized.

Results: The seven components of the MPS are DOE workers, nuclear weapons test participants, nuclear power plant workers, industrial radiographers, medical radiation workers, nuclear submariners, other U.S. Navy personnel, and radium dial painters. The MPS cohorts include tens of thousands of workers with elevated intakes of alpha particle emitters for which organ-specific doses are determined. Findings to date for chronic radiation exposure suggest that leukemia risk is lower than after acute exposure; lung cancer risk is much lower and there is little difference in risks between men and women; an increase in ischemic heart disease is yet to be seen; esophageal cancer is frequently elevated but not myelodysplastic syndrome; and Parkinson's disease may be associated with radiation exposure.

Conclusions: The MPS has provided provocative insights into the possible range of health effects following low-level chronic radiation exposure. When the 34 MPS cohorts are completed and combined, a powerful evaluation of radiation-effects will be possible. This final article in the MPS special issue summarizes the findings to date and the possibilities for the future. A National Center for Radiation Epidemiology and Biology is envisioned.

Introduction

The first international virtual symposium was held online on November 6, 2020, on the Study of One Million Radiation Workers and Veterans (MPS) on Low-Level Radiation Health Effects (Boice, Cohen, et al. 2019). The symposium was co-sponsored by Memorial Sloan Kettering Cancer Center (MSKCC), the National Council on Radiation Protection and Measurements (NCRP), and the Greater New York Chapter, the Baltimore-Washington Chapter, and the New Jersey Chapter of the Health Physics Society. The virtual workshop was attended by more than 300 individuals and included four presentations on the importance of radiation epidemiology to the missions of four U.S. government agencies by their representatives and 12 scientific presentations by collaborating members of the MPS research team, coupled with extensive question and answer sessions. A summary of the presentations, questions and answers, and conclusions, along with citations to other MPS publications, is reported here. Several presentations included preliminary results. Citations to published or submitted manuscripts referring to the formal results are provided to enable readers to seek additional details on specific cohorts, results, and conclusions.

Presentations

Welcome and introduction was presented by Dr. Kathryn D. Held (NCRP, and Massachusetts General Hospital/Harvard Medical School)

The critical evaluation of the health effects of low dose and low-dose rate radiation exposures in healthy U.S. populations being conducted through the MPS[1] is of great value for providing an enhanced understanding of the science needed for sound radiation protection policy and recommendations (Boice, Held, Shore et al. 2019). In partnership with numerous U.S. government agencies and other organizations, the NCRP has been involved in the MPS under the directorship of Dr. John Boice, for over a decade. NCRP is a Congressionally chartered, nongovernment, nonprofit organization that has the mission to support radiation protection by providing independent scientific analysis, information, and recommendations that represent the consensus of leading scientists. Hence, the NCRP has supported and fostered the conduct of the MPS epidemiology and dosimetry studies and helped disseminate the important findings of those scientific efforts through NCRP Reports (NCRP 2018a), Commentaries (NCRP 2020a), and other published works (Boice, Cohen et al. 2019, Boice et al. 2020; Boice, Cohen, Mumma, Howard et al. 2021), as well as at scientific meetings. This virtual symposium was a superb opportunity to further disseminate findings and information on the importance of the MPS.

A Million persons, a million dreams – overview of the MPS was presented by Dr. John D. Boice Jr. (NCRP, and Vanderbilt University School of Medicine) and Dr. Lawrence T. Dauer (MSKCC)

The MPS is designed to address the major unanswered question in radiation risk: What is the level of health effects when exposure is gradual over time and not delivered briefly (Boice, Cohen et al.et al. 2019; Boice and Dauer 2021)? Over a million healthy American workers and veterans are being studied currently to evaluate cancer and noncancer mortality following low-level low-linear energy transfer (LET) and high-LET exposure; rare cancers; intakes of radioactive elements; and differences in risks between women and men. As described more fully later in this report, the future research opportunities of the MPS will be substantially enriched in that for upwards of 800,000 workers and veterans, cancer incidence and clinical diagnoses of nonmalignant conditions, such as heart disease not leading to death, will become available within the year via linkages with files from the Centers for Medicare and Medicaid. Further, individual information will be available on important lifestyle factors such as smoking and on individual characteristics and conditions such as obesity, hypertension, diabetes, and on many hundreds of other health conditions (CMS 2018, 2021a, 2021b).

The MPS consists of seven categories of persons exposed to radiation from 1913 to the present (Table 1). Over 30 individual radiation cohorts are included in these broad groupings. The U.S. Department of Energy Health and Mortality study began over 40 years ago and is the source of upwards of 300,000 workers (Ellis et al. 2019). Over 25 years ago, the U.S. National Cancer Institute (NCI) collaborated with the U.S. Nuclear Regulatory Commission to effectively create a registry of radiation workers from which cohorts of nuclear power plant workers ($n = 135{,}193$) and industrial radiographers ($n = 123{,}556$) were drawn (Hagemeyer et al. 2018). For over 60 years, the U.S. Department of Defense collected data on aboveground nuclear weapons test participants ($n = 114{,}270$) (Boice et al. 2020). At the request of NCI in 1978, Landauer, Inc., a dosimetry service provider, began preserving their dosimetry databases which subsequently provided the exposure data used to identify a cohort of 109,019 medical workers as well as to identify supplementary workers at nuclear power plants and workers who were industrial radiographers (Yoder et al. 2019, 2021; NCRP 2020a). The U.S. Navy has recorded dosimetry information since the 1950s on nuclear submariners (~113,000) and nuclear shipyard workers (~96,000) (Mueller et al. 2020). The study of radium dial painters ($n = 3{,}276$) began in the 1920s and was recently reactivated (Martinez et al. 2021). The MPS is a U.S. national effort and relies on the active cooperation and support of federal agencies (US GAO 2017). The key to high-quality epidemiology is comprehensive dose reconstructions for individuals within each of the seven MPS components (NCRP 2018a; Dauer et al. 2018).

Table 1. Million Person Study cohorts and source of populations (NCRP 2018a; Boice, Cohen et al. 2019).

Cohort (source of cohort)	Reference	Number of workers [a]
Manhattan Project and nuclear facilities (DOE)	Ellis et al. (2019)	260,000
Atomic veterans (DOD)	Boice et al. (2020)	113,806
Nuclear power plant workers (NRC)	Boice, Cohen, Mumma, Hagemeyer et al. (2021)	135,193
Industrial radiographers (NRC, Landauer, Inc.)	NCRP (2018a), Mumma et al. (2019)	123,510
Medical radiation workers (Landauer, Inc.)	Boice, Cohen, Mumma, Howard et al. (2021)	>109,019
Nuclear submariners and other navy (US Navy)	Boice, Cohen et al. (2019)	>200,000
Radium dial workers (DOE)	Martinez et al. (2021)	3,276

[a]Numbers in the Tables may differ slightly as they relate to different analyses and follow-up.

The MPS vision is to provide broad scientific understanding of health effects following prolonged exposure to radiation. Such understanding will improve guidelines to protect workers and the public; improve input to compensation schemes for workers, veterans, and the public; provide guidance for policy and decision-makers; and provide evidence for or against the continued use of the linear non-threshold dose-response model in radiation protection to manage low doses.

Dosimetry is key to excellent epidemiology was presented by Dr. R. Craig Yoder (Landauer, Inc., retired)

Dosimetry enables the quantitative stratification of those workers receiving the most radiation exposure from those receiving the least. It permits epidemiologists to relate disease incidence with a radiological quantity and develop response functions. In the case of the MPS, dosimetry aims to estimate the mean or average absorbed dose to an organ or tissue (organ dose) usually abbreviated as D_T. This quantity cannot be measured and must be estimated using a combination of mathematical models and data from measurable quantities. Uncertainties arise from the differences between the real-world radiological exposure environments and the idealized conditions depicted in the models. Reconstructing workers' doses becomes an exercise of identifying the most appropriate models that predict organ dose and converting a large assortment of measured data into the inputs required by the models. Complexity arises from the large variety of exposure conditions that influence this process and changes in these conditions over the nearly 70 years of observation relevant to the MPS. Epidemiological analyses require organ doses to be expressed in both annual and lifetime accumulated values. Changes in the metrological methods and radiation safety quantities introduce issues affecting the comparability of lifetime data. Changes in regulatory requirements on radiation measurements and record-keeping, for example, monitoring to assess the maximally exposed part of the body, impart additional challenges to compiling lifetime organ absorbed dose values.

NCRP Report 178, Deriving Organ Doses and Their Uncertainty for Epidemiological Studies (NCRP 2018a; Dauer et al. 2018), provides a framework for the dose reconstruction process to be used for the various worker and atomic veteran cohorts forming the MPS. The key steps in this framework that influence the dosimetry are the construction of the pathways for internal and external exposure, and the selection of the irradiation geometries for external exposure and of the biokinetic models for internal exposure. Each cohort presents unique considerations regarding the types and energies of radiation leading to dose along with the approaches to radiological monitoring and availability of lifetime exposure data. Influences that create conditions leading to inhomogeneous irradiation of the body, such as leaded aprons and other shielding fixtures, present special challenges for tissues distributed widely in the body. Less scientific or technical difficulties arise from identifying the completeness of an individual's dose history and uncertainties affecting the ability to relate radiation safety monitoring results to a specific individual. These issues affect both internal and external dosimetry efforts and vary in importance depending on the specific worker cohort.

The recently published NCRP Commentary No. 30, Using Personal Monitoring Data to Derive Organ Doses for Medical Radiation Workers, with a Focus on Lung (Yoder et al. 2019, 2021; NCRP 2020a), details the external dosimetry process for medical workers. This cohort was primarily exposed to x and gamma rays and includes large numbers of women from which to examine possible sex-related differences in disease mortality or incidence, depending on the health effect of concern. NCRP Commentary No. 30 gives examples of both the technical and administrative challenges involved in providing annual doses that add up to a lifetime organ dose value.

Importance of radiation epidemiology: CDC perspective was presented by Dr. Armin Ansari (Centers for Disease Control and Prevention)

The practice of radiation protection is based on application of scientific data, together with consideration of ethical principles and societal factors, to provide for health and safety of workers and members of the public, including at-risk populations. During the last 100 years, we have learned a great deal about the biological effects of radiation on cellular and animal models. Radiation's mutagenic and carcinogenic effects are characterized in these experimental systems (NRC 2006; NCRP 2018b; UNSCEAR 2020). However, what informs our radiation protection and public heath practice concerning the long-term effects of human exposures to ionizing radiation comes primarily from epidemiologic data on cancer and noncancer diseases – most notably from the Life Span Study (LSS) of Japanese atomic bomb survivors. Except for occupational accidents, medical overexposures, or some acts of terrorism, most environmental exposures (including radon and technologically enhanced naturally occurring radioactive material [TENORM]), medical exposures, and occupational exposures (including nuclear workers and aircrews) are low doses accumulated from protracted low dose-rate exposures. Even in the aftermath of a major nuclear or radiological emergency, critical public health decisions in response and recovery (IAEA 2011; US EPA 2017; ICRP 2020) are informed by what we know about the risk of exposure at low doses and low dose-rates, and uncertainties of those estimates. These critical decisions include evacuation, relocation, the embargo of food and agricultural products, waste management, remediation, and long-term monitoring of exposed population. While direct observation of human health effects by epidemiologic means at low doses remains highly challenging, use of biologically based dose-response models can supplement epidemiological data and enhance the estimation of health effects (NCRP 2020b). Ultimately, while a multitude of factors can modulate the health effects of radiation exposure, we depend on reliable epidemiological studies to inform our radiation protection and public health practices (Ansari et al. 2019; Boice, Held, Shore et al. 2019).

Table 2. Absorbed dose to lung among the medical radiation workers by occupational category (Boice, Cohen, Mumma, Howard et al. (2021)).

		Absorbed dose to lung (mGy)			
Occupational category	Number of workers	Mean	Median	Min	Max
General radiology	48,837	11.6	9.1	4.5	313
Interventional radiology	44,498	3.0	2.1	0.3	65
Nuclear medicine and radiation oncology	9,597	50.8	39.0	7.5	1106
Other	6,087	38.6	20.9	7.5	1271
Total	109,019	13.0	6.8	0.3	1271

MPS Cohort: Medical worker study was presented by Dr. Lawrence T. Dauer (Memorial Sloan Kettering Cancer Center)

The MPS includes a cohort of U.S. workers exposed to radiation as a consequence of performing various medical procedures that involve the use of ionizing radiation (medical radiation workers). Among the different cohort groups studied as part of the MPS, the medical radiation worker group features the highest percentage of female subjects (49%) from which to examine adverse radiation effects that may vary according to sex, particularly the development of lung cancer (NCRP 2020a). NCRP Report No. 178 (NCRP 2018a) developed a framework for external radiation exposures along with tables and figures of conversion coefficients for relating measured approximations of the personal dose equivalent, $H_p(10)$ (the dose equivalent at a depth of 10 mm in the body expressed in units of mSv), to selected tissue and organ doses, D_T expressed in mGy (data essential for epidemiology studies). NCRP Commentary No. 30 (NCRP 2020a) has further and more specifically delineated considerations for deriving organ doses for the medical radiation worker cohort that depend on the development of radiation exposure scenarios to describe the radiological and physical conditions to which specific categories of medical workers may have been exposed. Dose-response analyses are based on organ-specific lung doses estimated for each worker based on his or her job-based exposure scenario and calendar years of monitoring. Four medical exposure scenarios were evaluated: general radiology characterized by low-energy x-ray exposure with no lead apron use; interventional radiologists or cardiologists (lead apron wearers); nuclear medicine personnel; and radiation oncologists (mainly nurses and technologists and some medical doctors) receiving high-energy gamma (photon) doses. Table 2 provides information on the lung doses for the four occupational categories. The lower doses among interventionalists are related to the use of lead aprons which appreciably diminished the absorbed dose to lung. The results of the study of U.S. medical radiation workers have recently become available (Boice, Cohen, Mumma, Howard et al. 2021). Overall, the MPS medical workers were at increased risk for lung cancer and risk was higher among men than women. There were no statistically significant radiation-associations with leukemia excluding chronic lymphocytic leukemia (CLL), ischemic heart disease (IHD) or other specific causes of death. A positive but not statistically significant dose response was found for Parkinson's disease.

Importance of radiation epidemiology: the U.S. Department of Energy perspective was presented by Dr. Joey Y. Zhou (Department of Energy)

The U.S. Department of Energy (US DOE) has supported all aspects of the MPS since the feasibility study in 2009–2010, funded by the Office of Science and now by the Office of Environment, Health, Safety & Security. Twenty-six percent (260,000) of the MPS study population comes from former U.S. DOE radiation workers (Boice, Cohen et al.et al. 2019). The U.S. DOE and its predecessors have a long history in conducting and supporting radiation epidemiological studies. Surveillance of radiation exposure and its health effects were implemented shortly after the Manhattan Project began in 1942. The Radiation Exposure Information and Reporting System (REIRS) was established in 1968 to serve as the central repository of occupational radiation exposure records (Hagemeyer et al. 2018; NRC 2020). The feasibility studies of using personnel, employment, medical, radiation exposure, and facility records to conduct epidemiologic mortality studies were completed in the 1960s while the health and mortality studies of radiation workers across U.S. DOE facilities were carried out from the early 1970s (Ellis et al. 2019). The Epidemiological Records Moratorium, "an agency-wide freeze" on the destruction of all epidemiological records, was issued in 1990. The Comprehensive Epidemiologic Data Resource (CEDR) also was created in 1990 to allow researchers to access data from the U.S. DOE epidemiological studies program (US DOE 2020). The CEDR became a major source of data used to extend the follow up U.S. DOE worker cohorts for the MPS. The U.S. DOE has data for over 650,000 former radiation workers and about 75,000 current radiation workers. The MPS can provide valuable research findings to improve U.S. DOE former worker medical screening and compensation programs, and to enhance the protection of current radiation workers.

MPS Cohort: Mortality among workers at the Los Alamos National Laboratory, 1943-2017 was presented by Dr. Sarah S. Cohen (EpidStrategies)

Los Alamos National Laboratory (LANL), established in 1942 during the Manhattan Project, continues operations today to solve national security challenges. This MPS study (Boice, Cohen, Mumma, Golden et al. 2021) includes 26,328 male and female workers who worked at LANL between 1943–1980 and were employed by LANL or Zia, the LANL maintenance contractor for the site. Vital status was determined through December 31, 2017. External radiation monitoring data were available from 1944 through 1990. The greatest potential for elevated doses from internal emitters arose from ^{238}Pu and ^{239}Pu. Doses to the 6,499 workers monitored for plutonium were based almost entirely on plutonium urinalyses for which there were over 158,000 (or 24.3 samples per worker on average). External doses also were received at facilities other than at LANL and were available from the U.S. DOE Radiation Exposure Monitoring System (REMS); U.S. Nuclear Regulatory Commission (NRC) REIRS; Landauer, Inc.; U.S. Navy Dosimetry System and other military databases; and the Nuclear Test Personnel Review Program. All dose estimates from photons, neutrons, tritium, ^{238}Pu, and ^{239}Pu were combined to obtain organ-specific doses received by each worker for each calendar year. Standardized Mortality Ratio (SMR) analyses were conducted as well as internal analyses using Cox proportional hazards models with adjustment for sex, educational attainment, and year of birth. Excess relative risks (ERR) were also estimated. The LANL population was 75% male and 81% White. At the end of vital status tracing, 60% had died, and only 32 (0.1%) were lost to follow-up. The presentation included SMRs for lung cancer and leukemia other than (CLL), as well as the associated hazard ratios (HRs) from the Cox model and ERRs. Full study results for the LANL population are available (Boice, Cohen, Mumma, Golden et al. 2021). Radiation was not found to increase the risk of leukemia or lung cancer. Esophageal cancer, however, was associated with radiation; bone cancer was linked to plutonium intakes, and there was a suggested radiation association for Parkinson's disease. Ischemic heart disease and cerebrovascular disease were not associated with radiation dose. More precise evaluations are planned with pooled analysis of workers with similar exposures such as at Rocky Flats (Boice, Cohen et al. 2019).

MPS Cohort: nuclear power plant workers and industrial radiographers was presented by Dr. Lawrence T. Dauer (Memorial Sloan Kettering Cancer Center)

The two largest cohorts within the MPS are the nuclear power plant workers and the industrial radiography workers. More than 500,000 workers have been employed in U.S. nuclear power plants since the first commercial production of electricity in 1957 (NCRP 2018a). The consistent reporting of annual worker doses required by the U.S. NRC for their licensees provides a high-quality dosimetry database that was redesigned in 1994 to facilitate epidemiologic study (Hagemeyer et al. 2018). Because workers' annual recorded dose in the nuclear industry has decreased over the years down to on the order of 2 mSv or less on average, only the workers at nuclear power plants first employed from 1957 through 1985 are included in the MPS. Cohort members were selected from databases available from the REIRS, which is maintained by the U.S. NRC, and supplemented with data available from Landauer, Inc. The number of nuclear power plant workers under study ($n = 135,193$) includes nearly three times the number of adult males and females over age 20 years at exposure than the study of Japanese atomic bomb survivors, and over seven times the number of adult males. Most radiation exposures were due to penetrating external gamma rays, although low-level neutron exposures were possible during some reactor work circumstances. There were no or negligible internal exposures. The results have been recently published (Boice, Cohen, Mumma, Hagemeyer et al. 2021). Occupational exposure to radiation over many years increased the risk of leukemia other than CLL as seen in the nuclear power plant workers and industrial radiographers (Tables 3, 4). Radiation associations were not found for either lung cancer or ischemic heart disease. An association with Parkinson's disease was suggested but was not statistically significant. There was no difference in the calculated radiation risk for lung cancer between men or women who worked at a nuclear power plant, although the estimates were imprecise. Mesothelioma and asbestosis were significantly elevated (Mumma et al. 2019).

Industrial radiography is an inspection method to detect fractures and other deficiencies in metallic and other dense materials by exploiting the penetrating ability of higher energy photons to create radiographic images, typically using film, of the defects. The period of greatest external exposure for radiographers is during the time the source (e.g., X-ray, ^{192}Ir and ^{60}Co) is outside its shielded container while being transported into and from the material being radiographed but also depends on the time required to achieve a radiographic image with the appropriate contrast to reveal any defects (NCRP 2018a). For the MPS, a cohort of 123,556 workers employed as industrial radiographers in the United States as early as 1940 was selected from records within the REIRS database and Landauer dosimetry database. To date, 26,000 of the industrial radiographers are known to have worked in naval shipyards. Preliminary epidemiology results (Table 3) show a significant excess of leukemia other than CLL (NCRP 2018b). A significant dose-response for lung cancer among males but not among females was seen (Boice, Ellis et al. 2019). Mesothelioma and

Table 3. Leukemia (excluding CLL) risk among MPS cohorts.

Cohort	Reference	Number of workers	ERR at 100 mGy (95% CI)
Nuclear power plant	Boice, Cohen, Mumma, Hagemeyer et al. (2021)	135,193	0.15 (0.00, 0.31)
Industrial radiographers	NCRP (2018b)	123,556	0.17 (−0.02, 0.35)
Medical radiation workers	Boice, Cohen, Mumma, Howard et al. (2021)	109,019	0.10 (−0.34, 0.54)
Atomic veterans	Boice et al. (2020)	114,270	−0.37 (−1.08, 0.33)
Mound [a]	Boice et al. (2014); NCRP (2018b)	4,954	0.04 (−0.37, 0.71)
Mallinckrodt	Golden et al. (2019)	2,514	−0.14 (−0. 60, 0.33)
Rocketdyne[a]	Boice et al. (2011); NCRP (2018b)	5,801	0.06 (−0.50, 1.13)
Los Alamos National Lab	Boice et al. (2021b)	26,328	−0.43 (−1.11, 0.24)

[a]Taken as the Hazards Ratio minus 1 at 100 mSv
ERR denotes Excess Relative Rate

Table 4. Absorbed dose to red bone marrow among MPS cohorts [a].

MPS cohort	Reference	Number of workers	Dose to red marrow (mGy)		
			Mean	Median	Max
Nuclear power plant	Boice, Cohen, Mumma, Hagemeyer et al. (2021)	135,193	37.9	19.7	953
Industrial radiographers	Boice, Ellis et al. (2019)	123,510	12.7	0.8	1267
Medical radiation workers	Boice, Cohen, Mumma, Howard et al. (2021)	109,019	12.9	6.9	1187
Atomic veterans	Boice et al. (2020)	113,806	6.2	2.3	953
Mound [b]	Boice et al. (2014); NCRP (2018b)	4,954	26.1 [b]	2.9	841
Mallinckrodt [b]	Golden et al. (2019)	2,514	35.9 [b]	13.0	1219
Rocketdyne [c]	Boice, Leggett et al. (2006); Boice, Leggett et al. (2011)	5,801	13.5 [c]	na	1000
Los Alamos National Laboratory [b]	Boice et al. (2021a); NCRP (2018b)	26,328	12.4 [b]	0.8	835
Tennessee Eastman Corporation	Boice et al. (2021b)	26,650	na	na	na

[a]Red bone marrow doses represent the latest study compilations and may differ slightly from published values.
[b]Dose weighting factor of 1 applied for plutonium, polonium and uranium dose.
[c]Personal dose equivalent in mSv
na denotes not available

asbestosis were significantly increased and linked to asbestos exposure in shipyards (Mumma et al. 2019).

Importance of radiation epidemiology: the U.S. Department of Defense perspective was presented by Dr. Paul K. Blake (Defense Threat Reduction Agency)

The U.S. Department of Defense (U.S. DoD) currently employs approximately three million military (active & reserve) and civilian workers. Approximately 70,000 U.S. DoD workers (2%) are annually monitored for ionizing radiation exposure (Blake and Komp 2014). U.S. DoD workers and their dependents can potentially be exposed to ionizing radiation for medical purposes and during nuclear war scenarios or operations other than war. The U.S. military was an early adopter of ionizing radiation for diagnostic purposes. Less than three years after Roentgen's discovery of the X-ray, both the U.S. Army and U.S. Navy were employing X-ray machines in the war with Spain (1898). However, U.S. DoD's existing occupational radiation exposure records programs arose during the Manhattan Project in World War II. These programs expanded with the U.S. development of nuclear weapons and nuclear power applications. The U.S. DoD radiation monitoring infrastructure includes three nationally accredited external personal radiation dosimetry programs (Army, Navy, and Air Force), a variety of internal personal monitoring programs, various environmental and food radiological analysis laboratories, and five radiation dose repositories with records on over two million individuals [Atomic Veterans (1945–1992), Army/National Guard, Navy/Marines, Air Force, and Operation Tomodachi Registry (OTR)] (Blake and Komp 2014). The OTR is unique in that it includes dependents, in addition to military and civilian adults (Marro 2014). U.S. DoD dose repositories also include Coast Guard and Merchant Marine exposures and non-US DoD visitor exposures. There have been numerous radiation epidemiology studies of these workers (Mueller et al. 2020). Most of these studies have been performed by external entities, including the ongoing MPS, which includes 114,270 male veterans who participated in seven U.S. atmospheric nuclear test series, and the first test at TRINITY (Boice et al. 2020). In summary, radiation epidemiology studies are important to U.S. DoD in (1) understanding the impact of ionizing radiation exposures on the health and safety of the U.S. DoD-affiliated population, (2) addressing the credibility of U.S. DoD's radiation safety programs, and (3) providing a technical basis for associated radiogenic disease compensation programs.

MPS Cohort: Nuclear weapons test participants was presented by Dr. Emily A. Caffrey (Radian Scientific, LLC) and Dr. John E. Till (Risk Assessment Corporation)

The nuclear weapons test participants form an important component of the MPS. These 114,270 military veterans are part of a larger group of approximately 235,000 individuals who took part in one or more atmospheric nuclear weapons tests at the Nevada Test Site (NTS) or the Pacific Proving Grounds (PPG) between 1945 and 1962 (Caldwell et al. 2016; Boice et al. 2020). The dosimetry for this cohort is of high quality (Coefficient of Variation (CV) of $\sim$ 0.5) due to the detailed historical records available to researchers, information provided about exposure rate fields, location of ships and units, and the availability of film badge dosimeter readings for about 20% of the participants (Till et al. 2014; Till, Beck, Aanenson et al. 2018). The mean estimated external dose for red bone marrow was 6 mGy (maximum of 108 mGy) based on a case-cohort study. Two-thirds of the sampled cohort received doses less than 5 mGy (Boice et al. 2020). The 65-year follow-up case-cohort epidemiological study concluded there was no evidence for increasing trends with radiation dose for leukemia (excluding CLL), myelodysplastic syndrome, multiple myeloma, ischemic heart disease, or cancers of the lung, prostate, breast, and brain (Boice et al. 2020). This presentation briefly reviewed the dosimetry and epidemiological results, with a focus on the radiological and non-radiological contributions of this component of the MPS. Insights into future low dose research gleaned from the nuclear weapons test participants cohort were discussed, along with ideas on how to combine radioepidemiological studies to expand our practical knowledge of low dose radiation effects for the purposes of radiation protection.

A collaborative study comparing biological dosimetry, dose reconstruction, and film badge readings was conducted among 12 veterans who had received doses in excess of 200 mGy and 12 matched control participants with no evidence of exposure (Simon et al. 2019). The range of doses for the full cohort used for this biodosimetry study are found in

Table 5. Lung cancer risk among MPS cohorts (Boice, Ellis et al. 2019).

Cohort	Reference	Number of workers	ERR at 100 mGy [a] (95% CI)
Nuclear power plant (NPP)	Boice, Cohen, Mumma, Hagemeyer et al. (2021)	135,193	−0.04 (−0.11, 0.02)
Industrial radiographers (IR)	Boice, Ellis et al. (2019)	123,556	0.13 (0.02, 0.23)
Medical radiation workers (MW)	Boice, Cohen, Mumma, Howard et al. (2021)	109,019	0.15 (0.02, 0.27)
NPP + IR + MW		367,722[b]	0.02 (−0.03, 0.07)
Atomic veterans	Boice et al. (2020)	114,270	0.04 (−0.11, 0.19)
Mound	Boice et al. (2014)	4,954	0.00 (−0.03, 0.04)[c]
Mallinckrodt	Golden et al. (2019)	2,514	−0.06 (−0.18, 0.06)
Rocketdyne	Boice et al. (2011)	5,801	−0.02 (−0.18, 0.17)[c]
Los Alamos National Laboratory	Boice et al. (2021a)	26,328	0.01 (−015, 0.17)
Tennessee Eastman Corporation	Boice et al. (2021b)	26,650	−0.08 (−0.17, 0.02)[c]
Rocky Flats	Preliminary	9,535	0.01 (−0.10, 0.11)

[a]Doses are at 100 mGy for Mallinckrodt, Los Alamos National Laboratory, Tennessee Eastman Corporation and Rocky Flats where a dose weighting factor of 1 was used for uranium and plutonium dose
[b]46 workers were in two cohorts and are counted only once in this pooled analysis.
[c]Taken as the Hazards Ratio minus 1 at 100 mSv, a dose weighting factor was not applied
ERR denotes Excess Relative Rate

Beck et al. (2017). The correlation between the three dosimetry methods was remarkable, especially considering it had been more than 60 years since the exposures to fallout from nuclear weapons tests had occurred. This radiobiology evaluation lends credibility to the estimated doses for the entire cohort. Further, reliable dose estimates were derived from the chromosome aberration-based retrospective biodosimetry technique used that might be applicable in other long-term dose reconstruction circumstances (McKenna et al. 2019). An inverse relationship between telomere length and dose estimates was unexpected but also might be pursued in other exposed populations.

The atomic veterans are a key component of the Million Person Study because they comprise a large, low dose cohort with high-quality dosimetry. There was no evidence that the low doses of radiation received by the veterans increased their risk of cancer or ischemic heart disease, although statistical variation was such that slight excesses could not be excluded with 95% confidence. Incidence rates of mesothelioma and prostate cancer were elevated, but these were not related to radiation exposure (Till, Beck, Boice et al. 2018; Boice et al. 2020). For the first time, many veterans are getting answers about what their radiation dose level means and its implications for their health. The MPS will build on this work, focusing on low doses in the range of 0–100 mGy above background, to continue to work toward shining a light on the effects of low doses of radiation.

Importance of radiation epidemiology: a NASA perspective was presented by Dr. Steve R. Blattnig (NASA Langley Research Center)

Radiation is one of the primary risk factors to human health that provides an obstacle to the safe exploration of space (NASA 2021). Radiation in space is of a different composition than most exposures on earth, consisting of high energy charged particles and resulting particle fragments from collisions with materials in spacecraft and the human body. Exposures can also be relatively high, with a mission to Mars estimated to be of the order of 300–450 mGy (870–1200 mSv) (Simonsen et al. 2020). Beyond low earth orbit, space radiation consists of continuous exposure to galactic cosmic rays (GCR) and sporadic eruptions from the sun that produce solar particle events. The health risks of primary concern are cancer, cardiovascular disease, and central nervous system diseases (NCRP 2014, 2020b). NASA's strategy to mitigate these effects currently consists of the development of permissible exposure limits that effectively limit the time individuals spend in space, as well as shielding from solar particle events (NCRP 2014). Because shielding from GCR is of limited effectiveness (Slaba et al. 2016, 2017), the Human Research Program is currently investigating the potential for biomedical countermeasures (NASA 2021; Werneth et al. 2020, 2021). A common theme of this risk mitigation strategy is a need to understand and quantify these health risks to counter them cost-effectively. Radiation epidemiology provides a primary basis to understand and quantify the response in humans and is integrated with animal and other experimental data to develop risk models for specific disease endpoints. Currently, the MPS is investigating lung cancer risks (Boice, Ellis et al. 2019) among women and men, and the risks of cognition dysfunction and dementia following intakes of radionuclides specifically for NASA (Boice 2017, 2019; NCRP 2019). However, all results that improve the understanding of radiation risks in humans can provide benefit to NASA's risk mitigation approach.

Lung cancer risk among men and women was presented by Dr. Ashley P. Golden (Oak Ridge Associated Universities)

There is conflicting evidence on the differences in radiation-related lung cancer risk between men and women. The lung cancer ERR estimated from Japanese atomic bomb survivors indicates that women are at nearly three times greater radiation risk than men on a relative scale (Ozasa et al. 2012; Cahoon et al. 2017). Recent results (Boice, Ellis et al. 2019) from five occupational cohorts within the MPS indicated little evidence that chronic or fractionated exposures increased the risk of lung cancer ($n = 403{,}067$ men and 50,679

Table 6. Absorbed dose to lung among MPS cohorts [a].

MPS cohort	Reference	Number of workers	Dose to lung (mGy)		
			Mean	Median	Max
Nuclear power plant	Boice, Cohen, Mumma, Hagemeyer et al. (2021)	135,193	43.2	22.4	1,085
Industrial radiographers	Boice, Ellis et al. (2019)	123,556	10.9	0.3	1,435
Medical radiation workers	Boice, Cohen, Mumma, Howard et al. (2021)	109,019	13.0	6.8	1,272
Atomic veterans	Boice et al. (2020)	113,806	6.2	2.3	972
Mound [b]	Boice et al. (2014)	4,954	98.7	10.2	17,478
Mallinckrodt [c]	Golden et al. (2019)	2,514	69.9	33.1	885
Rocketdyne [d]	Boice, Leggett et al. (2006); Boice et al. (2011)	5,801	19.0[d]	na	3,560
Los Alamos National Laboratory [c]	Boice et al. (2021a)	26,328	28.6[c]	0.9	16,811
Tennessee Eastman Corporation [c]	Boice et al. (2021b)	26,650	477 [c]	125.6	18,538

[a]Lung doses represent the latest study compilations and may differ slightly from published values. The atomic veteran doses, for example, are the ones used in the lagged analysis.
[b]Dose weighting factor of 1 applied for plutonium and polonium dose. Using a DWF of 20 results in a mean lung dose of 1.54 weighted-Gy.
[c]Dose weighting factor of 1 applied for plutonium and uranium.
[d]mSv presented, not mGy and dose weighting factor for uranium was 1
na denotes not available

women), and, not surprisingly, there was no evidence that radiation risk estimates were higher among women than men. Results for an additional four MPS cohorts (Tables 5, 6) also failed to uncover any significant differences in the lung cancer risks between men and women (Boice, Cohen, Mumma, Howard et al. 2021; Boice, Cohen, Mumma, Golden et al. 2021; Boice, Cohen, Mumma, Hagemeyer et al. 2021; Boice et al. 2021b). These studies are consistent with the absence of an association between fluoroscopic chest x-rays and lung cancer reported in the Canadian TB-Fluoroscopy Cohort Study and the Massachusetts TB-Fluoroscopy Study (Howe 1995; Brenner 2010). The reasons behind the conflicting evidence for lung cancer risks are not clear but might be related to different patterns of cigarette smoking among the Japanese bomb survivors, e.g., most male survivors smoked whereas most female survivors did not, and the ERR/Gy varied by number of cigarettes smoked and approached 0.00 for the heaviest smokers (Furukawa et al. 2010). Further, different population characteristics and circumstances may have influenced the radiation-related lung cancer risks: the Japanese survivors were exposed briefly in 1945 to the Hiroshima or Nagasaki atomic bombs and then lived in a war-torn country. The MPS cohorts were healthy U.S. radiation workers who were exposed chronically over many years during the course of their employment. Further, the 1945 Asian population has different characteristics than U.S. populations, and it is challenging to make comparisons, for example, to 'transfer risks' from a 1945 Asian population with different background rates for lung cancer to other populations.

NCRP Scientific Committee SC 1-27 is preparing a Commentary on the "Evaluation of Sex-Specific Differences in Lung Cancer Radiation Risks and Recommendations for Use in Transfer and Projection Models" which incorporates detailed evaluations of MPS cohort study results (NCRP 2021a). Since protection standards for astronauts currently are based on individual lifetime risk projections, any sex-specific difference in lung cancer risks limits the time women can spend in space (NASA 2021; NASEM 2021b).

Cognition and dementia following intakes of radionuclides was presented by Dr. John D. Boice Jr. (NCRP and Vanderbilt University School of Medicine) and Michael T. Mumma (International Epidemiology Institute and Vanderbilt University School of Medicine)

The American author and screenwriter, Ray Bradbury, once noted, 'It's not going to do any good to land on Mars if we're stupid' (NPR 2012). Relatively high cumulative doses to brain tissue from Galactic Cosmic Rays (GCR) are possible during a mission to Mars (Boice 2019; NCRP 2019; Simonsen et al. 2020). Galactic cosmic rays are a whole range of ions, including high-velocity ions (HZE particles) traveling near the speed of light through space. Animal studies have revealed early and late neurological disorders from relatively brief exposures to these high-velocity heavy ions (NCRP 2014, 2020b). These studies have raised concern about possible effects on astronauts that might impair performance so that the mission would not be completed or, conceivably, there might be a risk of Alzheimer's or dementia years after the voyage is over.

There are no human circumstances on earth that can approximate GCR exposures to brain tissue. In fact, there is little to no evidence in human studies that low-LET radiation (at doses below those used in radiotherapy) is associated with dementia or Alzheimer's disease. However, a recent study of Russian Mayak workers has suggested a link with Parkinson's disease following low-LET radiation (Azizova et al. 2020) which has increased interest in this area (Pasqual et al. 2021). Further, analyses within the MPS cohorts of populations exposed to low-LET radiation are consistent with the Mayak study (Tables 7, 8). As a possible, though imperfect, analogue to the high-LET portion of the GCR exposure, radium dial painters (Martinez et al. 2021; NCRP 2021c) and U.S. DOE workers with intakes of alpha-particle emitting radionuclides are being evaluated for dementia, Alzheimer's disease, Parkinson's disease, motor neuron disease and for cognitive impairment (Boice 2017, 2019). The internally deposited alpha emitters in radiation workers (Leggett et al. 2019) comprise a possible human analogue for high-LET GCR exposure to brain tissue in

Table 7. Parkinson's disease risk among MPS cohorts.

Cohort	Reference	Number of workers	ERR at 100 mGy (95% CI)
Nuclear power plant	Boice, Cohen, Mumma, Hagemeyer et al. (2021)	135,193	0.24 (−0.02, 0.50)
Industrial radiographers		123,556	0.24 (−0.02, 0.50)
Medical radiation workers	Boice, Cohen, Mumma, Howard et al. (2021)	109,019	0.17 (−0.20, 0.54)
Mound	Boice et al. (2014)	4,954	0.23 (−0.01,0.54)[a]
Mallinckrodt	Golden et al. (2019)	2,514	−0.06 (−0.18, 0.06)
Los Alamos National Lab	Boice et al. (2021b)	26,328	0.16 (−0.07, 0.40)

[a] Combined dementia, Alzheimer's, Parkinson's and motor neuron disease
ERR denotes Excess Relative Rate
A pooled analysis of the nuclear power plant workers, industrial radiographers, and medical radiation workers consisted of 367,722 after removing overlapping workers. The ERR at 100 mGy (95% CI) was 0.30 (0.08, 0.56).

space. All MPS cohorts now include Parkinson's disease as an outcome for dose-response evaluation and somewhat surprising the majority have been positive to date (Table 7). These epidemiologic studies are intended to provide another line of evidence to consider when making judgments for radiation protection guidance for flight crews on missions in space.

Interestingly, cigarette smoking is consistently seen to decrease the risk of Parkinson's disease (Mappin-Kasirer et al. 2020; Bloem et al. 2021). Preliminary results in MPS studies of atomic veterans, industrial radiographers, and nuclear power plant workers confirmed indirectly such negative associations. Surrogates of high smoking prevalence were taken as being an enlisted man for veterans and having only a high school education or less for workers. These persons had significantly low SMRs of dying from Parkinson's disease. In contrast, they had significantly high SMRs of dying from lung cancer. After linkage with the Medicare and Medicaid Services (CMS) files described below (CMS 2021b), we will have more detailed information on individual tobacco use to confirm these provocative associations.

Studies have begun of alpha particle (He nuclei) exposure to brain tissue and mortality from dementia, Alzheimer's, Parkinson's and motor neuron disease, as well as cognitive impairment that will be incorporating quantitative scores from neuropsychological testing available from Medicare and Medicaid Services (CMS) files and nursing home files (CMS 2018, 2021a). The cohorts of U.S. DOE workers being evaluated (Phase 1) include workers at Los Alamos National Laboratory, Mallinckrodt Chemical Works, Mound, Rocketdyne, Rocky Flats, Tennessee Eastman Corporation (TEC), Fernald, Middlesex, Portsmouth Gaseous Diffusion Plant, Paducah Gaseous Diffusion Plant, Oak Ridge National Laboratories (X-10, K-25, Y-12), Savannah River Site and Hanford. Of special note is the possibility that several thousand workers with plutonium burdens could be interviewed and administered the same or similar battery of cognition tests (COGNITION) taken by astronauts on the International Space Station today (Phase 2) (NCRP 2021b).

The strengths of this investigation are that the exposure is to humans, and not rodents; the exposure is from high-LET radiation at a low dose rate (over years and not minutes or weeks as in most of the animal experiments); the human exposure is to a mixed field of high-LET radiation and low-LET radiation (similar to exposures in space); the energy deposition is similar for a wide range of particle types and energies (cf, Brenner 1990; Zaider 1996; Hofmann et al. 2020); and human outcomes of interest can be directly evaluated, i.e., the occurrence of dementia and Alzheimer's disease as well as quantitative measures of cognitive impairment.

Low-energy alpha particles are a significant portion of the GCR exposure experienced inside a space vehicle on a long voyage. Much of the high-LET portion of radiation space exposure is actually from low-energy alpha particles and protons (NCRP 2014). While there are similarities between high-LET alpha-particle exposure to brain tissue and high-LET GCR exposure, there are also dissimilarities, e.g., GCR and alpha particles emitted from radionuclides may share the same LET values, but their track structures and energies are distinct. A critical component of this array of studies of high-LET exposure to brain tissue is the accurate assessment of individual worker doses. The next presentations by Caleigh Samuels (ORNL) and Sergei Tolmachev (USTUR) cover the comprehensive approaches being employed for dose reconstruction.

Brain dose estimates for alpha emitters at MPS sites was presented by Dr. Caleigh Samuels (Oak Ridge National Laboratory)

The MPS cohorts studied so far include tens of thousands of workers with elevated intakes of alpha particle emitters, primarily ^{238}Pu, ^{239}Pu, ^{241}Am, ^{226}Ra, ^{210}Po, and U isotopes. Some subsets of these alpha emitters generally dominate reconstructed doses to MPS cohorts from internal emitters and, for some sites, from all internal and external sources. In the course of the MPS, the brain has emerged as a tissue of concern due to potential cognitive effects from internal emitters, particularly alpha emitters (see the Presentation on "Cognition and dementia following intakes of radionuclides"). To this point, the biokinetic models applied in MPS dose reconstructions, which are generally the latest models of the International Commission on Radiological Protection (ICRP), do not explicitly address the brain. Instead, the brain is treated as a mass fraction of a pool of tissues called *Other* that represent the remainder of the body after the removal of tissues explicitly identified in the biokinetic model. We investigated the feasibility of revising the biokinetic models for radionuclides of concern to include the brain as an explicitly identified pool with parameter values

Table 8. Absorbed dose to brain among MPS cohorts[a].

MPS cohort	Reference	Number of workers	Dose to brain (mGy)		
			Mean	Median	Max
Nuclear power plant	Boice, Cohen, Mumma, Hagemeyer et al. (2021)	135,193	33.2	17.2	833.6
Industrial radiographers	Boice et al. (2019c)	123,510	10.3	0.7	1025
Medical radiation workers	Boice, Cohen, Mumma, Howard et al. (2021)	109,019	18.9	9.8	1077
Atomic veterans	Boice et al. (2020)	113,806	6.8	2.5	1058
Mound [b]	Boice et al. (2014)	4,954	26.1[b]	na	937
Mallinckrodt [c]	Golden et al. (2019)	2,514	38.8[c]	16.2	1151
Rocketdyne [b]	Boice, Leggett et al. (2006); Boice et al. (2011)	5,801	13.5 [b]	na	1000
Los Alamos National Laboratory [c]	Boice et al. (2021a)	26,328	11.7[c]	0.8	764.3
Tennessee Eastman Corporation	Boice et al. (2021b)	26,650	na	na	na

[a]Brain doses represent the latest study compilations and may differ slightly from published values.
[b]Personal dose equivalent, mSv
[c]Dose weighting factor of 1 applied for plutonium and uranium.
na denotes not available

derived from element-specific biokinetic data (Leggett et al. 2019; NCRP 2021b).

Case studies were performed for 17 radionuclides [^{52}Mn, ^{53}Mn, ^{54}Mn, ^{134}Cs, ^{194}Hg (vapor), ^{203}Hg (vapor), ^{207}Bi, ^{209}Pb, ^{210}Pb, ^{210}Po, ^{224}Ra, ^{226}Ra, ^{230}U, ^{234}U, ^{237}Pu, ^{239}Pu, ^{241}Am]. In addition to radionuclides frequently encountered in the MPS, the case studies include radioisotopes of elements for which brain kinetics have been examined in some detail (e.g., Mn, Hg, Pb). Injection dose coefficients (50-y equivalent doses to the brain following injection into blood of an adult) were calculated for each radionuclide using each of two versions of the ICRP's current systemic biokinetic model for that element for workers (ICRP 2016, 2017, 2019, 2021): the original version and a modified version differing only in the treatment of the brain. If the ICRP model contained an explicit brain region, the modified version of that model removed the compartment(s) representing the brain and included the brain as part of *Other*. If the ICRP model did not have an explicit brain pool but instead included the brain in *Other*, the modified version included an explicit brain region with kinetics based on best available brain-specific data (Leggett et al. 2019; NCRP 2021b). The comparison of dose coefficients for a given radionuclide was expressed as a ratio A:B, where A and B are the injection dose coefficients based on the versions of the model with and without an explicit brain region, respectively. The derived ratio A:B was <0.2 in two cases (^{224}Ra, ^{241}Am), in the range 0.5–2.0 in 12 cases (^{52}Mn, ^{54}Mn, ^{134}Cs, ^{203}Hg-vapor, ^{207}Bi, ^{209}Pb, ^{210}Po, ^{226}Ra, ^{230}U, ^{234}U, ^{237}Pu, ^{239}Pu), and in the range 3–5 in three cases (^{53}Mn, ^{194}Hg-vapor, ^{210}Pb). Thus, cases were found in the use of an explicit brain model that: (1) had little effect on the estimated dose to the brain, (2) resulted in a much lower dose estimate for brain or (3) resulted in a much higher dose to the brain (Leggett et al. 2019; NCRP 2021b). An interesting finding from these case studies and a broader review of the literature was that the brain usually has a much lower rate of uptake of elements per gram of tissue but a longer residence time than do most other studied soft tissues. Thus, an initially low uptake of a radionuclide by the brain should not be interpreted as indicating that the dose to the brain is substantially lower than that to most other tissues.

Radionuclide concentrations in brain segments: autopsy series was presented by Dr. Sergei Y. Tolmachev (Washington State University), Dana Wegge (University of Missouri), Dr. John Brockman (University of Missouri)

The United States Transuranium and Uranium Registries (USTUR) is a federal-grant-funded human tissue research program providing long-term study of actinide biokinetics in former nuclear workers with internal depositions of these elements. The USTUR conducts autopsies and performs radiochemical analyses of voluntarily donated tissue samples (Kathren and Tolmachev 2019; Tolmachev et al. 2019). The National Human Radiobiology Tissue Repository (NHRTR) holds all tissues donated to the USTUR, along with specimens acquired from the U.S. Radium Worker Studies (Rowland 1994; Martinez et al. 2021). The USTUR/NHRTR is a unique resource for retrospective analyses and distribution studies of plutonium, uranium, and americium, as well as radium and beryllium in the human body and in specific tissues and organs. This presentation provided preliminary results of plutonium (^{239}Pu), uranium (^{238}U), beryllium (^{9}Be), and radium (^{226}Ra) analyses in brain tissues from occupationally exposed individuals. Distributions of these elements among different segments of the brain and the impact of this distribution on brain dosimetry were discussed. This study was conducted in close collaboration with the MPS (Leggett et al. 2019; NCRP 2021b) and the University of Missouri Research Reactor Center.

Using alpha-spectrometry, ^{239}Pu activity concentrations were measured in the cerebellum and the right cerebrum lobe of six occupationally exposed male individuals. In cerebrum lobe samples, ^{239}Pu concentration was about two times higher than the average for other brain segments, suggesting non-uniform plutonium distribution between brain segments. The distribution of uranium, beryllium, and radium was studied in the brain of a female occupationally-exposed to ^{226}Ra. The concentrations of ^{238}U, ^{9}Be, and ^{226}Ra were measured with inductively coupled plasma mass-spectrometry (Thomas 2013) in the corpus callosum, the white and gray matter of the cerebrum lobes, the cerebellum, and brainstem segments of the brain. Preliminary results

Table 9. Ischemic heart disease risk among MPS cohorts.

Cohort	Reference	Number of deaths	ERR at 100 mGy (95% CI)
Nuclear power plant	Boice, Cohen, Mumma, Hagemeyer et al. (2021)	5,410	−0.01 (−0.06, 0.04)
Industrial radiographers	NCRP (2018b)	4,458	−0.03 (−0.08, 0.03)
Medical workers	Boice, Cohen, Mumma, Howard et al. (2021)	1,655	−0.10 (−0.27, 0.06)
Atomic veterans	Boice et al. (2020)	13,806	−0.01 (−0.12, 0.11)
Mound	NCRP (2018b)	221	−0.14 (−0.43, 0.14)
Mallinckrodt	Golden et al. (2019)	521	0.13 (−0.01, 0.28)
Los Alamos National Lab	Boice et al. (2021b)	3,043	−0.06 (−0.16, 0.04)

ERR denotes Excess Relative Rate

indicated that the highest ^{238}U concentration was measured in the cerebrum gray matter, while in other brain segments it was about 25% lower. High ^{9}Be concentrations were measured in the corpus callosum and, analogously with uranium, in the gray matter of the cerebrum lobe. In three other segments, beryllium concentrations were approximately 50% lower. In contrast to uranium and beryllium, radium accumulated preferentially in the white matter of the cerebral lobes. The ^{226}Ra concentration in the white matter was about four times higher than the average of all other brain segments. Acknowledging the limited number of cases studied, preliminary results suggest that plutonium, radium, beryllium, and, to a lesser extent, uranium are non-uniformly distributed in the human brain. This finding might have a direct impact on biokinetic modeling of internally-deposited radionuclides in the brain as well as on the assessment of radiation doses to the brain. Current systemic biokinetic models, recommended by the ICRP, assume a uniform distribution in the brain for any specific element.

A Million more dreams was presented by Dr. John D. Boice Jr. (NCRP and Vanderbilt University School of Medicine), Dr. Lawrence T. Dauer (Memorial Sloan Kettering Cancer Center), and Dr. Derek W. Jokisch (Francis Marion University and Oak Ridge National Laboratory)

The vision for the MPS includes a long-term follow-up of all exposure cohorts and an expansion of efforts in radiation biology. Currently, over 800,000 workers and veterans within 25 of the over 30 individual MPS cohorts have been followed for mortality, and in the next 2-3 years all will be. At regular 5 to 10 year intervals, there will be a new follow-up to update the mortality experience of the MPS and to conduct new dose-response evaluations.

The power of the MPS is in combining the similar datasets to make strong inferences about health effects from chronic, low-dose exposures (Zhang et al. 2014; Boice, Cohen et al. 2019; Boice, Ellis et al. 2019; Boice and Dauer 2021). NCRP SC1-27 has developed a methodology used to combine four available MPS cohorts for lifetime radiation risk projections for lung cancer (NCRP 2021a). The use of ensemble models to combine various diverse data sets within the MPS is being discussed (Simonsen and Slaba 2020). The MPS data will be harmonized as far as feasible while still taking account of differences in important variables in different cohorts, such as socioeconomic status. Combined studies will include workers with an intake of radionuclides (e.g., plutonium and uranium), and workers exposed primarily to gamma- and X-rays (nuclear power plant workers, industrial radiographers, medical radiation workers, atomic veterans). The evaluation of radiation-related ischemic heart disease is an important ongoing activity within the MPS (Tables 9, 10). The dose–response analyses will be organ-specific as the combining of all tumors together has limited biological meaning although it is important for radiation protection. Ways to combine organ-specific dose response relationships will be developed and enhanced. The MPS is a dynamic and evolving program of radiation studies. New inclusions this year are the study of nearly 113,000 U.S. Navy nuclear submariners starting with service on the Nautilus in 1954, the updated study of 3,276 radium dial painters, and the possible study of 97,000 nuclear shipyard workers. A large study of over 50,000 workers with measured values of cumulative neutron dose is being considered. While challenging, workers with measured bioassay samples of tritium could be considered (Little and Wakeford 2008). A new study of nearly 14,000 women who worked during WWII (1943-47) at the Tennessee Eastman Corporation (TEC) uranium processing plant is noteworthy in that the cohort has never been studied, lung doses from intakes of uranium are up to 1,000 mGy (Boice et al. 2021b), and the women are recognized by the public as 'the girls of the atomic city' (Kiernan 2013). This follow-up is in conjunction with an update of 18,000 men who worked at TEC during the same calendar years, which will facilitate sex-specific comparisons in lung cancer risk (Polednak and Frome 1981).

Also underway is a study revisiting a cohort of 3,276 radium dial painters (Rowland 1994; Fry 1998; Martinez et al. 2021). The last published follow-up was in 1980, and nearly 60% of the dial painters were alive at that time (Stebbings et al. 1984). In addition to updating the status of cohort members, the radiation dosimetry will be updated and enhanced. Updated biokinetic modeling will be applied and will be sex-specific and also specific to age throughout life. The biokinetics of radium progeny will be modeled independently and will include a sophisticated treatment of radon produced *in vivo* by the decay of radium. The latest nuclear decay and energy absorption data published by ICRP will be used to compute absorbed dose rates and annualized absorbed doses to targets of interest including the skeleton, red bone marrow, breast, brain, and lungs. A novel biological study of radium dial painter blood and

Table 10. Absorbed dose to heart among MPS cohorts[a].

			Dose to heart (mGy)		
MPS cohort	Reference	Number of workers	Mean	Median	Max
Nuclear power plant	Boice, Cohen, Mumma, Hagemeyer et al. (2021)	135,193	43.9	22.8	1105
Industrial radiographers	Boice et al. (2019c)	123,510	15.0	1.0	1504
Medical radiation workers	Boice, Cohen, Mumma, Howard et al. (2021)	109,019	14.6	8.0	1272
Atomic veterans	Boice et al. (2020)	113,806	6.1	2.2	953
Mound [b]	Boice et al. (2014)	4,954	24.0	2.4	630
Mallinckrodt [c]	Golden et al. (2019)	2,514	49.3 [c]	24.3	1345
Rocketdyne [d]	Boice, Leggett et al. (2006), Boice et al. (2011)	5,801	13.5 [d]	Na	1000
Los Alamos National Laboratory [c]	Boice et al. (2021a)	26,328	13.5 [c]	0.9	897
Tennessee Eastman Corporation	Boice et al. (2021b)	26,650	na	Na	na

[a]Heart doses represent the latest study compilations and may differ slightly from published values.
[b] Dose weighting factor of 1 applied for plutonium and polonium dose
[c]Dose weighting factor of 20 applied for plutonium and uranium
[d]Personal dose equivalent in mSv
na denotes not available

tissue is provided below as an example of what might be possible in MPS future investigations.

DNA methylation, neurodegenerative disorders, epigenetic age acceleration

A novel biological study is being considered to evaluate whether DNA methylation studies of radium dial painter blood can reveal associations with epigenetic age acceleration possibly associated with lower cognitive ability and brain vascular lesions. A key question to evaluate is whether such studies might predict radiation-related cognitive decline. Blood samples and smears are available from the radium dial painters and have been used previously to estimate radiation doses from alpha particles (Goans et al. 2019). DNA methylation studies and molecular signatures of aging have been linked in some studies as predictors of cognitive decline (Hillary et al. 2019; Nabais et al. 2021). Molecular signals of aging also have been associated with lifespan shortening, toxic exposures, and unfavorable health behaviors. Contrasts have been made between epigenetic age and chronological age. Persons with high-LET exposure to internal alpha emitters as well as low-LET external radiation may demonstrate epigenetic age acceleration (EAA) as estimated from genome-wide methylation evaluation of blood-derived DNA. These DNA methylation evaluations have found associations with lower cognitive ability and brain vascular lesions and appear to be predictors of mortality associated with different measures of brain health (Hillary et al. 2019; Nabais et al. 2021; Qin et al. 2021).

A National Center for Radiation Epidemiology and Biology (NCREB) is envisioned to provide continued support and guidance for addressing national needs. Currently, there is an infrastructure within the MPS collaborative teams that could be considered a *de facto* National Center for Radiation Epidemiology, but a radiation biology component will require a reinvigorated focus, expanded collaborators and substantial resources. We are encouraged by support from many agencies and the U.S. Congress (US GAO 2017). In FY20, for the third year in a row, the Senate Appropriations Bill included a line item to support 'the Epidemiologic Study of One Million U.S. Radiation Workers and Veterans …' (Boice, Held, Shore et al. 2019). Further, it is encouraging that the U.S. Congress has recently requested the National Academies to prepare a report entitled *Developing a Long-Term Strategy for Low-Dose Radiation Research in the United States* (NASEM 2021a).

The MPS will expand its role in training radiation scientists in epidemiology, statistics and dosimetry and will continue to offer opportunities for collaborative research as well as master and doctoral degrees. In addition to substantially improving knowledge of the potential health effects from low-level radiation exposures received gradually over time, the MPS will be able to make strong inferences regarding the adequacy and appropriateness of the LNT model as used in radiation protection.

The MPS builds upon 25 years of dreams, some sleepless nights, and heartfelt thanks and gratitude to the many hundreds who have made this vision a reality!

Virtual symposium wrap up was presented by Brian Quinn (MSKCC and Greater New York Chapter of the Health Physics Society)

This MPS virtual symposium, presentations, scientific discussions and preliminary results represented a tremendous collaboration of many different people and groups toward successful initial completion and ongoing follow-up and refinement of this important program of studies. The symposium was an opportunity not only to recognize the extraordinary individual and combined efforts but also to celebrate the results to date while envisioning the exciting work to come. It is always encouraging to see scientific progress while a diverse group of scientific experts openly acknowledges limitations to current knowledge, and then, they recognize the opportunity to address them in a meaningful way. Meetings such as these are important in part to spread the word about these projects to a diverse group of radiation scientists and protection professionals, but also to encourage and engage other individuals to be thinking about the big questions and gaps in knowledge that surround the ubiquitous radiation world we live in.

Questions and answers

Symposium attendees and panelists were encouraged to provide written questions to speakers throughout the presentations for a fruitful dialogue and discussion on related topics. Several responses were provided by the panelists during the symposium. The submitted questions were combined and compiled below by category along with answers/responses.

Department of Defense radiation studies

Has the USS Ronald Reagan study been published?

Yes. In response to the Consolidated Appropriations Act, 2014, U.S. DoD published a final report for the Congressional Defense Committees concerning personnel radiation exposures while serving on the United States Ship (USS) RONALD REAGAN (CVN 76) during Operation Tomodachi in 2011 (US DOD 2014). The report is available to the public (see https://www.health.mil/Reference-Center/Reports/2014/06/19/Radiation-Exposure-Report. . . . .)

Has the *in utero* group study been published?

Yes, see the report by Conlin et al. (2013) entitled 'Outcomes among pregnant women included in the Operation Tomodachi Registry.'

Space radiation exposures

What are examples of radiation countermeasures being considered for extended space missions?

For the unlikely yet still possible occurrence of acute radiation syndrome from an extremely large solar particle event there are FDA-approved-countermeasures that were developed for terrestrial applications that can be used (Carnell 2020). For late effects such as cancer, cardiovascular diseases and neurodegenerative diseases there aren't any biomedical countermeasures that have been established to be effective (Werneth et al. 2021). However, aspirin is currently being investigated because of its apparent effectiveness in reducing background risk of gastrointestinal as well as other cancers (Werneth et al. 2020; Zhang et al. 2021).

How accurate have we become at calculating and measuring radiation doses from the various forms of space radiation which vary in type, linear energy transfer, and energy over fairly wide ranges?

While some gaps remain in the understanding of the physics, we can predict the exposures from GCR quite accurately (Norbury et al. 2019) with the major remaining uncertainties in the biological response (Simonsen et al. 2020). The capability of predicting solar particle events is still largely lacking but calculating exposures after the fact is quite feasible (Mertens and Slaba 2019). Measurement capability has also improved significantly over the last decade to measure dose equivalent and even the energy deposition spectra quite well (Kroupa et al. 2015; Stoffel et al. 2016)

Is NASA flying dosimeters on the Mars rover missions in anticipation of manned expeditions?

Yes, the Radiation Assessment Detector flown on the Curiosity rover has measurements in transit to Mars (Zeitlin et al. 2013) as well as on the surface (Zeitlin et al. 2019).

Confounding and effect measure modification in the MPS

How is smoking as a confounder for lung cancer being handled by the MPS? Is COPD being analyzed as a possible test for smoking effects?

The MPS includes over 30 individual cohort studies and the approach to addressing tobacco use as a potential confounder is addressed in various ways. In some of the cohorts, comprehensive interviews with workers had been conducted, such as at LANL, and this information is being used directly and indirectly to confirm that education was a good surrogate for tobacco use (Mahoney and Wilkinson 1987; Boice et al. 2021a). For some studies, we were able to sample workers and ask for information on smoking which was used to confirm our use of pay type (white collar/blue collar) as a surrogate measure for tobacco use (Boice, Cohen et al. 2006). In other studies, smoking histories were available on medical questionnaires administered during workers' employment and we have abstracted and continue to abstract this information from archival records (Petersen et al. 1990; Dupree et al. 1995).

A broad objective for each MPS cohort is to develop an understanding of possible differences in smoking habits by the level of radiation exposure. As is often the case in epidemiologic studies, without direct information on cigarette smoking use for individual workers, a measure of socioeconomic status (SES) based on employment job categories, salary, education, or other demographic characteristics is used and assumed as an adequate adjustment for smoking and other lifestyle habits. For our studies of atomic veterans, rank (enlisted vs officer) is used; and for other studies, job category, pay type (white collar/blue collar) or education are most often used. In some studies, SES assignments for individuals were based on area-wide educational levels obtained from census-block group residential histories (Cohen et al. 2018; Boice, Cohen, Mumma, Hagemeyer et al. 2021).

Chronic obstructive pulmonary disease (COPD) and nonmalignant respiratory diseases are being evaluated as possible indicators of smoking histories. Elevated SMRs of these conditions related to cigarette smoking would suggest an increased prevalence of tobacco use compared with the general population. Internal cohort analyses would suggest the possible confounding effect of tobacco use if a dose response were revealed, i.e., for conditions not known to be related to tobacco use. Further, another statistical approach includes the joint modeling of COPD and lung cancer mortality as an indirect approach to control for unmeasured confounding of tobacco use (Richardson and Wing 2011).

Are there differences between the LSS and MPS female cohorts on smoking prevalence?

In the LSS, the majority of female adult survivors were nonsmokers. The differences in smoking prevalence vary by the MPS cohorts of female workers. In the MPS, the majority of most medical radiation workers (e.g., radiologists and interventional fluoroscopists) were also nonsmokers. For women employed during WWII, smoking prevalence might be as high as 20-25% (Mahoney and Wilkinson 1987; Dupree et al. 1995). Information on smoking histories is an important aspect of the MPS and additional efforts continue to obtain as many individual histories of tobacco use as available in archival records.

How are potential exposures to other chemicals such as asbestos, beryllium, or stable heavy metals as confounders being handled by the MPS?

Each cohort is evaluated on a case-by-case basis. Asbestos exposure has been evaluated among nuclear weapons test participants who served in the Navy (Till, Beck, Boice et al. 2018), among workers in the nuclear power industry, and among industrial radiographers, including those who worked at naval shipyards (Mumma et al. 2019). In contrast to the challenges in identifying health effects at very low levels of radiation, the signals associated with asbestos and asbestosis are large, often greater than 5-fold. For example, among nuclear power plant workers the SMR for asbestos-related mesothelioma and pleural cancer was 5.6, whereas for leukemia, a radiation-related malignancy, the SMR was 1.06 (Boice, Cohen, Mumma, Hagemeyer et al. 2021). Beryllium is an issue regarding several of the U.S. DOE facilities (US DOE 2019). U.S. DOE has a comprehensive program for beryllium evaluations which provides information on levels of potential exposures. Increases in berylliosis, though based on small numbers, have been identified in studies of LANL (Stefaniak et al. 2003; Boice et al. 2021a) and Rocky Flats workers (Viet et al 2000). Further, within the USTUR, we have the opportunity for quantitative assessments of tissue levels of beryllium. Several radiation workers with high exposures to beryllium have donated their bodies to science, and development of biokinetic models of systemic distribution specific to levels of exposure to beryllium (high versus low exposure) is feasible (Tolmachev et al. 2019).

Is ethnicity (and potential for higher rates of certain cancers in some ethnic groups) being considered for cohorts in the MPS?

Yes, this is considered to the extent that we have ethnicity available within the occupational records. For example, the LANL population included statistically meaningful numbers of Hispanic workers, and separate analyses identified increased rates for certain diseases in comparison with the white population. However, there was no correlation of any of these diseases with radiation dose (Boice et al. 2021a). It is standard procedure to obtain all available information on race/ethnicity for all workers, including Asian, Black, Hispanic, Native American, and White. It will only be when the MPS cohorts are combined, however, that more precise evaluations of radiation health effects by ethnicity are possible.

Epidemiological methods

How does MPS analyze uncertainty in dosimetry or modeling (e.g., sensitivity analysis, using regression calibration or Monte Carlo maximum likelihood techniques)? What confidence intervals are reported for dose-response analyses?

While most of the current uncertainty analyses for MPS have included sensitivity analyses, more rigorous, comprehensive uncertainty analyses are yet to be conducted within the MPS. However, they are high priority and soon will be initiated. We plan to apply, to the extent feasible, the new approach published by Stram et al. (2021) for uncertainty analyses among Mayak workers where there appeared to be little if any effect for photon exposures, but a meaningful effect for the plutonium intakes. For our recent paper on LANL workers, an uncertainty analysis was conducted following the methods outlined by Gilbert and colleagues (Gilbert and Fix 1996; Gilbert et al. 1996; Gilbert 1998). However, given that there were few changes in the point estimates or confidence limits, we presented the dose-response confidence intervals from the primary analyses. We are also looking at innovative ways to apply ensemble models with regard to the uncertainties associated with particular parameter inputs (Simonsen and Slaba 2020). For sensitivity analyses, we often change the parameter values for the dosimetric models for intakes of radionuclides. Among the TEC men and women, the primary exposure came from the inhalation of uranium dust associated with the operation of calutrons. There were four key parameters in the dosimetry model, including particle size and breathing rate. Minimum and maximum realistic values for these parameters were assumed and the dose–response analyses for lung cancer were conducted to test whether there was any appreciable difference in the dose response. As each one of over 30 individual cohorts comprising the MPS provides a different scenario for exposure uncertainty, individual approaches are taken. Another example would be in the study of medical radiation workers where the wearing of a lead apron was the primary determinant of dose, followed by orientation and assumption of incident energy. Various combinations of the parameter values in the preferred model – extreme, minimum, and maximum – were considered and evaluated regarding their effect on the various dose-response relationships.

What is the statistical power for individual cohorts and for the entire pooled studies for the MPS likely to be? Can this be compared to the next largest study?

The statistical power of the MPS to reveal health effects associated with low-level radiation experienced gradually over time is exceptionally large since power depends on sample size and broad distribution of organ-specific doses. As we complete and combine various cohorts, we can either detect underlying health effects or, equally important, exclude relatively low-risk levels with 95% confidence. For example, we have recently combined three studies of low-LET radiation exposure received gradually over time and evaluated lung cancer risk. The combined nuclear power plant worker and industrial radiographer studies had been published (Boice, Ellis et al. 2019), and recently the study of medical radiation workers has been included. A study of nearly 400,000 workers indicated a radiation association that was not statistically significant (ERR per 100 mGy of 0.02; 95% CI -0.03, 0.07) which indicates that an excess relative risk above 0.07 at 100 mGy can be excluded with 95% confidence. As the additional 600,000 MPS workers are included in the combined analysis, much more precision will be achieved.

How does length of follow up on the MPS play into the significance of any sex-dependent risk estimates (e.g., for lung cancer)?

The statistical power to uncover any underlying health effects greatly increases when the follow-up is long and when workers have lived to ages later in life when the background rates of lung cancer are high. While the time-dependent analyses evaluate risk among the workers as they age up to about 95 y, the power of the analysis depends on the number of workers reaching 'old age'. Currently, there are six MPS cohorts which include males and females for whom sex-specific lung cancer risks have been evaluated and are being incorporated into pooled and combined analyses (Boice, Ellis et al. 2019; NCRP 2021a). Comprehensive follow-up has been conducted through 2011 for all cohorts and has been or is being updated for all cohorts through 2017-2019. The percentage of workers in each U.S. DOE worker cohort who have died is usually over 50% and the lengths of follow-up are over 40 y on average. The cohorts that began after WW II have shorter follow-up times. The percentage of workers who have died and the mean duration follow-up are for LANL 60% and 45 y, TEC women 89% and 51 y, Mound workers 51% and 40 y, nuclear power plant workers 22% and 30 y, industrial radiographers 17% and 22 y, and medical radiation workers 10.5% and 25 y, respectively. Updated tracing will increase the percentage who have died as well as the mean duration of follow-up. More importantly, the follow-up occurs later in life when cancer rates increase. The medical radiation worker cohort includes more recent workers, shorter mean follow-up and fewer percentages of deaths. Because nearly 40% of the cohort was over age 65 y at last follow-up and since mortality rates increase markedly among the elderly, the new follow-up will improve the power of the analyses to reveal any differences in sex-specific lung cancer risk estimates.

While the MPS individual cohort results to date appear to show that there is little evidence for lung (and other) cancer risks from fractionated, low dose, low dose-rate exposures, will a quantitative dose threshold analysis be performed for the pooled MPS study?

Most MPS cohorts find evidence of radiation effects; for example, among nuclear power plant workers and industrial radiographers, there were clear excesses of radiation-associated leukemia (Tables 3, 4). Also, while sex-specific differences in lung cancer radiation risks are not apparent, several cohorts find significant associations such as the male medical radiation workers and the male industrial radiographers. Several cohorts reveal significant dose-response relationships for esophageal cancer, as does the Mayak study (Sokolnikov et al. 2015), suggesting the possibility of confounding by alcohol. When the pooled results are complete, hopefully in the next two years, we will be able to evaluate radiation risks in a variety of ways. First, we will be able to report the statistically significant findings and the various model fits for individual site-specific risks. Model fits include linear and quadratic fits. Second, if significant findings are not apparent, we will be able to present the level of radiation risks that could be excluded with 95% confidence. Third, we hope to be able to incorporate biologically-based models in a way to improve the estimate of risk in the low-dose domain (NCRP 2020b; Preston et al. 2021; Mi and Norman 2020). As is reported in several non-MPS studies, we would present dose-response relationships showing the lowest range of doses where a significant response is apparent. Again, if there is a relatively flat or even negative response in the low-dose domain, we will present confidence bounds indicating the lowest level of radiation risk that could be excluded with confidence.

Will the MPS also evaluate the possibility of hormetic effects of low doses of radiation? While it appears that the MPS typically uses a < 5 mSv or < 10 mSv referent group for dose–response evaluations, does this introduce a negative bias in that the possibility that low doses might have a hormetic effect on cancer incidence? Is it possible to also compare to results derived using non-radiation workers as a referent group?

We often use a <5 mSv or <10 mSv category as a "referent group" to make categorical comparisons, that is, visual graphics for the reader to accompany the model dose-response fits, usually linear but also quadratic, as opposed to just presenting model fits, that is, the regression model lines. Our analyses do not have a specific referent or a specific category for risks among the non- or low-exposed. Many readers appreciate having some point of reference, although arbitrary. What happens, though, is that we can change the appearance of the point estimates and confidence intervals just by changing the referent group and the categories. However, the plot of the dose response, whether it's a linear or quadratic model, always remains the same because it is not dependent on our chosen referent level for graphical comparison. The most valid analyses are comparing radiation workers to radiation workers over various levels of dose category. What is often seen in occupational studies is that the non-radiation workers (those who are not badged) are often appreciably different from the radiation workers in terms of SES and associated demographic and lifestyle factors which are not measured and thus we are not able to control. We found this to be a concern in our study of Rocketdyne and Mound workers (Boice, Cohen et al. 2006; Boice et al. 2014). The non-radiation workers were evaluated and presented but we did not use them in the internal dose–response categories because their patterns of mortality were so completely different compared with those of the very low-dose radiation workers. So, we continue to choose our comparisons to be 'like to like', radiation worker to radiation worker. Also, when we evaluate various cut-off points for dose response to identify if possible the lowest dose range for which a radiation effect (positive or negative) may be apparent, if there was evidence for adaptive response given the fractionated and often daily exposure to low doses, we would be able to pick that up and analyses to do so certainly will be conducted.

What is the potential overlap of the MPS Medical Worker cohort with the National Cancer Institute United States Radiologic Technologists (USRT) study?

The MPS medical radiation worker cohort (Boice, Cohen, Mumma, Howard et al. 2021) is quite different from the U.S. Radiologic Technologists (USRT) study (Simon et al. 2006) in design, calendar years of selection, male/female percentages, and completeness of recorded monitoring data. (1) The USRT study includes only radiologic technologists, whereas the U.S. medical radiation worker study includes not only technologists, but also radiologists, interventional fluoroscopists, radiation oncologists, and nuclear medicine physicians. Also included were nuclear pharmacists, medical and radiation physicists, nurses, veterinarians, chiropractors, dentists and allied healthcare support workers monitored for radiation in similar environments. (2) The selections of the cohorts were very different. The USRT cohort was selected from the computerized files of the American Association of Radiological Technologists (ARRT) and covered the years 1926–1980 (Boice et al. 1992). The U.S. medical radiation worker study was selected from the computerized Landauer® dosimetry database and covered the years 1965–1994. Thus, 65% of the workers in the MPS medical worker study were first monitored after 1980, that is, the last year for USRT inclusion, and could not be in the USRT study. (3) The USRT cohort included only questionnaire respondents who were alive in 1980; thus, anyone dying before 1980 was excluded. The U.S. medical worker study was not too dissimilar in this regard but did exclude anyone who died before 1977. (4) The initial linkage of the USRT study roster against the Landauer® dosimetry database found matches for only 19% of the cohort (Boice et al. 1992). The U.S. medical radiation worker cohort selection is based entirely on long-duration of measurement coverage within the Landauer® dosimetry database. The mean duration of monitored personal dosimetry records was 27 y with few gaps in coverage from first to last monitoring. (5) The USRT study includes approximately 25% males in contrast with 51% within the U.S. medical worker study. Specifically, the USRT study consists of 106,068 radiologic technologists, including 80,180 women and 25, 888 men (Velazquez-Kronen et al. 2020). The MPS medical worker study consists of 109,019 workers, including 53,801 women and 55,218 men. So, while there likely is some overlap between the two studies, the design, calendar years of selection, male/female percentages, and completeness of recorded monitoring data coverage within the Landauer® dosimetry database indicate that any overlap is small.

What MPS cohorts have the potential for radiation dose from internal emitters? Will there be enough data in the pooled results for meaningful comparisons between external and internal exposures on epidemiologic outcomes?

One of the unique strengths of the MPS is the number of cohorts with detailed information on internal emitters so that organ-specific estimates of radiation dose can be made. These cohorts include the U.S. DOE workers at LANL, Rocky Flats, Hanford, Savannah River, Linde, Middlesex, Mallinckrodt, Mound, TEC, Y-12, Fernald, gaseous diffusion plants (K-25, Portsmouth and Paducah) (Boice, Cohen et al. 2019), workers at Rocketdyne (Boice, Cohen et al. 2006; Boice, Leggett et al. 2006; Boice et al. 2011), and the radium dial painters (Martinez et al. 2021). Also, there will be sufficient information to contrast radiation effects from the intakes of radionuclides with external exposures. This is one of the important goals of the pooled analyses for which meaningful comparisons between external and internal exposures are possible. And it is not only the large numbers of workers with intakes of radionuclides that are being evaluated that is noteworthy, but also the comprehensive, and time-consuming dose reconstructions to obtain individual organ-specific estimates of dose from the intakes of radionuclides, in large part alpha particle emitters. Comparisons are being made now within the individual cohorts, but study sizes are not large enough to be definitive. Such comparisons, for example, have been done for the workers at Rocketdyne, Mound, and Los Alamos National Laboratory. The wealth of data for future comparisons and combined analyses will include the workers at Hanford, Rocky Flats, Savannah River, Mallinckrodt, Linde and gaseous diffusion plants where meaningful organ doses from the intakes of radionuclides and from external gamma radiation can be contrasted. In more basic analyses, risk estimates from cohorts with only low-LET exposures, such as industrial radiographers, nuclear power plant workers, medical radiation workers, and nuclear submariners can be contrasted with estimates from cohorts with primarily high-LET doses such as the TEC cohort and the radium dial painters. Although methodically challenging, there are substantial numbers of workers with neutron exposures that will be evaluated, since there are no human populations that have been able to evaluate possible effectiveness of neutrons in causing cancers or place upper bounds on possible equivalent dose estimates. Finally, quite a number of the workers had intakes of tritium and possible health effects will be evaluated, although this is challenging because of the higher concomitant doses from gamma and other radiation (Boice, Cohen et al. 2019). The complexity of these comparisons is that gamma radiation often dominates the organ-specific doses, and few studies include workers with only high-LET exposures, only neutron exposures, or only tritium exposures. The TEC study, however, is mainly high-LET exposure to the lung from uranium dust as external exposures were minimal. Overall, we are optimistic that meaningful comparisons can be made when the large numbers of workers with intakes of radionuclides are pooled, organ-specific doses are estimated based on the latest biokinetic models, and contrasts are made with low-LET exposures.

Are flight crews classified as radiation workers and could they be included in MPS?

NCRP has recognized that the cockpit and cabin crews of commercial aviation are occupational groups that may receive the highest average annual effective doses of all occupationally exposed workers (NCRP 2018c). These commercial airline crews and the pilots of cargo and corporate aircraft, however, are not normally monitored. Nonetheless, effective doses are estimated based on calculations of solar and GCR exposures for various flight routes and altitudes (NCRP 2018c). Because this exposure to naturally occurring sources of radiation results from occupational activities, the ICRP judged that these workers should be considered occupationally- exposed and managed as such. CDC and OSHA also recognize flight crews as occupational-exposed workers (Waters et al. 2000; FAA/OSHA 2000; Grajewski et al. 2011; CDC 2017; OSHA 2021). NCRP has not taken an official position at this time. NRC generally does not regulate naturally occurring sources of radiation, and it is left to individual states (Agreement States) to do so. NASA provides radiation safety standards for astronauts that currently differ from terrestrially-based standards for workers (NCRP 2014; NASEM 2021b). Commercial flight crews are currently being comprehensively studied by NIOSH (Yong et al. 2014). Conceivably, such studies could be included in the MPS, or pooled in the future, and would certainly add value given the unique exposures received at high altitudes. Because such workers are already being studied, there is no need to initiate new efforts within the MPS.

Is there an MPS cohort for criticality accident-exposed individuals?

No. Any worker who received >250 mSv in a given calendar year is excluded because such an exposure would not be consistent with the overall goal to evaluate low-level and low-dose rate exposure over time (Boice, Cohen et al. 2019). Nonetheless, the MPS can identify the excluded workers with both high-dose rate exposures and those who were exposed during criticality accidents and survived. Thanks for the question; it is an interesting idea for the future.

Is there an MPS cohort group for uranium miners (either U.S. or international)?

No. MPS does not currently include uranium miners or populations exposed to indoor radon. Our primary focus has been following workers and veterans exposed to low-level radiation at low-dose rates so that improved estimates of health consequences can be made with increased validity and precision for today's populations who are exposed in occupational, medical, and environmental circumstances. There are notable comprehensive studies of underground miners in Germany and Europe (Richardson et al. 2021) that have extended the compilation of underground miner cohorts that was done some 20 years ago at the National Cancer Institute and comprehensively evaluated in the BEIR VI report (NRC 1999).

How does the current global COVID-19 pandemic affect the epidemiological study? Does it skew health effects data or change the characteristics of the cohorts to unhealthy?

COVID-19 will not affect the MPS epidemiological studies currently being conducted. All our investigations provide the cause of death mortality information through or before Dec 31, 2018, i.e., before the start of the pandemic. It is of note that NCRP, NCI and NIAID recently held a workshop to evaluate the rationale behind using low-dose radiotherapy (0.3–1.5 Gy) as a treatment for severe COVID-19 respiratory disease (Prasanna et al. 2020). Without refuting or endorsing low-dose radiotherapy, general guidance to clinicians and researchers was provided. Others, however, come down a bit stronger and find little justification and potential long-term risk from such radiation treatments (Kirsch et al. 2020).

In addition to utilizing the Centers for Medicare and Medicaid Services (CMS) data to investigate neurocognitive effects, are there more general plans to use that data to analyze other nonmalignant effects (e.g., cardiovascular effects, metabolic syndromes, etc.)?

Yes. When we link all MPS study participants to CMS data, we will obtain all medical diagnosis and procedure codes associated with Medicare and Medicaid claims (CMS 2018). Having this information will allow us to investigate not only the CNS conditions of import but also cancer incidence, less fatal cancers, and a broad range of other nonmalignant diseases and medical conditions including depression, cataracts, cardiovascular and cerebrovascular disease, and renal failure.

The CNS evaluations include cognitive function scores within the Minimum Data Set of Nursing Home admissions (CMS 2021a). A unique aspect of the nationwide CMS data is the possibility to obtain smoking information among all workers (CMS 2021b). After the initial evaluations are completed the opportunities for evaluated radiation-disease associations among the 600,000 workers and veterans for whom CMS linkage is possible are limited only by the imagination of the investigators.

How diverse (age, race, gender, economic status, etc.) is the MPS? Is it currently a population representative of the workforce or of the United States overall? Will any results need to be adjusted to be representative of the U.S.?

The MPS is diverse and includes healthy workers and veterans of all races, SES levels, geographic residence, years of birth (calendar years) and sex, but not ages. The MPS comprises adults and a few pre-teens or teenagers (11 to 21 y) but no children (ages 0 to 10 y). Over 250,000 adult women are being studied, including the 'girls of the atomic city', and there are about 1,000 pre-teen or teenage radium dial painters. The SES levels cover all walks of life: blue collar and white collar, enlisted men and officers, elementary school education and advanced Ph.D. degrees (think Los Alamos and Oppenheimer). Most previous occupational studies have focused on White males, but the MPS is much more diverse. While we are unaware of any studies showing racial differences in effects of ionizing radiation, even though differences in the background rates for disease can be substantial, potential differences in radiation response will be evaluated especially for Blacks where nuclear workers appear notably healthier than the U.S. population average (Wartenberg et al. 2001). Hispanics at LANL have been studied independently: increased mortality rates were seen for diabetes and certain cancers but there was no difference in radiation response (Boice et al. 2021a). Workers who are Asian or Native American are identified but the numbers are not large enough to date for independent analyses. Exposed workers spanned nearly a century, from the radium dial painters in the 1920s to medical workers in the 2010s. Age (young, middle age, elderly), sex, and race (White, Black), SES (education; white-collar, blue-collar) specific cancer and noncancer risks will be evaluated once the million individual workers in the over 30 MPS cohorts are combined. For some of the future analyses, which could involve upwards of a billion data variables per individual (when including the CMS and Nursing Home linkages), cluster computing (perhaps with the use of supercomputers) would likely be of inestimable value.

While the MPS is currently for United States workers, are there any plans or possibilities for extending to other international radiation worker cohorts?

There are no current plans for extending the MPS to include other international radiation worker cohorts. One of the largest international cohorts, the INWORKS study (Thierry-Chef et al. 2015), already includes 5 U.S. cohorts of which 3 (ORNL, SRS and Hanford) overlap with the MPS. Idaho National Laboratory (INL) and Portsmouth Naval Shipyard are included in INWORKS but not in the MPS. The previous 15-Country study (Cardis et al. 2007) included four U.S. cohorts: ORNL, Hanford, a cohort of nuclear power plant workers, and INL. Only INL is not included in the MPS. The 3-Country study (Cardis et al. 1995) included 3 MPS cohorts: Hanford, Rocky Flats and ORNL. Once we have completed our first evaluation and pooling of all the more than 30 cohorts, any unique opportunity for collaboration and comparisons with our international colleagues, including the Canadian and UK worker studies, would be encouraged.

How much do animal models play into the MPS research at moderate dose-rate effects?

Animal models are essential for assigning dose weighting factors (DWF) for alpha-particle emitters, tritium, and thermal and fast neutrons. Animal models can also provide guidance as to the shape of dose–response functions that might be considered for organ-specific evaluations, e.g., linear, quadratic, threshold, over a broad range of doses. Biologically-based models including animal and cellular models will play an important future role for the application of the MPS research. NCRP has completed a recent report addressing the integration of radiation biology and epidemiology (NCRP 2020b; Preston et al. 2021). The committee concluded that populations such as the MPS will provide important epidemiologic data for which biologically-based models can be applied to enhance and improve the estimation of radiation risks in the low-dose domain. Biologically-based models are not limited only to carcinogenesis, e.g., adverse outcome pathways for space radiation-induced neurological diseases, e.g., Alzheimer's, are being evaluated (Mi and Norman 2020). Animal experiments play a critical role in assessing space-radiation induced cognitive impairment (Britten et al. 2021).

What is the current speculation about the basis for the difference in relative risk for lung cancer in women compared to the atomic bomb survivor study? Is it all *healthy worker effect* or has that been controlled for? Is it dose rate?

The cohort populations in the MPS are quite different from the LSS population of Japanese atomic bomb survivors. Perhaps the most important difference that might partially explain the absence of a sex-specific difference in the MPS cohorts is that the exposure in Japan was brief, perhaps on the order of a second, whereas the exposure in the MPS cohorts was gradual over time and over a period of years (Boice, Cohen et al. 2019). Animal studies have provided evidence that for low-LET radiation the risk at cumulative dose comparisons is lower for prolonged exposures (Dauer et al. 2010). Another difference is the general health of the exposed populations and environmental conditions. The Japanese survivors were exposed in 1945 and had to live in a war-torn country under conditions of deprivation, infections, and malnutrition (Boice, Cohen et al. 2019). The MPS cohorts started as very healthy men and women working in the United States. The sex-specific comparisons in Japan are made between women who had a low prevalence of smoking in contrast to the men who had a very high prevalence. For the comparisons in the MPS, some of the female cohorts were in large part nonsmokers and this is particularly apparent among the medical radiation workers which included radiologists and fluoroscopists, and other physicians and health-conscious individuals. Perhaps the strongest evidence for an absence of sex-specific differences in lung cancer risk is among the population of TEC workers. Here, approximately 30,000 workers, 58% men and 42% women, were evaluated and over 90% have died. The estimated lung doses for the inhalation of uranium reached 0.5 Gy (DWF = 1) and the weighted dose (DWF = 20) up to 10 Gy. Information at the time of employment on tobacco use was obtained from questionnaires and is being used in the adjustment process. There is probably a combination of factors influencing the differences in the findings from the Japanese survivor study and the MPS. These include: dose rate, race (and the underlying differences in background rates of cancer between Japanese and Americans), unmeasured effects of deprivation, uncontrolled effects of cigarette smoking (there is a peculiar ERR response in the LSS by numbers of cigarettes smoked per day, i.e., no radiation risk among the heaviest smokers), and a Japanese cohort exposed during one period of time during a few seconds in August 1945. Further, it is important to note that it is not

Table 11. Standardized mortality ratios (SMRs) and number of breast cancers for males and females in 8 MPS cohorts (consisting of 545,771workers and veterans) by quality of exposure.

MPS cohort	Number of Workers	Male SMR (number of breast cancers)	Female SMR (number of breast cancers)
High-linear energy transfer (LET)[a] exposures			
Mound (Boice et al. 2014)	4,954	0 (0)	1.00 (41)
LANL (Boice et al. 2021b)	26,328	0.59 (3)	1.11 (192)
TEC (Boice et al. 2021b)	26,650	0.31 (1)	0.92 (414)
Rocketdyne (Boice et al. 2011)	5,801	0 (0)	0.96 (8)
Low-linear energy transfer (LET)[b] exposures			
Nuclear power plant (Boice et al. 2021c)	135,193	0.35 (4)	1.15 (37)
Industrial radiographers	123,556	0.91 (7)	0.94 (35)
Medical workers (Boice, Cohen, Mumma, Howard et al. 2021)	109,019	1.04 (5)	0.79 (311)
Atomic veterans (Boice et al. 2020)	114,270	1.17 (29)	Na
Total	545,771	0.91 (49)	0.91 (1,038)

[a]High-LET exposures consist primarily of internal intakes of radionuclides and neutrons.
[b]Low-LET exposures consist primarily of external photon exposures, mainly gamma rays but also x-rays.
na denotes not applicable.

only the occupational cohorts within the MPS that fail to reveal differences in lung cancer rates between women and men, but also results from the pooling of over 20 comprehensive international studies of indoor radon, including nonsmokers (Darby et al. 2006; Cheng et al. 2021), and from studies of tuberculosis patients in two countries where the total lung doses from repeated chest fluoroscopies reached exceptionally high levels (i.e., up to 4 Gy) (Howe 1995; Brenner 2010; Boice, Ellis et al. 2019).

When evaluating risk differences between males and females for lung cancer, were there any differences in the risk of breast cancer?

Yes. We will conduct sex-specific internal dose-response comparisons for breast and other cancers among the MPS cohorts, similar to what is being done for lung cancer. However, one of the challenges for breast cancer comparisons is that male breast cancer is rare; thus, the power of detecting a radiation association or any sex-specific differences may only be possible from the combination of the MPS cohorts. Another issue may be the relatively low doses to the breast in comparison with the higher doses to the lung, especially for workers who had inhaled radionuclides. The TEC workers, for example, had a substantial lung dose related to the inhalation of uranium dust but the dose from the minimal external exposures was negligible to the breast. Table 11 shows the SMRs and numbers of breast cancers for males and females in 8 MPS cohorts consisting of over 545,000 workers. While the number of male workers is very large, the number of male breast cancers is small, only 49, in contrast to 1,038 breast cancers among females. When compared with the general population, the total SMRs for males and females are 0.91, showing that both males and females show a healthy worker effect that indicates a 9% lower chance of dying from breast cancer.

Are there any plans to compare individuals working at boiling water reactors to those working at pressurized water reactors? Are there differences in exposure versus megawatt of the unit?

These are interesting questions. We had not thought about comparing individuals who worked at boiling water reactors with those working at pressurized water reactors per se as we would not anticipate any difference in regard to the effectiveness of radiation to cause a health effect. It is interesting, though, to consider tabulating the cumulative exposure levels for workers per MW. The NRC publishes detailed information on the annual doses by reactor characteristics, but not by cumulative doses.

Dosimetry for the MPS cohorts

What tissue weighting factor is being used for tritium for MPS cohort studies (e.g., LANL, etc.)?

In our evaluations, the computed dose from intakes of tritium is assumed to be a whole-body dose and not tied to a specific tissue weighting factor, so the weighting factor would be 1. For the analysis, we assume different dose weighting factors from 1 to 3 (NCRP 2018d).

What dosimetry conversion factors are used in the MPS (ICRP Report 74 or ICRP Report 116)?

As noted in NCRP Report 178 (NCRP 2018a), dose coefficients relating personal dose equivalent to organ dose for photons are derived primarily from ICRP Report 116 (ICRP 2010) and for neutrons are derived from Tables C1-C30 of ICRP Report 116 (ICRP 2010) and Table A.42 of ICRP Report 74 (ICRP 1996).

How are minimum detectable levels (MDL) for dosimetry being handled by the MPS?

Both NCRP Report 178 (NCRP 2018a) (for all cohorts) and NCRP Commentary 30 (NCRP 2020a) provide specific guidance for handling issues associated with MDL practices over time. An example on how this is done in practice is found in the study of medical radiation workers (Boice, Cohen, Mumma, Howard et al. 2021). Adjustments for 'missed dose' were done by increasing each annual dose to a minimum value of 0.4 mSv and adding 0.2 mSv for annual doses that were between 0.4 and 1.0 mSv. Such adjustments for doses below the MDL of the recording dosimeter did not influence the dose responses reported in this study.

How are occupational doses over several different cohorts during the working life of an individual handled by the MPS and what cohort are you assigned to?

When creating a worker's dose history, we use occupational doses from several data sources, including dosimetry records from the facility in which the worker was employed, the U.S. DOE REMS database, the NRC REIRS database, dose records from a private dosimetry company (Landauer), and military service (DTRA, Army, Navy and Air Force) dose records. A manuscript we published in May 2006 entitled 'A comprehensive dose reconstruction methodology for former Rocketdyne/atomics international radiation workers' (Boice, Leggett et al. 2006) provides an overview on how we use multiple radiation databases to reconstruct a worker's full career dose history. Workers employed at multiple U.S. DOE facilities would be included in each facility-specific cohort, combining the doses they received at other facilities in their full dose profile. Some workers had been employed at as many as five different facilities. When the cohorts are eventually combined, a worker will be counted only once.

How does the Landauer database overlap with REIRS? Are there any indications of the incompleteness of individual dose records in REIRS and can the Landauer database be used to fill in any dosimetry data gaps?

In our manuscript on nuclear power plant workers, we indicate the overlap between the REIRS dataset and the Landauer dataset (Boice, Cohen, Mumma, Hagemeyer et al. 2021). For the 135,193 nuclear power plant workers in the study cohort, 16.6% were in both datasets, 82.6% in REIRS only, and 0.8% in Landauer only. One of the benefits of the Landauer database is that there is the ability to provide information for workers that may not be included in REIRS or were included but with incomplete coverage (mainly for the very early workers for which voluntary retrospective reports to REIRS were not as complete as for workers in later years) (Hagemeyer et al. 2018).

What are the dose ranges observed in the MPS cohorts to date?

Table 6 provides information on absorbed dose to the lung for nine MPS cohorts completed to date. Information is presented on nearly 550,000 workers with individual dosimetry determinations. The mean doses to lung range from a low of 6.4 mGy among the atomic veterans to highs of 477 and 508 weighted-mGy among TEC and Mallinckrodt workers, respectively. The maximum doses to lung range from 970 mGy among the atomic veterans to highs of 16–18 weighted-Gy among TEC, Mallinckrodt and LANL workers. These absorbed dose distributions differ from the personal dose equivalent distributions reported in NCRP Report 178 (2018a), as expected. Years of effort were needed to convert the badge and personal dosimetry reading into organ doses accounting for energies and orientation, for incorporating organ doses from the intakes of radionuclides, and for adding career doses for individual workers who were employed at multiple facilities. The dose ranges for red bone marrow, brain, and heart are presented in Tables 4, 8, and 10.

The dose distribution of MPS workers and veterans contains more high dose persons, that is, > 50 mSv, than the Japanese study of atomic bomb survivors, that is, $\sim$180,000 adults compared with 25,035 children and adults (Boice, Cohen et al. 2019). The dosimetry for the LSS also continues to be revised with significant organ-dose modifications possible once the question of higher neutron dose and gamma dose to organs closer to the body surface than the colon is resolved (Cordova and Cullings 2019). Further a new set of hybrid phantoms has been developed based on the Japanese 1945 population with improved anatomical and age definitions than the stylized phantoms currently used (Griffin et al. 2019). The hybrid phantoms apparently result in different doses than DS02 and impact the organ-specific radiation risk estimates. The importance of dosimetry is again reinforced as the cornerstone for accurate and valid epidemiologic research.

Radiation and cognitive effects

When reviewing the CMS data, are any particular confounding variables being factored into the analysis?

Ongoing research will incorporate CMS data based on traditional Medicare claims and Minimum Data Set (MDS) nursing home assessments (CMS 2018, 2021a). These linkages will provide information on Alzheimer's disease and related dementias (ADRD) as well as cognitive function scores (Thomas et al. 2017; Nikpay et al. 2021). All U.S. radiation workers and veterans who were alive anytime from 1999 through 2020 and eligible for Medicare and/or Medicaid benefits, for example, aged 65 years or over, will have linkages requested for CMS claims and prescription drug data. Approximately 600,000 MPS cohort members will be submitted for linkage. The pivotal comparisons will be between estimated radiation dose to the brain from high-LET alpha particles, in the presence of dose from low-LET photons, and the incidence of ADRD, with particular emphasis on cognitive function scores that are available within the CMS MDS. An important evaluation will be the odds of specialized geriatric care as an evaluation of cohort members likely to avail themselves of such facilities and receive care. Additionally, to account for or evaluate the determinants of ADRD and associated cognition scores, analyses will be conducted based on age, sex, race, marital status; income to the extent possibly available; education; and residence (urban vs. rural). The history of any long-term nursing home stays since 2001 will also be evaluated based on nursing home assessments. Studies conducted at Vanderbilt University have revealed that persons with ADRD compared with persons without such diagnoses are more likely to have visited a geriatric specialist, to be older, female, less likely to be currently married, have a low household income, have only a high school or less education, and a history of residence in a nursing home. No differences were identified by race or rural residency (Nikpay et al. 2021).

Is there evidence on the latency period for dementia development after radiation exposure? How do we know that it would develop within 2–3 years which is necessary to reach Mars?

NASA divides potential nervous system effects into two different categories: acute inflight and late degenerative (NASA 2021). The latter category would encompass typical diseases like Alzheimer's. As discussed in the Presentation on "Cognition and dementia following intakes of radionuclides", current low-dose radiation epidemiology has not yet established the existence of an effect and less so estimates of latency potential latency. However, conventional forms of dementia seem unlikely to progress rapidly enough to be an inflight risk. Short latency cognitive impairment has been suggested mainly by animal experiments (Cekanaviciute et al. 2018). However, impairment has, at least in some cases, also been shown to occur after durations of a substantial portion of the lifespan of the animal (Rabin et al. 2014). In summary, the existence of cognitive impairments from space radiation-like exposures has not been established in humans, extrapolations of latencies from short-lived animal models are questionable, and the link between potential inflight effects and late degenerative ones is largely unknown; the MPS will provide the first data in humans from relevant high-LET exposures that may be able to help address these questions.

How similar, microdosimetrically, is radiation track distribution and energy from alpha particles to that from galactic cosmic radiation?

The track structure of an alpha particle is significantly less complex than for a full GCR exposure. However, in a realistic scenario for the human body inside a vehicle, low-energy alpha particles and protons make up a substantial portion of the exposure (Slaba et al. 2016, see in particular figure 10) so in that sense, these exposures aren't any less relevant than any other laboratory experiment that uses a single beam. Zaider (1996) compares the energy deposition spectra of alpha emitters with higher energy heavy ion beams that have been commonly used in experiments. The difference in the track structure is that the low energy alpha particles will have comparatively very short ranges with narrower track widths that have more concentrated energy deposition. Additionally, on a more macroscopic scale, GCR exposures are relatively homogenous throughout the brain and other internal organs (Slaba et al. 2016), but it is still an open question how uniformly internal emitters are distributed throughout the brain (see the Presentation on "Radionuclide concentrations in brain segments: autopsy series").

Could the non-uniform distribution of certain radionuclides in the brain be related to inhomogeneous blood flow to these areas?

It seems likely that nonuniform blood flow to different regions of the brain (Bentourkia et al. 2000) is an important contributing factor to the nonuniform distributions of some radionuclides in the brain. However, we do not have any compelling evidence for this. Perhaps a more important contributing factor to the distribution of radionuclides in the brain is the distribution of essential elements in the brain, as radioisotopes of those essential elements and chemically similar elements may follow the mode of entry of the essential elements and distribute in the same way as the essential elements.

Has USTUR used alpha autoradiography and neutron-induced radiography of thin brain sections to study distribution of alpha emitters and specifically plutonium in brain tissue?

We did not use these techniques. However, in collaboration with Northwestern University and Advanced Photon Source facility at Argonne National Laboratory we applied synchrotron-based X-ray fluorescence microscopy (Chen et al. 2015) to study plutonium distribution using 5 μm-thickness paraffin-embedded sections of the brainstem, cortex, hippocampus, thalamus, and cerebellum. Unfortunately, we were not able to detect plutonium due to the insufficient limit of the detection. We are planning to apply a recently developed, and available in-house, ionizing-radiation quantum imaging detector (iQID) to image brain sections. The iQID is a digital alternative of conventional autoradiography (Miller et al. 2015) and was successfully used at the USTUR to study ^{241}Am distribution in bones (Tabatadze et al. 2019).

General information requests related to the MPS

Where can one find the link to the Centers for Disease Control and Prevention (CDC) Radiation Epidemiology for Public Health Decision Making set of training videos?

See https://www.cdc.gov/nceh/radiation/emergencies/radiation-epidemiology.htm, which provides an overview of important considerations of radiation epidemiology, describes what distinguishes a well-designed or reliable study from an unreliable or a flawed study, explores how the results of epidemiologic studies may be misused or misrepresented, and discusses the impact of such studies on creating public policy and evidence-based health practices.

Please provide the website details for the Comprehensive Epidemiologic Data Repository (CEDR) managed by the U.S. Department of Energy Oak Ridge Institute for Science and Education (ORISE)?

https://oriseapps.orau.gov/cedr/

Conclusions

The First International Virtual Symposium on the MPS provided a sweeping overview of nearly 25 years of epidemiologic and dosimetric research. The dynamic nature of the program of investigations is evident. Not only were new results for the seven components which the MPS comprises presented in one forum, for the first time, but a broad landscape for the future was painted. We live in a world of low-level radiation exposure from medical, occupational, environmental, dietary, and lifestyle factors experienced over a lifetime. The MPS compilation of radiation-exposed populations, from the radium dial painters of the 1920s to the Girls of the Atomic City in the 1940s to medical radiation workers in the 2000s, form the basis for new insights and risk assessments of relevance today and tomorrow. The scientific results will inform policy and decision-makers responsible for the safety of society without unduly restricting the substantial benefits from the use of radiation. It is envisioned that sound scientific data will help in the communication and understanding of the real and not the perceived risks following gradual low-level radiation exposures received over time.

The symposium participants would like to thank those who attended the virtual format and provided such excellent questions that we, with due diligence, attempted to respond to in an informative way. The questions provided us with new ideas for future work as well as clarifying and expanding upon issues raised during the presentations. The partnering of radiation protection organizations and societies with government agencies, universities, and private enterprises has made for an exciting and synergistic forum for ideas and future opportunities. Stay tuned! A future symposium may very well be sponsored by the National Center for Radiation Epidemiology and Biology.

Disclosure statement

Most of the data used to reconstruct doses for the medical radiation worker cohort arose from measurements made by Landauer, Inc. and its predecessors over a period in excess of 50 years. Dr. R. Craig Yoder, a former long-term employee of Landauer, Inc and now retired, was chair of the NCRP Scientific Committee that recently completed Commentary 30 on dosimetry guidance for medical workers (NCRP 2020). He contributed to the assembling, evaluation and interpretation of the recorded doses used in the analyses as well as assuring the historical accuracy of both technical and administrative data, His participation does not reflect any endorsement of the commercial offerings of Landauer by the NCRP, and he received no compensation from Landauer with regard to any aspect of the research reported herein. Further, he attests that his former associations had no influence on the scientific accuracy, or any other aspect of the work reported here. The other authors report no conflicts of interest. The authors alone are responsible for the content and writing of the paper.

Notes

1. also known as the Million U.S. Workers and Veterans Study (MWS), and earlier as the Atomic and Nuclear Energy Worker (ANEW) Study

Acknowledgments

The authors appreciate all who attended this first MPS virtual symposium, especially those who participated by asking questions and joining in the discussions. The authors appreciate the radioanalytical brain autopsy results provided by Dr. John Brockman and Dana Wegge from the University of Missouri. The authors appreciate the technical support graciously provided by Burk and Associates, whose team (lead by Brett Burk) donated their time and expertise to ensure a flawless virtual symposium. The conclusions and opinions in this report are those solely of the authors. Its publication does not imply endorsement by the National Council on Radiation Protection and Measurements, Vanderbilt University, Memorial Sloan Kettering Cancer Center, the Health Physics Society or any of the acknowledged agencies. Specifically, the views expressed are those of the authors and do not reflect the official policy or position of the DOE, CDC, DOD, NASA, or the US Government.

Disclosure statement

No potential conflict of interest was reported by the author(s).

Funding

NCRP acknowledges the financial support of the MPS-related work presented in this symposium from U.S. DOE [grants # DE-AU0000042 and DE-AU0000046], NASA [grants # 80NSSC17M0016 and 80NSSC19M0161], Naval Sea Systems Command [contract # N00024-17-C-4322], U.S. NRC (grant # NRC-HQ-60-14-G-0011), and the CDC [grant # 5NUE1EH001315].

Study of Low-Dose Health Effects for nearly twenty years, providing analytic support as well as coauthoring numerous publications.

ORCID

John D. Boice Jr. http://orcid.org/0000-0002-8755-1299
Brian Quinn http://orcid.org/0000-0002-1664-7437
Isaf Al-Nabulsi http://orcid.org/0000-0002-6234-3071
Armin Ansari http://orcid.org/0000-0002-3752-8488
Emily A. Caffrey http://orcid.org/0000-0002-3530-099X
Sarah S. Cohen http://orcid.org/0000-0003-0421-1983
Ashley P. Golden http://orcid.org/0000-0002-9430-7702
Kathryn D. Held http://orcid.org/0000-0003-0918-294X
Derek W. Jokisch http://orcid.org/0000-0002-1567-5268
Michael T. Mumma http://orcid.org/0000-0001-7506-8710
Caleigh Samuels http://orcid.org/0000-0002-7854-0029
John E. Till http://orcid.org/0000-0003-3666-8806
Sergei Y. Tolmachev http://orcid.org/0000-0003-0077-106X
R. Craig Yoder http://orcid.org/0000-0002-5132-967X
Lawrence T. Dauer http://orcid.org/0000-0002-5629-8462

References

Ansari A, Kleinhans K, Boice JD. Jr. 2019. Potential health effects of low dose radiation and what it means to the practice of radiation protection. J Radiol Prot. 39(4):E9–E13.

Azizova TV, Bannikova MV, Grigoryeva ES, Rybkina VL, Hamada N. 2020. Occupational exposure to chronic ionizing radiation increases risk of Parkinson's disease incidence in Russian Mayak workers. Int J Epidemiol. 49(2):435–447.

Beck HL, Till JE, Grogan HA, Aanenson JW, Mohler HJ, Mohler SS, Voillequé PG. 2017. Red bone marrow and male breast doses for a cohort of atomic veterans. Radiat Res. 187(2):221–228.

Bentourkia M, Bol A, Ivanoiu A, Labar D, Sibomana M, Coppens A, Michel C, Cosnard G, De Volder AG. 2000. Comparison of regional cerebral blood flow and glucose metabolism in the normal brain: effect of aging. J Neurol Sci. 181(1-2):19–28.

Blake PK, Komp GR. 2014. Radiation exposure of U.S. military individuals. Health Phys. 106(2):272–278.

Bloem BR, Okun MS, Klein C. 2021. Parkinson's disease. Lancet. 397(10291):2284–2303.

Boice JD, Cohen SS, Mumma MT, Chen H, Golden AP, Beck HL, Till JE. 2020. Mortality among U.S. military participants at eight above-ground nuclear weapons test series. Int J Radiat Biol. :1–22.

Boice JD, Cohen SS, Mumma MT, Dupree Ellis ED, Eckerman KF, Leggett RW, Boecker B, Brill A, Henderson B. 2006. Mortality among radiation workers at Rocketdyne (Atomics International), 1948-1999. Radiat Res. 166(1 Pt 1):98–115.

Boice JD, Cohen SS, Mumma MT, Ellis ED, Cragle DL, Eckerman KF, Wallace PW, Chadda B, Sonderman JS, Wiggs LD, et al. 2014. Mortality among Mound workers exposed to polonium-210 and other sources of radiation, 1944-1979. Radiat Res. 181(2):208–228.

Boice JD, Held KD, Shore RE. 2019. Radiation epidemiology and health effects following low-level radiation exposure. J Radiol Prot. 39(4): S14–S27.

Boice JD, Jr, Cohen SS, Mumma MT, Ellis ED, Eckerman KF, Leggett RW, Boecker BB, Brill AB, Henderson BE. 2011. Updated mortality analysis of radiation workers at Rocketdyne (Atomics International), 1948-2008. Radiat Res. 176(2):244–258.

Boice JD, Jr, Cohen SS, Mumma MT, Ellis ED. 2019. The Million Person Study, whence it came and why. Int J Radiat Biol. 1–14. doi: 10.1080/09553002.2019.1589015.

Boice JD, Jr, Cohen SS, Mumma MT, Golden AP, Howard S, Girardi DJ, Ellis ED, Bellamy M, Dauer LT, Eckerman KF, et al. 2021b. Mortality among Tennessee Eastman Corporation uranium processing workers. Int J Radiat Biol. 1943–2017. [submitted]

Boice JD, Jr, Cohen SS, Mumma MT, Golden AP, Howard SC, Girardi DJ, Dupree Ellis ED, Bellamy M, Dauer LT, Samuels C, Eckerman KF, et al. 2021a. Mortality among workers at the Los Alamos National Laboratory. Int J Radiat Biol. 1–28. doi: 10.1080/09553002.2021.1917784.

Boice JD, Jr, Cohen SS, Mumma MT, Hagemeyer DA, Chen H, Golden AP, Yoder RC, Dauer LT. 2021. Mortality from leukemia, cancer and heart disease among U.S. nuclear power plant workers, 1957-2011. Int J Radiat Biol. [in press].

Boice JD, Jr, Cohen SS, Mumma MT, Howard SC, Yoder RC, Dauer LT. 2021. Mortality among medical radiation workers in the United States, 1965-2016. Int J Radiat Biol. [in press].

Boice JD, Jr, Dauer LT. 2021. Million Person Study of Low-Dose Radiation Health Effects. Trans Am Nucl Soc. (in press).

Boice JD, Jr, Ellis ED, Golden AP, Zablotska LB, Mumma MT, Cohen SS. 2019. Sex-specific lung cancer risk among radiation workers in the Million Person Study and among TB fluoroscopy patients. Int J Radiat Biol. 1–12. doi: 10.1080/09553002.2018.1547441.

Boice JD, Jr, Leggett RW, Ellis ED, Wallace PW, Mumma M, Cohen SS, Brill AB, Chadda B, Boecker BB, Yoder RC, Eckerman KF. 2006. A comprehensive dose reconstruction methodology for former Rocketdyne/Atomics International radiation workers. Health Phys. 90(5):409–430.

Boice JD, Jr, Mandel JS, Doody MM, Yoder RC, McGowan R. 1992. A health survey of radiologic technologists. Cancer. 69(2):586–598.

Boice JD. Jr. 2017. Space: the final Frontier-Research Relevant to Mars. Health Phys. 112(4):392–397.

Boice JD. Jr. 2019. Relevance of the million person study to research needs for NASA and space exploration. Int J Radiat Biol. 1–9. doi: 10.1080/09553002.2019.1589020.

Brenner AV. 2010. Lung cancer mortality after exposure to fractionated ionizing radiation in a cohort of Massachusetts tuberculosis patients. Meeting Report. New Developments and Future Directions in Radiation Research. American Statistical Association Conference on Radiation and Health. Vail, Colorado, June 15–18, 2008. Radiat Res. 173(3):397.

Brenner DJ. 1990. The microdosimetry of radon daughters and its significance. Radiat Prot Dosimet. 31(1-4):399–403.

Britten RA, Wellman LL, Sanford LD. 2021. Progressive increase in the complexity and translatability of rodent testing to assess space-radiation induced cognitive impairment. Neurosci Biobehav Rev. 126: 159–174.

Cahoon EK, Preston DL, Pierce DA, Grant E, Brenner AV, Mabuchi K, Utada M, Ozasa K. 2017. Lung, laryngeal and other respiratory cancer incidence among Japanese atomic bomb survivors: an updated analysis from 1958 through 2009. Radiat Res. 187(5): 538–548.

Caldwell GG, Zack MM, Mumma MT, Falk H, Heath CW, Till JE, Chen H, Boice JD. 2016. Mortality among military participants at the 1957 PLUMBBOB nuclear weapons test series and from leukemia among participants at the SMOKY test. J Radiol Prot. 36(3): 474–489.

Cardis E, Gilbert ES, Carpenter L, Howe G, Kato I, Armstrong BK, Beral V, Cowper G, Douglas A, Fix J, et al. 1995. Effects of low doses and low dose rates of external ionizing radiation: cancer mortality among nuclear industry workers in three countries. Radiat Res. 142(2):117–132.

Cardis E, Vrijheid M, Blettner M, Gilbert E, Hakama M, Hill C, Howe G, Kaldor J, Muirhead CR, Schubauer-Berigan M, et al. 2007. The

15-country collaborative study of cancer risk among radiation workers in the nuclear industry: estimates of radiation-related cancer risks. Radiat Res. 167(4):396–416.

Carnell LS. 2020. Spaceflight medical countermeasures: a strategic approach for mitigating effects from solar particle events. Int J Radiat Biol. 1–7. doi: 10.1080/09553002.2020.1820603.

CDC 2017. Centers for Disease Control and Prevention. Aircrew safety & health: cosmic radiation. National Institute for Occupational Safety and Health. CDC. [accessed 2021 August 1]. [https://www.cdc.gov/niosh/topics/aircrew/cosmicionizingradiation.html.].

Cekanaviciute E, Rosi S, Costes SV. 2018. Central nervous system responses to simulated galactic cosmic rays. IJMS. 19(11):3669.

Chen S, Paunesku T, Yuan Y, Jin Q, Hornberger B, Flachenecker C, Lai B, Brister K, Jacobsen C, Woloschak G, et al. 2015. The bionanoprobe: synchrotron-based hard x-ray fluorescence microscopy for 2D/3D trace element mapping. Micros Today. 23(3):26–29.

Cheng ES, Egger S, Hughes S, Weber M, Steinberg J, Rahman B, Worth H, Ruano-Ravina A, Rawstorne P, Yu XQ. 2021. Systematic review and meta-analysis of residential radon and lung cancer in never-smokers. Eur Respir Rev. 30(159):200230. [accessed 2021 March 22]. [https://err.ersjournals.com/content/errev/30/159/200230.full.pdf.].

CMS 2018. Centers for Medicare and Medicaid Services. Long-Term Care Facility Resident Assessment Instrument 3.0 User's Manual. Version 1.16. October 2018. (accessed August 31, 2021). [https://downloads.cms.gov/files/1-MDS-30-RAI-Manual-v1-16-October-1-2018.pdf.].

CMS 2021a. Centers for Medicare and Medicaid Serves. Minimum Data Set 3.0 Public Reports. (accessed August 31, 2021). [https://www.cms.gov/Research-Statistics-Data-and-Systems/Computer-Data-and-Systems/Minimum-Data-Set-3-0-Public-Reports/index.html.].

CMS 2021b. Centers for Medicare and Medicaid Serves. Chronic Conditions Data Warehouse. Tobacco use. (accessed August 31, 2021). [https://www2.ccwdata.org/web/guest/condition-categories.].

Cohen SS, Mumma MT, Ellis ED, Boice JD. Jr 2018. Validating the use of census data on education as a measure of socioeconomic status in an occupational cohort. Int J Radiat Biol. 1–10. doi: 10.1080/09553002.2018.1549758.

Conlin AMS, Sevick CJ, Bukowinski AT, Crum-Cianflone NF. 2013. Outcomes among pregnant women included in the Operation Tomodachi registry. [accessed 2021 August 1]. [https://apps.dtic.mil/dtic/tr/fulltext/u2/a592154.pdf.].

Cordova KA, Cullings HM. 2019. Assessing the relative biological effectiveness of neutrons across organs of varying depth among the atomic bomb survivors. Radiat Res. 192(4):380–387.

Darby S, Hill D, Deo H, Auvinen A, Barros-Dios JM, Baysson H, Bochicchio F, Falk R, Farchi S, Figueiras A, et al. 2006. Residential radon and lung cancer–detailed results of a collaborative analysis of individual data on 7148 persons with lung cancer and 14,208 persons without lung cancer from 13 epidemiologic studies in. Europe. Scand J Work Environ Health. 32(Suppl 1):1–83.

Dauer LT, Bouville A, Toohey RE, Boice JD, Jr, Beck HL, Eckerman KF, Hagemeyer D, Leggett RW, Mumma MT, Napier B, et al. 2018. Dosimetry and uncertainty approaches for the million-worker study of radiation workers and veterans: overview of the recommendations in NCRP Report No. 178. Int J Radiat Biol. 1–10. doi: 10.1080/09553002.2018.1536299.

Dauer LT, Brooks A, Hoel D, Morgan W, Stram D, Tran P. 2010. Review and evaluation of updated research on the health effects and risks associated with low dose ionizing radiation. Radiat Protect Dosim. 140(2):103–136.

Dupree EA, Watkins JP, Ingle JN, Wallace PW, West CM, Tankersley WG. 1995. Uranium dust exposure and lung cancer risk in four uranium processing operations. Epidemiology. 6(4):370–375.

Ellis ED, Girardi D, Golden AP, Wallace PW, Phillips J, Cragle DL. 2019. Historical perspective on the Department of Energy mortality studies: Focus on the collection and storage of individual worker data. Int J Radiat Biol. 1–8. Nov 29.

FAA/OSHA 2000. Federal Aviation Administration/Occupational Safety and Health Agency. Application of OSHA's requirements to employees on aircraft in operations. Aviation safety and health team (First Report). [accessed 2021 May 14]. [https://www.faa.gov/about/initiatives/ashp/media/faa-osha-report.pdf.].

Fry SA. 1998. Studies of U.S. radium dial workers: an epidemiological classic. Radiat Res. 150(5):S21–S29.

Furukawa K, Preston DL, Lönn S, Funamoto S, Yonehara S, Matsuo T, Egawa H, Tokuoka S, Ozasa K, Kasagi F, et al. 2010. Radiation and smoking effects on lung cancer incidence among atomic bomb survivors. Radiat Res. 174(1):72–82.

Gilbert ES, Fix JJ, Baumgartner WV. 1996. An approach to evaluating bias and uncertainty in estimates of external dose obtained from personal dosimeters. Health Phys. 70(3):336–345.

Gilbert ES, Fix JJ. 1996. Laboratory measurement error in external dose estimates and its effects on dose-response analyses of Hanford worker mortality data. PNL-11289. Richland (WA): Pacific Northwest Laboratory. [accessed 2021 March 20]. [https://www.osti.gov/servlets/purl/379945.].

Gilbert ES. 1998. Accounting for errors in dose estimates used in studies of workers exposed to external radiation. Health Phys. 74(1):22–29.

Goans RE, Toohey RE, Iddins CJ, McComish SL, Tolmachev SY, Dainiak N. 2019. The Pseudo-Pelger Huët cell as a retrospective dosimeter: analysis of a radium dial painter cohort. Health Phys. 117(2):143–148.

Golden AP, Ellis ED, Cohen SS, Mumma MT, Leggett RW, Wallace PW, Girardi DJ, Watkins JP, Shore RE, Boice JD, Jr. 2019. Update mortality analysis of the Mallinckrodt uranium processing workers, 1942–2012. Int J Radiat Biol. 1–21 doi: 10.1080/09553002.2019.1569773.

Grajewski B, Waters MA, Yong LC, Tseng CY, Zivkovich Z, Cassinelli RT. 2011. Airline pilot cosmic radiation and circadian disruption exposure assessment from logbooks and company records. Ann Occup Hyg. 55(5):465–475.

Griffin K, Paulbeck C, Bolch W, Cullings H, Egbert S, Funamoto S, Sato T, Endo A, Hertel N, Lee C. 2019. Dosimetric impact of a new computational voxel phantom series for the Japanese atomic bomb survivors: children and adults. Radiat Res. 191(4):369–379. Dosimetric Impact of a New Computational Voxel Phantom Series for the Japanese Atomic Bomb Survivors: Children and Adults.

Hagemeyer D, Nichols G, Mumma MT, Boice JD, Brock TA. 2018. 50 years of the Radiation Exposure Information and Reporting System. Int J Radiat Biol. 1–4. doi: 10.1080/09553002.2018.1540896.

Hillary RF, Stevenson AJ, Cox SR, McCartney DL, Harris SE, Seeboth A, Higham J, Sproul D, Taylor AM, Redmond P, et al. 2019. An epigenetic predictor of death captures multi-modal measures of brain health. Mol Psychiatry. doi: 10.1038/s41380-019-0616-9.

Hofmann W, Li WB, Friedland W, Miller BW, Madas B, Bardiès M, Balásházy I. 2020. Internal microdosimetry of alpha-emitting radionuclides. Radiat Environ Biophys. 59(1):29–62.

Howe GR. 1995. Lung cancer mortality between 1950 and 1987 after exposure to fractionated moderate-dose-rate ionizing radiation in the Canadian fluoroscopy cohort study and a comparison with lung cancer mortality in the atomic bomb survivors study. Radiat Res. 142(3):295–304.

IAEA 2011. International Atomic Energy Agency. Criteria for use in preparedness and response for a nuclear or radiological emergency, IAEA Safety Standards Series No. GSG-2. Vienna (AT): IAEA.

ICRP 2021. International Commission on Radiological Protection. Occupational Intakes of Radionuclides.] Part 5. [in press].

ICRP. International Commission on Radiological Protection. 1996. Conversion coefficients for use in radiological protection against external radiation. ICRP Publication 74. Ann ICRP. 26 (3-4)

ICRP. International Commission on Radiological Protection. 2010. Conversion coefficients for radiological protection quantities for external radiation exposures. ICRP Publication 116. Ann ICRP. 40(2-5)

ICRP. International Commission on Radiological Protection. 2016. Occupational intakes of radionuclides: Part 2. ICRP Publication 134. Ann ICRP. 45:3–4.

ICRP. International Commission on Radiological Protection. 2017. Occupational intakes of radionuclides: Part 3. ICRP Publication 137. Ann ICRP. 46:3–4.

ICRP. International Commission on Radiological Protection. 2019. Occupational intakes of radionuclides: Part 4. ICRP Publication 141. Ann ICRP. 48(2-3)

ICRP. International Commission on Radiological Protection. 2020. Radiological protection of people and the environment in the event of a large nuclear accident, ICRP Publication 146. Ann ICRP. 49(4)

Kathren RL, Tolmachev SY. 2019. The United States Transuranium and Uranium Registries (USTUR): a five-decade follow-up of plutonium and uranium workers. Health Phys. 117(2):118–132.

Kiernan D. 2013. The girls of atomic city: the untold story of the women who helped win World War II. New York (NY): Atria Books.

Kirsch DG, Diehn M, Cucinotta FA, Weichselbaum R. 2020. Lack of supporting data make the risks of a clinical trial of radiation therapy as a treatment for COVID-19 pneumonia unacceptable. Radiother Oncol. 147:217–220.

Kroupa M, Bahadori A, Campbell-Ricketts T, Empl A, Hoang SM, Idarraga-Munoz J, Rios R, Semones E, Stoffle N, Tlustos L, et al. 2015. A semiconductor radiation imaging pixel detector for space radiation dosimetry. Life Sci Space Res (Amst)). 6:69–78.

Leggett RW, Tolmachev SY, Boice JD. Jr 2019. Potential improvements in brain dose estimates for internal emitters. Int J Radiat Biol. 1–13. Dec 4.

Little MP, Wakeford R. 2008. Systematic review of epidemiological studies of exposure to tritium. J Radiol Prot. 28(1):9–32.

Mahoney MC, Wilkinson GS. 1987. Smoking patterns among Los Alamos National Laboratory employees. Report LA-10650. Los Alamos (NM): Los Alamos National Laboratory. [accessed 2021 March 20]. [https://www.osti.gov/biblio/6403953-smoking-patternsamong-los-alamos-national-laboratory-employees.].

Mappin-Kasirer B, Pan H, Lewington S, Kizza J, Gray R, Clarke R, Peto R. 2020. Tobacco smoking and the risk of Parkinson disease: A 65-year follow-up of 30,000 male British doctors. Neurology. 94(20): e2132–e2138.

Marro R, McKenzie-Carter M, Rademacher S, Knappmiller K, Ranellone R, Case D, Dunavant J, Miles T. 2014. Radiation dose assessments for fleet-based individuals in Operation Tomodachi, Rev. 1, DTRA-TR-12-041 (R1). Fort Belvoir (VA): Defense Threat Reduction Agency. [accessed 2021 August 1] [https://www.hsdl.org/?view&did=759522.]

Martinez NE, Jokisch DW, Dauer LT, Eckerman KF, Goans RE, Brockman JD, Tolmachev SY, Avtandilashvili M, Mumma MT, Boice JD, Jr, et al. 2021. Radium dial workers: back to the future. Int J Radiat Biol. 1–19. doi: 10.1080/09553002.2021.1917785.

McKenna MJ, Robinson E, Taylor L, Tompkins C, Cornforth MN, Simon SL, Bailey SM. 2019. Chromosome translocations, inversions and telomere length for retrospective biodosimetry on exposed U.S. atomic veterans. Radiat Res. 191(4):311–322.

Mertens CJ, Slaba TC. 2019. Characterization of solar energetic particle radiation dose to astronaut crew on deep-space exploration missions. Space Weather. 17(12):1650–1658.

Mi K, Norman R. 2020. An adverse outcome pathway for potential space radiation induced neurological diseases. NASA/TP-2020-220443. Hampton (VA): Langley Research Center [accessed 2021 August 1]. [https://ntrs.nasa.gov/api/citations/20200001144/downloads/20200001144.pdf.].

Miller BW, Frost SH, Frayo SL, Kenoyer AL, Santos E, Jones JC, Green DJ, Hamlin DK, Wilbur DS, Fisher DR, et al. 2015. Quantitative single-particle digital autoradiography with α-particle emitters for targeted radionuclide therapy using the iQID camera. Med Phys. 42(7): 4094–4105.

Mueller TJ, Weishar TM, Hallworth JM, Draper DA. 2020. Occupational radiation exposure from U.S. naval nuclear plants and their support facilities. Washington (DC): Naval Nuclear Propulsion Program. [accessed 2021 May 18]. [https://www.energy.gov/sites/default/files/2020/07/f77/NT-20-2.pdf.].

Mumma MT, Sirko JL, Boice JD, Jr, Blot WJ. 2019. Mesothelioma mortality within two radiation monitored occupational cohorts. Int J Radiat Biol. 1–9. doi: 10.1080/09553002.2019.1642540.

Nabais MF, Alzheimer's Disease Neuroimaging Initiative, Laws SM, Lin T, Vallerga CL, Armstrong NJ, Blair IP, Kwok JB, Mather KA, Mellick GD, Sachdev PS, et al. 2021. Meta-analysis of genome-wide DNA methylation identifies shared associations across neurodegenerative disorders. Genome Biol. 22(1):90.

NASA 2021. National Aeronautics and Space Administration. Human research roadmap. A risk reduction strategy for human space exploration. [accessed 2021 August 1]. [https://humanresearchroadmap.nasa.gov/.].

NASEM 2021a. National Academies of Sciences, Engineering, and Medicine. Developing a Long-Term Strategy for Low-Dose Radiation Research in the United States. [accessed 2021 August 5]. [https://www.nationalacademies.org/our-work/developing-a-long-term-strategy-for-low-dose-radiation-research-in-the-united-states.]

NASEM 2021b. National Academies of Sciences, Engineering, and Medicine. Space Radiation and Astronaut Health: Managing and Communicating Cancer Risks. Washington, DC: The National Academies Press. [accessed 2021 August 5] [https://doi.org/10.17226/26155]

NCRP 2014. National Council on Radiation Protection and Measurements. Radiation protection for space activities: supplement to previous recommendations. NCRP Commentary 23. Bethesda (MD): NCRP.

NCRP 2018a. National Council on Radiation Protection and Measurements. Deriving organ doses and their uncertainty for epidemiologic studies (with a focus on the one million US workers and veteran study of low-dose radiation health effects). NCRP Report 178. Bethesda (MD): NCRP.

NCRP 2018b. National Council on Radiation Protection and Measurements. Implications of recent epidemiologic studies for the linear-nonthreshold model and radiation protection. NCRP Commentary 27. Bethesda (MD): NCRP.

NCRP 2018c. National Council on Radiation Protection and Measurements. Management of exposure to ionizing radiation: radiation protection guidance for the United States. NCRP Report 180. Bethesda (MD): NCRP.

NCRP 2018d. National Council on Radiation Protection and Measurements. Evaluation of the relative effectiveness of low-energy photons and electrons in inducing cancer in humans. NCRP Report 181. Bethesda (MD): NCRP.

NCRP 2019. National Council on Radiation Protection and Measurements. Radiation exposures in space and the potential for central nervous system effects: phase II. NCRP Report 183. Bethesda (MD): NCRP.

NCRP 2020a. National Council on Radiation Protection and Measurements. Using personal monitoring data to derive organ doses for medical radiation workers, with a focus on lung. NCRP Commentary 30. Bethesda (MD): NCRP.

NCRP 2020b. National Council on Radiation Protection and Measurements. Approaches for integrating information from radiation biology and epidemiology to enhance low-dose health risk assessment. NCRP Report 186. Bethesda (MD): NCRP.

NCRP 2021a. National Council on Radiation Protection and Measurements. Scientific Committee 1-27: evaluation of sex-specific differences in lung cancer radiation risks and recommendations for use in transfer and projection models. [accessed 2021 Mar 17]. [https://ncrponline.org/program-areas/sc-1-27-evaluation-of-sex-specific-differences-in-lung-cancer-radiation-risks-and-recommendations-for-use-in-transfer-and-projection-models/.].

NCRP 2021b. National Council on Radiation Protection and Measurements. Development of kinetic and anatomical models for brain dosimetry for internally deposited radionuclides. NCRP SC 6–12. Bethesda (MD): NCRP. [accessed 2021 May 18]. [https://ncrponline.org/program-areas/sc-6-12/.].

NCRP 2021c. National Council on Radiation Protection and Measurements. Methods and models for estimating organ doses from intakes of radium. NCRP SC 6-13. Bethesda (MD): NCRP.

[accessed 2021 Aug 24]. [https://ncrponline.org/program-areas/sc-6-13/.].

Nikpay S, Keohane LM, Cheng A, Braun K, Buntin M, Lipworth L, Stevenson D. 2021. Utilization of specialized geriatric care among Medicare beneficiaries with Alzheimer's disease and related dementia: an observational analysis. J Gen Intern Med. doi: 10.1007/s11606-020-06460-3.

Norbury JW, Slaba TC, Aghara S, Badavi FF, Blattnig SR, Clowdsley MS, Heilbronn LH, Lee K, Maung KM, Mertens CJ, et al. 2019. Advances in space radiation physics and transport at NASA. Life Sci Space Res (Amst)). 22:98–124.

NPR 2012. National Public Radio. Obituaries: Ray Bradbury: 'It's Lack That Gives Us Inspiration.' [accessed 2021 May 18]. [https://www.npr.org/2012/06/08/154524695/ray-bradbury-its-lack-that-gives-us-inspiration.].

NRC 1999. National Research Council. Health effects of exposure to radon: BEIR VI. Washington (DC): The National Academies Press. [accessed May 18, 2021]. [https://doi.org/10.17226/5499].

NRC 2006. National Research Council. Health risks from exposure to low levels of ionizing radiation: BEIR VII Phase 2. Washington (DC): The National Academies Press.

NRC 2020. Nuclear Regulatory Commission. Radiation exposure information and reporting system (REIRS) for radiation workers. Washington (DC): NRC. [accessed 2021 May 18]. [www.reirs.com.].

OSHA 2021. Occupational Safety and Health Administration. Ionizing radiation. United States Department of Labor. Washington (DC): OSHA. [accessed 2021 May 18]. [www.osha.gov/ionizing-radiation.].

Ozasa K, Shimizu Y, Suyama A, Kasagi F, Soda M, Grant EJ, Sakata R, Sugiyama H, Kodama K. 2012. Studies of the mortality of atomic bomb survivors, Report 14, 1950-2003: an overview of cancer and noncancer diseases. Radiat Res. 177(3):229–243.

Pasqual E, Boussin F, Bazyka D, Nordenskjold A, Yamada M, Ozasa K, Pazzaglia S, Roy L, Thierry-Chef I, de Vathaire F, et al. 2021. Cognitive effects of low dose of ionizing radiation - Lessons learned and research gaps from epidemiological and biological studies. Environ Int. 147:106295.

Petersen GR, Gilbert ES, Buchanan JA, Stevens RG. 1990. A case-cohort study of lung cancer, ionizing radiation, and tobacco smoking among males at the Hanford Site. Health Phys. 58(1):3–11.

Polednak AP, Frome EL. 1981. Mortality among men employed between 1943 and 1947 at a uranium-processing plant. J Occup Med. 23(3):169–178.

Prasanna PG, Woloschak GE, DiCarlo AL, Buchsbaum JC, Schaue D, Chakravarti A, Cucinotta FA, Formenti SC, Guha C, Hu DJ, et al. 2020. Low-dose radiation therapy (LDRT) for COVID-19: benefits or risks? Radiat Res. 194(5):452–464.

Preston RJ, Rühm W, Azzam EI, Boice JD, Bouffler S, Held KD, Little MP, Shore RE, Shuryak I, Weil MM. 2021. Adverse outcome pathways, key events, and radiation risk assessment. Int J Radiat Biol. 97(6):804–814.

Qin N, Li Z, Song N, Wilson CL, Easton J, Mulder H, Plyler E, Neale G, Walker E, Zhou X, et al. 2021. Epigenetic age acceleration and chronic health conditions among adult survivors of childhood cancer. J Natl Cancer Inst. 113(5):597–605.

Rabin BM, Shukitt-Hale B, Carrihill-Knoll KL. 2014. Effects of age on the disruption of cognitive performance by exposure to space radiation. JBBS. 04 (07):297–307.

Richardson DB, Rage E, Demers PA, Do MT, DeBono N, Fenske N, Deffner V, Kreuzer M, Samet J, Wiggins C, et al. 2021. Mortality among uranium miners in North America and Europe: the Pooled Uranium Miners Analysis (PUMA). Int J Epidemiol. 50(2):633–643.

Richardson DB, Wing S. 2011. Evidence of confounding by smoking of associations between radiation and lung cancer mortality among workers at the Savannah River Site. Am J Ind Med. 54(6):421–437.

Rowland RE. 1994. Radium in humans. a review of U.S. studies. Report ANL/ER-3. Argonne (IL): Argonne National Laboratory [accessed 2021 May 18]. [https://publications.anl.gov/anlpubs/1994/11/16311.pdf.].

Simon SL, Bailey SM, Beck HL, Boice JD, Bouville A, Brill AB, Cornforth MN, Inskip PD, McKenna MJ, Mumma MT, et al. 2019. Estimation of radiation doses to US military test participants from nuclear testing: a comparison of historical film-badge measurements, dose reconstruction and retrospective biodosimetry. Radiat Res. 191(4):297–310.

Simon SL, Weinstock RM, Doody MM, Neton J, Wenzl T, Stewart P, Mohan AK, Yoder RC, Hauptmann M, Freedman DM, et al. 2006. Estimating historical radiation doses to a cohort of U.S. radiologic technologists. Radiat Res. 166(1 Pt 2):174–192.

Simonsen LC, Slaba TC, Guida P, Rusek A. 2020. NASA's first ground-based galactic cosmic ray simulator: enabling a new era in space radiobiology research. PLoS Biol. 18(5):e3000669.

Simonsen LC, Slaba TC. 2020. Ensemble methodologies for astronaut cancer risk assessment in the face of large uncertainties. NASA TP 2020-5008710. [accessed 2021 Mar 16]. [https://www.biorxiv.org/content/10.1101/2021.01.29.428854v1.].

Slaba TC, Bahadori AA, Reddell BD, Singleterry RC, Clowdsley MS, Blattnig SR. 2017. Optimal shielding thickness for galactic cosmic ray environments. Life Sci Space Res (Amst)). 12:1–15.

Slaba TC, Blattnig SR, Norbury JW, Rusek A, La Tessa C. 2016. Reference field specification and preliminary beam selection strategy for accelerator-based GCR simulation. Life Sci Space Res (Amst)). 8: 52–67.

Sokolnikov M, Preston D, Gilbert E, Schonfeld S, Koshurnikova N. 2015. Radiation effects on mortality from solid cancers other than lung, liver, and bone cancer in the Mayak worker cohort: 1948-2008. PLoS One. 10(2):e0117784.

Stebbings JH, Lucas HF, Stehney AF. 1984. Mortality from cancers of major sites in female radium dial workers. Am J Ind Med. 5(6): 435–459.

Stefaniak AB, Weaver VM, Cadorette M, Puckett LG, Schwartz BS, Wiggs LD, Jankowski MD, Breysse PN. 2003. Summary of historical beryllium uses and airborne concentration levels at Los Alamos National Laboratory. Appl Occup Environ Hyg. 18(9):708–715.

Stoffel N, Gaza R, Nounu H, Lee K, Bahadori A. 2016. Comparison of passive and active exploration flight test 1 radiation detector measurements with trapped proton and vehicle shielding model calculations. NASA-TP-218599. Washington (DC): NASA. [accessed 2021 May 18]. [https://krex.k-state.edu/dspace/bitstream/handle/2097/34588/TP-2016-218599.pdf?sequence=1&isAllowed=y.].

Stram DO, Sokolnikov M, Napier BA, Vostrotin VV, Efimov A, Preston DL. 2021. Lung cancer in the Mayak workers cohort: risk estimation and uncertainty analysis. Radiat Res. 195(4):334–346.

Tabatadze G, Miller BW, Tolmachev SY. 2019. Mapping ^{241}Am spatial distribution within anatomical bone structures using digital autoradiography. Health Phys. 117(2):179–186.

Thierry-Chef I, INWORKS Consortium, Richardson DB, Daniels RD, Gillies M, Hamra GB, Haylock R, Kesminiene A, Laurier D, Leuraud K, Moissonnier M, et al. 2015. Dose estimation for a study of nuclear workers in France, the United Kingdom and the United States of America: methods for the International Nuclear Workers Study (INWORKS). Radiat Res. 183(6):632–642.

Thomas KS, Dosa D, Wysocki A, Mor V. 2017. The Minimum Data Set 3.0 Cognitive Function Scale. Med Care. 55(9):e68–e72.

Thomas R. 2013. Practical guide to ICP-MS: a tutorial for beginners. Boca Raton (FL): CRC Press.

Till J, Beck H, Boice JD, Jr, Mohler H, Mumma M, Aanenson J, Grogan H. 2018. Asbestos exposure and mesothelioma mortality among atomic veterans. Int J Radiat Biol. 1–5.

Till JE, Beck HL, Aanenson JW, Grogan HA, Mohler HJ, Mohler SS, Voillequé PG. 2014. Military participants at U.S. atmospheric nuclear weapons testing-methodology for estimating dose and uncertainty. Radiat Res. 181(5):471–484.

Till JE, Beck HL, Aanenson JW, Grogan HA, Mohler HJ, Mohler SS, Voillequé PG. 2018. Dosimetry associated with veterans who participated in nuclear weapons testing. Int J Radiat Biol. 1–9.

Tolmachev SY, Swint MJ, Bistline RW, McClellan RO, McInroy JF, Kathren RL, Filipy RE, Toohey RE. 2019. USTUR special sessions roundtable: United States Transuranium and Uranium Registries

(USTUR): a five-decade follow-up of plutonium and uranium workers. Health Phys. 117(2):211–222.

UNSCEAR 2020. United Nations Scientific Committee on the Effects of Atomic Radiation. UNSCEAR 2019 Report to the General Assembly, with scientific annexes A and B. New York (NY): United Nations. [accessed 2021 January 10]. [www.unscear.org/unscear/en/publications/2019.html.].

US DoD 2014. US Department of Defense. Final report to the Congressional Defense Committees in response to the Joint Explanatory Statement accompanying the Department of Defense Appropriations Act (p. 90, "Radiation Exposure"). [accessed 2021 Jul 27]. [https://www.health.mil/Reference-Center/Reports/2014/06/19/Radiation-Exposure-Report.].

US DOE 2019. US Department of Energy. Former worker medical screening program report. 2019 Annual Report. Washington (DC): Office of Environment, Health, Safety and Security. [accessed 2021 March 20]. [https://www.energy.gov/sites/prod/files/2020/06/f75/2019-FW-Medical-Screening-Program-Annual-Report.pdf.].

US DOE 2020. US Department of Energy. Comprehensive Epidemiologic Data Resource: CEDR Epidemiologic Databases. Oak Ridge (TN): Oak Ridge Institute for Science and Education. [accessed 2021 May 18]. [https://oriseapps.orau.gov/cedr/.].

US EPA 2017. US Environmental Protection Agency. PAG Manual: protective action guides and planning guidance for radiological incidents. EPA-400/R-17/001. Washington (DC): US EPA.

US GAO 2017. US Government Accountability Office. Low-dose radiation: interagency collaboration on planning research could improve information on health effects. GAO-17-546. Washington (DC): GAO. [accessed 2021 Mar 16]. [https://www.gao.gov/products/gao-17-546.].

Velazquez-Kronen R, Gilbert ES, Linet MS, Moysich KB, Freudenheim JL, Wactawski-Wende J, Simon SL, Cahoon EK, Alexander BH, Doody MM, et al. 2020. 2020. Lung cancer mortality associated with protracted low-dose occupational radiation exposures and smoking behaviors in U.S. radiologic technologists, 1983-2012. Int J Cancer. 147(11):3130–3138.

Viet SM, Torma-Krajewski J, Rogers J. 2000. Chronic beryllium disease and beryllium sensitization at Rocky Flats: a case-control study. AIHAJ. 61(2):244–254.

Wartenberg D, Brown S, Mohr S, Cragle D, Friedlander B. 2001. Are African-American nuclear workers at lower mortality risk than Caucasians? J Occup Environ Med. 43(10):861–871.

Waters M, Bloom TF, Grajewski B. 2000. The NIOSH/FAA Working Women's Health Study: evaluation of the cosmic-radiation exposures of flight attendants. Health Phys. 79(5):553–559.

Werneth CM, Slaba TC, Blattnig SR, Huff JL, Norman RB. 2020. A methodology for investigating the impact of medical countermeasures on the risk of exposure induced death. Life Sci Space Res. 25: 71–102.

Werneth CM, Slaba TC, Simonsen LC. 2021. Medical countermeasure requirements for meeting permissible radiation exposure limits in space. NASA/TP–20210009708. [accessed 2021 July 27]. [https://ntrs.nasa.gov/api/citations/20210009708/downloads/NASA-TP-20210009708.pdf.].

Yoder CR, Balter S, Boice JD, Jr, Grogan H, Mumma M, Rothenberg LN, Passmore C, Vetter RJ, Dauer LT. 2021. Using personal monitoring data to derive organ doses for medical radiation workers in the Million Person Study – considerations regarding NCRP Commentary no. 30. J Radiol Prot. 41(1):118–128.

Yoder RC, Dauer L, Balter S, Boice JD, Jr, Grogan H, Mumma M, Passmore CN, Rothenberg LN, Vetter RJ. 2019. Dosimetry for the study of medical radiation workers with a focus on the mean absorbed dose to the lung, brain and other organs. Int J Radiat Biol. 1–12. doi: 10.1080/09553002.2018.1549756.

Yong LC, Pinkerton LE, Yiin JH, Anderson JL, Deddens JA. 2014. Mortality among a cohort of U.S. commercial airline cockpit crew. Am J Ind Med. 57(8):906–914.

Zaider M. 1996. Microdosimetric-based risk factors for radiation received in space activities during a trip to Mars. Health Phys. 70(6):845–851.

Zeitlin C, Hassler DM, Cucinotta FA, Ehresmann B, Wimmer-Schweingruber RF, Brinza DE, Kang S, Weigle G, Böttcher S, Böhm E, et al. 2013. Measurements of energetic particle radiation in transit to Mars on the Mars Science Laboratory. Science. 340(6136): 1080–1084.

Zeitlin C, Hassler DM, Ehresmann B, Rafkin SCR, Guo J, Wimmer-Schweingruber RF, Berger T, Matthiä D. 2019. Measurements of radiation quality factor on Mars with the Mars Science Laboratory Radiation Assessment Detector. Life Sci Space Res (Amst)). 22: 89–97.

Zhang J, Stram DO, Cohen SS, Pawel D, Sesso H, Boice J. 2014. Non-cancer mortality in two radiation worker cohorts (PS7-70). 21st International Meeting - Conference on Radiation Health; September; Las Vegas (NV).

Zhang Y, Chan AT, Meyerhardt JA, Giovannucci EL. 2021. Timing of aspirin use in colorectal cancer chemoprevention: a prospective cohort study. J Natl Cancer Inst. 113(7):841–851.

[illegible]

Index

Note: **Bold** page numbers refer to tables; *italic* page numbers refer to figures and page numbers followed by "n" denote endnotes.